Fox

Fundamentals of Human Physiology

ON TRACK . . .

Various types of interactives and quizzing keep you motivated and on track in mastering the key concepts.

- **Animations** Access to over 100 animations of key physiological processes will help you visualize and comprehend important concepts depicted in the text. The animations even include quiz questions to help ensure that you are retaining the information.
- **Test Yourself** Take a chapter quiz at the *Fundamentals of Human Physiology* ARIS site to gauge your mastery of chapter content. Each quiz is specially constructed to test your comprehension of key concepts. You can even e-mail your quiz results to your professor!
- **Learning Activities** In addition to interactive online quizzing and animations, each chapter offers relevant case study presentations, text images, vocabulary flash cards, and other activities designed to reinforce learning.

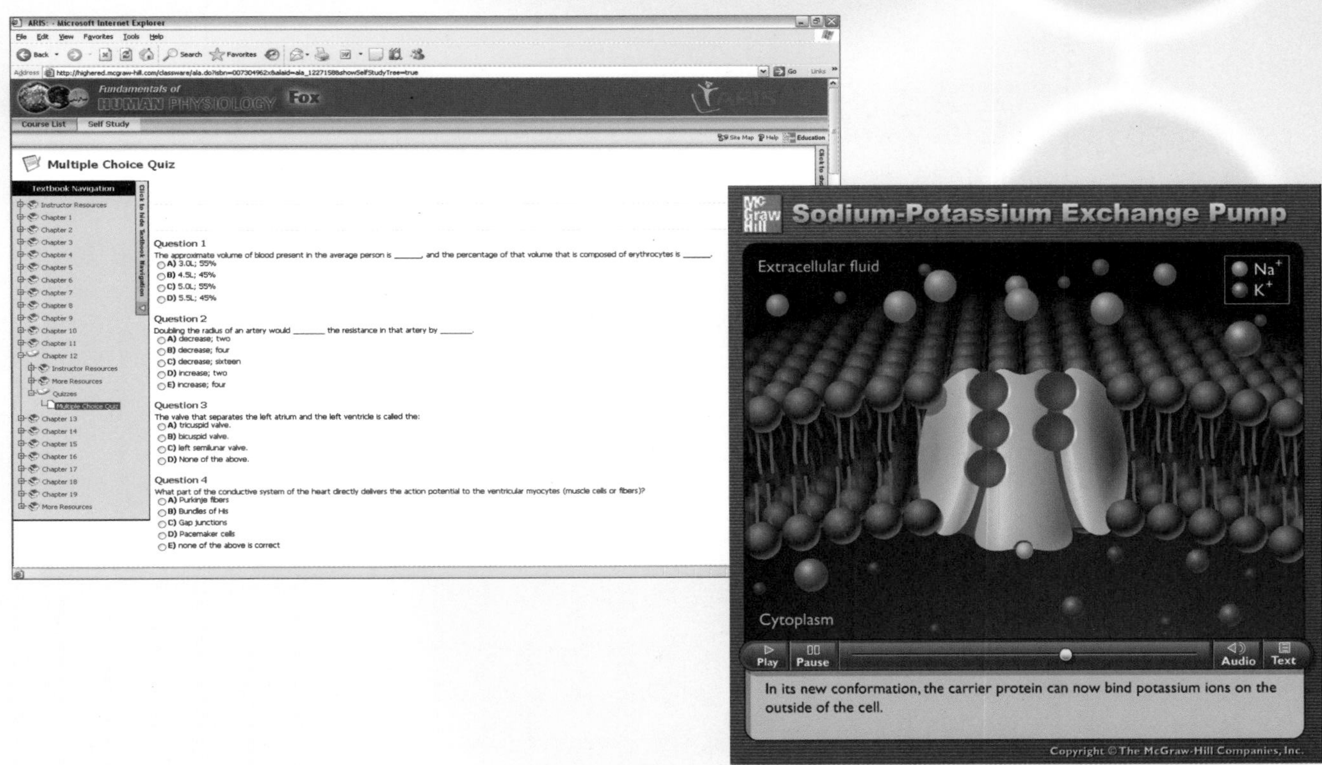

Fundamentals of HUMAN PHYSIOLOGY

Stuart Ira Fox
PIERCE COLLEGE

FUNDAMENTALS OF HUMAN PHYSIOLOGY

Published by McGraw-Hill, a business unit of The McGraw-Hill Companies, Inc., 1221 Avenue of the Americas, New York, NY 10020.

Printed in China

3 4 5 6 7 8 9 0 CTP/CTP 11 10

ISBN 978–0–07–340349–6
MHID 0–07–340349–0

Publisher: *Michelle Watnick*
Vice-President New Product Launches: *Michael Lange*
Executive Editor: *Colin H. Wheatley*
Developmental Editor: *Kathleen R. Loewenberg*
Marketing Manager: *Lynn M. Breithaupt*
Project Manager: *April R. Southwood*
Senior Production Supervisor: *Sherry L. Kane*
Senior Media Project Manager: *Tammy Juran*
Senior Coordinator of Freelance Design: *Michelle D. Whitaker*
Cover/Interior Designer: *Elise Lansdon*
(USE) Cover Image: *(top and bottom right): © Dr. Don Fawcett/Getty Images; (middle): © Manfred Kage/Peter Arnold, Inc.; (bottom): © Comstock Images/PictureQuest (RF); background: © Jason Reed/Getty Images (RF)*
Senior Photo Research Coordinator: *John C. Leland*
Photo Research: *Mary Reeg*
Supplement Coordinator: *Mary Jane Lampe*
Compositor: *Argosy Publishing, Inc.*
Typeface: *10/12 Times Roman*
Printer: *CTPS*

The credits section for this book begins on page 429 and is considered an extension of the copyright page.

Library of Congress Cataloging-in-Publication Data

Fox, Stuart Ira.
Fundamentals of human physiology / Stuart Ira Fox. – 1st ed.
p. cm.
Includes index.
ISBN 978–0–07–340349–6 — ISBN 0–07–340349–0 (hard copy : alk. paper)
1. Human physiology. I. Title.

QP34.F76 2009
612—dc22

2007043505

www.mhhe.com

This book is dedicated to the students who will use it, in the hope that it will help ease their long and often arduous journey toward their goals.

About the Author

Stuart Ira Fox earned a Ph.D. in human physiology from the Department of Physiology, School of Medicine, at the University of Southern California, after earning degrees at the University of California at Los Angeles (UCLA); California State University, Los Angeles; and UC Santa Barbara. He has spent most of his professional life teaching at Los Angeles City College; California State University, Northridge; and Pierce College, where he has won numerous teaching awards, including several Golden Apples. Stuart has authored thirty-five editions of seven textbooks, which are used worldwide and have been translated into several languages. When not engaged in professional activities, he likes to hike, fly fish, and cross-country ski in the Sierra Nevada Mountains.

Brief Contents

Contents

8 Endocrine System 186

9 Muscle Physiology 215

10 Blood and Circulation 240

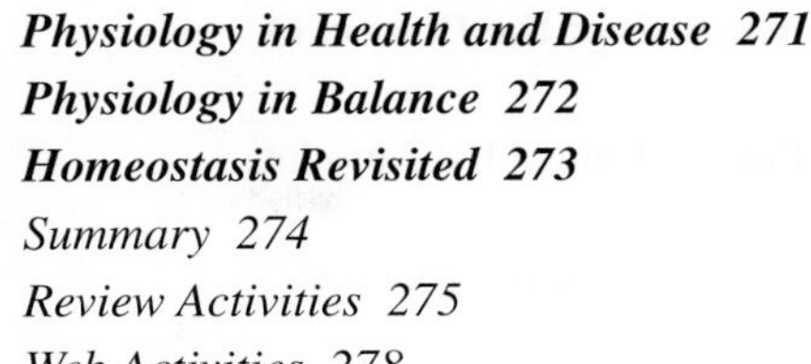

11 Immune System 279

12 Respiratory System 297

Preface

Human physiology is one of the most interesting subjects to study—almost everyone wants to learn some aspects of body function and related medical applications. And most students in a human physiology course are highly motivated to succeed, because the course is prerequisite for entry into many health professions. Despite this, in my years teaching human physiology and communication with others, I've learned that this subject is uniquely challenging for both students and instructors due to the nature of the subject and the varied levels of academic preparation of the students.

Some students take this course early in their college careers, having had few, if any, prerequisites. Others may be biology or physiology majors, may have varied exposures to related fields, or may even be in graduate or professional programs. Students with diverse backgrounds and interests are often in the same classroom, presenting challenges that are heightened by the nature of the subject, which emphasizes an understanding of concepts rather than rote memorization of facts, using terms and concepts from biology, anatomy, chemistry, and physics.

The differences among students are more profound in some colleges than in others, where human physiology courses may be offered at both lower- and upper-division levels. Nevertheless, students who share a common goal—to enter health professions—often find themselves in the same human physiology course despite their varied academic backgrounds. Instructors are set with the difficult task of choosing the most appropriate text for their own course level and classes.

Why Write Another Human Physiology Book?

The first thing you may notice about *Fundamentals of Human Physiology* is that it's shorter than other physiology texts; this was achieved by eliminating material extraneous to the immediate needs of lower-division students. A glance at the contents will then reveal that it's written in a more informal manner than most physiology texts, contains a number of unique, student-friendly features, and minimizes mathematical concepts. Students will find this textbook very readable, interesting, and relevant to their professional goals. It provides the physiological information required for entering nursing and related allied health professions, but offers less depth than larger texts. *Fundamentals of Human Physiology* emphasizes homeostasis and the medical applications of basic physiological concepts. This emphasis is strikingly evident in the many features of the text described in the next section.

"I feel that the author hit the nail on the head in his writing style on a text geared for non-majors. It's simplistic, yet direct and to the point. I think that any college student, regardless of their science background experience, would appreciate this method of composition. Dr. Fox's style is very student friendly."

—Susan Mounce, Eastern Illinois University

What Sets This Book Apart?

Proven Writing Style

Stuart Fox has been teaching and authoring textbooks for over 30 years. His in-the-classroom experience and honed writing skills have put him at the very top of the list of respected and talented authors of human physiology textbooks. Stuart's clear, concise explanations are presented in logical steps and include relevant examples and analogies. He avoids unnecessary chatter and keeps the coverage at precisely the right depth for students just being introduced to this discipline. Reviewers of Stuart's material continuously concur that he is a master writer:

"This will be an excellent textbook for the human physiology course and for pre-med undergraduates. Besides having many innovative and attractive characteristics, this textbook is written in a very easy and attractive way. The author's writing style is definitely one of the book's most exciting features. Chapters two and four read almost like a novel—effortless!"

—Daewoo Lee, Ohio University

"The readability of the fundamentals text is good, clear, and concise. I especially like the explanation of depolarization and repolarization of membranes."

—Linda Collins, University of Tennessee at Chattanooga

"I would give this chapter an A. In terms of organization, depth, and usefulness in a Human Physiology class, this is the finest endocrine physiology chapter that I have ever read."

—Kip McGilliard, Eastern Illinois University

Unique Features

Homeostasis Emphasis

Each chapter opens with a paragraph summary of the relationship of the chapter contents to the overall theme of homeostasis. This theme is flagged with an icon (different for each chapter) of an athletic performance requiring balance, evoking the common function of homeostatic mechanisms: maintaining balance of the internal environment.

> "Homeostasis is absolutely essential as the central theme for understanding physiology. This is an excellent choice as an opening and closing theme for every chapter."
>
> —Jon Hunter, Texas A&M University

Clinical Investigation

A clinical investigation is presented on the same chapter opener page as the homeostasis paragraph, illustrating what can happen if homeostasis is not maintained. The clinical investigation increases student interest in the chapter because it demonstrates the medical relevance of physiological concepts. This is presented as a mystery to be solved, much like the medical mysteries that open many television series and dramas. Students are expected to be baffled by the Clinical Investigation at first, and then pleased as they progress through the chapter and see how their initial confusion is supplanted by knowledge and understanding.

Chemistry Refresher

This textbook assumes no prior scientific knowledge, although most students will have had some previous exposure to these subjects. Thus, the chemistry refresher section in chapter 1 can be used to learn this fundamental information for the first time, to refresh a student's memory of selected information, or as a reference to be used as needed during a student's progress through the textbook.

> "In chapter one, I particularly like the chemistry review, it is succinct and well-defined so that students who need it know where to find it, and those that don't can easily skip it. I was also impressed by Stuart's description of trans fats; I have struggled for years to give my students a cogent explanation and he did it in a few sentences!"
>
> *—Laurel Roberts, University of Pittsburgh*

Clinical Investigation Clues

Clues for understanding the Clinical Investigation in the chapter opener are found within each chapter. These clues are located immediately following physiological information relevant to the investigation, and ask students to use the information they just learned from the text to help solve the medical mystery. This activity provides an enjoyable way for students to engage in active learning and analytical thinking, and thus gain a deeper understanding of physiological concepts.

Pyramid Paragraphs

The occasional pyramid paragraphs (indicated by a pyramid icon) are a unique feature of this textbook. The pyramid concept involves layering advanced knowledge on a foundation of more basic information and concepts. The pyramid paragraphs serve several functions: (1) they signal students that this is material that's particularly fundamental for understanding concepts covered later in this book; (2) they help students realize that basic material is important for their understanding of the subject and thus for their career goals; and (3) they alert students to their progress as they proceed through the course and textbook, providing positive feedback for their studies.

> The description of cardiac muscle in this section will serve as a basis for a later discussion of the nature of the heartbeat and the electrocardiogram, and for the regulation of heart function in the next chapter. Similarly, the description of smooth muscles in this section will provide a basis for aspects of the physiology of the digestive, urinary, and reproductive systems that will be discussed in later chapters. Thus, this section on cardiac and smooth muscle will provide part of a sturdy foundation for your knowledge pyramid in the physiology of the body systems.

> "The pyramid paragraphs are a very good feature. I think students really need this. I do something similar with my lectures. Especially with this discipline, the concepts build one topic onto the next, and I constantly try to remind my students of the cumulative nature of physiology."
>
> *—Todd McBride, California State University*

Clinical Applications

Clinical Applications are boxed paragraphs that provide examples of how the physiological concepts in the immediately preceding text are medically relevant. These help reinforce the physiological concepts by demonstrating their relevance and importance to the career goals of the students.

FYI Footnotes

These footnotes are for fun. FYI stands for "For Your Information," a title that suggests that they're asides, snippets of interesting information that make the textbook more enjoyable to read and the subject of physiology more alive. They present items of historical interest (Where does the name "Ringer" from "Ringer's solutions" come from?) as well as assorted items of general interest (Does a dog see the same colors we see?). Students can take a breather when they get to an FYI footnote; this isn't test information.

Check Points

Following every major section within each chapter are Check Point questions. These provide "reality checks" for students to assess their understanding and memory of the material just read. A student who has difficulty with a particular question can look up the answer in the material immediately preceding the Check Point, and review it again before reading on.

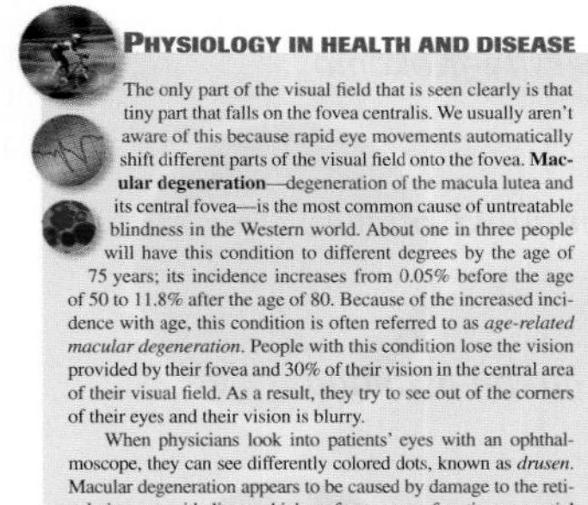

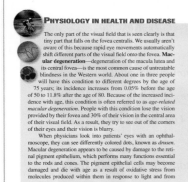

PHYSIOLOGY IN HEALTH AND DISEASE

The only part of the visual field that is seen clearly is that tiny part that falls on the fovea centralis. We usually aren't aware of this because rapid eye movements automatically shift different parts of the visual field onto the fovea. **Macular degeneration**—degeneration of the macula lutea and its central fovea—is the most common cause of untreatable blindness in the Western world. About one in three people will have this condition to different degrees by the age of 75 years; its incidence increases from 0.05% before the age of 50 to 11.8% after the age of 80. Because of the increased incidence with age, this condition is often referred to as *age-related macular degeneration*. People with this condition lose the vision provided by their fovea and 30% of their vision in the central area of their visual field. As a result, they try to see out of the corners of their eyes and their vision is blurry.

When physicians look into patients' eyes with an ophthalmoscope, they can see differently colored dots, known as *drusen*. Macular degeneration appears to be caused by damage to the retinal pigment epithelium, which performs many functions essential to the rods and cones. The pigment epithelial cells may become damaged and die with age as a result of oxidative stress from molecules produced within them in response to light and from smoking (a clear risk factor). Certain genetic mutations have also been associated with an increased risk of macular degeneration. You can't do anything about the genes you inherit, but you can stop smoking, wear protective sunglasses when necessary, and take antioxidants such as zinc, beta carotene, and vitamins C and E.

The deterioration of the retina in people with macular degeneration sometimes causes blood vessels to grow into the retina from the underlying layer called the choroid. The growth of new blood vessels to a tissue is known as **neovascularization** (growth of blood vessels in general is called *angiogenesis*). Although it occurs in only about 10% of people with macular degeneration, this process is responsible for most of the cases of legal blindness caused by this disease. Neovascularization also occurs in cancer and provides increased blood flow to growing tumors. Knowing this, scientists have developed drugs to block the growth of new blood vessels as a cancer treatment. These drugs are antibodies that block the action of *vascular endothelial growth factor* (*VEGF*), a paracrine regulator (chapter 3) that stimulates the growth of blood vessels to the tumor. These antibodies against VEGF, developed as a treatment for cancer, have now also been successfully used to treat the neovascularization that contributes to blindness in macular degeneration.

CHECK POINT

1. Describe the reactions that interchange the phosphate group between ATP and creatine during rest and exercise, and explain the significance of phosphocreatine.
2. When do muscles perform anaerobic respiration? What are the benefits and costs of this metabolic pathway?
3. How do muscles obtain energy at rest and following 2 minutes of moderate exercise? Which molecules are used for this process?
4. Define the oxygen debt and explain the reasons it exists.
5. Distinguish between the different types of skeletal muscle fibers, and explain how these fibers are affected by exercise training.

Physiology in Balance

At the end of each chapter is a page entitled Physiology in Balance, which associates the material covered in the chapter with concepts related to other body systems and other chapters. This feature helps students to understand how the physiological mechanisms in the chapter work together with those of other body systems to maintain homeostasis.

Homeostasis Revisited Tables

Also unique to this text are tables that revisit the physiological mechanisms discussed in the chapter by organizing them around the components of negative feedback loops. This helps students understand how these mechanisms relate to the common theme of homeostasis, thereby fostering a better understanding of physiological processes. These tables, together with the Physiology in Balance pages and the chapter-opening material on homeostasis, provide students with a homeostasis framework for their study of physiology.

Homeostasis Revisited

Topic	Set Point	Integrating Center	Sensors	Effectors
How homeostasis of arterial oxygen and carbon dioxide is maintained by the peripheral chemoreceptors when a person hypoventilates	Plasma pH	Respiratory control center in the medulla oblongata	Aortic and carotid bodies	Somatic motor neurons stimulate the diaphragm and other respiratory muscles to contract, causing breathing.
How homeostasis of arterial oxygen and carbon dioxide is maintained by the central chemoreceptors when a person hypoventilates	pH of brain interstitial fluid	Respiratory control center in the medulla oblongata	Central chemoreceptors in the medulla oblongata	Somatic motor neurons stimulate the diaphragm and other respiratory muscles to contract, causing breathing.
How the Bohr effect helps sustain aerobic respiration by exercising skeletal muscles	Oxygen available for aerobic respiration in muscle fibers.	Red blood cells	Hemoglobin molecules	The pH of exercising skeletal muscles decreases, reducing the affinity of hemoglobin for oxygen, thereby increasing oxygen unloading to the muscle fibers.
How the regulation of breathing operates to partially restore homeostasis of arterial pH when a person has a metabolic acidosis	Arterial pH	Respiratory control center in medulla oblongata	Aortic and carotid body chemoreceptors	Somatic motor neurons to respiratory muscles stimulate hyperventilation, which induces a respiratory alkalosis to partially offset the metabolic acidosis.
How 2,3-BPG in the RBCs helps sustain aerobic respiration when a person goes to a high altitude	Oxygen available for aerobic respiration in muscle fibers.	Red blood cells	Enzyme that promotes production of 2,3-BPG	Reduced oxyhemoglobin stimulates the production of 2,3-BPG, which reduces the affinity of hemoglobin for oxygen, thereby increasing the unloading of oxygen to the tissues.

Review Activities

Each chapter ends with Review Activities, a section divided into three parts. The first part consists of objective questions, with answers in Appendix 3. The second part contains essay questions that test understanding of the physiological concepts presented in the chapter. The third part presents essay questions that require more analytical thinking. Answers to the essay questions are provided in the online *Instructor's Guide* accompanying this textbook.

Acknowledgments

The professors who reviewed this book in manuscript form collectively brought many decades of teaching and research to this task, and for this I wish to thank them and assure readers that any errors or typos that remain are solely my own fault. Also, this book would not exist, nor have the features it does, without the creativity and labor of many people at McGraw-Hill. In particular, I would especially like to thank Kathy Loewenberg (developmental editor), April Southwood (project manager), Colin Wheatley (sponsoring editor), Michelle Whitaker (design coordinator), Lynn Breithaupt (marketing manager), and Michelle Watnick (publisher) for all they have done.

Board of Advisors

Faith Vruggink, *Kellogg Community College*

Laurel Roberts, *University of Pittsburgh*

Sandra Kreiling, *Harper College*

Denise Yordy, *Community College of Rhode Island*

Marie Kelly-Worden, *Ball State University*

Michael Finkler, *Indiana University, Kokomo*

Reviewers

William D. Blaker, *Furman University*

Linda T. Collins, *University of Tennessee–Chattanooga*

J. Helen Cronenberger, *University of Texas–San Antonio*

Jerome Dempsey, *University of Wisconsin–Madison*

Mark R. Eichinger, *Luther College*

Carmen D. Eilertson, *Georgia State University*

Michael S. Finkler, *Indiana University–Kokomo*

Judy Jiang, *Triton College*

Suzanne Kempke, *Ivy Tech Community College*

Sandra Kreiling, *Harper College*

John Leatherland, *University of Guelph–Canada*

Daewoo Lee, *Ohio University*

Todd A. McBride, *California State University–Bakersfield*

Kip L. McGilliard, *Eastern Illinois University*

Jeanne Mitchell, *Truman State University*

Susan E. Mounce, *Eastern Illinois University*

William Reddan, *University of Wisconsin–Madison*

Laurel Roberts, *University of Pittsburgh*

Faith Vruggink, *Kellogg Community College*

Crista Wagner, *California State University–San Marcos*

R. Douglas Watson, *University of Alabama–Birmingham*

Denise M. Yordy, *Community College of Rhode Island*

Teaching and Learning Supplements

McGraw-Hill offers various tools and technology products to support *Fundamentals of Human Physiology*. Instructors can obtain teaching aids by calling the McGraw-Hill Customer Service Department at 1–800–338–3987, visiting our Anatomy & Physiology catalog at www.mhhe.com/ap or contacting their local McGraw-Hill sales representative.

ARIS Text Website

Instructors will find a complete electronic homework and course management system on the ARIS website that accompanies this textbook. They can create and share course materials with colleagues, edit questions, import their own content, and create announcements and/or due dates for assignments—all with just a few clicks of the mouse. ARIS also offers automatic grading and reporting of easy-to-assign homework, quizzing, and testing.

Students will appreciate the animations, practice quizzes, case studies, downloadable content for portable players, helpful Internet links and more—a whole semester's worth of study help!

Visit www.mhhe.com/foxfundamentals and start benefiting today!

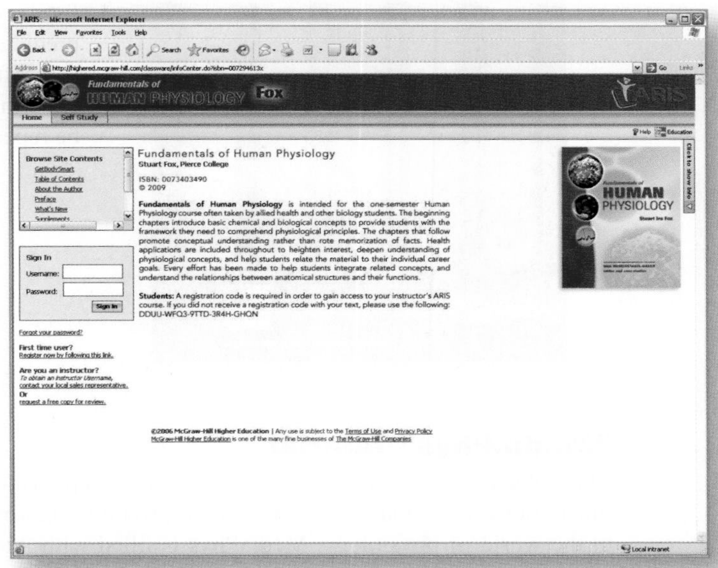

Complete Set of Electronic Images and Assets for Instructors

Instructors, build instructional materials wherever, whenever, and however you want!

Part of the ARIS website, the digital library contains assets such as photos, artwork, animations, PowerPoints, and other media resources that can be used to create customized lectures, visually enhance tests and quizzes, and design compelling course websites or attractive printed support materials. All assets are copyrighted by McGraw-Hill Higher Education but can be used by instructors for classroom purposes. The visual resources in this collection include:

- **Art** Full-color digital files of all illustrations in the book can be readily incorporated into lecture presentations, exams, or custom-made classroom materials. In addition, all files are pre-inserted into blank PowerPoint slides for ease of lecture preparation.
- **Photos** The photo collection contains digital files of photographs from the text, which can be reproduced for multiple classroom uses.
- **Tables** Every table in the text has been saved in electronic form for use in classroom presentations and/or quizzes.
- **Animations** Numerous full-color animations illustrating important physiological processes are also provided. Harness the visual impact of concepts in motion by importing these files into classroom presentations or online course materials.
- **Lecture Outlines** Specially prepared custom outlines for each chapter are offered in easy-to-use PowerPoint slides.

Onine Student Study Guide

Written by an experienced instructor who teaches students with a broad range of preparation, this study guide uses a unique format to present questions. Bypassing the typical multiple-choice or true/false styles, these intersting questions are worded to provoke active thought and yet are still easy to grade. Each chapter also includes a fun corssword puzzle to aid in retention of terminology and concepts.

Computerized Test Bank Online

A comprehensive bank of test questions is provided within a computerized test bank powered by McGraw-Hill's flexible electronic testing program EZ Test Online. EZ Test Online allows instructors to create and access paper or online tests and quizzes in an easy to use program anywhere, at any time without installing the testing software. Now, with EZ Test Online, instructors can select questions from multiple McGraw-Hill test banks or author their own, and then either print the test for paper distribution or administer the test online. Visit www.eztestonline.com to learn more about creating and managing tests, online scoring and reporting, and support resources.

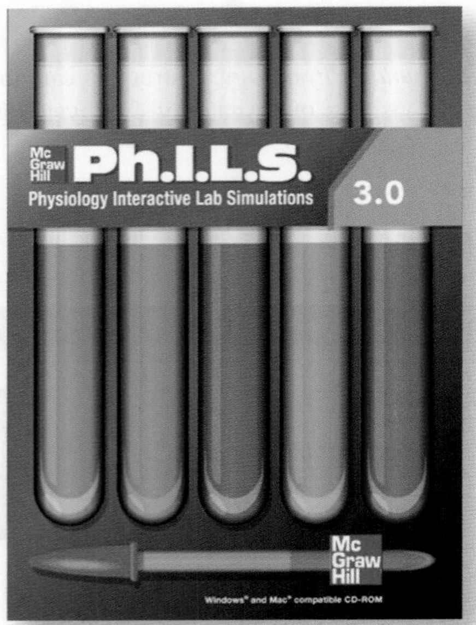

Physiology Interactive Lab Simulations (Ph.I.L.S.) 3.0

This unique student study tool is the perfect way to reinforce key physiology concepts with powerful lab experiments. Created by Dr. Phil Stephens of Villanova University, the program offers 37 laboratory simulations that may be used to supplement or substitute for wet labs. Students can adjust variables, view outcomes, make predictions, draw conclusions, and print lab reports. The easy-to-use software offers the flexibility to change the parameters of the lab experiment—there is no limit to the amount of times an experiment can be repeated.

"MediaPhys" Tutorial

This physiology study aid offers detailed explanations, high-quality illustrations, and amazing animations to provide a thorough introduction to the world of physiology. MediaPhys is filled with interactive activities and quizzes to help reinforce physiology concepts that are often difficult to understand.

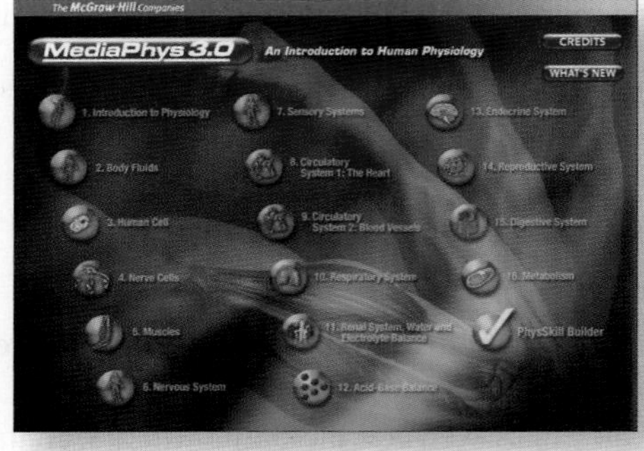

Anatomy & Physiology | REVEALED® 2.0

Anatomy & Physiology | REVEALED® 2.0 is a unique multimedia study aid designed to help students learn and review human anatomy using digital cadaver specimens. Dissections, animatins, imaging, and self-tests all work together as an exceptional tool for the study of structure and function. Version 2.0 now includes all body systems, histology material, and expanded physiology content. Visit www.mhhe.com/aprevealed for more information.

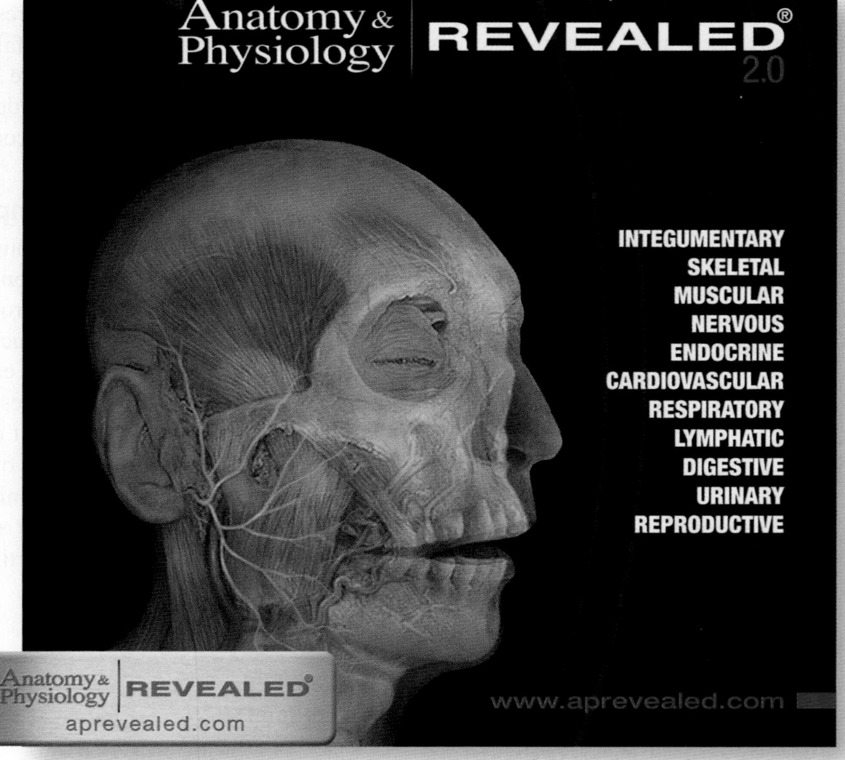

Electronic Books

If you, or your students, are ready for an alternative version of the traditional textbook, McGraw-Hill and VitalSource have partnered to introduce innovative and inexpensive electronic textbooks. By purchasing E-books from McGraw-Hill and VitalSource, students can save as much as 50% on selected titles delivered on the ***most advanced*** E-book platform available, VitalSource Bookshelf.

E-books from McGraw-Hill and VitalSource are *smart, interactive, searchable, and portable*. VitalSource Bookshelf comes with a powerful suite of built-in tools that allow detailed searching, highlighting, note taking, and student-to-student or instructor-to-student note sharing. In addition, the media-rich E-book for *Fundamentals of Human Physiology* integrates relevant animations and videos into the textbook content for a true multimedia learning experience.

E-books from McGraw-Hill and VitalSource will help students study smarter and quickly find the information they need while saving money. Instructors: contact your McGraw-Hill sales representative to discuss E-book packaging options.

e-Instruction with CPS

The Classroom Performance System (CPS) is an interactive system that allows the instructor to administer in-class questions electronically. Students answer questions via hand-held remote control keypads (clickers), and their individual responses are logged into a grade book. Aggregated responses can be displayed in graphical form. Using this immediate feedback, the instructor can quickly determine if students understand the lecture topic, or if more clarification is needed. CPS promotes student participation, class productivity, and individual student confidence and accountability. Specially designed questions for e-Instruction to accompany *Fundamentals of Human Physiology* are provided through the book's ARIS website.

Course Delivery Systems

In addition to McGraw-Hill's ARIS course management options, instructors can also design and control their course content with help from our partners WebCT, Blackboard, Top-Class, and eCollege. Course cartridges containing website content, online testing, and powerful student tracking features are readily available for use within these or any other HTML-based course management platforms.

Guided Tour

A Focus on Homeostasis Opens and Closes Every Chapter

Each chapter of *Fundamentals of Human Physiology* opens with a reader-friendly overview of the relationship of homeostasis to the chapter content. The introduction describes the mechanisms needed to maintain the balance of homeostasis.

But if homeostasis is not maintained . . .

Clinical Investigation

A Clinical Investigation is also presented on the chapter-opening page and illustrates what can happen if homeostasis does not remain in balance. The case study demonstrates the medical relevance of physiological concepts and is described as a puzzle to be solved, with clues coming later in the chapter.

Clinical Investigation Clues are carefully placed next to content relevant to the opening case study. This educational activity provides an opportunity for students to engage in active learning and analytical thinking.

Homeostasis Revisited Tables

Fundamentals of Human Physiology is the first physiology text to use tables that organize the physiological mechanism discussed in the chapter around the components of negative feedback loops. This approach helps students understand how these mechanisms relate to the common theme of homeostasis.

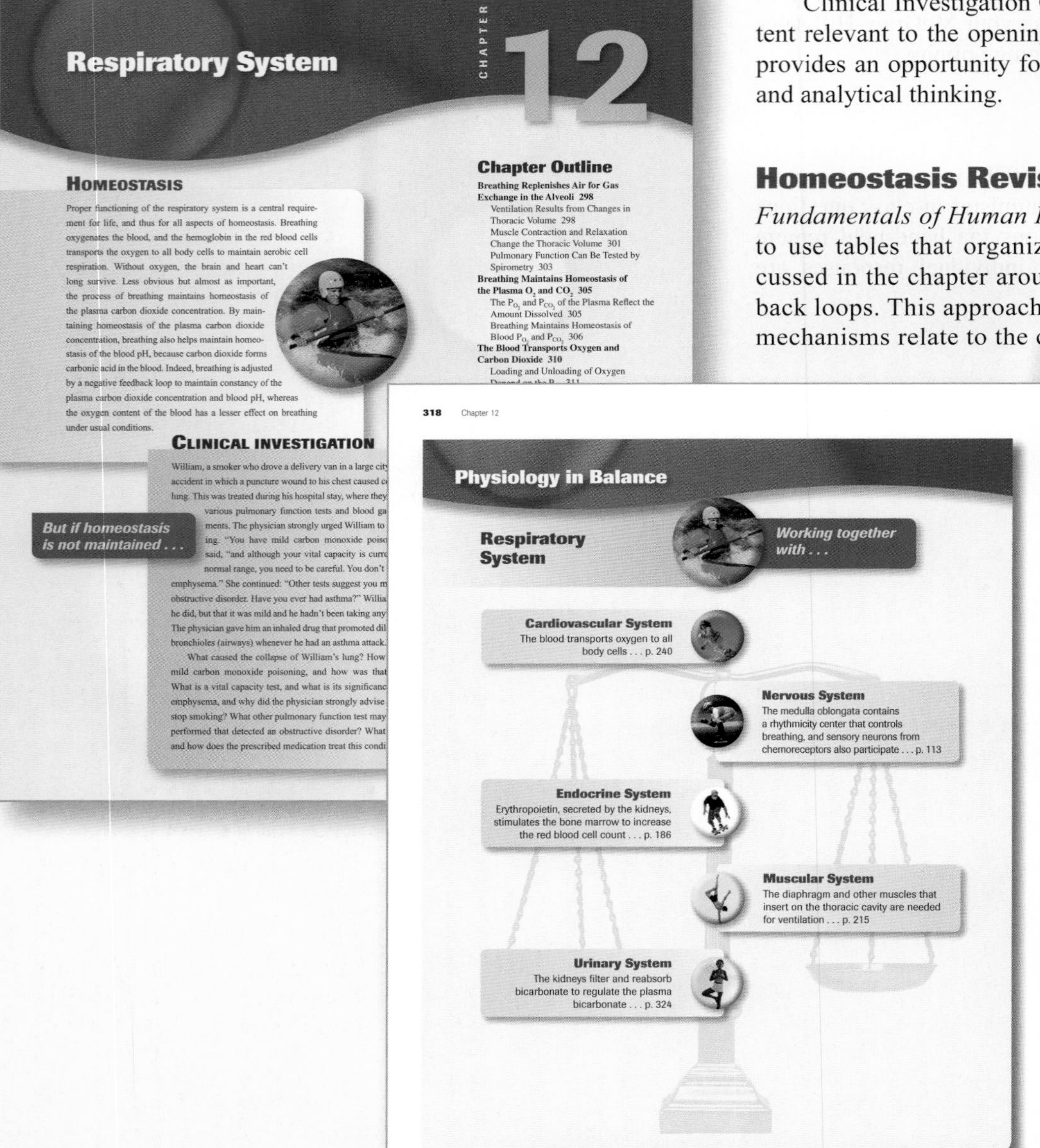

Respiratory System

CHAPTER 12

HOMEOSTASIS

Proper functioning of the respiratory system is a central requirement for life, and thus for all aspects of homeostasis. Breathing oxygenates the blood, and the hemoglobin in the red blood cells transports the oxygen to all body cells to maintain aerobic cell respiration. Without oxygen, the brain and heart can't long survive. Less obvious but almost as important, the process of breathing maintains homeostasis of the plasma carbon dioxide concentration. By maintaining homeostasis of the plasma carbon dioxide concentration, breathing also helps maintain homeostasis of the blood pH, because carbon dioxide forms carbonic acid in the blood. Indeed, breathing is adjusted by a negative feedback loop to maintain constancy of the plasma carbon dioxide concentration and blood pH, whereas the oxygen content of the blood has a lesser effect on breathing under usual conditions.

Chapter Outline

CLINICAL INVESTIGATION

William, a smoker who drove a delivery van in a large cit[y] ... accident in which a puncture wound to his chest caused c[...] lung. This was treated during his hospital stay, where they ... various pulmonary function tests and blood ga[...] ments. The physician strongly urged William to ... ing. "You have mild carbon monoxide poiso[...] said, "and although your vital capacity is curr[...] normal range, you need to be careful. You don't ... emphysema." She continued: "Other tests suggest you m[...] obstructive disorder. Have you ever had asthma?" Willia[...] he did, but that it was mild and he hadn't been taking any ... The physician gave him an inhaled drug that promoted dil[...] bronchioles (airways) whenever he had an asthma attack.

What caused the collapse of William's lung? How ... mild carbon monoxide poisoning, and how was that ... What is a vital capacity test, and what is its significanc[e] ... emphysema, and why did the physician strongly advise ... stop smoking? What other pulmonary function test may ... performed that detected an obstructive disorder? What ... and how does the prescribed medication treat this condi[...]

Physiology in Balance

Respiratory System

Working together with . . .

Cardiovascular System
The blood transports oxygen to all body cells . . . p. 240

Nervous System
The medulla oblongata contains a rhythmicity center that controls breathing, and sensory neurons from chemoreceptors also participate . . . p. 113

Endocrine System
Erythropoietin, secreted by the kidneys, stimulates the bone marrow to increase the red blood cell count . . . p. 186

Muscular System
The diaphragm and other muscles that insert on the thoracic cavity are needed for ventilation . . . p. 215

Urinary System
The kidneys filter and reabsorb bicarbonate to regulate the plasma bicarbonate . . . p. 324

Physiology in Balance

At the end of each chapter is a page entitled Physiology in Balance, which associates the material covered in the chapter with concepts related to other body systems and other chapters. This feature helps students to understand how the physiological mechanisms in that chapter work together with those of other body systems to maintain homeostasis.

Truly Innovative Pedagogy Based On Years of Teaching Experience

Pyramid Paragraphs

The pyramid concept involves layering advanced knowledge on a foundation of more basic information and concepts. The Pyramid Paragraphs alert students to material that's fundamental for understanding concepts covered later in the text and also serve as excellent study tools. Written in a "coaching" style, the paragraphs offer positive feedback to the readers, while alerting them to important textual content.

"I think the Chemistry Refresher is excellent! Dr. Fox approaches chemistry in exactly the same conceptual frame that I use. It's almost like he read my mind. The chapter on Cells is also very clear and presented in an engaging fashion. The illustrations are clean, simple, and not cluttered with superfluous labels. I really think this is a wonderful textbook."

—Faith Vruggink, Kellogg Community College

Chemistry Refresher

Unique to all human physiology textbooks, the Chemistry Refresher in chapter 1 can be used to learn fundamental information for the first time, to refresh a student's memory of selected information, or as a reference during a student's progression through the textbook.

Chemistry Refresher

Basic Chemical Concepts Used in Physiology

The study of physiology utilizes some concepts and terminology from chemistry. This is needed in order to explain much of cellular function, which is the basis of the physiology of organs and systems. The major types of chemical bonds, together with acids, bases, and the pH scale, are chemical concepts that are particularly fundamental to an understanding of body function.

Physiology is a science that can't be separated from anatomy, since structure and function are clearly interrelated. Also, we can't talk about how the body works without using some basic terminology and concepts of chemistry and biology. You may have had courses in these subjects; if so, you're ahead of the game. Even if you've had recent courses in these subjects, chances are that you could use a little refresher. Perhaps you can skip this section for now, returning to aspects of it as the need arises later. Feel free to use this section as you (or your instructor) think best.

An **atom** is the smallest unit of the chemical elements. The **nucleus** at the center of each atom contains two types of particles—**protons**, which are positively charged, and **neutrons**, which have mass but are not charged. The mass of a proton is equal to the mass of a neutron, and the sum of the protons and neutrons in an atom is equal to the **atomic mass** of the atom. Carbon, for example, has 6 neutrons and 6 protons in its nucleus, and so has an atomic mass of 12.

The number of protons in the nucleus is the **atomic number** of the atom. For example, carbon has 6 protons and so has an atomic number of 6. Outside of the atomic nucleus are negatively charged particles called **electrons**. The number of electrons in an atom is equal to the number of protons in its nucleus, and so the positive and negative charges are balanced. An *isotope* of an atom has the same atomic number but a different number of neutrons, and thus a different atomic mass. The different isotopes of a particular atom constitute a *chemical element*. Hydrogen, for example, is an element that contains 3 isotopes: an atom with only a proton in the nucleus (the most common isotope), an atom with 1 proton and 1 neutron, and a radioactive isotope (called *tritium*) that has 1 proton and 2 neutrons.

Although diagrams of atoms often show electrons as particles orbiting the nucleus, like planets orbiting the sun, this mental picture isn't quite accurate. An electron can occupy any position within a particular energy "shell." Each shell can hold up to a maximum number of electrons, and the shells are filled from the innermost shell outward. The first shell, closest to the nucleus, can contain up to 2 electrons. If an atom has more than 2 electrons (as they all do, except hydrogen and helium), the additional electrons must be located in shells farther from the nucleus. The second shell can contain a maximum of 8 electrons. A carbon atom (with an atomic number of 6) has 2 electrons in the first shell and 4 electrons in the second shell (fig. 1.6).

A carbon atom would need 4 more electrons to reach the maximum of 8 electrons for its outer shell. As we will soon see, this allows the carbon atom to share electrons with other atoms as it forms chemical bonds with them. The electrons in the outermost shell are known as **valence electrons**, and it is always the valence electrons that participate in chemical reactions.

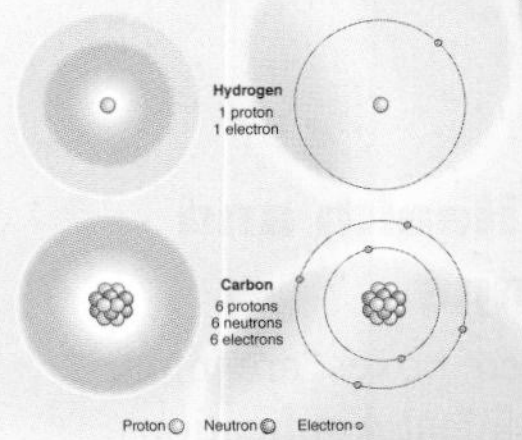

FIGURE 1.6 Diagrams of the hydrogen and carbon atoms. The electron shells on the left are represented by shaded spheres indicating probable positions of the electrons. The shells on the right are represented by concentric circles.

7

Check Points

At the end of every major heading within each chapter are Check Point questions. These provide "reality checks" for students to assess their understanding and memory of the material they just read. If they have difficulty with a particular question, they can look up the answer in the material immediately preceding the Check Point, and review it again before reading on.

CHECK POINT

1. Describe the reactions that interchange the phosphate group between ATP and creatine during rest and exercise, and explain the significance of phosphocreatine.
2. When do muscles perform anaerobic respiration? What are the benefits and costs of this metabolic pathway?
3. How do muscles obtain energy at rest and following 2 minutes of moderate exercise? Which molecules are used for this process?
4. Define the oxygen debt and explain the reasons it exists.
5. Distinguish between the different types of skeletal muscle fibers, and explain how these fibers are affected by exercise training.

"I do think my students will benefit from a refresher on chemistry. Many of them are taking chemistry concurrently with physiology; some come in completely without any background in chemistry."

—Denise Yordy, Community College of Rhode Island

Guided Tour

FYI Footnotes

FYI stands for "For Your Information," a title that suggests that these footnotes are asides—snippets of interesting information that make the textbook more enjoyable to read and the subject of physiology more fun. They include items of historical interest (Where does the name "Ringer" from "Ringer's solutions" come from?) as well as items of general interest (Does a dog see the same colors we see?).

FYI [5]Scientists have recently shown that there are stem cells near the middle of a hair follicle, located in a small bulge. These can divide to produce cells that migrate down to the region of the follicle that generates the hair shaft; thus, absence of those stem cells produces baldness. Also, there are stem cells here that differentiate into melanocytes, which are the cells that give hair its color. Graying of hair results from the loss of these melanocyte stem cells.

Clinical Content Adds Relevance

Clinical Applications

Clinical Applications reinforce the physiological concepts in the immediately preceding text by demonstrating their relevance and importance to the students' career goals.

CLINICAL APPLICATIONS

The **appendix** is a short, thin outpouching from the cecum (fig. 14.13). It doesn't appear to have a digestive function, but contains numerous lymphatic nodules that may offer some immune protection. However, it's most famous because it can become dangerously inflamed, a condition called **appendicitis.** This is often detected in its later stages by pain in the lower right quadrant of the abdomen. If the appendix ruptures, the infection can spread throughout the abdominal cavity, causing inflammation of the *peritoneum* (the membranes that line the abdominal cavity and cover its organs), or *peritonitis*. This dangerous event can be prevented by an *appendectomy*, which is the surgical removal of the appendix.

Physiology in Health and Disease readings

These clinically based boxes are more extensive than the Clinical Applications, and generally involve medical issues that encompass topics and concepts covered in several sections of the chapter.

PHYSIOLOGY IN HEALTH AND DISEASE

The only part of the visual field that is seen clearly is that tiny part that falls on the fovea centralis. We usually aren't aware of this because rapid eye movements automatically shift different parts of the visual field onto the fovea. **Macular degeneration**—degeneration of the macula lutea and its central fovea—is the most common cause of untreatable blindness in the Western world. About one in three people will have this condition to different degrees by the age of 75 years; its incidence increases from 0.05% before the age of 50 to 11.8% after the age of 80. Because of the increased incidence with age, this condition is often referred to as *age-related macular degeneration*. People with this condition lose the vision provided by their fovea and 30% of their vision in the central area of their visual field. As a result, they try to see out of the corners of their eyes and their vision is blurry.

When physicians look into patients' eyes with an ophthalmoscope, they can see differently colored dots, known as *drusen*. Macular degeneration appears to be caused by damage to the retinal pigment epithelium, which performs many functions essential to the rods and cones. The pigment epithelial cells may become damaged and die with age as a result of oxidative stress from molecules produced within them in response to light and from smoking (a clear risk factor). Certain genetic mutations have also been associated with an increased risk of macular degeneration. You can't do anything about the genes you inherit, but you can stop smoking, wear protective sunglasses when necessary, and take antioxidants such as zinc, beta carotene, and vitamins C and E.

The deterioration of the retina in people with macular degeneration sometimes causes blood vessels to grow into the retina from the underlying layer called the choroid. The growth of new blood vessels to a tissue is known as **neovascularization** (growth of blood vessels in general is called *angiogenesis*). Although it occurs in only about 10% of people with macular degeneration, this process is responsible for most of the cases of legal blindness caused by this disease. Neovascularization also occurs in cancer and provides increased blood flow to growing tumors. Knowing this, scientists have developed drugs to block the growth of new blood vessels as a cancer treatment. These drugs are antibodies that block the action of *vascular endothelial growth factor* (*VEGF*), a paracrine regulator (chapter 3) that stimulates the growth of blood vessels to the tumor. These antibodies against VEGF, developed as a treatment for cancer, have now also been successfully used to treat the neovascularization that contributes to blindness in macular degeneration.

Vivid Art Program Clarifies Concepts

Stepped-Out Figures

Many of the figures in *Fundamentals of Human Physiology* present stepped-out views to aid student comprehension by focusing attention on one part of the figure at a time, in the correct sequence.

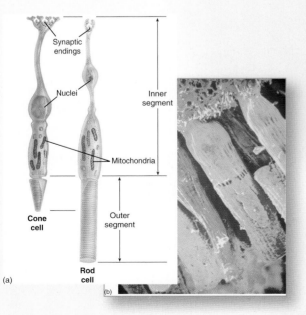

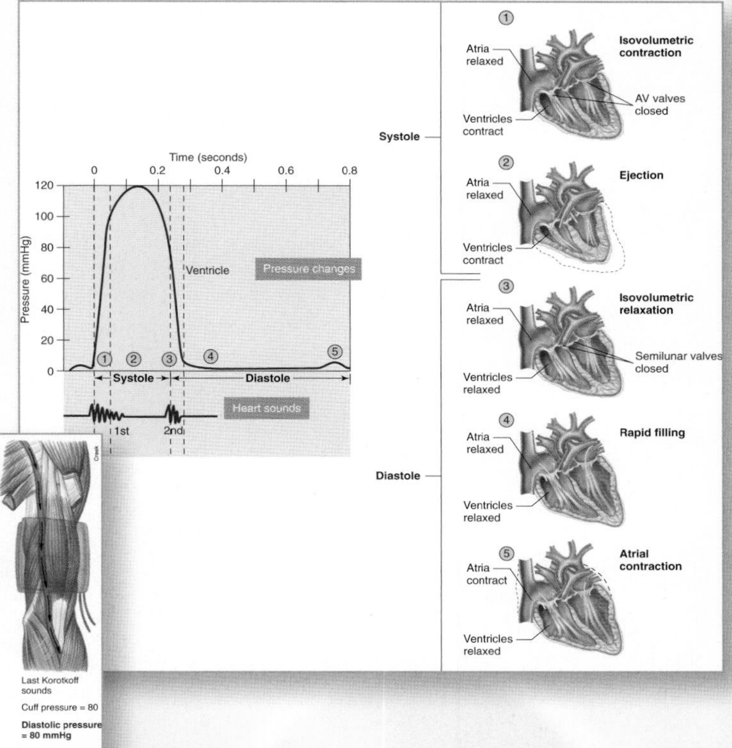

No sounds
Cuff pressure = 140
First Korotkoff sounds
Cuff pressure = 120
Systolic pressure = 120 mmHg
Sounds at every systole
Cuff pressure = 100
Last Korotkoff sounds
Cuff pressure = 80
Diastolic pressure = 80 mmHg
Blood pressure = 120/80

Micrographs Paired with Illustrations

Art renderings and scanning electron micrographs each bring different advantages to an art program, which is why it's best to display both side-by-side when possible. *Fundamentals of Human Physiology* often uses this approach to allow study of labeled detail in comparison with actual depictions.

Synaptic endings
Nuclei
Inner segment
Mitochondria
Outer segment
Cone cell
Rod cell
(a)
(b)

Macroscopic to Microscopic Views

Progressive views of key structures emphasize relationships and help students visualize the figure as a whole.

Consistent Use of Color

Throughout the entire art program of *Fundamentals of Human Physiology*, key structures are beautifully rendered in consistent colors, making it easier for students to interpret each figure while improving the clarity of the overall presentation.

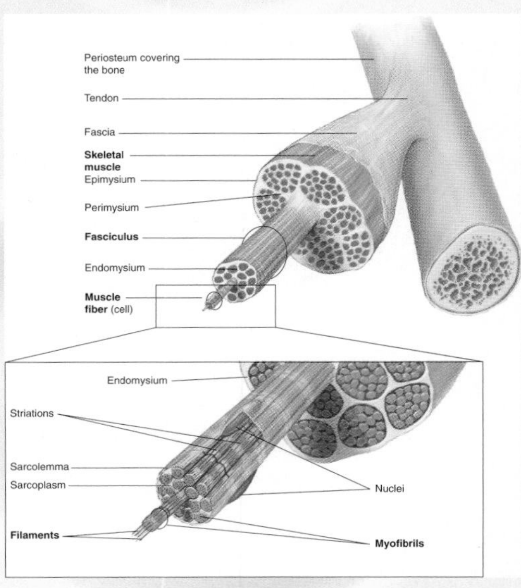

CHAPTER 1

Introduction to Human Physiology

Chapter Outline

Homeostasis

Most of the properties of the internal environment of the body—its temperature, the concentration of molecules in the blood, the blood pressure, and many others—are kept in a state of dynamic constancy. This is called homeostasis, and it is maintained by regulatory mechanisms that operate by negative feedback mechanisms. In these, sensors detect changes and effectors work to counteract those changes. Most negative feedback corrections involve the nervous and endocrine systems.

Carbohydrates, lipids, proteins, and nucleic acids perform many important physiological functions. Carbohydrates and lipids are the major sources of energy for the body cells. Proteins can be used for energy, but they serve many other, highly specialized, functions. Larger forms of these molecules are produced from their smaller subunits and serve for energy storage. The larger molecules, in turn, can be broken down into their subunits for the energy requirements of the body cells.

Clinical Investigation

But if homeostasis is not maintained . . .

Laura decided to lose weight quickly and went on a very strict diet that was low in carbohydrates and calories and emphasized eating protein and fat. She succeeded in losing weight in her first four weeks, but noticed that she felt weak, lacked energy, and seemed to be thirsty all the time. She went to a doctor, who told her to check her urine with a dipstick test for ketone bodies. The physician suggested that Laura add some complex carbohydrates (polysaccharides) to her diet.

How did the diet promote a reduction in body fat? What are ketone bodies, and why did the physician ask Laura to check her urine for them? What might account for her weakness and thirst? What are polysaccharides, and how would eating them help relieve her symptoms?

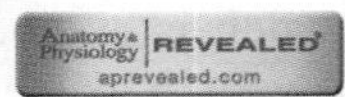

A virtual cadaver dissection experience

The Science of Physiology

Human physiology is the scientific study of normal body function. Scientists apply the scientific method to learn the cause-and-effect mechanisms of how cells, tissues, organs, and body systems work when they are in a healthy state. An understanding of physiological processes and mechanisms is basic to all health professions.

You are constructing a pyramid of new knowledge and understanding. The bottom layer supports all of the other layers above it, so it's critical that this foundation be as strong as you can make it. When studying a science, the bottom layer, the foundation of the pyramid you are constructing, is the "basics." Here, the term *basic* does not mean "simple"; the term refers to the foundational nature of the information. Actually, the "basics" are often more demanding to learn than the more advanced information. In constructing your pyramid of physiological knowledge and understanding, your effort to build a strong foundation will more than pay for itself as you add layers, and as you use this knowledge and understanding throughout your student and professional life.

What Is Physiology?

Think of a theatrical production, perhaps a Shakespearean play or a movie. If you memorize only the names of the parts, their backgrounds, and the way each relates to the others, you would be doing something akin to studying anatomy. If you then watch the play or movie to see what each part actually says and does, and how each interacts with the others in the service of the play as a whole, you would be doing something more akin to studying physiology. This analogy is a little unfair to anatomy, in that the study of anatomy involves so many parts (the names of bones, muscles, vessels, and so on) that it is often best to study it separately from physiology. However, the study of physiology is the payoff for the work of learning the anatomical parts, because it is in the study of physiology that you learn what the parts do and how they interact in the service of the entire organism—you.

Physiology is the study of normal biological function, from cell to tissue, tissue to organ, and organ to system, as well as how the organism as a whole accomplishes particular tasks essential to life. In the study of physiology, the emphasis is on *mechanisms*—on exactly how a particular function is performed. For example, how is the heartbeat produced, and how is the heart rate regulated? How do we digest dietary fat? How is the blood sugar concentration regulated? The answers to these questions contain cause-and-effect sequences. Thus, in this physiology textbook you will see many flowcharts (with arrows showing cause and effect) and numbered lists of the steps of particular processes.

Human physiology is a part of the more general science of mammalian physiology, which is a subcategory of vertebrate physiology. The physiology of both vertebrates and invertebrates is part of the more general science of animal physiology. The study of how particular body functions are accomplished in different vertebrate and invertebrate groups is known as comparative physiology.

The physiological mechanisms that we study in human physiology are generally not unique to humans, and may not even be unique to mammals in general. Indeed, they were always discovered first in nonhuman mammals (such as mice) or other vertebrates (such as fish, frogs, or reptiles), or perhaps even in more distantly related organisms. Since it is much more difficult to study physiological mechanisms in humans (for ethical reasons), we often do not know how these mechanisms work in humans as well as we know how they work in mice or laboratory rats. Despite this, we must do our best to understand how physiological mechanisms operate in humans.

Why is the study of human physiology so important? Although it is certainly interesting and important in its own right, the study of human physiology gains special significance because it is the major scientific foundation of medicine and other health applications. This relationship is recognized by the Nobel Prize committee, which awards prizes in a category called "Physiology or Medicine." We must understand how a cell, tissue, organ, or system normally functions (its normal physiology) before we can understand how the physiological processes become altered in disease or injury, which is a related science known as *pathophysiology*.

How Should You Study Physiology?

The science of physiology is not a recital of facts, but an activity that people do (usually in laboratories) that generates observations regarding the mechanisms of body function. These observations are generally presented in textbooks as facts, but scientists understand that they are only as valid as the data on which they are based, and that they are always subject to modifications as a result of future experiments. However, a student needs to become familiar with our current understanding of human physiology in order to understand future advances, and in order to understand present and future health applications of physiology.

Why are you studying physiology? Odds are, it's because you have to. Most students who take lower-division college courses in human physiology do so because it's a prerequisite for entering heath-related professions. Actually, this course is perhaps the most important prerequisite: professional schools have found that success in a human physiology course is a very good predictor for success in the health profession curriculum and practice. This should provide you with a strong motivation for performing well in the human physiology course.

So, what are good techniques for studying human physiology? If you look around in the classroom, you'll see students who are active participants in the class. They ask relevant questions during the class, see the professor during office hours, rewrite (or type) their lecture notes between classes, and so on. Rewriting and reorganizing lecture notes, perhaps incorporating additional information from the textbook readings, is a particularly effective study technique. Students who are active learners, who take ownership of their learning experience, generally do well in any course. However, this is particularly true in a human physiology course, where the emphasis is on knowledge of processes and understanding of concepts.

Although there is much new information to learn, you'll do best if you don't try to memorize facts, but rather try to understand the steps in cause-and-effect processes. Physiology is the study of *how* something happens—how your heart rate increases when you stand up quickly, for example, or how your blood sugar goes back down after you eat a candy bar. These are like stories that you are asked to tell from memory. Just as you wouldn't try to memorize a story sentence by sentence, you shouldn't try to memorize a physiological process by memorizing the facts in isolation. Instead, fit the steps of a process together in a flowchart or outline form so that they tell a story.

Here is a good reality check: if you can't explain a physiological process to someone else, you don't understand it. If you don't have a study partner, an imaginary one will do. However, it's best to also tell the story in writing. That's the function of essay questions. There are essay-type questions at the end of every major heading in the chapters of this book, to allow you to test yourself before you continue. You may not want to write out the answers to all of these, but it would benefit you to write out those that most pertain to the emphasis of your class. This writing, together with the rewriting and reorganizing of your lecture notes, can help to make you a more active, and thus a more successful, student.

There are other aids for studying provided by this textbook. Review activities at the end of each chapter provide ways to test yourself by answering both objective and essay questions. Also, the website for this book (www.mhhe.com/foxfundamentals) offers a student study guide and many additional useful resources. However, your active participation in learning the lecture and textbook should remain your primary study emphasis; other resources are supplemental.

Above all, have fun! Human physiology is the study of how your body works: what could be more interesting and practical? Apply what you learn in class to aspects of your life, watch how your understanding of medical TV shows improves, and help your relatives and friends better understand their own bodies and medical conditions. Your human physiology course may not only be the most important prerequisite for your intended health profession, it is also one of the most interesting and enjoyable subjects anyone could have the good fortune to study!

CHECK POINT

1. What is physiology? What specifically are human physiology and pathophysiology?
2. Why study physiology? What are the best techniques for studying human physiology?

The Theme of Physiology Is Homeostasis

The physiological processes and regulatory mechanisms that we study in physiology exist for a common purpose: maintaining homeostasis, or constancy of the body's internal environment. When changes start to occur, sensory information evokes physiological responses that act to defend the internal environment against the changes. Thus, much of physiology involves the study of physiological mechanisms that maintain homeostasis.

When you study anatomy, you learn the structure of tissues, how the tissues form organs, and how organs compose the body systems. You can learn these facts by making lists, constructing study cards, and using other techniques that aid the memorization of a large number of facts. However, when you study physiology, you are not so much memorizing the parts as learning what the parts do and—very importantly—the purpose of their actions. When you understand the "why" as well as the "how," the study of physiology becomes easier and much more meaningful. And, amazing as it may seem, the "purpose" of almost any particular physiological process—the "why" of the process—can usually be stated in two words: maintaining homeostasis.

Homeostasis may be defined as the *dynamic constancy of the internal environment*. The terms *dynamic* and *constant* seem contradictory. However, this definition of homeostasis refers to the active, dynamic physiological control processes that must fight changes in order to maintain relatively constant conditions within our body. The internal environment here refers to such conditions as deep body temperature, blood volume and the concentrations of different molecules, blood pressure, and many others.

The study of physiology is largely a process of answering "how" questions, such as, How is the heart rate regulated? As we delve into a particular physiological process, which can comprise many steps and involve several different organs or systems, we might lose our perspective. It's much easier to keep our bearings if we remember the theme of homeostasis. For example, one nerve (a parasympathetic nerve) slows the heart rate and another nerve (a sympathetic nerve) increases the heart rate. If we remember the theme of homeostasis, then it becomes clear that the heart rate "should" increase in response to a fall in blood pressure (because a faster heart rate raises the blood pressure, thereby defending against a fall in blood pressure). The theme of homeostasis answers the "why" question: Why should the sympathetic nerve be stimulated, and the parasympathetic nerve be inhibited, when blood pressure falls? The specific physiological responses make sense: we don't have to memorize each change by rote.

CLINICAL APPLICATIONS

When we go to a physician for a physical exam, the nurse or physician will sample our internal environment to see if we are maintaining homeostasis. They usually measure our internal body temperature, then our blood pressure, and they may order various blood tests. They expect the body temperature, blood pressure, and measurements of the blood concentration of various molecules to be within a normal range. If one or more of these measurements is outside of the normal range, it indicates that homeostasis is not being maintained, and we are not healthy. The results of tests such as these, combined with clinical observations, may allow medical personnel to determine which physiological mechanisms are not operating properly.

Homeostasis Is Maintained by Negative Feedback Loops

Suppose you have a room thermostat that you set at 70° F—this is the **set point**. If the room gets sufficiently cooler than this, a **sensor** will detect that deviation from the set point. The sensor will indirectly activate a heater, which has the effect of correcting this deviation from the set point. The heater thus serves as an **effector** in this mechanism for defending the set point. If you had a fancier room thermostat that controlled an air conditioner as well as a heater, the sensor might send its information to an **integrating center**, which in turn controls both effectors—the heater and the air conditioner. In that case, the air conditioner would turn off and the heater would turn on to correct a fall in room temperature from the set point. The opposite would occur if the room got significantly hotter than the set point.

The relationship just described, where a sensor is activated by a deviation from a set point and an effector responds to oppose that deviation, is called a **negative feedback loop**. "Negative" describes the action of effectors to oppose the deviation. "Loop" describes the circular nature of the cause-and-effect sequence; the effector "feeds back" on the initial deviation (the stimulus, which is the first step in the sequence) to oppose it and help maintain homeostasis. In this textbook, the last step in a negative feedback loop will be shown by a dashed arrow with a negative sign to indicate the opposing action of the effectors on the initial stimulus (fig. 1.1, *top*). The deviated measurement and its negative feedback correction can be charted over time as shown in figure 1.1, *bottom*. It's important to also realize that a deviation in the opposite direction must also be corrected by a negative feedback loop in order to maintain homeostasis (fig. 1.2).

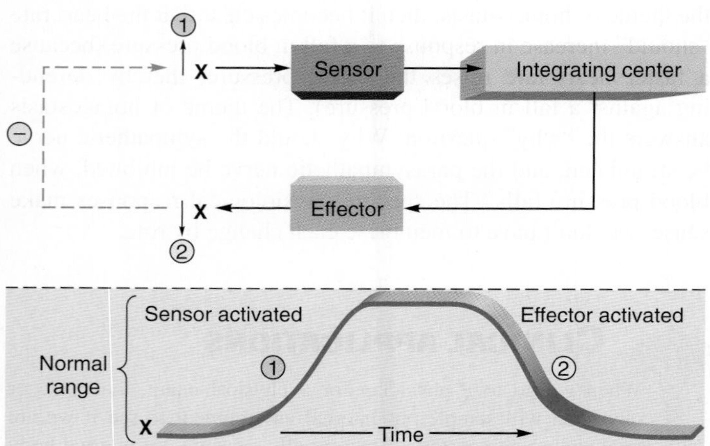

FIGURE 1.1 A rise in some factor of the internal environment (↑X) is detected by a sensor. This information is relayed to an integrating center, which causes an effector to produce a change in the opposite direction (↓X). The initial deviation is thus reversed, completing a negative feedback loop (shown by the dashed arrow and negative sign). The numbers indicate the sequence of changes.

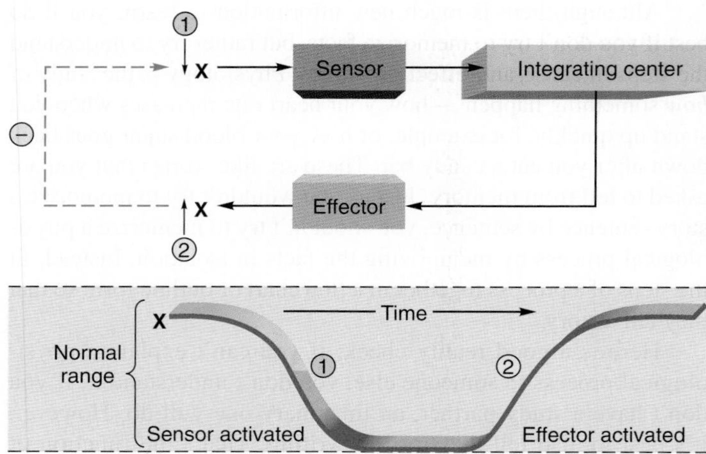

FIGURE 1.2 A fall in some factor of the internal environment (↓X) is detected by a sensor. Compare this negative feedback loop with that shown in fig. 1.1.

Our physiological mechanisms maintain homeostasis by operating through negative feedback loops. In our body, the integrating center is often (but not always) a particular area of the brain. Sensors are cells (often neurons or gland cells) that respond to specific stimuli, and effectors are usually muscles and glands. For example, in order to maintain homeostasis of body temperature, we have neurons that act as temperature sensors in our brain. The integrating center is also located there. In response to a fall in deep body temperature, effectors (such as shivering skeletal muscles) act to raise the body temperature back to the set point. In response to a rise in body temperature above the set point, effectors (including sweat glands) are activated to cool the body temperature back to the set point. In this way, a normal body temperature is maintained by opposing negative feedback loops (fig. 1.3). Notice that in figure 1.3 there is a normal range; constancy is dynamic and relative, not absolute.

Shivering skeletal muscles and sweat glands are **antagonistic effectors**. Both act through negative feedback loops, but they

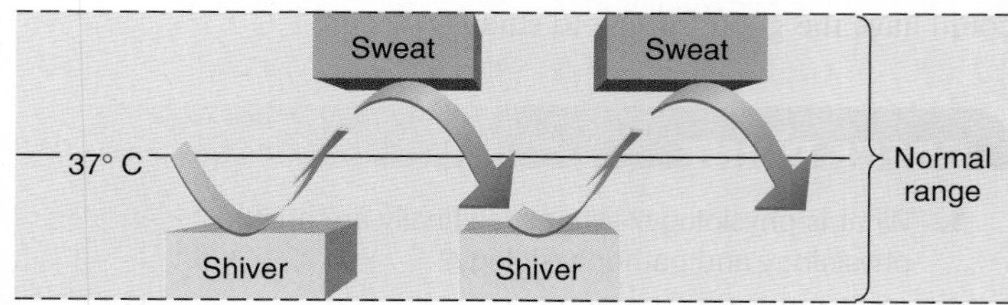

FIGURE 1.3 How body temperature is maintained within the normal range. The body temperature normally has a set point of 37° C. This is maintained, in part, by two antagonistic mechanisms—shivering and sweating. Shivering is induced when the body temperature falls too low, and it gradually subsides as the temperature rises. Sweating occurs when the body temperature is too high, and it diminishes as the temperature falls. Most aspects of the internal environment are regulated by the antagonistic actions of different effector mechanisms.

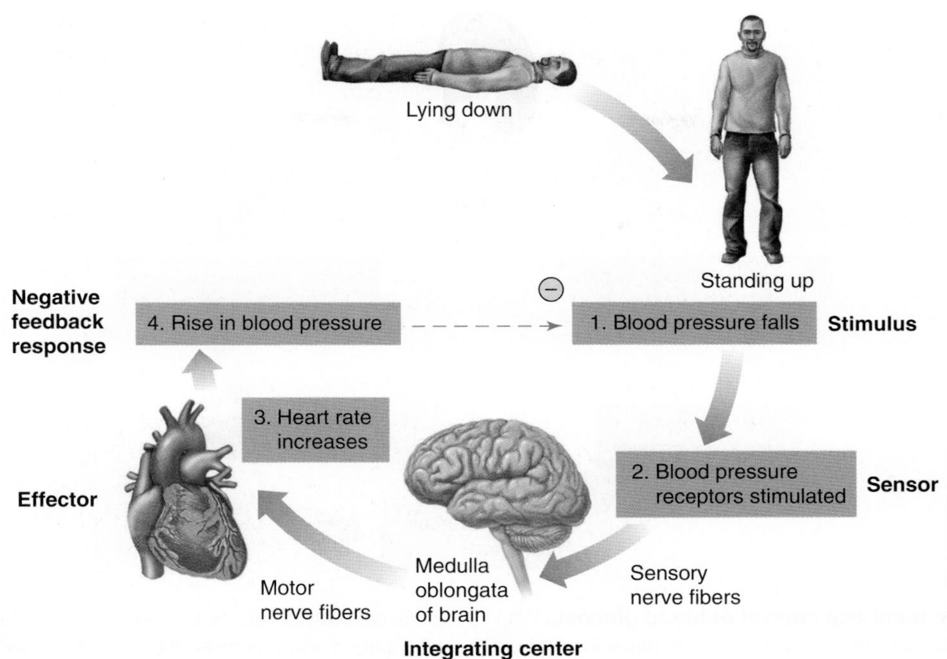

FIGURE 1.4 Negative feedback control of blood pressure. This negative feedback loop helps correct a fall in blood pressure by stimulating an increased heart rate. In this negative feedback loop, sensory and motor nerve fibers convey information between the sensor, integrating center, and effector.

have opposing effects. Thus, their negative feedback loops are activated by opposite stimuli: we shiver in response to a fall in body temperature and sweat in response to a rise in body temperature. Antagonistic effectors afford better, tighter control than would be possible if we had only one type of effector. For example, constant body temperature would not be very well maintained if we shivered when it was cold and simply didn't shiver when it was hot. Most aspects of the internal environment are maintained in a state of homeostasis by numerous antagonistic negative feedback loops.

Neural and Endocrine Regulation of Homeostasis

Let's follow a typical negative feedback loop involving neural control. Suppose that there is a fall in blood pressure—for example, because a person who was lying down stands up quickly. The fall in blood pressure below the set point is the stimulus that activates the sensors, which are blood pressure receptors (baroreceptors). These stimulate sensory neurons that convey this information to an integrating center in the brain (in the medulla oblongata). This integrating center then causes motor (autonomic) neurons that slow the heart rate (parasympathetic neurons) to become less active, while neurons that cause a faster heart rate (sympathetic neurons) become more active. As a result, the heart, which is the effector in this story, increases its rate of beat. This helps raise the blood pressure back to the set point, thereby completing the negative feedback loop (fig. 1.4).

Now let's look at a negative feedback loop that involves a hormone. Suppose that a person eats a candy bar, raising the blood glucose (sugar) concentration higher than the set point. The change is detected by sensors, which are cells sensitive to the blood glucose concentration. This story may be confusing because it's so simple: the sensor cells are also the integrating center and the effector. They are the cells located in clusters (called islets) within the pancreas that secrete the hormone insulin when the blood glucose concentration rises. Insulin then stimulates certain tissues (primarily skeletal muscles, liver, and adipose tissue) to take glucose out of the blood. This lowers the blood glucose concentration, completing the negative feedback loop (fig. 1.5). Conversely, when a person fasts and the blood glucose concentration starts to fall, insulin secretion is decreased.

It's not important at this time for you to learn the negative feedback loops depicted in figures 1.4 and 1.5, as they will be presented again more completely later in this book. These negative feedback loops are shown here only to illustrate how negative feedback loops operate to maintain homeostasis. Although a negative feedback loop may involve many steps and the cooperation of different organs and systems, these examples show that the effects are easy to understand. If you know the direction of the initial change from the set point (the stimulus, or first step, of the negative feedback loop), then you know that the effectors must cause a change in the opposite direction in order to maintain homeostasis.

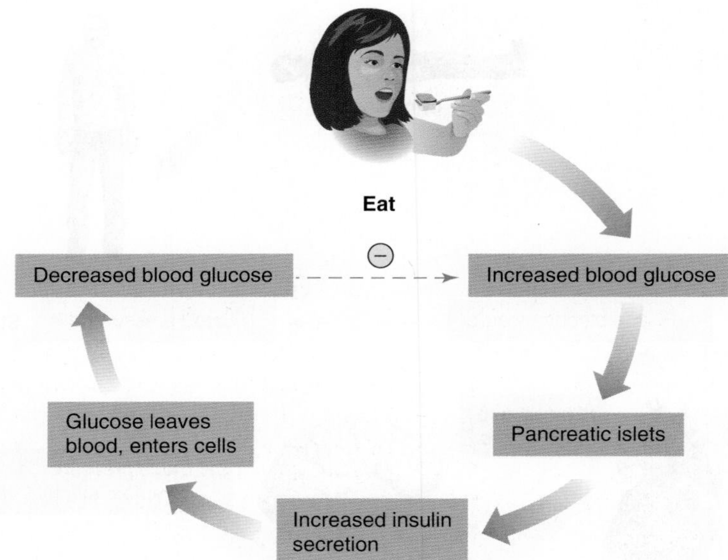

FIGURE 1.5 Negative feedback control of blood glucose. When someone eats carbohydrates (sugars or starch), the blood glucose concentration rises. This stimulates the secretion of the hormone insulin from the pancreatic islets (also called the islets of Langerhans). Insulin stimulates blood glucose to enter cells (primarily of skeletal muscles), so that the elevated blood glucose is brought back down to the level it was at before the carbohydrates were eaten.

Clinical Investigation Clues

Remember that Laura ate a low-carbohydrate, low-calorie diet.

- How might the diet affect her blood glucose concentration?
- How would this influence her level of insulin secretion?
- Given that high insulin promotes fat storage, and low insulin promotes fat breakdown, what might account for her weight loss?

Positive Feedback Amplifies Changes

Constancy requires that a change be compensated by an opposing (negative) change. Thus, homeostasis can be maintained only by negative feedback mechanisms. **Positive feedback** mechanisms operate to amplify changes. In response to a small change in a particular direction, a positive feedback loop would cause the change to become greater and greater, like an avalanche.

There are a few cases where positive feedback loops operate in the body. For example, damage to a blood vessel initiates a change in the blood that starts a positive feedback cascade, where one clotting factor activates another. This positive feedback mechanism produces a blood clot, which helps stem the loss of blood from the damaged vessel. Because the clot prevents blood loss, it helps maintain homeostasis of blood volume.

Other examples of positive feedback operate in the reproductive system. For example, a positive feedback effect of hormones operating between the ovary and the pituitary gland culminate in the explosive event of ovulation (extrusion of an egg cell from the ovary). Another positive feedback mechanism operates between the uterus and the pituitary of a pregnant woman, culminating in the forceful contractions of the uterus during labor and delivery. These examples help the ovary and uterus do their jobs, which is not to maintain homeostasis but to reproduce.

Check Point

1. What is the definition of homeostasis, and what is the significance of homeostasis in physiology and medicine?
2. Describe the components of a negative feedback loop. Use an example to show how a negative feedback loop helps maintain homeostasis.
3. What is a positive feedback loop? Can positive feedback loops act to maintain homeostasis? Explain, and give an example of a positive feedback loop acting in the body.

Chemistry Refresher

Basic Chemical Concepts Used in Physiology

The study of physiology utilizes some concepts and terminology from chemistry. This is needed in order to explain much of cellular function, which is the basis of the physiology of organs and systems. The major types of chemical bonds, together with acids, bases, and the pH scale, are chemical concepts that are particularly fundamental to an understanding of body function.

Physiology is a science that can't be separated from anatomy, since structure and function are clearly interrelated. Also, we can't talk about how the body works without using some basic terminology and concepts of chemistry and biology. You may have had courses in these subjects; if so, you're ahead of the game. Even if you've had recent courses in these subjects, chances are that you could use a little refresher. Perhaps you can skip this section for now, returning to aspects of it as the need arises later. Feel free to use this section as you (or your instructor) think best.

An **atom** is the smallest unit of the chemical elements. The **nucleus** at the center of each atom contains two types of particles—**protons**, which are positively charged, and **neutrons**, which have mass but are not charged. The mass of a proton is equal to the mass of a neutron, and the sum of the protons and neutrons in an atom is equal to the **atomic mass** of the atom. Carbon, for example, has 6 neutrons and 6 protons in its nucleus, and so has an atomic mass of 12.

The number of protons in the nucleus is the **atomic number** of the atom. For example, carbon has 6 protons and so has an atomic number of 6. Outside of the atomic nucleus are negatively charged particles called **electrons**. The number of electrons in an atom is equal to the number of protons in its nucleus, and so the positive and negative charges are balanced. An *isotope* of an atom has the same atomic number but a different number of neutrons, and thus a different atomic mass. The different isotopes of a particular atom constitute a *chemical element*. Hydrogen, for example, is an element that contains 3 isotopes: an atom with only a proton in the nucleus (the most common isotope), an atom with 1 proton and 1 neutron, and a radioactive isotope (called *tritium*) that has 1 proton and 2 neutrons.

Although diagrams of atoms often show electrons as particles orbiting the nucleus, like planets orbiting the sun, this mental picture isn't quite accurate. An electron can occupy any position within a particular energy "shell." Each shell can hold up to a maximum number of electrons, and the shells are filled from the innermost shell outward. The first shell, closest to the nucleus, can contain up to 2 electrons. If an atom has more than 2 electrons (as they all do, except hydrogen and helium), the additional electrons must be located in shells farther from the nucleus. The second shell can contain a maximum of 8 electrons. A carbon atom (with an atomic number of 6) has 2 electrons in the first shell and 4 electrons in the second shell (fig. 1.6).

A carbon atom would need 4 more electrons to reach the maximum of 8 electrons for its outer shell. As we will soon see, this allows the carbon atom to share electrons with other atoms as it forms chemical bonds with them. The electrons in the outermost shell are known as **valence electrons**, and it is always the valence electrons that participate in chemical reactions.

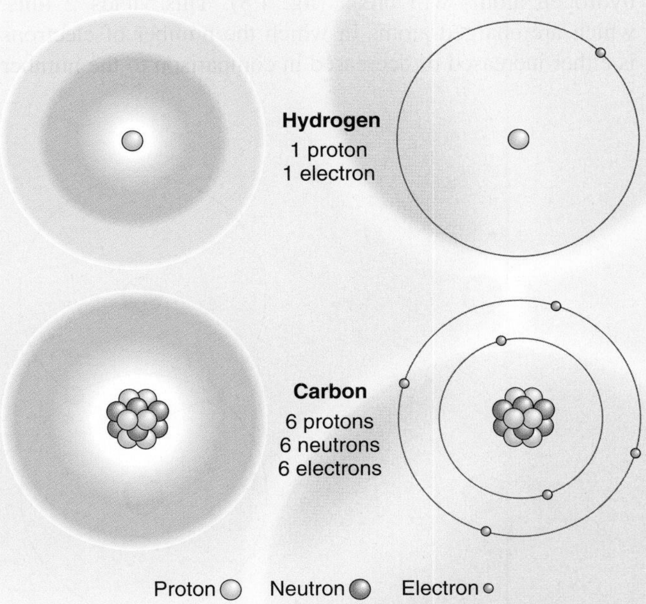

FIGURE 1.6 Diagrams of the hydrogen and carbon atoms. The electron shells on the left are represented by shaded spheres indicating probable positions of the electrons. The shells on the right are represented by concentric circles.

The Strongest Bonds Are Covalent Bonds

Covalent bonds are those formed when atoms share their valence electrons. Covalent bonds are the strongest type of chemical bond, but not all covalent bonds are equally strong. The strongest type of covalent bond occurs when two atoms share their valence electrons equally, as occurs when 2 atoms of the same element bond together (such as O_2 and H_2—fig. 1.7). Molecules formed by covalent bonds in which electrons are equally shared are **nonpolar molecules**, because they are symmetrically charged—they lack plus and minus sides, or poles. In contrast to this, molecules formed by covalent bonds in which the electrons are not equally shared are **polar molecules**—they have a positive and a negative pole. The atoms of oxygen, nitrogen, and phosphorous are notorious bullies, in that they don't share electrons well when they form covalent bonds with other atoms. Instead, they have a strong tendency to pull electrons over to their side of the molecule, so that molecules formed with oxygen, nitrogen, or phosphorous tend to be polar.

Water molecules are an excellent example of polar molecules. Water is the most abundant molecule in the body, and its polar nature is of particular importance. When the oxygen atom shares electrons with 2 hydrogen atoms to make a water molecule, the electrons spend more time at the oxygen side of the water molecule than at the hydrogen side. As a result, the oxygen pole of the molecule is negatively charged in comparison to the hydrogen pole, which can be indicated as positive (fig. 1.8). Since the electrons are unequally shared, there are times when the oxygen atom's bond to one of the hydrogen atoms will break (fig. 1.8). This yields 2 **ions**, which are charged atoms in which the number of electrons is either increased or decreased in comparison to the number of protons. In the case of a water molecule that dissociates, this yields 2 ions: a hydrogen ion (H^+), which is a hydrogen atom minus its electron (in other words, just a proton), and a *hydroxide ion* (OH^-), which has the electron that was swiped from the hydrogen atom.

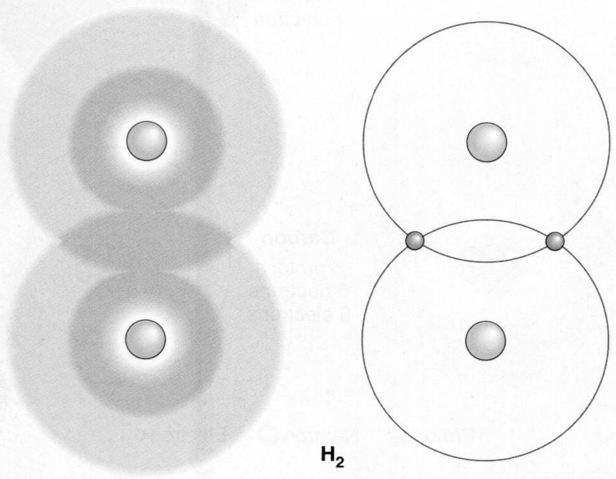

FIGURE 1.7 A hydrogen molecule showing the covalent bonds between hydrogen atoms. In this molecule, these are formed by the equal sharing of electrons.

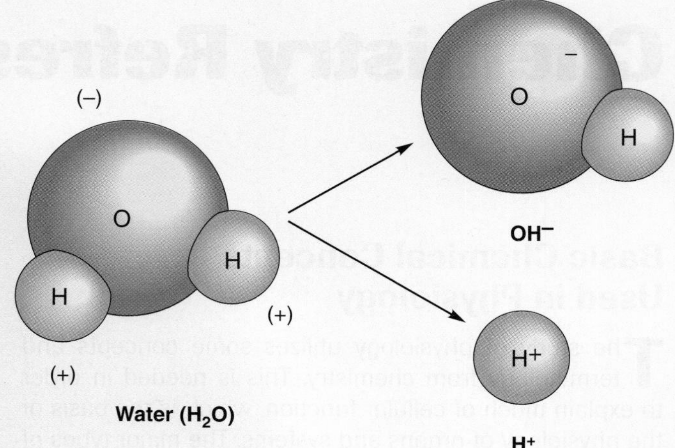

FIGURE 1.8 Water is a polar molecule. Notice that the oxygen side of the molecule is negative, whereas the hydrogen side is positive. Polar covalent bonds are weaker than nonpolar covalent bonds. As a result, some water molecules ionize to form a hydroxyl ion (OH^-) and hydrogen ion (H^+).

Ionic Bonds Are Weaker

Ionic bonds are those that are formed between atoms when one or more valence electrons are completely transferred from one atom to another. This produces 2 ions—a positively charged ion (called a **cation**) that is missing electrons, and a negatively charged ion (called an **anion**) that has gained the extra electrons. The cation and anion then stick together by electrical attraction to form an **ionic compound**. The most important of these in physiology is sodium chloride (NaCl)—common table salt. In the formation of this ionic compound, the lone electron in sodium's outer shell is transferred to the outer shell of the chlorine atom, creating a sodium ion (Na^+) and chloride ion (Cl^-). In solid form, table salt is actually composed of the two ions stuck together as an ionic compound.

When NaCl is dissolved in water, the 2 ions come apart. Each becomes surrounded by water molecules, which form hydration spheres around each ion (fig. 1.9). This is possible because the negative poles of the water molecules are attracted to the Na^+, and the positive poles of the water molecules are attracted to the Cl^-. Then, other water molecules can surround the first layer, because the positive pole of a water molecule is attracted to the negative pole of another. In this way, water is an excellent *solvent*: it can dissolve compounds that are polar, with a negative and positive

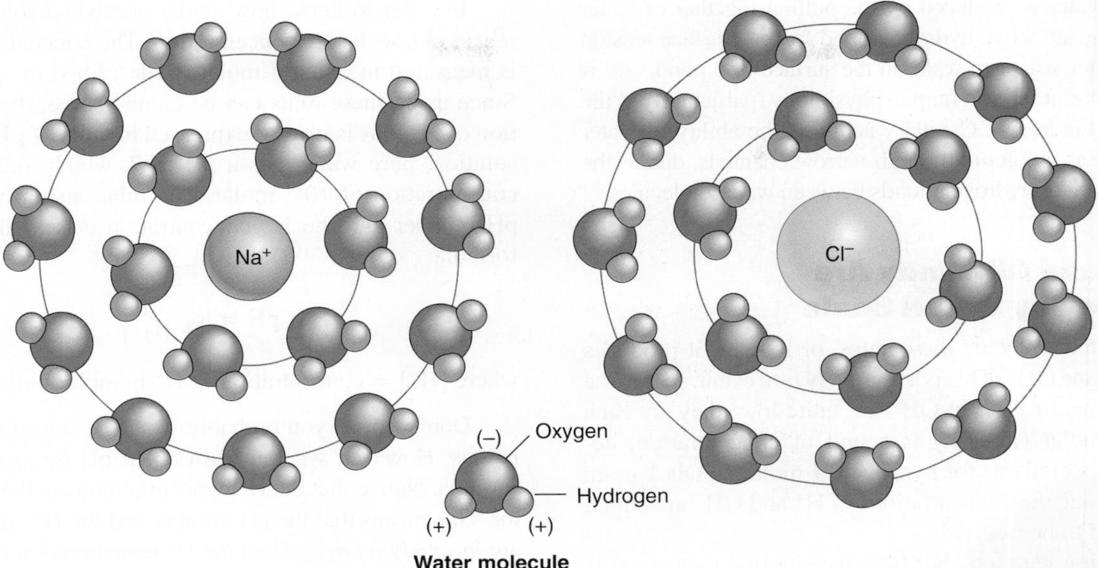

FIGURE 1.9 How NaCl dissolves in water. The negatively charged oxygen-ends of water molecules are attracted to the positively charged Na^+, whereas the positively charged hydrogen-ends of water molecules are attracted to the negatively charged Cl^-. Other water molecules are attracted to this first concentric layer of water, forming hydration spheres around the sodium and chloride ions.

charge. Therefore, water (a polar solvent) can dissolve ionic compounds and other polar molecules.

"**Like dissolves like**" is a statement that summarizes this very important concept. Polar compounds are soluble (can dissolve in) polar solvents, such as water. If you want to dissolve a nonpolar molecule (such as fat, described later), you would have to use a nonpolar solvent (such as alcohol, benzene, or paint thinner). This is why oil and water don't mix, and why you have to use paint thinner to dilute oil-based paint. Those molecules that are polar, and can thus dissolve in water, are called **hydrophilic**; those that are nonpolar cannot dissolve in water and actually do their best to avoid contact with water; they are **hydrophobic**. The division of molecules into polar and nonpolar, hydrophilic and hydrophobic is a concept fundamental to the way that certain physiological processes operate in the body.

Hydrogen Bonds Are Even Weaker

When a hydrogen atom forms a bond with an atom of oxygen or nitrogen, the hydrogen gains a slight positive charge as the oxygen or nitrogen atom pulls the shared electron closer to its side of the molecule, which thereby gains a slight negative charge. When the same thing happens to another, different hydrogen atom bonded to an oxygen or nitrogen atom that's nearby, the positive hydrogen of one pair will have a weak attraction for the negative oxygen or nitrogen of the pair nearby. This weak attraction is called a **hydrogen bond**. Although each hydrogen bond is weak, the sum of their attractive force is largely responsible for holding the three-dimensional shape of proteins, which fold and bend in ways that are stabilized by hydrogen bonds. Also, hydrogen bonds are responsible for joining the two strands of the DNA double helix together (as will be described later in this chapter).

Hydrogen bonds also form between water molecules (fig. 1.10). The hydrogen bonds that form between water molecules is responsible for many of the important properties of water. These include *surface tension* and *capillary action*. Surface

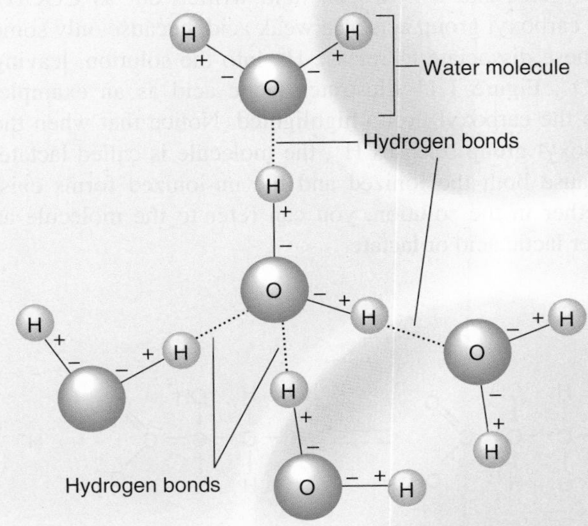

FIGURE 1.10 Hydrogen bonds between water molecules. The oxygen atoms of water molecules are weakly joined together by the attraction of the electronegative oxygen for the positively charged hydrogen. These weak bonds are called hydrogen bonds.

tension of water is produced by the pulling together of water molecules by attractive hydrogen bond forces. Surface tension allows a water strider to walk on the surface of a pond, and is also an important force in human physiology (particularly in the function of the lungs). Capillary action is the ability of water to be pulled as a column through narrow channels, due to the attractive force of hydrogen bonds between water molecules.

Acidity and Alkalinity Are Measured by the pH Scale

When a water molecule dissociates, or ionizes, it produces one H^+ and one OH^-. This is a relatively rare event, so that the concentrations of H^+ and OH^- are quite low: they are each only 10^{-7} molar (molarity is a unit of concentration, described in appendix 1; for hydrogen, 1 molar equals 1 gram per liter). Since the concentrations of H^+ and OH^- are equal, water is said to be **neutral**.

A solution with a higher H^+ concentration than water is said to be **acidic**; one with a lower H^+ concentration is **basic**, or **alkaline**. An **acid** is defined as a molecule that can release protons (H^+) into solution. A **base** is a compound that combines with H^+ and thereby removes it from solution; it is a "proton acceptor." For example, NaOH is a base because it ionizes to release hydroxide (OH^-) ions that can bind to H^+ and form H_2O.

Hydrochloric acid (HCl) is a very strong acid because it almost completely dissociates to release H^+ and Cl^-, thereby greatly increasing the H^+ concentration (acidity) of the solution. Organic acids, such as amino acids, fatty acids, lactic acid, citric acid, and acetic acid, all have a **carboxyl group** (or **carboxylic acid group**), consisting of a carbon, 2 oxygens, and a hydrogen (and written out as COOH). The carboxyl group acts as a weak acid, because only some of these dissociate to release H^+ into the solution, leaving COO^-. Figure 1.11 illustrates lactic acid as an example, with the carboxyl group highlighted. Notice that when the carboxyl group releases H^+, the molecule is called lactate. Because both the ionized and the un-ionized forms exist together in the solution, you can refer to the molecule as either lactic acid or lactate.

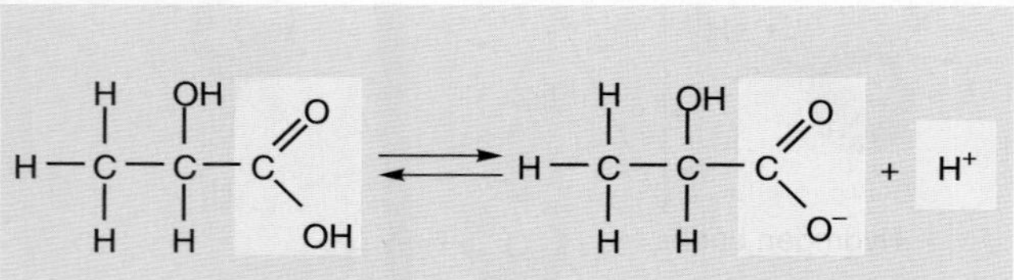

Lactic acid ⇌ Lactate + H^+

FIGURE 1.11 The carboxyl group of an organic acid. This group can ionize to yield a free proton, which is a hydrogen ion (H^+). This process is shown for lactic acid, with the double arrows indicating that the reaction is reversible.

In order to know how acidic or basic a solution is, we need to know its H^+ concentration. The concentration of H^+ is measured in units of molarity (described in appendix 1). Since using these units can be cumbersome, the concentration of H^+ ions is instead expressed in units of **pH**. A neutral solution, pure water, has a pH of 7, which indicates a H^+ concentration of 10^{-7} molar. The relationship between the pH number and the H^+ concentration is described by this formula:

$$\text{pH} = \log \frac{1}{[H^+]}$$

where $[H^+]$ = concentration of H^+ in molar units.

Don't worry: you probably won't have to do any calculations. However, we can learn about pH by examining the formula. Notice that the H^+ concentration is in the denominator. This means that the pH number and the H^+ concentration are *inversely related*: when the H^+ concentration goes up, the pH goes down. Conversely, when the H^+ concentration goes down, the pH goes up. Now look at the formula again, and notice the "log," or logarithmic, relationship. This indicates that if the H^+ concentration goes up 10 times, the pH number goes down by 1. For example, water has a pH of 7 and lemon juice has a pH of about 5. So a solution with 10 times the H^+ concentration of water would have a pH of 6, and one with 100 times the H^+ concentration of water would have a pH of 5, like lemon juice.[1]

Solutions with a pH lower than 7 are therefore acidic, and solutions with a pH greater than 7 are basic, or alkaline. Later in this book we will discuss blood pH, and you will see that normal blood pH is in the range of 7.35 to 7.45. Blood pH lower than the normal range is a condition called *acidosis*, and blood pH above the normal range is *alkalosis*. Notice that blood can be acidotic but not acidic; a blood pH of 7.15 is acidotic, because it's to the acid side of normal, but it's not acidic because its pH is greater than 7.0.

CHECK POINT

1. Describe the nature and relative strength of covalent, ionic, and hydrogen bonds.
2. Distinguish between polar and nonpolar molecules, and give examples.
3. Describe the ability of water to act as a solvent.
4. What is meant by the terms *acid* and *base, acidic* and *basic*? How do these terms relate to the pH scale?

FYI [1]Gastric juice (in the stomach) has a pH of 2. Can you figure out how much greater its H^+ concentration is compared to water? (Just keep multiplying by 10 every time you go down a pH number: gastric juice, at pH 2, has 100,000 times the H^+ concentration of water).

Carbohydrates and Lipids Have Different Characteristics

Carbohydrates and lipids are both major sources of energy for the body. Although they both consist of carbon, hydrogen, and oxygen, they have different ratios of these atoms, are constructed differently, and have different physical properties. Carbohydrates and lipids are each large families of molecules, and each contains different subcategories that have distinct structures and uses in the body.

The remaining sections of this chapter continue the background review of chemistry. However, these sections relate to organic molecules so fundamental to many aspects of physiology that the border between what's chemistry and what's physiology can be very blurred. Given this, the sections on organic molecules that follow can be used as you and your instructor think best.

Organic molecules are those that contain the element carbon, generally bonded to hydrogen and oxygen. Because of the number of valence electrons in the outermost shell, each carbon atom can form 4 bonds with other atoms. Each hydrogen atom can form only 1 bond, and each oxygen atom can form 2 bonds. A great variety of molecules can be constructed using these rules. Carbohydrates share a common theme in the ratio of these atoms in the molecule. Lipids are more structurally variable, but share a common theme of being primarily nonpolar molecules.

Monosaccharides, Disaccharides, and Polysaccharides Are Subcategories of Carbohydrates

All carbohydrates share the same ratio of carbon, hydrogen, and oxygen atoms described by their name: "carbo" (carbon) and "hydrate" (water, H_2O). The general formula for a carbohydrate is thus $C_nH_{2n}O_n$. In this formula, *n* is some number. Whatever the number, there is going to be the same number of carbon atoms as oxygen atoms, and twice the number of hydrogen atoms as either carbon or oxygen.

The subcategory of carbohydrates called **monosaccharides** is the simplest: they can't be broken down any smaller and still be carbohydrates. Monosaccharides (literally, "1 sugar") are also called *simple sugars*: the most important for our study of human physiology are *glucose*, *galactose*, and *fructose*. All three of these monosaccharides happen to contain 6 carbons, and so all three have the same formula: $C_6H_{12}O_6$. They are *structural isomers* of each other, differing in the way their atoms are put together (fig. 1.12). Glucose is the monosaccharide we will encounter most often in our study of human physiology, because it is the blood sugar, and is required by the brain for energy.

Two monosaccharides can be joined together by a covalent bond to form a double sugar, or **disaccharide**. Common disaccharides include *sucrose* (glucose bonded to fructose; this is table sugar), *lactose* (glucose bonded to galactose; this is milk sugar), and *maltose* (glucose bonded to glucose). When many monosaccharides are covalently bonded together, a **polysaccharide** is produced. The polysaccharide of major significance that will be encountered in this textbook is *glycogen*. Glycogen has been called "animal starch." It is essentially the same as plant starch, consisting of hundreds of glucose molecules bonded together, but it has more branching than plant starch (fig. 1.13). Glycogen is an energy storage molecule, formed primarily within the cells of skeletal muscles and the liver.

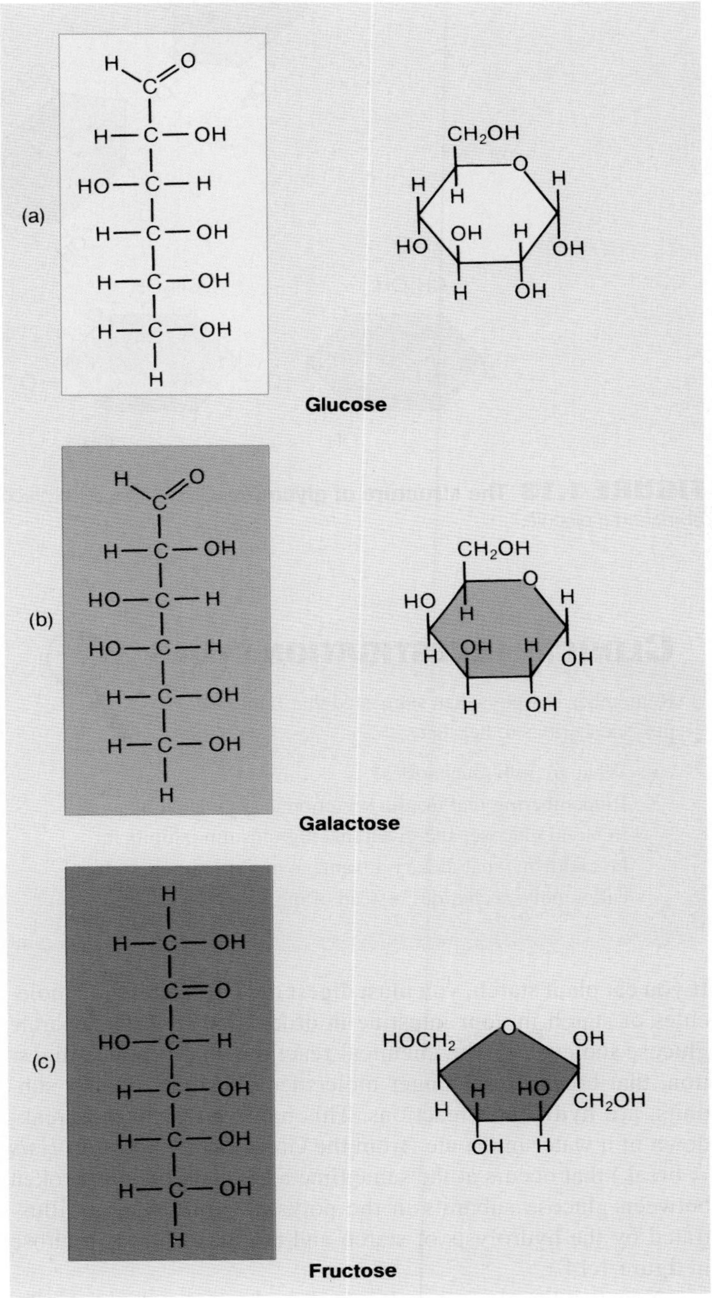

FIGURE 1.12 Structural formulas for three hexose sugars. These are (*a*) glucose, (*b*) galactose, and (*c*) fructose. All three have the same ratio of atoms—$C_6H_{12}O_6$. The representations on the left more clearly show the atoms in each molecule, while the ring structures on the right more accurately reflect the way these atoms are arranged.

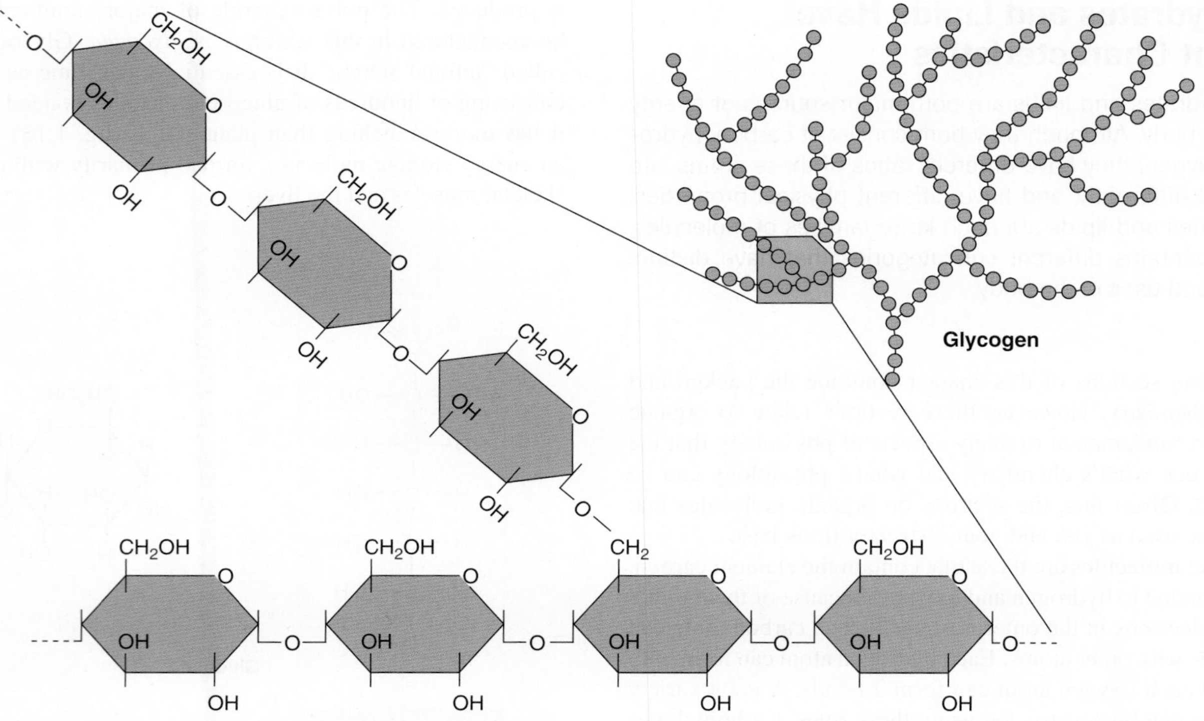

FIGURE 1.13 The structure of glycogen. Glycogen is a polysaccharide composed of glucose subunits joined together to form a large, highly branched molecule.

CLINICAL INVESTIGATION CLUES

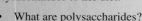

Remember, the physician suggested that Laura add polysaccharides to her diet.

- What are polysaccharides?
- Remembering that insulin secretion is stimulated by a rise in blood glucose, and given that high insulin inhibits fat breakdown, what dietary advantage might Laura have by eating polysaccharides instead of mono- or disaccharides?

If you eat plant starch, you must digest the polysaccharide molecules of starch in your intestine in order to absorb the separate glucose molecules. The chemical reactions that produce digestion, that break down longer molecules into their smaller subunits, are **hydrolysis** reactions. This name refers to the breakdown of a water molecule (from the Greek *hydro* = water; *lysis* = break) that occurs at the same time a covalent bond is broken between glucose subunits in the polysaccharide. This is illustrated by the hydrolysis of starch and the disaccharide maltose in figure 1.14.

Now that the glucose molecules have been absorbed from the intestine into your blood, they can travel to the liver and skeletal muscles. The cells of these organs can take the glucose molecules out of the blood and use them for a variety of purposes. They can break glucose down further for energy (as described in chapter 2), or they can store the carbohydrates in the form of glycogen. In order to do that, these cells must bond the glucose molecules together. This reaction requires that a hydrogen atom (H) be removed from one glucose molecule and a hydroxyl (OH) be removed from another. Then a bond can form between the oxygen atom of one glucose and the carbon atom of the other glucose molecule (fig. 1.15). In this process, the hydrogen and hydroxyl are also joined together to make HOH, or water. Because water is formed, the type of chemical reaction that builds a larger molecule out of smaller subunits is called **dehydration synthesis**.

Actually, both hydrolysis and dehydration synthesis reactions require the participation of specific enzymes (described shortly), and because of this they can be carefully regulated. These reactions are important, and so it's worth repeating: *breakdown reactions occur by hydrolysis; building reactions occur by dehydration synthesis*. This is not only true for carbohydrates, but it's also true for lipids, proteins, and other categories of molecules, as will be described.

Let's get back to the glycogen we just made from blood glucose. Suppose that you now fast or go on a strict diet. Your blood glucose concentration will start to fall, a change that must be corrected to maintain homeostasis. Sensors will detect the fall in blood glucose and activate negative feedback loops to correct that change. As a result, the liver's stored glycogen will be hydrolyzed into free glucose molecules, which can be secreted into the blood. Also, fat stored in adipose cells (and formed by dehydration synthesis reactions) will be hydrolyzed to release its subunits into the blood, as described in the next section.

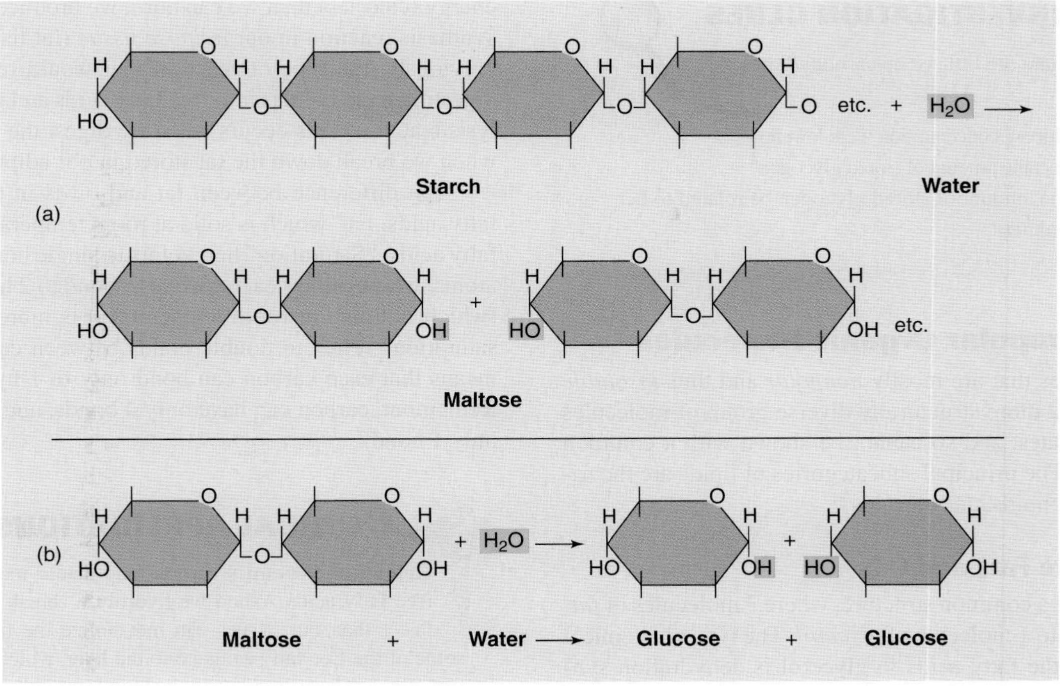

(a) Starch + Water (H_2O) → Maltose + Maltose etc.

(b) Maltose + Water (H_2O) → Glucose + Glucose

FIGURE 1.14 The hydrolysis of starch. The polysaccharide is first hydrolyzed into (*a*) disaccharides (maltose) and then into (*b*) monosaccharides (glucose). Notice that as the covalent bond between the subunits breaks, a molecule of water is split. In this way, the hydrogen atom and hydroxyl group from the water are added to the ends of the released subunits.

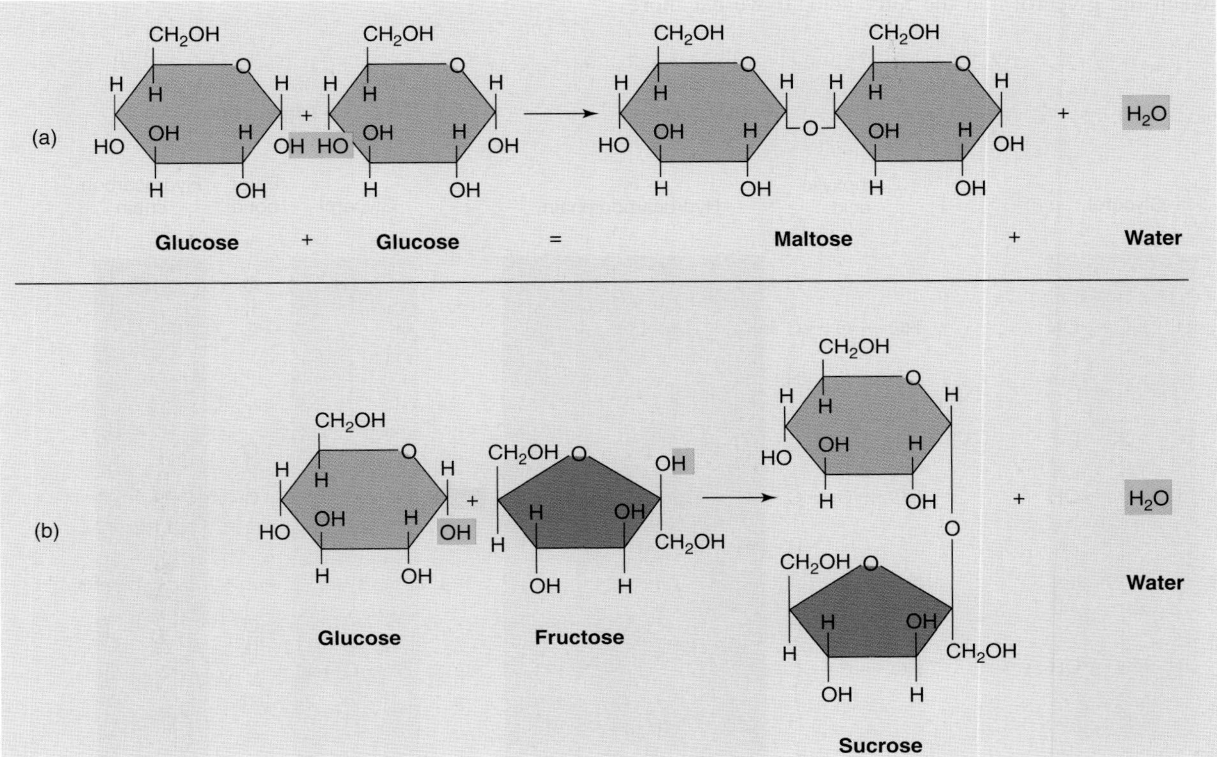

FIGURE 1.15 Dehydration synthesis of disaccharides. The two disaccharides formed here are (*a*) maltose and (*b*) sucrose (table sugar). Notice that a molecule of water is produced as the disaccharides are formed.

Clinical Investigation Clues

Remember that Laura ate little or no carbohydrates and felt weak.

- If her blood glucose concentration were low, how would this affect the amount of stored glycogen?
- How might the amount of stored glycogen be related to her feeling of weakness?

Lipids Are Nonpolar Organic Molecules

Lipids are molecules that are mostly *nonpolar* and thus *insoluble in water*. They are a more structurally diverse group of molecules than the carbohydrates, and so cannot be shown with a common chemical formula. The principal subcategories of lipids are the triglycerides, phospholipids, and steroids.

Triglycerides Are Fats and Oils

Triglycerides share a common structure, where 3 molecules of *fatty acids* are bonded to 1 molecule of *glycerol*. The type of chemical reaction that joins the fatty acids to glycerol is dehydration synthesis, as shown in figure 1.16. Notice that glycerol is a 3-carbon-long molecule, and each carbon atom has a hydroxyl group that can participate in the reaction. Each fatty acid is a long chain of carbons and hydrogens, with a carboxyl (carboxylic acid) group on one end. The hydrogen from the carboxyl group bonds with the hydroxyl group from glycerol to produce water, as a bond is formed between the glycerol and the fatty acid. If we eat more energy (calories) than we can burn, we produce fat by dehydration synthesis reaction in our *adipose tissue* (fat tissue). As mentioned previously, the type of reaction that would reverse this process, digesting triglycerides into free fatty acids and glycerol, is a hydrolysis reaction. This occurs when we digest the fat we eat, and also when we break down the fat stored in our adipose tissue.

The difference between fat and oil is in the structure of the fatty acids. Fat, which is solid at room temperatures, has saturated fatty acids. "Saturation" here refers to single bonds between carbon atoms, which allows each carbon to bond to 2 hydrogen atoms. Oil (which is liquid at room temperature) is more unsaturated. "Unsaturation" refers to double bonds between carbon atoms, which means that each carbon can bond only to 1 hydrogen (fig. 1.17). Remember, carbon can have only 4 bonds, and hydrogen can have only 1 bond.

Clinical Applications

Ketone bodies are 4-carbon-long acidic molecules derived from free fatty acids. When we hydrolyze (break down) our stored fat faster than our tissues can metabolize the fatty acids for energy, some of the free fatty acids go to the liver, which forms ketone bodies from them. These circulate in the blood (providing energy for the body) and some "spill over" into the urine. Ketone bodies in the urine can be tested by a simple dipstick method. Indeed, there are strict weight-reducing diets that suggest that dieters test their urine for ketone bodies to see if they are breaking down fat. A dangerous level of ketone bodies in the blood, which lowers the blood pH, can occur in uncontrolled type 1 diabetes mellitus (discussed in chapter 8).

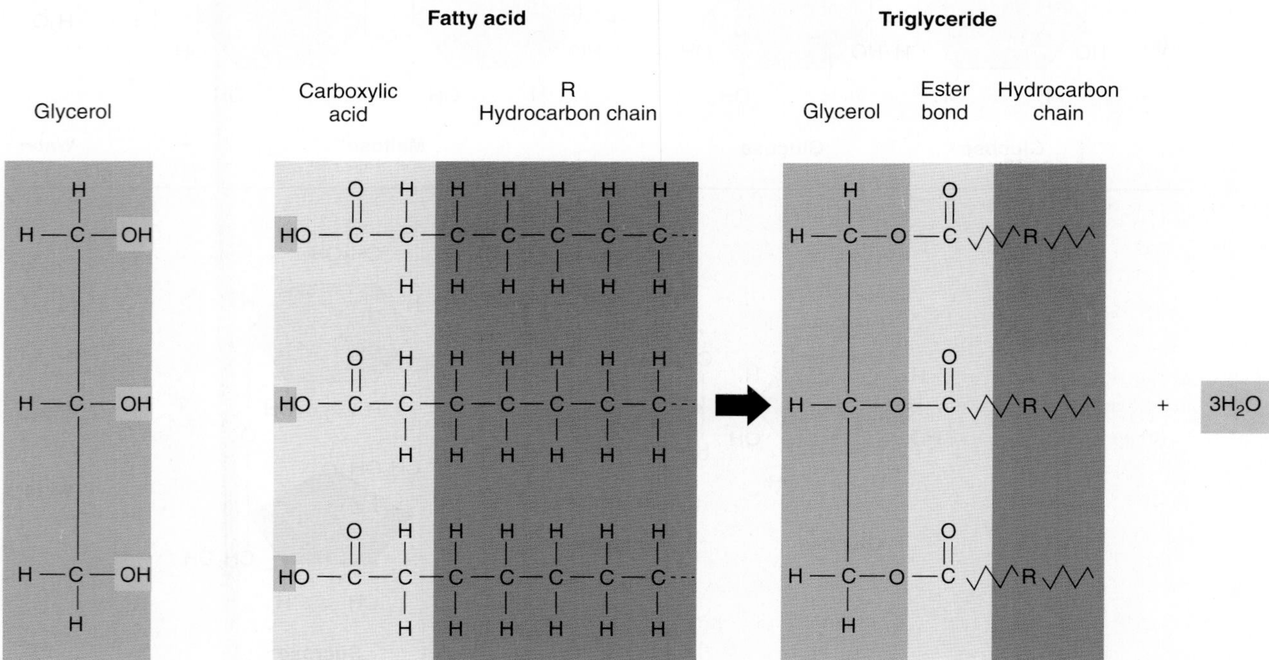

FIGURE 1.16 The formation of a triglyceride molecule from glycerol and 3 fatty acids by dehydration synthesis reactions. A molecule of water is produced as a bond forms between each fatty acid and the glycerol. Sawtooth lines represent hydrocarbon chains, which are symbolized by an *R*.

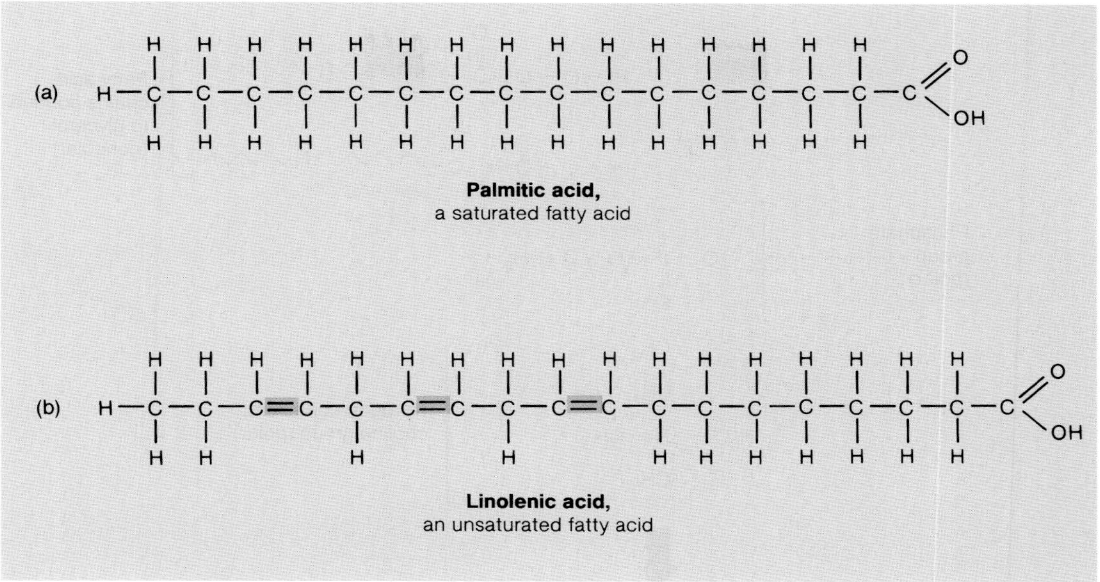

(a) **Palmitic acid,**
a saturated fatty acid

(b) **Linolenic acid,**
an unsaturated fatty acid

FIGURE 1.17 Structural formulas for fatty acids. (*a*) The formula for saturated fatty acids and (*b*) the formula for unsaturated fatty acids. Double bonds, which are points of unsaturation, are highlighted in yellow.

CLINICAL INVESTIGATION CLUES

Remember that Laura was weak and thirsty, and the physician asked that she test her urine for ketone bodies.

- What are ketone bodies, and how would a diet cause more to be produced?
- Given that excretion of ketone bodies in the urine also results in polyuria (excess urination), what might account for Laura's thirstiness?
- How would adding some complex carbohydrates to her diet reduce her production of ketone bodies (ketosis)?

Phospholipids Have a Split Personality

Although lipids, by definition, are nonpolar, phospholipids are a type of lipid that only partially fits that description. Phospholipids don't dissolve in water, and so are mostly nonpolar, but part of a phospholipid molecule actually is polar.

The most common type of phospholipid has a 3-carbon-long glycerol bonded to 2 fatty acids, and so starts out much like the nonpolar triglycerides. However, the third carbon is bonded to a phosphate group, which is highly polar (a phosphate ion has the formula PO_4^{3-}). If the phosphate group is then bonded to a nitrogen-containing organic molecule called *choline*, the phospholipid molecule is known as **lecithin** (or *phosphatidylcholine*). Figure 1.18 shows a detailed structure of lecithin and the common simplified structure. Notice that in the simplified structure of a phospholipid, the polar portion of the molecule is shown as a sphere or circle, and the nonpolar portion is shown using two sawtoothed lines to indicate the two fatty acid chains of carbons and hydrogens.[2]

The split personality of phospholipids enables them to "duck their heads" into the surface of water (that is, the polar portions of the molecule, but not the rest, can be immersed in the water). This breaks up the surface of the water, thereby lowering its surface tension. The ability of phospholipids to lower surface tension, or act as a *surfactant*, is of critical importance in our lungs, preventing them from collapsing (as described in chapter 12). Also, the split personality of phospholipids—part polar and part nonpolar—enables them to form the major structure of cell (plasma) membranes (chapter 2).

Steroids Are Derived from Cholesterol

Cholesterol is generally regarded as something bad for the body, because high blood cholesterol promotes heart disease and stroke (chapter 10). However, our body produces cholesterol, and it is needed as a precursor (parent molecule) for **steroid hormones** (fig. 1.19). Steroid hormones are produced by the gonads—ovaries produce *estradiol*, and testes produce *testosterone*. Also, steroid hormones are produced by the adrenal cortex, which is the outer region of the adrenal glands. These are

FYI [2]The word *choline* may ring a bell, if you had a previous biology course. Choline is part of the molecule acetylcholine, which is an important neurotransmitter (as will be discussed in chapter 4). For that reason, some people thought that eating lecithin (as pills or in egg yolk) would improve a person's memory. Unfortunately, it doesn't seem to work.

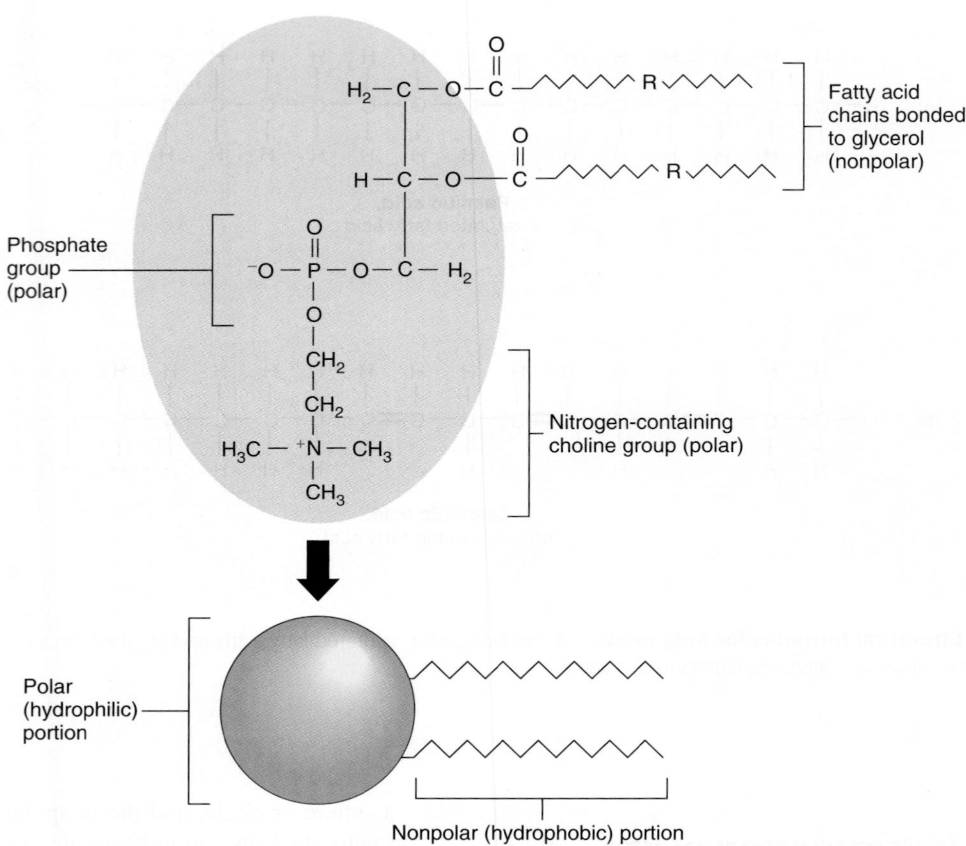

FIGURE 1.18 The structure of lecithin. Lecithin is also called phosphatidylcholine, where choline is the nitrogen-containing portion of the molecule. The detailed structure of the phospholipid (*top*) is usually shown in simplified form (*bottom*), where the circle represents the polar portion and the sawtoothed lines the nonpolar portion of the molecule.

collectively termed *corticosteroids*, and the best-known of these is *cortisol* (*hydrocortisone*).

Notice that the steroid structure is shown as rings—three 6-carbon rings and one 5-carbon ring. In chemistry, the corners of the rings are understood to contain carbon atoms, and the lines connecting the corners represent the bonds between the carbon atoms. Also, notice how similar these structures look to our eyes. Fortunately, our tissues have no difficulty telling them apart, as everyone knows from the different effects that estradiol and testosterone produce in women and men, respectively.

CHECK POINT

1. Give examples of monosaccharides, disaccharides, and polysaccharides, and describe the chemical reactions that allow monosaccharides to be formed from polysaccharides and vice versa.
2. What is the common characteristic of lipids? Identify the major subcategories of lipids, and give examples of each.
3. Draw the simplified structure of a phospholipid and label the polar and nonpolar parts. What functional benefits does this "split personality" provide?

Proteins and Nucleic Acids

Proteins and nucleic acids (DNA and RNA) are very large molecules, or macromolecules. Proteins are formed from subunits known as amino acids, while the nucleic acid subunits are called nucleotides. The complex three-dimensional structure of proteins allows enzyme proteins to function as catalysts, making specific reactions go faster. Complementary base pairing of nucleotides is fundamental to the function of DNA and RNA.

There is an enormous diversity of protein structure, made possible by the fact that proteins are made from 20 different building blocks, the amino acids. Because of their diverse structures, proteins have a great many different functions in cells and in the body. Some proteins are structural: *collagen*, for example, is a fibrous protein that provides strength in connective tissues (tendons and ligaments are composed predominantly of collagen). Many proteins are *enzymes*, which function as biological catalysts to cause specific chemical reactions to go faster. *Antibodies* are proteins, as are *receptors* on cell surfaces for regulatory molecules and *transport carriers* in the plasma (cell) membrane that move substances into or out of the cells. These and other types of molecules have **specificity of function**, a property of

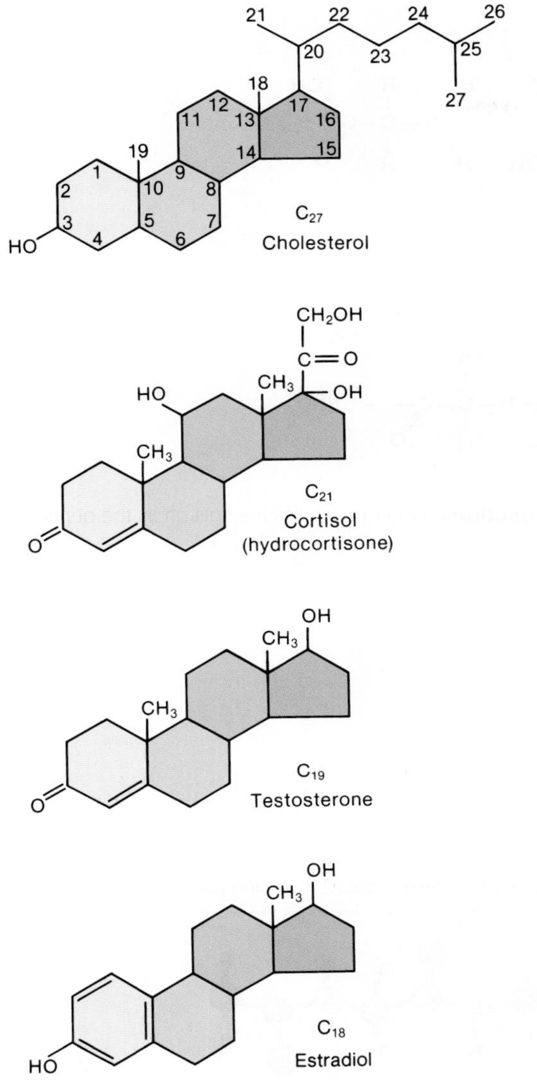

FIGURE 1.19 Cholesterol and some of the steroid hormones derived from cholesterol. The steroid hormones are secreted by the gonads and the adrenal cortex. The numbers on the cholesterol molecule indicate positions of carbon atoms.

proteins because of their great diversity of structure. Protein structure is coded by the genes (DNA), using RNA as the intermediary (as will be discussed in chapter 2).

Structure of Proteins

Each **amino acid**, as implied by its name, has an *amino group* (NH_2) at one end and a carboxyl group at the other end of the molecule (fig. 1.20). There are about 20 different amino acids, and each differs from the other in the part called the *functional group* (abbreviated *R*—think of it as the "rest of the molecule"). Each particular protein has a definite sequence of amino acids (coded by the genes) that are joined together by dehydration synthesis reactions. When water molecules are formed, bonds also form between the amino acids (fig. 1.21). These bonds are called *peptide bonds*,

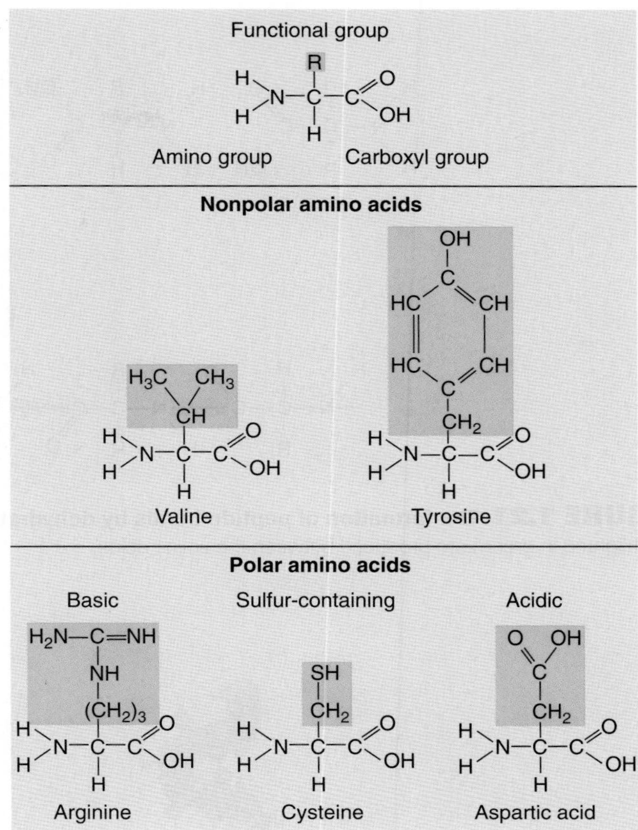

FIGURE 1.20 Representative amino acids. The figure depicts different types of functional (R) groups. Each amino acid differs from other amino acids in the number and arrangement of its functional groups.

and the chain of amino acids formed this way is a **polypeptide**. When the polypeptide gets large enough (about 100 amino acids long), it's called a **protein**.

The specific sequence of amino acids in a protein is called its **primary structure** (fig. 1.22). Because of interactions between amino acids in the polypeptide chain, the chain can twist into a helical shape. This is the **secondary structure** of a protein. Then, the protein can bend and twist into a complex three-dimensional shape, which is the **tertiary structure** of the protein. Finally, some proteins are composed of more than one polypeptide chain joined together—the hemoglobin protein in red blood cells, for example, is composed of 4 polypeptide chains. This is referred to as the **quaternary structure** of the protein.

The specific shape of a protein is often the key to its specific function. For example, a particular regulatory molecule (like a hormone or neurotransmitter) fits into a pocket within the three-dimensional shape of its specific receptor protein; a molecule to be transported across a plasma membrane (such as glucose) fits into a complementary-shaped pocket in its specific transport protein; and a molecule that reacts with another in a reaction catalyzed by an enzyme fits into its specific slot in the enzyme protein. Since the primary structure (amino acid sequence) of a protein is determined by the genetic code, and since the higher-order structure of the protein is a result of its amino acid sequence, you can see how genes can regulate the activities of these types of proteins, and through them regulate the cell.

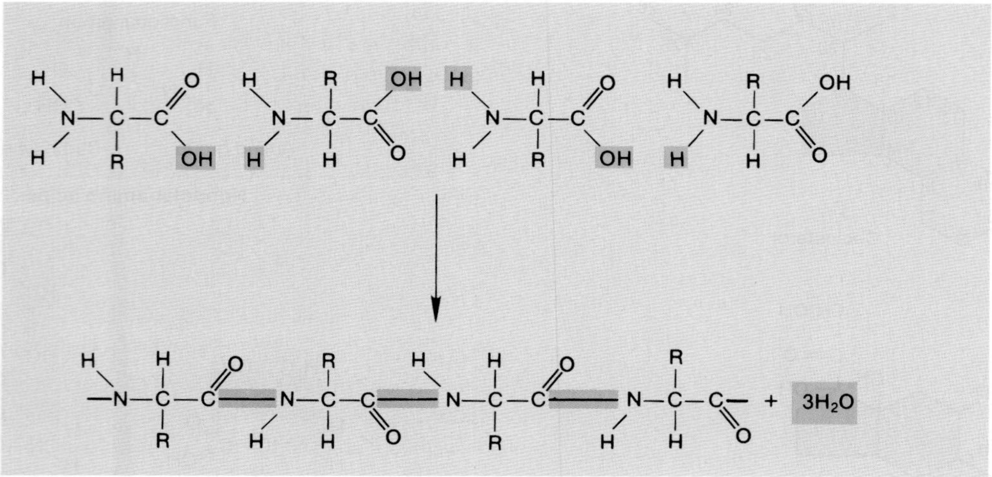

FIGURE 1.21 The formation of peptide bonds by dehydration synthesis reactions. Water molecules are split off as the peptide bonds (highlighted in green) are produced between the amino acids.

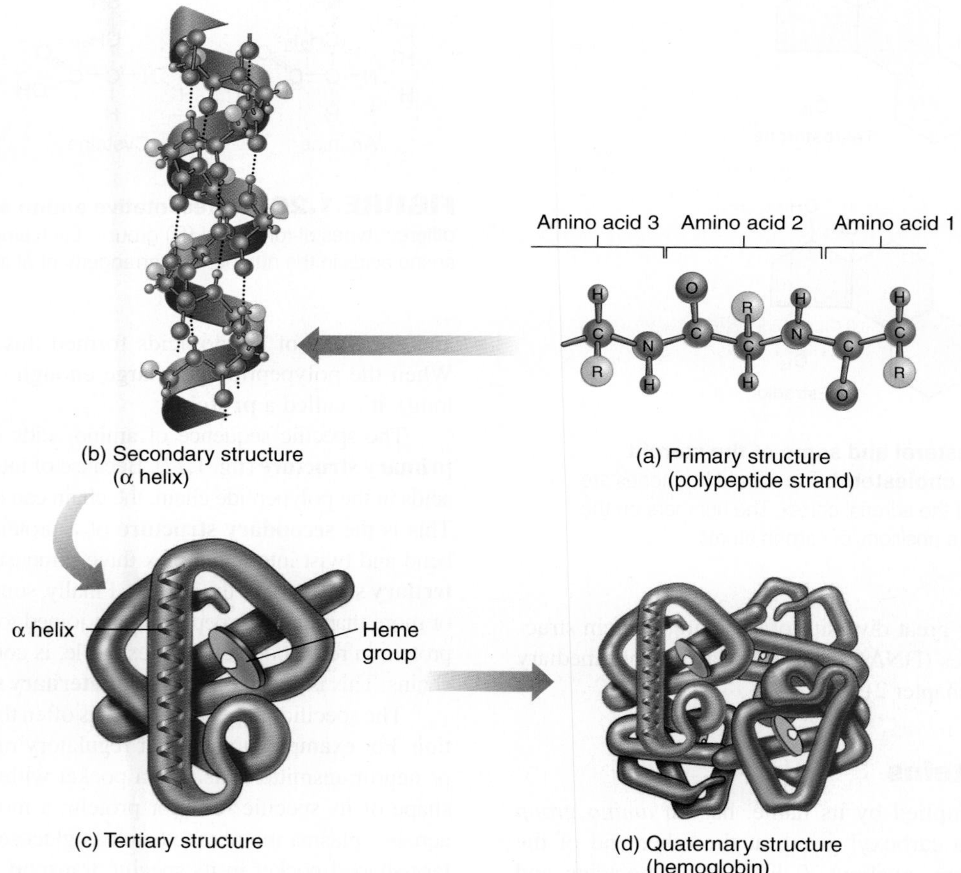

FIGURE 1.22 The structure of proteins. *(a)* The primary structure refers to the sequence of amino acids in the polypeptide chain. The secondary structure refers to the conformation of the chain created by hydrogen bonding between amino acids; this can be an alpha helix *(b)*. The tertiary structure *(c)* is the three-dimensional structure of the protein. The formation of a protein by the bonding together of two or more polypeptide chains is the quaternary structure *(d)* of the protein. Hemoglobin, the protein in red blood cells that carries oxygen, is used here as an example.

Enzyme Proteins Are Highly Specific Catalysts

Catalysts are substances that *increase the rate of chemical reactions* without being changed by the reaction, and without changing the nature of the reaction. They do this by making it easier for the molecules to react, because they lower the energy a molecule needs in order to participate in the reaction. This, described as *lowering the activation energy* for the reaction, is like lowering the price of admission; more molecules can join in the fun and participate in the reaction. As a result, the reaction goes faster than it would if the catalyst were absent.

Enzymes are proteins that function as catalysts. Because each enzyme protein has a specific tertiary structure, it is very specific as to the reaction it catalyzes. This can be understood by examining figure 1.23, which depicts the **lock-and-key** mental model of enzyme function. The molecules that will react, called the *substrates* of the enzyme, can fit into specific pockets in the shape of the enzyme proteins. These pockets are the **active sites** of the enzyme, and each different enzyme will have differently shaped active sites. Once the substrates have docked in the active sites of the enzyme, a chemical reaction can occur whereby the substrates are converted into the *products* of the reaction. The products dissociate from the enzyme, so the enzyme hasn't been changed by the reaction and can work over again many times with new substrate molecules.

Although the lock-and-key model is somewhat of a simplification, it allows us to understand some important properties of enzymes. One property is *specificity*—one enzyme can catalyze the conversion of only certain, specific substrates into products. Because there are thousands of different enzyme-catalyzed reactions in the body, there must be thousands of different enzymes. This affords a fine degree of control as to which reactions are allowed to occur at particular times, because if the enzyme is not present and active, the reaction will be negligibly slow. Thus, the metabolism of cells can be adjusted to follow different possible pathways by regulating the production and activation of enzymes.

Since the activity of an enzyme depends on the shape of the active site, which in turn depends on the tertiary structure of the protein, anything that bends the enzyme out of shape will interfere with its activity. For example, heating an enzyme too much may permanently change the protein's tertiary structure (a change called *denaturation*) and prevent the enzyme from functioning. Also, changes in pH can influence the hydrogen bonding that holds a protein in its specific tertiary structure. Different enzymes are adapted to work best at a particular pH—this is their **pH optimum**. When the pH is either increased or decreased from the pH optimum, enzyme activity declines (fig. 1.24).

The names of the enzymes given in figure 1.24 follow different patterns. Pepsin (a protein-digesting enzyme in gastric juice) and trypsin (a protein-digesting enzyme in pancreatic juice) were named according to an older convention. The newer and more useful enzyme names tell the action of the enzyme and end with the

Clinical Applications

When tissue cells die and disintegrate due to diseases, they release some cellular enzymes into the extracellular fluid. These get washed away in the blood, where they are not normally found and are not active. However, their enzyme activity can be measured in a test tube by adding the appropriate substrate, and this is commonly done in clinical laboratories to help detect damage to specific organs. For example, damage to the prostate gland might result in elevation of blood *acid phosphatase*; damage to the heart may elevate *creatine phosphokinase* (*CPK*), *lactate dehydrogenase* (*LDH*), or other enzymes abbreviated *AST* and *ALT*; and damage to the pancreas may result in elevated levels of *pancreatic amylase* in the blood.

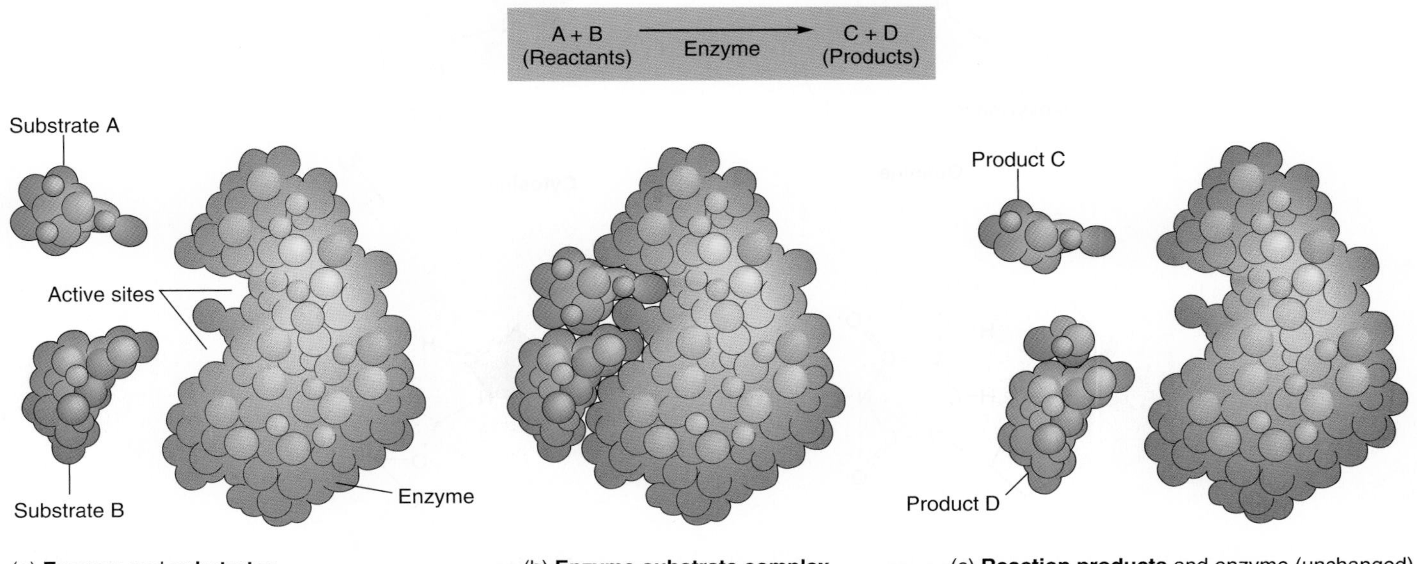

FIGURE 1.23 The lock-and-key model of enzyme action. (*a*) Substrates A and B fit into active sites in the enzyme, forming an enzyme-substrate complex. (*b*) This complex then dissociates (*c*), releasing the products of the reaction and the free enzyme.

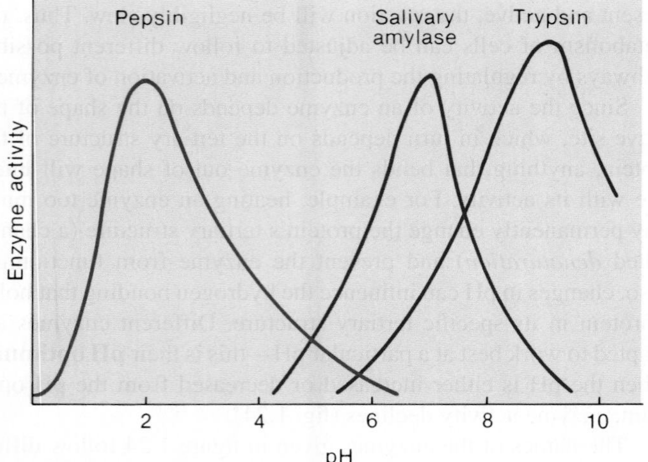

FIGURE 1.24 The effect of pH on the activity of three digestive enzymes. Salivary amylase is found in saliva, which has a pH close to neutral; pepsin is found in acidic gastric juice, and trypsin is found in alkaline pancreatic juice.

suffix *-ase*. For example, *hydrolases* promote hydrolysis reactions, *phosphatases* remove phosphate groups, and *synthetases* catalyze dehydration synthesis reactions. The names of many enzymes tell the substrate and the action of the enzyme (such as *lactate dehydrogenase*), or the product and the action of the enzyme (such as *glycogen synthetase*). The third enzyme named in figure 1.24—*salivary amylase*—is an enzyme in saliva that digests starch.

Structure of Nucleic Acids

The nucleic acids are **DNA (deoxyribonucleic acid)** and **RNA (ribonucleic acid)**. These are very long molecules composed of subunits called **nucleotides**. Each nucleotide consists of three parts: a phosphate, a 5-carbon sugar molecule (either *deoxyribose* or *ribose*, which gives the molecule its name), and a group called a **nitrogenous base** (or just *base*). There are 4 nitrogenous bases in DNA: *adenine*, *guanine*, *cytosine*, and *thymine*, producing 4 different DNA nucleotides (fig. 1.25).

Phosphate groups can link nucleotides together, producing a strand with a "sugar-phosphate backbone" that has the

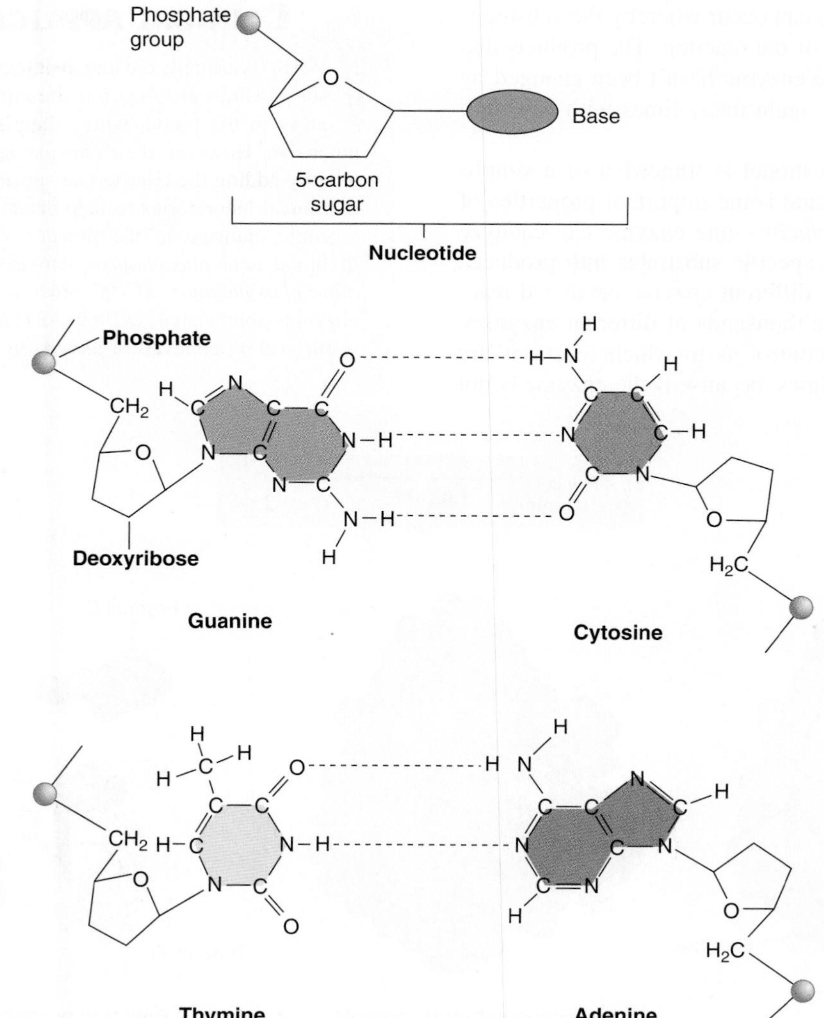

FIGURE 1.25 DNA nucleotides. The general structure of a nucleotide is depicted at the top of the figure. The DNA bases guanine and cytosine are shown on one DNA strand, and the DNA bases cytosine and thymine are shown on another DNA strand. The two DNA strands are joined by hydrogen bonds (dashed lines) between complementary bases.

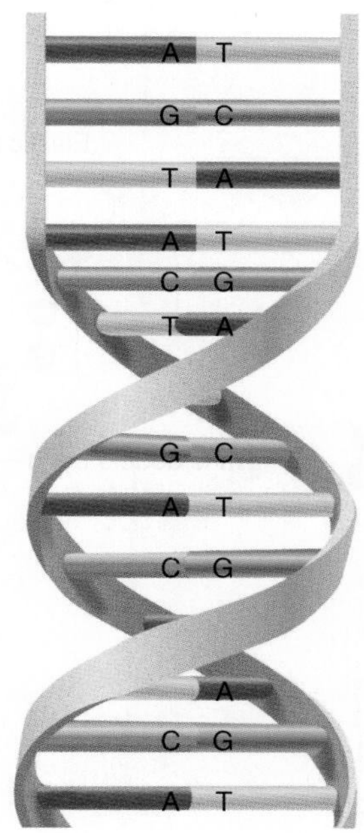

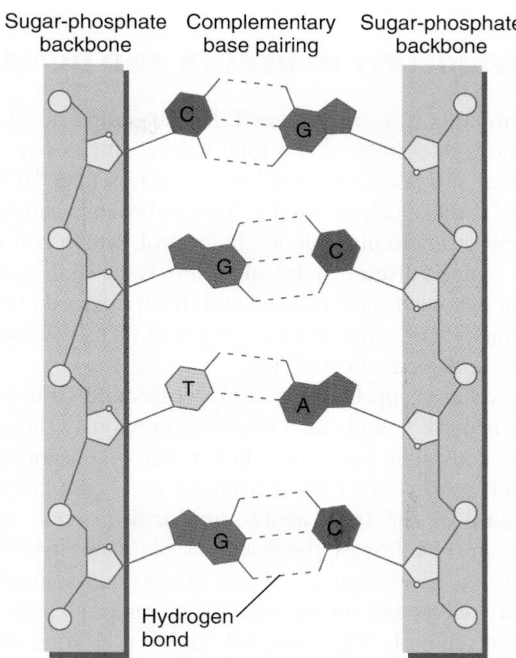

FIGURE 1.26 The double-helix structure of DNA. The two strands are held together by hydrogen bonds between complementary bases in each strand.

bases protruding to one side (fig. 1.25). In DNA, 1 sugar-phosphate strand is joined to another to form the famous **double helix** (fig. 1.26). The two DNA strands of the double helix are joined by weak hydrogen bonds between their bases. Not just any 2 bases can bond in this way. According to the **law of complementary base pairing**, *adenine can bond only with thymine*, and *guanine can bond only with cytosine*. Thus, if you know the sequence of bases on 1 strand, you can tell the sequence of bases on the complementary strand. However, it is the sequence of bases along 1 DNA strand that forms the genetic code.[3]

RNA also consists of long chains of nucleotides that are joined together (through dehydration synthesis) by sugar-phosphate bonds. However, RNA differs from DNA in that (1) the sugar is ribose, rather than deoxyribose; (2) its bases are *adenine*, *guanine*, *cytosine*, and *uracil* instead of thymine (fig 1.27); and (3) it is single-stranded rather than double-stranded. The RNA nucleotides can bond by complementary base pairing with the DNA bases on one of the strands (uracil in RNA bonds to adenine in DNA), so that the RNA molecule is a complementary copy of a portion of the DNA strand. The RNA is like a blueprint, which directs the synthesis of a specific protein according to the instructions coded in the base sequence of DNA (a process described in chapter 2).

DNA nucleotides contain | **RNA nucleotides contain**

Deoxyribose instead of **Ribose**

Thymine instead of **Uracil**

FIGURE 1.27 Differences between the nucleotides and sugars in DNA and RNA. DNA has deoxyribose and thymine; RNA has ribose and uracil. The other three bases are the same in DNA and RNA.

CHECK POINT

1. Why do proteins have such a great diversity of structure and function? Give examples of some of the different functional types of proteins we have in the body.
2. Describe how peptide bonds are formed, and identify the different orders of protein structure.
3. What are the substrates of an enzyme, and how do enzymes catalyze specific reactions that convert substrates into products?
4. Identify the subunits of DNA and RNA, and describe how the structures of DNA and RNA differ from each other.

FYI [3]The total human genome (all of the genes in a cell) consists of more than 3 billion base pairs that would be over a meter long if the DNA molecules were unraveled and stretched out.

Physiology in Health and Disease

Health authorities recommend that fat should not make up more than 30% of the total caloric intake per day, and that saturated fat should make up less than 10% of the total daily caloric intake. This is because saturated fat may promote high blood cholesterol, which is a risk factor in heart disease and stroke. Here is the saturated fat content of some foods: rapeseed oil (6%); olive oil (14%); margarine (17%); chicken fat (31%); beef fat (52%); butter fat (66%); and coconut oil (77%).

But the relationship between triglycerides and cardiovascular health is more complicated than was previously thought. Those fatty acids that have their first point of unsaturation (double bond) at the third carbon from the end—and are thus called **omega-3** or **n-3 fatty acids**—appear to provide some protection against cardiovascular disease. Cold-water fish (such as salmon, trout, and herring) and fish oil are rich sources of omega-3 fatty acids, and on this basis it seems prudent to eat fish or fish oil regularly. However, fish high on the food chain (such as salmon and tuna) can also concentrate mercury, which is a toxic pollutant. So it's wise to become educated in which fish are lowest in mercury and highest in omega-3 fatty acids (example: wild salmon have lower mercury levels than farmed salmon), and to eat accordingly.

There is yet another complication. Vegetable oils (which are unsaturated and liquid at room temperature) can be made into solids (such as margarine) by hydrogenating them. This results in the formation of **trans fatty acids**, in which the hydrogen atoms on each of 2 double-bonded carbons are on opposite sides of the fatty acid molecule (fig. 1.28). This is different from naturally occurring fatty acids, called *cis fatty acids*, in which the hydrogen atoms are on the same side (fig. 1.28). The resulting *trans fats* are used in almost all commercially prepared fried and baked foods. Since trans fats appear to increase the risk of cardiovascular disease, the Food and Drug Administration (FDA) now requires that trans fats be listed on food labels.

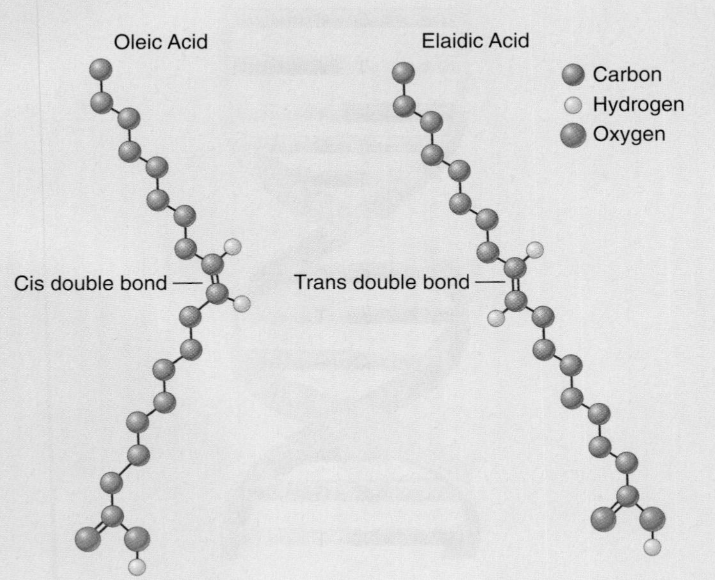

FIGURE 1.28 The structure of cis and trans fatty acids. Oleic acid is a naturally occurring fatty acid with 1 double bond. Notice that both hydrogen atoms (yellow) on the carbons that share this double bond are on the same side of the molecule—this is called the *cis* configuration. The *cis* configuration makes this naturally occurring fatty acid bend. The fatty acid on the right is the same size and also has 1 double bond, but its hydrogens here are on opposite sides of the molecule, known as the *trans* configuration. This makes the fatty acid stay straight, more like a saturated fatty acid. Note that only these hydrogens and the ones on the carboxyl groups (*bottom*) are shown. Those carbons that are joined by single bonds are also each bonded to 2 hydrogen atoms, but those hydrogens are not illustrated.

Physiology in Balance

Introduction to Human Physiology

Working together with . . .

Cardiovascular System

Homeostasis of blood pressure is maintained . . . p. 240

Respiratory System

Homeostasis of blood CO_2 and pH are maintained . . . p. 297

Urinary System

Homeostasis of blood potassium is maintained . . . p. 324

Endocrine System

Homeostasis of blood glucose is maintained . . . p. 186

Digestive System

Homeostasis of blood cholesterol is maintained . . . p. 348

Summary

The Science of Physiology 2

- Human physiology is the study of how our bodies normally function, and much of its information is derived from studying animals other than humans; pathophysiology is the study of the physiology of disease states.
- The science of physiology is best studied actively.

The Theme of Physiology Is Homeostasis 3

- Homeostasis is the dynamic constancy of the internal environment, whereby changes are maintained within limits by negative feedback loops.
- In negative feedback loops, deviations from a set point are detected by a sensor, which relays this information to an integrating center, which then activates an effector that works to correct that deviation.
- Positive feedback mechanisms amplify changes; these operate during the formation of a blood clot, and in the reproductive system for ovulation and labor contractions.

Basic Chemical Concepts Used in Physiology 7

- The valence electrons in an atom are those in its outermost shell; it is these electrons that participate in chemical reactions.
- The strongest bonds are covalent bonds, where atoms bond by sharing valence electrons; ionic bonds are weaker, and hydrogen bonds are weaker still.
- Molecules that are polar—usually because of bonds to oxygen or nitrogen—have positive and negative sides, and ions are charged because they either have extra electrons or are missing electrons.
- "Like dissolves like"; polar solvents such as water can dissolve polar compounds; nonpolar solvents are required to dissolve nonpolar molecules.
- Acids are molecules that donate H^+ (protons) to a solution; bases remove H^+ from the solution.
- Solutions with a higher H^+ concentration than water (a pH lower than 7) are acidic; those with a H^+ concentration lower than water (a pH greater than 7) are basic, or alkaline.

Carbohydrates and Lipids Have Different Characteristics 11

- Monosaccharides (including glucose, fructose, and galactose) can be combined by dehydration synthesis reactions into disaccharides and long molecules of polysaccharides.
- The polysaccharide glycogen, produced in the cells of the liver and skeletal muscles, can be broken down by hydrolysis reactions into its glucose subunits.
- Lipids are nonpolar organic molecules, divided into three major subclasses: triglycerides (fats and oils), phospholipids, and steroids.
- Triglycerides are stored in adipose tissue; they're formed by dehydration synthesis of fatty acids and glycerol.
- Fatty acids can have a double bond between 2 carbon atoms, in which case each of those carbons can bond only to 1 hydrogen; this is referred to as a point of unsaturation, and oils generally are more unsaturated than fats.
- Phospholipids, such as lecithin, have polar and nonpolar sides; this dual nature permits phospholipids to have a surfactant action (lowering the surface tension of water) and to form the major structure of the plasma (cell) membrane.
- Steroids are lipids with a unique ring structure, and examples include the steroid hormones of the ovaries (such as estradiol), the testes (testosterone), and the adrenal cortex (cortisol).

Proteins and Nucleic Acids 16

- Proteins are large polypeptides, formed by dehydration synthesis reactions that join amino acids together through peptide bonds.
- The polypeptide chain of a specific sequence of amino acids (the primary structure of the protein) is twisted in a helical shape (the secondary protein structure), and then bent and folded in a complex, three-dimensional shape (the tertiary protein structure).
- Enzymes are proteins that function as catalysts, which speed up reactions without being changed by the reaction or changing the nature of the reaction.
- Each enzyme catalyzes only a very specific reaction; this is because of the unique tertiary structure of the enzyme protein.
- The substrates of the enzyme dock into pockets in the tertiary structure of the protein called the active sites; this allows the substrates to react at a lower energy level than if the reaction occurred without the enzyme catalyst.
- DNA and RNA are macromolecules produced by dehydration synthesis reactions of nucleotides; this forms long chains, where the deoxyribose or ribose sugars are joined together through bonds to phosphate groups.
- Dehydration systhesis of nucleotides produces a molecule with a sugar-phosphate "backbone," from which the nucleotide bases protrude; the DNA bases are adenine, guanine, cytosine, and thymine.
- DNA is a double helix containing 2 sugar-phosphate strands joined together by hydrogen bonds between the complementary bases of each: guanine on one strand can bind only to cytosine (and vice versa) on the other; adenine on one strand can bind only to thymine (and vice versa) on the other.
- RNA is single-stranded, it contains the sugar ribose, and its bases are adenine, guanine, cytosine, and uracil (instead of thymine); it is formed by the pairing of RNA nucleotides to complementary bases on one strand of DNA.

Review Activities

Objective Questions: Test Your Knowledge

1. Which of the following statements regarding homeostasis is true?
   a. It refers to a state of absolute constancy of the internal environment.
   b. It requires the action of positive feedback loops.
   c. Effectors act to correct deviations from a set point.
   d. An integrating center activates a sensor.
2. Breathing adds oxygen to the blood and removes carbon dioxide. Therefore, which of the following deviations from a set point would you expect to stimulate breathing?
   a. A rise in blood oxygen and a rise in blood carbon dioxide
   b. A fall in blood oxygen and a fall in blood carbon dioxide
   c. A rise in blood oxygen and a fall in blood carbon dioxide
   d. A fall in blood oxygen and a rise in blood carbon dioxide
3. Contractions of the uterus during labor are stimulated by the hormone oxytocin, and the secretion of this hormone from the pituitary is stimulated by contractions of the uterus.

Which statement best describes the interaction between the pituitary gland and uterus in this example?
a. The uterus is the effector in a negative feedback loop.
b. The pituitary is the effector in a negative feedback loop.
c. The pituitary is the integrating center in a negative feedback loop.
d. The pituitary and uterus are interacting in a positive feedback loop.

4. Which of the following statements about an atom and an ion is false?
a. The atomic weight is the sum of the protons and neutrons.
b. The atomic number is equal to the number of protons.
c. An ion has the same number of electrons as it has protons.
d. Only the electrons in the outermost shell participate in chemical reactions.

5. Chemical bonds in which electrons are unequally shared between 2 atoms are
a. polar covalent bonds.
b. nonpolar covalent bonds.
c. ionic bonds.
d. hydrogen bonds.

6. Which of the following statements about water is true?
a. Water is a nonpolar molecule.
b. Water molecules sometimes dissociate into H^+ and OH^-.
c. Water is a good solvent for triglycerides and steroids.
d. Water molecules repel each other.

7. Organic acids, such as lactic acid,
a. can dissociate and donate H^+ to a solution.
b. have a carboxyl group.
c. can lower the pH of a solution.
d. all of these statements are true.

8. Solution *X* has a pH of 10, and solution *Y* has a pH of 7. Which of the following statements about these 2 solutions is true?
a. Solution *X* is basic, while solution *Y* is acidic.
b. Solution *Y* has three times the H^+ concentration of solution *X*.
c. Solution *X* has one-thousandth the H^+ concentration of solution *Y*.
d. Solution *X* is acidic, while solution *Y* is neutral.

9. Which of the following molecules is a monosaccharide?
a. Glycogen
b. Glucose
c. Lecithin
d. Sucrose

10. Which of the following is an example of a hydrolysis reaction?
a. The conversion of glycogen to glucose
b. The conversion of glucose and fructose to sucrose
c. The conversion of amino acids into a protein
d. The conversion of fatty acids and glycerol into triglycerides

11. Which of the following carbohydrates is found in the blood?
a. Glycogen
b. Sucrose
c. Lactose
d. Glucose

12. Which of the following statements about lipids is true?
a. Adipose tissue stores lecithin for energy.
b. Lipids have the general formula $C_nH_{2n}O_n$.
c. Lipids are mostly nonpolar organic molecules.
d. Lipids are soluble in water.

13. Which of the following is a subcategory of lipids that releases free fatty acids into the blood for energy when we are fasting?
a. Steroids
b. Triglycerides
c. Phospholipids
d. Maltose

14. Which of the following is a subcategory of lipids that can partially be immersed in water?
a. Steroids
b. Triglycerides
c. Phospholipids
d. Maltose

15. What are the subunits of proteins, and what are the chemical bonds that join them together?
a. Monosaccharides; glycosidic
b. Fatty acids; ester
c. Nucleotides; sugar-phosphate
d. Amino acids; peptide

16. Which of the following statements about enzymes is false?
a. Enzymes continue to work better the higher the temperature.
b. Each enzyme works best at a particular pH.
c. Enzymes convert their substrates into products.
d. Enzyme activity depends on the tertiary structure of the enzyme protein.

17. Which of the following statements regarding a molecule named RNA polymerase is true?
a. It is an enzyme.
b. It forms RNA.
c. It catalyzes a dehydration synthesis reaction.
d. All of these statements are true.

18. In a double-stranded DNA molecule, a guanine base on one strand will hydrogen bond to
a. a thymine base on the other strand.
b. a uracil base on the other strand.
c. a cytosine base on the other strand.
d. an adenine base on the other strand.

19. Which of the following statements about RNA is false?
a. RNA has ribose as its sugar.
b. RNA is double-stranded.
c. RNA has uracil instead of thymine.
d. RNA is made as a complementary copy of a portion of the DNA.

Essay Questions 1: *Test Your Understanding*

1. Define homeostasis and describe how negative feedback loops in general help maintain homeostasis.
2. Explain the negative feedback loop involved in the way that a thermostat maintains a constant room temperature, and compare this to how the body maintains a constant temperature.
3. Using only the information in this chapter, describe how regulation of the heart rate can operate to maintain homeostasis of blood pressure.
4. Describe the characteristics of a positive feedback loop, and give two examples of this in the body.
5. Explain the meaning of the expression "Like dissolves like," and give examples of polar and nonpolar compounds.
6. Distinguish between covalent bonds and ionic bonds.
7. Define the terms *acid* and *acidic*, and explain the pH scale.

8. Describe how monosaccharides and polysaccharides are related, and explain how they are interconverted in the body.
9. Define lipids, and describe the characteristics of the three major lipid subcategories.
10. Describe the structure of phospholipids and explain how their structure relates to their functions.
11. Explain, in terms of dehydration synthesis and hydrolysis reactions, how triglycerides are stored and how they can be used for energy.
12. Describe the levels of protein structure and explain why proteins can have a great diversity of structures.
13. Describe the chemical nature of enzymes and how enzymes catalyze reactions.
14. Explain the meaning of the pH optimum of enzymes, and why enzymes have a pH optimum.
15. Describe the structure of a nucleotide, and distinguish between DNA and RNA nucleotides.
16. Describe the law of complementary base pairing for DNA, and explain how that relates to the structure of DNA.
17. How does the structure of RNA compare with the structure of DNA, and how is the sequence of bases in RNA determined?

Essay Questions 2: *Test Your Analytical Ability*

1. Suppose you ate a candy bar. Using just the information in this chapter, explain how insulin secretion will help maintain homeostasis of your blood glucose concentration.
2. Given that if you hold your breath, your blood carbon dioxide concentration will rise, explain a negative feedback mechanism that could maintain homeostasis of your blood carbon dioxide concentration.
3. Suppose that you came across a person suffering from a rapid fall in blood pressure, perhaps because of blood loss. If you felt the pulse, would you expect it to be fast or slow? Explain how this could occur.
4. Suppose you poured vinegar and oil into a flask. What would happen? Now, imagine that you put in a pinch of lecithin. Where would you expect to see the lecithin? Explain your answer.
5. Suppose there was a drug that had about the same size and shape as a particular substrate, and could fit into the active site of a specific enzyme, but the drug couldn't participate in the reaction. Explain how this drug would affect the reaction catalyzed by that enzyme.
6. Trace, using the reactions of hydrolysis and dehydration synthesis, how a glucose molecule that is part of the starch in a potato could become part of a glycogen molecule in your liver.
7. Given that the DNA of genes makes molecules of RNA, and that different RNA molecules code for different proteins (including enzyme proteins), explain how the activation of different genes could change the metabolism of the cell.
8. Glucose travels in your blood dissolved in the blood plasma (the liquid portion of the blood), whereas cholesterol and triglycerides travel bound to protein carriers in the blood. Explain the reason for this difference.
9. Describe the digestion of a bite of bread once you've swallowed it and it reaches your stomach. (Hint: think of the enzymes involved and their pH optima.)
10. Normal blood pH is in the range of 7.35 to 7.45. Suppose that the blood contained too many ketone bodies or lactic acid. How would this affect the blood pH? What would be needed to bring the pH back to normal? Explain.

Web Activities

ARIS™ **For additional study tools, go to: www.aris.mhhe.com. Click on your course topic and then this textbook's author/title. Register once for a semester's worth of interactive activities and practice quizzing to help boost your grade!**

From Cells to Systems

CHAPTER

Chapter Outline

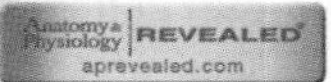

A virtual cadaver dissection experience

Homeostasis

Cells are alive and have structures within them—called organelles—that carry out the functions required for life. Their structural and functional characteristics organize cells into tissues, organs, and systems. Like a conductor regulating many members of an orchestra, the nervous and endocrine systems regulate the different tissues, organs, and systems so that homeostasis is maintained.

The genome (all the genes) determines the proteome (all of the proteins) in a person. Some of these proteins are enzymes, and it is the activity of the enzymes in each cell that determines its metabolism. All cells have metabolic pathways that provide energy-rich ATP, required for all types of cellular work. Glucose is particularly important and can be used to provide ATP through two related pathways, lactic acid fermentation and aerobic respiration. Homeostasis is maintained by the different body systems cooperating in these cellular activities.

Clinical Investigation

But if homeostasis is not maintained . . .

Eric had hepatomegaly (an enlarged liver) when he was just a few months old, and blood tests revealed that he also had hypoglycemia (low blood glucose concentration). They treated the hypoglycemia by giving Eric glucose intravenously. Physicians took a biopsy (living tissue sample) from his liver and found an abnormally large amount of glycogen within the cells and a deficiency in the enzyme glucose 6-phosphatase. The diagnosis was confirmed several months later by a newly developed test for a mutated gene, which was made possible by gene cloning and biotechnology. Now that Eric is a 5-year-old child, he eats uncooked cornstarch to prevent a recurrence of the hypoglycemia. Because his parents were told that Eric could not metabolize fructose or galactose, they give him a special diet.

What could have caused the hepatomegaly and hypoglycemia, and how would eating uncooked cornstarch help prevent this? Why can't Eric metabolize fructose and galactose? What is the relationship between a mutated gene and an enzyme deficiency? What's the function of glucose 6-phosphatase, and how does this relate to the liver?

The Cell Is the Basic Unit of Body Structure and Function

Our tissues, organs, and systems are composed of cells, which perform the functions required by the body. The plasma membrane is the highly functional cell structure that separates the intracellular from the extracellular environments. Cilia and flagella are structures capable of movement that may be associated with the plasma membrane.

Learning any science is like building a pyramid, with one layer on top of another and the entire structure dependent on the strength of the foundation. Chapter 1 contains a chemistry refresher that can be used as you and your instructor think best. Similarly, this chapter contains information that you may have learned previously, if you had a course in introductory biology. To be sure that your foundation is secure, it may be wise to review the cell biology in this chapter, or perhaps learn some of this information for the first time. You can also refer back to specific topics in these two chapters later as needed. The material in chapter 2, like the chemistry reviewed in chapter 1, is background information—part of the foundation for the pyramid you are constructing—in your study of physiology.

Most people think that size reflects complexity—that the bigger something is, the more parts it can contain, and so the more complex its functions can be. Although this thinking may be valid generally, it can ambush us when we consider cells. So tiny that they cannot be seen without a microscope (except for a human egg cell, which is about the size of a period), each cell is a living entity unto itself, capable of metabolism and most of reproduction. Yet the cells of our body are specialized to perform the great diversity of tasks required of our tissues, organs, and systems.

Although cells are small, each contains parts that are smaller still, which are needed to perform the specialized functions of the cell. The smaller, subcellular structures within a cell are known generally as **organelles**. The size and shape of cells, and the organelles they contain, vary greatly because the tasks that they perform vary in different organs. However, most cells contain the structures illustrated in figure 2.1.

Each cell is surrounded by a very thin structure called a **plasma** (or **cell**) **membrane**. The term *membrane* can refer to any relatively thin structure, and is often confusingly used to name structures that have very different compositions. The plasma (or cell) membrane separates the fluid outside the cell (the **extracellular** environment, or compartment), from the fluid and its contents inside the cell (the **intracellular** environment, or compartment). The term *plasma* is used because the extracellular fluid is either blood plasma (the

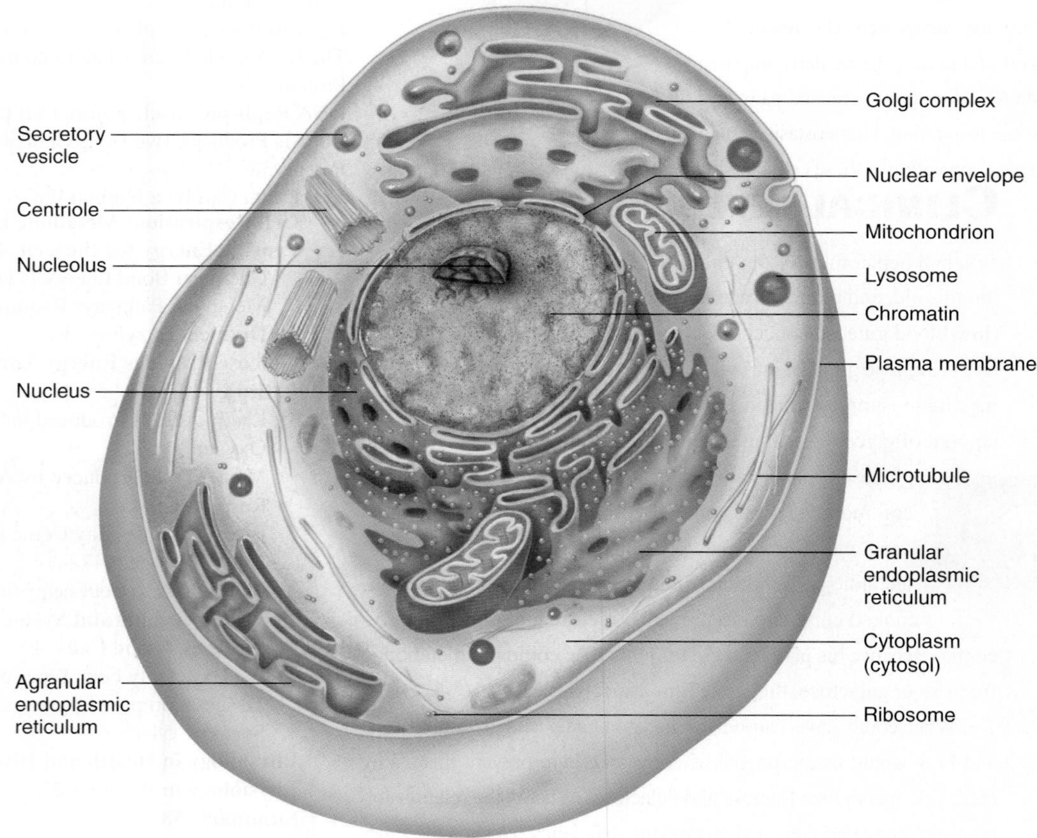

FIGURE 2.1 A generalized human cell showing the principal organelles. Since most cells of the body are highly specialized, they have structures that differ from those shown here.

liquid portion of the blood) or derived from blood plasma. The term *plasma membrane* is preferred to "cell membrane," because membranes also surround some of the intracellular organelles; for that reason, the term *plasma membrane* will be used throughout the rest of this text.

The part of the cell located inside the plasma membrane but outside the nucleus is the **cytoplasm**. The cytoplasm contains numerous organelles, some of which are surrounded by their own membranes. The term **cytosol** refers specifically to the fluid portion of the cytoplasm, exclusive of the organelles. The **nucleus** is the largest organelle and contains the genetic information in the DNA.

Plasma Membrane Structure Allows for Complex Functions

The body is composed mostly of water; approximately 67% of the total body water is within the intracellular compartment, while the remainder is in the extracellular compartment. Since the plasma membrane separates these two body compartments, it must be composed of a material that is not soluble in water—lipids. Lipids, however, are hydrophobic (discussed in chapter 1), so how can they be in an aqueous (watery) environment? Phospholipids, which are part polar and part nonpolar (chapter 1), provide the solution. The polar portion of the phospholipid molecule can face water on the inside or outside of the cell. A *double layer of phospholipids* is thus arranged so that the polar portions face water on both sides of the membrane, while the nonpolar portions face each other in the center of the plasma membrane (fig. 2.2).

The double phospholipid membrane is more liquid than solid. Within the phospholipid "sea" are icebergs—proteins. Some membrane proteins span the thickness of the entire membrane, while others are only partially embedded in the phospholipids. The mental model of membrane structure depicted in figure 2.2 has been called the **fluid-mosaic model**. Functions of membrane proteins include acting as enzymes, transport carriers, receptors for regulatory molecules, and many others.

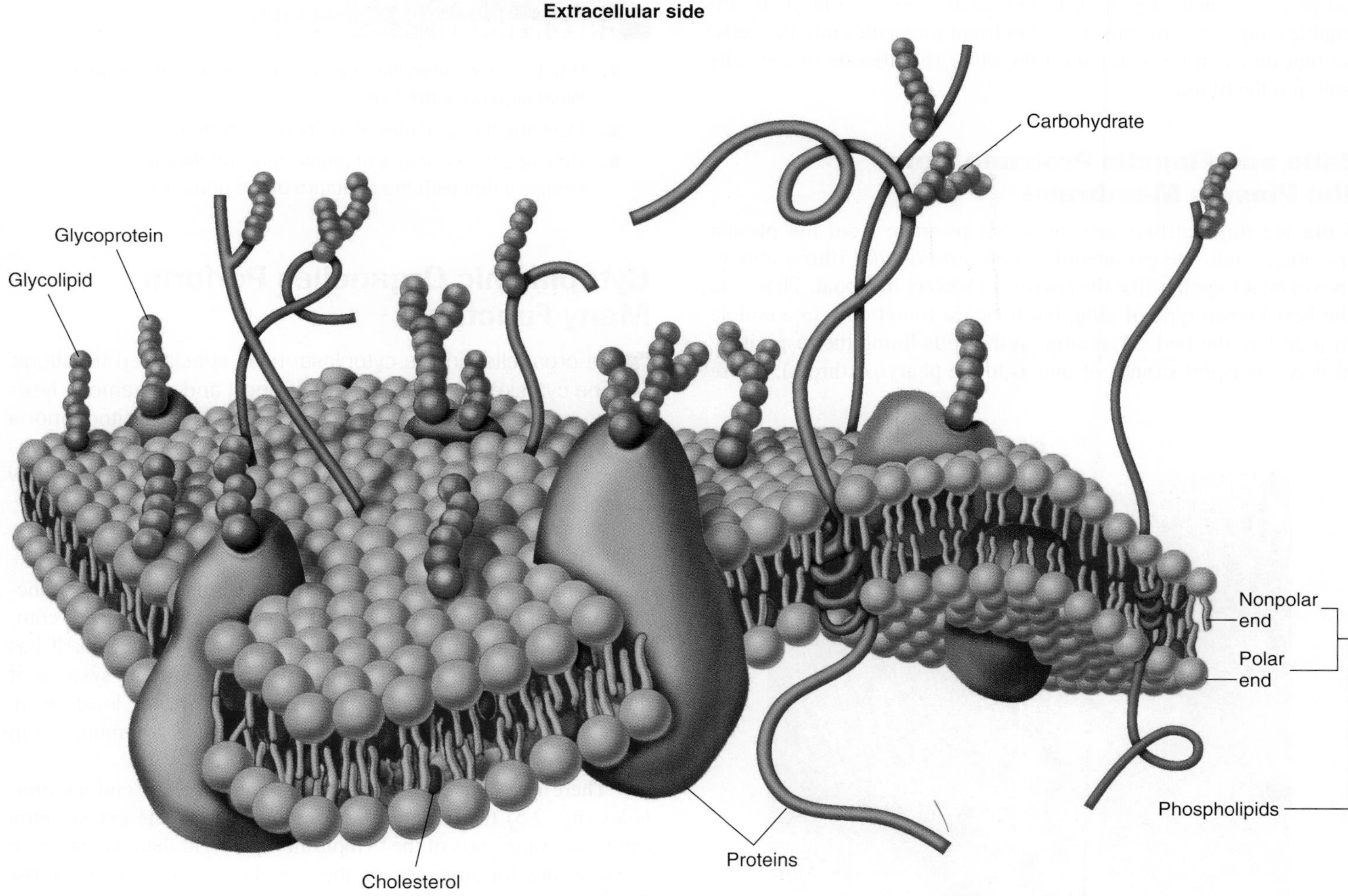

FIGURE 2.2 The fluid-mosaic model of the plasma membrane. The membrane consists of a double layer of phospholipids, with the polar regions (shown by spheres) oriented outward and the nonpolar hydrocarbons (wavy lines) oriented toward the center. Proteins may completely or partially span the membrane. Carbohydrates are attached to the outer surface.

Microvilli Increase the Surface Area of the Plasma Membrane

The plasma membrane is the barrier that a molecule in the extracellular fluid must cross to enter the cell. Thus, anything that increases the surface area of this membrane will allow faster transport, in the sense that more molecules could pass per given time period. Consider, for example, a theme park that had only one side with an entrance facing a parking lot, compared to a theme park that had two sides with entrances facing parking lots. With more area containing entrances, more people could enter the park every minute in the second example. A cell with more plasma membrane, containing entrances for more molecules (by means of structures discussed in chapter 3), would have a faster rate of transport.

Microvilli are tiny folds of the plasma membrane found on the apical surface (the surfaces facing the lumen, or cavity) of the cells that line the small intestine, and in some other parts of the body. These folds are extremely tiny—too small to be seen with an ordinary (light) microscope; you need a powerful electron microscope to see them (fig. 2.3). In the small intestine, the microvilli enable a rapid rate of transport of nutrient molecules into the cells, so that they can be secreted on the other (basal) side of the cells and into the blood.

Cilia and Flagella Protrude from the Plasma Membrane

Cilia are tiny hairlike structures that protrude from the plasma membrane into the extracellular fluid. *Motile cilia* (those able to move) beat together like the strokes of rowers in a boat. These are the best-known type of cilia, but they are found only in a couple of places in the body. The cilia on the cells lining the respiratory airways transport strands of mucus to the pharynx (throat), where the mucus can be swallowed or expectorated. Cilia on the cells that line the uterine tubes beat to produce currents that draw the ovum (egg cell) into the tube and then move the ovum along toward the uterus. Almost every cell in the body has a single, nonmotile *primary cilium*. The function of the primary cilium is more mysterious, but scientists think that it generally has a sensory function.

Cilia are composed of *microtubules*—thin cylinders formed from proteins—and are surrounded by a part of the plasma membrane. There are 9 pairs of microtubules arranged around the circumference of the cilium. In motile cilia, there is also a pair of microtubules in the center, producing an arrangement known as "9 + 2" (fig. 2.4). The nonmotile cilia do not have the central pair of microtubules, so their arrangement is "9 + 0."

Sperm are the only cells in the body that have flagella. A flagellum is a single, whiplike structure that propels the sperm through its environment. Like the motile cilia, a flagellum has the "9 + 2" arrangement of microtubules. The subject of sperm motility and function is considered with the reproductive system in chapter 15.

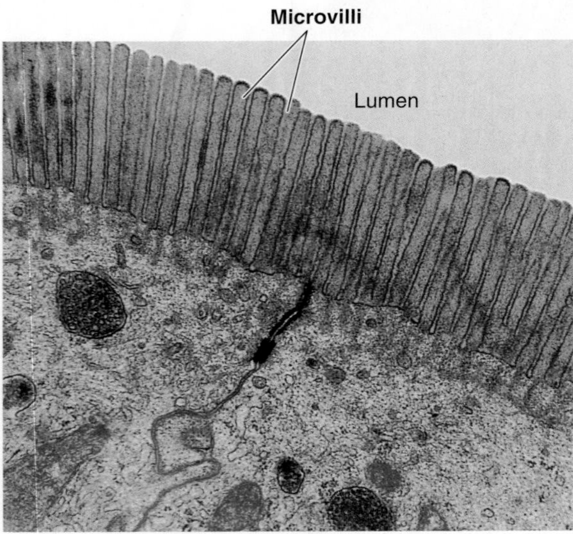

FIGURE 2.3 Microvilli in the small intestine. Microvilli are seen in this colorized electron micrograph. The microvilli face the lumen, or cavity, of the small intestine.

CHECK POINT

1. What are the intracellular and extracellular compartments? What separates the two?
2. Describe the structure of the plasma membrane.
3. Describe the structure of motile cilia and flagella, and compare that with the structure of the primary cilium.

Cytoplasmic Organelles Perform Many Functions

The organelles in the cytoplasm have specialized functions. The cytoskeleton provides for support and movement; lysosomes remove molecules and other organelles; mitochondria supply cellular energy; ribosomes are factories for protein synthesis; and the endoplasmic reticulum and Golgi complex allow molecules to be packaged into vesicles.

A cell, like the whole body, must perform many different functions at the same time, and to do this it requires specialized components. Like the different organs of the body, the cell itself has little organs—the organelles. And—like the organs and systems of the body—the organelles of a cell must be in specific locations in order to perform their functions, and often must communicate in some way with each other.

There is a latticework of protein *microfilaments* and *microtubules* (fig. 2.5) that function as a **cytoskeleton**. The cytoskeleton can make some parts of the cytoplasm more rigid than others, serve as an anchor for structures in the cytoplasm or protruding across the plasma membrane, and also serve as a transport system within a cell. This transport function is provided by other proteins associated with the cytoskeleton, which can move organelles along "tracks" of cytoskeleton from one part of the cell to another. Also, contractile proteins (like those in our muscles) associated with the cytoskeleton

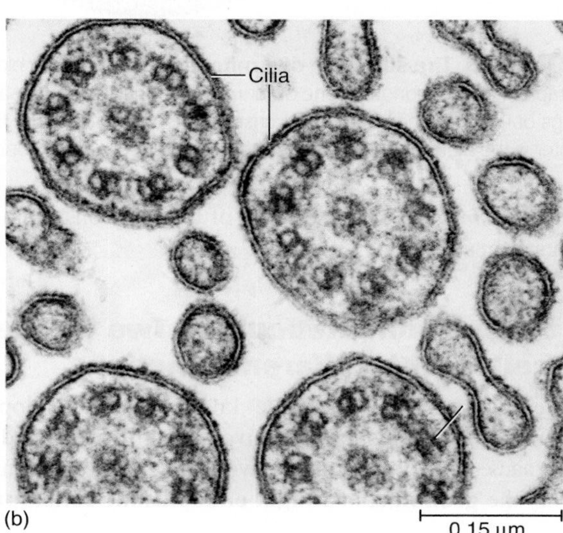

FIGURE 2.4 Cilia, as seen with the electron microscope. (*a*) Scanning electron micrograph of cilia on the epithelium lining the trachea; (*b*) transmission electron micrograph of a cross section of cilia, showing the "9 + 2" arrangement of microtubules within each cilium.

can produce movement of the cell, allowing regions of the plasma membrane to pinch in to form pockets (discussed in chapter 3) or certain white blood cells to move like an amoeba. A living cell is much like a miniature city, with different regions, specialization of tasks, and much movement and commerce between its parts.

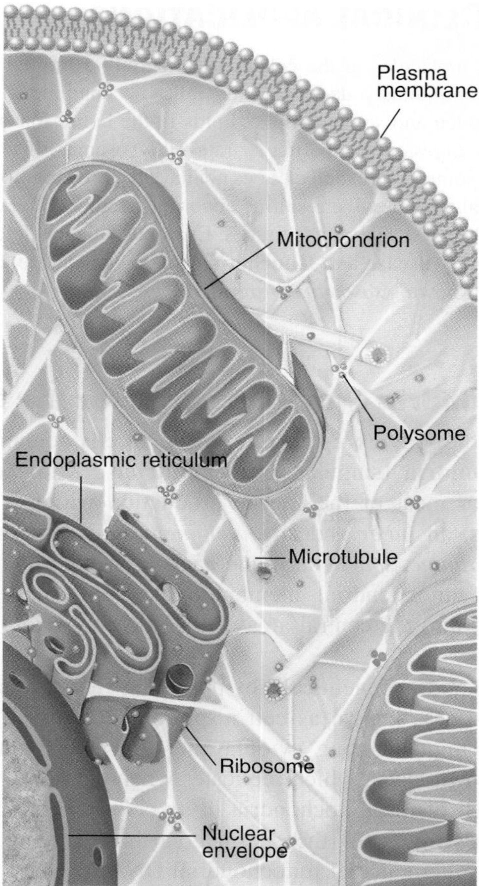

FIGURE 2.5 The formation of the cytoskeleton by microtubules. Microtubules are also important in the motility (movement) of the cell, and movement of materials within the cell.

Lysosomes Contain Digestive Enzymes

Lysosomes are organelles, surrounded by membranes, which contain more than 60 different digestive enzymes. When a cell eats by phagocytosis, or takes in part of its extracellular environment by pinching in a small portion of its plasma membrane (a process called endocytosis, discussed in chapter 3), it forms a membranous bag (a *vesicle*) containing molecules from the extracellular environment. This vesicle then fuses with the lysosome, so that the digestive enzymes of the lysosome can work on the molecules within the vesicle. In this capacity, the lysosome works like the "stomach" of a cell.

Lysosomes can also destroy other organelles. This is a normal process—called **autophagy** ("self-eating")—required for the turnover of old organelles and for terminating the function of certain organelles. Lysosomes have been called "suicide bags," because if their membrane should break, the leakage of digestive enzymes into the cytoplasm would kill the cell. This happens naturally at times, in a process called **programmed cell death**, or **apoptosis**. Apoptosis is required for the organ remodeling that occurs during embryonic development, for example, and for the normal turnover of certain cells in many organs. This process will be described in more detail in a later section.

Clinical Applications

Most molecules in the cell persist for only a limited time. They are continuously destroyed and must be continuously replaced. Glycogen and some complex lipids in the brain, for example, are normally digested at a particular rate by lysosomes. If a person, because of some genetic defect, does not have the proper amount of these lysosomal enzymes, the resulting abnormal accumulation of glycogen and lipids could destroy the tissues. Examples of such defects include **Tay Sach's disease** and **Gaucher's disease**, which are examples of the forty known *lysosomal storage diseases*.

Mitochondria Provide Energy for the Cell

All cells in the body, except mature red blood cells, contain hundreds of **mitochondria** (singular, **mitochondrion**). Mitochondria are the sites for most of the energy production within a cell. Two membranes, an *inner* and an *outer membrane*, separated by an *intramembranous space*, surround each mitochondrion (fig. 2.6). The inner mitochondrial membrane has folds, called *cristae*, that project into the central area of the mitochondrion, known as the *matrix*.

Mitochondria benefit cells by providing energy in an oxygen-utilizing process known as cell respiration (discussed in a later section).[1] Unique among cytoplasmic organelles, mitochondria have their own DNA, which replicates when the mitochondria reproduce. All of the mitochondria in a person's body came from the mother's egg cell (the sperm contribute essentially none), and so the inheritance of mitochondrial DNA is strictly mother to child. This is important in tracking certain diseases inherited in the mitochondrial DNA.

Ribosomes Are the Protein Factories

Proteins are produced by dehydration synthesis reactions of amino acids to form peptide bonds, and the specific sequence of amino acids in a particular protein—the primary structure of the protein (chapter 1)—is determined by information contained in a type of RNA known as *messenger RNA* (discussed later). However, in order for the messenger RNA code to be read, it must enter into a **ribosome**. For that reason, ribosomes have been described as "protein factories."

Each ribosome is very tiny (about 25 nanometers); some are free in the cytoplasm, and some are on the surface of an organelle described next, the endoplasmic reticulum. In an electron microscope ribosomes look like dark dots, but actually each is composed of two subunits made of protein and a type of RNA called *ribosomal RNA*. Contrary to what most scientists expected, it now appears that the ribosomal RNA functions as enzymes (called *ribozymes*) to catalyze some of the reactions required for protein synthesis.

FYI [1]Although the use of oxygen in cell respiration is required for energy production, oxygen can have a dark side. Mitochondria can produce molecules known as superoxide radicals that provoke an "oxidative stress," which may contribute to various diseases and aging. So, although the presence of mitochondria is essential for our lives, it is not without costs.

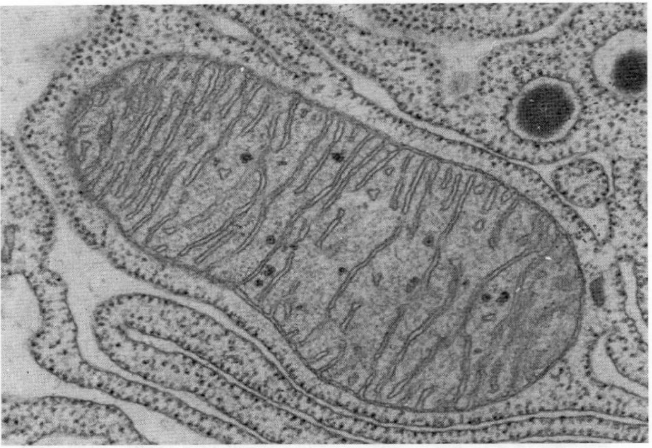

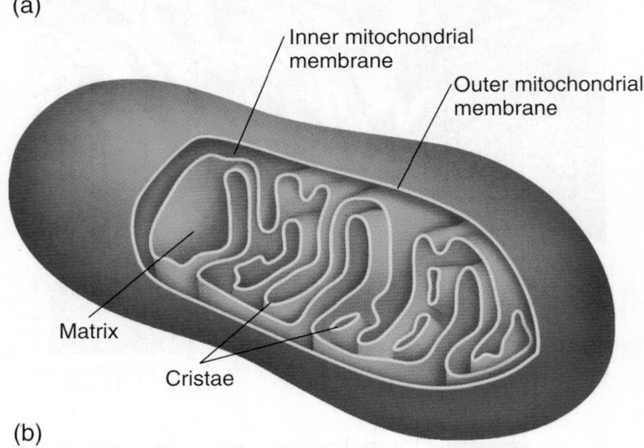

FIGURE 2.6 The structure of a mitochondrion. (*a*) An electron micrograph of a mitochondrion. The outer mitochondrial membrane and the infoldings of the inner membrane—the cristae—are clearly seen. The fluid in the center is the matrix. (*b*) A diagram of the structure of a mitochondrion.

The Endoplasmic Reticulum: Two Related Organelles with Different Functions

Most cells contain an organelle known as the endoplasmic reticulum, which forms an extensive system of flattened membranous sacs. There are actually two types, with different functions: (1) the **granular**, or **rough**, **endoplasmic reticulum**; and

Clinical Applications

The agranular endoplasmic reticulum in liver cells contains enzymes used for the inactivation of steroid hormones and many drugs. When people take certain drugs (such as alcohol and phenobarbital) for a long period of time, increasingly large doses of these compounds are required to achieve the effect produced initially. This phenomenon, called **tolerance**, is accompanied by growth of the agranular endoplasmic reticulum, and thus an increase in the amount of enzymes that work to inactivate these drugs.

(2) the **agranular**, or **smooth**, **endoplasmic reticulum** (fig. 2.7). The granular, or rough, endoplasmic reticulum has ribosomes on its surface and functions in protein synthesis and secretion. The agranular, or smooth, endoplasmic reticulum serves different functions in different cells. In some cells, such as in striated muscle cells, it serves as a site for the storage of calcium ions (Ca^{2+}). In other cells, the smooth endoplasmic reticulum contains particular enzymes that function to metabolize biologically active molecules.

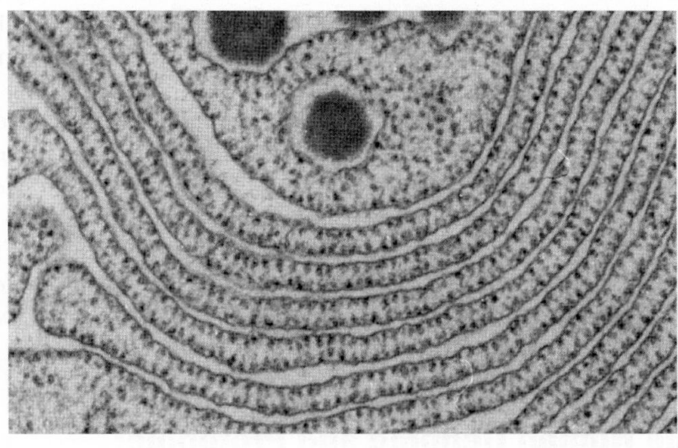
(a)

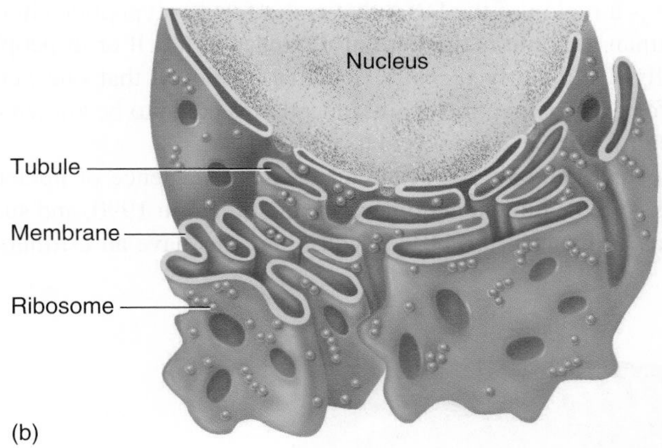

(b)

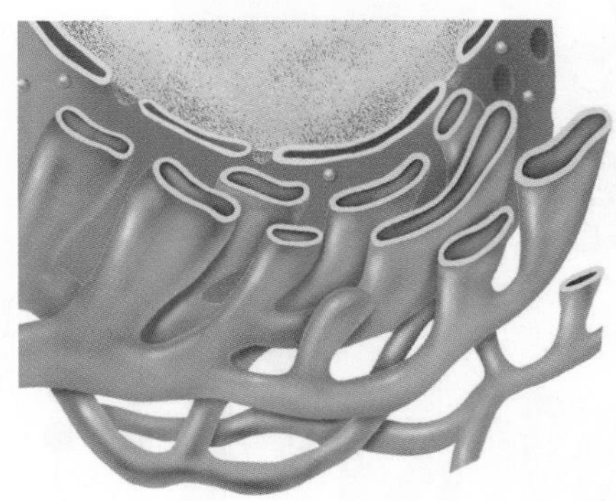
(c)

FIGURE 2.7 The endoplasmic reticulum. *(a)* An electron micrograph of a granular endoplasmic reticulum (about 100,000×). The granular endoplasmic reticulum *(b)* has ribosomes attached to its surface, whereas the agranular endoplasmic reticulum *(c)* lacks ribosomes.

The Golgi Complex Forms Different Vesicles

The **Golgi complex**, also called the **Golgi apparatus**, consists of several flattened sacs, stacked like pancakes (fig. 2.8). The sacs are hollow, and their cavities are known as *cisternae*. One side of the stack faces the rough endoplasmic reticulum, which buds off tiny vesicles containing molecules it produces. These vesicles fuse with the first sac of the Golgi complex, which may then modify the product and pass it along to the next sac, until the last sac buds off a vesicle containing the modified product. This vesicle enters the cytoplasm, and then may have different fates depending on the nature of the product it contains. For example, those vesicles that

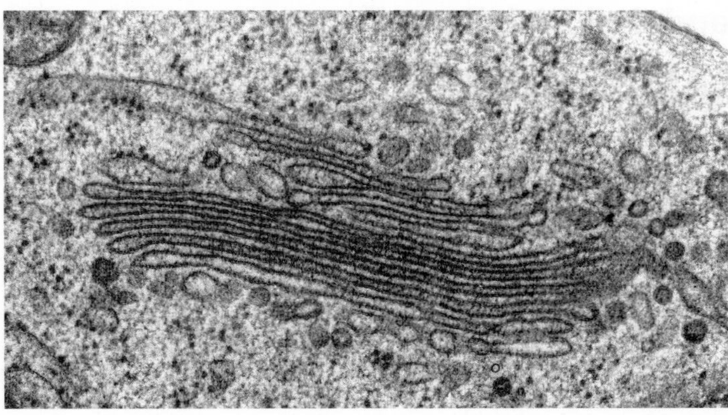
(a)

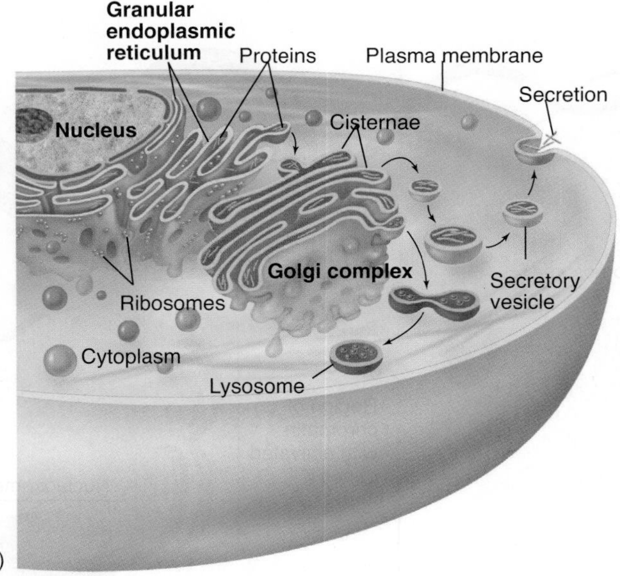

(b)

FIGURE 2.8 The Golgi complex. *(a)* An electron micrograph of a Golgi complex. Notice the formation of vesicles at the ends of some of the flattened sacs. *(b)* An illustration of the processing of proteins by the granular endosplasmic reticulum and Golgi complex.

contain lysosomal enzymes become lysosomes; those that contain other types of products may become storage granules. Still other vesicles will contain products that the cell must secrete, and those vesicles will fuse with the plasma membrane so that their contents can be released into the extracellular environment.

CHECK POINT

1. What is the cytoskeleton, and what functions does it perform?
2. Describe the structure and function of mitochondria.
3. Describe the composition of ribosomes and their function.
4. Distinguish between the structure and function of the granular and agranular endoplasmic reticulum.
5. Describe the structure and function of the Golgi complex.

Cell Nucleus, DNA Function, and Cell Division

Within the nucleus, DNA in the form of chromatin has a code that is transcribed into RNA, and this code is later translated into the primary structure of proteins. Before a cell can divide, its DNA must replicate so that identical copies can be passed to each daughter cell. DNA transcription, RNA translation, and DNA replication are dependent on the complementary pairing of nucleotide bases.

Most people have heard of chromosomes, and most who have had a biology course have seen photographs of them. Yet this widespread knowledge is often accompanied by some misconceptions. A common one is that the chromosomes are the DNA, and the form of the genetic material inside cells. This is misleading on two counts: (1) chromosomes contain important regulatory proteins as well as DNA; and (2) the chromosomes that are seen and counted under a microscope are the short, stubby form of DNA that is packaged so that it can be moved into the daughter cells during cell division. The form of DNA and protein that serves as the functioning genetic material within the nuclei of nondividing cells, known as **chromatin**, is long and threadlike. The way the chromatin is spooled onto proteins to form *nucleosome* particles, and the relationship between DNA, chromatin, and chromosomes, is depicted in figure 2.9.

The functioning DNA in the nucleus of nondividing cells directs the synthesis of RNA, which must then leave the nucleus by passing through its surrounding membrane, termed the *nuclear envelope*. The nuclear envelope is composed of an inner and an outer membrane fused together by special protein channels, the *nuclear pore complexes*, which form openings known as *nuclear pores* (fig. 2.10). The RNA made within the nucleus can pass through the nuclear pores to enter the cytoplasm.

The Human Genome and Proteome: A Success Story Still Unfolding

A **gene** is a region of the DNA that codes for a polypeptide chain. The human **genome** is the total of all genes in a cell or in people generally. Until fairly recently, scientists believed that one gene codes for one polypeptide chain; indeed, this came to be known as the "central dogma" of molecular biology.

An international effort to determine the sequence of nucleotide bases in the human genome was undertaken in 1990, and success was announced in 2001. Because humans have an estimated

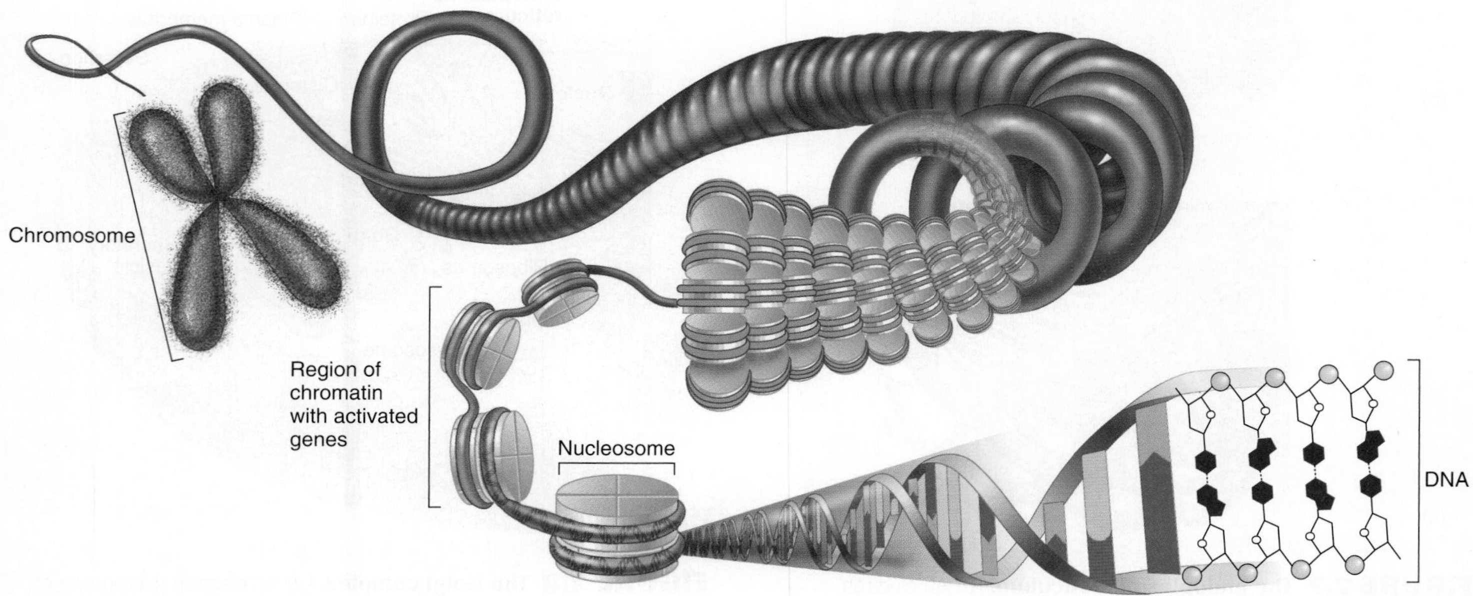

FIGURE 2.9 The structure of chromatin. Part of the DNA is wound around complexes of histone proteins, forming particles known as nucleosomes.

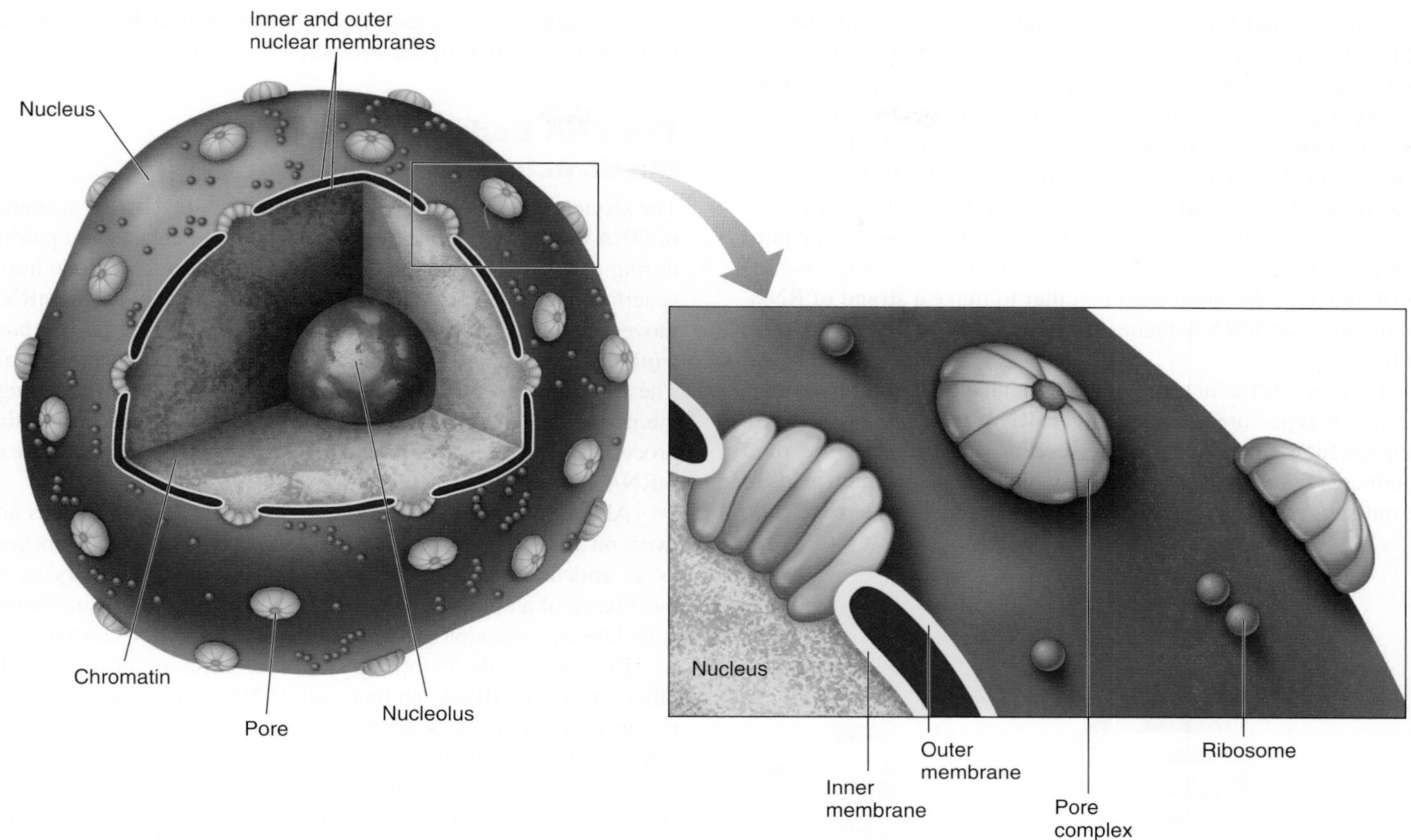

FIGURE 2.10 The structure of the cell nucleus. Illustration of a nucleus, showing the nuclear envelope and nuclear pores as well as other structures associated with the nucleus.

100,000 or so different types of proteins, scientists generally believed that we must have about that many different genes. Consequently, most scientists were surprised when it turned out that there are "only" about 25,000 genes in the human genome.

The term **proteome** has been coined to refer to all of the different types of proteins produced by the human genome. Since the proteome is much larger than the genome, it must mean that one gene can code for a number of different polypeptide chains. There are several ways that this is accomplished, affording many sites for potential regulation. Academic and biotechnology research laboratories are currently busy trying to identify all of the proteins in the human proteome, the remaining portions of the genome still undiscovered, and the ways in which the genome and proteome are regulated.

The numbers of genes in the genome, let alone proteins in the proteome, can seem overwhelming. However, scientists estimate that only about 300 genes are active in a given cell at a given time. Genes that are functional in a particular cell can be temporarily shut off and later turned back on. Such alteration of genetic expression is key to how certain physiological processes are regulated. Also, many genes are permanently inactivated in each specialized cell type. For example, a gene specific for a muscle cell may be silenced in an epithelial cell. The process of cell specialization, or **differentiation**, by which the different tissues and organs are formed, requires that certain genes be permanently turned on while others are turned off.

Clinical Applications

The differentiated cells of an adult are derived, or "stem from," those of the embryo. Early **embryonic stem cells** can become any cell in the body—they are said to be *totipotent*. As development proceeds, most genes are silenced as cells become more differentiated. **Adult stem cells** can differentiate into a range of specific cell types, but are not totipotent. Adult stem cells in the bone marrow can form the different types of blood cells, for example, and neural stem cells can differentiate into neurons and glial cells. Many scientists hope that stem cells grown in tissue culture might someday be used to grow transplantable tissues and organs.

DNA Acts as a Template for the Synthesis of RNA

A gene is typically a region of DNA several thousand nucleotide bases long. In order for the gene to determine the sequence of amino acids in a protein—the protein's primary structure (chapter 1)—the DNA of the gene must first direct the synthesis of a complementary RNA molecule. This DNA-directed synthesis of RNA is known as **genetic transcription**.

In genetic transcription, an enzyme called *RNA polymerase* breaks the weak hydrogen bonds between DNA bases in a part of the double helix containing a code for "start." Only one of the two DNA strands will be used to direct the synthesis of RNA (fig. 2.11). RNA nucleotides around this DNA strand form hydrogen bonds with their complementary DNA bases. An RNA nucleotide containing adenine will bond to the DNA base thymine; an RNA nucleotide containing uracil will bond to the DNA base adenine; and cytosine and guanine containing nucleotide bases will bond according to the law of complementary base pairing (chapter 1). The ribose sugars and phosphates of the ribonucleotides are joined together to make a strand of RNA, and the finished RNA detaches from the DNA so that it can leave the nucleus.

There are different types of RNA that have different functions. The major types are (1) **messenger RNA (mRNA)**, which is a complementary copy of a gene, coding for the structure of a protein; (2) **transfer RNA (tRNA)**, which transfers amino acids to the messenger RNA, helping to translate the nucleotide base code into an amino acid sequence; and (3) **ribosomal RNA (rRNA)**, which makes up part of the structure of ribosomes.[2]

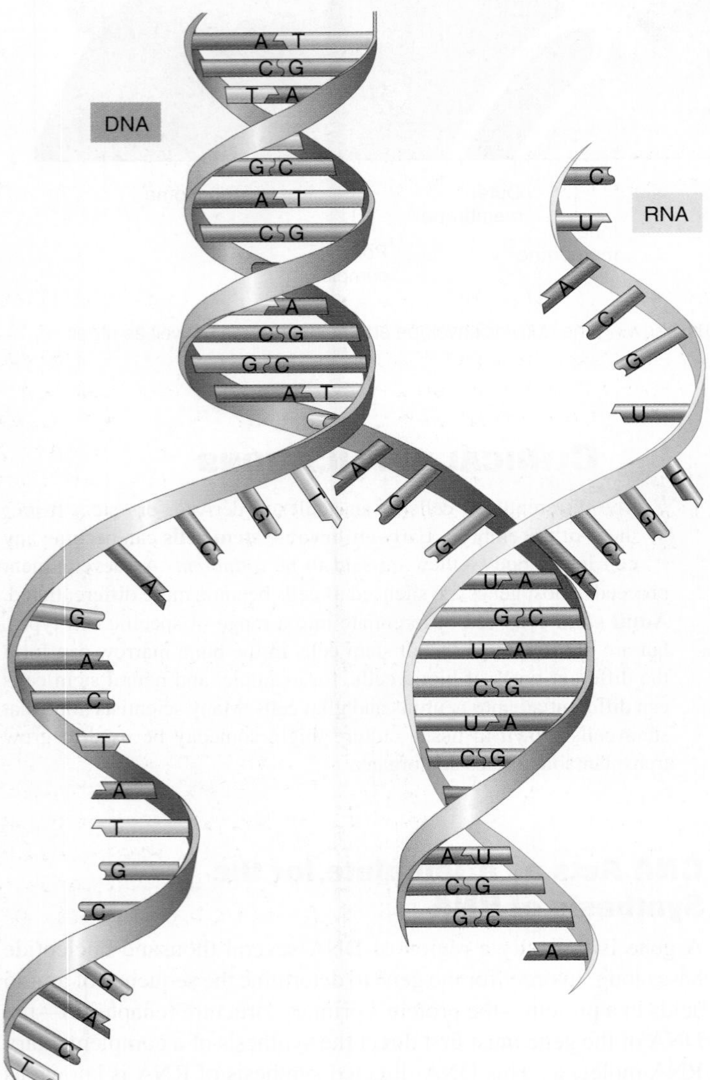

FIGURE 2.11 RNA synthesis (transcription). Notice that only one of the two DNA strands is used to form a single-stranded molecule of RNA.

The RNA Code Is Translated into the Structure of Proteins

The sequence of bases in the mRNA is determined by the sequence of DNA bases in the gene, because of complementary base pairing during genetic transcription. Translation of the mRNA code into a specific sequence of amino acids in a protein occurs as the mRNA moves through ribosomes. Every three mRNA bases, or *base triplet*, is a code word—called a **codon**—for a specific amino acid. These codons are "read" in sequence as the mRNA moves through the ribosome. The growing polypeptide chain formed during this process has a sequence of amino acids specified by the sequence of mRNA codons (fig. 2.12).

Although transfer RNA (tRNA) is single-stranded, it bends and twists on itself. One end of this structure has a base triplet that functions as an **anticodon**, with three nucleotide bases complementary to the three bases of an mRNA codon. There are different tRNA molecules, with different anticodons, for each of the different mRNA codons.

Enzymes in the cytoplasm join a specific amino acid to the other end of the tRNA, so that each tRNA with a particular anticodon base triplet is bonded to a particular amino acid. As the mRNA moves through a ribosome, each mRNA codon bonds by complementary base pairing to only a specific tRNA, which carries a specific amino acid. When the second tRNA binds to the second codon, the amino acid it carries forms a peptide bond with the first amino acid. A third tRNA then binds to the third codon, bringing a third amino acid to form a peptide bond with the second amino acid. In this way, a polypeptide grows as new amino acids are added to its growing tip (fig. 2.13).

CLINICAL INVESTIGATION CLUES

Remember that Eric was found to have a mutated gene and a defective enzyme.

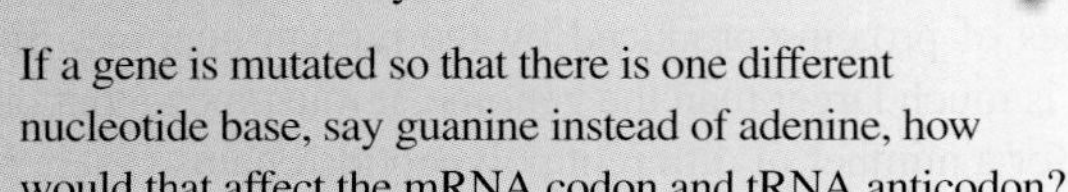

- If a gene is mutated so that there is one different nucleotide base, say guanine instead of adenine, how would that affect the mRNA codon and tRNA anticodon?
- Remembering that enzymes are proteins, how could such a change influence the ability of the enzyme to function?

DNA Replicates Itself Before Cell Division

When a cell is going to divide, the DNA must make a copy of itself so that one copy can later be distributed to each daughter cell. The ability of DNA to copy itself is called **DNA replication**. When

FYI [2]There is also a pre-mRNA—a larger RNA from which segments are cut and spliced together to make mRNA. In the continuing saga of scientific discovery, a new process known as RNA interference has recently been uncovered. This involves small RNA strands known as short interfering RNA (siRNA) and microRNA (miRNA), which interfere with the ability of specific mRNA molecules to produce proteins by suppressing the mRNA expression or promoting its destruction. Scientists hope for future medical applications involving siRNA and miRNA.

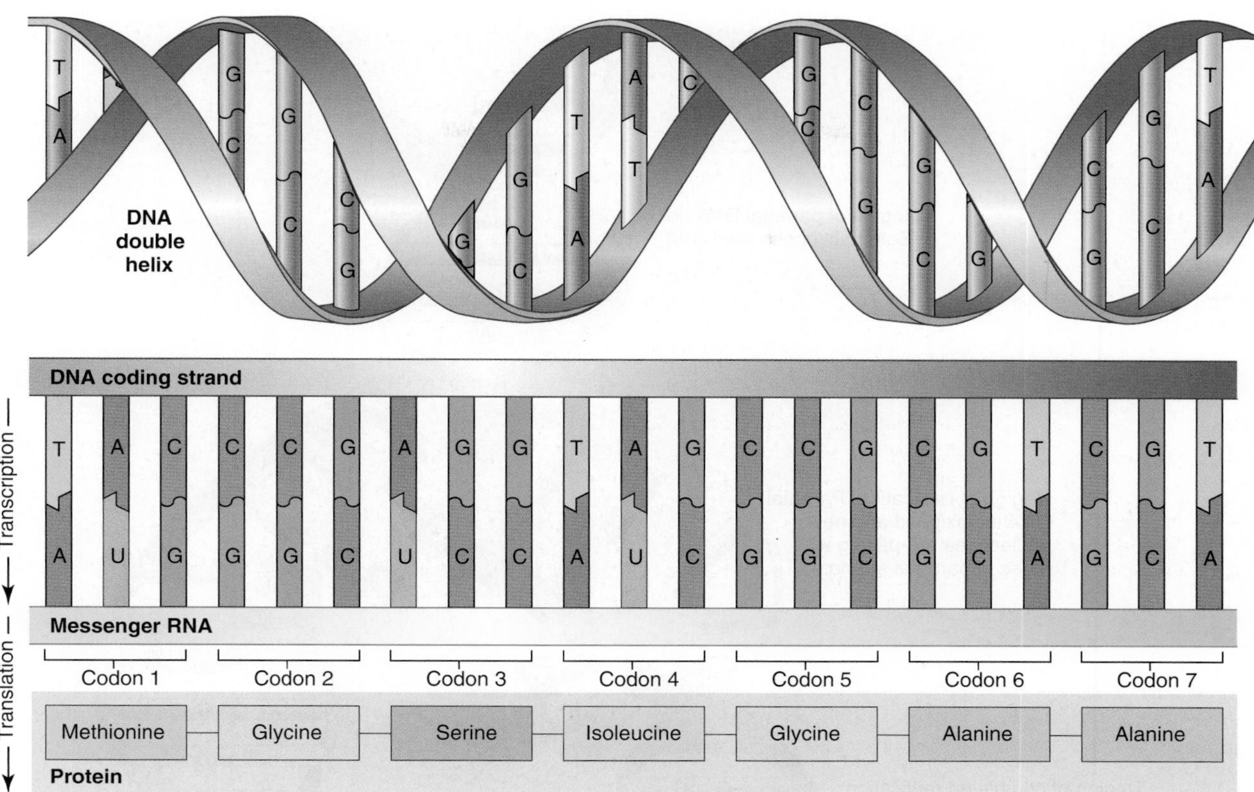

FIGURE 2.12 Transcription and translation. The genetic code is first transcribed into base triplets (codons) in mRNA and then translated into a specific sequence of amino acids in a polypeptide.

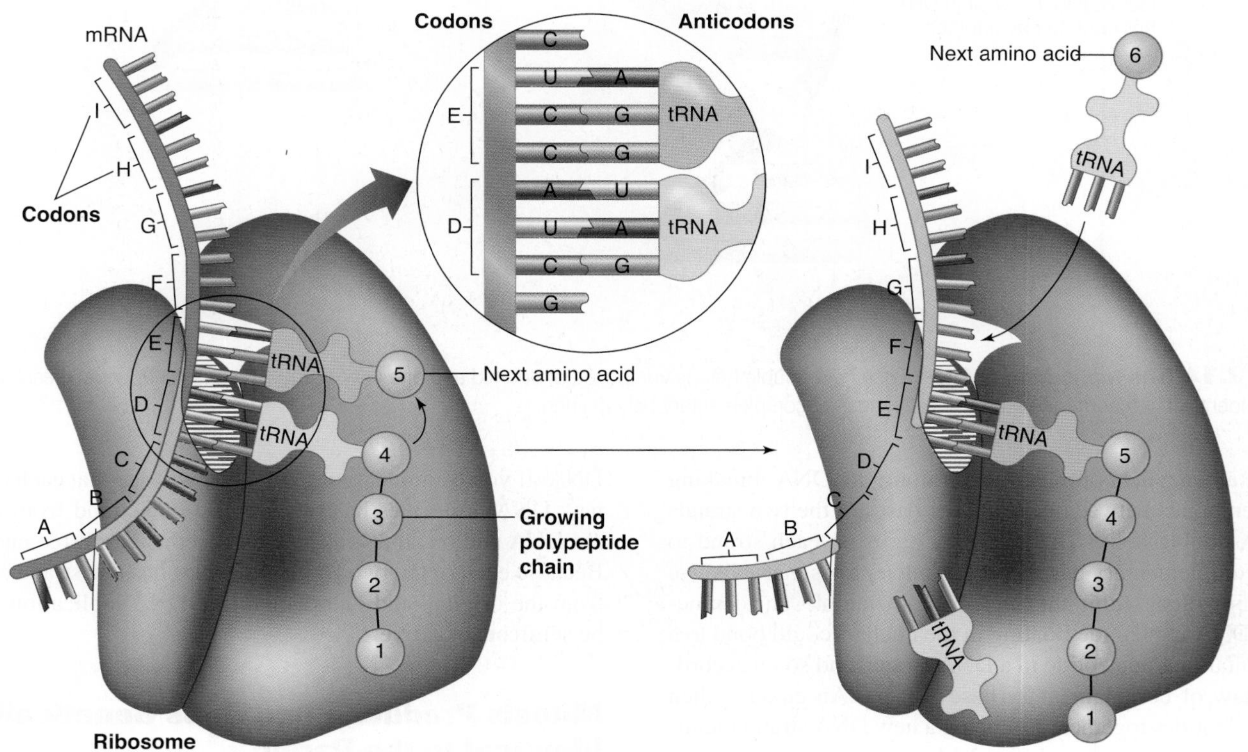

FIGURE 2.13 The translation of messenger RNA (mRNA). As the anticodon of each new aminoacyl-tRNA bonds with a codon on the mRNA, new amino acids are joined to the growing tip of the polypeptide chain.

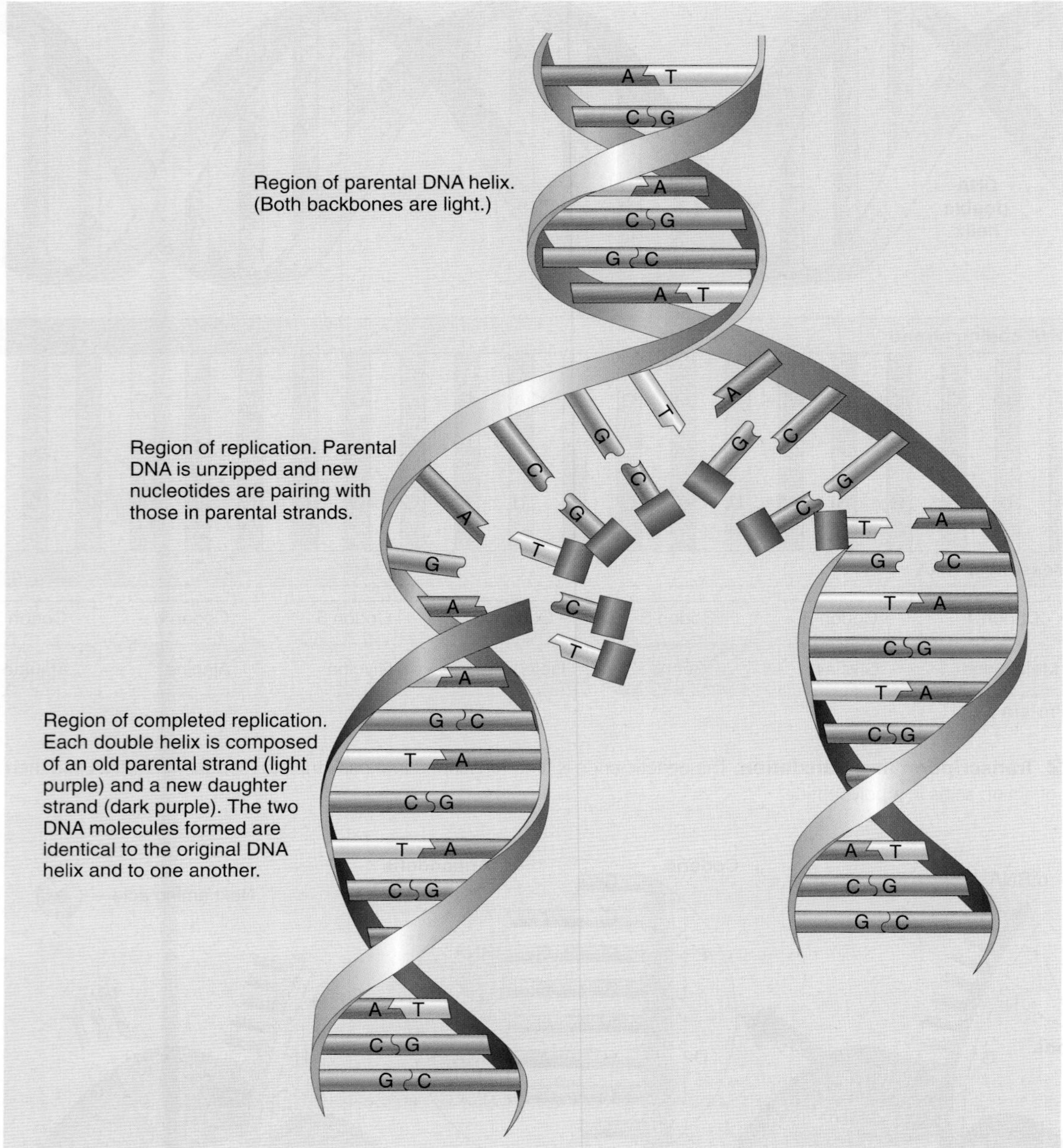

Region of parental DNA helix. (Both backbones are light.)
Region of replication. Parental DNA is unzipped and new nucleotides are pairing with those in parental strands.
Region of completed replication. Each double helix is composed of an old parental strand (light purple) and a new daughter strand (dark purple). The two DNA molecules formed are identical to the original DNA helix and to one another.

FIGURE 2.14 The replication of DNA. Each new double helix is composed of one old and one new strand. The base sequence of each of the new molecules is identical to that of the parent DNA because of complementary base pairing.

this occurs, an enzyme complex moves along the DNA, breaking the hydrogen bonds between each pair of bases in the two strands of the DNA double helix. This frees the bases on each strand to bond with complementary bases of DNA nucleotides that—like an alphabet soup—float within the nucleus. For example, an adenine-containing DNA nucleotide floating in the nucleus could bond to a thymine-containing nucleotide in a DNA strand, and so on according to the law of complementary base pairing. An enzyme then joins the nucleotides together, producing a new DNA strand identical to the former partner strand (fig. 2.14).

Therefore, when DNA replicates, it produces two DNA molecules identical in base sequence to each other and to the original DNA. If you examine figure 2.14, you will see that each of the two new DNA molecules has two strands, one strand from the original DNA and one that was just put together from free nucleotides. Because each of the two DNA molecules has one strand conserved from the original and one that is new, DNA replication is said to be **semiconservative**.

Mitosis Produces Two Cells Genetically Identical to the Parent

A cell is potentially immortal if it and its progeny continue to divide, because when a cell divides, its daughter cells have the same identity as the parent. Cell division that produces such identical daughter cells is known as **mitosis**. This is the type of cell division that allows a fertilized egg to grow into a baby, that allows organs to grow, and that allows cells to be replenished in a tissue. For example, you get a brand new epidermis (with new cells) approximately every 2 weeks, and a brand new stomach lining every 2 to 3 days. There is also another type of cell division, termed *meiosis*, that only occurs in the gonads (ovaries and testes) during the formation of gametes (egg cells and sperm); meiosis is discussed together with the reproductive system (chapter 15).

Clinical Applications

Growth of an organ because of an increase in the number of its cells, due to mitotic cell division, is called **hyperplasia**. A well-known example of this is the growth of the prostate in older men, a condition called *benign prostatic hyperplasia* (*BPH*) that interferes with urination. Growth of an organ because its cells get larger (rather than more numerous through cell division) is called **hypertrophy**. A good example is the *compensatory hypertrophy* that occurs in skeletal muscles when a person exercises with weights. Compensatory hypertrophy can also occur in the heart if its workload increases. The opposite of hypertrophy is **atrophy**, when an organ gets smaller because its cells get smaller.

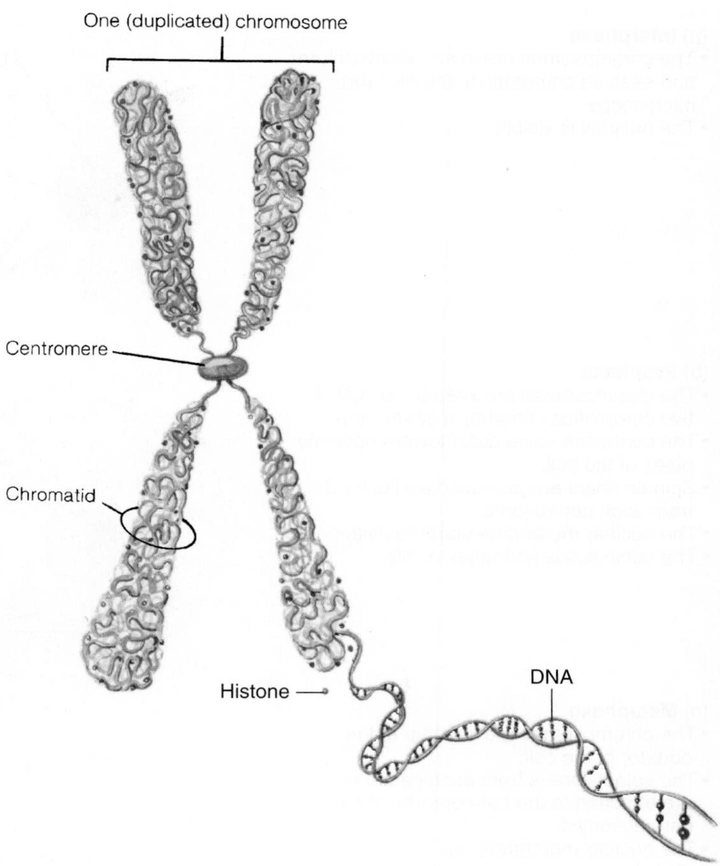

FIGURE 2.15 The structure of a chromosome after DNA replication. At this stage, a chromosome consists of two identical strands, or chromatids.

If a cell is going to divide, it replicates its DNA during a phase of mitosis called *interphase*. After DNA replication, the chromatin begins to condense into a more compacted form. The chromatin then becomes converted into the shorter, stubbier forms that can be seen under the microscope and identified as **chromosomes** (fig. 2.15); these can be seen during the next phase of mitosis, *prophase*. The chromosomes line up at the equator of the cell during *metaphase*. Meanwhile, organelles called *centrioles* have formed *spindle fibers* that attach to a structure, the *centromere*, that binds the duplicated chromosomes together. Next, at *anaphase*, the duplicated chromosomes are pulled to opposite poles of the cell by the spindle fibers. Finally, at *telophase*, the cell cytoplasm divides to form two cells as the chromosomes change back to their extended form once again (fig. 2.16).

Clinical Investigation Clues

Remember that Eric had hepatomegaly (an enlarged liver) as a baby. The liver biopsy revealed that the cells contained an unusually large amount of glycogen.

- If the liver was enlarged because of an increased number of cells, what term could be used to describe the type of growth?
- The large amount of stored glycogen made the liver cells larger. What term describes this type of growth?
- What is the chemical nature of glycogen (see chapter 1)?

Cell Death Can Be a Normal Physiological Process

Cell death can occur as part of a normal physiological process, or it may occur for abnormal (pathological) reasons. Pathological cell death, leading to tissue death, is called **necrosis**. The process of normal physiological cell death is known as **apoptosis** (from a Greek word describing the shedding of leaves from a tree). Apoptosis is a necessary, physiologically important process, as recognized in 2002 when its discoverers were awarded a Nobel Prize in Physiology or Medicine. As an example of the importance of apoptosis, remember that the cells that line the stomach, and those in the epidermis of the skin, are frequently renewed; mitosis occurs because apoptosis occurs, and both ensure that we have a fresh layer of cells facing the harsh conditions inside our stomach and outside our skin.

Apoptosis can be triggered by chemical signals that come from either the outside or the inside of the cells. Extracellular death signals are specific molecules released by other cells; intracellular death signals can be triggered by damage to DNA or by oxidative stress in the cell. These cause certain molecules to leak out of mitochondria and activate the executioners of the cell, enzymes called *caspases*. Caspase action leads to fragmentation of the DNA

(a) **Interphase**
- The chromosomes are in an extended form and seen as chromatin in the electron microscope.
- The nucleus is visible

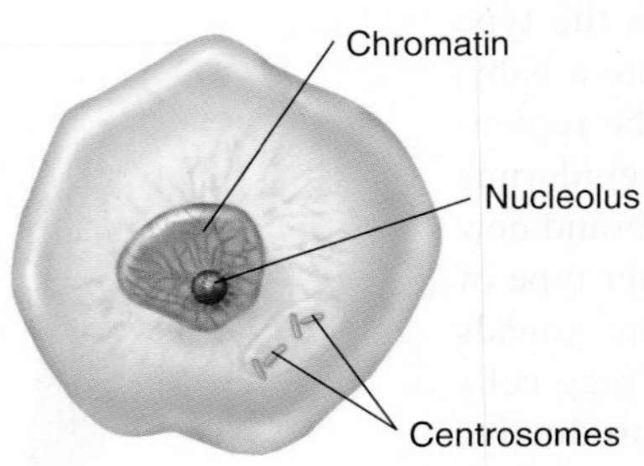

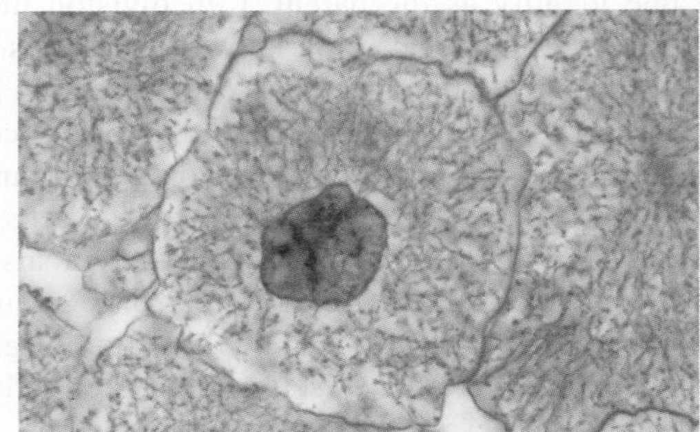

(b) **Prophase**
- The chromosomes are seen to consist of two chromatids joined by a centromere.
- The centrioles move apart toward opposite poles of the cell.
- Spindle fibers are produced and extend from each centrosome.
- The nuclear membrane starts to disappear.
- The nucleolus is no longer visible.

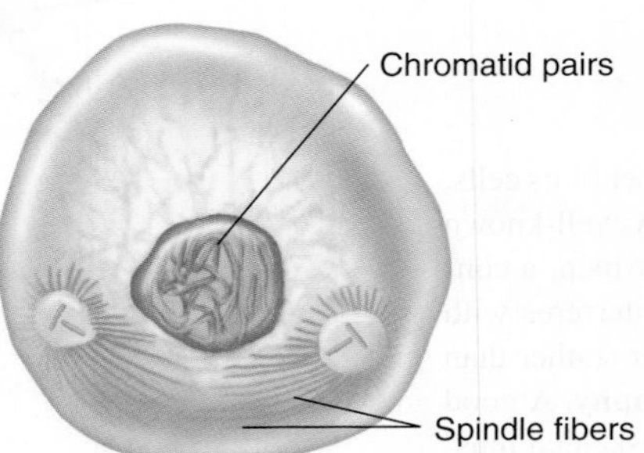

(c) **Metaphase**
- The chromosomes are lined up at the equator of the cell.
- The spindle fibers from each centriole are attached to the centromeres of the chromosomes.
- The nuclear membrane has disappeared.

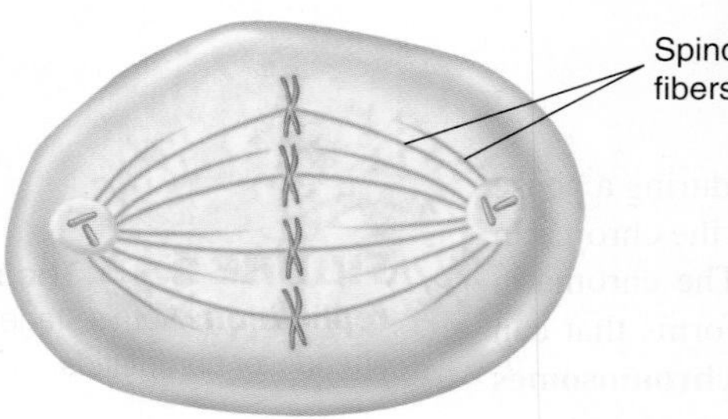

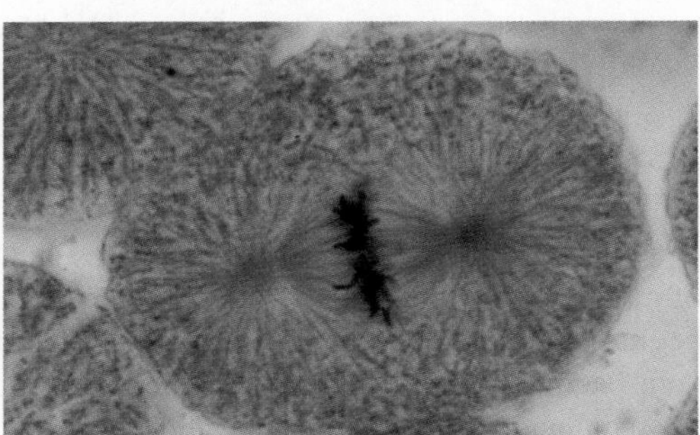

(d) **Anaphase**
- The centromere split, and the sister chromatids separate as each is pulled to an opposite pole.

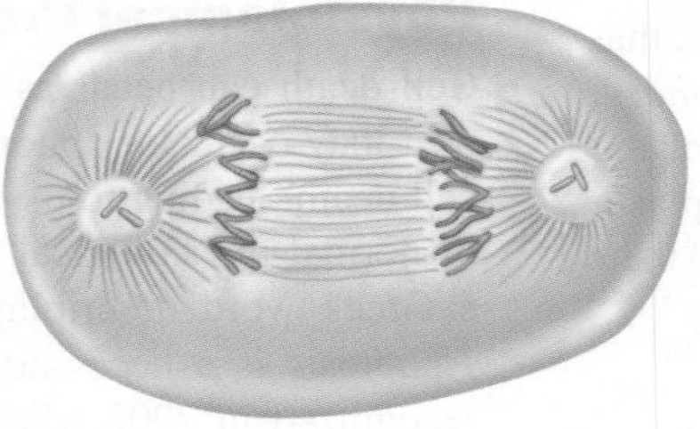

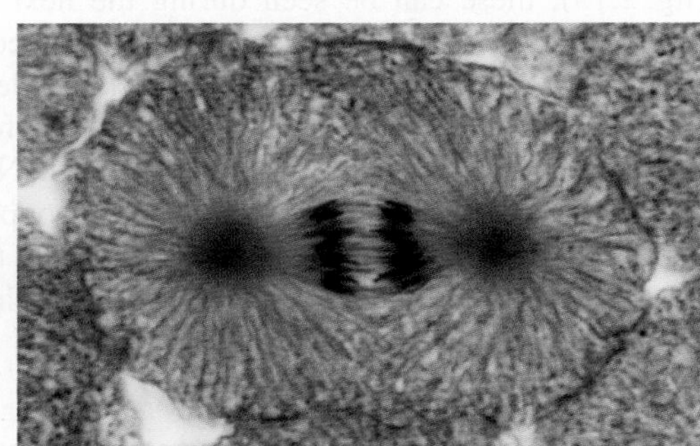

(e) **Telophase**
- The chromosomes become longer, thinner, and less distinct.
- New nuclear membranes form.
- The nucleolus reappears.
- Cell division is nearly complete.

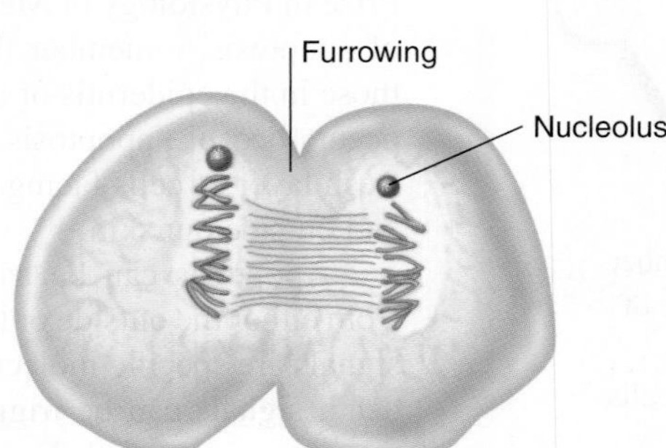

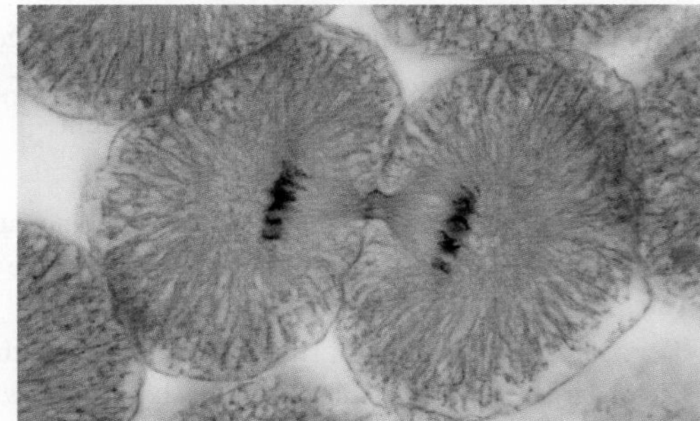

FIGURE 2.16 The stages of mitosis. The events that occur in each stage are indicated in the figure.

and death of the cell. This is a programmed cell death, which is necessary for normal growth, remodeling of organs, and many other physiological processes.

CHECK POINT

1. Describe the relationship between DNA, nucleosomes, chromatin, and chromosomes.
2. Describe what is meant by the terms *genome* and *proteome*; also explain the meaning of the terms *differentiation* and *stem cell*.
3. Explain how mRNA and tRNA function to produce a specific protein according to the instructions in a gene.
4. Describe what can happen to a protein after it's formed.
5. Describe how DNA is replicated and why this replication is called semiconservative.
6. Identify the phases and the major characteristics of each phase of mitosis, and explain the roles of mitosis in the body.
7. Define apoptosis and explain why this process is needed.

Cell Respiration: Metabolic Pathways That Provide Energy for the Cell

The term *metabolic pathway* refers to a sequence of chemical changes produced by a network of enzymes. *Cell respiration* refers to the metabolic pathways that provide energy for the cell. In cell respiration, the chemical bond energy of molecules is released in a stepwise manner and a portion of that energy is captured in the chemical bonds that hold the last phosphate groups of ATP. ATP then functions as the universal energy carrier of the cell.

This topic involves chemical concepts, so it might be useful to refer back to parts of chapter 1 as you go. Don't let the chemistry intimidate you; there's nothing here that you cannot understand with the aid of the information presented in these first two chapters. And think of how important this is for you; it relates to how every aspect of your body functions (since organ functions require food energy), how body weight can be gained or lost, and all the medical implications of metabolism and body weight. Let that motivate you to carefully lay this important foundation for your later study of physiology. This topic is interesting, important, and fun to learn—so let's get to it.

Chemical Bond Energy Is Used to Make ATP

Like all other animals, we must eat to live. The molecules in food serve a variety of functions, but one of the major functions of food is to provide the body with the energy it needs to function. Breathing, the beating of the heart, contractions of other muscles, and critical transport processes (described in chapter 3) are only some of the types of energy-requiring activities performed by the body cells. But what is the nature of the energy in food molecules, and where did that energy come from?

"Energy can neither be created nor destroyed" is a well-known description of the law of physics called the *first law of thermodynamics*. So, if the bread we eat has energy, the energy had to come from somewhere else. Plants use light energy from the sun to convert CO_2 and H_2O into glucose ($C_6H_{12}O_6$), a process known as *photosynthesis*. Since energy cannot be created or destroyed, the energy must still be there, in the form of chemical bond energy. Scientists say that the glucose has more *free energy* than the CO_2 and H_2O (fig. 2.17). Free energy is the energy available to do work.

The energy present in foods (such as starch formed from glucose by plants, or molecules produced by animals that ate the plants) can be measured by combusting the foods into molecules of CO_2 and H_2O. When the chemical bonds that were formed using light energy are broken, their energy must be released (because energy can't be destroyed). The energy released by combustion is in the form of heat. Heat is measured in calorie units; 1 **calorie** is the amount of heat required to raise the temperature of 1 cubic centimeter of water by 1 degree Celsius. This unit is too small to be convenient when measuring the energy in food. Consequently, we use **kilocalories** (or **Calories** spelled with a capital letter) to measure the energy value of food.

In our body cells, we break down molecules of glucose and others derived from our food in small steps, so that the energy will be released in smaller packets. A portion of that energy will indeed escape as heat; that's unavoidable, according to another law of physics called the *second law of thermodynamics*. However, a portion will be captured as usable (free) energy in the form of a chemical bond. In other words, the energy-releasing (*exergonic*) process of breaking down glucose (or another molecule) into

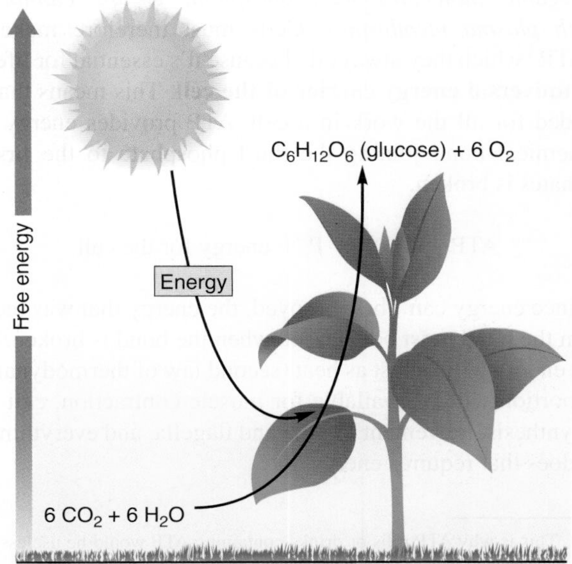

FIGURE 2.17 A simplified diagram of photosynthesis. Some of the sun's radiant energy is captured by plants and used to produce glucose from carbon dioxide and water. As the product of this endergonic reaction, glucose has more free energy than the initial reactants.

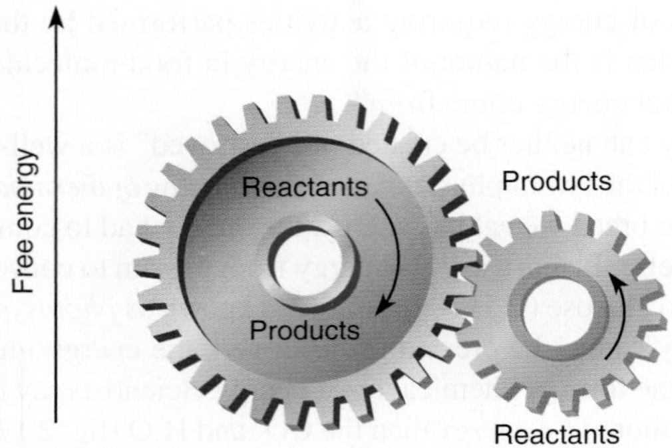

FIGURE 2.18 A model of the coupling of exergonic and endergonic reactions. The reactants of the exergonic reaction (represented by the larger gear) have more free energy than the products of the endergonic reaction because the coupling is not 100% efficient—some energy is lost as heat.

smaller components will drive an energy-requiring (*endergonic*) reaction where a chemical bond is formed. This is analogous to the way that the turning of one gear causes the turning of another (fig. 2.18).

The energy released by the breakdown of food molecules is used to directly drive one endergonic reaction (fig. 2.19). This is the addition of an *inorganic phosphate* group (P_i) to a molecule of *adenosine diphosphate* (*ADP*) to form a molecule of **adenosine triphosphate** (**ATP**). The bond joining the third phosphate to the other two must, according to first law, contain some of the energy released by the breakdown of food molecules. Biologically, the major benefit of eating is to provide our cells with the energy to make ATP.

Organic molecules with phosphate groups cannot pass through plasma membranes. Cells must therefore make their own ATP,[3] which they always do because it's essential for life. ATP is the **universal energy carrier of the cell**. This means that ATP is needed for all the work in a cell. ATP provides energy when the chemical bond joining the third phosphate to the first two phosphates is broken.

$$\text{ATP} \rightarrow \text{ADP} + \text{P}_i + \text{energy for the cell}$$

Since energy can't be destroyed, the energy that was required to form the bond must be released when the bond is broken. Some of that energy will be lost as heat (second law of thermodynamics), but a portion will be available for muscle contraction, cell transport, synthesis, movement of cilia and flagella, and everything else a cell does that requires energy.

FYI [3]This is why ATP pills or drinks containing ATP would be useless. These would merely fortify the feces; the ATP can't pass through the plasma membrane of the intestinal cells and enter the blood. But can we inject ATP directly into the blood before a sports event to get an energy burst? No, because it can't pass through the plasma membrane of our body cells either, and so it can't leave the blood. If we tried intravenous ATP, it would be filtered by the kidneys and simply enrich the urine. So it looks like we'll have to keep on eating.

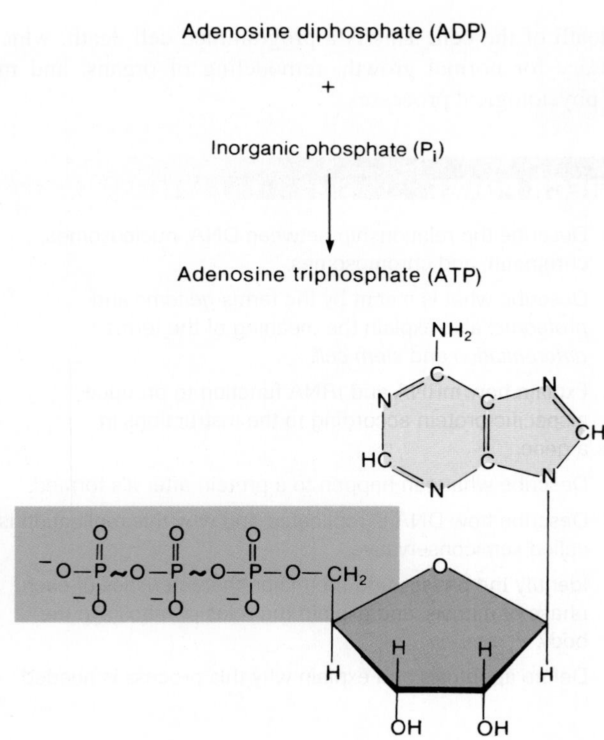

FIGURE 2.19 The formation and structure of adenosine triphosphate (ATP). ATP is the universal energy carrier of the cell. High-energy bonds are indicated by a squiggle (~).

Metabolic Pathways Require Cooperation of Different Enzymes

The term **metabolism** refers to all the chemical changes in our cells. These chemical changes can be relatively small, such as the conversion of glucose into fructose (two related monosaccharides described in chapter 1; see fig. 1.12); others can be much more extensive. Sequential chemical changes, produced by cooperating enzymes, are called **metabolic pathways**.

As described in chapter 1, an enzyme catalyzes a chemical reaction where the reactants—called the *substrates* of the enzyme—are converted into the *products* of the reaction. By catalyzing the reaction, an enzyme makes the reaction go faster than it would if the enzyme were absent. However, from the viewpoint of a cell, the reaction doesn't go sufficiently fast to matter if the enzyme is absent. So, from the viewpoint of a cell, the presence or absence of an enzyme can work like a switch; the reaction goes if the enzyme is present, and doesn't go if the enzyme is absent.

Enzymes must cooperate with each other to produce a metabolic pathway. For example, suppose molecule A is the substrate of one enzyme and B is its product. A second enzyme then converts B into C, and so on (fig. 2.20). This can involve quite a number of enzymes that convert an initial substrate into a final product of the metabolic pathway.

Many metabolic pathways have branch points (fig. 2.21). At a branch point, two or more different enzymes use the same molecule as their substrate. In fact, this is quite common, so

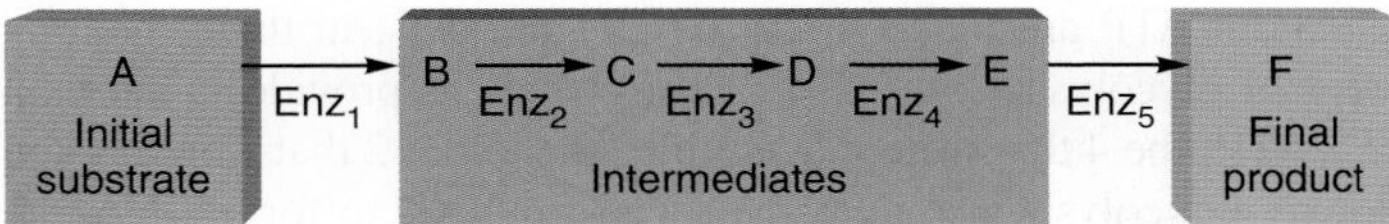

FIGURE 2.20 The general pattern of a metabolic pathway. In metabolic pathways, the product of one enzyme becomes the substrate of the next.

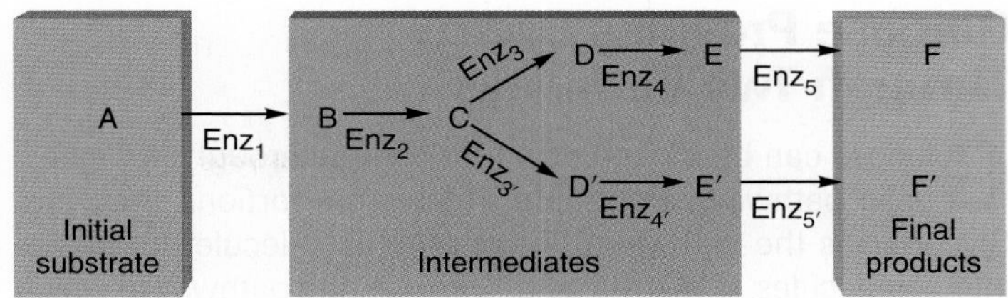

FIGURE 2.21 A branched metabolic pathway. Two or more different enzymes can work on the same substrate at the branch point of the pathway, catalyzing two or more different reactions.

that one molecule can be used for a number of different purposes and form different products. This allows genes to control the metabolism of the cell. For example, the gene coding for enzyme D in figure 2.21 may be inactivated, whereas the gene for enzyme D′ is active; this would switch metabolism away from product F and toward product F′. Like switching railroad tracks for an oncoming train, gene activation/inactivation and enzyme activation/inactivation can control the direction taken by a metabolic pathway.

Cell respiration refers to the metabolic pathways that result in the production of ATP, and thus that provide energy for the cell. In cell respiration, an initial substrate, such as glucose, fatty acids, or amino acids, is converted through numerous enzymatically controlled reactions into smaller final products. Energy is released at particular steps (fig. 2.22), and the total amount released in this controlled fashion is the same as when the glucose is completely combusted. However, in the cell, some of the energy released when chemical bonds are broken is used to power the production of ATP.

CHECK POINT

1. Describe how the first and second laws of thermodynamics apply to the way that energy is obtained and used in the body.
2. How is ATP produced, and what is its significance? Why can't we take ATP directly instead of eating food?
3. What is a metabolic pathway? What is the significance of branched metabolic pathways?

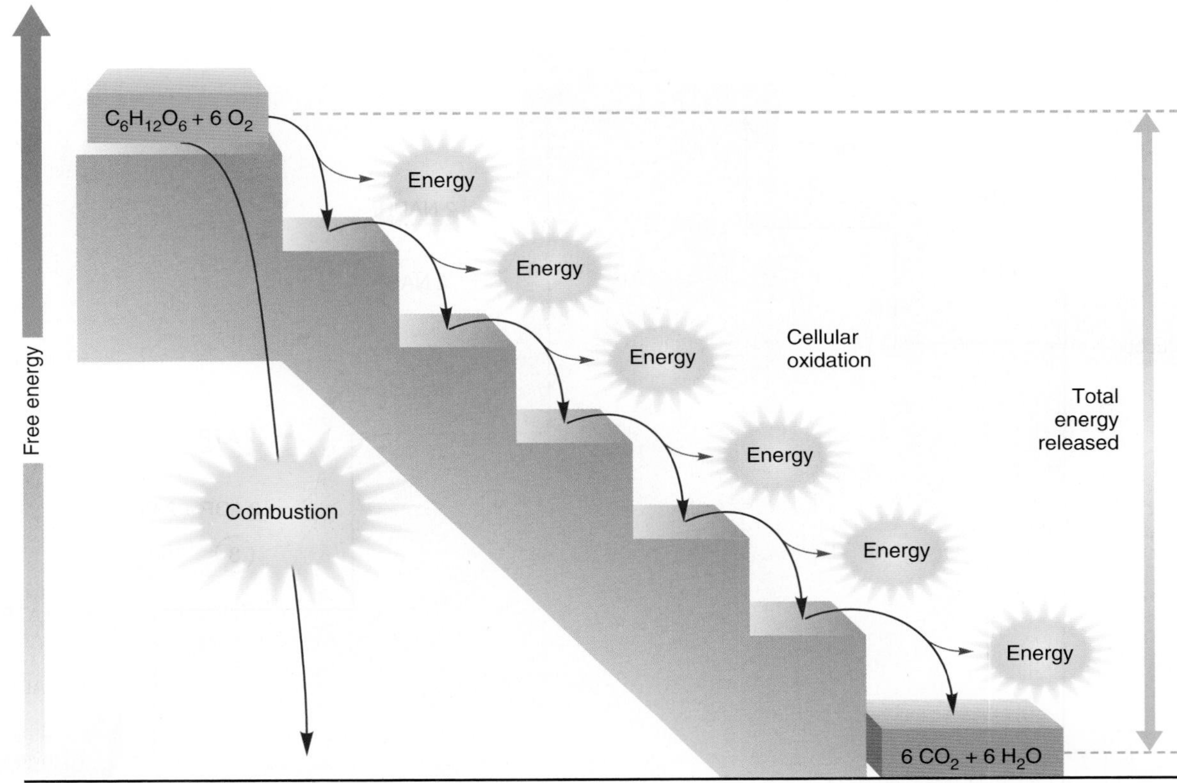

FIGURE 2.22 A comparison of combustion and cell respiration. Since glucose contains more energy than 6 separate molecules each of carbon dioxide and water, the combustion of glucose is an exergonic process. The same amount of energy is released when glucose is broken down stepwise within the cell. Each step represents an intermediate compound in the aerobic respiration of glucose.

Glucose Provides Energy Through Two Pathways

Glucose can be broken down for energy through two metabolic pathways that share a common portion: glycolysis. Glycolysis is the pathway that converts 1 molecule of glucose into 2 molecules of pyruvic acid. Through one pathway in which oxygen doesn't participate, pyruvic acid is converted into lactic acid. In aerobic respiration, the pyruvic acid is completely oxidized into CO_2 and H_2O. Aerobic respiration requires oxygen and produces more ATP than the lactic acid pathway.

The first step in the cell respiration of glucose is the stepwise conversion of glucose (a 6-carbon-long molecule) into 2 molecules of *pyruvic acid* (each 3-carbons-long). This metabolic pathway is termed **glycolysis** (from the Greek *glyksis*, referring to sugar, and *lysis*, referring to breaking). In this metabolic pathway, one ATP is initially converted into ADP, so that its phosphate can be added to glucose to form glucose 6-phosphate. Since organic molecules with phosphate groups can't pass through the plasma membrane, this traps the glucose within the cell. Then, the glucose is converted into its isomer, fructose, and another ATP is used to add a phosphate to form the molecule fructose 1,6-biphosphate (fig. 2.23). Since we've lost 2 ATP in this process, we can think of this loss as an energy investment that must be made up before we can reap an energy profit.

Four ATP are made, from 4 ADP and 4 P_i, in the remaining steps of glycolysis (fig. 2.23). Thus, glycolysis provides a net gain of 2 ATP (the 4 that were made minus the initial 2 that were broken down). Glycolysis also adds hydrogens to NAD to form 2 NADH. *NAD* refers to *nicotinamide adenine dinucleotide*, a vitamin B_3 (niacin) derivative that helps transfer hydrogen atoms and electrons from one reaction to another. NADH actually transfers 2 hydrogens. This is because it combines with 1 hydrogen atom and the electron from another; this leaves a free proton (H^+), which tags along and recombines with its electron when the pair of hydrogen atoms is transferred. This will become important in the next stages of the cell respiration of glucose. A more detailed accounting of the specific reactions involved in glycolysis is provided in appendix 2.

Clinical Investigation Clues

Remember that Eric cannot metabolize fructose or galactose.

- What are fructose and galactose, and what is their relationship to glucose (see chapter 1)?
- What foods are rich in fructose or galactose (see chapter 1)?
- Given that Eric has a defective glucose 6-phosphatase, what does his nutritional problem tell you about the way that fructose and galactose are normally metabolized?

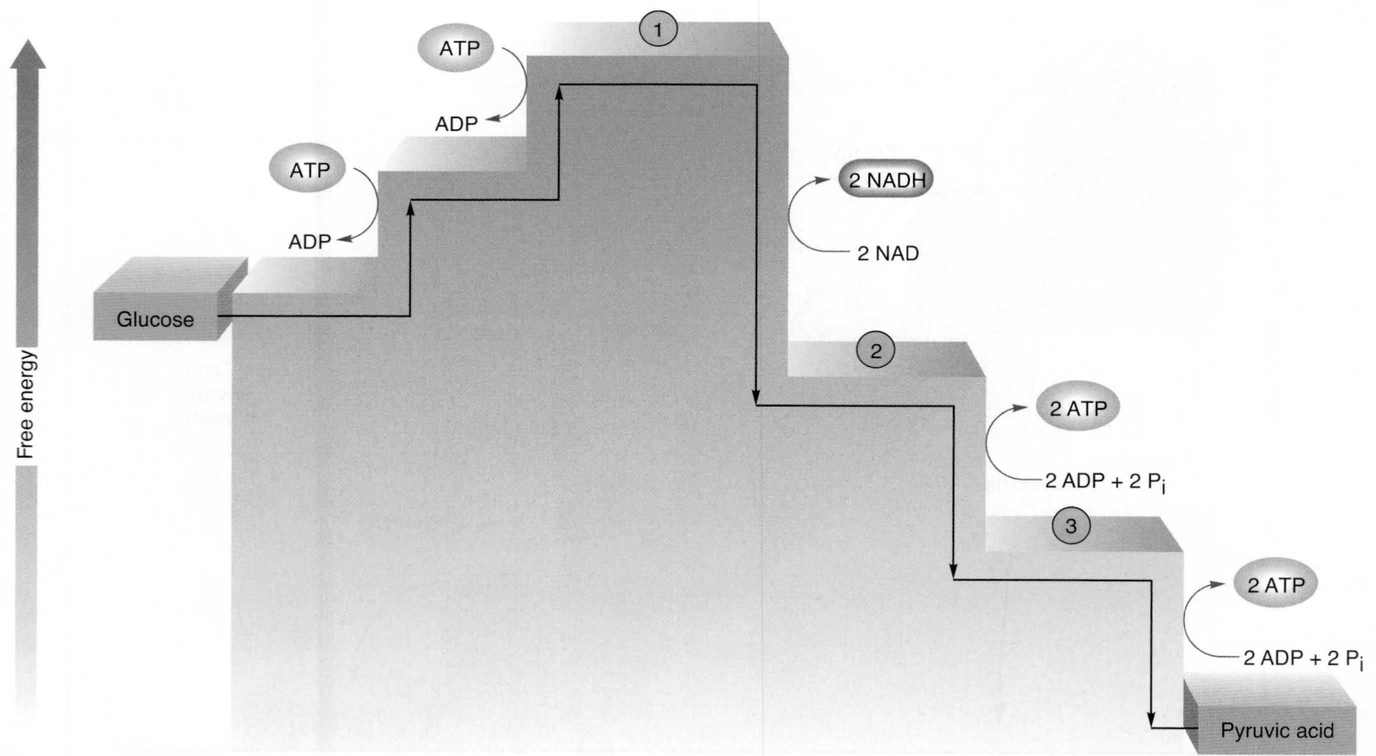

FIGURE 2.23 The energy expenditure and gain in glycolysis. Notice that there is a "net profit" of 2 ATP and 2 NADH for every molecule of glucose that enters the glycolytic pathway. Molecules listed by number are (*1*) fructose 1,6-biphosphate, (*2*) 1,3-biphosphoglyceric acid, and (*3*) 3-phosphoglyceric acid (see appendix 2).

Lactic Acid Is Produced in the Absence of Oxygen

Mitochondria, the organelles required for aerobic respiration (using oxygen), are absent in mature red blood cells, which therefore must produce ATP without oxygen. This is ironic but logical, given that each red blood cell carries almost a billion molecules of oxygen for use by other body cells. Skeletal muscle cells must also produce ATP without oxygen when their metabolic demand during exercise increases more than the supply of oxygen delivered to them in the blood. This generally occurs during the first minute or so of exercise, before the cardiovascular system has made the adjustments necessary to provide the needed increase in blood flow.

In these cases, glycolysis provides ATP for the cells. However, remember that glycolysis also results in the conversion of NAD to NADH. In order for glycolysis to continue, the NADH must be converted back again to NAD by the transfer of hydrogens to another molecule. In **aerobic respiration**, 2 hydrogen atoms are eventually given to oxygen, forming water (H_2O). In the absence of oxygen, the 2 hydrogens are transferred to pyruvic acid, forming **lactic acid** (fig. 2.24).

The metabolic pathway that converts glucose to lactic acid is an *anaerobic* pathway, because oxygen isn't involved. Many physiologists call this pathway **anaerobic respiration**. However, biologists generally prefer to call this process **lactic acid fermentation**, because this pathway is analogous to the one that yeast cells use to produce alcohol (ethanol). Whatever terms are used, the lactic acid pathway results in the net gain per glucose molecule of only those 2 ATP that were made in glycolysis.

Most ATP Is Produced by Aerobic Cell Respiration

If cells (other than red blood cells) have oxygen available, they will produce their ATP by aerobic cell respiration. When glucose is respired aerobically, it first undergoes glycolysis and is converted into 2 molecules of pyruvic acid, as was described in the preceding section. In aerobic respiration, the pyruvic acid enters into the matrix of a mitochondrion and has a carbon dioxide (CO_2) removed to form a 2-carbon-long molecule (fig. 2.25). This combines with a transfer molecule called *coenzyme A*. The resulting molecule is *acetyl coenzyme A* (abbreviated *acetyl CoA*). The "acetyl" in the name refers to the 2-carbon-long derivative of pyruvic acid, because it's related to acetic acid (from vinegar), which is 2 carbons long.

The coenzyme A transfers the 2-carbon-long molecule to an enzyme that combines it with a 4-carbon-long molecule already in the mitochondria, forming a 6-carbon-long molecule called *citric acid* (fig. 2.26). This begins a cyclic metabolic pathway; the 6-carbon-long citric acid has a carbon dioxide removed to produce a 5-carbon-long molecule (α-ketoglutaric acid), which then has

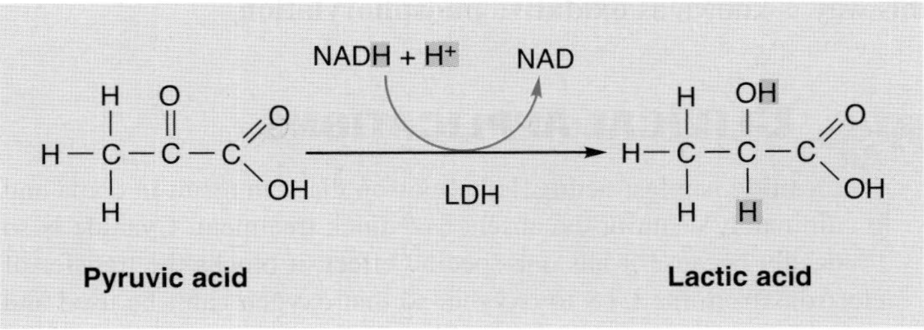

FIGURE 2.24 The formation of lactic acid. The addition of 2 hydrogen atoms (colored boxes) from NADH to pyruvic acid produces lactic acid and NAD. This reaction is catalyzed by lactic acid dehydrogenase (LDH) and is reversible under the proper conditions.

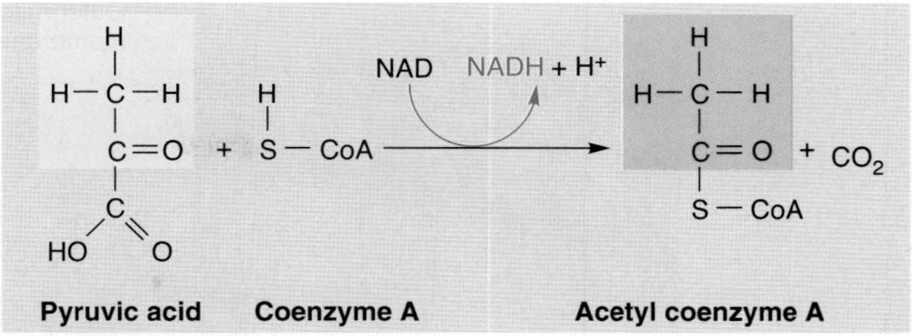

FIGURE 2.25 The formation of acetyl coenzyme A in aerobic respiration. Notice that NAD is converted into NADH in this process.

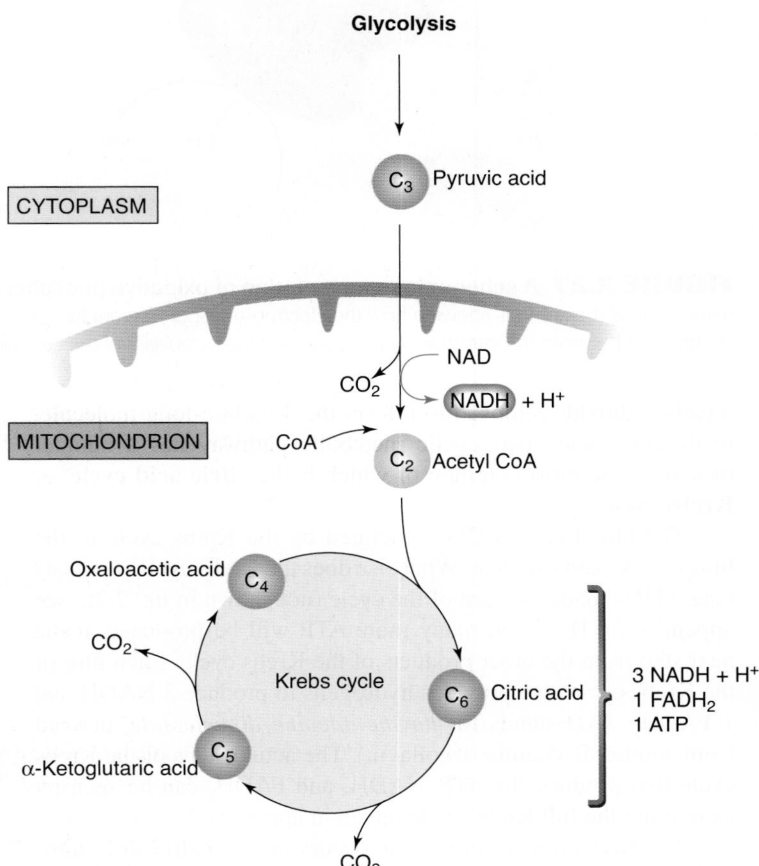

FIGURE 2.26 A simplified diagram of the Krebs cycle. This diagram shows how the original 4-carbon-long oxaloacetic acid is regenerated at the end of the cyclic pathway. Only the numbers of carbon atoms in the Krebs cycle intermediates are shown; the numbers of hydrogens and oxygens are not accounted for in this simplified scheme. The complete Krebs cycle is shown in appendix 2.

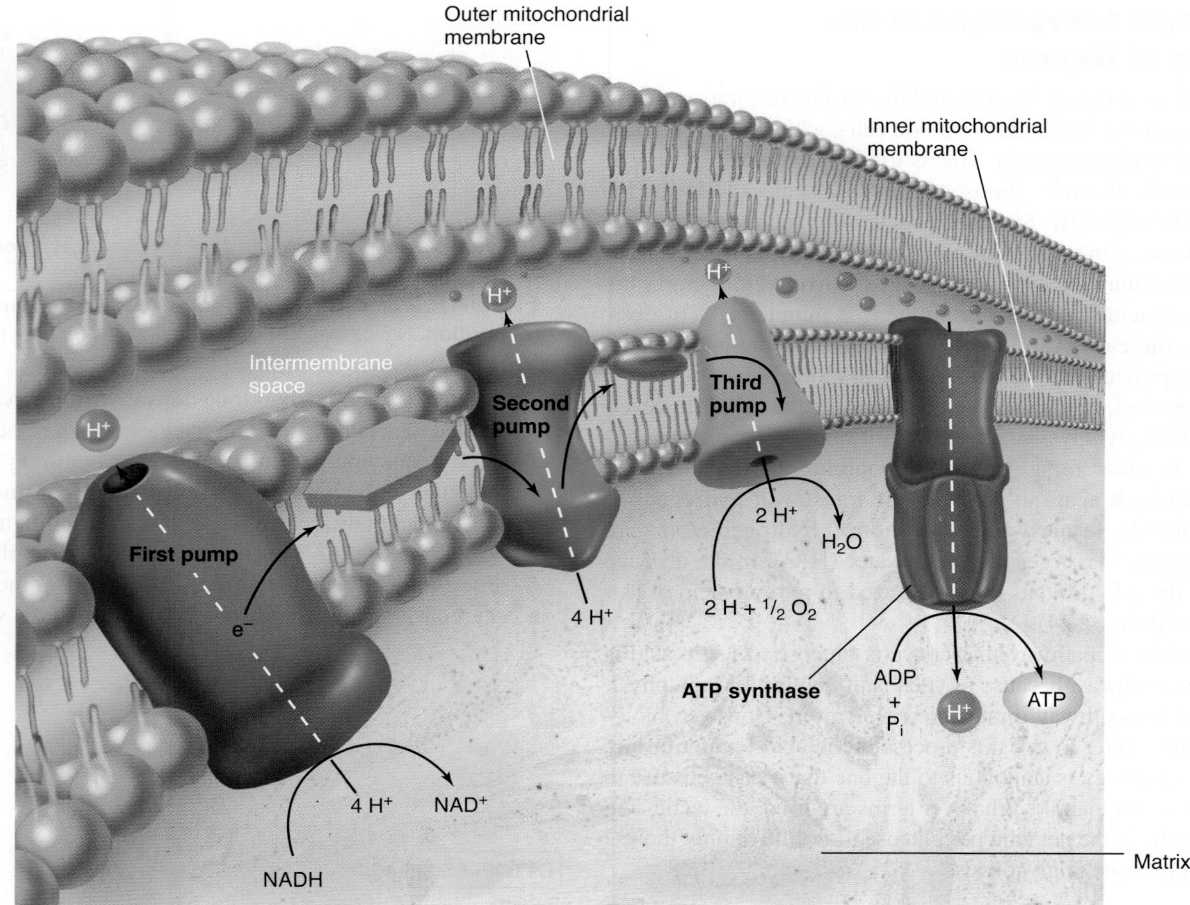

FIGURE 2.27 A schematic representation of oxidative phosphorylation. The matrix and the compartment between the inner and outer mitochondrial membranes showing how the electron-transport system functions to pump H^+ from the matrix to the intermembrane space. This results in a steep H^+ gradient between the intermembrane space and the cytoplasm of the cell. The diffusion of H^+ through ATP synthase results in the production of ATP.

a carbon dioxide removed to reform the 4-carbon-long molecule, oxaloacetic acid. This cyclic metabolic pathway has a number of names, the most common of which is the **citric acid cycle**, or **Krebs cycle**.

The blood carries CO_2 generated by the Krebs cycle to the lungs, so we can exhale it. What else does the Krebs cycle produce? One ATP is made per turn of the cycle (not shown in fig. 2.26; see appendix 2). However, many more ATP will be produced at the next step from the other products of the Krebs cycle. Each turn of the Krebs cycle also provides hydrogens to produce 3 NADH and 1 $FADH_2$. *FAD* stands for *flavine adenine dinucleotide*, derived from another B vitamin (riboflavin). The actual steps of the Krebs cycle that produce the ATP, NADH, and $FADH_2$ can be seen by examining the full Krebs cycle, given in appendix 2.

The Krebs (citric acid) cycle occurs in the matrix of a mitochondrion. The next act of the story takes place in the cristae, the foldings of the inner mitochondrial membrane (see fig. 2.6). This is where the molecules of the **electron-transport system (ETS)** are located. The molecules of the ETS are organized into three pumps that take electrons from the hydrogen atoms provided by NADH and $FADH_2$; the electrons are passed in sequence from one pump to the next, like a bucket brigade (fig. 2.27). This electron transport powers the pumping of protons (H^+), so that the proton (H^+) that goes with each electron of a hydrogen atom is moved into the intramembranous space between the outer and inner mitochondrial membranes. Then, at the very last step, *oxygen serves as the final electron acceptor* from the ETS. Two protons (H^+) join their electrons in the oxygen atom to form water (H_2O). During this last step, H^+ moves from the intramembranous space into the matrix and powers an enzyme, *ATP synthase*, that catalyzes the conversion of ADP and P_i into ATP (fig. 2.27). The production of ATP in this way is known as **oxidative phosphorylation**.

Clinical Applications

Cyanide is a fast-acting, lethal poison that can result in coma and ultimately death in the absence of quick treatment. Cyanide is so deadly because it has one specific effect: it blocks the transfer of electrons from the ETS to oxygen, so that oxygen can't be used and water isn't formed. As a result, oxidative phosphorylation stops. All of the medical problems associated with cyanide poisoning, leading up to death, stem from inadequate amounts of ATP being available to the brain, heart, and other organs.

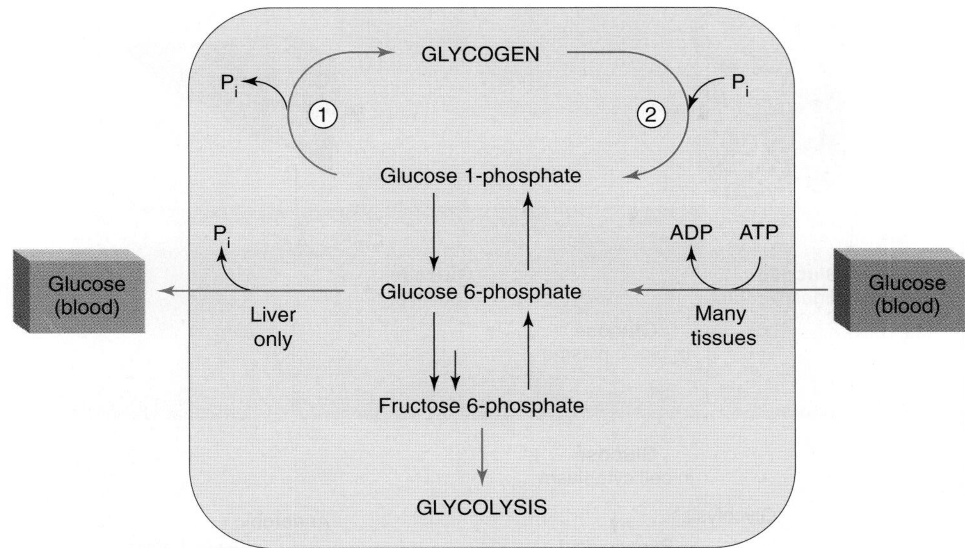

FIGURE 2.28 Glycogenesis and glycogenolysis. Blood glucose entering tissue cells is phosphorylated to glucose 6-phosphate. This intermediate can be metabolized for energy in glycolysis, or it can be converted to glycogen (*1*) in a process called *glycogenesis*. Glycogen represents a storage form of carbohydrates that can be used as a source for new glucose 6-phosphate (*2*) in a process called *glycogenolysis*. The liver contains an enzyme that removes the phosphate from glucose 6-phosphate.

The ETS and oxidative phosphorylation produce 34 to 36 ATP per glucose. Because glycolysis itself produces 2 ATP per glucose, the aerobic cell respiration of glucose produces a total of 36 to 38 ATP within the mitochondrion. However, there are inefficiencies involved in moving the ATP out of the mitochondrion and into the cytoplasm, where it is needed to power the work of the cell. Thus, most scientists currently believe that 1 glucose molecule that is aerobically respired provides the cell cytoplasm with about 30 to 32 ATP.

Regardless of how the accounting is performed, cells certainly get much more ATP from aerobic respiration than from the lactic acid pathway. This explains why our skeletal and cardiac muscles rely on the lactic acid pathway only when they are forced to do so by the lack of oxygen. As soon as sufficient oxygen becomes available, our cells switch to aerobic cell respiration.[4] The switch to aerobic respiration explains the "second wind" that we get after a minute or so of exercise, when our cardiovascular system has made the adjustments necessary to deliver sufficient oxygen.

Blood Glucose May Come from Food or from Liver Glycogen

When we eat carbohydrates, we digest them in the small intestine into monosaccharides (chapter 1), which are then transported across the intestinal lining into the blood. As a result, there is a rise in the blood glucose concentration. The blood from the intestine goes first to the liver, in a vessel called the *hepatic portal vein*. If the blood glucose concentration is high after a meal, the liver can remove a certain amount of glucose and convert it into the polysaccharide glycogen (chapter 1). Blood later carries glucose to skeletal muscles, which can also take glucose from the blood and store it as glycogen. (The hormone insulin is required for the movement of blood glucose into skeletal muscle and liver cells.)

If we don't eat for some time, the blood glucose concentration starts to fall. This could be very dangerous, because *the brain gets its energy only from the aerobic respiration of blood glucose*. It doesn't store glycogen, and (unlike most other organs) it can't use molecules other than glucose very well for energy. If we aren't absorbing glucose from the food in our intestine, the blood glucose must come from somewhere else or we'll die. Fortunately, the liver (and only the liver) can secrete glucose into the blood. The liver can obtain this glucose by hydrolysis of its stored glycogen, a process called **glycogenolysis**.

You may wonder why the skeletal muscles, which collectively store much more glycogen than the liver, can't secrete glucose into the blood. The reason is that, when glycogen is broken down, it forms glucose 6-phosphate. Remember, organic molecules with phosphate groups can't pass through a plasma membrane; the glucose 6-phosphate is trapped inside the cell. The liver, and only the liver, has the enzyme *glucose 6-phosphatase*, which can remove the phosphate group to produce free glucose that can enter the blood (fig. 2.28).

Thus, blood glucose may come from either the intestine or the liver. The cells throughout our body can remove glucose from the blood and use it for either the lactic acid pathway or aerobic respiration (fig. 2.29) to derive energy for the production of ATP. The ATP is then used to power all of the work of the cell.

FYI [4]The French wine industry asked Louis Pasteur (1822–1895) why wine turns sour. In his investigations, he was the first to distinguish between aerobic respiration and anaerobic fermentation. Microbes ferment when they lack oxygen. Pasteur discovered that fermentation stops when microbes are supplied with oxygen. This sensible response of microbes (and of human cells) to oxygen is widely known as the *Pasteur effect*.

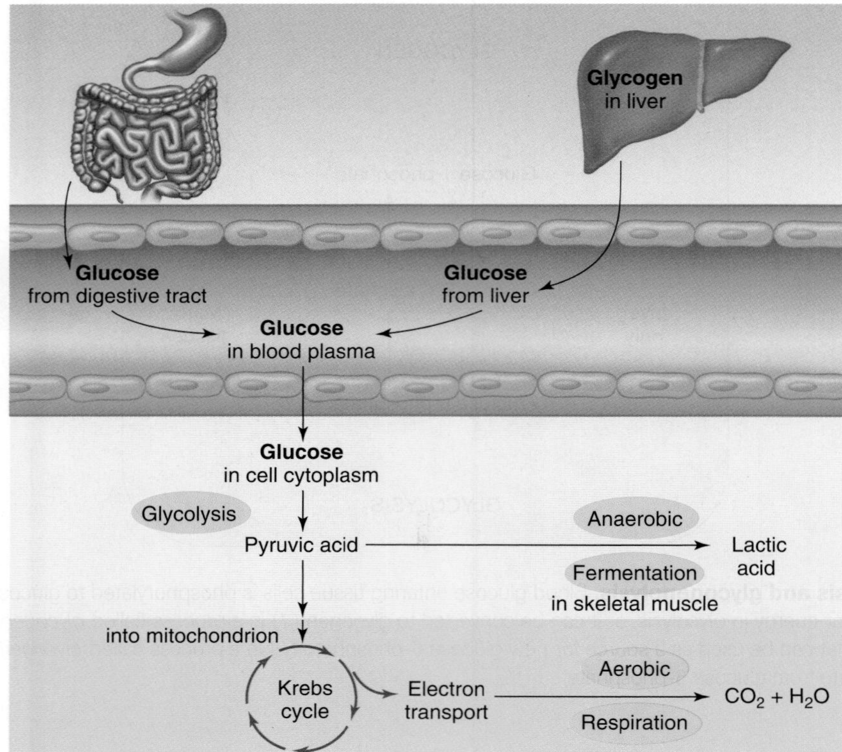

FIGURE 2.29 Overview of energy metabolism using blood glucose. Blood glucose may be obtained from food via the digestive tract, or the liver may produce it from stored glycogen. Plasma glucose enters the cytoplasm of cells, where it can be used for energy by either anaerobic lactic acid fermentation or aerobic cell respiration.

CLINICAL INVESTIGATION CLUES

Remember that Eric had excessive glycogen stored in his liver cells, and a tendency to get hypoglycemia (low blood glucose) that he countered by eating cornstarch.

- What is the function of glucose 6-phosphatase in liver cells, and how would a defect in this enzyme cause the accumulation of excessive glycogen?
- How would a defective hepatic (liver) glucose 6-phosphatase result in hypoglycemia, particularly during fasting?
- How would eating cornstarch help to prevent the hypoglycemia? (See chapter 1 for a discussion of starch and its digestion.)

Cells Can Also Get Energy from Fat and Protein

In the study of physiology, there is a lot of emphasis on glucose metabolism; skeletal muscles regularly ferment glucose to lactic acid during certain stages of exercise, and the brain obtains essentially all of its energy by the aerobic respiration of glucose. Fortunately for the brain, the liver and resting skeletal muscles prefer to get their energy from the aerobic respiration of fatty acids, thereby saving the blood glucose for the brain.

Adipose tissue stores fat (triglyceride; see chapter 1) in large droplets. When this fat is to be used for energy, a lipase enzyme hydrolyzes (breaks down) a triglyceride molecule into its component parts of fatty acids and glycerol (chapter 1; see fig. 1.16). These can then circulate in the blood and be taken into cells for aerobic cell respiration. When a fatty acid is used for energy, the hydrocarbon chain is converted into acetyl CoA molecules, which can start Krebs cycles (fig. 2.30). A 20-carbon-long fatty acid, for example, can be converted into 10 acetyl CoA molecules and start 10 Krebs cycles. In this way, a typical fatty acid can liberate enough energy to produce more than 100 molecules of ATP! Glycerol can also be used for energy (by being converted into pyruvic acid), but it releases much less energy (forming less ATP) than the fatty acids.

We don't store proteins in our body for energy; the proteins all have other functions. However, we can derive energy from amino acids that may come from food proteins or from the breakdown of our muscle proteins. Proteins can be hydrolyzed into amino acids, which may then be aerobically respired for energy by their conversion into pyruvic acid, acetyl CoA, or one of the Krebs cycle acids (fig. 2.30). However, this requires the removal of the amine group from the amino acid (chapter 1; see fig. 1.20). If the amine group enters the blood, it will form ammonia (NH_3), a toxic compound. Fortunately, the liver converts ammonia into urea, a less toxic molecule that can be eliminated in the urine.

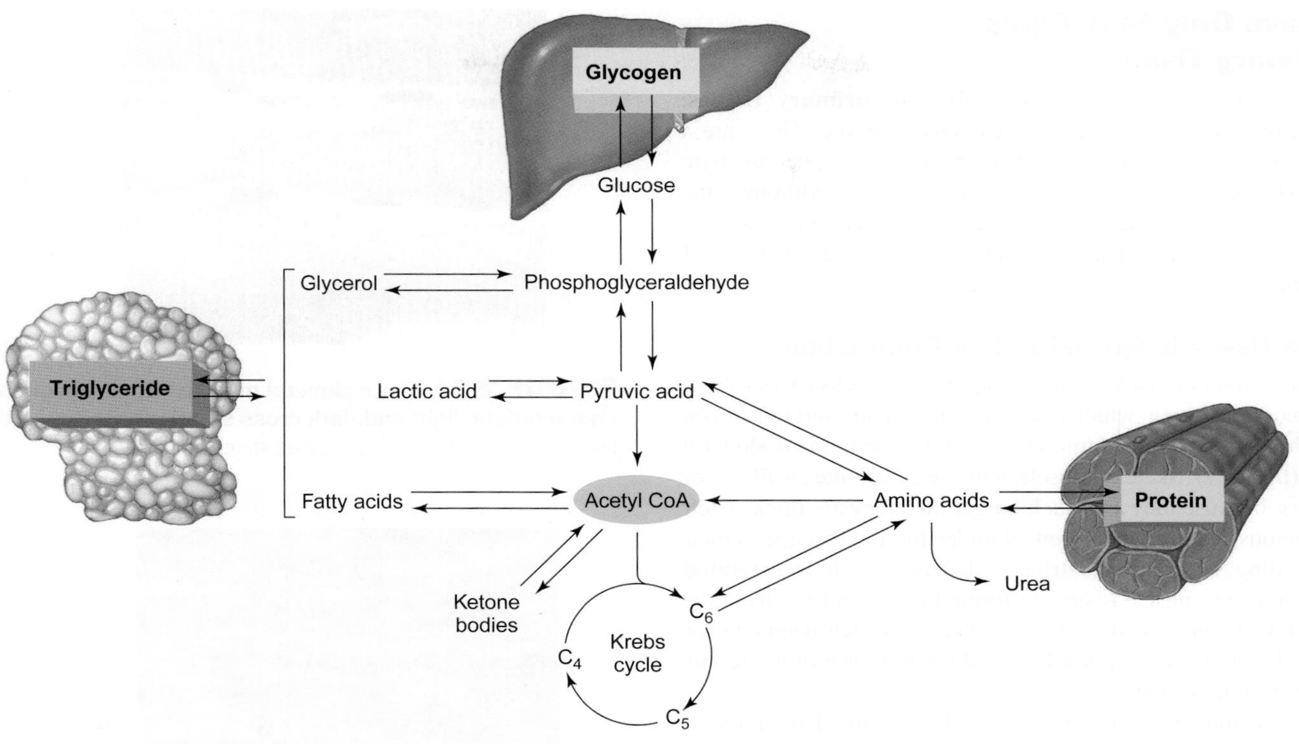

FIGURE 2.30 The interconversion of glycogen, fat, and protein. These simplified metabolic pathways show how glycogen, fat, and protein can be interconverted. Note that while most reactions are reversible, the reaction from pyruvic acid to acetyl CoA is not. This is because a CO_2 is removed in the process. (Only plants, in a phase of photosynthesis called the dark reaction, can use CO_2 to produce glucose.)

CHECK POINT

1. Describe the metabolic pathway that produces lactic acid. Under what circumstances, and where, does this occur in the body?
2. Describe the fate of pyruvic acid in the aerobic respiration of glucose. Explain the role of oxygen in this process.
3. Compare the amount of ATP made by the anaerobic fermentation and aerobic respiration of glucose. How is most of the ATP made in aerobic respiration?
4. Describe how glycogen can be used for energy, and explain why the liver is the only organ that can secrete glucose into the blood.
5. Explain how fat and protein can be used by cells for energy.

The Body Organs and Systems Are Composed of Specialized Tissue Cells

Cells that are specialized for different functions are grouped together to form the tissues of the body. There are 4 primary tissues: muscle, nervous, epithelial, and connective tissue. Organs are composed of different tissues that contribute to each organ's physiology, and particular organs cooperate with each other to produce the body systems: the nervous, endocrine, digestive, urinary, reproductive, immune, muscular, skeletal, and integumentary systems.

Here is more background information, this time relating to anatomical topics you may have learned previously in an introductory biology or anatomy course. If so, you may want to look these topics over to refresh your memory, or perhaps learn certain ones for the first time. As with the previously presented information in chemistry and cell biology, you can use this basic anatomical refresher as you and your instructor think best to build your knowledge pyramid as you study physiology.

Organs are composed of tissues consisting of highly specialized, or *differentiated*, cells. This contrasts with the cells of a very early embryo, which are so *undifferentiated* (unspecialized) that they can become any of the body tissues. At a particular embryonic stage, these undifferentiated cells are also known as *embryonic stem cells*, as will be discussed in conjunction with the reproductive system in chapter 15. As the embryo develops, its cells become increasingly specialized along different lines—they become more differentiated. In the adult, many organs retain small populations of cells that are less differentiated, and thus more able to divide and differentiate into the specialized cells of the organs. These less differentiated cells are known as *adult stem cells*.[5]

We Have Only Four Types of Primary Tissues

The entire body is composed of only four **primary tissues**: *nervous*, *muscular*, *epithelial*, and *connective* tissue. There are a number of subcategories of each primary tissue type, but all share certain characteristics that group them together. Although the study of the tissues is more a topic of anatomy than of physiology, a familiarity with the tissues is essential for an understanding of body function.

Muscle Tissue Is Specialized for Contraction

The contraction of muscle tissue occurs when its cells try (successfully or not) to shorten, which causes the muscle to exert tension on its attachments. The most familiar type of muscle tissue is **skeletal muscle** (fig. 2.31). Skeletal muscle cells are called *muscle fibers*, or *myofibers*, because they are much longer than they are thick. Their most obvious feature when viewed under the microscope is their cross banding, or striations (stripes); skeletal muscle is a **striated muscle**. Skeletal muscle fibers are formed in the embryo from the end-to-end fusion of cells called *myoblasts*, which accounts for the great length of many myofibers and the presence of numerous nuclei in each myofiber.

Skeletal muscles generally attach (by means of tendons) at each of their ends to bones; when a skeletal muscle contracts and shortens, it moves the skeleton at a joint. Since we can voluntarily contract our skeletal muscles, they are also known as *voluntary muscles*. However, the voluntary nature of skeletal muscle contraction relates to the types of nerves that innervate (stimulate) these muscles, not to any particular property of the muscles themselves.

Cardiac muscle (fig. 2.32) is the type of muscle tissue found only in the heart. Like skeletal muscle, cardiac muscle is striated; because of this, both skeletal and cardiac muscles contract by means of similar processes (discussed in chapter 9). However, unlike skeletal muscle, cardiac muscle is not under voluntary control. **Smooth muscle** (fig. 2.33) and cardiac muscle are *involuntary muscles* because of their type of neural regulation. However, unlike cardiac muscle, the smooth muscle cells are not striated and so contract by a different means. Smooth muscles are found in the walls of many organs, including blood vessels, the gastrointestinal tract, urinary bladder, and other organs.

Nervous Tissue Regulates the Body Tissues, Organs, and Systems

Nerve tissue consists of cells called **neurons**, which are the principal regulating cells of the nervous system. Neurons produce and conduct electrochemical nerve impulses and release chemical neurotransmitter molecules (described in chapter 4) that regulate other neurons, muscles, and glands. The nervous system consists of neurons and **supporting cells**, which are also called *neuroglial cells* (or just *glial cells*) when they are in the brain and spinal cord. The functions of the neuroglial cells are also described in chapter 4.

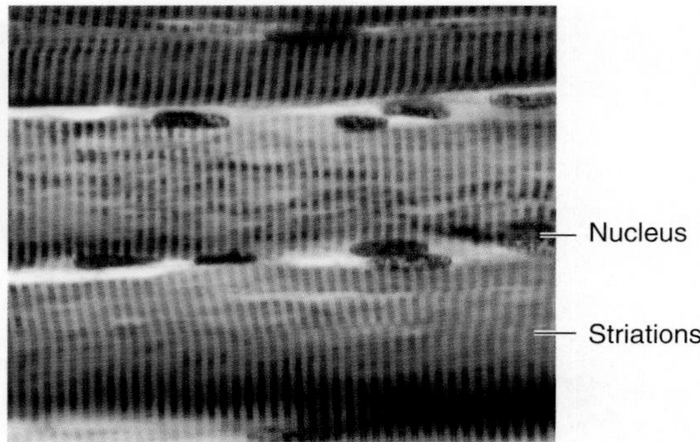

FIGURE 2.31 Three skeletal muscle fibers showing the characteristic light and dark cross striations. Because of this feature, skeletal muscle is also called striated muscle.

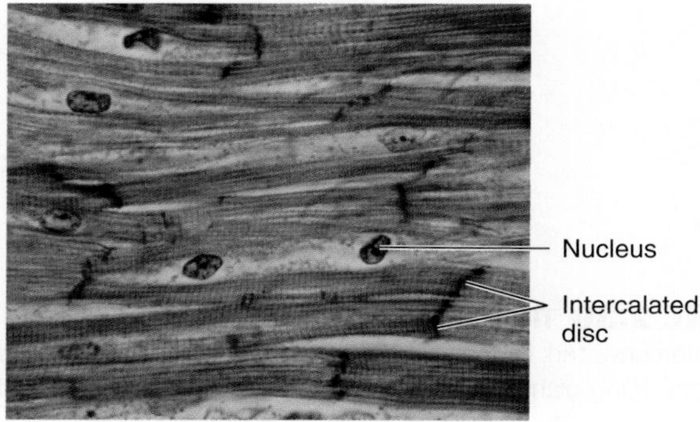

FIGURE 2.32 Human cardiac muscle. Notice the striated appearance and dark-staining intercalated discs, which are characteristic of cardiac muscle.

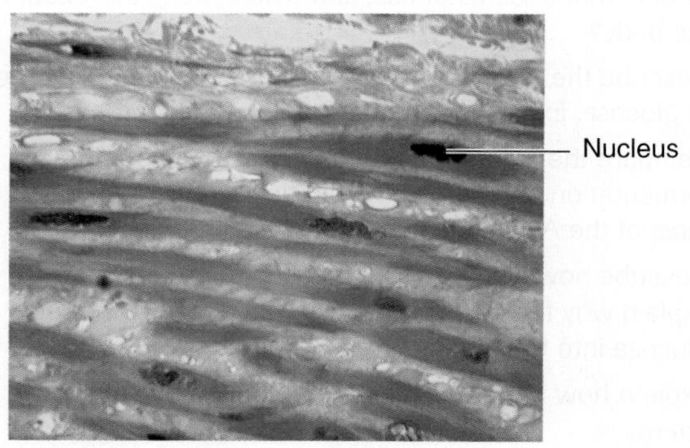

FIGURE 2.33 A photomicrograph of smooth muscle cells. Notice that these cells contain single, centrally located nuclei and lack striations.

FYI [5]Scientists have recently shown that there are stem cells near the middle of a hair follicle, located in a small bulge. These can divide to produce cells that migrate down to the region of the follicle that generates the hair shaft; thus, absence of those stem cells produces baldness. Also, there are stem cells here that differentiate into melanocytes, which are the cells that give hair its color. Graying of hair results from the loss of these melanocyte stem cells.

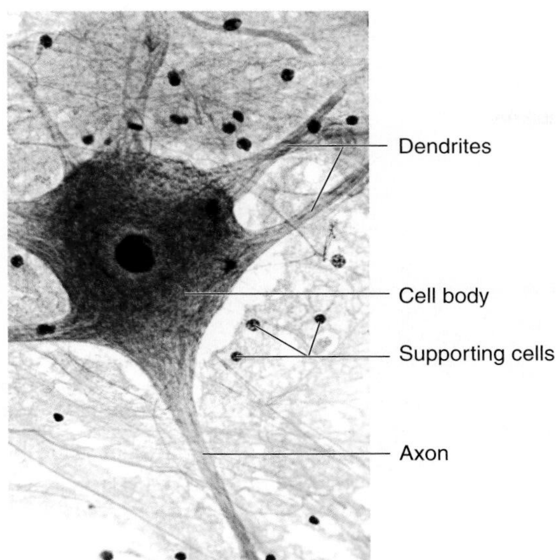

FIGURE 2.34 A photomicrograph of nerve tissue. A single neuron and numerous smaller supporting cells can be seen.

Each neuron has three principal parts (fig. 2.34). The **cell body** is the part of the neuron that contains the nucleus; it serves as the manufacturing center of the neuron. There are two categories of cytoplasmic processes that extend from the cell body. The term *process* in biology refers to a thin extension from whatever is being discussed. **Dendrites** are shorter, branched processes that receive stimulation and conduct it toward the cell body. The **axon** is a single, usually longer extension that conducts impulses away from the cell body. Although the axon leaves the cell body as a single process, it can branch extensively near its terminal.

Epithelial Tissue Forms Membranes and Glands

Epithelial tissue includes **epithelial membranes** and **glands**, which are derived from those membranes. Epithelial membranes are thin structures consisting of epithelial cells that are tightly joined together. *Epithelial membranes cover all body surfaces*, and so serve as a barrier (among other functions) between the tissues underneath and the outside environment. The junctions between epithelial cells prevent substances from easily passing between the cells, a logical property of a tissue barrier. The nature of the junctions and the ways in which molecules may get across epithelial membranes will be considered in chapter 3.

Epithelial membranes are attached to the underlying connective tissue by a gluelike layer, consisting of proteins and polysaccharides, called a *basement membrane* (remember, a membrane can be any thin structure). If the epithelial membrane is only one cell layer thick—so all of the cells are on the basement membrane—it's a **simple epithelial membrane** (fig. 2.35). These can be classified further by cell shape. The cells of a *squamous epithelium* are squat (flat); those of a *cuboidal epithelium* are roughly cube-shaped; and those of a *columnar epithelium* are taller than they are wide. If the membrane is a number of layers thick, with only the bottom layer

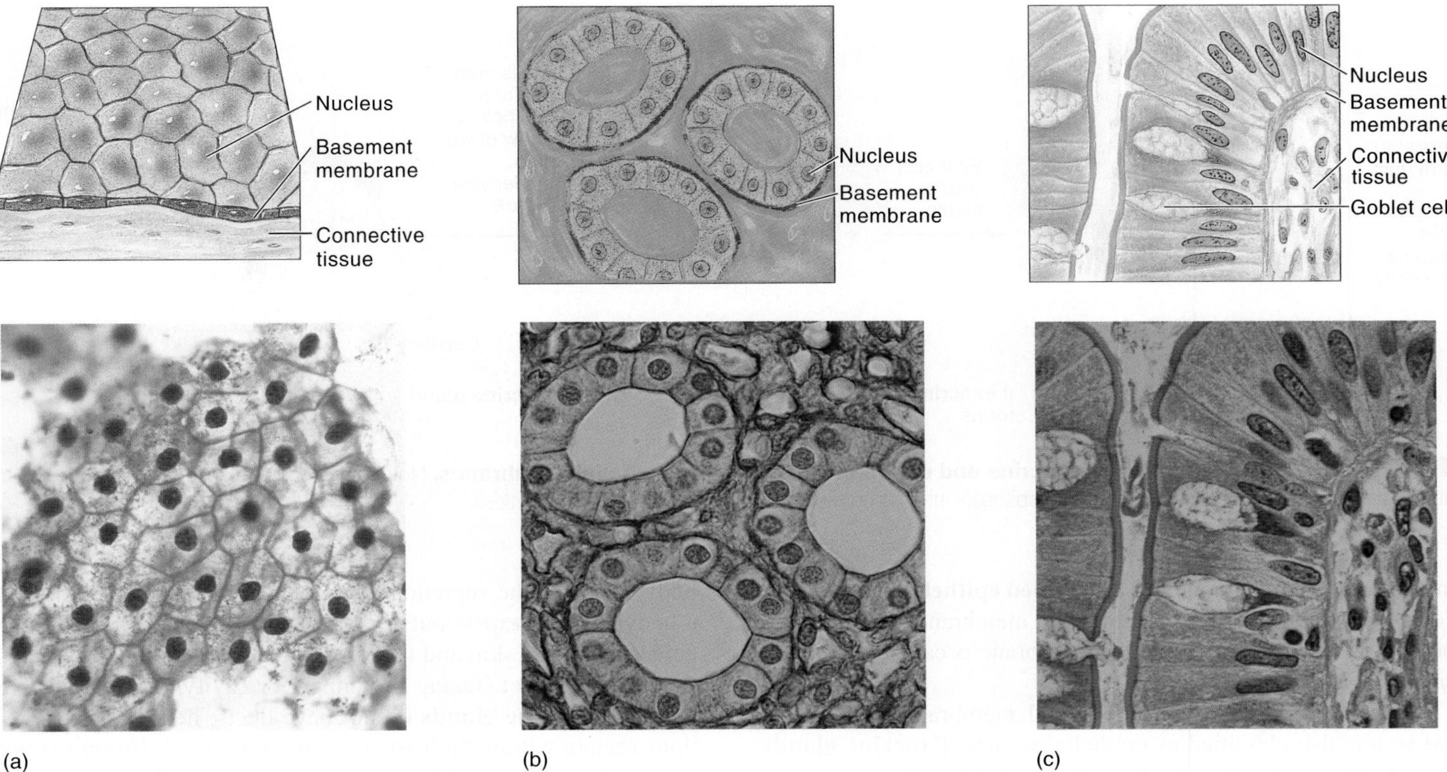

FIGURE 2.35 Different types of simple epithelial membranes. (*a*) Simple squamous, (*b*) simple cuboidal, and (*c*) simple columnar epithelial membranes. The tissue beneath each membrane is connective tissue.

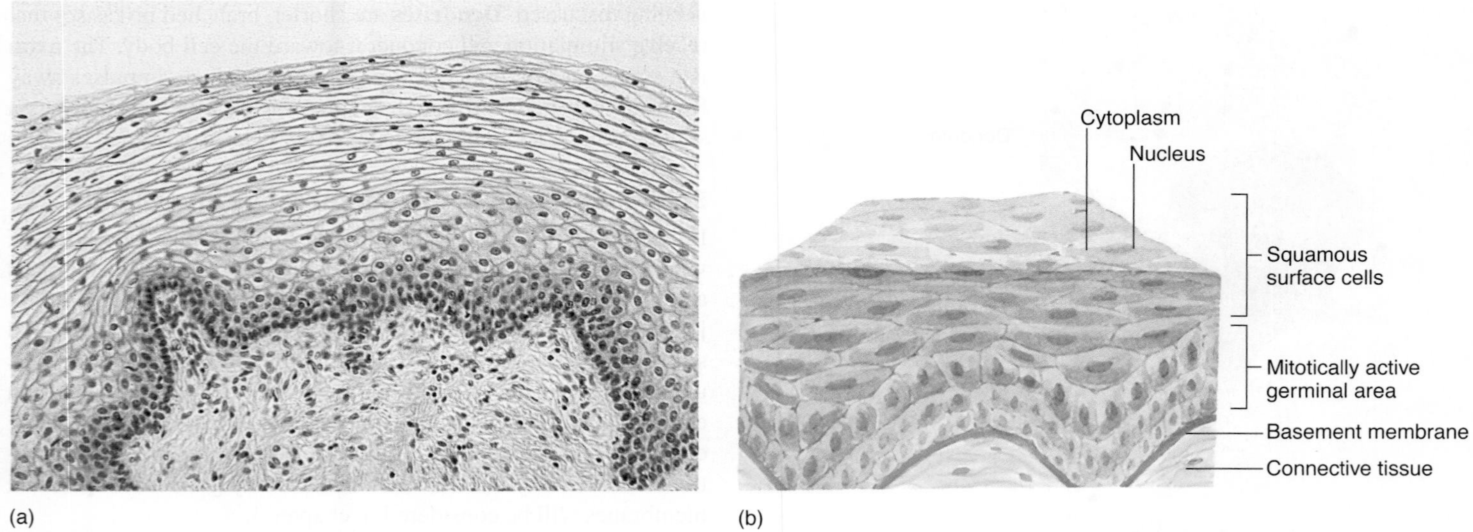

FIGURE 2.36 A stratified squamous nonkeratinized epithelial membrane. This is a photomicrograph (*a*) and illustration (*b*) of the epithelial lining of the vagina.

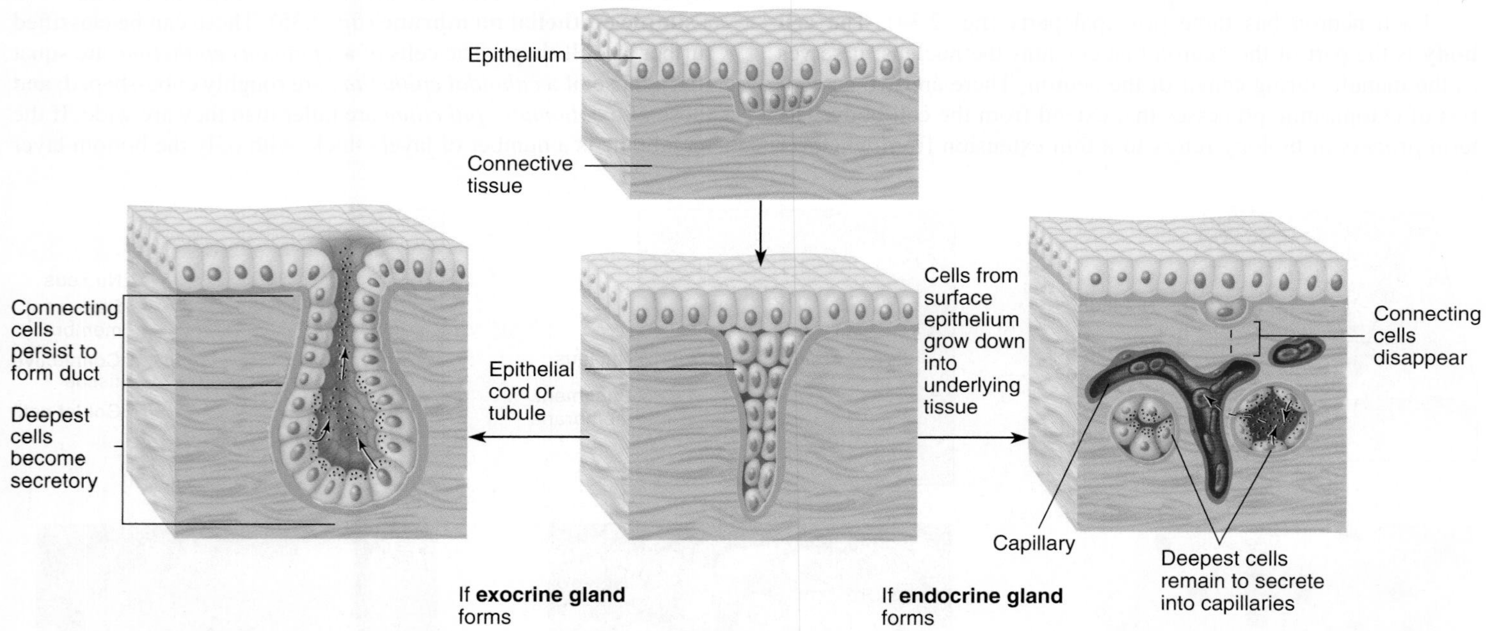

FIGURE 2.37 The formation of exocrine and endocrine glands from epithelial membranes. Note that exocrine glands retain a duct that can carry their secretion to the surface of the epithelial membrane, whereas endocrine glands are ductless.

on the basement membrane, it's a **stratified epithelial membrane.** Cell shapes vary in a stratified epithelial membrane; when cells in the upper layer are squamous, the membrane is called a *stratified squamous epithelium* (fig. 2.36).

Glands are derived from epithelial membranes (fig. 2.37), and so are also classified as epithelial tissues. **Exocrine glands** have ducts to carry the product of the gland to the surface of an epithelial membrane. Since epithelial membranes cover all body surfaces, the secretion is thereby carried to the outside of a body surface (*exo* = outside). This includes the outside of the epidermis of the skin and the outside of the epithelium that lines the digestive tract (facing the lumen, or cavity, of the digestive tract). **Endocrine glands** do not have ducts; hence, their secretions remain within the body and enter the blood. *Hormones* are the regulatory molecules secreted by endocrine glands into the blood (chapter 8).

Clinical Applications

Basement membranes consist primarily of a structural protein known as *collagen*. The specific type of collagen in basement membranes is known as *collagen IV*, a large protein assembled from 6 different polypeptide chains coded by 6 different genes. **Alport's syndrome** is a genetic disorder of the collagen subunits. This leads to their degradation and can cause a variety of problems, including kidney failure. **Goodpasture's syndrome** is a disease caused when a person's own immune system makes antibodies against the basement membrane. When basement membranes are attacked in this way, a person may develop lung and kidney impairment.

Connective Tissues Have Abundant Extracellular Material

Connective tissues are characterized by the presence of abundant extracellular material, which composes the *matrix* between connective tissue cells. The relatively great distance that separates the connective tissue cells makes the presence of an extracellular matrix possible. Connective tissues are grouped according to the nature of the extracellular material in the matrix.

In **connective tissue proper**, the extracellular matrix consists of protein fibers (primarily collagen) embedded in a gel-like *ground substance*. In *loose connective tissue* (also called *areolar tissue*), the protein fibers are spread apart in the ground substance (fig. 2.38). In *dense connective tissue*, the fibers are more densely packed; for example, in a tendon the collagen fibers are packed tightly together in parallel groups (fig. 2.39). The cells of connective tissue proper are in the ground substance; the principal cells are *fibroblasts*, which secrete the extracellular matrix. A type of adult stem cell, called a *mesenchymal cell*, is also present to help renew the population of fibroblasts and other connective tissue cells. *Adipose tissue* (fig. 2.40) is a specialized type of loose connective tissue. Each adipose cell, or adipocyte, has its cytoplasm stretched across a large globule of fat. Enzymes in the cytoplasm of the adipocyte can hydrolyze the fat when we need to use our stored fat for energy.

Cartilage is a specialized connective tissue in which the protein that composes the ground substance of the extracellular matrix is tougher and more resilient than that of connective tissue proper. **Bone** is another specialized connective tissue, which is stronger than cartilage because its extracellular matrix is hardened by calcium phosphate crystals. Because the extracellular matrix is calcified, the bone cells (*osteocytes*), trapped within little spaces (*lacunae*) in the calcified matrix, can't be kept alive by diffusion of oxygen and nutrients through the matrix. Instead, the osteocytes

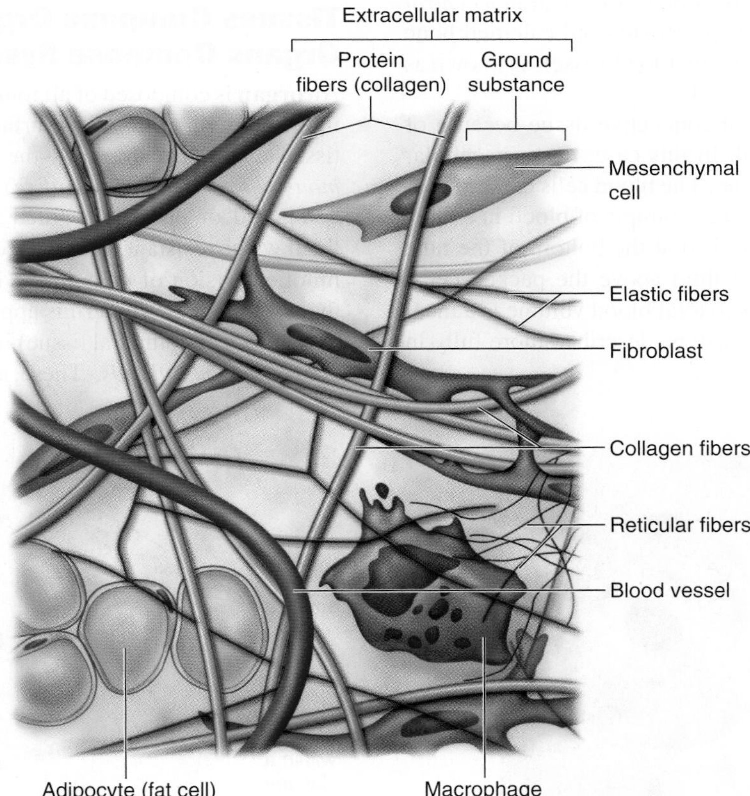

FIGURE 2.38 Loose connective tissue. This illustration shows the cells and protein fibers characteristic of connective tissue proper. The ground substance is the extracellular background material, against which the different protein fibers can be seen. The macrophage is a phagocytic connective tissue cell, which can be derived from monocytes (a type of white blood cell).

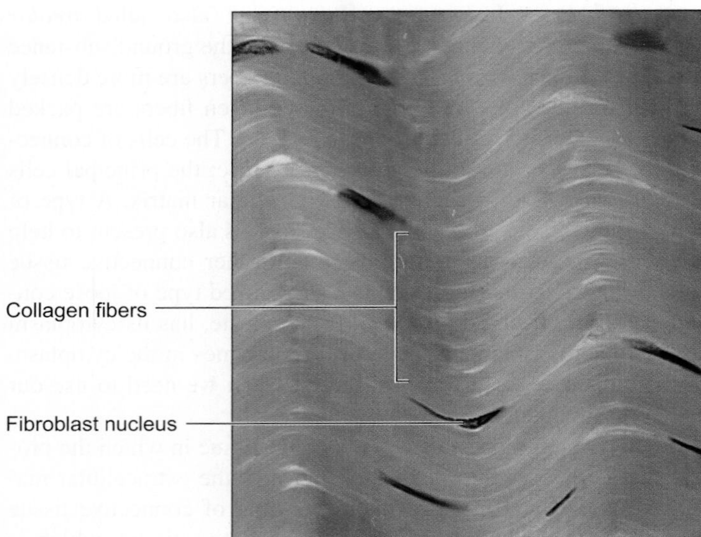

FIGURE 2.39 Dense connective tissue. In this photomicrograph, the collagen fibers in a tendon are packaged densely into parallel groups. The ground substance is in the tiny spaces between the collagen fibers.

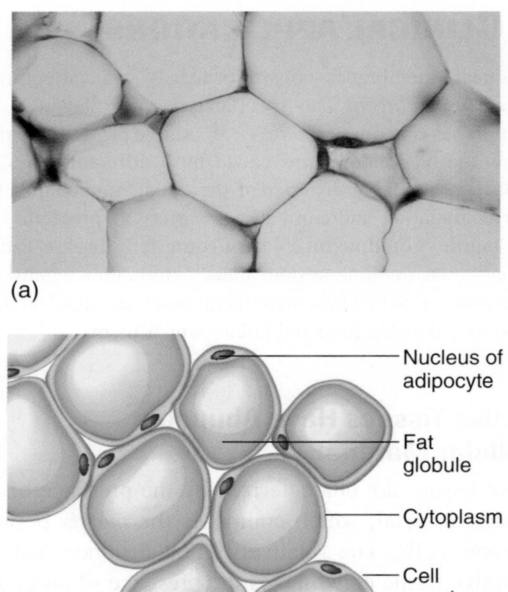

FIGURE 2.40 Adipose tissue. Each adipocyte contains a large, central globule of fat surrounded by the cytoplasm of the adipocyte. (*a*) Photomicrograph and (*b*) illustration adipose tissue.

are kept alive by little canals (*canaliculi*) that extend from the lacunae, through the calcified matrix, to blood vessels within a *central canal*. The circular arrangement of osteocytes and calcified bone matrix around a central canal containing blood vessels is known as an *osteon*, or *haversian system* (fig. 2.41).

Blood is classified as a special connective tissue because of its abundant extracellular material. In this case, the extracellular material is a fluid—the blood *plasma*. The blood cells can be separated from the plasma by centrifuging a sample of blood in a tube. When this is done, the cells are packed at the bottom of the tube and the plasma is a straw-colored fluid above the packed cells. The ratio of packed red blood cells to total blood volume is a measurement called the *hematocrit*, as will be described more fully in chapter 10.

Tissues Compose Organs, and Organs Compose Systems

An **organ** is composed of all four primary tissues. The largest organ in the body (in terms of its surface area) is the skin (fig. 2.42). The tissue exposed to the air—the **epidermis**—is a *stratified squamous keratinized* (or *cornified*) *epithelium*. The upper layers are composed of dead cells filled with the protein *keratin*, making them water-resistant. These are continuously shed and replaced by mitotic division of cells in the bottom layer of the epidermis, so that we get a new epidermis approximately every 2 weeks. Glands (a category of epithelial tissue) in the skin include the *sweat glands* and *sebaceous glands*. These are exocrine glands because their

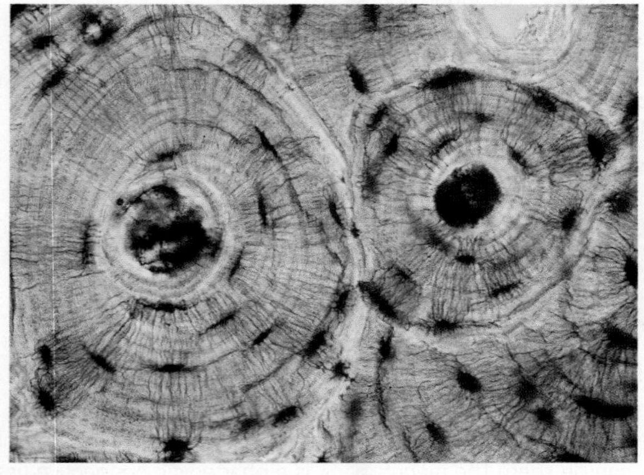

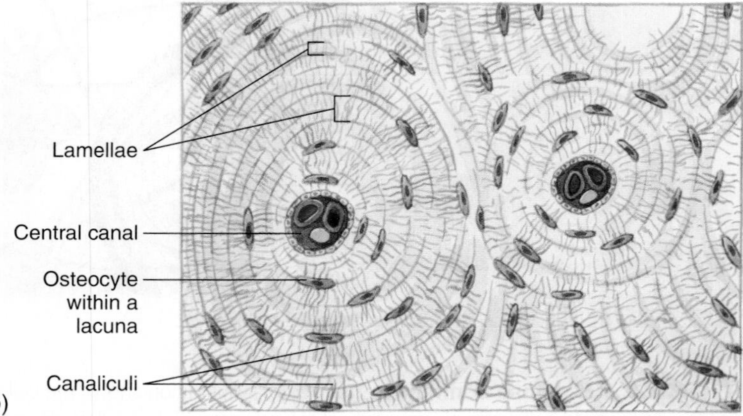

FIGURE 2.41 The structure of bone. (*a*) A photomicrograph showing osteons (haversian systems), and (*b*) a diagram of osteons. Within each central canal, an artery (red), a vein (blue), and a nerve (yellow) is illustrated.

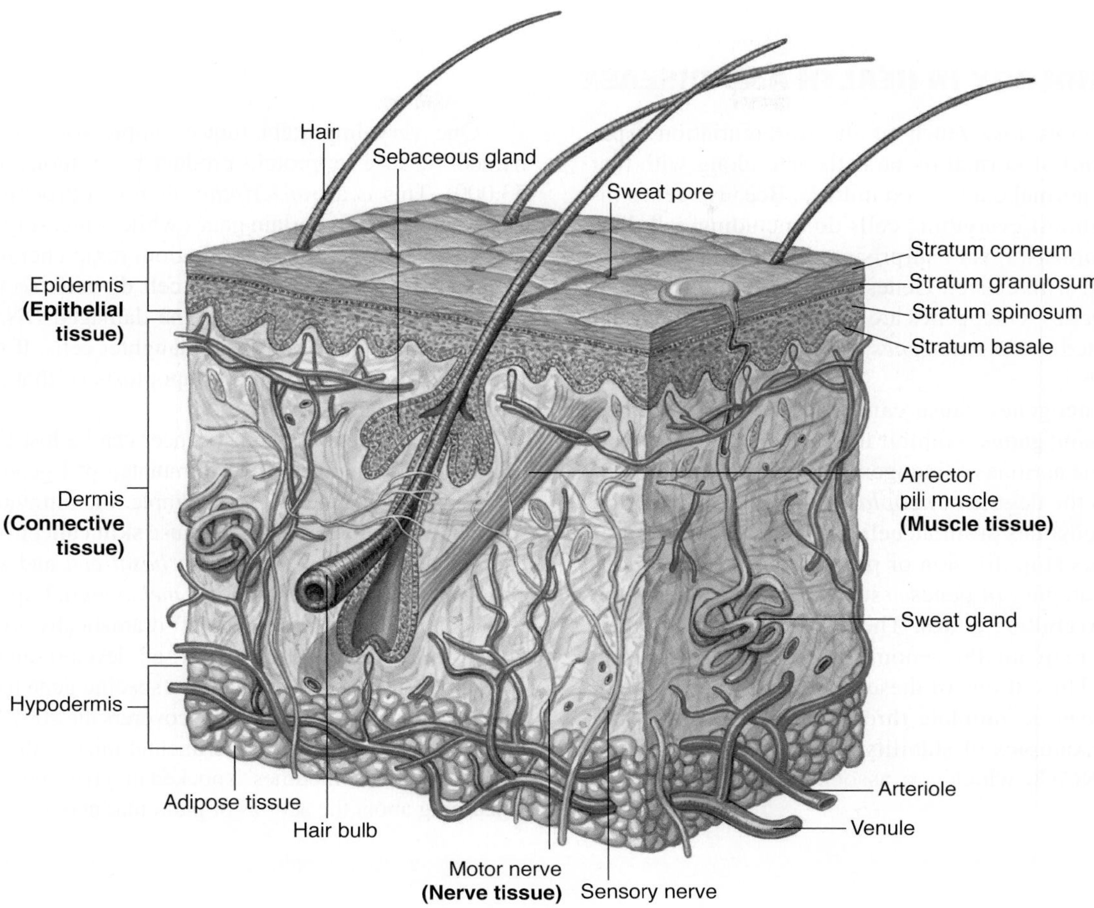

FIGURE 2.42 A diagram of the skin. The skin is an organ that contains all four types of the primary tissues.

watery (sweat) or oily (*sebum*) secretions are channeled through ducts to the surface of the epidermis.

Because cells of an epithelial membrane are tightly joined together, there is no room in an epithelial membrane for blood vessels (which are organs in their own right) and nerves. These enter the skin in the connective tissue **dermis**, located beneath the epidermis. Adipose tissue, another connective tissue, is just below the dermis in the **hypodermis**; the fat here is sometimes called *subcutaneous fat*. Muscle tissue is also found in the skin, as the *arrector pili* muscles. These smooth muscles attach to hair follicles and to the dermis; when they contract, they may cause hair at the back of the neck to stand up. More generally, these are our "goosebump" muscles.

Organs located in different regions of the body that perform related functions are grouped into **systems**. The nervous and endocrine systems help regulate the other body systems, and our regulatory mechanisms coordinate the activity of the different organs of a system in order to maintain homeostasis.

CHECK POINT

1. Define the terms *undifferentiated* and *differentiated*, and explain the significance of adult stem cells.
2. Describe the characteristics of the different types of muscle tissue.
3. Describe the structure and location of epithelial membranes.
4. Identify the parts of a neuron and their functions.
5. What characteristic do all connective tissues share in common? Describe the different types of connective tissues.
6. Identify the different tissues in a typical organ, the skin.

Physiology in Health and Disease

Cancer cells lose much of the differentiation (specialization) of normal tissue cells and, along with that loss, the normal controls on mitosis. Because genes influence almost everything cells do, including cell division by mitosis, it isn't surprising that certain genes are associated with cancer. Genes that contribute to cancer are **oncogenes**. Oncogenes are mutated forms of normal genes, called *proto-oncogenes*, that function in normal, healthy cells.

Whereas oncogenes cause cancer, other genes—called **tumor suppressor genes**—inhibit its development. The balance between the actions of oncogenes and tumor suppressor genes regulates the degree of *neoplasia* (new cell production by mitosis); oncogenes promote cell division, whereas tumor suppressor genes stop division or promote cell death (apoptosis). A third category of genes associated with cancer is the **stability**, or **caretaker**, **genes**. These help repair the small errors that can creep into the genome during DNA replication as cells divide. Thus, if one of these genes becomes mutated, genetic errors can accumulate throughout the genome and cause cancer. Examples of stability gene mutations include *BRCA1* and *BRCA2*, which are associated with hereditary **breast cancer**.

One very important tumor suppressor gene is **p53** (so-named because its protein product has a molecular weight of 53,000). This is a *transcription factor:* a protein that binds to DNA and activates certain genes while repressing others. When DNA is damaged—due to radiation, toxic chemicals, or other cellular stresses—p53 can stop cell division so that the DNA can be repaired. This prevents the damaged DNA from being replicated and passed on to the daughter cells. If the DNA cannot be repaired, p53 promotes apoptosis so that the cell won't replicate the damaged DNA.

These protections against cancer can be lost if the p53 gene itself becomes mutated. Indeed, mutated p53 genes are found in over 50% of all cancers. For example, the ultraviolet portion of sunlight can damage DNA and cause **skin cancer** if the p53 gene has mutated. This occurs in both *basal cell* and *squamous cell carcinomas* (but apparently not in *melanoma*). Experimental mice with their p53 genes "knocked out" dramatically reveal the importance of p53 in preventing cancer: all develop tumors. *Knockout mice* are strains of mice in which a specific gene has been inactivated, a technique that won its discoverers the 2007 Nobel Prize in Physiology or Medicine. Experimental mice with different genes knocked out (or sometimes "knocked in") have proved very useful in learning about the actions of particular genes.

Physiology in Balance

From Cells to Systems

Working together with . . .

Endocrine System

Homeostasis of blood glucose is maintained . . . p. 186

Respiratory System

Homeostasis of blood O_2 and CO_2 is maintained . . . p. 297

Digestive System

Digestion of carbohydrates supplies the blood with glucose . . . p. 348

Muscular System

Lactic acid is produced by muscles during exercise . . . p. 215

Nervous System

The brain aerobically respires blood glucose . . . p. 113

Urinary System

The kidneys reabsorb blood glucose . . . p. 324

Summary

The Cell Is the Basic Unit of Structure and Function 28

- The plasma membrane separates the intracellular from the extracellular environments.
- The plasma membrane consists of a double layer of phospholipids and proteins, described by the fluid-mosaic model of the membrane.
- Microvilli increase the surface area of the plasma membrane in certain body locations; this increases the rate of transport across the plasma membrane.
- Flagella and motile cilia have specialized functions and consist of a "9+2" arrangement of microtubules.

Cytoplasmic Organelles Perform Many Functions 30

- Lysosomes are organelles that contain many digestive enzymes.
- Mitochondria are the sites where aerobic cell respiration occurs, and where most of the ATP is produced.
- Ribosomes are the protein factories, where messenger RNA is translated into proteins.
- The rough (granular) endoplasmic reticulum is where proteins are produced that are destined for secretion; the smooth (agranular) endoplasmic reticulum has diverse functions, including storage of Ca^{2+} in some cells and enzymatic reactions in others.
- The Golgi complex is a system of membranous sacs that receives products in vesicles budded from the rough endoplasmic reticulum, modifies the products, and then releases the modified products within vesicles into the cytoplasm.

Cell Nucleus, DNA Function, and Cell Division 34

- A gene is a region of DNA that codes for a polypeptide chain; the genome is the total genes in a cell; the proteome is the total number of different proteins in a cell.
- DNA directs the synthesis of RNA through complementary base pairing with one of the 2 DNA strands; RNA synthesis is called genetic transcription.
- Three bases in messenger RNA (mRNA) compose a codon that specifies a particular amino acid; a transfer RNA (tRNA), which carries the amino acid, has 3 bases that form an anticodon that binds to the codon for protein synthesis, or genetic translation.
- Before a cell divides, its DNA replicates itself when its 2 strands separate and new partner DNA strands are formed; replication is semiconservative, because each DNA consists of 1 strand from an "old" DNA and 1 new strand, identical to the original DNA because of complementary base pairing.
- When a cell divides by mitosis, its DNA replicates to form duplicate chromosomes, so that one set can be distributed to each of the daughter cells; mitotic cell division progresses through stages identified as interphase, prophase, metaphase, anaphase, and telophase.
- Cell death can occur abnormally in necrosis, or as a normal, physiological process called apoptosis, which is necessary for the continual replacement of worn-out cells.

Cell Respiration: Metabolic Pathways That Provide Energy for the Cell 41

- Chemical bond energy in food molecules, such as glucose, is partially transferred into the bond joining the last phosphate in ATP, the universal energy carrier of the cell.
- Many enzymes can cooperate to produce a metabolic pathway, in which sequential chemical changes occur; the metabolic pathways that provide energy for ATP production in cells are known as cell respiration.

Glucose Provides Energy Through Two Pathways 44

- The conversion of glucose into 2 molecules of pyruvic acid, accompanied by the net production of 2 ATP, is called glycolysis; in the absence of oxygen, a pathway called lactic acid fermentation results in the conversion of pyruvic acid into lactic acid.
- When oxygen is available, respiration is aerobic; the pyruvic acid enters the matrix of a mitochondrion and is converted into acetyl Coenzyme A, which starts a citric acid (Krebs) cycle.
- The Krebs cycle produces an ATP, as well as NADH and $FADH_2$, which donate hydrogens to the electron transport system in the cristae; oxygen is the final electron acceptor (becoming part of a water molecule), and much ATP is produced by electron transport in the process of oxidative phosphorylation.
- Blood glucose may come from the glucose that passes across the wall of the intestine, or from glucose secreted by the liver; the liver can obtain glucose by hydrolysis (breakdown) of its stored glycogen, and is the only organ that can secrete glucose into the blood.
- Cells can also obtain energy for ATP production by the aerobic cell respiration of fatty acids (which each produce many ATP molecules) and glycerol from fat, and by the aerobic cell respiration of amino acids from proteins.

The Body Organs and Systems Are Composed of Specialized Tissue Cells 49

- Muscle tissue is specialized for contraction; we have two types of striated muscles—skeletal muscle and cardiac muscle. Smooth muscle, found in the walls of many internal organs, is not striated.
- Nervous tissue contains neurons, which are the principal regulating cells of the nervous system; a neuron consists of a cell body and two types of processes, dendrites and an axon.
- Epithelial tissue forms membranes and glands; epithelial membranes may be simple or stratified, and they have junctional complexes that keep their cells tightly together; glands may be exocrine (with ducts) or endocrine (without ducts).
- Connective tissues have abundant extracellular material, which is the matrix of connective tissues; these tissues are very diverse in structure, ranging from connective tissue proper to the specialized connective tissues of cartilage, bone, and blood.
- An organ consists of all four primary tissues, and organs in different body locations that perform related functions are grouped into the body systems.

Review Activities

Objective Questions: Test Your Knowledge

1. Which of the following statements regarding the plasma membrane is true?
   a. It is composed of a layer of phospholipids sandwiched between two layers of proteins.
   b. It is composed of a double layer of phospholipids, with the polar groups of each layer facing each other in the center of the membrane.
   c. It is composed of a double layer of phospholipids, with the nonpolar groups facing each other in the center of the membrane.
   d. There is a layer of proteins sandwiched between a double layer of phospholipids.
2. Autophagy and apoptosis are associated with which organelle?
   a. Lysosomes
   b. Rough endoplasmic reticulum
   c. Smooth endoplasmic reticulum
   d. Golgi complex
   e. Mitochondria
3. Which of the following statements about DNA is false?
   a. Mitochondria contain DNA.
   b. The DNA doesn't form visible chromosomes when the cell isn't going to divide.
   c. DNA exits the nucleus through the nuclear pores.
   d. Chromatin contains regions where the DNA is spooled around proteins that compose nucleosome particles.
4. A codon and an anticodon are three complementary bases on
   a. DNA and mRNA, respectively.
   b. mRNA and tRNA, respectively.
   c. tRNA and rRNA, respectively.
   d. rRNA that pair with three amino acids in a protein.
5. Which of the following statements about the genome is false?
   a. The human genome contains approximately 25,000 genes.
   b. The genome codes for the proteome.
   c. We have the same number of different proteins as we have different genes.
   d. In any cell, about 300 different genes are active at any one time.
6. In DNA replication,
   a. the DNA strands separate, and each strand serves as a template for the production of a complementary strand.
   b. after replication, both strands of one DNA are new and both strands of the other DNA are from the original molecule.
   c. DNA synthesis is said to be "conservative."
   d. DNA synthesis is called genetic transcription.
7. Which of the following statements about the chemical bond energy in food molecules is true?
   a. It is all converted into heat energy.
   b. It is all captured in the chemical bond joining the third phosphate group to ATP.
   c. It is used to directly power the energy-requiring processes of the cell.
   d. A portion of the energy escapes as heat, and a portion is captured in the chemical bond joining the third phosphate group to ATP.
8. Which of the following statements about glycolysis is true?
   a. In glycolysis, glucose is converted into lactic acid.
   b. It is a metabolic pathway requiring the action of many enzymes.
   c. Glycolysis results in a net gain of 4 ATP.
   d. Glycolysis occurs in the mitochondria.
9. Which of the following statements about aerobic cell respiration is true?
   a. Acetyl Coenzyme A is formed.
   b. A Krebs cycle (citric acid cycle) occurs.
   c. One ATP, 3 NADH, and 1 $FADH_2$ are produced per Krebs cycle.
   d. All of these statements are true.
10. The production of most of the ATP in a typical cell occurs
    a. by the process of oxidative phosphorylation.
    b. in the intramembranous space of a mitochondrion.
    c. during the Krebs cycle.
    d. during glycolysis.
11. Which of the following correctly describes the role of oxygen in aerobic respiration?
    a. It combines with carbon to form carbon dioxide.
    b. It is the final electron acceptor from the electron transport system.
    c. It combines with hydrogen to form water during the Krebs cycle.
    d. It is required for the conversion of glucose to pyruvic acid.
12. The lactic acid pathway
    a. produces more ATP per glucose than aerobic cell respiration.
    b. occurs in the mitochondria.
    c. can start with either glucose or fatty acids.
    d. is also called lactic acid fermentation.
13. Glucose derived from stored glycogen
    a. starts as glucose 6-phosphate.
    b. is produced by the process of glycogenolysis.
    c. can be produced and secreted only by the liver.
    d. All of these statements are true.
14. The primary tissue characterized by junctional complexes that join cells tightly together and a basement membrane is
    a. muscle tissue.
    b. epithelial tissue.
    c. connective tissue.
    d. nervous tissue.
15. The primary tissue characterized by the presence of an abundant extracellular matrix is
    a. muscle tissue.
    b. epithelial tissue.
    c. connective tissue.
    d. nervous tissue.
16. Which of the following statements about muscle tissue is false?
    a. Skeletal muscle contains intercalated discs.
    b. Cardiac muscle and smooth muscle are involuntary muscles.
    c. Skeletal muscle and cardiac muscle are striated muscles.
    d. Skeletal muscle fibers are derived from the end-to-end fusion of myoblast cells.
17. Which of the following statements about glands is false?
    a. All glands are epithelial tissue.
    b. Endocrine glands secrete hormones into the blood.
    c. Exocrine glands lack ducts.
    d. Exocrine glands secrete to the outside of epithelial membranes.

18. Skeletal muscles produce lactic acid
    a. at all times, regardless of the availability of oxygen.
    b. at the beginning of exercise.
    c. whenever they metabolize fatty acids.
    d. by the addition of hydrogens from pyruvic acid to NAD.

19. Organic molecules with phosphate groups, such as ATP or glucose 6-phosphate,
    a. can always pass through plasma membranes.
    b. can sometimes pass through plasma membranes.
    c. can never pass through plasma membranes.
    d. are produced in the extracellular fluid.

20. The brain gets its energy almost exclusively from the
    a. aerobic respiration of blood glucose.
    b. anaerobic respiration of blood glucose.
    c. aerobic respiration of glucose derived from stored glycogen.
    d. anaerobic respiration of glucose derived from stored glycogen.

Essay Questions 1: *Test Your Understanding*

1. Describe the composition of the plasma membrane, and explain why the term *plasma* is used to name this membrane.
2. Distinguish between microvilli, cilia, and flagella, in terms of their structures and functions.
3. What is apoptosis? What organelles are involved in this process, and what is its physiological significance?
4. Distinguish between chromatin and chromosomes.
5. What is a gene? Describe the relationship between the genome and the proteome.
6. Explain the meaning of the term *differentiated* in terms of the tissue types and the genes that are active in a particular cell. What are "adult stem cells," and what functions do they perform?
7. Explain how genes are transcribed into codons in messenger RNA.
8. Explain how the codons in messenger RNA are translated into the sequence of amino acids in a protein.
9. When does DNA replication occur? Explain the process of DNA replication.
10. What are the functions of mitosis? Describe the events that occur during the phases of mitosis.
11. What is the physiological significance of apoptosis?
12. Explain the role of ATP in the energy flow of a cell, and how its formation and breakdown allows it to play this role.
13. Identify the initial substrate and final products of glycolysis, and how many ATP are gained in this process.
14. When and where does lactic acid production occur in the body, and how does it relate to glycolysis?
15. How does the aerobic respiration of glucose relate to glycolysis? Identify the cellular location and products of the Krebs (citric acid) cycle.
16. Where is the electron transport system located, and what is its function and significance? Explain why most cells will perform aerobic respiration if they are given a supply of oxygen.
17. The liver is the only organ that can use its stored glycogen to secrete glucose into the blood. Explain why this is true.
18. Describe how cells can use fatty acids and amino acids for energy.
19. Identify the characteristics of epithelial tissue and contrast these with the characteristics of connective tissue. Explain why blood is considered a type of connective tissue.
20. Describe the similarities and differences between skeletal muscle and cardiac muscle.

Essay Questions 2: *Test Your Analytical Ability*

1. Describe the structure and characteristics of phospholipids that allows them to form a double layer in a plasma membrane. How would this phospholipid structure influence the ability of ions to pass through the membrane? How would this compare with the ability of steroid hormones to pass through the membrane?
2. What advantages might be provided by the microvilli in the simple columnar epithelium of the small intestine? When microvilli are observed in other parts of the body, such as tubules in the kidneys, what might you infer from their presence?
3. Explain how the presence of somewhat undifferentiated adult stem cells in an organ can be related to the ongoing apoptosis of some of the organ's cells.
4. Suppose there is a particular metabolic pathway that has a branch point. Remembering that enzymes are proteins, explain how the direction taken by this pathway can be influenced by gene activation and inactivation.
5. Suppose that a DNA is made using nucleotides containing radioactive deoxyribose sugars. Then, the DNA is allowed to replicate using only nonradioactive nucleotides. How do you think the radioactivity of the daughter DNA would compare to the parent DNA? Explain.
6. Suppose a slice of pie contains 300 kilocalories. If you eat this slice of pie, will you get 300 kilocalories worth of chemical bond energy in ATP? Explain.
7. Lactic acid, produced from glucose by skeletal muscles, can travel in the blood to the liver, where it can be converted back to glucose. Using just the information in this chapter, explain how that could be accomplished.
8. After exercising, much of the lactic acid is aerobically respired for energy; the extra oxygen required contributes to the "oxygen debt" following exercise. Describe the steps by which lactic acid can be used for energy.
9. Red blood cells lack mitochondria, whereas cardiac muscle cells are very rich in mitochondria. What do these observations suggest regarding the physiology of red blood cells and cardiac muscle cells?
10. There are more kilocalories in a gram of fat than in a gram of carbohydrates, and we get more ATP energy from fatty acids than from glucose. Explain the metabolic pathways involved, providing only the level of detail presented in this chapter.
11. Describe the distinction between voluntary and involuntary muscle, and provide reasons why this may not be a very accurate distinction.
12. Using the dermis and epidermis of the skin as examples, explain how the characteristics of connective tissue and epithelial tissue can complement each other.

Web Activities

For additional study tools, go to: www.aris.mhhe.com. Click on your course topic and then this textbook's author/title. Register once for a semester's worth of interactive activities and practice quizzing to help boost your grade!

CHAPTER 3

Interactions Between Cells and Their Environment

Homeostasis

All the cells in the body must interact with their external environment, the extracellular fluid in which they live. They obtain oxygen and nutrients and eliminate waste products, so the extracellular fluid must be circulated and refreshed. Homeostasis of the extracellular fluid requires the proper functioning of many organs, including the heart and blood vessels, lungs, intestine, liver, and kidneys.

Membrane transport processes allow the cells of all organs to function, and are needed for the normal functioning of those organs charged with maintaining homeostasis of the extracellular fluid. The lungs allow oxygen and carbon dioxide to pass between the air and blood; the liver produces and secretes many molecules into the blood, and transports toxins and waste products from the blood into the bile for elimination in the feces; the intestine absorbs the products of food digestion into the blood; and transport processes in the kidneys are required for the elimination of certain waste products in the urine, as well as for the retention of molecules required by the body.

But if homeostasis is not maintained . . .

Clinical Investigation

Jill, a 50-year-old woman, went to the doctor because she felt weak and dehydrated and was alarmed that she was getting puffy, a condition the physician called edema (excessive interstitial, or tissue, fluid). The doctor performed a urine test that revealed a significant amount of glucose, and further tests confirmed the diagnosis of diabetes mellitus. Blood tests demonstrated an abnormally low plasma protein concentration, and more alarming, other tests, including a liver biopsy, revealed that Jill had cirrhosis of the liver. The doctor explained that the liver produces most of the plasma proteins, and a low plasma protein concentration could account for Jill's edema.

What is the relationship between Jill's symptoms (of weakness, dehydration, and edema), her lab results of glucose in the urine and a low plasma protein concentration, and her diagnosis of diabetes mellitus and cirrhosis?

Chapter Outline

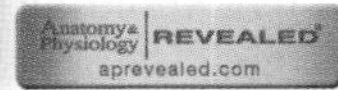

A virtual cadaver dissection experience

The Plasma Membrane Can Move Many Molecules Together

Related cell processes that move many molecules at the same time into a cell include phagocytosis, pinocytosis, and receptor-mediated endocytosis. Through these processes, substances from the extracellular fluid are brought into a cell enclosed in a vacuole or vesicle. Exocytosis allows a cell to secrete products contained in intracellular vesicles, by fusion of those vesicles with the plasma membrane.

Although some of this information is covered in introductory biology courses, the topic of plasma membrane transport is genuinely a physiology topic. It's fundamental to understanding the physiology of neurons, muscle cells, and indeed the cells in all organs. It's worth emphasizing at this point that the physiology you might think is most relevant to health applications—blood and guts physiology—is based on the function of cells. And to a large degree, cell function and its broader physiological implications are based on the processes that occur at the plasma membrane.

The plasma membrane separates the intracellular from the extracellular environments, or compartments. The fluid in the extracellular compartment is divided between the *blood plasma*, which circulates within blood vessels, and the **interstitial (tissue) fluid**, which immediately surrounds the tissue cells. The interstitial fluid is derived from blood plasma and returns to it, so it's more like a lively lake with inlets and outlets than like a stagnant pond. All tissue cells continuously remove molecules from this interstitial fluid and release molecules into it, and such activities help maintain homeostasis of the blood plasma and thus the health of the body.

A cell can move many molecules at the same time from the extracellular fluid into the cell, as long as these molecules are encased in a bubble of membrane derived from the plasma membrane. The membrane bubble is called a **vacuole** when it's large and a **vesicle** when it's small. In order for the extracellular molecules encased within the membrane bubbles of vacuoles and vesicles to enter the cytoplasm of cells, these molecules must pass through the vacuole or vesicle membrane that surrounds them. This may require the digestion of the molecules to smaller components, as may occur when a vacuole fuses with a lysosome (the organelle that contains digestive enzymes, as described in chapter 2). The processes that move many molecules together contrasts with those that move single molecules across the plasma membrane directly into or out of the cytoplasm, which will be described in later sections of this chapter.

Phagocytosis Is Cell Eating

Some body cells are able to move like an amoeba. They perform this *amoeboid motion* by extending **pseudopods** (literally, "false feet")—portions of their cytoplasm that can reach out and pull the cell along—that allow the cell to crawl and thereby move independently. Cells that can form pseudopods may use them to surround an article of "food"—such as a bacterium—and fuse their pseudopods to form a vacuole surrounding the engulfed "food" (fig. 3.1). This process is called **phagocytosis**.

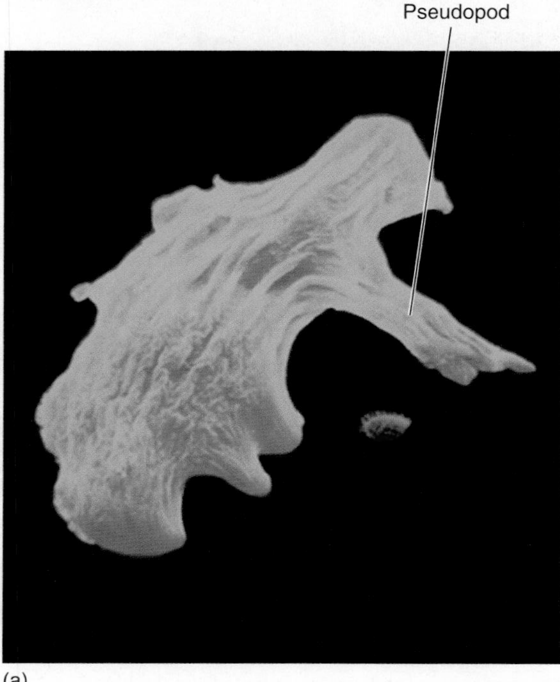

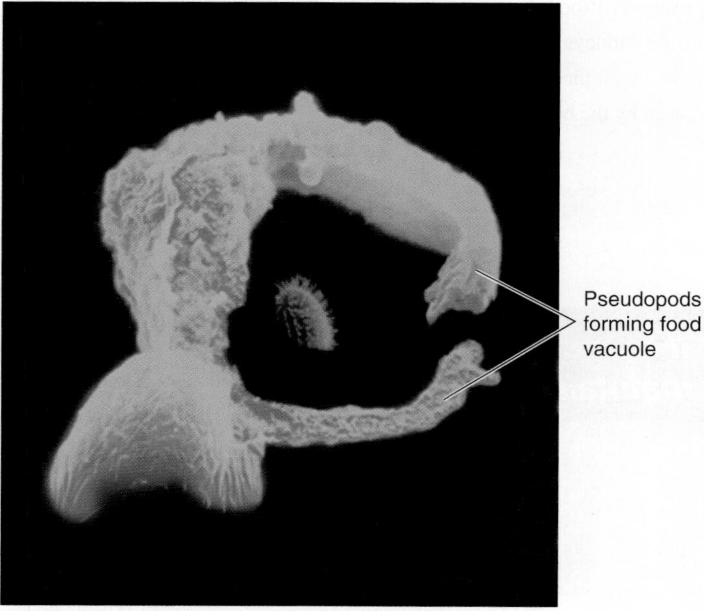

FIGURE 3.1 Scanning electron micrographs of phagocytosis. (*a*) The formation of pseudopods and (*b*) the entrapment of the prey within a food vacuole.

The bacterium is now within a vacuole inside the *phagocytic cell* (the cell that can perform phagocytosis). This vacuole will fuse with a lysosome, so that the enzymes in the lysosome will be able to digest the bacterium.

Phagocytic cells that perform amoeboid motion include white blood cells called *neutrophils*, and to a lesser degree white blood cells called *monocytes*. Monocytes are notable because, when they enter connective tissues, they can transform into *macrophages* (literally, "big eaters").[1] There are also some amoeboid phagocytic cells that are characteristic of only particular organs; these include the *microglia* of the brain, which are related to the monocytes and macrophages.

In addition to the phagocytic cells that can wander around by amoeboid movement, there are also **fixed phagocytes**. The term refers to the fact that these phagocytic cells are immobile (fixed); they are stuck in small blood channels called sinusoids. Examples include the *Kupffer cells* in the liver, as well as fixed phagocytic cells in the sinusoids of the spleen and lymph nodes. These cells cannot move by amoeboid motion, but they can extend pseudopods that surround passing toxins, debris, and pathogens (disease-causing organisms, such as bacteria) in the blood or lymph.

Endocytosis Produces Vesicles Containing Extracellular Materials

In phagocytosis, extracellular materials are enclosed by pseudopods that extend out from the cell. In **endocytosis**, the plasma membrane *invaginates* inward; that is, it forms an inward pocket or channel. The plasma membrane then fuses together at the invagination, forming a vesicle that, at first, is still attached to the inside of the plasma membrane. Finally, the vesicle buds off from the plasma membrane, so that it enters the cell with its contents of extracellular material.

When the invaginated epithelium forms a narrow channel that nonselectively takes in extracellular fluid, the type of endocytosis is called **pinocytosis** (fig. 3.2). Pinocytosis produces tiny vesicles that contain extracellular fluid and molecules such as proteins that could not enter the cell by other means. Most cells in the body are capable of pinocytosis.

In contrast to pinocytosis, only certain cells in some organs can perform **receptor-mediated endocytosis**. The "receptor" here is a protein within the plasma membrane that can very selectively bind to a particular molecule in the extracellular fluid. The subject of receptor proteins will be described in more detail near the end of this chapter. For now, it is important to understand that the interaction between the receptor proteins stuck in the plasma membrane and particular extracellular molecules is highly selective, much like the specific binding of an enzyme protein and its substrate (discussed in chapter 1). In receptor-mediated endocytosis, the binding of extracellular molecules with their membrane receptors stimulates the plasma membrane at that region to invaginate, forming a small pocket. The pocket then pinches off to form a small vesicle as the plasma membrane fuses at the invagination (fig. 3.3). This highly selective form of endocytosis takes specific molecules from the extracellular fluid into the cell.

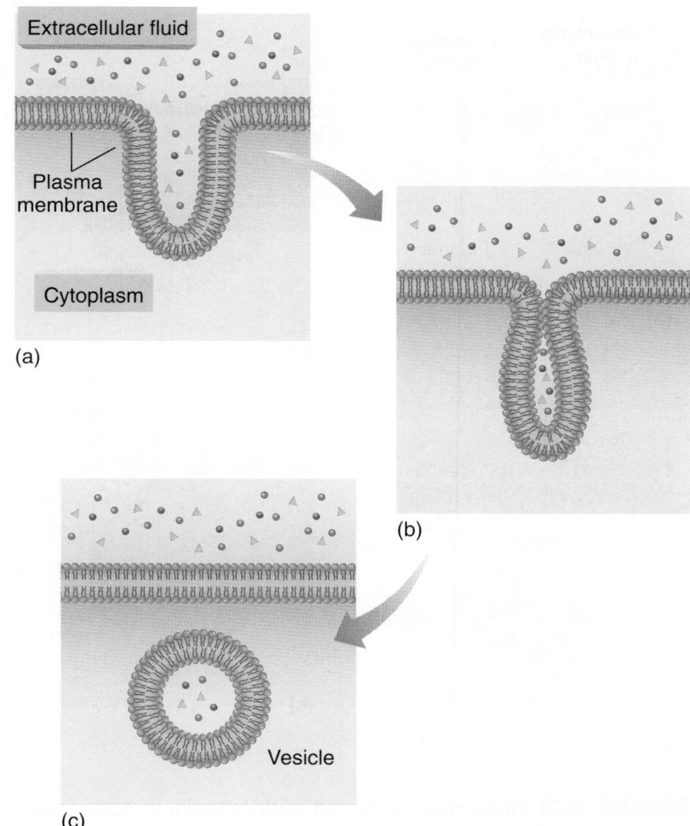

FIGURE 3.2 Pinocytosis. Cells can take in many molecules at the same time through pinocytosis. This is a nonspecific process in which (*a*) the plasma membrane furrows inward and then (*b*) pinches off to form (*c*) an intracellular vesicle containing molecules from the extracellular fluid.

There is a famous example of receptor-mediated endocytosis that results in the uptake of cholesterol from the blood into liver cells. Cholesterol is a lipid, therefore nonpolar and insoluble in water (chapter 1). Thus, cholesterol (like other lipids) must travel in the blood attached to *carrier proteins* in the plasma. The carrier proteins are like tiny cruise ships, allowing the cholesterol (and other lipids) to travel in the blood without getting wet (without actually dissolving in the plasma). Depending on the type of carrier protein, cholesterol could be ferried to an artery wall (the "bad cholesterol"; see chapter 10), or to the liver (the "good cholesterol").

FYI [1]When you get a cut in your skin, the bacteria that enter the dermis are met by macrophages that wander in the loose connective tissues. If they can't handle the job alone, other phagocytic cells from the blood are recruited by chemical attractants, which lure neutrophils to the infected sites like the aroma of a barbecue can attract passing neighbors. Neutrophils arrive first, squeezing out of small vessels in the area. Monocytes follow, become transformed into macrophages, and finish the job of destroying the bacteria. However, the digestive enzymes released from their lysosomes also kill the phagocytic cells. The dead, liquified white blood cells form pus.

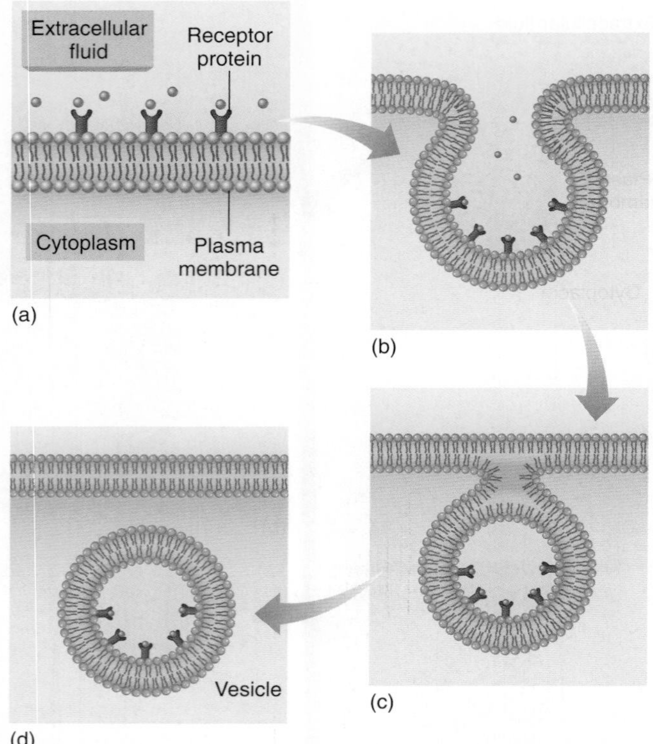

FIGURE 3.3 Receptor-mediated endocytosis. In this process, (*a*) the plasma membrane has specific protein receptors for particular molecules in the extracellular fluid. (*b*) When the molecules bind to their receptor proteins, (*c*) endocytosis occurs. This results in (*d*) the formation of an intracellular vesicle containing the specfic molecules that could bind to the receptor proteins.

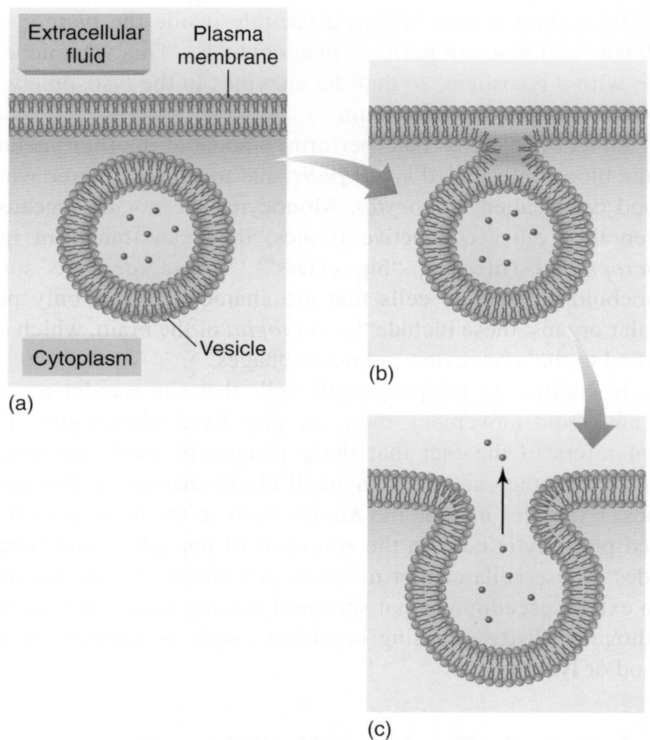

FIGURE 3.4 Exocytosis. In the process of exocytosis, (*a*) an intracellular vesicle (*b*) fuses with the plasma membrane to (*c*) release its contents into the extracellular fluid. This process allows cells to secrete their products and also adds new membrane to the plasma membrane.

Clinical Applications

The carrier proteins that ferry cholesterol through the tiny blood channels (sinusoids) in the liver can bind to the receptor proteins in the plasma membrane of the liver cells and trigger endocytosis. The endocytotic vesicles formed in this way contain these combinations of lipids and proteins, or *lipoproteins*, which the liver cells can then metabolize. The liver cells convert cholesterol into bile acids, which are excreted via the bile into the intestine (and eventually the feces). Thus, the receptor-mediated endocytosis of cholesterol bound to its carrier proteins into liver cells helps lower blood cholesterol, and thereby helps protect against *atherosclerosis*, a major cause of **heart disease** and **stroke** (discussed in chapter 10). People who (because of a mutation) do not produce the specific receptor proteins required for this receptor-mediated endocytosis can develop cardiovascular disease at an early age.

Exocytosis Releases Cell Products into the Extracellular Fluid

Exocytosis is a process in which an intracellular vesicle fuses with the plasma membrane so that its contents can be released in the extracellular fluid (fig. 3.4). This is essentially the reverse of endocytosis. Since the vesicle that undergoes exocytosis contains proteins or other products of the cell, these products can be released into the extracellular fluid when the vesicle membrane fuses with the plasma membrane. The vesicles were budded off from the Golgi complex (chapter 2; see fig. 2.8). Examples of exocytosis include the exocrine gland release of digestive enzymes from the pancreas; the release of hormones from endocrine glands; and the release of neurotransmitter molecules from the axon terminals of neurons.

Notice that endocytosis removes small portions of plasma membrane as the plasma membrane invaginates to form a vesicle, and that exocytosis adds new membrane to the plasma membrane through fusion with an intracellular vesicle. Exocytosis can thus help replenish plasma membrane lost by endocytosis. However, this two-way trafficking in plasma membrane material allows for something else as well: the two-way trafficking of specific membrane proteins. For example, the membrane surrounding an intracellular vesicle can contain receptor proteins for the hormone insulin; when the vesicle fuses with the plasma membrane, the insulin receptors will then be stuck into the plasma membrane on the surface of the cell, so that the cell can respond to insulin (chapter 8). At other times, the receptors can be removed from the plasma membrane when it invaginates and forms an intracellular vesicle, as in endocytosis. Similarly, membrane proteins that allow the passage of water through the plasma membrane (osmosis, discussed shortly) can either be inserted into or removed from the plasma membrane by exocytosis and endocytosis, respectively (see fig. 3.14). This is an important aspect of kidney function (chapter 13).

CHECK POINT

1. Describe the process of phagocytosis, and identify some of the types of phagocytic cells in the body.
2. How does endocytosis differ from phagocytosis? How does pinocytosis differ from receptor-mediated endocytosis?
3. Describe the process of exocytosis, and explain how it is used to secrete cellular products and to reverse the loss of membrane material (and specific membrane proteins) from the plasma membrane.

The Plasma Membrane Can Transport Individual Molecules

The plasma membrane is selectively permeable, allowing certain molecules to pass through but not others. Passive membrane transport occurs by diffusion, where substances move from higher to lower concentrations. Active membrane transport requires the expenditure of ATP energy to move substances from lower to higher concentration.

The plasma membrane is the barrier that separates the intracellular and extracellular compartments of the body. As a barrier, it prevents many cellular molecules and ions from leaking out into the extracellular fluid. Preventing cellular proteins, organic molecules with phosphate groups (such as ATP; see chapter 2), and other substances from escaping into the extracellular fluid is critically important for the life of the cell. Similarly, cells must allow only particular molecules and ions to enter from the extracellular fluid. The plasma membrane is thus described as **selectively permeable**: it allows only certain molecules and ions to penetrate (*permeate*) easily from one side to the other. The plasma membrane is less permeable to other substances, and completely *impermeable* (not permeable) to still others.

Net Diffusion Occurs from Higher to Lower Concentrations

Molecules and ions dissolved in water are called *solutes*; the water is the *solvent*. All of the substances in a *solution* (the solvent and its dissolved solutes) are in a constant state of random motion because of their heat (thermal) energy. If the solution is hotter, the random motion will be faster. Since the water and solutes are buzzing around randomly, they are just as likely to move left as right, up as down. However, if one region of the solution has a higher concentration of a particular substance than another, more of that substance will move by chance in the direction of the lower concentration than in the direction of the higher concentration.

There is a terminology used to describe this process. A difference in concentration between two regions of a solution is described as a *concentration difference* or **concentration gradient**. Since random motion caused by heat energy makes the solutes and water molecules more diffusely (evenly) spread out, the process is called **diffusion**. The random motion of solute and solvent in a solution will cause more of these substances to go from their higher to their lower concentrations—will cause them to go *down their concentration gradients*—than in the opposite direction. This net movement can be referred to as **net diffusion** (fig. 3.5). Net diffusion stops when the concentrations are equal in all parts of the solution, although random motion of solute and solvent molecules still occurs.

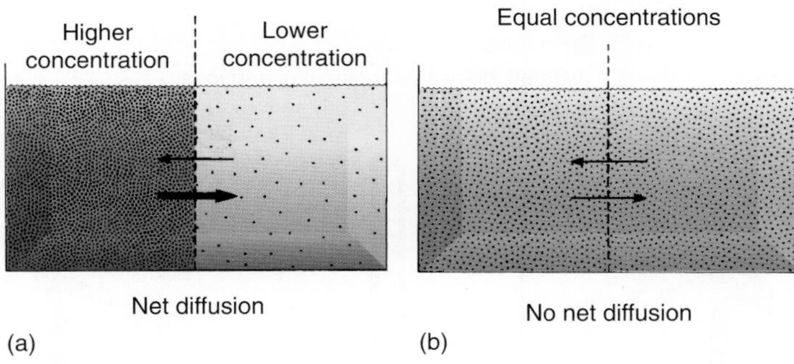

FIGURE 3.5 Diffusion of a solute. (*a*) Net diffusion occurs when there is a concentration difference (or concentration gradient) between two regions of a solution, provided that the membrane separating these regions is permeable to the diffusing substance. (*b*) Diffusion tends to equalize the concentrations of these regions, and thus to eliminate the concentration differences.

A membrane may cause one region of a solution to be more concentrated in a particular substance than another. For example, a membrane may be very permeable to water but not at all permeable to a particular solute, so that the solute can build up to a high concentration on one side of the membrane and remain at a low concentration on the other side. In the case of the plasma membrane, the permeability to that solute might be regulated by physiological mechanisms. If something caused the plasma membrane to suddenly become permeable to that solute, net diffusion would then occur in the direction of the lower solute concentration.

CLINICAL APPLICATIONS

Dialysis is a technique that uses an artificial (plastic) *semipermeable membrane* to separate molecules based on their sizes. The dialysis membrane has pores that allow water and small solutes to permeate but not larger molecules, such as proteins (fig. 3.6). In **hemodialysis** (dialysis of blood), the plasma proteins stay on one side of the dialysis membrane while small waste products can diffuse through the pores in the membrane into the surrounding fluid (which is discarded). By contrast, there is no net diffusion of glucose out of the blood because glucose is added at the same concentration to the surrounding fluid, preventing net diffusion. The kidneys normally also eliminate wastes while retaining plasma proteins, glucose, and other needed molecules, and so hemodialysis must be performed on patients with kidney failure.

Membrane Transport Processes Are Passive or Active

There may be a higher concentration of a particular molecule or ion on one side of the plasma membrane than on the other side, higher either in the extracellular fluid or the intracellular fluid. In other words, there may be a concentration difference, or concentration

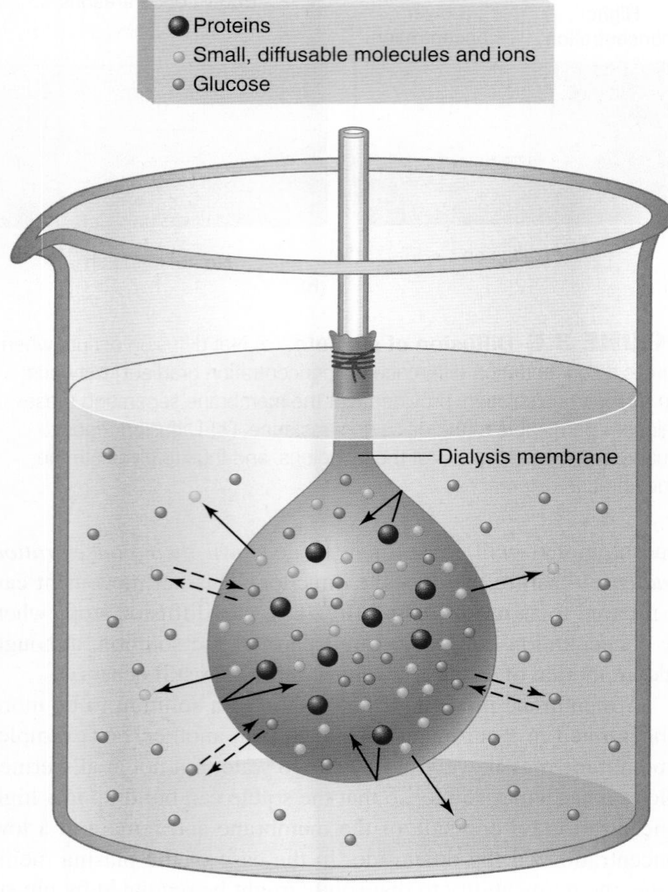

FIGURE 3.6 Diffusion through a dialysis membrane. A dialysis membrane is an artificially semipermeable membrane with tiny pores of a certain size. Proteins inside the dialysis bag are too large to get through the pores (bent arrows), but the small, diffusible molecules and ions are able to fit through the pores and diffuse (solid, straight arrows) from higher to lower concentration out of the bag and into the surrounding fluid. Glucose can also fit through the pores, but since it is present at the same concentration outside of the bag, there is no net diffusion (double dashed arrows).

gradient, across the plasma membrane. If the plasma membrane is permeable to that molecule or ion, it will undergo net diffusion from higher to lower concentration. This occurs because of the thermal energy of the molecules and their resulting random motion, not because of the expenditure of cellular ATP energy. Because ATP is not needed, membrane transport processes in which molecules or ions move down their concentration gradients are called **passive transport** processes.

Passive transport processes are the result of net diffusion, and so the rate of passive transport depends on:

1. *The "steepness" of the concentration gradient.* When the difference in concentration between the two sides of the membrane is greater for a molecule or ion, its rate of net diffusion will be faster. This is easily understood by looking at the extreme case: when there is zero concentration difference between the two sides of the plasma membrane (when there is no concentration gradient), there will be no passive transport.
2. *The permeability of the plasma membrane.* When there is a specific concentration gradient for a particular molecule or ion, the rate of its net diffusion across the plasma membrane depends on the degree to which the membrane is permeable to it. Again, this is easily understood by examining an extreme case: if the membrane is impermeable to a particular molecule or ion, there can be no passive transport. This is physiologically very important: with a given concentration gradient, the rate of passive transport of a particular molecule or ion depends entirely on the permeability of the plasma membrane.
3. *The temperature of the solution.* The warmer the solution, the faster diffusion occurs. However, since body temperature is relatively constant (at 37° C), this factor isn't generally significant in human physiology.
4. *The surface area of the membrane.* The greater the surface area of plasma membrane, the faster the rate of net diffusion into or out of the cell. This is a structural feature of a cell (such as the presence of microvilli), not a physiological adjustment that can be made from one moment to the next.

There are different ways that a molecule or ion can get across a plasma membrane by passive transport. These different passive transport processes depend on the chemical nature of the diffusing substance. On this basis, there are three ways that molecules or ions can cross the plasma membrane by passive transport (fig. 3.7):

1. *Nonpolar molecules can diffuse across the double phospholipid layers of the plasma membrane.* Nonpolar molecules cannot dissolve in water, but they are soluble in nonpolar solvents (chapter 1). The nonpolar portions of the membrane phospholipids permit nonpolar molecules to cross through the plasma membrane. Thus, nonpolar molecules such as oxygen, carbon dioxide, and steroid hormones can easily pass through any plasma membrane.
2. *Inorganic ions and water molecules can cross the plasma membrane through specific protein channels in the membrane.* The protein channels go all the way through the plasma membrane and are hollow, so that they are like tiny waterways that the polar inorganic ions (such as Na^+ and K^+) and water molecules can use to cross the otherwise nonpolar phospholipid barrier.
3. *Small organic molecules such as glucose may use specific carrier proteins to cross the plasma membrane.* Glucose is small, as organic molecules go, but it is too large to fit through the protein channels that permit inorganic ions and water molecules to cross. So, many body cells have membrane proteins that function as carriers, to transport glucose and similar molecules across the membrane. When the direction of net movement is from higher to lower concentration, this is a type of passive transport called *facilitated diffusion.*

There are important differences between protein channels and protein carriers. **Protein channels (channel proteins)** don't move, although they may have "gates" that can open and close. They simply provide an aqueous (watery) passageway for the water-soluble

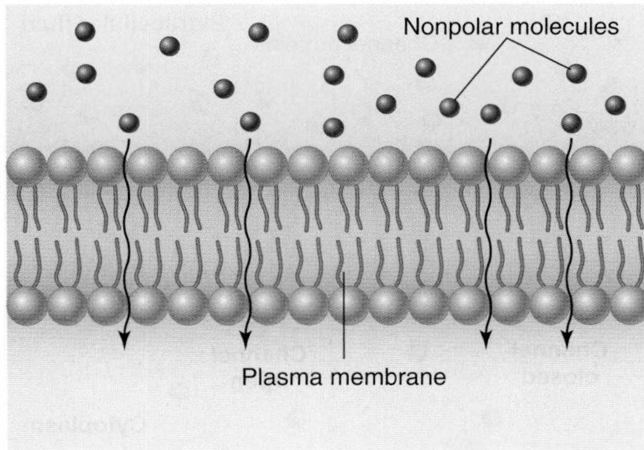

(a)

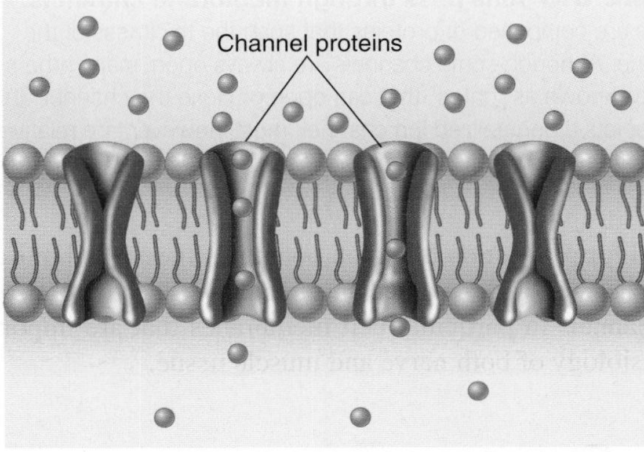

(b)

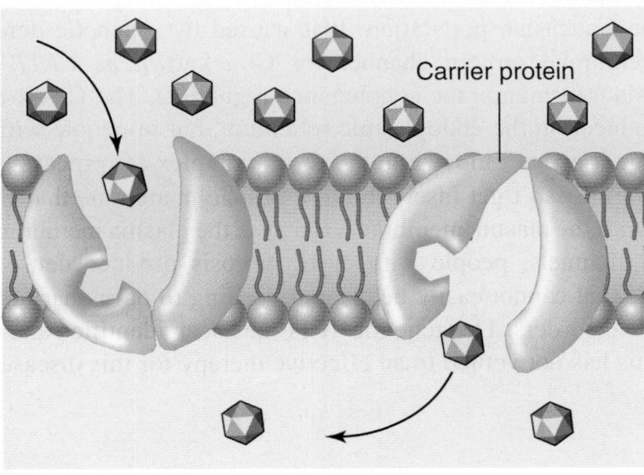

(c)

FIGURE 3.7 Three types of passive transport. (*a*) Nonpolar molecules can move by simple diffusion through the double phospholipid layers of the plasma membrane. (*b*) Inorganic ions and water molecules can move by simple diffusion through protein channels in the plasma membrane. (*c*) Small organic molecules, such as glucose, can move by facilitated diffusion through the plasma membrane using protein carriers.

inorganic ions (and water molecules themselves) to cross the otherwise impermeable plasma membrane. **Carriers proteins**, by contrast, first bind in a specific fashion to the molecule they will transport, and then move in some way to transport the molecule to the other side of the membrane. The characteristics of *carrier-mediated transport* will be discussed in a later section of this chapter.

The net diffusion of nonpolar molecules across the double phospholipid layers, and the net diffusion of ions and water through protein channels in the plasma membrane (numbers 1 and 2 in the previous list), can be called **simple diffusion**. This is in contrast to the net diffusion of molecules through the membrane by means of protein carriers, which is known as **facilitated diffusion**. Thus, passive transport processes include simple diffusion and facilitated diffusion.

Active transport is the transport of molecules and ions against their concentration gradients—that is, movement across the plasma membrane that is "uphill," from the region of lower to the region of higher concentration. Since the direction of transport is against the concentration gradient, cellular ATP energy must be spent in active transport. The breakdown of ATP is used to power the movement of particular carrier proteins in the plasma membrane. Thus, both facilitated diffusion and active transport require carrier proteins. In the case of active transport, the carriers move molecules and ions "uphill," and so active transport carriers are often referred to as **pumps**. Active transport and facilitated diffusion are discussed more in a later section.

CHECK POINT

1. Describe diffusion, and explain how the rate of diffusion across a membrane is affected by temperature, concentration gradient, membrane surface area, and membrane permeability.
2. Identify the types of passive transport, and which substances are transported by each of these types.
3. Distinguish between passive and active transport.

Passive Transport Can Occur by Simple Diffusion

Passive transport by simple diffusion includes the net diffusion across the plasma membrane of nonpolar molecules through the double phospholipid layers, and the net diffusion of ions and water molecules through specific protein channels. Inorganic ions can cross the plasma membrane through channels that may be opened or closed. The net diffusion of water across a membrane, called osmosis, occurs from the solution that is more dilute to the solution that has the higher solute concentration.

As described in the previous section, passive transport processes include simple diffusion and facilitated diffusion. Facilitated diffusion is the net diffusion of molecules across the plasma membrane from higher to lower concentration by means of specific membrane carrier proteins. This will be discussed in the next major section. Simple diffusion is the net diffusion of molecules and ions across the plasma membrane without using protein carriers. The ways that this can be accomplished depend on the nature of the diffusing molecule and ion.

Nonpolar Molecules Diffuse Across the Phospholipid Layers of the Membrane

The plasma membrane consists primarily of a double layer of phospholipids arranged so that the nonpolar portions of the phospholipid molecules form the interior of the membrane (chapter 2; see fig. 2.2). This nonpolar core of the plasma membrane presents a barrier that polar ions and molecules cannot penetrate. However, nonpolar molecules—such as oxygen, carbon dioxide, and steroid hormones—can diffuse easily through the nonpolar core of the membrane. Thus, just like walls present no barriers to movie ghosts, plasma membranes are no barriers to nonpolar molecules.

Cells consume oxygen in aerobic cell respiration (because they convert oxygen into water) and produce carbon dioxide at the same time. Thus, there is always a lower concentration of oxygen inside cells than in the extracellular fluid, and a lower concentration of carbon dioxide in the extracellular fluid than inside the cells. Since the plasma membrane is permeable to nonpolar oxygen and carbon dioxide, these gases will move by simple diffusion down their concentration gradients. As a result, there is net diffusion of oxygen into cells and of carbon dioxide out of cells (fig. 3.8).

Inorganic Ions Diffuse Through Protein Channels in the Plasma Membrane

Inorganic ions—Na^+, K^+, Ca^{2+}, and Cl^-, among others—are too polar to pass through the phospholipid layers of the plasma membrane. Instead, these ions move through protein channels that are relatively specific for each ion. Some of these channels are always open, but some are **gated channels**—channels that can be opened or closed by portions of the polypeptide chains that compose the channels (fig. 3.9).

For example, there are channels for K^+ that are always open—sometimes referred to as *leakage channels*—and other K^+ channels that are gated. By contrast, all of the channels for Na^+ are gated. The opening and closing of gated Na^+ and K^+ channels are required for production of action potentials (nerve and muscle impulses), as will be explained in chapter 4. There are also gated Ca^{2+} channels in particular cell membranes that are important in the physiology of both nerve and muscle tissue.

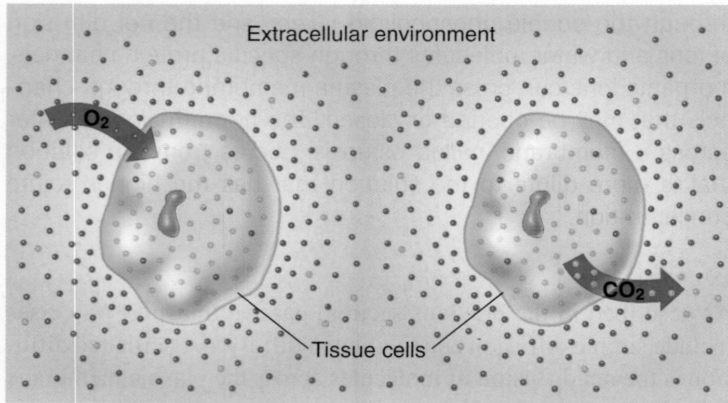

FIGURE 3.8 Gas exchange occurs by diffusion. The colored dots, which represent oxygen and carbon dioxide molecules, indicate relative concentrations inside the cell and in the extracellular environment. Gas exchange between the intracellular and extracellular compartments thus occurs by diffusion.

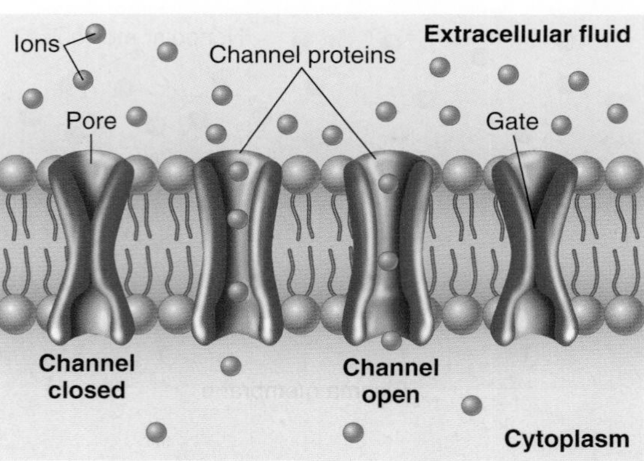

FIGURE 3.9 Ions pass through membrane channels. These channels are composed of proteins that span the thickness of the membrane. Although some channels are always open, many others have structures known as "gates" that can open or close the channel. This figure depicts a generalized ion channel; most, however, are relatively selective–they allow only particular ions to pass.

Clinical Applications

Cystic fibrosis occurs with a frequency of 1 in 2,500 births among the Caucasian population. It is caused by a genetic defect in a membrane protein channel for Cl^-, known as *CFTR* (cystic fibrosis transmembrane conductance regulator). The CFTR channel is produced in the endoplasmic reticulum, but in people with cystic fibrosis it doesn't move into the Golgi complex for processing. As a result, it doesn't get inserted into a vesicle membrane that can later fuse with the plasma membrane. Because the plasma membrane lacks these channels, people with cystic fibrosis produce dense, heavy mucus that cannot easily be cleared, leading to pulmonary and pancreatic disorders. The gene for CFTR has been identified and cloned, but this has not yet led to an effective therapy for this disease.

Osmosis Is the Net Diffusion of Water Through a Membrane

Osmosis is defined as the net diffusion of the solvent (water) through a membrane. In order for osmosis to occur, two conditions must be met (fig. 3.10):

1. *There must be a concentration difference across the membrane.* The solution on one side of the membrane must have a higher concentration of solute—and thus a lower concentration of water—than the other side.
2. *The membrane must be more permeable to water than to the solute.* The membrane is described as semipermeable; its relatively low permeability to the solute prevents the net diffusion of solute from equalizing the concentrations

of the solutions on each side of the membrane. Because the concentration of solute is higher on one side of the membrane than the other, and because the membrane is permeable to water, osmosis can occur.

The membrane can be made of plastic, like the dialysis membrane described earlier in this chapter. Suppose a bag composed of dialysis membrane is partially filled with a solution of sucrose (a disaccharide; see chapter 1) at a concentration of 4 grams per liter (g/L), and this bag is inserted in a beaker that contains 2 g/L of sucrose. Therefore, there is a concentration difference across the membrane. For osmosis to occur, the membrane must be permeable to water but not to sucrose. Osmosis will occur from the more dilute (watery) solution in the beaker into the less dilute solution in the bag (fig. 3.11). If the bag is able to expand and hold all the water that undergoes a net diffusion, osmosis will stop when the concentrations are equal on both sides of the membrane.

Osmosis is passive transport; the water is diffusing from its higher to its lower concentration. Pure water has the highest water concentration, while a solution containing solute molecules has a lower water concentration (see fig. 3.10). Common language can be confusing when discussing osmosis, because we don't usually talk about water concentration. We say, "This is a concentrated sugar solution"; we don't usually say, "This solution has a low water concentration." We can say that the 4 g/L sucrose solution has a higher concentration than the 2 g/L sucrose solution, but we must understand when discussing osmosis that the 2 g/L solution has the higher water concentration. Thus, osmosis occurs from the 2 g/L solution to the 4 g/L solution by the simple diffusion of water, making osmosis an example of passive transport.

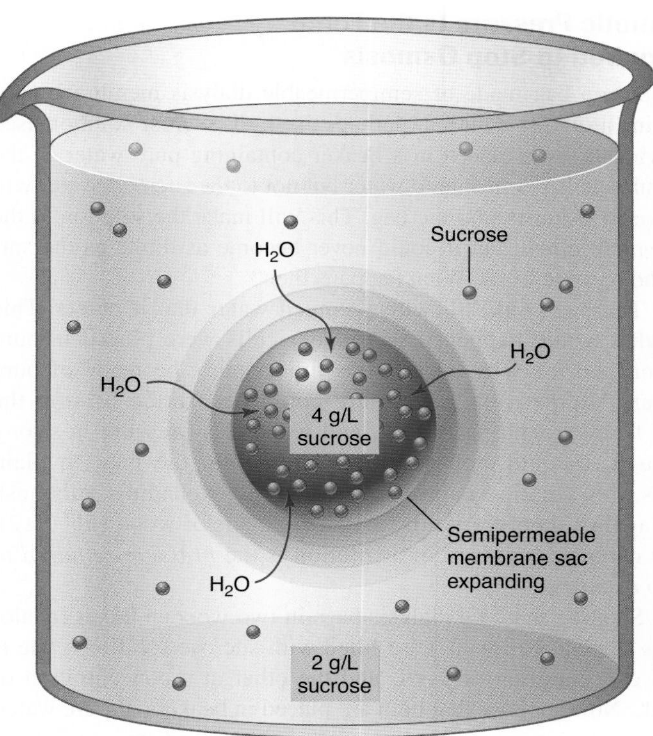

FIGURE 3.11 Osmosis occurs from the more dilute to the less dilute solution. The bag is composed of an artificial semipermeable membrane that is permeable to water but not to sucrose. In that case, the water will undergo a net diffusion across the membrane (osmosis) from the more dilute solution outside the bag to the less dilute solution (with a higher sucrose concentration) inside the bag.

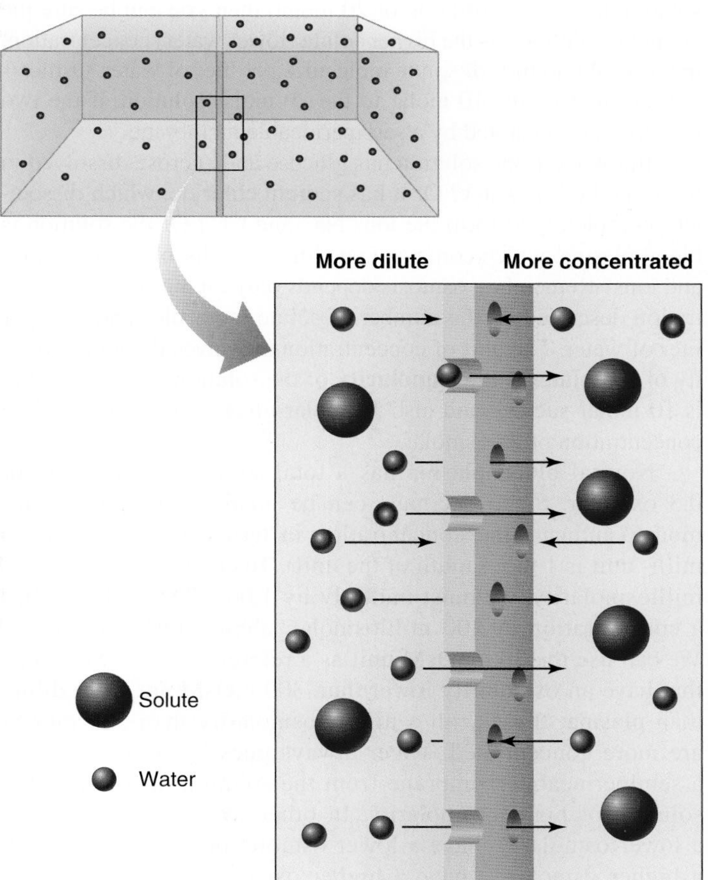

FIGURE 3.10 A model of osmosis. The diagram illustrates the net movement of water from the solution of lesser solute concentration (higher water concentration) to the solution of greater solute concentration (lower water concentration).

CLINICAL INVESTIGATION CLUES

Remember that Jill had edema (excessive tissue fluid) and her plasma protein concentration was low, due to her cirrhosis. Normally there is a higher concentration of proteins in the blood plasma than in the interstitial fluid, because proteins (unlike other plasma solutes and water) can't filter out from the blood plasma across vessel walls into the interstitial fluid.

- What is the relationship between the liver cirrhosis and the low plasma protein concentration?
- Given that the wall of tiny blood capillaries acts like a semipermeable membrane separating the plasma from the interstitial fluid, how could a low plasma protein concentration cause edema?

Osmotic Pressure Is the Force Required to Stop Osmosis

Suppose a bag made of semipermeable dialysis membrane containing a sucrose solution (perhaps the 4 g/L sucrose solution used previously) was placed in a beaker containing pure water. If the membrane is permeable to water but not to the sucrose, water will move by osmosis into the bag. This will make the solution in the bag more dilute, but it could never become as dilute as the surrounding pure water. What happens then?

The bag could fill with so much water that it bursts. This is what would happen if one of our cells were placed in pure water, and it's why intravenous fluid cannot be made of pure water. Alternatively, a pressure could be exerted to stop the bag from expanding further, and when this pressure is strong enough, it could stop further osmosis into the bag. In plant cells, this pressure is produced when the expanding cells push out against their cell walls, which are made of wood (fig. 3.12). The **osmotic pressure** of a solution is the *pressure required to stop osmosis*.

Suppose that two dialysis bags in two wooden boxes (analogous to plant cell walls) are filled with sucrose solutions; one is at a concentration of 2 g/L and the other at a concentration of 4 g/L. Now suppose that both are placed in beakers of pure water. What pressure must be exerted to stop osmosis into each bag? We don't care here about the actual pressure; what is important here is that the pressure required to stop osmosis—the osmotic pressure—will be twice as great for the 4 g/L solution as for the 2 g/L solution (fig. 3.12).

Since animal cells, including our own, don't have cell walls outside the plasma membrane, they will burst when enough water enters them.[2] The expression of osmotic pressure will be used in this human physiology text simply to indicate the solute concentration and thus the *tendency of a solution to take in water by osmosis*. If one solution has a higher solute concentration than another, it has a higher osmotic pressure. If a semipermeable membrane separates both solutions, water will diffuse from the solution of lower to the solution of higher osmotic pressure.

Osmolarity Describes Solution Concentrations

If there are two solutions separated by a semipermeable membrane, osmosis will occur from the more dilute to the more concentrated solution. Thus, in order to know if osmosis will occur between two solutions separated by a membrane, and in which direction, we need to know their concentrations. In figure 3.12, the concentrations are given in grams per liter (g/L). This works because both solutions have the same solute, sucrose. But what if the solutions have different solutes? For example, what if one solution contains 4 g/L of sucrose and the other solution has 3 g/L of glucose (and the membrane is impermeable to both solutes)?

It turns out that water will undergo net diffusion from the side of lower to the side of higher solute concentration, based on the *number of solute molecules or ions per liter*, not based on the weight of the solutes. Glucose is a monosaccharide, while sucrose is a disaccharide; 1 molecule of glucose weighs less than 1 molecule of sucrose. Therefore, it takes more glucose molecules to make the 3 g/L solution than it takes sucrose molecules to make the 4 g/L solution. The 3 g/L glucose solution has more solute molecules per liter than does the 4 g/L sucrose solution, and so water will undergo a net diffusion from the more dilute 4 g/L sucrose to the less dilute 3 g/L glucose solution.

In this example, expressing concentration in terms of grams per liter of solute is misleading and confusing. It is better to express concentration in terms of solute molecules or ions dissolved in a liter of solution, which is known as the *molarity* of the solution (the calculations for molarity units are described in appendix 1). Here, we will simply use the molarity units as a way to indicate the concentration of solutions when describing osmosis. If you know that a sucrose solution has a concentration of, say 10 molar, and a glucose solution has a concentration of 20 molar, then you can be sure the 20 molar solution has the higher solute (lower water) concentration, in terms of numbers of solute molecules per liter of water. Osmosis will occur from the 10 molar to the 20 molar solution, if the two solutions are separated by a semipermeable membrane.

But what if one solution has glucose and sucrose dissolved in the same liter of water? Or it has sodium chloride, which dissociates completely to form the ions Na^+ and Cl^-? Or the solution is blood plasma, which contains many different dissolved molecules and ions? Remember, osmosis depends only on the solute concentration described by the number of solute molecules and ions per liter of water. The unit of concentration that gives the total molarity of all solutes is the **osmolarity** of the solution. A solution that is 10 molar sucrose and also 10 molar glucose has a total solute concentration of 20 osmolar.

Normal blood plasma has a total solute concentration of 0.3 osmolar. Since decimals can be misread, it's more common to indicate the concentration in terms of milli- units (a milli- unit is 1 thousandth of the unit). To convert osmolarity to milliosmolarity, we must multiply by 1,000. Thus, plasma has a concentration of 300 milliosmolar, abbreviated *300 mOsM*. We can use the 300 mOsM unit as a reference value. Solutions that have an osmolarity lower than 300 mOsM are more dilute than plasma; those with a higher osmolarity than 300 mOsM are more concentrated. Water always goes by osmosis across a semipermeable membrane from the solution of lower to the solution of higher osmolarity. In other words, solutions with a lower osmolarity have a lower osmotic pressure; those with a higher osmolarity have a higher osmotic pressure. When a semipermeable membrane separates two solutions, osmosis will be from the more dilute, lower osmolarity, lower osmotic pressure solution into the less dilute, higher osmolarity, higher osmotic pressure solution.

FYI [2]Water sprayed on celery and cucumbers in the supermarket makes them crispy. When plants are watered, limp leaves and stems become *turgid*, or rigid because of hydrostatic pressure. This is because water enters the cells by osmosis, pressing the plasma membranes hard against their surrounding cell walls. Animal cells, including our own, don't have cell walls, so water doesn't make us crispy. Instead, pure water given intravenously is deadly, because it causes our cells to explode.

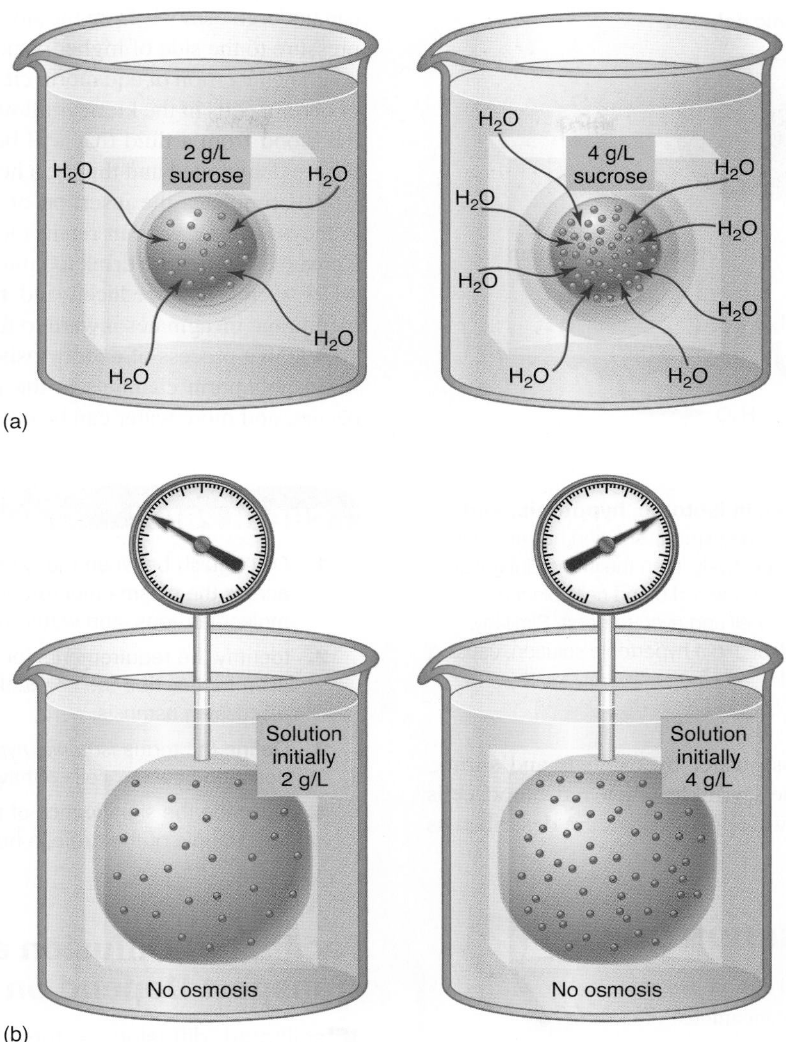

FIGURE 3.12 Osmosis and osmotic pressure. Bags composed of an artificial semipermeable membrane, permeable to water but not to sucrose, are within solid but permeable boxes (like the cell walls of plants) suspended in beakers of pure water. (*a*) Each bag expands as osmosis occurs, but the 4 g/L sucrose solution bag expands more rapidly than the 2 g/L sucrose solution bag. (*b*) Each bag has filled its box and presses hard against it. The pressure produced by the pressing of the bag against the box walls reaches a point where it can stop further osmosis. This is the osmotic pressure, and is twice as great for the 4 g/L solution as for the 2 g/L solution.

Tonicity Describes How Solutions Compare to Blood Plasma

If one of our red blood cells is placed in a beaker of normal plasma, it will neither swell with water nor shrink because of loss of water. This is how it behaves as it circulates in our blood. Because osmosis—the net diffusion of water across the plasma membrane, in this case—doesn't occur, we conclude that the cytoplasm of the red blood cell and the surrounding plasma have the same osmolarity and osmotic pressure. We can describe these solutions as being **isotonic** (*iso* = same; *tonic* = strength) to each other.

If there is no osmosis between any two solutions separated by a semipermeable membrane, we can say that the two solutions are isotonic to each other. However, if a solution is called "isotonic" and no reference solution is specified, the reference is assumed to be blood plasma. Thus, if a vial is labeled "isotonic saline," it's understood to contain a salt solution isotonic to blood plasma. This can be given intravenously or to exposed tissues without fear of causing osmotic damage to the body cells. Isotonic saline—also called *normal saline*—is at a concentration of 0.9 g NaCl per 100 milliliters (ml) of water. Isotonic glucose—also called *5% dextrose*—is at a concentration of 5 g/100 ml. Normal saline and 5% dextrose must each have a total solute concentration of 300 mOsM, because they are isotonic to blood plasma, which is 300 mOsM.

A solution with a lower osmolarity and osmotic pressure than plasma (or the cytoplasm of cells) is a **hypotonic** solution. Since a hypotonic solution is more dilute (watery) than plasma, cells placed in a hypotonic solution will gain water, swell, and perhaps burst (fig. 3.13). Conversely, a solution with a higher osmolarity and osmotic pressure than plasma is a **hypertonic solution**. A cell

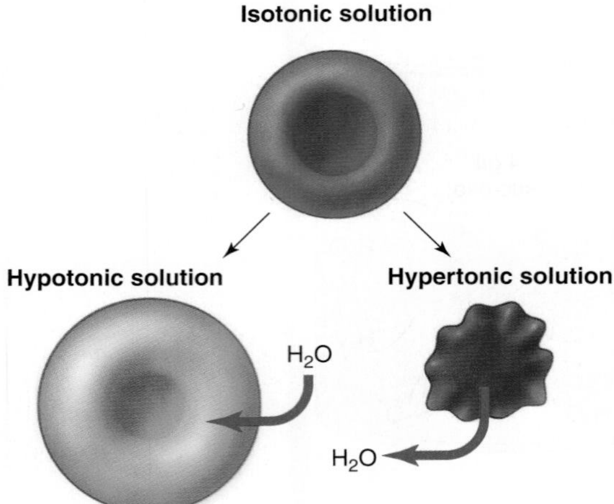

FIGURE 3.13 Red blood cells in isotonic, hypotonic, and hypertonic solutions. In each case, the external solution has an equal, lower, or higher osmotic pressure, respectively, than the intracellular fluid. As a result, water moves by osmosis into the red blood cells placed in hypotonic solutions, causing them to swell and even to burst. Similarly, water moves out of red blood cells placed in a hypertonic solution, causing them to shrink and become crenated.

in a hypertonic solution will lose water by osmosis and shrink. Because their plasma membranes wrinkle when red blood cells shrink, such cells that have lost water by osmosis are described as *crenated* (fig. 3.13).

CLINICAL INVESTIGATION CLUES

Remember that Jill felt weak and dehydrated, and that her urine contained a significant amount of glucose.

- Given that glucose is a solute not normally present in the urine, how would the added glucose affect the osmolarity and osmotic pressure of the urine?
- Given that the fluid to become urine is separated by a semipermeable membrane from the blood plasma, how would the changed urine osmolarity affect the amount of water excreted in the urine?
- How could these changes produce dehydration and weakness?

Aquaporin Channels Permit Osmosis Across Plasma Membranes

The simple diffusion of water molecules across the plasma membrane occurs very slowly unless the membrane has specific protein channels, called **aquaporin channels**. The aquaporin channel proteins are produced in the endoplasmic reticulum and inserted by the Golgi apparatus into the membrane of intracellular vesicles. Under proper stimulation, the vesicles can fuse with the plasma membrane, as if they were going to release their contents in the process of exocytosis (fig. 3.14). Once inserted into the plasma membrane, water can undergo osmosis from the side of lower osmolarity and osmotic pressure to the side of higher osmolarity and osmotic pressure.

The insertion of aquaporin channels into the plasma membrane of certain cells in the kidneys allows water to move by osmosis into the blood from a fluid that will become urine (chapter 13). When we are dehydrated and thirsty, a hormone (*antidiuretic hormone*, or *ADH*) stimulates the insertion of these aquaporin channels in the kidneys, so that we can retain more water and excrete less urine. Conversely, when we drink too much water and are overly hydrated, ADH secretion is reduced and the reverse occurs. The plasma membrane invaginates inward to form a pouch, which pinches off a vesicle in a process of endocytosis (fig. 3.14). As a result, there are fewer aquaporin channels in the plasma membrane, less osmosis occurs, and more water can be excreted in the urine.

CHECK POINT

1. Distinguish between the ways in which simple diffusion across the plasma membrane occurs for nonpolar molecules, ions, and water molecules.
2. Identify the requirements for osmosis, and the terms used to describe the concentrations of solutions when discussing osmosis.
3. Define the terms *isotonic*, *hypotonic*, and *hypertonic*, and describe how our cells behave in each of these solutions.
4. Describe the significance of aquaporin channels, and give an example of their role in human physiology.

Facilitated Diffusion and Active Transport Depend on Carrier Proteins

Facilitated diffusion is the downhill, passive transport of molecules using protein carriers in the plasma membrane. Like simple diffusion, there is net diffusion from higher to lower concentrations. Active transport is the uphill transport of molecules and ions from lower to higher concentration, using membrane protein carriers powered by the breakdown of ATP. Carrier proteins for both facilitated diffusion and active transport have specificity and can reach a transport maximum when they are saturated.

Transport across the plasma membrane using protein carriers is known as **carrier-mediated transport** and includes both facilitated diffusion and active transport. Membrane carrier proteins are different from the protein channels for inorganic ions discussed in the last section. Channel proteins may be gated (opened or closed by polypeptide "gates") and specific for particular ions, but the channel proteins themselves don't move. Channel proteins simply provide aqueous passageways for the net diffusion of water-soluble ions through the otherwise nonpolar phospholipid layers of the membrane. By contrast, carrier proteins move to transport small, polar organic molecules (such as glucose) across the membrane.

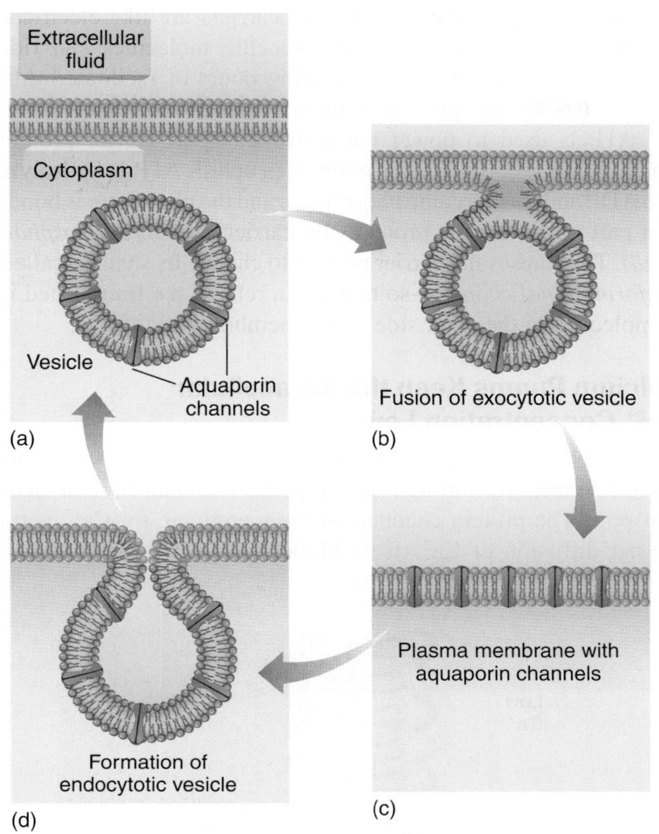

FIGURE 3.14 Insertion and removal of aquaporin channels. (*a*) The plasma membrane lacks aquaporin channels, but there are aquaporin channels in the membrane of an intracellular vesicle, which can (*b*) fuse with the plasma membrane. This results in (*c*) the insertion of aquaporin channels into the plasma membrane, so that the membrane can be permeable to water for osmosis. Under other conditions, (*d*) the aquaporin channels can be removed from the plasma membrane by the formation of an endocytotic vesicle containing the channels.

Carrier proteins have two characteristics that help distinguish them from channel proteins:

1. *Carrier proteins are highly specific.* A carrier protein for glucose will only transport glucose, for example. This is similar to the specificity of enzyme proteins for their substrates (chapter 1), and for similar reasons: the carrier protein's tertiary structure (shape) has a pocket that only the transported molecule can fit. The transported molecule fits in and binds to that pocket like a substrate molecule fits in and binds to the active site of an enzyme.
2. *Carrier proteins can become saturated and reach a transport maximum* (fig. 3.15). There is only a certain number of carrier proteins in the plasma membrane of a cell. When all of the carrier proteins are working (when they are saturated), the maximum transport rate will be reached (the transport maximum).

The concept of saturation and a transport maximum can be understood by thinking of carriers as buses. Let's suppose a bus can hold 40 people, and one bus arrives at a bus stop each hour. If there are 10 people waiting at the bus stop, the transport rate is 10 people per hour. If 20 people wait, it's 20 people per hour; if 30 wait, it's 30 people per hour. If 40, 50, or 1,000 people are waiting, the transport rate is still 40 people per hour, because that's all the bus can hold (no standing is allowed!). The bus is saturated at 40 people, and its transport maximum is 40 people per hour.[3]

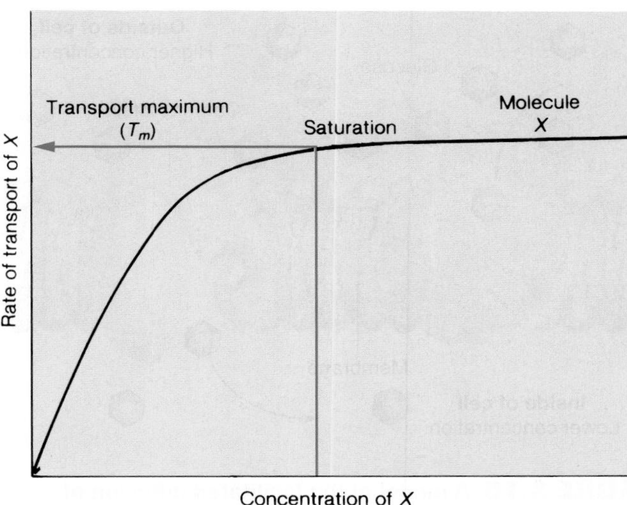

FIGURE 3.15 Saturation and transport maximum of carrier proteins. When all carriers are busy transporting molecules, they are said to be saturated. At this point, increasing the concentration of molecules to be transported (*X*) does not further increase the rate of transport.

CLINICAL INVESTIGATION CLUES

Remember that Jill had glucose in her urine and was later tested to confirm a diagnosis of diabetes mellitus.

- Would a blood test have shown that Jill had hyperglycemia (high blood glucose) when her urine tested positive for glucose?
- What is the relationship between hyperglycemia and glycosuria (glucose in the urine)?

Facilitated Diffusion Uses Carriers to Transport Molecules Passively

Physiological mechanisms, including action of the hormone insulin, maintain homeostasis of the blood glucose concentration. Keeping blood glucose within a normal range of concentrations is important, because glucose enters tissue cells passively from the blood. That is,

FYI [3]Glucose from the blood gets filtered out into small tubes (tubules) in the kidneys, so the filtrate that will become urine contains glucose. Normally there are enough carrier proteins (seats on the bus) in those tubules to transport all of the filtered glucose back into the blood. However, if a person has high blood glucose (*hyperglycemia*), the carriers can become saturated and glucose molecules that don't get a seat can't wait for the next bus. Instead, they go with the flow and are excreted in the urine (*glycosuria*). This is how diabetics can have sugar in the urine.

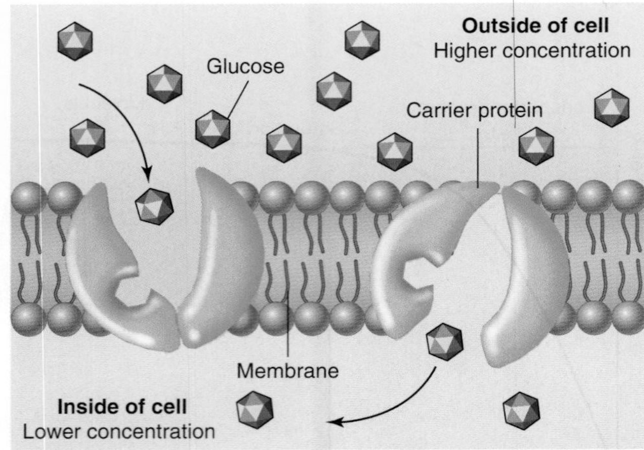

FIGURE 3.16 A model of the facilitated diffusion of glucose. A carrier—with characteristics of specificity and saturation—is required for this transport, which occurs from the blood into cells such as muscle, liver, and fat cells. This is passive transport because the net movement is to the region of lower concentrations, and ATP is not required.

glucose diffuses from the higher concentration in the blood to the lower concentration inside cells, using carrier proteins to get across the plasma membrane (fig. 3.16). If channel proteins for inorganic ions are like private doorways, the carrier proteins in facilitated diffusion are more like personal (specific) revolving doors, pushed by the heat energy of the transported molecule.

This is passive transport, and so the rate at which glucose enters the cells depends on the steepness of the concentration gradient. If the blood glucose concentration falls, glucose will enter cells at a slower rate. Because the neurons of the brain get almost all of their energy from the aerobic cell respiration of blood glucose, a fall in blood glucose (*hypoglycemia*) can cause a person to feel weak and dizzy, or even to lose consciousness.

Because homeostasis of the blood glucose concentration is normally maintained, the number of glucose carriers in the plasma membrane primarily determines the rate of glucose entry into cells. This is like increasing or decreasing the number of seats on a bus, in the analogy used previously. Insulin acts to increase the number of carrier proteins inserted into the plasma membrane of skeletal muscle cells, for example, so that the muscles can remove more glucose from the blood. When more glucose enters the muscle cells by facilitated diffusion, the blood glucose concentration is lowered.

Active Transport Carriers Pump Against the Concentration Gradient

Active transport is the membrane transport of molecules and ions against their concentration gradients; it is "uphill" transport, from the region of lower concentration to the region of higher concentration. This is why active transport carriers are called *pumps*. Because this goes against the direction of net diffusion due to heat energy, the cell must expend metabolic energy, in the form of ATP, for active transport. Active transport carriers are like electrically powered up-escalators (that only specific molecules can ride), rather than like the personal revolving doors of facilitated diffusion carriers that are pushed by the transported molecules.

ATP is used to power the active transport carrier proteins, which pump specific ions or molecules uphill. ATP is hydrolyzed into ADP and P_i (inorganic phosphate), and the phosphate is bonded to a part of the carrier protein (the carrier protein is *phosphorylated*). This causes the carrier protein to change its shape—called a *conformational change*—so that it can release the transported ion or molecule on the other side of the membrane (fig. 3.17).

Calcium Pumps Keep the Intracellular Ca^{2+} Concentration Low

The membrane transport of calcium ions (Ca^{2+}) involves both passive transport, through gated protein channels, and active transport. The protein channels in the membrane for Ca^{2+} permit the net diffusion of Ca^{2+} from higher to lower concentration, if

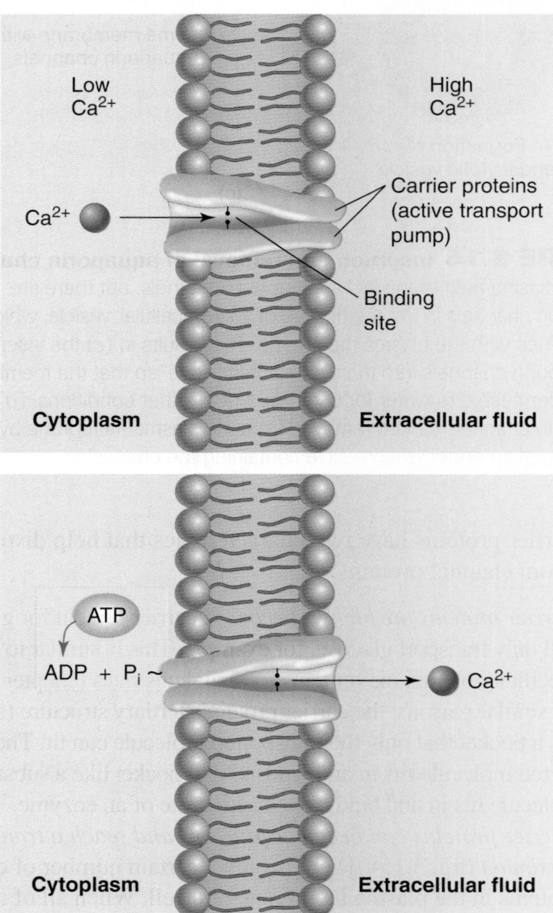

FIGURE 3.17 An active transport carrier works like a pump. The active transport carrier protein undergoes a conformational (shape) change when it gets phosphorylated (binds to a phosphate group by ATP). This allows the carrier to pump from a lower to a higher concentration. In this example, the carrier is a calcium pump, transporting Ca^{2+} up its concentration gradient out of the cell cytoplasm.

the channel gates are open. These gates are generally closed until they are opened by specific stimuli. The active transport of Ca^{2+} uses **Ca^{2+} pumps** (fig. 3.17) to transport Ca^{2+} from inside the cell (where the concentration of Ca^{2+} is lower) to the extracellular fluid (where the concentration of Ca^{2+} is higher).

Because of the active transport Ca^{2+} pumps, there is always a difference in concentration of Ca^{2+} across the plasma membrane. This requires metabolic energy, and so we might wonder why our cells would spend ATP energy pumping Ca^{2+}. They do this to maintain the concentration gradient, so that Ca^{2+} will stream into the cell by net diffusion when the gated membrane Ca^{2+} channels are opened. As we will see in later chapters, this diffusion of Ca^{2+} into the cytoplasm serves as an important physiological stimulus. It stimulates axon terminals to release neurotransmitter chemicals, for example, and it stimulates muscles to contract.

The Na^+/K^+ (ATPase) Pumps Transport Na^+ and K^+ in Opposite Directions

Another type of active transport carriers, found in the plasma membrane of all cells, are the **Na^+/K^+ (ATPase) pumps**. Each Na^+/K^+ (ATPase) pump transports Na^+ and K^+ at the same time, but in opposite directions: Na^+ is pumped out of the cell, and K^+ is pumped into the cell.[4] Actually, 3 Na^+ ions are pumped out for every 2 K^+ ions pumped in (fig. 3.18). This pumping requires the breakdown of ATP, which is a reaction catalyzed by the enzymatic activity of the pump protein (indicated by the term *ATPase*). Because of the activity of active transport pumps, there is a difference in the concentration of Na^+, K^+, Ca^{2+}, and other ions across the plasma membrane (fig. 3.19).

The Na^+/K^+ (ATPase) pumps are constantly working in all cells, expending metabolic energy. Because of this, you might suppose that this activity is important, and you would be right. The Na^+/K^+ (ATPase) pumps keep the Na^+ concentration higher outside the cells and the K^+ concentration higher inside the cells. These concentration gradients for Na^+ and K^+ are critical for the production of electrical impulses in nerve and muscle cells (chapter 4), including the muscle cells of the heart (chapter 10).

The active transport of Na^+ and K^+, and the concentration gradients that result from this pumping activity, are also important for other physiological processes. One of these is the **cotransport** of Na^+ and glucose across the plasma membrane (fig. 3.20) of epithelial cells in the kidney tubules and small

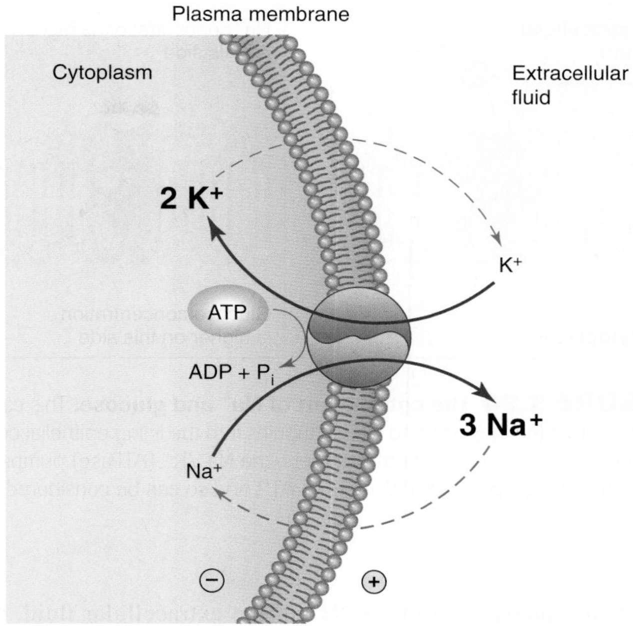

FIGURE 3.18 The exchange of intracellular Na^+ for K^+ by the Na^+/K^+ pump. The active transport carrier itself is an ATPase that breaks down ATP for energy. Dashed arrows indicate the direction of passive transport (diffusion); solid arrows indicate the direction of active transport. Because 3 Na^+ are pumped out for every 2 K^+ pumped in, the action of the Na^+/K^+ (ATPase) pumps help to produce a difference in charge, or potential difference, across the membrane.

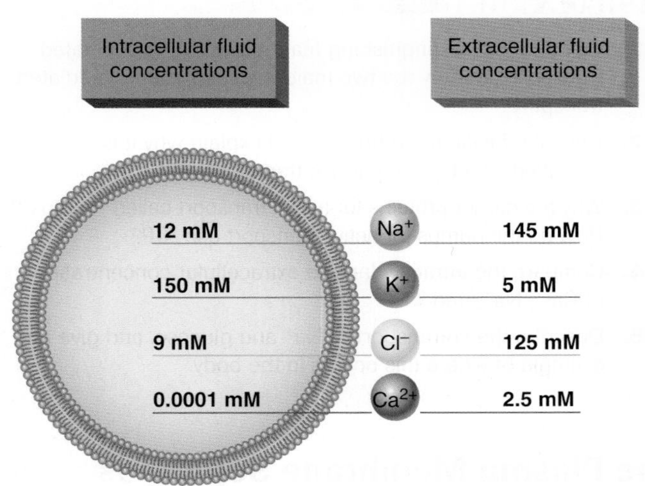

FIGURE 3.19 Ion concentrations in the intracellular and extracellular fluids. The differences in the concentrations of these ions result from the permeability properties of the plasma membrane and the constant activity of active transport ion pumps that work to counter the diffusion gradients.

intestine. In this case, a particular carrier protein transports Na^+ and glucose together. The Na^+ moves from the extracellular fluid (where the Na^+ concentration is higher) to the cytoplasm (where the Na^+ concentration is lower). The same carrier

FYI [4]This is hard, calorie-consuming work. The active transport Na^+/K^+ (ATPase) pumps consume enough energy (calories) to account for about 12% of the basal metabolic rate (BMR). The BMR is the body's energy consumption at rest (without exercising or digesting food). Considering that you also have to keep breathing and your heart beating, among other activities, the energy spent pumping Na^+ is an impressive fraction of the BMR. So if anyone ever accuses you of goofing off, you can explain that you are actually quite busy exercising your pumps.

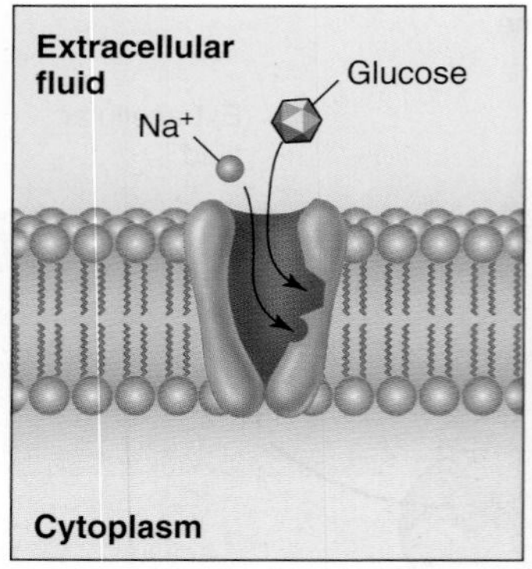
Extracellular fluid
Glucose
Na+
Cytoplasm
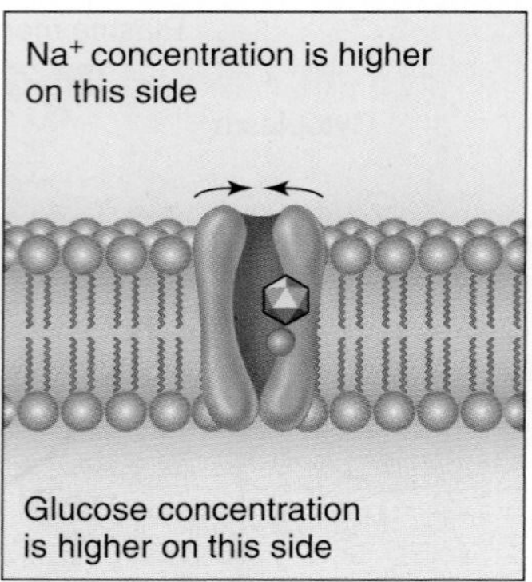
Na+ concentration is higher on this side
Glucose concentration is higher on this side
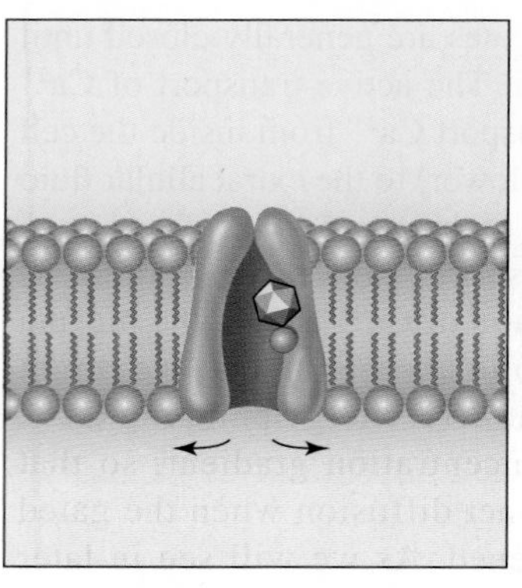
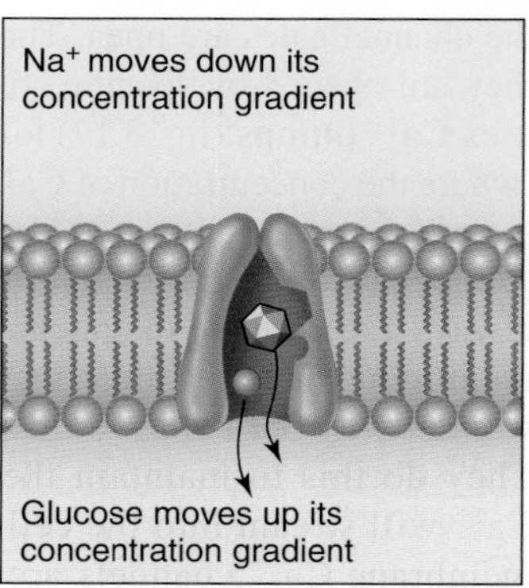
Na+ moves down its concentration gradient
Glucose moves up its concentration gradient

FIGURE 3.20 The cotransport of Na^+ and glucose. This carrier protein transports Na^+ and glucose at the same time, moving them from the lumen of the intestine and kidney tubules into the lining epithelial cells. This cotransport requires a lower intracellular concentration of Na^+, which is dependent on the action of other carriers, the Na^+/K^+ (ATPase) pumps. Since ATP is needed to power the Na^+/K^+ (ATPase) pumps, the cotransport of Na^+ and glucose depends indirectly on ATP, and so can be considered a type of active transport.

binds to glucose as well as Na^+ in the extracellular fluid, and so moves the glucose into the cell from a lower to a higher concentration. This cotransport is possible because the Na^+/K^+ (ATPase) pumps operate elsewhere in the plasma membrane to lower the intracellular Na^+ concentration. As a result, glucose and Na^+ can be absorbed together from the small intestine into the blood, and reabsorbed from the kidney filtrate (destined to become urine) into the blood.

CHECK POINT

1. What are the distinguishing features of carrier-mediated transport? Identify the two major types of carrier-mediated transport.
2. Describe facilitated diffusion, and explain why it is classified as a type of passive transport.
3. Why are carrier proteins for active transport called "pumps"? How do the pumps for active transport use ATP?
4. Compare the intracellular and extracellular concentration of Ca^{2+}, Na^+, and K^+.
5. Describe the cotransport of Na^+ and glucose, and give an example of where this occurs in the body.

The Plasma Membrane Separates Charges, Producing a Membrane Potential

The plasma membrane separates charges between the extracellular fluid and the cytoplasm. This produces a difference in charge, or potential difference, across the plasma membrane measured in millivolt units. The inside of the membrane is the negative pole, and the magnitude of the voltage in a resting cell is primarily determined by the difference in K^+ concentrations across the plasma membrane.

The topic of the membrane potential is presented in this chapter, because the electrical properties of cells result from plasma membrane function. However, it could have instead been presented at the beginning of chapter 4, because the membrane potential is fundamental to the physiology of neurons and muscle cells, which are "excitable cells" able to change their membrane potential when they produce impulses. So, to learn how the nervous and muscular systems work—topics central to the study of human physiology—you first need to understand the information presented in this section. Get ready to add another layer to your knowledge pyramid in physiology.

The selective permeability of the plasma membrane gives rise to an unequal distribution of ions between the extracellular fluid and the cytoplasm. For example, cellular proteins and organic phosphates (such as ATP) cannot penetrate the plasma membrane and so are trapped within the cell. These molecules have a negative charge in the cytoplasm, and so can be called *fixed anions* ("fixed" means they can't move; "anions" refers to negatively charged ions). Small, inorganic ions can penetrate the plasma membrane if their membrane channels are open (remember, many ion channels are gated).

Because the permeability of the plasma membrane to the different inorganic ions varies, and because there are active transport pumps for the different ions, there are differences in ion concentrations across the plasma membrane (see fig. 3.19). These differences in inorganic ion concentrations between the inside and outside of the cell, and the presence of fixed anions within the cell, produce a difference in charges across the plasma membrane. The difference in charge, or **potential difference**, across the plasma membrane is called the **membrane potential**.

The membrane potential is measured in units of voltage, and voltage is a measure of the *difference* in charge between two places. A voltmeter used to measure voltage must thus have two "leads," or wires. If you placed both leads inside the cell (or both outside

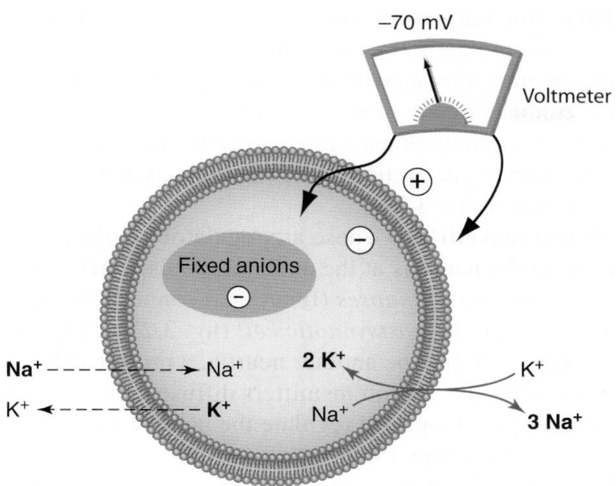

FIGURE 3.21 The resting membrane potential. The permeability of the plasma membrane to different inorganic ions (primarily K^+), its lack of permeability to intracellular fixed anions (proteins and organic phosphates), and the actions of active transport ion pumps—primarily the Na^+/K^+ (ATPase) pumps—produces a separation of charges across the membrane. This is the membrane potential, and in neurons at rest (not producing impulses) it averages −70 mV. The negative sign indicates that the inside of the cell is the negative pole.

the cell), the voltmeter would read "zero" because there would be no difference in charge between the two leads. If you instead put one lead inside the cell and one outside, you would record a voltage, because there is a difference in charge between these two locations. This difference in charge (created by the permeability properties and active transport pumps of the plasma membrane) is the membrane potential, and it measures about −70 millivolts (mV) in a typical neuron (fig. 3.21). The membrane potential in other cell types can have somewhat different values; in certain cardiac muscle cells, for example, it is −85 mV. The negative sign in front of the voltage indicates that the inside of the cell is the negative pole.[5] (It is useful to know the polarity of the inside of the cell, as will become evident when we examine nerve impulses in chapter 4.)

All cells have a membrane potential, because all have fixed anions, and all have the distribution of ions indicated in figure 3.19. However, only neurons and muscle cells can change their membrane potential (produce "impulses") when stimulated. Neurons and muscle cells are described as being *excitable*. Because these excitable cells can change their membrane potential when stimulated, the membrane potential before stimulation is called the **resting membrane potential**.

The voltage of the resting membrane potential is most greatly influenced by the difference in concentration of K^+ across the plasma membrane. This is because, of all the ions, K^+ can cross the resting plasma membrane most easily. The membrane is most permeable to K^+ because it contains two kinds of K^+ channels: one is gated, and one is not gated. At rest (not producing impulses), the gated channels are closed but the ungated channels are open—this is why they are referred to as *leakage channels*. Since K^+ is more highly concentrated inside the cell, it diffuses out of the cell through the leakage channels. By contrast, Na^+ has a higher concentration outside of the cell and so would diffuse into the cell. However, this diffusion is restricted because the plasma membrane contains only gated channels for Na^+, and these channels are closed in the resting membrane. Despite this, some Na^+ does diffuse into the cell because the gated Na^+ channels can "flicker" open occasionally.

The diffusion of K^+ out of and Na^+ into the cell would eventually change the concentrations of these ions in the intracellular and extracellular fluids, and thus would change the resting membrane potential. However, this is prevented by the constant activity of the Na^+/K^+ (ATPase) pumps. The pumps repair the leaks and then some; because each pump transports 3 Na^+ ions out of the cell for every 2 K^+ it transports in (see fig. 3.18), it contributes to the unequal distribution of charges and thus to the resting membrane potential. The factors that contribute to the resting membrane potential are summarized in figure 3.22.

CLINICAL APPLICATIONS

Because the plasma membrane is most permeable to K^+, changes in the K^+ concentration of the blood plasma (and thus the extracellular fluid) are particularly dangerous. A rise in the blood plasma concentration of K^+—called **hyperkalemia**—reduces the concentration gradient for the diffusion of K^+ out of the cells. This increases the K^+ concentration within the cells, thereby making the inside of the cells less negatively charged. In other words, the resting membrane potential is decreased, a change called *depolarization*. Depolarization of the heart is particularly dangerous, as it could cause abnormal rhythms.[6] For this reason, blood electrolytes (ions) are carefully monitored in patients with heart or kidney disease.

CHECK POINT

1. What is a potential difference, and what is the role of the plasma membrane in establishing a membrane potential?
2. Explain why there is no voltage if two leads of a voltmeter are placed outside (or inside) a cell.
3. What does the negative sign in front of the resting membrane potential indicate?
4. Explain why the resting membrane potential is particularly sensitive to the blood concentration of K^+.

FYI [5]The cell is like a tiny battery, with a voltage of a little under 0.1 volt. The poles of a battery are designated "+" and "−", a convention invented by Benjamin Franklin, who helped to put electricity on a scientific footing while serving as a founding father of the United States (among his other accomplishments).

FYI [6]Lethal injections of animals "put to sleep" and of condemned convicts often contain potassium chloride (KCl). The KCl ionizes, releasing K^+ to produce hyperkalemia. This depolarizes the heart and produces arrhythmias (abnormal heart rhythms), leading to death.

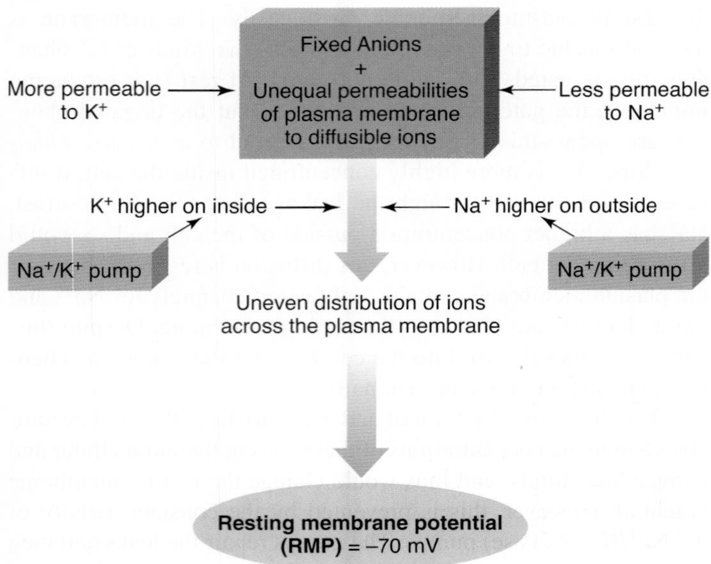

FIGURE 3.22 Factors contributing to the resting membrane potential. The resting membrane potential results from the presence of negatively charged ions (anions) that can't diffuse out of the cell (the fixed anions); the greater permeability of the resting plasma membrane to K^+ than to Na^+; and the activity of the Na^+/K^+ (ATPase) pumps, which pump out 3 Na^+ for every 2 K^+ pumped into the cell.

Regulatory Molecules from the Extracellular Fluid Influence Cell Activities

Regulatory molecules, including hormones and neurotransmitters, influence the activities of their target cells. In order to do this, the regulatory molecules must bind to specific receptor proteins. Polar regulatory molecules are water-soluble and bind to receptors located in the plasma membrane; nonpolar regulatory molecules bind to receptors located inside the target cells.

The physiology of the individual cells in an organ must be coordinated and regulated so that the organ can maintain homeostasis of the organism (you). This requires chemical signals sent from one tissue to another in an organ, and from one organ to another in the body. These chemical regulators will be discussed individually in later chapters; however, because the signals must interact with the plasma membrane surrounding cells in order to affect the cells, the basic concepts are presented here for you to add to your foundation of knowledge.

The activities of all cells are influenced by a variety of regulatory molecules released into the extracellular fluid by other cells. The cells that respond to a particular regulatory molecule are known as the regulator's *target cells*. There are different types of regulatory molecules, including:

1. **Paracrine regulators**. These are regulatory molecules released into the extracellular fluid by different tissue cells belonging to the same organ as the target cells. For example, the epithelial lining of blood vessels (the endothelium) releases a variety of paracrine regulators that can stimulate relaxation or contraction of the smooth muscle layer of the same vessel (fig. 3.23*a*).
2. **Neurotransmitters**. These are regulatory molecules released by neurons at the axon terminals, where the terminals make *synapses* (functional connections) with another cell, the *postsynaptic cell* (fig. 3.23*b*). The postsynaptic cell may be another neuron, a muscle cell, or a gland cell. The neurotransmitters diffuse across a very narrow synaptic gap to regulate the postsynaptic cell (as discussed in chapter 4).
3. **Hormones**. Hormones are regulatory molecules secreted into the blood by endocrine glands. The blood carries hormones to every cell in the body, but only the target cells for a particular hormone can respond (fig. 3.23*c*). Endocrine regulation is described more fully in chapter 8.

Target cells must have **receptor proteins** to be able to respond to a particular regulatory molecule. Receptor proteins, like enzyme proteins and membrane carrier proteins, have *specificity*. For example, the receptor protein for estradiol (the major estrogen, or female hormone) will not respond to testosterone (the major androgen, or male hormone), or even to progesterone, a different female hormone. The specificity of receptor proteins allows a given cell to respond in different ways to many different regulatory molecules.

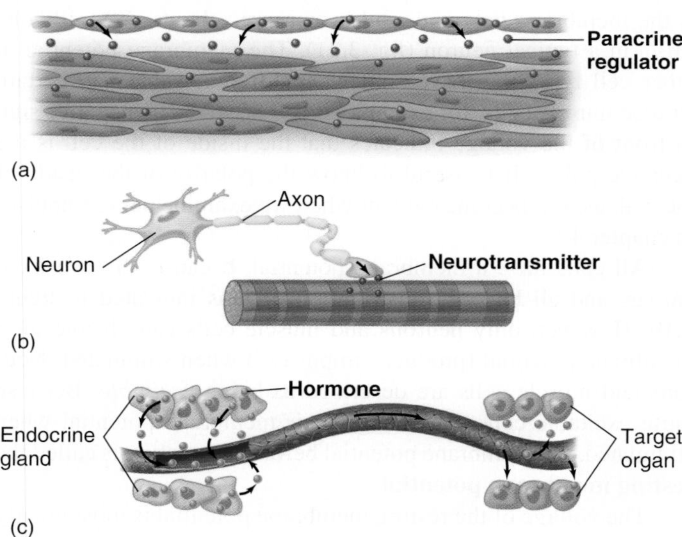

FIGURE 3.23 Types of regulatory molecules. (*a*) Paracrine regulators are released by one tissue (such as epithelial cells) and act on another tissue (such as the smooth muscle shown here) in the same organ. (*b*) Neurotransmitters are released by axon terminals at synapses and regulate the cells located across the synaptic gap. (*c*) Hormones are secreted into the blood by endocrine glands and are carried by the blood to the cells of target organs.

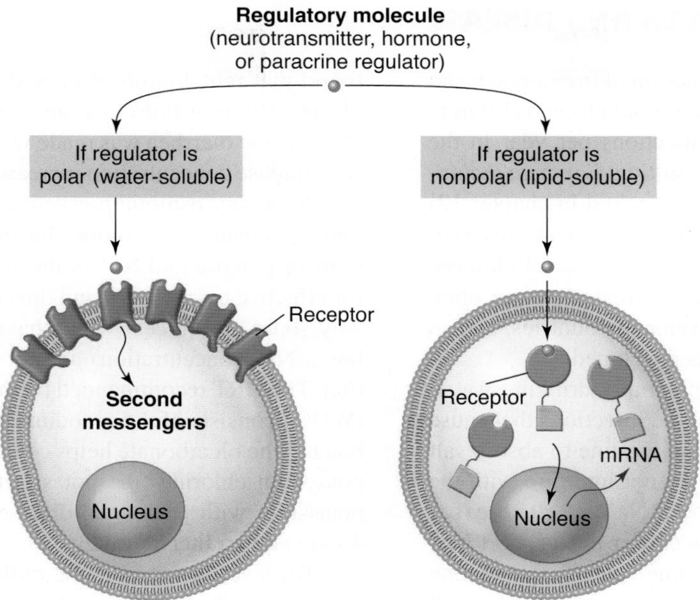

FIGURE 3.24 How regulatory molecules influence their target cells. Polar, water-soluble regulatory molecules bind to their receptors on the outer surface of the plasma membrane. When this occurs, intracellular second messengers act as middlemen to carry out the effects of the regulatory molecules inside the target cells. If the regulatory molecule is nonpolar and lipid-soluble, it enters its target cells and binds to intracellular receptor proteins. This leads to activation of genes, production of messenger RNA (mRNA), and thus the synthesis of specific proteins in the target cell.

If a regulatory molecule is polar, it is water-soluble and so can travel dissolved in the blood plasma. However, when the polar regulatory molecule arrives at its target cell, it cannot dissolve into the nonpolar double phospholipid layers of the plasma membrane. In general, polar regulatory molecules *do not enter* their target cells. Because polar regulatory molecules don't enter their target cells, they must bind to receptor proteins that are built into the plasma membrane (fig. 3.24). The binding of the regulatory molecule to its membrane receptor then triggers a sequence of intracellular events. Because the binding of the regulator to its receptor occurs at the plasma membrane, but must affect affairs deep in the cell, intracellular middlemen—known as *second messengers*—are required. Examples of polar regulatory molecules include the hormones insulin and epinephrine (adrenalin), and the neurotransmitter acetylcholine.

By contrast, nonpolar regulatory molecules are not water-soluble and so cannot dissolve in the blood plasma (they instead bind to carrier proteins, which act like cruise ships to ferry the regulators through the blood). When they arrive at their target cells, the nonpolar regulatory molecules leave their carrier proteins and pass through the plasma membrane. They can easily do this because they are lipid-soluble and so can penetrate the double phospholipid layers of the membrane. The receptor proteins for nonpolar regulatory molecules are thus located inside the cell (fig. 3.24). In general, this leads to the activation of specific genes, synthesis of messenger RNA, and thus the production of specific proteins in response to the regulator. Examples of nonpolar regulatory molecules include steroid hormones of the gonads, such as estradiol, progesterone, and testosterone; and thyroxine from the thyroid gland.

CHECK POINT

1. What are the distinctions between paracrine regulators, neurotransmitters, and hormones?
2. What are the characteristics of receptor proteins, and how do these compare with enzyme proteins and membrane carrier proteins?
3. Identify the locations of receptor proteins in the target cells for polar and nonpolar regulatory molecules, and explain why these locations are different.

Physiology in Health and Disease

Acute gastroenteritis (inflammation of the stomach and intestines) is a common affliction of infants and children, causing about 200,000 hospitalizations per year in the United States. Worldwide, the resulting *diarrhea, malnutrition*, and *metabolic acidosis* (discussed in chapter 12) that results from different causes of gastroenteritis produce approximately 4 million deaths per year of children under the age of 4 years. Intravenous treatments are often not possible, especially in underdeveloped countries, and so **oral rehydration therapy** (**ORT**) was developed.

The diarrhea cannot be treated by simply drinking water, or even water with added salt, because the infections that cause diarrhea interfere with the ability of the intestine to absorb salt and water. However, these infections do not interfere with the membrane protein carrier that cotransports Na^+ and glucose (see fig. 3.20). Glucose is required for this carrier to transport Na^+ (and vice versa) from the intestinal lumen across the plasma membrane of epithelial cells. When Na^+ enters an epithelial cell and then leaves the cell to enter the blood, water follows the Na^+ by osmosis. This is because Na^+ is the major extracellular solute, and increased extracellular Na^+ exerts an osmotic pressure that draws water from the intestinal lumen.

The cotransport of sodium and glucose by the intestine was an accidental discovery. In the late 1940s medical personnel found that rehydration of patients improved when they added glucose (for nutrition) to a salt solution. If they added too much glucose, the diarrhea was made worse. This is because an excessive glucose concentration increases the osmolarity of the drink, and the higher osmotic pressure draws water into the intestinal lumen, causing an osmotic diarrhea. The molarity concentrations of glucose and Na^+ in the solution should be about equal for effective cotransport, and thus effective rehydration—this is why sodas and juices (which have too high a glucose and too low a Na^+ concentration) are not effective drinks for rehydration. The ORT recommended by the World Health Organization (WHO) consists of 3.5 g sodium chloride; 2.5 g sodium bicarbonate (the bicarbonate helps counter metabolic acidosis); 1.5 g potassium chloride (the potassium counters the loss in blood potassium with prolonged diarrhea); and 20 g of glucose, all dissolved in a liter of water.

The lives of more than a million small children a year are saved by oral rehydration therapy. The effectiveness of ORT in treating the dehydration and other consequences of diarrhea is particularly impressive in the treatment of *cholera*. Without treatment, mortality from cholera is greater than 50%; with oral rehydration therapy, mortality is reduced to less than 10%.

Physiology in Balance

Interactions Between Cells and Their Environment

Working together with . . .

Endocrine System

Homeostasis of blood glucose is needed for the facilitated diffusion of glucose into cells . . . p. 186

Respiratory System

Diffusion of O_2 and CO_2 between air and blood is needed for homeostasis of blood gases . . . p. 297

Digestive System

Absorption of food molecules into the blood requires transport through the plasma membrane of epithelial cells . . . p. 348

Cardiovascular System

The heart pumps blood through the body, thereby circulating the extracellular fluid compartment . . . p. 240

Urinary System

Membrane transport processes move molecules from the fluid that will become urine into the blood . . . p. 324

Summary

The Plasma Membrane Can Move Many Molecules Together 62

- Phagocytosis involves extension of pseudopods that fuse to form a vacuole.
- Endocytosis includes pinocytosis and receptor-mediated endocytosis.
- Pinocytosis involves a nonspecific furrowing of the plasma membrane to produce a vesicle containing extracellular fluid.
- Receptor-mediated endocytosis requires the binding of extracellular molecules with very specific receptor proteins in the plasma membrane, and the invagination of the plasma membrane to produce a vesicle.
- Exocytosis occurs when an intracellular vesicle fuses with the plasma membrane; this can release cellular products and add new membrane to the plasma membrane.

The Plasma Membrane Can Transport Individual Molecules 65

- Net diffusion occurs because thermal (heat) energy produces random motion of molecules and ions, resulting in net movement from higher to lower concentrations.
- Passive membrane transport results from net diffusion; its rate depends on the steepness of the concentration gradient and the permeability of the plasma membrane.
- Nonpolar molecules diffuse through the phospholipid layers of the membrane; inorganic ions diffuse through protein channels through the membrane; these "simple diffusion" processes contrast with diffusion by means of carrier proteins, called facilitated diffusion.
- Active transport also requires carriers, but the carriers pump the transported molecule or ion against its concentration gradient, from lower to higher concentration; this pumping action requires the breakdown of ATP.

Passive Transport Can Occur by Simple Diffusion 67

- Oxygen, carbon dioxide, and nonpolar organic molecules such as steroid hormones can diffuse across the double phospholipid layers of the plasma membrane.
- Inorganic ions diffuse through fairly specific protein channels in the plasma membrane, many of which have gates that can open or close the channels.
- Osmosis, the net diffusion of water across a membrane, requires that a concentration gradient exist across the membrane and that the membrane be more permeable to water than to a solute.
- The more concentrated a solution is, in terms of the numbers of solute molecules and ions per liter of water, the higher its osmolarity and its tendency to draw water by osmosis.
- The osmotic pressure of a solution is directly related to its solute concentration, and so indicates the tendency of a solution to take in water.
- Solutions can be described as isotonic, hypotonic, and hypertonic, depending on their osmolarity and osmotic pressure in comparison to a reference solution, usually blood plasma.
- Water is able to move by osmosis through the plasma membrane with the aid of aquaporin channels that may be inserted into the membrane.

Facilitated Diffusion and Active Transport Depend on Carrier Proteins 72

- Carrier proteins are highly specific as to what they transport, and they can become saturated; at saturation, the transport rate reaches a maximum value.
- Facilitated diffusion is passive transport from higher to lower concentrations; for example, there are carrier proteins that move glucose from the blood plasma into skeletal muscle cells by facilitated diffusion.
- Active transport carriers are called pumps, because they pump molecules and ions uphill, against their diffusion gradients.
- The Ca^{2+} pump actively transports Ca^{2+} out of cells, so that the intracellular Ca^{2+} concentration remains lower than the extracellular concentration.
- The Na^+/K^+ (ATPase) pumps actively transport 3 Na^+ out of the cell for every 2 K^+ they transport into the cell.
- As a result of the action of Na^+/K^+ (ATPase) pumps, the concentration of Na^+ outside of cells is kept higher than inside of cells; and the concentration of K^+ inside cells is kept higher than outside of cells.

The Plasma Membrane Separates Charges, Producing a Membrane Potential 76

- Cytoplasmic proteins and organic phosphates (such as ATP) cannot leave the cell; they are fixed anions (negatively charged ions).
- The plasma membrane at rest (not producing impulses) is most permeable to K^+, because it contains open, ungated channels for K^+; it is much less permeable to Na^+, because for Na^+ the resting plasma membrane contains only closed, gated channels.
- Because of the selective permeability of the plasma membrane to ions, and because of the action of the Na^+/K^+ pumps, all cells have an unequal distribution of charges between the inside and the outside of the plasma membrane; this is a potential difference, and because it's caused by the plasma membrane, it's called a membrane potential.
- A typical resting neuron (one not producing impulses) has a resting membrane potential of about -70 millivolts (mV); the negative sign indicates that the inside of the cell is the negative pole.
- The resting membrane potential is affected more by changes in the extracellular concentration of K^+ than of Na^+, because the membrane is more permeable to K^+.

Regulatory Molecules from the Extracellular Fluid Influence Cell Activities 78

- Paracrine regulators are regulatory molecules released by one tissue that regulate another tissue of the same organ.
- Neurotransmitters are regulatory molecules released by axon terminals where they make a functional connection, or synapse, with a postsynaptic cell.
- Hormones are regulatory molecules secreted into the blood by endocrine glands; hormones travel through the blood to regulate the activities of their target cells.
- Polar regulatory molecules bind to receptor proteins located on the plasma membrane; this activates second messengers inside the target cell, which produce the effects of the regulatory molecule on the target cell.
- Nonpolar regulatory molecules pass through the plasma membrane and bind to receptor proteins within the target cell; this activates genes, producing messenger RNA (mRNA) that codes for the production of specific proteins.

Review Activities

Objective Questions: Test Your Knowledge

1. Which of the following phagocytic cells are fixed in position and cannot move by amoeboid motion?
   a. Kupffer cells in the liver
   b. Microglia in the brain
   c. Neutrophils in the blood and connective tissues
   d. Macrophages in the connective tissues

2. The uptake of cholesterol from the blood into cells of the liver or artery wall is an example of
   a. pinocytosis.
   b. exocytosis.
   c. receptor-mediated endocytosis.
   d. phagocytosis.

3. When an intracellular vesicle joins with the plasma membrane, the process is termed
   a. pinocytosis.
   b. exocytosis.
   c. receptor-mediated endocytosis.
   d. phagocytosis.

4. The rate of net diffusion across the plasma membrane depends on
   a. the steepness of the concentration gradient.
   b. the permeability of the plasma membrane.
   c. the presence of active transport pumps.
   d. Both "a" and "b" are true.
   e. Both "b" and "c" are true.

5. Which of the following statements about passive transport is true?
   a. Nonpolar molecules diffuse through protein channels in the membrane.
   b. Inorganic ions diffuse through carrier proteins.
   c. Glucose diffuses through the membrane using carrier proteins.
   d. ATP breakdown is required.

6. In passive membrane transport, the direction of net movement is always
   a. from higher to lower concentration.
   b. from the inside of the cell to the outside.
   c. from the outside of the cell to the inside.
   d. from lower to higher concentration.

7. Which of the following statements about osmosis is true?
   a. It is an example of active transport.
   b. It involves the net diffusion of solute across a membrane.
   c. It occurs only if the plasma membrane is more permeable to solute than to solvent.
   d. It occurs only if the solution on one side of the membrane is more concentrated than the other side.

8. The solution with the higher osmotic pressure is the one with
   a. the higher number of grams per liter of solute.
   b. the solute with the greater molecular weight.
   c. the solution with the greater number of solute molecules and ions per liter.
   d. the solution with the lower molar concentration.

9. Which of the following statements about an isotonic solution is false?
   a. It has an osmolarity of about 300 mOsM.
   b. A cell placed in it will crenate.
   c. It has an osmolarity equal to that of plasma.
   d. A cell placed in it neither gains nor loses water.

10. When two solutions are separated by a semipermeable membrane, osmosis will occur
    a. from the solution of lower to the solution of higher osmotic pressure.
    b. from the solution of higher to the solution of lower osmotic pressure.
    c. from the solution of higher to the solution of lower osmolarity.
    d. from the solution of lower to the solution of higher water concentration.

11. A transport maximum is a characteristic of which type of membrane transport?
    a. Diffusion of nonpolar molecules through the double phospholipid layers
    b. Diffusion of inorganic ions and water through protein channels
    c. Passive and active transport using carrier proteins
    d. All of these

12. Which of the following is an example of facilitated diffusion?
    a. The movement of Na^+ through gated membrane channel proteins
    b. The movement of oxygen through the membrane
    c. The movement of Na^+ from the inside to the outside of a cell
    d. The movement of glucose from the blood plasma into cells

13. Which of the following statements about the membrane transport of Ca^{2+} is true?
    a. Ca^{2+} can diffuse into cells through ion channel proteins.
    b. Ca^{2+} is transported out of cells by a Ca^{2+} pump.
    c. The intracellular Ca^{2+} concentration is kept much lower than the extracellular concentration by active transport.
    d. All of these are true.

14. Which of the following statements about the Na^+/K^+ (ATPase) pumps is true?
    a. They participate in facilitated diffusion.
    b. They require the breakdown of ATP.
    c. They pump Na^+ into the cell and K^+ out of the cell.
    d. They transport 3 K^+ for every 2 Na^+ ions.

15. The term *fixed anions* refers to the
    a. organic phosphates and proteins within the cell.
    b. positively charged ions within the cell.
    c. chloride and other negatively charged ions outside of the cell.
    d. Na^+, K^+, and other positively charged ions outside of the cell.

16. The major ion in the extracellular fluid, and the solute which contributes most to the osmotic pressure of plasma, is:
    a. Ca^{2+}
    b. K^+
    c. Cl^-
    d. Na^+

17. The resting membrane (one not producing impulses) is most permeable to which ion?
    a. Ca^{2+}
    b. K^+
    c. Cl^-
    d. Na^+

18. Which of the following statements about the resting membrane potential is false?
    a. It is very sensitive to the concentration of K^+ in the extracellular fluid.
    b. It is approximately 70 millivolts in a typical neuron.
    c. The outside of the membrane is the negative pole.
    d. It can be measured by voltmeter leads placed outside and inside the cell.
19. The resting membrane potential is produced by
    a. activity of the Na^+/K^+ (ATPase) pumps.
    b. fixed anions within the cell.
    c. the permeability of the plasma membrane to inorganic ions, chiefly K^+.
    d. all of these.
20. Regulatory molecules that are secreted into the blood and act on distant cells are
    a. hormones.
    b. paracrine regulators.
    c. neurotransmitters.
    d. all of these.
21. Which of the following statements about nonpolar regulatory molecules is false?
    a. They bind to intracellular receptors.
    b. They activate genes.
    c. They require the action of second messengers.
    d. They stimulate the synthesis of mRNA and protein.

Essay Questions 1: *Test Your Understanding*

1. What is receptor-mediated endocytosis, and how does it differ from pinocytosis? Give an example of receptor-mediated endocytosis.
2. What is phagocytosis? Give examples of different phagocytic cells and state where they are found.
3. Compare exocytosis with endocytosis, and explain how these two processes can complement each other.
4. Define diffusion and describe the factors that affect the rate of diffusion across a plasma membrane.
5. Describe three different ways that substances can move by passive transport across a plasma membrane, and give examples.
6. Explain how active transport differs from passive transport.
7. Describe how Na^+ can travel across the plasma membrane, and compare this with the ways that K^+ can cross the membrane.
8. Define osmosis and state the conditions needed for osmosis to occur.
9. Which solution has the higher osmotic pressure: one with a solute concentration of 100 mOsM, or one with a concentration of 200 mOsM? Explain why this is so.
10. Explain why solutions given intravenously are isotonic. Describe what happens to cells in hypotonic and hypertonic solutions.
11. Describe the characteristic of carrier-mediated transport.
12. Explain how facilitated diffusion and active transport are similar and how they are different.
13. Describe the action of the Ca^{2+} pumps and the physiological significance of these pumps.
14. Describe the action of the Na^+/K^+ (ATPase) pumps. How do these pumps affect the intracellular and extracellular concentrations of Na^+ and K^+, and how do they affect the membrane potential?
15. Describe the cotransport of Na^+ and glucose, and identify where in the body it occurs.
16. What is a membrane potential? What factors contribute to the membrane potential, and how is it measured?
17. What is the significance of the sign in front of the membrane potential voltage? What is the significance of the word *resting* in the term *resting membrane potential*?
18. What are the differences between paracrine regulators, hormones, and neurotransmitters?
19. Identify the cellular location of the receptor proteins for polar and nonpolar regulatory molecules. Explain the reasons for this difference.
20. Which regulatory molecules require second messengers in order to control their target cells? Explain why this is so.

Essay Questions 2: *Test Your Analytical Ability*

1. Explain the significance of receptor-mediated endocytosis in the regulation of blood cholesterol, and explain how this relates to heart disease.
2. Receptor proteins for insulin, and aquaporin channel proteins for water, can be inserted into the plasma membrane when needed and removed when not needed. Explain how this relates to endocytosis and exocytosis.
3. Water moves from the less concentrated to the more concentrated solution. Does this mean that osmosis is active transport? Explain.
4. A normal saline solution containing 9 g NaCl per 100 ml, and a glucose solution containing 5 g glucose per 100 ml, are isotonic to each other and isotonic to blood plasma. Explain (without using mathematics) why this statement is true.
5. Suppose substance X is present at a higher concentration outside of cells, and that after a time its concentration inside cells increases. What must you know to tell if substance X is diffusing through the phospholipid layers of the membrane, diffusing through protein channels, or diffusing through the membrane using carrier proteins?
6. Suppose you poison a cell with cyanide, so that its ability to produce ATP is inhibited (chapter 2). Which types of membrane transport would stop, and which would still occur? Explain.
7. If a cell is poisoned with cyanide (see question 6), what will happen to the intracellular and extracellular concentrations of Na^+ and K^+ over time? What will happen to the resting membrane potential? Explain.
8. Pouring K^+ (in the form of KCl) outside of cells will depolarize them. Pouring Na^+ (in the form of NaCl) outside of cells has less of an effect on the membrane potential, but draws water out of the cell by osmosis. Explain why these statements are true.
9. If two solutions are separated by a semipermeable membrane, water moves by osmosis from the solution of lower to the one of higher osmotic pressure. Yet air in the atmosphere always moves from higher to lower pressure areas (producing wind). How can this seeming contradiction be explained?
10. When you drink lots of water, you produce more urine. This is because the secretion of ADH is inhibited. Conversely, you produce less urine when you are dehydrated because ADH secretion is stimulated. Using only the information in this chapter, explain how this is accomplished.
11. More of the water you drink can be absorbed through the intestine into the blood if the water contains the correct proportions of both Na^+ and glucose. What physiological mechanism can account for this?
12. All cells have a resting membrane potential. What, then, must be true for all cells?
13. What would happen to the resting membrane potential if the plasma membrane suddenly became permeable to Na^+ or Ca^{2+}? What

would happen if it suddenly became permeable to Cl^- (when the concentration of Cl^- is higher outside of the cell)? Explain.

14. The steroid and thyroid hormones are nonpolar, whereas insulin is a polar compound (a polypeptide). The steroids and thyroid hormones can be taken orally (as a pill), whereas insulin cannot (it has to be injected). What properties of the steroid and thyroid hormones allow them to be taken orally?
15. A second messenger in the action of some polar regulatory molecules is cyclic adenosine monophosphate (cAMP). If you injected cAMP into a target cell, it would produce the same effect as the regulatory molecule. However, eating cAMP, or injecting it into the blood, would have no effect. Why would injection of cAMP into target cells have the same effect as the regulatory molecule? Why would eating or injecting cAMP into the blood be ineffective? (Hint: cAMP is an organic phosphate molecule).

Web Activities

ARIS™ **For additional study tools, go to: www.aris.mhhe.com. Click on your course topic and then this textbook's author/title. Register once for a semester's worth of interactive activities and practice quizzing to help boost your grade!**

CHAPTER 4

Nervous System: Neurons and Synapses

Chapter Outline

HOMEOSTASIS

Motor neurons stimulate the effector organs—the muscles (skeletal, cardiac, and smooth muscles) and glands (exocrine and endocrine glands). Activity of motor neurons is stimulated or inhibited by association neurons located in the brain and spinal cord, which can produce willed, conscious actions in some cases and unconscious reflexes in others. Activity of the association and motor neurons is influenced by sensory information brought to the nervous system by way of sensory neurons. In order for neurons to communicate with other neurons and with effector cells, they must produce impulses that convey information, conduct these impulses over long distances, and then transmit information to other cells. These functions are required so that the body can respond to changes in the internal environment. The activity of the nervous system is critical for a great many of the physiological mechanisms that operate by way of negative feedback loops to maintain homeostasis.

CLINICAL INVESTIGATION

But if homeostasis is not maintained . . .

Lisa, a 36-year-old woman, complained of extraordinary muscle weakness after eating mussels gathered by a friend from the local seashore. A paramedic who came on the scene mentioned that this was probably the problem, because there was a red tide and the mussels contained saxitoxin, a poison. He said Lisa was lucky she hadn't eaten more than she did, because saxitoxin causes potentially fatal paralytic shellfish poisoning. However, upon further examination he saw that her eyelid drooped, and she mentioned that she had a recent Botox treatment. The paramedic wondered if that might have caused the problems, but he became more confused when he found prednisone (a corticoid drug used for inflammation) and a bottle of Prostigmin (an inhibitor of acetylcholinesterase) in her purse. When she was questioned, Lisa explained that she had myasthenia gravis and used these drugs to treat it.

How are skeletal muscles affected by myasthenia gravis, Botox treatment, and eating mussels with saxitoxin? Could any of these have caused Lisa's symptoms? What benefit would an inhibitor of acetylcholinesterase have in treating myasthenia gravis?

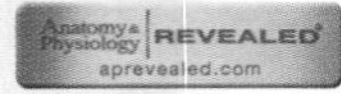

A virtual cadaver dissection experience

Nervous System Function Depends on Neurons and Supporting Cells

The nervous system is divided into the central nervous system (CNS) and peripheral nervous system (PNS). Neurons are classified in different ways depending on their structure and function, and there are also different types of supporting cells, called neuroglia, in the CNS. The functions of the nervous system are based on the physiology of neurons and supporting cells.

There are two body systems—the endocrine and nervous systems—that regulate one another and the other body systems as well. Regulation by the nervous system is so pervasive that we can't discuss the physiology of any organ system without frequent references to its neural regulation. Neural regulation of an organ allows sensors to communicate with an integrating center (often the CNS), and the integrating center to activate effectors (chapter 1) that help adjust the organ's function to maintain homeostasis. Thus, when you learn the way neurons produce impulses and communicate with other cells, you are building an important foundation for your knowledge pyramid in human physiology.

Neurons (described in chapter 2) consist of three major parts: the *cell body*, which contains the nucleus; the *dendrites*, which function to receive stimulation; and the *axon*, which conducts impulses away from the cell body (fig. 4.1). An insulating material called a **myelin sheath** surrounds many axons (fig. 4.2). Notice that the myelin sheath does not surround the *axon hillock*, which is the initial segment of the axon. The myelin sheath is produced by certain supporting cells, which will be described shortly.

There are two major divisions of the nervous system. The **central nervous system (CNS)** consists of the brain and spinal cord. The **peripheral nervous system (PNS)** consists of all the other parts of the nervous system, specifically nerves and ganglia. **Nerves** are bundles of axons in the PNS that enter either the brain or the spinal cord; **ganglia** are bundles of neuron cell bodies in the PNS. Their counterparts in the CNS are **tracts** (collections of axons in the CNS) and **nuclei** (groupings of neuron cell bodies in the CNS).

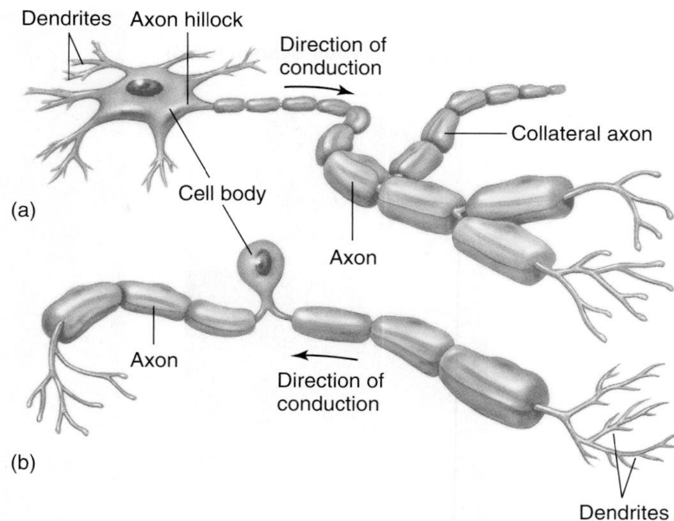

FIGURE 4.1 The structure of two kinds of neurons. (*a*) A motor neuron and (*b*) a sensory neuron are depicted here.

There Are Different Types of Neurons and Supporting Cells

Neurons are classified by their structure and function. Functionally, neurons in the PNS are either **sensory neurons** or **motor neurons** (fig. 4.3). Sensory neurons conduct impulses into the CNS (and thus are *afferent neurons*), whereas motor neurons conduct impulses out of the CNS (and thus are *efferent neurons*). Motor

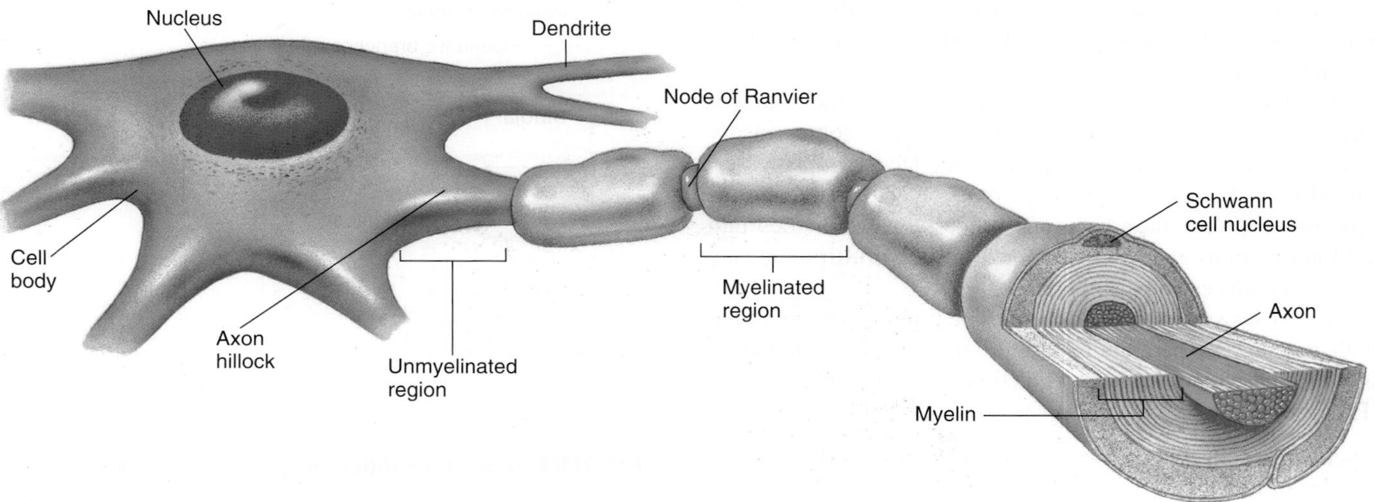

FIGURE 4.2 Parts of a neuron. The axon of this neuron is wrapped by Schwann cells, which form a myelin sheath.

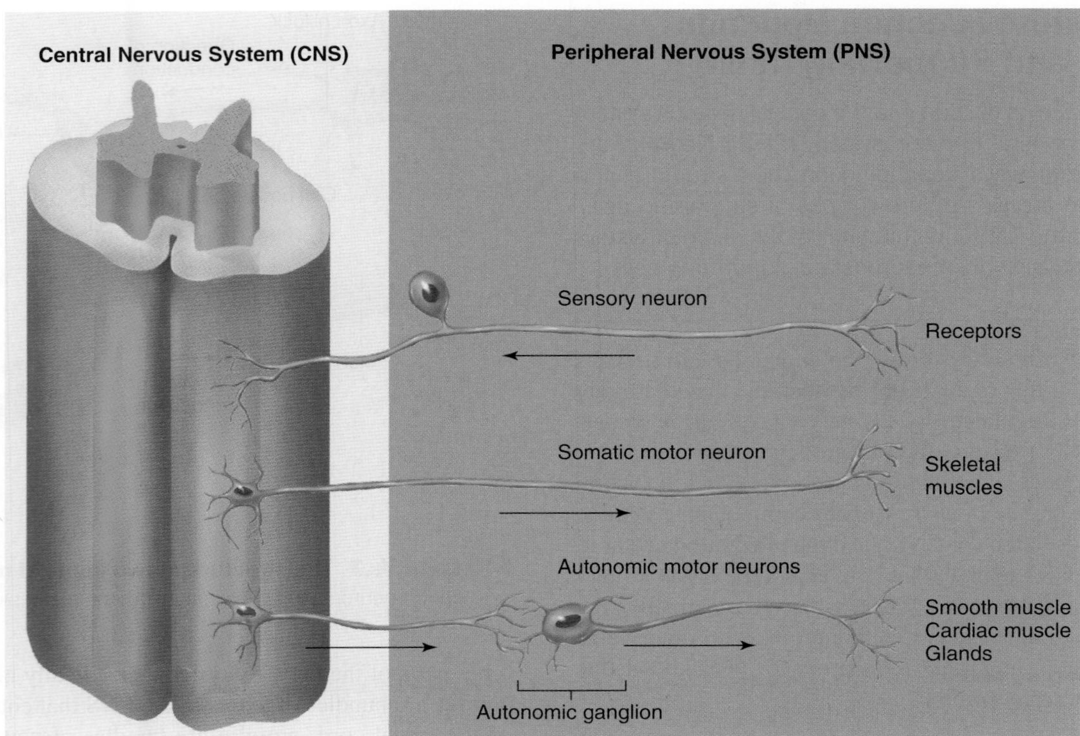

FIGURE 4.3 The relationship between the CNS and PNS. Sensory and motor neurons of the peripheral nervous system carry information into and out of, respectively, the central nervous system (brain and spinal cord).

neurons are divided into those that stimulate skeletal muscles (*somatic motor neurons*), and those that stimulate smooth muscles, cardiac muscles, and glands (*autonomic motor neurons*). Autonomic neurons are subdivided into *sympathetic* and *parasympathetic* divisions. Skeletal muscles are under voluntary control because they're innervated by somatic motor neurons, while the effectors innervated by autonomic motor neurons—smooth muscle, cardiac muscle, and glands—are involuntary. Neurons with their dendrites, cell body, and axon located entirely within the CNS are known as **association neurons**, because they can receive sensory information, integrate and analyze information, and stimulate motor neurons.

Structurally, neurons are classified according to their number of extensions, or processes (fig. 4.4). For example, some sensory neurons in the eye and ear are **bipolar**: they have two processes, one dendrite and one axon. However, most other sensory neurons are **pseudounipolar**: they have one process, but it divides into two. Motor neurons and association neurons are **multipolar**: they have many dendrites and a single axon.

The nervous system contains about five times more supporting cells than neurons. There are several different types of supporting cells, which have many important functions. The supporting cells in the PNS include:

1. **Schwann cells**, which form myelin sheaths around many axons of the PNS.
2. **Satellite cells**, which support neuron cell bodies in the ganglia of the PNS.

The supporting cells of the CNS are known as **neuroglial cells** (also called *neuroglia*, *glial cells*, or simply *glia*). There are several types of neuroglial cells (fig. 4.5):

1. **Oligodendrocytes**, which form myelin sheaths around many axons of the CNS.
2. **Astrocytes**, which have extensions that surround blood capillaries and extensions in close proximity to axons at

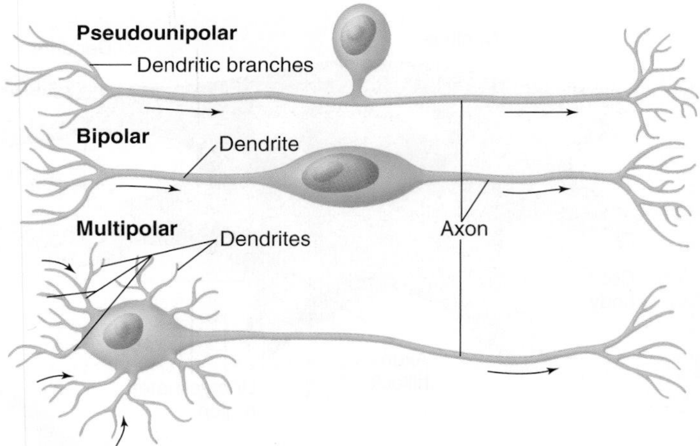

FIGURE 4.4 Three different types of neurons. Pseudounipolar neurons, which are sensory, have one process that splits. Bipolar neurons, found in the eye and ear, have two processes. Multipolar neurons, which are motor and association neurons, have many dendrites and one axon.

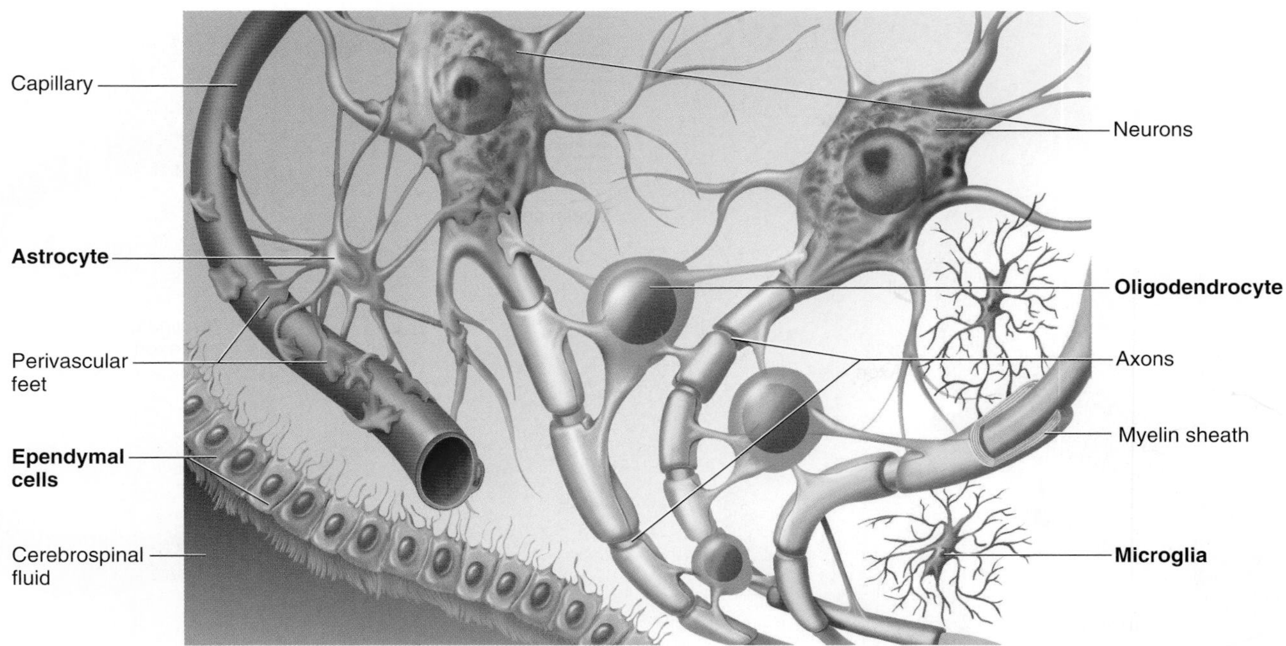

FIGURE 4.5 The different types of neuroglial cells. Myelin sheaths around axons are formed in the CNS by oligodendrocytes. Astrocytes have extensions that surround both blood capillaries and neurons. Microglia are phagocytic, and ependymal cells line the brain ventricles and central canal of the spinal cord.

synapses; astrocytes have diverse functions that support the activity of neurons.

3. **Microglia**, which are phagocytic cells within the nervous system (chapter 3).
4. **Ependyma**, which is an epithelial layer that lines the ventricles (cavities of the brain) and the central canal of the spinal cord.

The blood capillaries (the thinnest blood vessels) in the brain, unlike those of most other organs, don't have pores in their walls. As a result, there is no filtration of fluid out of the capillaries of the brain, the way there is in other organs. How then can the neurons of the brain obtain nutrients and regulatory molecules from the blood? These must move through the cytoplasm of the cells that compose the capillary walls (the endothelial cells) by diffusion, active transport, endocytosis, and exocytosis (described in chapter 3), which can be highly selective processes. This selectivity as to which molecules can pass from the blood to the brain is known as the **blood-brain barrier**. The selectivity of the blood-brain barrier is of great significance physiologically and medically, as it determines which regulatory molecules or drugs can influence the CNS. There is evidence that astrocytes induce many aspects of the blood-brain barrier, including the formation of tight junctions between endothelial cells, and the production of specific carrier proteins and ion channels.

Myelin Sheaths Are Formed by Schwann Cells and Oligodendrocytes

Your nervous system contains axons that have myelin sheaths, called *myelinated axons*, and those that don't, which are described as *unmyelinated axons*. In the PNS, Schwann cells produce the myelin sheaths; in the CNS, the myelin sheaths are produced by oligodendrocytes.

In the PNS, Schwann cells surround both myelinated and unmyelinated axons, forming a **sheath of Schwann**, or **neurilemma**. The sheath of Schwann is believed to aid the regeneration of damaged axons in the PNS by forming a "regeneration tube" that guides a growing axon to its proper destination. The difference between the myelinated and unmyelinated axons in the PNS is that, for myelinated axons, the Schwann cells wrap themselves around the axon (fig. 4.6). As a Schwann cell wraps itself around the axon, its cytoplasm gets squeezed to the outside (fig. 4.7). This results in successive wrappings of Schwann cell plasma membrane surrounding the axon, much like the wrapping of electrical tape around a wire. It is these overlapping layers of plasma membrane that compose the myelin sheath.

If you were to use electrical tape to insulate a wire, you would wrap the tape at an angle so that there would be no gaps in the insulation. This isn't how the Schwann cells do it. Instead, each Schwann cell wraps itself over and over again around the same section of axon; then a different Schwann cell wraps itself around the next section, and so on. Thus, there are gaps in the myelin sheath. These gaps are known as the **nodes of Ranvier** (see fig. 4.2), and they serve a very important function in the conduction of impulses by myelinated axons.

In the CNS, oligodendrocytes form myelin sheaths in much the same way, except that an oligodendrocyte has several extensions that can form myelin sheaths around several axons (fig. 4.8). Because of this, there isn't a continuous sheath around axons in the CNS the way there is a continuous sheath of Schwann around axons in the PNS. This may be an important reason why CNS axons don't generally regenerate after they're damaged. However, regeneration of CNS axons is also inhibited by molecules produced by oligodendrocytes and astrocytes.

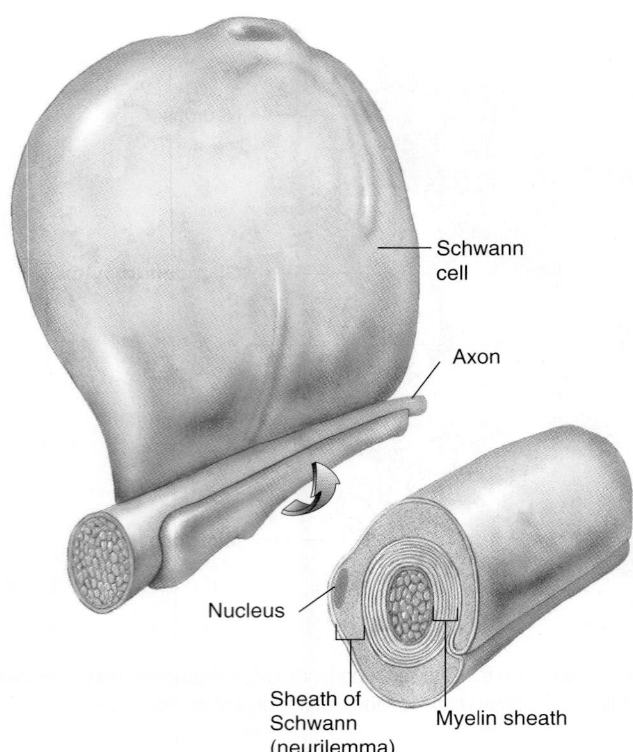

FIGURE 4.6 The formation of a myelin sheath around a peripheral axon. The myelin sheath is formed by successive wrappings of the Schwann cell membranes, leaving most of the Schwann cell cytoplasm outside the myelin. The sheath of Schwann is thus external to the myelin sheath.

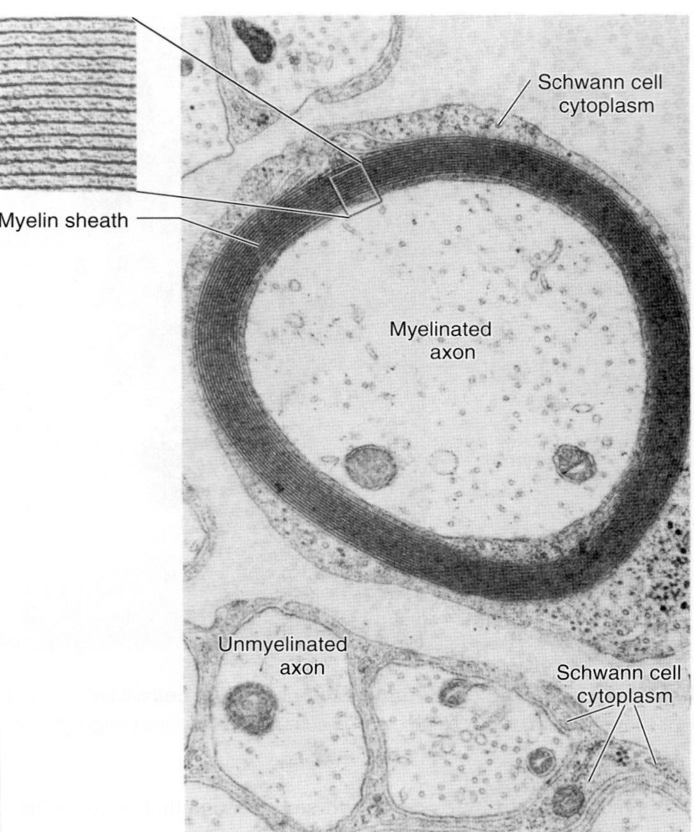

FIGURE 4.7 An electron micrograph of unmyelinated and myelinated axons. Notice that myelinated axons have Schwann cell cytoplasm to the outside of their myelin sheath, and that Schwann cell cytoplasm also surrounds unmyelinated axons.

The areas of the CNS that contain mostly axons have a white color because of the phospholipids of plasma membranes, which make up myelin sheaths. Thus, the **white matter** of the CNS contains axon tracts. Because there are no myelin sheaths surrounding the cell bodies and dendrites, areas of the CNS that contain cell bodies and dendrites are gray in color, and compose the **gray matter**. In the brain, gray matter is on the surface with white matter underneath; in the spinal cord the order is reversed, with white matter on the outside and gray matter forming a butterfly-shaped area in the center of the spinal cord (chapter 5).

CLINICAL APPLICATIONS

Multiple sclerosis (MS) is a neurological disease usually diagnosed in people between the ages of 20 and 40. It is a chronic, degenerating, remitting, and relapsing disease that progressively destroys the myelin sheaths of neurons in multiple areas of the CNS. Initially, lesions form on the myelin sheaths and soon develop into hardened *scleroses,* or scars (from the Greek *sklerosis* = hardened). Destruction of the myelin sheaths prohibits the normal conduction of impulses, resulting in a progressive loss of functions. Because myelin degeneration is widespread and affects different areas of the nervous system in different people, MS has a wider variety of symptoms than any other neurological disease.

CHECK POINT

1. Distinguish between sensory neurons and motor neurons, and between the two major types of motor neurons. How do these relate to the terms *afferent* and *efferent*?
2. Compare the structures, functions, and locations of bipolar, pseudounipolar, and multipolar neurons.
3. What is the blood-brain barrier, and what is its significance?
4. Describe how myelin sheaths and nodes of Ranvier are formed in the PNS, and the structure and significance of the sheath of Schwann.
5. Explain how myelin sheaths are formed in the CNS, and describe the different compositions of white and gray matter in the CNS.

Axons Produce Action Potentials

Action potentials are nerve impulses produced by axons. This occurs in response to a depolarization stimulus that causes changes in the permeability of the plasma membrane to first Na^+ and then K^+. Action potentials are all-or-none events that cannot summate, and that are conducted along the axon without decreasing in amplitude.

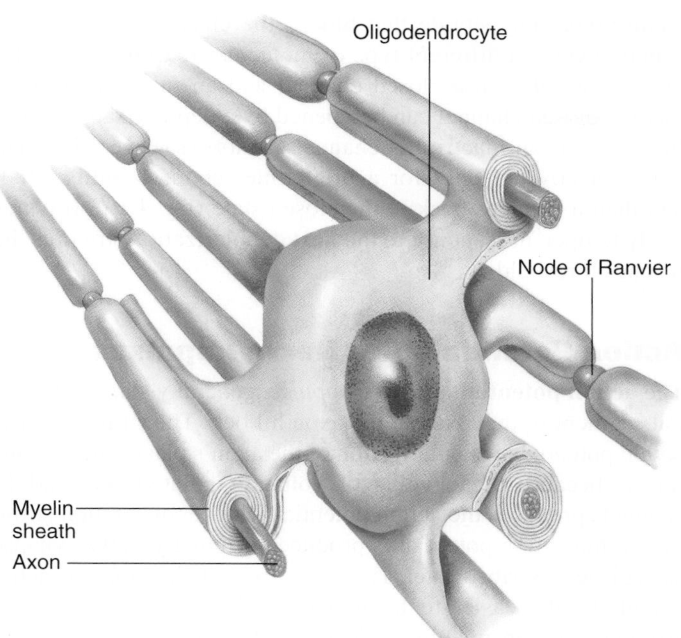

FIGURE 4.8 The formation of myelin sheaths in the CNS by an oligodendrocyte. One oligodendrocyte forms myelin sheaths around several axons.

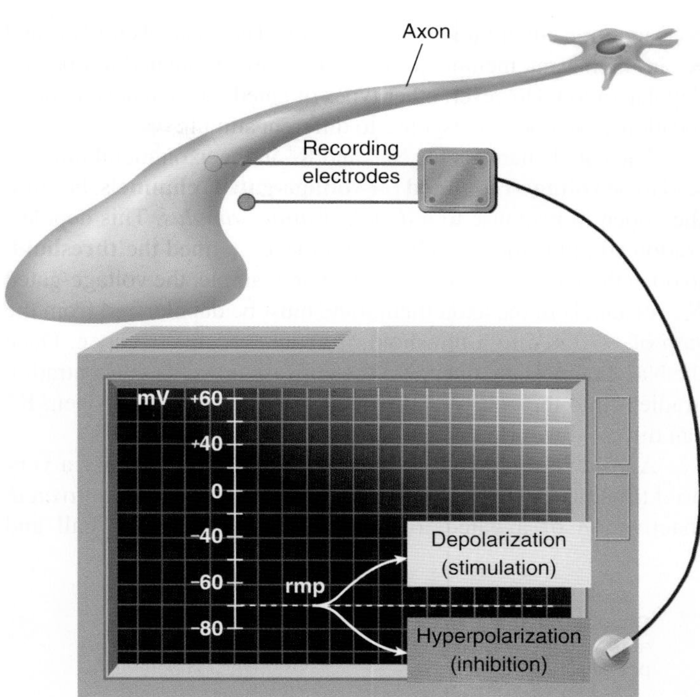

FIGURE 4.9 Depolarization and hyperpolarization. Depolarization is seen as a deflection upward from the resting membrane potential (rmp), whereas hyperpolarization is seen as a deflection downward from the rmp. The oscilloscope screen depicted here measures membrane potential in millivolts (mV).

Neurons and muscle cells are classified as "excitable," meaning that they can respond to a stimulus by producing impulses, which are action potentials. In neurons, these action potentials are the way information is carried over long distances by axons. In muscle cells, action potentials are the signals that trigger contraction. This is true of all types of muscles, including cardiac muscle and smooth muscle. So, when you learn the nature of action potentials, you are not only laying a foundation for the understanding of neural and skeletal muscle physiology, you are also building a foundation for the later study of the cardiovascular system, digestive system, and all of the other body systems.

As described in chapter 3, all cells maintain a membrane potential, or difference in charge across the plasma membrane. When this is a membrane potential in a neuron or muscle cell that is not producing impulses, it's called a *resting membrane potential* (*rmp*). The amount of the potential difference (charge difference) is measured in voltage, and a sign is placed in front of the voltage number to indicate which pole (negative or positive) is the inside of the cell. A typical resting membrane potential for mammalian neurons is −70 mV (fig. 4.9).

Measurement of the membrane potential requires two wires, or leads—one placed inside the plasma membrane and one placed outside. These leads can be connected to an oscilloscope, which indicates the voltage and quickly graphs rapid changes in the membrane potential. If the membrane potential decreases (goes toward zero voltage), the change in membrane potential is called a **depolarization**. If the membrane potential increases (goes to a higher voltage that is even more negative on the inside), the change is called a **hyperpolarization**. As illustrated in figure 4.9, the oscilloscope shows a depolarization as a deflection upward, and a hyperpolarization as a deflection downward, from the resting membrane potential.

Gated Ion Channels in the Axon Are Voltage-Regulated

The resting membrane potential, and the ability of neuron and muscle cells to change their membrane potential in response to stimulation, depend on the permeability of the plasma membrane to inorganic ions, chiefly Na^+ and K^+. As described in chapter 3, the plasma membrane contains protein channels for inorganic ions that may be gated or ungated. The membrane contains only gated channels for Na^+, but contains both gated and ungated channels for K^+. Because the gated Na^+ channels are closed in the resting membrane, but the ungated K^+ channels are open, the resting membrane potential is most greatly influenced by the difference in K^+ concentration between the inside and outside of the cell.

The resting membrane potential also requires the activity of the Na^+/K^+ (ATPase) pumps, which work constantly to pump Na^+ out of and K^+ into the cell, against their concentration gradients. In the absence of stimulation, the resting membrane potential will be maintained for as long as the cell is alive and has ATP to power the pumps. However, when neuron or muscle cells are stimulated, specific gated ion channels open, allowing the ions to diffuse from higher to lower concentration either into or out of the cell. The channels for Na^+ and K^+ in the plasma membrane of the axon are stimulated to open by depolarization. However, other types of gated ion channels, in other locations, can open in response to different stimuli.

The gated channels for Na^+ and K^+ in the axon membrane are said to be **voltage-regulated** (or **voltage-gated**) **channels**, because they open in response to a *depolarization stimulus*. This depolarization stimulus must reach a certain level, termed the **threshold**, to open the voltage-gated channels. For example, the voltage-gated Na^+ channels in the axon membrane must be depolarized from the rmp of −70 mV to a threshold of about −55 mV to open. Once the Na^+ channels are open, Na^+ can diffuse down its concentration gradient into the cell. When voltage-gated K^+ channels open, K^+ can diffuse down its concentration gradient out of the cell.

A voltage-regulated ion channel stays open for only a very brief time (less than a millisecond) before it becomes *inactivated*. Inactivation of a gated ion channel may involve a "ball and chain" type polypeptide that blocks the channel (fig. 4.10), or it may involve a different type of molecular rearrangement. The difference between a closed channel and an inactivated one is that the closed channel can be opened by a depolarization stimulus, whereas an inactivated channel cannot. However, the channel is inactivated only for a brief time (about 1 millisecond), and then it changes back to the closed state (fig. 4.10). It is now ready to open in response to the next depolarization stimulus that reaches threshold.

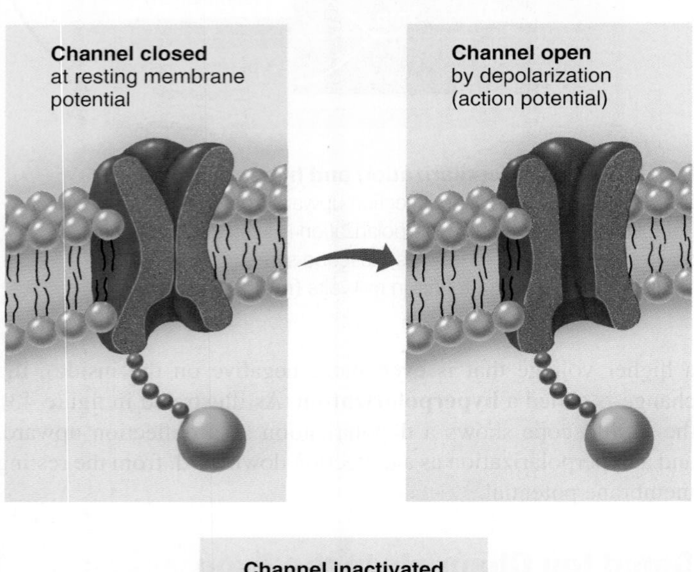

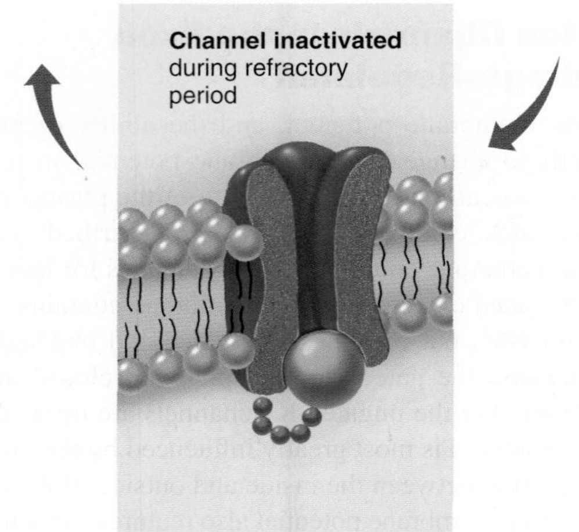

FIGURE 4.10 A model of a voltage-gated ion channel. The channel is closed at the resting membrane potential but opens in response to a threshold level of depolarization. This permits the diffusion of ions required for action potentials. After a brief period of time, the channel is inactivated by the "ball and chain" portion of a polypeptide chain.

Action Potentials Are Nerve Impulses

The **action potential**, or *nerve impulse*, generally starts at the initial segment of the axon, at the axon hillock. The stimulus for an action potential is depolarization of the axon's plasma membrane; this is because the opening of voltage-regulated Na^+ and K^+ channels produces the action potential. The depolarization stimulus for the action potential is produced originally at the synapse, as will be described later. However, to start the story of action potentials, it's easier to begin with a depolarization produced artificially by an imaginary device that injects positive charges into the axon hillock (fig. 4.11).

If the depolarization is not sufficient to reach threshold (about −55 mV), the voltage-regulated ion channels remain closed and no action potential will be produced. The depolarization will gradually decay (fig. 4.11), disappearing after it travels a very short distance

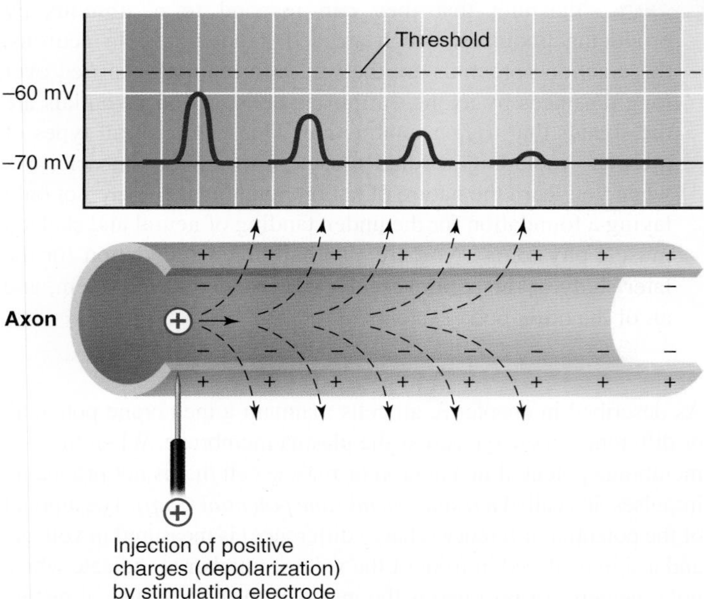

FIGURE 4.11 Cable properties of an axon. The cable properties of an axon are the properties that permit it to conduct potential changes over distances. If a stimulating electrode injects positive charges and produces a depolarization (*blue*) at one point in the axon, the depolarization will quickly dissipate if it doesn't trigger an action potential. The decreasing amplitude of the depolarization is due to leakage of charges through the axon membrane (dashed *arrows*). This results in a poor ability of the axon to conduct changes in potential over distances.

of about 1–2 millimeters (mm). Analogous to the conduction of electricity by a wire, the ability of a potential change to spread along the axon in this way is referred to as the *cable properties* of the axon. The spread of a potential change over distance is short because charges can leak through the plasma membrane, and because there is internal resistance of the axon cytoplasm to the spread of charges. As a result, an axon conducts charges very poorly compared to the conduction of electricity by a wire.

Sequential Opening and Closing of Na^+ and K^+ Channels Produces an Action Potential

The story is very different if the depolarization stimulus reaches threshold. Once this occurs, the depolarization stimulates a sequence of changes in the voltage-gated Na^+ and K^+ channels. These changes occur in the following sequence (fig. 4.12):

1. Voltage-gated Na^+ channels open, and Na^+ diffuses down its concentration gradient *into* the axon. The entry of Na^+ causes the membrane potential at that small region of axon plasma membrane to become even more depolarized. This is a positive feedback effect, causing the membrane potential to shoot past zero and actually reverse polarity (reaching about +30 mV at its peak) at the stimulated region of the axon.
2. The voltage-gated Na^+ channels then become inactivated, and at the same time the voltage-gated K^+ channels open. This allows K^+ to diffuse down its concentration gradient *out of* the axon. As a result, the membrane potential is brought back to the rmp of −70 mV.

Step 1 results in the entry of some Na^+ into the axon; step 2 results in the loss of some K^+ from the axon. These sequential changes in ion movements produce a sequence of change in the membrane potential. First, there is a rapid change from the rmp of −70 mV to +30 mV; this is referred to as *depolarization*. Second, there is a reversal of the membrane potential back to the rmp; this is called *repolarization* (fig. 4.13). The changes in the voltage-gated Na^+ and K^+ channels, the ion movements that result, and the change in the membrane potential produced by these ion movements constitute the action potential. Notice that these steps do not directly use ATP. However, as previously discussed, ATP is required for active transport by the Na^+/K^+ (ATPase) pumps to maintain the concentration gradients needed for these events. Also, ATP is used by the pumps to restore the concentration of Na^+ and K^+ after the action potential is completed.

CLINICAL INVESTIGATION CLUES

Remember that Lisa ate mussels gathered during a red tide. She may have ingested saxitoxin, a poison with a very specific action: it blocks voltage-gated Na^+ channels.

- What effect would saxitoxin's action have on the production of action potentials?
- Given the mechanism of saxitoxin action, how would this influence muscle contraction?
- How do these effects relate to Lisa's symptoms?

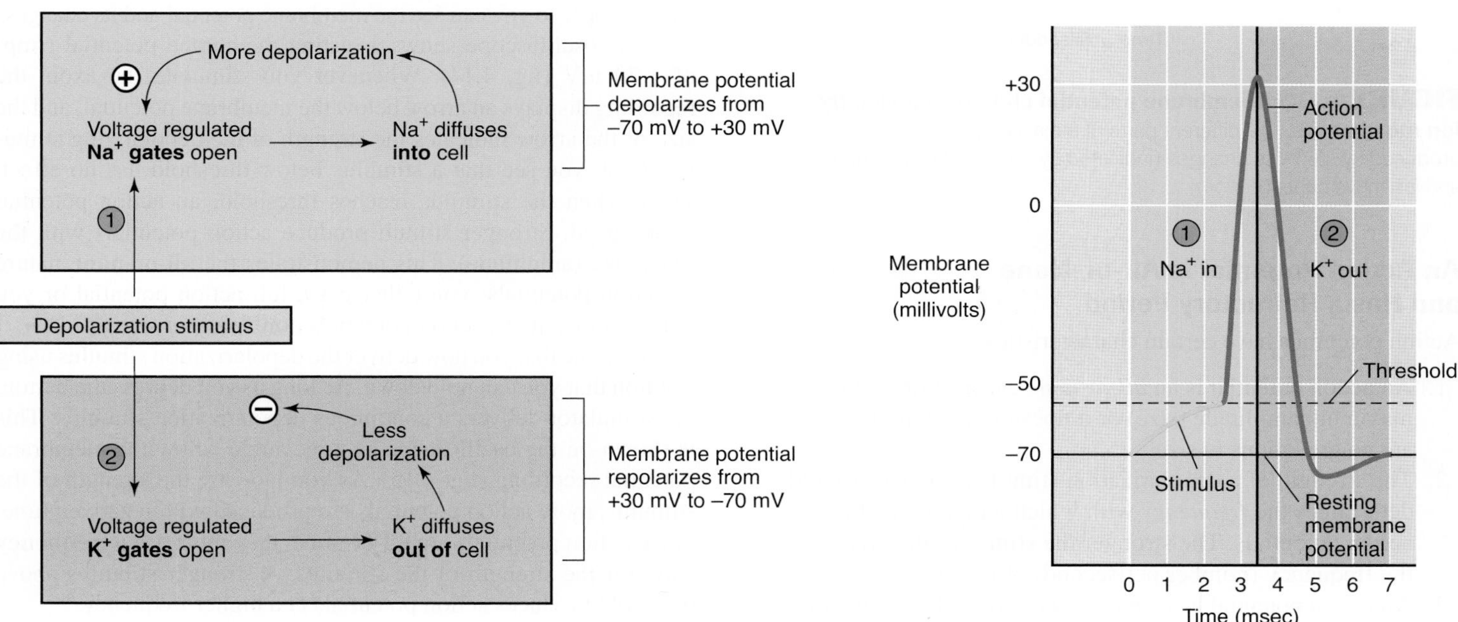

FIGURE 4.12 Depolarization of an axon affects Na^+ and K^+ diffusion in sequence. (*1*) Na^+ gates open and Na^+ diffuses into the cell. (*2*) After a brief period, K^+ gates open and K^+ diffuses out of the cell. An inward diffusion of Na^+ causes further depolarization, which in turn causes further opening of Na^+ gates in a positive feedback (+) fashion. The opening of K^+ gates and outward diffusion of K^+ makes the inside of the cell more negative, and thus has a negative feedback effect (−) on the initial depolarization. Step 1 produces the rising phase of the action potential, and step 2 produces the falling phase.

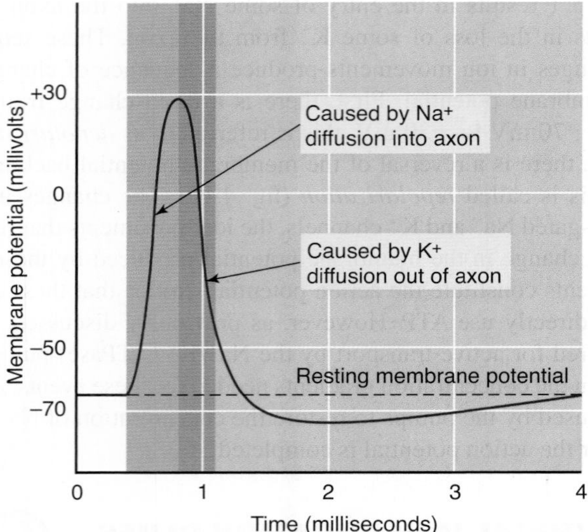

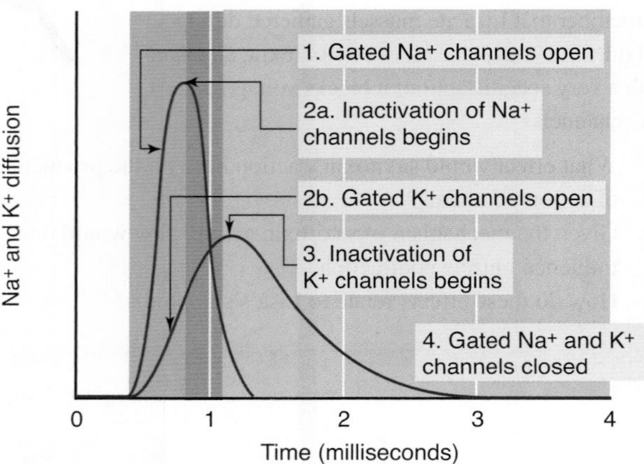

FIGURE 4.13 Membrane potential changes caused by ion movements. The different parts of the action potential (*top*) are produced by stepwise changes (indicated by numbers) in ion channels and ion movements (*bottom*).

An Action Potential Is All-or-None and Has a Refractory Period

Action potentials have certain characteristics:

1. An action potential is an *all-or-none* event. It fires to its maximum amplitude (size) or it doesn't fire at all; there are no gradations in size.
2. The strength of a depolarization stimulus above threshold determines the *frequency* with which a membrane fires action potentials. The stronger the stimulus, the greater the frequency (number per second) of action potentials.
3. An action potential has a *refractory period*. This is the time when the plasma membrane cannot produce a second action potential after producing a first one. The refractory period prevents action potentials from adding together or running together continuously.

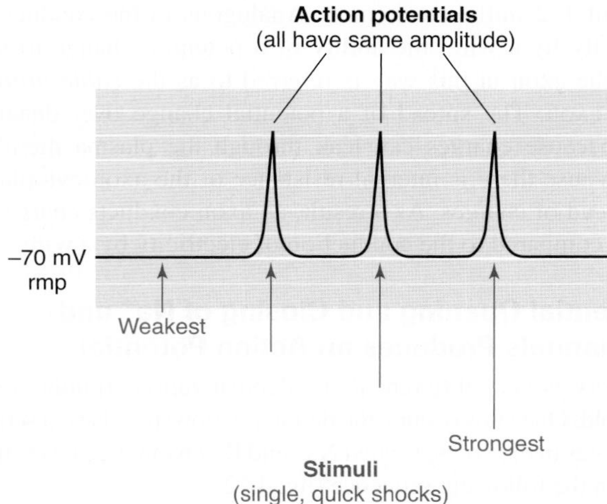

FIGURE 4.14 The all-or-none law of action potentials. A single, quick shock delivered to an axon can serve as a depolarizing stimulus. If the stimulus is below threshold, no action potential is produced by the axon. Once the stimulus has reached threshold, a full action potential is produced. Any greater stimulus does not produce greater action potentials. Thus, action potentials are not graded (varied); they are all-or-none.

Imagine that you are injecting positive charges into the axon hillock to depolarize it. You have a toggle switch you can depress; whenever it clicks, a brief depolarization stimulus is delivered to the axon. Further, you have a knob you can turn to increase the strength of the depolarization stimulus. The axon hillock also has recording electrodes (one inside the axon and one outside) connected to an oscilloscope, so you can see the membrane potential and its changes.

The oscilloscope shows a resting membrane potential (rmp) of −70 mV (fig. 4.14). Whenever you stimulate the axon, the recording displays an arrow below the membrane potential, and the size of the arrow indicates the strength of the depolarizing stimulus. First, you see that a stimulus below threshold has no effect. Then, when the stimulus reaches threshold, an action potential is produced. Stronger stimuli produce action potentials with the same size (amplitude). This demonstrates the **all-or-none** nature of action potentials: you either get a full action potential or you don't get one at all; action potentials don't vary in size.

Imagine that you now deliver the depolarization stimulus using a button that you can hold down. As long as you depress the button, the stimulator delivers a continuous depolarization stimulus. This is shown on the oscilloscope as a rectangle below the membrane potential recording (fig. 4.15). As you increase the strength of the stimulus, more action potentials are produced within a given time. Since action potentials are all-or-none, they must use a **frequency code** for the strength of the stimulus. A stronger stimulus above threshold produces action potentials at a higher frequency.

Finally, action potentials have a **refractory period** (fig. 4.16). This means that the plasma membrane of the stimulated region of the axon is refractory (cannot respond) to another stimulus for a period of time. The *absolute refractory period* (when the membrane

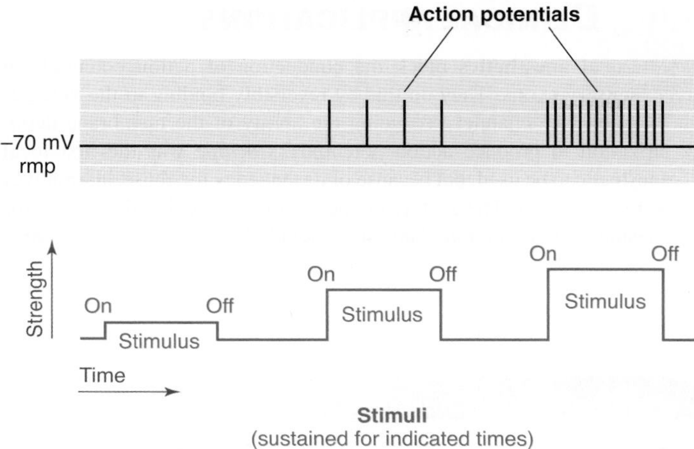

FIGURE 4.15 The effect of stimulus strength on action-potential frequency. Stimuli that are sustained for a period of time are given to an axon. In the first case, the stimulus is weaker than required to reach threshold, and no action potentials are produced. In the second case, a stronger stimulus is delivered, which causes the production of a few action potentials while the stimulus is sustained. In the last case, an even stronger stimulus produces a greater number of action potentials in the same time period. This demonstrates that stimulus strength is coded by the frequency (rather than the amplitude) of action potentials.

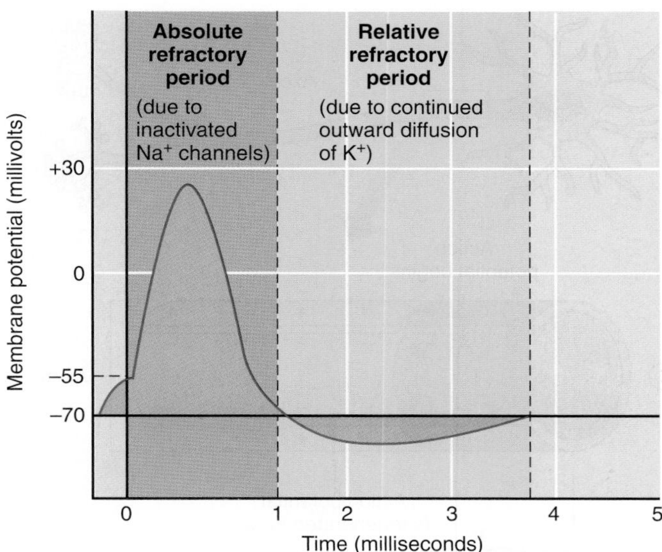

FIGURE 4.16 Absolute and relative refractory periods. While a segment of axon is producing an action potential, the membrane is absolutely or relatively resistant (refractory) to further stimulation.

absolutely cannot respond to a second stimulus) occurs during the production of the action potential, and results from the inactivation of the Na^+ channels. The *relative refractory period* occurs immediately following the action potential, when the continued outward diffusion of K^+ causes the membrane potential to overshoot the rmp and actually hyperpolarize for a brief time (fig. 4.16). A second stimulus would have to be unusually strong to overcome this and trigger an action potential.

Because of the refractory period:

1. Action potentials cannot run together and summate. They remain separate, all-or-none events that can code for the strength of a stimulus by their frequency of firing.
2. Action potentials cannot backtrack. An action potential produced in the axon hillock can go toward the axon terminals, but not backward. An action potential that arrives at the axon terminal cannot reverse direction, because the membrane behind it is still in its refractory period.

Action Potentials Are Regenerated Along the Axon

When the cable properties of axons were discussed, it was mentioned that axons are poor conductors (see fig. 4.11). If the axon had to conduct impulses by their cable properties, the way a wire conducts electricity, the axon couldn't be much longer than 1 mm. Our nervous systems have axons that are a meter or more in length, so it seems evident that they must conduct action potentials differently. In fact, they don't exactly conduct the action potential; rather, they regenerate it along the length of the axon. As a result, action potentials are **conducted without decrement**—they don't decrease in amplitude as they are regenerated along the axon from the axon hillock to the axon terminals.

Remember, an action potential is actually a stepwise opening of voltage-gated Na^+ and K^+ channels at a small patch of axon membrane that has been depolarized to threshold. This produces a stepwise inward diffusion of Na^+ (causing further depolarization) followed by an outward diffusion of K^+ (causing repolarization). It's also important to realize that an action potential is a *local event*—it occurs only in the tiny stimulated region of the axon membrane, not in the rest of the axon. The reversal of the membrane potential (to +30 mV at the "top" of the action potential) is produced only at the stimulated region of axon membrane. Farther along the axon, the plasma membrane is still maintaining its resting membrane potential of −70 mV.

At this instant, there is a potential difference between the stimulated region of the axon and the adjacent region that has not yet been stimulated. This results in movements of charges that can depolarize the next region along the axon. In an *unmyelinated axon*, this movement of charges can bring the very next adjacent region of plasma membrane to a threshold depolarization. Thus, in an unmyelinated axon, this next adjacent region of the axon will be stimulated to reproduce the action potential (fig. 4.17). These events repeat from the first segment of axon to the last.[1]

FYI [1]Conduction of action potentials is much like conduction of the "wave" by spectators in a stadium. Just as a spectator doesn't actually move from one seat to another, the action potential doesn't actually move down the axon. Each spectator stands up and raises her arms and then sits down as the next spectator stands up. Each patch of axon membrane depolarizes and repolarizes, and stimulates the next patch to depolarize and repolarize. We talk about action potential conduction, but our mental image should be more like the traveling wave in a stadium.

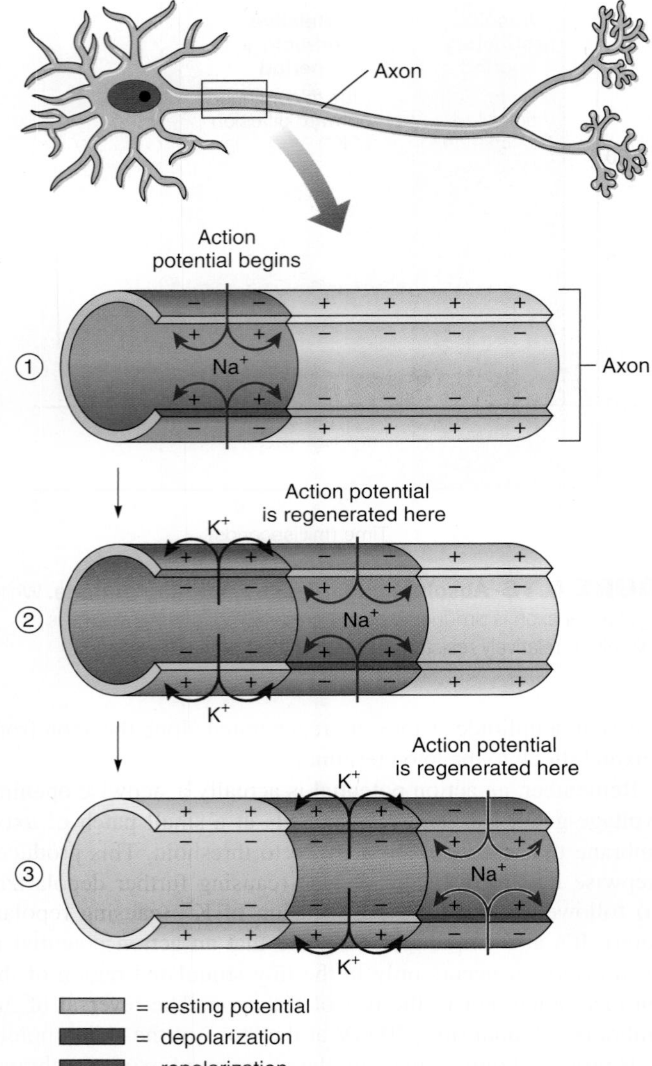

FIGURE 4.17 The conduction of action potentials in an unmyelinated axon. Each action potential "injects" positive charges that spread to adjacent regions. The region that has just produced an action potential is refractory. The next region, not having been stimulated previously, is partially depolarized. As a result, its voltage-regulated Na^+ gates open and the process is repeated. Successive segments of the axon thereby regenerate, or "conduct," the action potential.

Conduction of action potentials in a *myelinated axon* is similar, but differs in that a myelinated axon can produce action potentials only at the nodes of Ranvier (fig. 4.18). This is called **saltatory conduction**, because the action potentials seem to jump from node to node. Each node must be depolarized to threshold by an action potential produced at the previous node. Thus, a myelinated axon reproduces the action potential fewer times than would an equal length of unmyelinated axon. Because conduction of charges from one node to the next by cable properties is faster than the production of an action potential, saltatory conduction of action potentials by myelinated axons is faster than the conduction of action potentials by unmyelinated axons.

CLINICAL APPLICATIONS

Local anesthetics block the conduction of action potentials in sensory axons. They do this by reversibly binding to the voltage-gated Na^+ channels, reducing the ability of the membrane depolarization to produce action potentials. *Cocaine* was the first local anesthetic to be used, but because of its tendency for abuse, alternatives were developed. The first synthetic analog of cocaine developed for anesthesia was *procaine*; later ones include *lidocaine* and *tetracaine*.

CHECK POINT

1. In a stepwise manner, describe the changes that occur in the voltage-gated Na^+ channels at the axon membrane when it is depolarized to threshold. Correlate this with the changes in membrane potential that occur.
2. In a stepwise manner, describe the changes that occur in the voltage-gated K^+ channels at the axon membrane when an action potential is produced, and correlate these with the changes in membrane potential that occur.
3. Explain the all-or-none nature of action potentials, and how the strength of a stimulus is coded by action potential frequency.
4. What is the refractory period? What is its significance?

Neurons Regulate Other Cells at Synapses

In most synapses, a chemical neurotransmitter is released by the axon terminals. This neurotransmitter changes the permeability of the postsynaptic membrane to specific ions, causing the postsynaptic cell to become depolarized or hyperpolarized. Depolarization can then stimulate the production of action potentials, whereas hyperpolarization inhibits action potential production by the postsynaptic cell.

An axon terminal makes a functional connection, or **synapse**, with another cell, called a **postsynaptic cell**. If the postsynaptic cell is another neuron, the synapse is generally with a dendrite or the cell body of the postsynaptic neuron. However, the axons of motor neurons make synapses with muscle or gland cells. In the most common type of synapse, there is a tiny space, the **synaptic cleft** (or **synaptic gap**), separating the axon terminal from the postsynaptic cell (fig. 4.19).

When action potentials depolarize the plasma membrane of the axon terminal, voltage-gated Ca^{2+} channels open. This permits Ca^{2+} to diffuse down its concentration gradient into the cytoplasm, where it stimulates the release of a **chemical neurotransmitter**. Chemical neurotransmitter molecules are stored in **synaptic vesicles**, and the entry of Ca^{2+} stimulates the *exocytosis* of these vesicles so that the neurotransmitter can be released into the synaptic cleft (fig. 4.20). The greater the frequency of action

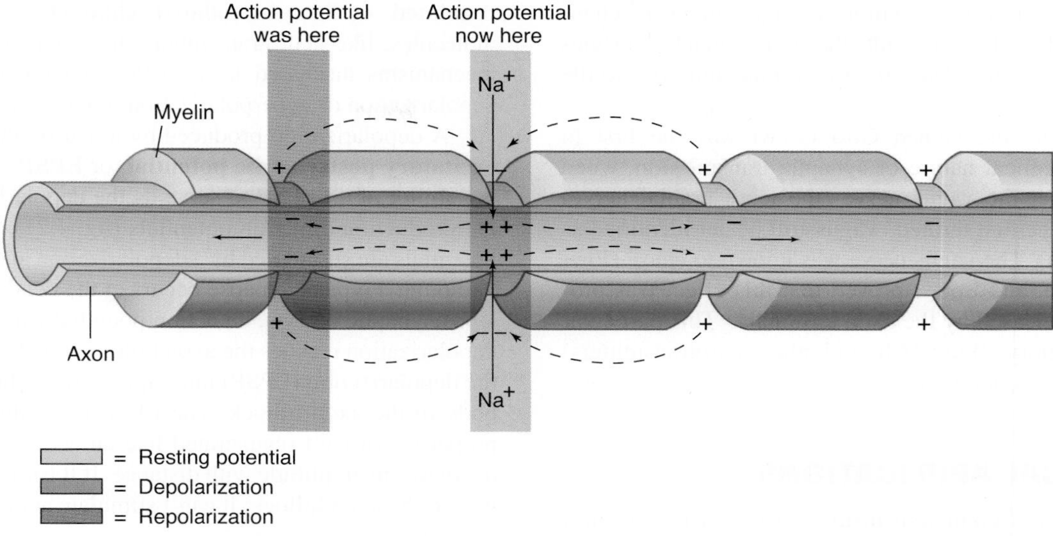

FIGURE 4.18 The conduction of a nerve impulse in a myelinated axon. Since the myelin sheath prevents inward Na^+ current, action potentials can be produced only at gaps in the myelin sheath called the nodes of Ranvier. This "leaping" of the action potential from node to node is known as saltatory conduction.

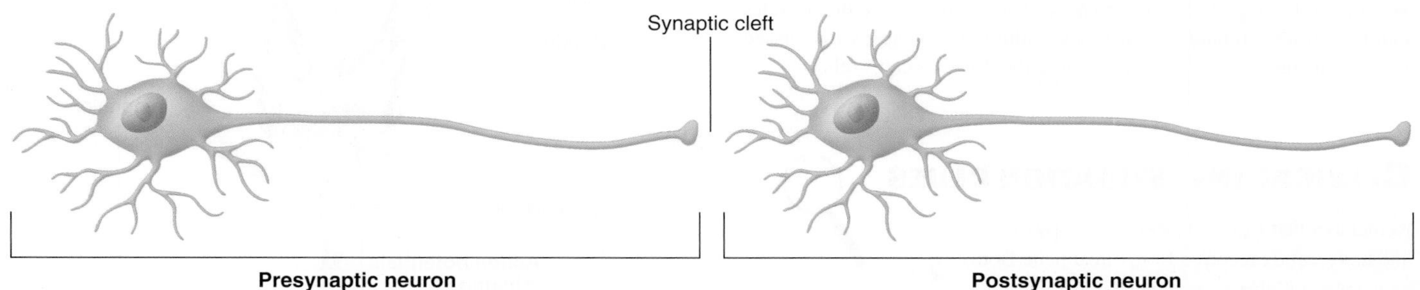

FIGURE 4.19 Presynaptic and postsynaptic neurons. When one neuron makes a synapse with another, the first is the presynaptic neuron and the second is the postsynaptic neuron.

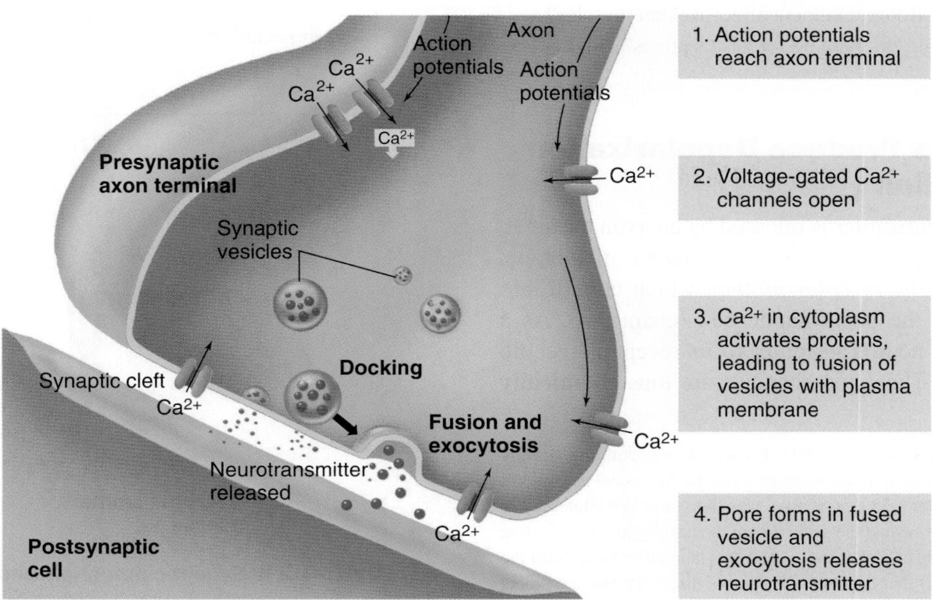

FIGURE 4.20 The release of neurotransmitter. Action potentials, by opening Ca^{2+} channels, stimulate the fusion of docked synaptic vesicles with the cell membrane of the axon terminals. This leads to exocytosis and the release of neurotransmitter, as indicated in the numbered steps.

potentials at the axon terminal, the more voltage-gated Ca^{2+} channels will be opened, and, as a result, the more synaptic vesicles will undergo exocytosis to release their neurotransmitters into the synaptic cleft.

An Austrian scientist named Otto Loewi was the first to demonstrate the chemical nature of synaptic transmission, when he demonstrated that the vagus nerve (the tenth cranial nerve) releases a chemical he first termed "vagusstoff," later identified as **acetylcholine** (**ACh**). The vagus nerve was known to cause slowing of the heart rate, and Loewi proved that this effect was produced by the ACh released by the axon terminals.[2] This particular example is an inhibitory effect; ACh (and other neurotransmitters) can also produce excitatory effects.

CLINICAL APPLICATIONS

Tetanus toxin and **botulinum toxin** are bacterial products that cause paralysis by preventing neurotransmission. These neurotoxins function as protein-digesting enzymes, digesting molecules that allow synaptic vesicles to fuse with the plasma membrane of the axon terminals for exocytosis of neurotransmitters. Botulinum toxin prevents the release of ACh, causing *flaccid paralysis* (where the muscles can't contract). Tetanus toxin blocks inhibitory synapses (discussed later), causing *spastic paralysis* (where the muscles can't relax).

CLINICAL INVESTIGATION CLUES

Remember that Lisa had ptosis (a droopy eyelid) and had recently received Botox treatment. Botox contains botulinum toxin, a poison that interferes with the ability of synaptic vesicles to dock and undergo exocytosis. Ptosis is a common adverse reaction to Botox treatment.

- Given that ACh is the neurotransmitter that stimulates skeletal muscles, how would Botox treatment affect skeletal muscles?
- What other reasons might there be for Lisa's ptosis?

Neurotransmitters Produce Depolarization or Hyperpolarization

When a particular neurotransmitter is released by an axon terminal, it diffuses across the synaptic gap to the postsynaptic membrane. There, it binds to its specific *receptor protein*, which is built into the plasma membrane of the postsynaptic cell (chapter 3). As a result of the binding of a neurotransmitter to its receptor, specific gated ion channels open. These ion channels are thus **chemically regulated**, or **ligand-regulated**, **channels**. (Ligands are smaller molecules, like neurotransmitters that bind to proteins.) Through mechanisms discussed later in this chapter, this leads to either depolarization or hyperpolarization of the postsynaptic membrane.

A depolarization produced by a neurotransmitter is called an **excitatory postsynaptic potential**, or **EPSP**. This name refers to the ability of an EPSP to serve as the depolarization stimulus for the production of action potentials (fig. 4.21). (Remember, action potentials are stimulated by a depolarization at or above threshold.) Although a typical synapse occurs at the dendrites or cell body of a postsynaptic neuron, an action potential isn't produced until the depolarization reaches the axon hillock (fig. 4.22). This means that the depolarization (EPSP) must spread from the dendrites and cell body to the axon hillock. The EPSP is conducted only by cable properties (it isn't regenerated like an action potential), and so it decreases in amplitude with distance. If it isn't at threshold when it gets to the axon hillock, it can't stimulate action potentials.

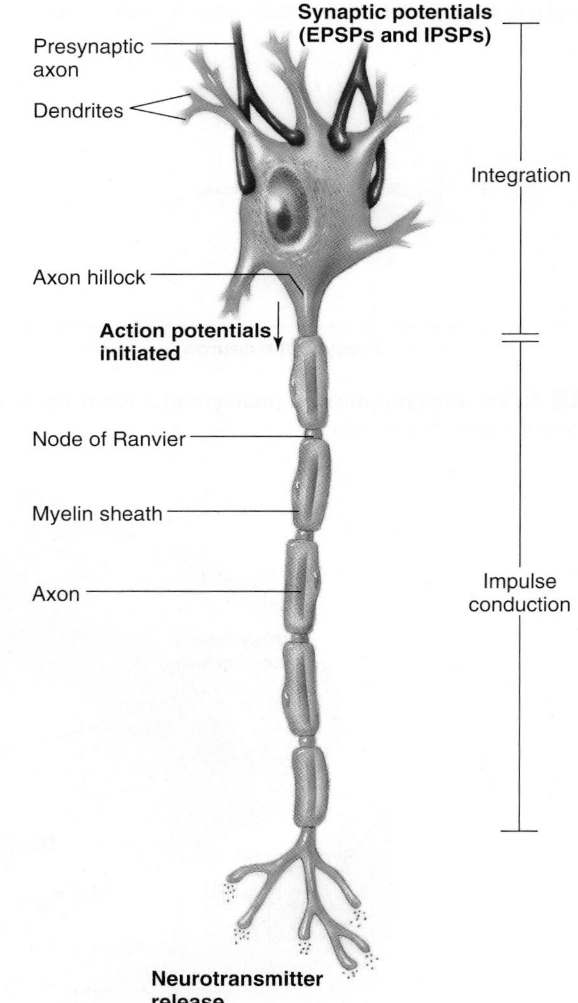

FIGURE 4.21 Synaptic potentials and action potentials in a neuron. The dendrites and cell body are specialized to receive stimulation and produce synaptic potentials (EPSPs and IPSPs). Action potentials (nerve impulses) are produced first at the axon hillock and conducted along the axon.

FYI [2]On the evening of Easter Saturday, 1921, Loewi had a dream and sleepily wrote down some notes, but at 6 a.m. when he looked at his scrawl he couldn't read his own handwriting. He later said that Sunday was the "most desperate day in my whole scientific life." Luckily, he had the same dream the next night, and this time didn't take any chances. He awoke at 3 a.m. and went to his laboratory to perform the experiment (completed by 5 a.m.) that eventually led to his discovery and the Nobel Prize. Loewi won the Nobel Prize in Physiology or Medicine in 1936, but was forced to use his Nobel Prize money to flee the Nazis in 1938; he became a professor at New York University in 1940.

Other neurotransmitters produce a hyperpolarization of the postsynaptic membrane, called an **inhibitory postsynaptic potential (IPSP)**. The hyperpolarization is inhibitory because a bigger depolarization is then required to reach threshold. Like an EPSP, an IPSP decreases in amplitude as it travels toward the axon hillock.

Synaptic potentials (EPSPs and IPSPs) have these characteristics:

1. *Synaptic potentials decrease in amplitude with distance*, because they aren't regenerated. This follows from the way they're produced from the opening of chemically regulated, rather than voltage-regulated, channels. They can only be produced at the postsynaptic membrane where neurotransmitter chemicals bind to their membrane receptor proteins.
2. *Synaptic potentials are graded*, not all-or-none. Fewer neurotransmitter molecules released (from fewer synaptic vesicles undergoing exocytosis) produce a smaller depolarization; more produce a larger depolarization.
3. *Synaptic potentials do not have a refractory period and can summate.* Because EPSPs don't have refractory periods, they can "ride piggyback" on each other and produce a larger summated depolarization. Alternatively, an EPSP and IPSP can interact in such a way that the hyperpolarization subtracts from the depolarization.

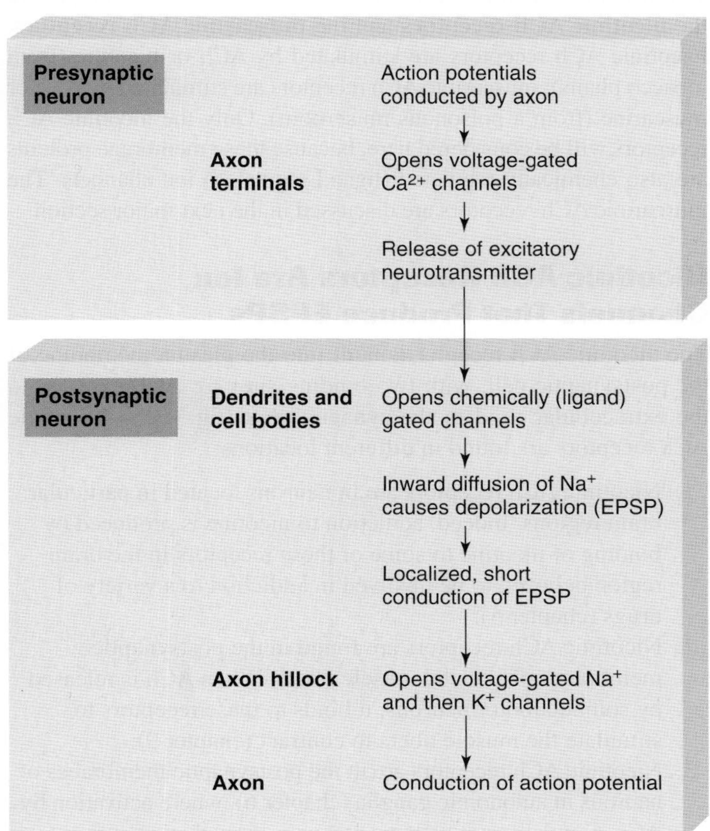

FIGURE 4.22 The sequence of events in synaptic transmission. The different regions of the postsynaptic neuron are specialized, with ligand (chemically) gated channels located in the dendrites and cell body, and voltage-gated channels located in the axon.

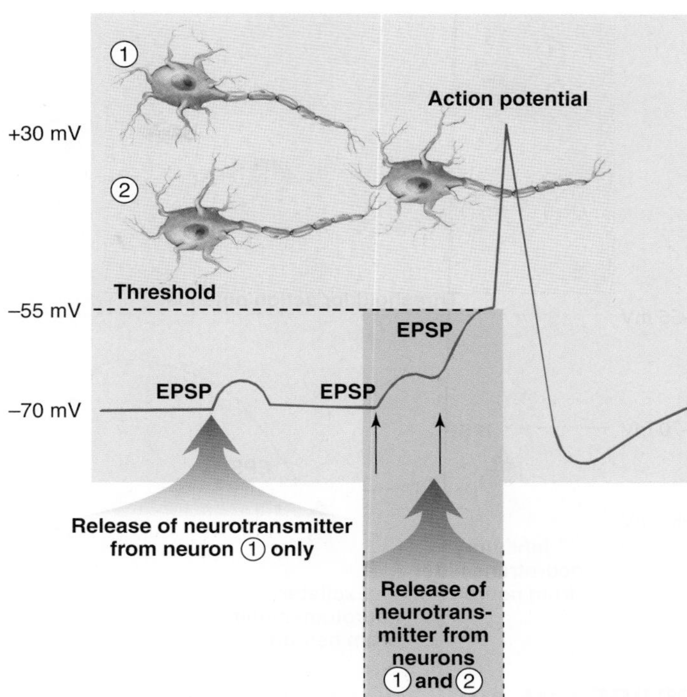

FIGURE 4.23 Spatial summation. When only one presynaptic neuron releases excitatory neurotransmitter, the EPSP produced may not be sufficiently strong to stimulate action potentials in the postsynaptic neuron. When more than one presynaptic neuron produces EPSPs at the same time, however, the EPSPs can summate at the axon hillock to produce action potentials.

Summation of Synaptic Potentials Integrates Information

Multipolar neurons in the CNS have extensively branched dendritic trees, so they can receive many presynaptic inputs. Suppose a particular neuron receives synaptic inputs from two presynaptic neurons. If only one of these presynaptic neuron releases an excitatory neurotransmitter and produces an EPSP, the depolarization may not reach threshold by the time it arrives at the axon hillock. However, if both presynaptic neurons release excitatory neurotransmitters at the same time, two different EPSPs will be produced at different locations in the postsynaptic neuron. These can add together at the axon hillock to produce a stronger depolarization able to stimulate action potentials in the postsynaptic neuron (fig. 4.23). Summation of EPSPs produced in different locations in the postsynaptic neuron is called **spatial summation**.[3]

Many of these presynaptic inputs can also be inhibitory. That is, the synapses can release inhibitory neurotransmitters that cause

FYI [3]As hard as it is to picture, motor neurons in the spinal cord can receive as many as 1,000 presynaptic axon terminals. This is necessary because the cell body of the motor neuron is much larger than the axons from the association neurons that synapse with it. So, a great amount of spatial summation of EPSPs is needed for the motor neuron to reach threshold and fire action potentials to stimulate muscle fibers to contract.

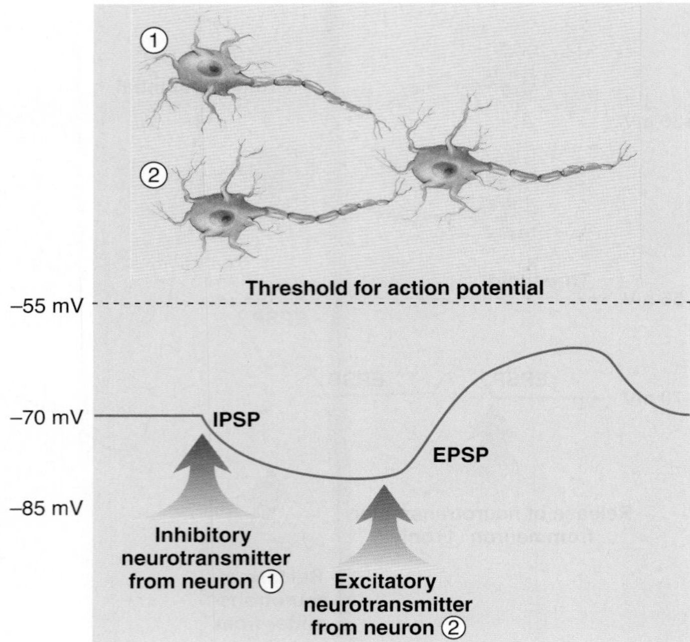

FIGURE 4.24 Postsynaptic inhibition. An inhibitory postsynaptic potential (IPSP) makes the inside of the postsynaptic membrane more negative than the resting potential—it hyperpolarizes the membrane. Therefore, excitatory postsynaptic potentials (EPSPs), which are depolarizations, must be stronger to reach the threshold required to generate action potentials at the axon hillock.

hyperpolarizations (IPSPs). In this case, the IPSPs can summate in a negative manner with the EPSPs. This decreases the strength of the depolarization, reducing the frequency of action potentials produced by the postsynaptic neuron or even inhibiting it completely (fig. 4.24). The IPSPs are said to produce **postsynaptic inhibition**. Inhibition is extremely important in the nervous system; for example, without it we would lose motor control as muscles became inappropriately stimulated in spastic paralysis.

CHECK POINT

1. Describe how axon terminals release chemical neurotransmitters.
2. Explain how and where EPSPs are produced, and how they differ from action potentials.
3. What are IPSPs, and what effect do they have in the nervous system?

Some Neurotransmitter Receptors Are Also Ion Channels

There is a family of neurotransmitter receptor proteins that are also ion channels. When the neurotransmitter binds to these receptors, an ion channel opens in the receptor protein. If the channel permits Na^+ to enter the cell, a depolarization (EPSP) is produced. If the channel allows Cl^- to enter, a hyperpolarization (IPSP) is produced. These chemically regulated (or ligand-regulated) channels include certain excitatory receptors for acetylcholine and receptors for GABA, an inhibitory neurotransmitter.

Now it may seem that we're getting down to some serious, detailed physiology. Actually, these are very "basic" physiological concepts, if we use the term *basic* as defined in an early chapter to mean "fundamental." How regulatory molecules, including neurotransmitters, signal cells to behave is fundamental to the physiology of all organs. So, as always in this text, remember: there is nothing here that you cannot understand, and the fundamental importance of this information is an exciting opportunity to add to the knowledge pyramid you are constructing.

The neurotransmitter acetylcholine (ACh) is released by diverse neurons; at some synapses, it is excitatory (producing EPSPs); at others, it's inhibitory (producing IPSPs). Some of the differences in these effects can be attributed to the existence of two kinds of ACh receptor proteins in the postsynaptic membranes of different cells. Somewhat illogically, these two receptor types for ACh are named according to specific poisons that (along with ACh) bind to and stimulate the receptors. Accordingly, the two ACh receptor types are the **nicotinic ACh receptors** and the **muscarinic ACh receptors**. Nicotinic ACh receptors are stimulated by ACh or nicotine (from tobacco plants); muscarinic ACh receptors are stimulated by ACh or muscarine (from a poisonous mushroom). Only the nicotinic ACh receptors will be considered here, because these membrane proteins are also chemically regulated (ligand-regulated) ion channels. The muscarinic ACh receptors are discussed in the next major section.

Nicotinic ACh Receptors Are Ion Channels That Produce EPSPs

The nicotinic ACh receptor is built into the plasma membrane of the postsynaptic cell, with two binding sites for ACh exposed to the extracellular fluid in the synaptic cleft (fig. 4.25). Nicotinic ACh receptors are found in different locations:

1. Nicotinic ACh receptors are in neurons located in particular brain regions. Indeed, addiction to nicotine is produced by binding of nicotine to some of these receptors in the brain region believed to be involved in addiction to a variety of drugs (chapter 5).
2. Nicotinic ACh receptors are found in the postsynaptic membranes of skeletal muscle fibers. When ACh is released by somatic motor neurons, it binds to these receptors to stimulate the muscle fibers to contract (chapter 9).
3. Nicotinic ACh receptors are in the postsynaptic membranes of neurons in autonomic ganglia (chapter 6), where activation by ACh stimulates sympathetic and parasympathetic neurons.

The nicotinic ACh receptor protein is composed of 5 polypeptide chains that surround an ion channel, which is closed until the receptor binds to 2 ACh molecules. When that occurs, the channel opens and Na^+ and K^+ diffuse at the same time in opposite

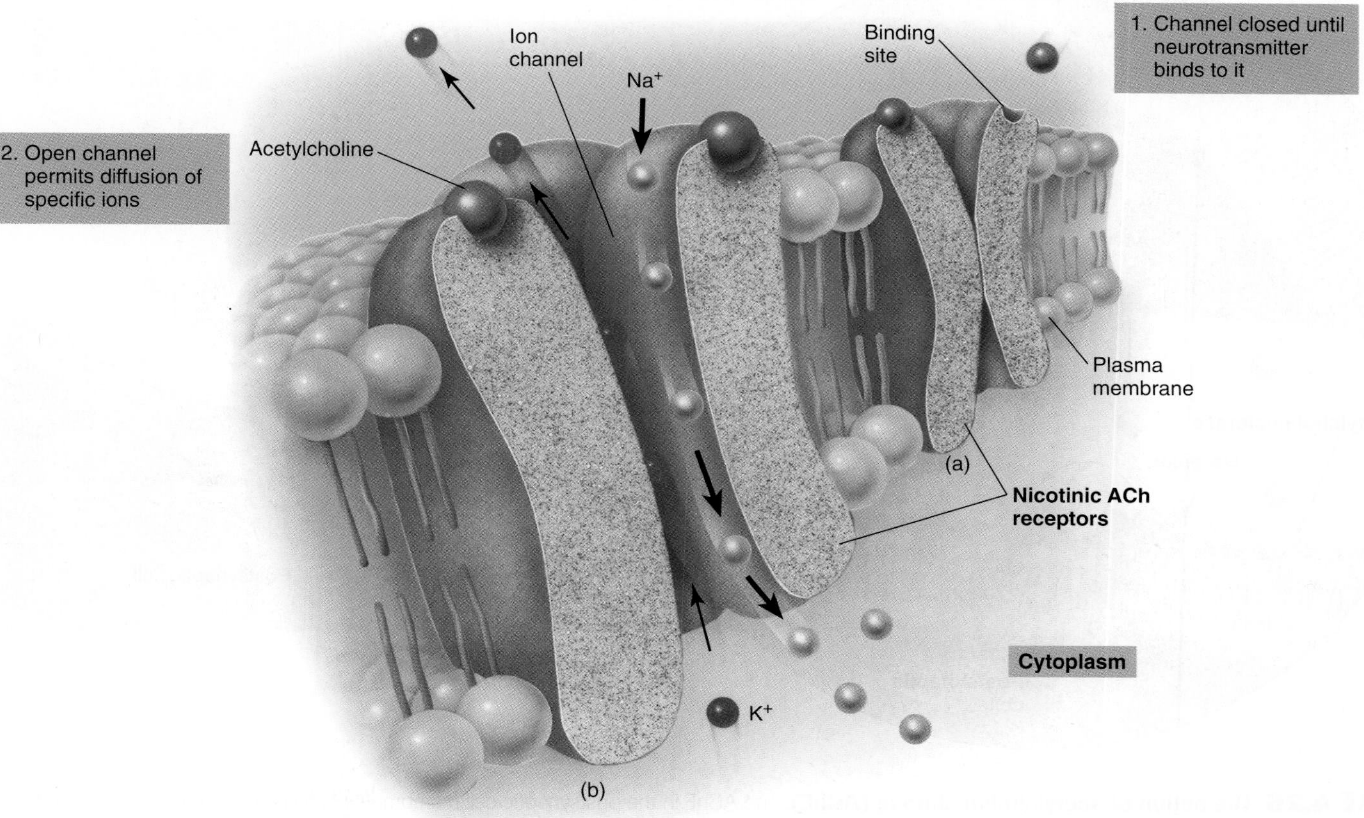

FIGURE 4.25 Nicotinic acetylcholine (ACh) receptors also function as ion channels. The nicotinic acetylcholine receptor contains a channel that is closed (*a*) until the receptor binds to ACh. (*b*) Na^+ and K^+ diffuse simultaneously, and in opposite directions, through the open ion channels. The electrochemical gradient for Na^+ is greater than for K^+, so that the effect of the inward diffusion of Na^+ predominates, resulting in a depolarization known as an excitatory postsynaptic potential (EPSP).

CLINICAL APPLICATIONS

The drug **curare** was derived from the resin of a South American tree and originally used for poison darts. It causes *flaccid paralysis* (where the muscles can't contract) by blocking the binding of ACh to its nicotinic receptors in skeletal muscles. Because of this, curare is sometimes used clinically to promote muscle relaxation. People with **myasthenia gravis** similarly can have flaccid paralysis due to antibodies, produced by their own immune systems, that block their muscle nicotinic ACh receptors. Diseases like myasthenia gravis that are caused by the patient's own immune system are known as *autoimmune diseases* (chapter 11).

directions down their concentration gradients (fig. 4.25). It might seem that these ion movements would cancel each other out—but that doesn't occur, because the *electrochemical gradient* (sum of the electrical and concentration gradients) for Na^+ is greater than for K^+. Thus, there is greater movement of Na^+ into, than of K^+ out of, the postsynaptic membrane. The predominant effect of Na^+ entry causes a depolarization, the EPSP. Thus, nicotinic ACh receptors are always excitatory.

The ACh molecules bind to their receptors only for a brief time and then they release. They may bind to a receptor again, to produce another EPSP, or they may instead bind to an enzyme present in the postsynaptic membrane. This enzyme, **acetylcholinesterase (AChE)**, breaks down ACh (fig. 4.26) and so stops the stimulation. Because of AChE, the presynaptic axon must continue to release additional ACh if the stimulation of the postsynaptic membrane is to continue.

CLINICAL APPLICATIONS

Nerve gas exerts its odious effects by inhibiting AChE in skeletal muscles. Because ACh is not degraded, it can continue to bind to its nicotinic receptor proteins and stimulate skeletal muscle contraction. This produces *spastic paralysis*. Clinically, inhibitors of AChE (*cholinesterase inhibitors*), such as neostigmine, are used to enhance the ability of ACh to stimulate muscle contraction when a patient has a disorder of neuromuscular transmission like *myasthenia gravis*.

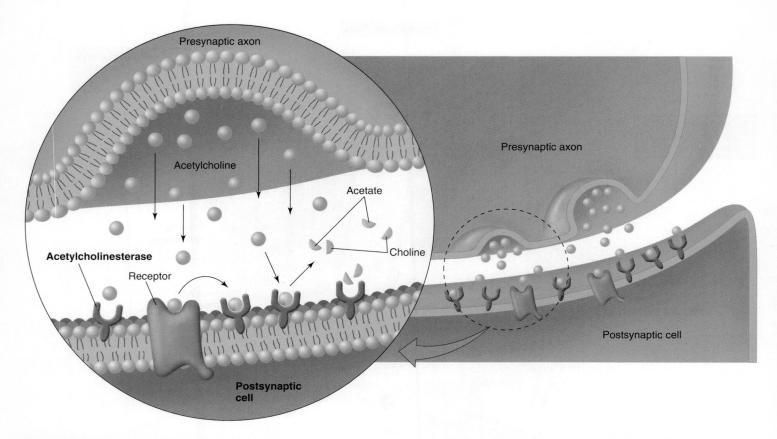

FIGURE 4.26 The action of acetylcholinesterase (AChE). The AChE in the postsynaptic cell membrane inactivates the ACh released into the synaptic cleft. This prevents continued stimulation of the postsynaptic cell unless more ACh is released by the axon.

CLINICAL INVESTIGATION CLUES

Remember that Lisa had myasthenia gravis, which can cause ptosis (droopy eyelid). She was treating her myasthenia gravis with prednisone, a corticoid drug that inhibits inflammation, and Prostigmin, a cholinesterase inhibitor (inhibitor of acetylcholinesterase).

- By what mechanism might a person with myasthenia gravis develop ptosis?
- What benefit might an anti-inflammatory drug like prednisone provide in treating myasthenia gravis?
- How would a cholinesterase inhibitor benefit a person with myasthenia gravis?

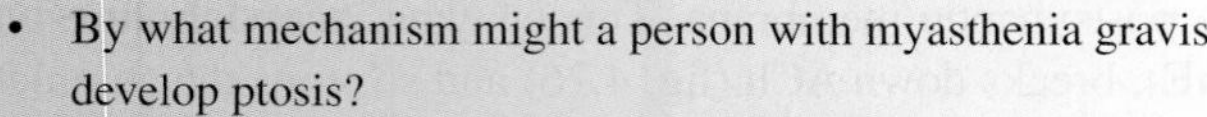

GABA Receptors Are Ion Channels That Produce IPSPs

Many other neurotransmitter chemicals were discovered after acetylcholine, some of which are excitatory and some inhibitory. **Gamma-aminobutyric acid (GABA)** is an important inhibitory neurotransmitter in the CNS. Its receptor is a member of the same structural family as the nicotinic ACh receptors, and works in a similar manner. When the neurotransmitter molecules of GABA bind to the receptor, an ion channel through the receptor opens. This chemically regulated channel is specific for *chloride* ion (*Cl^-*), which is negatively charged. The Cl^- diffuses down its concentration gradient into the postsynaptic cell (fig. 4.27). When this occurs, the postsynaptic membrane becomes hyperpolarized and an IPSP is produced.

Inhibition produced by GABA (and by a related neurotransmitter, *glycine*) is very important for the control of muscles by the CNS. For example, a deficiency of GABA-releasing neurons is responsible for the uncontrolled movements of people with *Huntington's disease*, a neurodegenerative disorder caused by a genetic defect. Also, people convulsing with muscular spasms are often given drugs that promote the ability of GABA to open Cl^- channels and promote the inhibition of spinal motor neurons.

CLINICAL APPLICATIONS

Benzodiazepines are drugs that increase the ability of GABA to activate its receptors in the brain and spinal cord. The intravenous infusion of benzodiazepines can inhibit the muscular spasms in seizures resulting from epilepsy or from drug overdose and poisons, because benzodiazepines promote GABA-induced inhibition of the spinal motor neurons (those that stimulate skeletal muscles to contract). The ability to fall asleep requires activity of GABA-releasing neurons (chapter 5), and so benzodiazepines such as *Valium* are given orally to treat anxiety and sleeplessness.

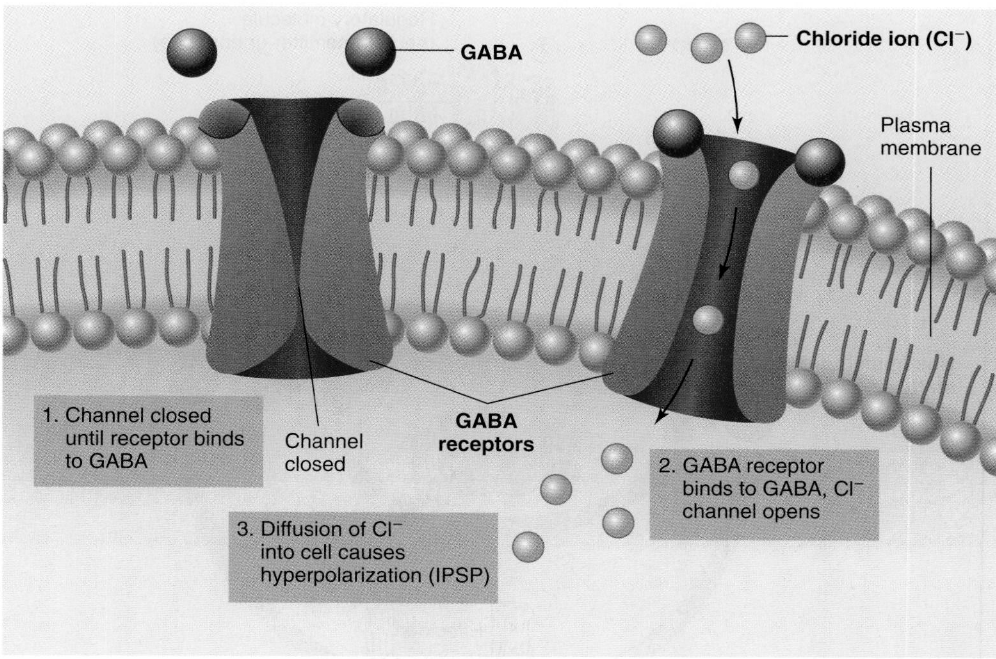

FIGURE 4.27 GABA receptors contain a chloride channel. When GABA (gamma-aminobutyric acid) binds to its receptor, a chloride ion (Cl^-) channel opens through the receptor. This permits the inward diffusion of Cl^-, resulting in the hyperpolarization on an IPSP.

CHECK POINT

1. Describe the structure of the nicotinic ACh receptors and how they also function as ion channels to produce EPSPs.
2. Explain how ACh is inactivated after it is released into the synaptic gap.
3. Describe how GABA produces IPSPs, and explain the significance of this neurotransmitter.

Some Neurotransmitters Activate G-Protein-Coupled Receptors

There are synapses where the neurotransmitter binds to a receptor protein that is separate from the ion channel. In order for the neurotransmitter to open (or close) the ion channel, a "middleman" must go from the activated receptor to the ion channel. This role is played by G-proteins. Such G-protein-coupled receptors include the muscarinic ACh receptor and receptors for the monoamine neurotransmitters.

The synapses described in the previous section are relatively simple, because the receptor is also the chemically gated ion channel. Nicotinic ACh receptors produce a depolarization (EPSP) in all synapses in which they are found, and thus are always excitatory. GABA receptors open to permit the entry of Cl^-, and so produce a hyperpolarization (IPSP). These receptors contrast with another type of receptors that are not ion channels, but spatially separated in the plasma membrane from the chemically gated ion channels they regulate. In order for these receptors to act on the ion channels, a "middleman" is required that can shuttle between these two different proteins.

The middleman is a complex of three smaller proteins called the **G-proteins**. The name refers to their regulation by guanosine nucleotides—GDP and GTP (fig. 4.28). The G-proteins are designated *alpha*, *beta*, and *gamma* (α, β, and γ). When a regulatory molecule, such as a neurotransmitter or hormone, binds to its G-protein-coupled receptor, these 3 subunits dissociate from the receptor. Also, the α subunit dissociates from the $\beta\gamma$ subunits, which stick together. Either the α subunit, or the $\beta\gamma$ complex, then moves through the plasma membrane to an effector molecule, which is an ion channel (for neurotransmitter function) or an enzyme. This can open (or close) the ion channel, or activate an enzyme in the plasma membrane that was previously inactive. There are about 400 to 500 different G-protein-coupled receptors for neurotransmitters and hormones, and in 1994 the discoverers of the G-proteins were awarded a Nobel Prize in Physiology or Medicine.

Muscarinic ACh Receptors Cause Excitation or Inhibition

The benefit for having the effector protein (such as a chemically regulated ion channel) separated from the receptor and operated by G-proteins is the greater flexibility this affords. This flexibility is illustrated by the action of ACh in synapses that have muscarinic ACh receptors. In some cases, the G-proteins cause depolarization and excitation; in other cases, the G-proteins cause hyperpolarization and inhibition. Muscarinic ACh receptors are located in postsynaptic

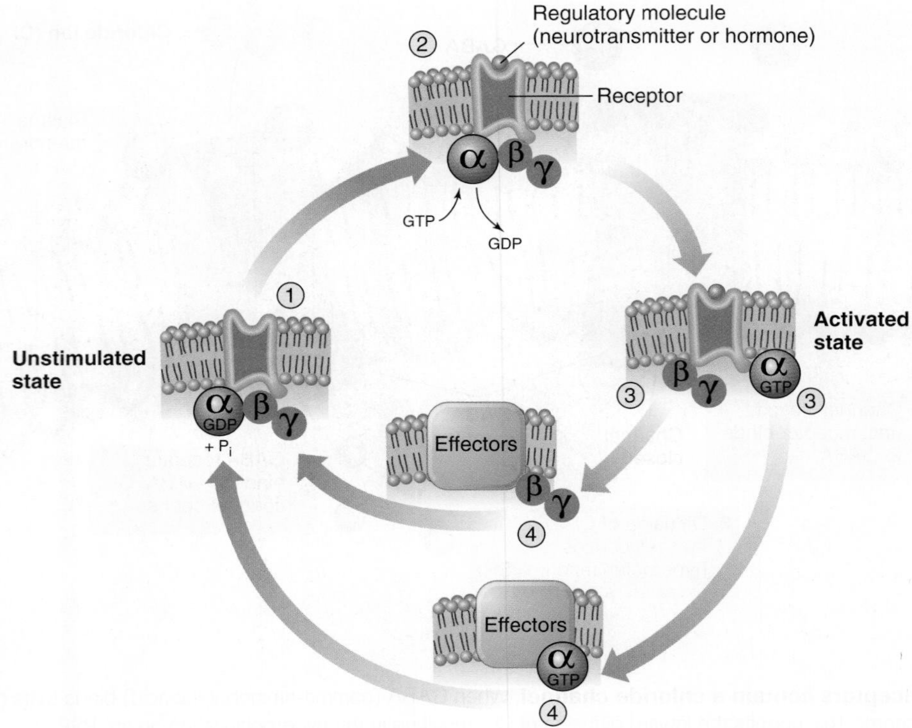

FIGURE 4.28 G-proteins allow receptors to communicate with effectors. The 3 G-proteins (alpha, beta, and gamma—α, β, and γ) associate with certain receptors in the plasma membrane (*1*). When the receptor binds to a regulatory molecule, GDP is exchanged for GTP (*2*). This forms an activated state (*3*), allowing the α subunit to dissociate from the βγ subunits, which stick together. These subunits move through the plasma membrane to activate effector molecules, which may be ion channels that produce EPSPs or IPSPs, or enzymes in the membrane.

membranes of some neurons in the CNS, and in smooth muscle and gland cells innervated by parasympathetic axons.

For example, ACh binding to muscarinic ACh receptors has an excitatory effect at synapses where parasympathetic neurons stimulate the smooth muscle of the gastrointestinal tract and the iris of the eyes. In order to do this, the G-proteins released from the muscarinic receptors here promote depolarizations (EPSPs). This will be described in conjunction with the autonomic nervous system in chapter 6.

However, the parasympathetic axons in the vagus nerves innervating the heart also release ACh, but the effects of ACh in the heart produce inhibition. Inhibition results because the βγ complex that dissociates from the muscarinic ACh receptor here causes gated K^+ channels to open (fig. 4.29). This allows K^+ to diffuse out of cells in the pacemaker region of the heart, making the inside of the cells more negative. As a result, it takes a longer time for these cells to become depolarized and produce an action potential. This is the way the "vagusstoff" (ACh) discovered by Otto Loewi works to slow the heart rate.

Dopamine, Norepinephrine, and Serotonin Are Monoamines

There are many other neurotransmitters in the nervous system that bind to G-protein-coupled receptors. Examples include the **monoamines**, an important family of regulatory molecules. Some monoamines are:

1. **Dopamine**, **norepinephrine**, and **epinephrine**. In general, epinephrine functions as a hormone (secreted by the adrenal medulla gland; it's also called *adrenaline*). Norepinephrine functions as both a neurotransmitter and a hormone, and dopamine is only a neurotransmitter. Dopamine, norepinephrine, and epinephrine are in the same subfamily of monoamines, called **catecholamines**, and are derived from the amino acid tyrosine.
2. **Serotonin**. Serotonin is an important neurotransmitter of certain neurons in the brain. It's a monoamine but not a catecholamine, because it's derived from the amino acid tryptophan rather than tyrosine.

The receptors for most monoamine neurotransmitters are G-protein-coupled receptors. When a monoamine neurotransmitter binds to its receptor, the α subunit dissociates and moves through the plasma membrane. The effector activated by the G-protein in this case is an enzyme, *adenylate cyclase*. This enzyme catalyzes the conversion of ATP into **cyclic AMP (cAMP)**, which is a *second messenger* (chapter 3). Increased production of cAMP in the postsynaptic cell indirectly opens ion channels, thereby stimulating the cell (fig. 4.30). Cyclic AMP has different effects in different cells, but it always works by one action: it activates a specific enzyme in the cytoplasm that was previously inactive. This enzyme is called *protein kinase*, a name that refers to its ability to add phosphate groups to other proteins. Through this action, a wide variety of effects can be produced.

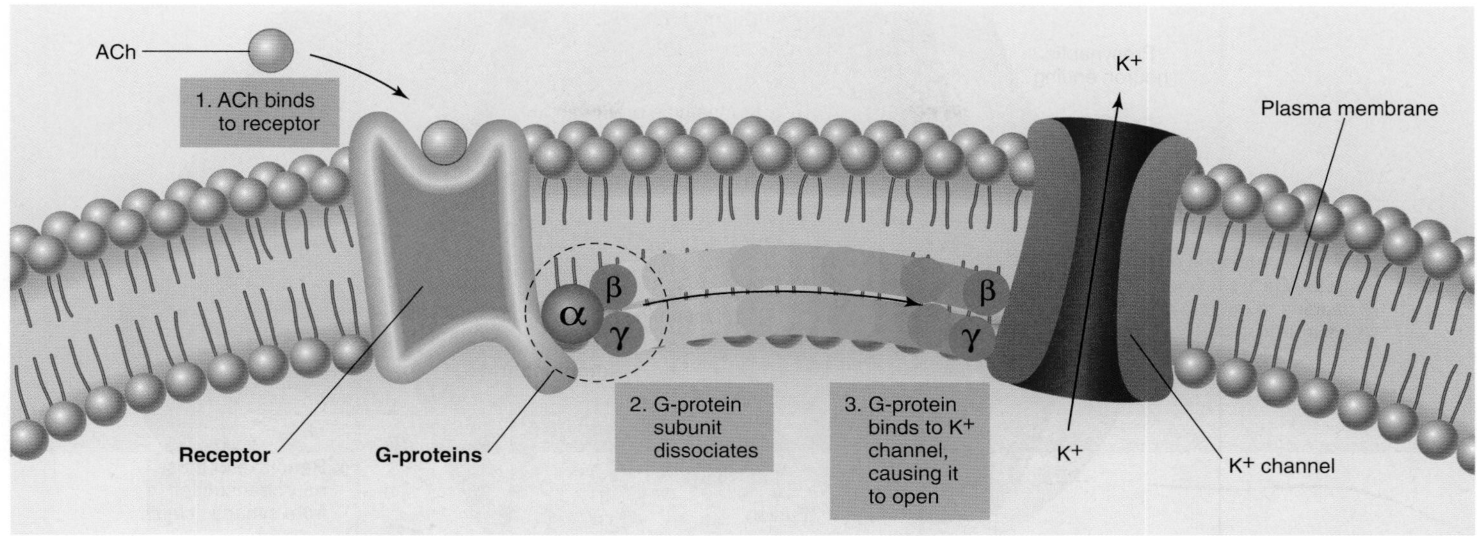

FIGURE 4.29 How ACh makes the heart rate slower. ACh binds to muscarinic ACh receptors in the pacemaker cells of the heart. This receptor works by means of G-proteins to cause the opening of chemically gated K^+ channels in the plasma membrane. This allows K^+ to diffuse out of the cells, making them more negative inside. As a result, it takes longer for the cells to depolarize sufficiently to produce an action potential, and because of this it takes them longer to produce a contraction.

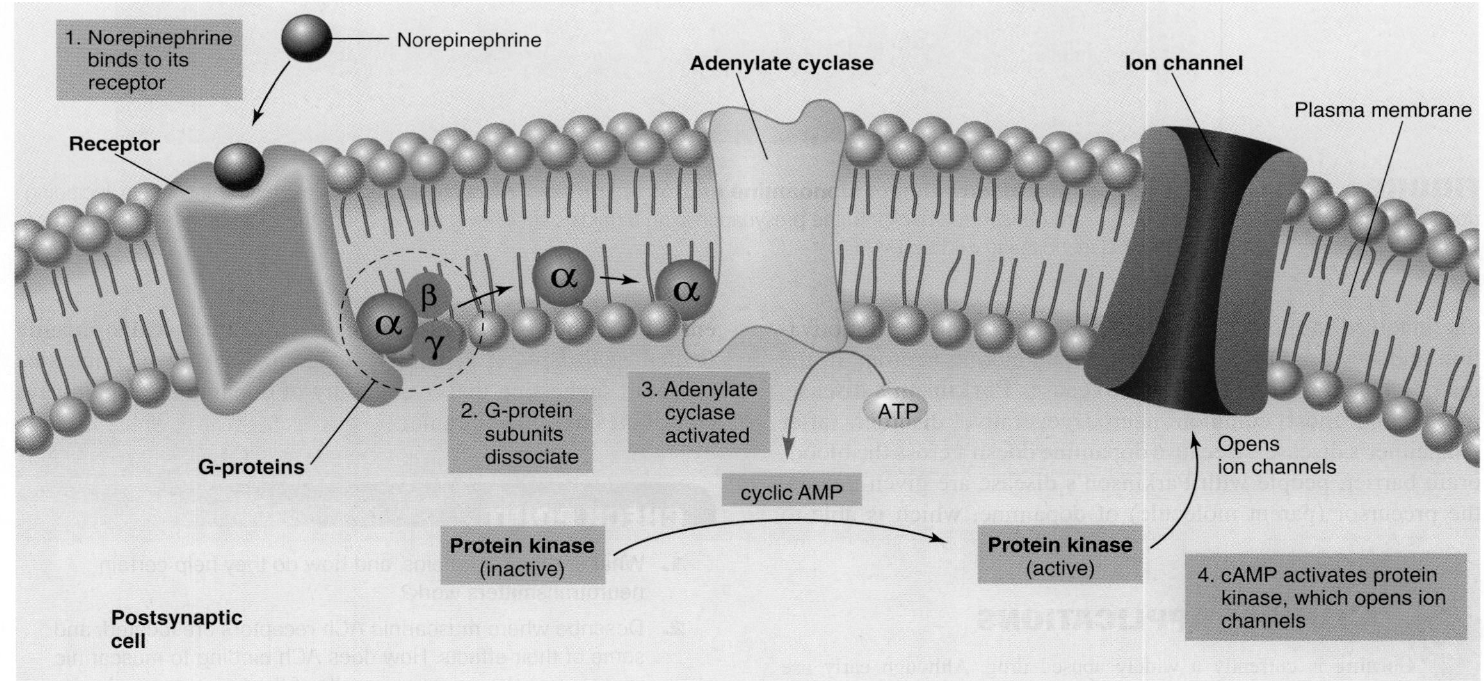

FIGURE 4.30 Norepinephrine action requires G-proteins. The binding of norepinephrine to its receptor causes the dissociation of G-proteins. Binding of the alpha G-protein subunit to the enzyme adenylate cyclase activates this enzyme, leading to the production of cyclic AMP. Cyclic AMP, in turn, activates protein kinase, which can open ion channels and produce other effects.

Just as ACh has to be broken down (by AChE) after it is released into the synaptic gap, the monoamine neurotransmitters in the synaptic gap must also be inactivated to achieve control over synaptic transmission. In the case of the monoamines, inactivation is generally a two-step process: (1) the monoamines in the synaptic gap are transported back into the presynaptic axons; and then (2) the monoamines are broken down by an enzyme called *monoamine oxidase* (*MAO*), located within the presynaptic axon (fig. 4.31).

Neurons that release dopamine are referred to as *dopaminergic neurons*. These form two neural pathways in the brain,

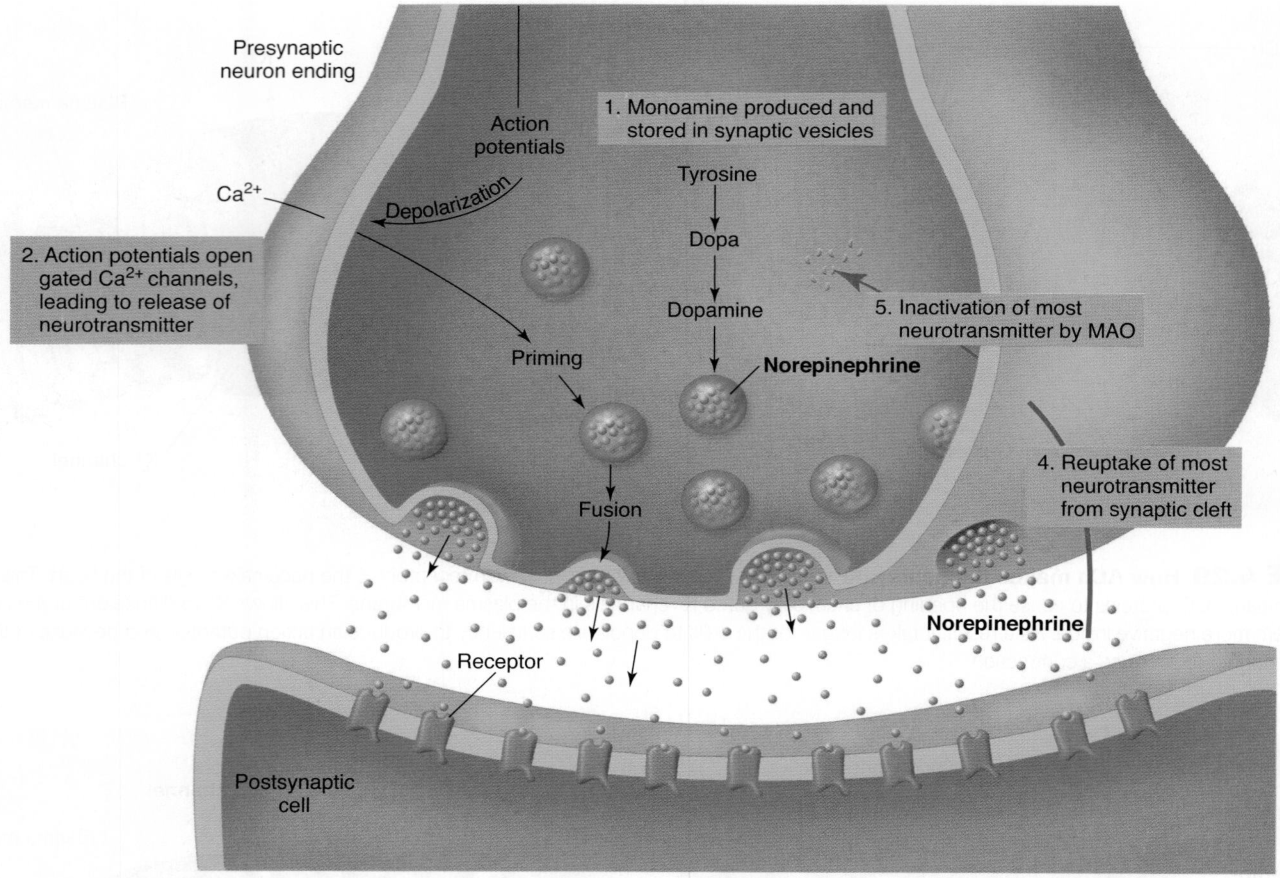

FIGURE 4.31 Release, reuptake, and inactivation of monoamine neurotransmitters. Most of the monoamine neurotransmitters, including dopamine, norepinephrine, and serotonin, are transported back into the presynaptic axon terminals after being released into the synaptic gap. They are then degraded and inactivated by an enzyme, monoamine oxidase (MAO).

one involved in motor control and the other involved in motivation and reward. Degeneration of dopaminergic neurons in the motor areas of the brain (chapter 5) causes **Parkinson's disease**, the second most common neurodegenerative disorder (after Alzheimer's disease). Because dopamine doesn't cross the blood-brain barrier, people with Parkinson's disease are given L-dopa, the precursor (parent molecule) of dopamine, which is able to enter the brain. By contrast, drugs used to treat **schizophrenia** (called *neuroleptics*) generally antagonize a type of dopamine receptor, suggesting that overactivity of dopaminergic pathways contributes to schizophrenia.

CLINICAL APPLICATIONS

Cocaine is currently a widely abused drug. Although early use may produce feelings of euphoria, continued use leads to social withdrawal, depression, dependence, and serious cardiovascular and renal damage that can cause heart and kidney failure. The numerous effects of cocaine in the CNS appear to be produced by one primary mechanism: cocaine binds to the reuptake transporter for dopamine (and other monoamines) and thereby blocks reuptake of dopamine into the presynaptic axon terminals. This results in overstimulation of neural pathways that use dopamine and other monoamines as neurotransmitters.

CHECK POINT

1. What are the G-proteins, and how do they help certain neurotransmitters work?
2. Describe where muscarinic ACh receptors are located, and some of their effects. How does ACh binding to muscarinic receptors in the pacemaker cells of the heart cause slowing of the heart rate?
3. Identify the monoamine neurotransmitters and describe how they act on the postsynaptic membrane.
4. Explain how monoamine neurotransmitters are inactivated after they are released into the synaptic gap.

Physiology in Health and Disease

Mood disorders—including **major depressive disorder** and **panic disorder**—are treated pharmaceutically with drugs that affect the monoamine neurotransmitters. The monoamine neurotransmitters include dopamine, norepinephrine, and serotonin; the drugs used to treat depression and panic attacks include **monoamine oxidase (MAO) inhibitors**, **selective serotonin reuptake inhibitors (SSRIs)**, and the **tricyclic antidepressants**. We need to know how monoamines work in synaptic transmission in order to understand how these drugs affect the brain.

Recall that the monoamine neurotransmitters are released by exocytosis into the synaptic gap, bind to their receptor proteins, and then detach. They can again bind to the receptor proteins (and produce another EPSP), but the free neurotransmitters in the synaptic gap can also be transported back into the presynaptic axon terminals. Transport carriers in the plasma membrane of the presynaptic terminals bind to the monoamine neurotransmitters in a relatively specific fashion to accomplish this. Once the neurotransmitters are in the presynaptic axon terminals, they are degraded and inactivated by the monoamine oxidase (MAO) enzyme (fig. 4.31).

The SSRIs specifically inhibit the transporters for serotonin, so that less serotonin is taken out of the synaptic gap and more is left in the synapse to stimulate the postsynaptic neuron. Common examples of SSRIs include escitalopram (*Lexapro*), fluoxetine (*Prozac*), paroxetine (*Paxil*), and sertraline (*Zoloft*). The tricyclic antidepressants are an older group of pharmaceuticals that have a less definitive mechanism of action. They are believed to be effective in treating depression because they interfere with the reuptake of norepinephrine and serotonin into the presynaptic axon terminals. However, the tricyclic antidepressants also have antihistamine and anticholinergic (anti-ACh) effects.

The MAO inhibitors reduce activity of monoamine oxidase, and so interfere with the ability of the presynaptic axon terminals to break down the monoamines (fig. 4.31). With more of these neurotransmitters building up in the presynaptic axon terminals, there is a reduced transport of them out of the synaptic cleft. As a result, there will be increased stimulation at synapses that use serotonin, norepinephrine, and dopamine as neurotransmitters. This helps treat depression and panic attacks, but can produce side effects. The most dangerous of these is hypertensive crisis (a dangerous elevation of blood pressure), possibly as a result of increased norepinephrine action. Eating foods rich in *tyramine*, a monoamine metabolized by the MAO enzyme, may trigger adverse reactions in people using MAO inhibitors. Foods rich in tyramine include many cheeses, wine, aged or cured meats, nuts, and soy sauce.[4] The tyramine derived from these foods can also produce severe migraine headaches in susceptible people.

FYI [4]The *cheese syndrome* is the name given to the interaction between MAO inhibitors and tyramine-rich foods. The name comes from the history of the discovery: a British pharmacist noticed that his wife (who had been taking an MAO inhibitor) developed severe headaches whenever she ate cheese. In alternate universes it might be called the salami or the sauerkraut syndrome.

Physiology in Balance

Nervous System: Neurons and Synapses

Working together with . . .

Endocrine System

The brain regulates the pituitary gland . . . p. 186

Respiratory System

Reflex control of breathing requires sensory and motor neurons . . . p. 297

Digestive System

Neurons coordinate activities of the intestines . . . p. 348

Cardiovascular System

A neural reflex helps maintain homeostasis of blood pressure . . . p. 240

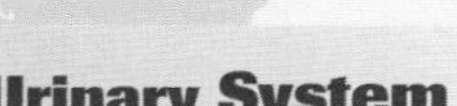

Urinary System

Urination is regulated by a neural reflex . . . p. 324

Summary

Nervous System Function Depends on Neurons and Supporting Cells 87

- Sensory (afferent) neurons transmit impulses into the central nervous system (CNS); motor (efferent) neurons transmit impulses out of the CNS to stimulate muscles and glands.
- Most sensory neurons are pseudounipolar; most motor neurons and association neurons are multipolar.
- Schwann cells produce myelin sheaths in the peripheral nervous system (PNS); oligodendrocytes produce myelin sheaths in the CNS.
- Myelinated axons have breaks in the myelin sheath every 1 to 2 mm; these are the nodes of Ranvier.

Axons Produce Action Potentials 90

- A decrease in the resting membrane potential (rmp), indicated by a movement upward from the rmp on the oscilloscope screen, is called depolarization.
- An increase in the membrane potential, indicated by a downward deflection below the rmp on the oscilloscope, is called hyperpolarization.
- Action potentials are based on voltage-regulated (or voltage-gated) Na^+ and K^+ channels, which are opened sequentially in response to a depolarization stimulus that reaches a threshold level.
- Stepwise opening, inactivation, and closing of first the voltage-regulated Na^+ channels and then the voltage-regulated K^+ channels produce an action potential.
- Na^+ diffusing into the axon produces the depolarization phase of the action potential, where the membrane potential reaches +30 mV.
- K^+ diffusion out of the axon produces the repolarization phase of the action potential, where the membrane potential eventually returns to the rmp.
- Action potentials are all-or-none; they don't vary in size.
- The stronger the depolarization stimulus above threshold, the greater the frequency of action potentials.
- The plasma membrane is in a refractory period when it is producing an action potential; therefore action potentials cannot run together or summate.
- An action potential at one segment of an unmyelinated axon depolarizes the very next region to threshold, so that the action potential is regenerated all along the axon.
- In a myelinated axon, the action potential is regenerated only at the nodes of Ranvier; this saltatory conduction is faster than conduction in an unmyelinated axon.
- Action potentials are conducted without decrement; because they are regenerated, action potentials are the same size at the axon terminal as at the axon hillock.

Neurons Regulate Other Cells at Synapses 96

- The axon terminal contains synaptic vesicles with chemical neurotransmitter molecules; fusion of these vesicles with the plasma membrane results in the release of neurotransmitters by exocytosis into the synaptic cleft.
- The release of neurotransmitters is stimulated by action potentials, which cause the opening of voltage-regulated Ca^{2+} channels in the axon terminals.
- Acetylcholine (ACh) was the first neurotransmitter to be discovered.
- Neurotransmitters bind to receptor proteins in the postsynaptic membrane and promote the opening of chemically regulated (or ligand-regulated) ion channels.
- A depolarization produced by chemically regulated channels is an excitatory postsynaptic potential (EPSP), which serves as the normal depolarization stimulus for action potentials
- A hyperpolarization produced by chemically regulated channels is an inhibitory postsynaptic potential (IPSP).
- EPSPs and IPSPs are graded; they decrease in amplitude with distance; they don't have refractory periods; and they can summate.
- EPSPs can add together by spatial summation to produce larger depolarizations, or an IPSP can reduce the depolarization produced by an EPSP in postsynaptic inhibition.

Some Neurotransmitter Receptors Are Also Ion Channels 100

- Nicotinic ACh receptors, found in the postsynaptic membrane of many neurons and skeletal muscle cells, are stimulated by ACh and also by nicotine.
- When ACh binds to the nicotinic ACh receptors, a chemically regulated channel opens in the receptor, permitting Na^+ to diffuse in and K^+ to diffuse out of the cell; this causes an EPSP.
- ACh is broken down by the enzyme acetylcholinesterase (AChE), thereby stopping the action of ACh in the synapse.
- GABA (gamma-aminobutyric acid) is an inhibitory neurotransmitter in the CNS, important in the control of motor neurons and sleep.
- When GABA binds to its receptor, it causes the opening of a channel for chloride ions (Cl^-) through the receptor, resulting in the inward diffusion of Cl^- that produces an IPSP.

Some Neurotransmitters Activate G-Protein-Coupled Receptors 103

- There are 3 small G-proteins, alpha (α), beta (β), and gamma (γ), associated with the receptors for some neurotransmitters; these are necessary when the receptor protein is a different molecule from the ion channel protein.
- When a neurotransmitter binds to its G-protein-coupled receptor, the G-proteins dissociate from the receptor, and the alpha protein dissociates from the other two, which stick together.
- The α protein, or the $\beta\gamma$ complex, then moves through the plasma membrane to bind to a chemically regulated ion channel; this either opens or closes the channel.
- Muscarinic ACh receptors are G-protein-coupled receptors that bind to either ACh or muscarine; they're found in the CNS and also in cardiac muscle and smooth muscle.
- In some tissues, ACh causes depolarization and excitation as a result of the dissociation of G-proteins.
- In other cases, ACh binds to muscarinic ACh receptors to cause inhibition; this is the case in the heart, where ACh promotes a slowing of the heart rate as a result of the $\beta\gamma$ complex causing the opening of K^+ channels.
- Monoamine neurotransmitters (dopamine, norepinephrine, and serotonin) generally stimulate G-protein-coupled receptors.
- In the action of the monoamines, the G-proteins can cause activation of an enzyme known as adenylate cyclase, which catalyzes the conversion of ATP in the cell to cyclic AMP.
- Cyclic AMP is a second messenger in the postsynaptic cell, where it activates the enzyme protein kinase and indirectly causes the opening of chemically regulated ion channels.

- Monoamine neurotransmitters are eliminated from the synapse in a two-stage process: first, they are transported back into the presynaptic axon terminals; then, they are degraded and inactivated by an enzyme, monoamine oxidase.

Review Activities

Objective Questions: Test Your Knowledge

1. Which of the following statements about the myelin sheath is true?
   a. All neurons are myelinated.
   b. The myelin sheath surrounds dendrites and axons.
   c. Schwann cells form myelin sheaths in the PNS.
   d. Astrocytes form myelin sheaths in the CNS.
2. Sensory neurons are
   a. pseudounipolar.
   b. part of the PNS.
   c. afferent neurons.
   d. all of these.
3. Which of the following is needed for the formation of the blood-brain barrier?
   a. Oligodendrocytes
   b. Astrocytes
   c. Schwann cells
   d. Microglia
4. Which of the following statements regarding the nodes of Ranvier is false?
   a. They are found in myelinated axons of the PNS only.
   b. They are required for impulse conduction by myelinated axons.
   c. They are gaps in the myelin sheath.
   d. They are involved in saltatory conduction.
5. A change in the resting membrane potential that results in the inside of the cell becoming less negative is a _________; this is indicated by a(n) _________ deflection of the oscilloscope recording.
   a. depolarization; downward
   b. depolarization; upward
   c. hyperpolarization; downward
   d. hyperpolarization; upward
6. The voltage-gated Na^+ channels are closed until the membrane is
   a. depolarized to a threshold level.
   b. depolarized by any amount.
   c. hyperpolarized to a threshold level.
   d. hyperpolarized by any amount.
7. After voltage-gated ion channels are opened,
   a. they stay open until the stimulus is removed.
   b. they enter the closed state until the stimulus is removed.
   c. they enter an inactivated state and stay that way until the stimulus is removed.
   d. they enter an inactivated state for a brief period, then enter the closed state.
8. The first action potential in a neuron is usually produced at the
   a. dendrites.
   b. cell body.
   c. axon hillock.
   d. axon terminals.
9. Which of the following statements regarding the depolarization phase of the action potential is false?
   a. Na^+ diffuses out of the axon.
   b. The membrane potential shoots to +30 mV.
   c. A depolarization stimulus evokes a greater depolarization.
   d. This phase is ended by inactivation of the Na^+ channels.
10. Which of the following statements regarding the repolarization phase of the action potential is false?
    a. The membrane potential eventually returns to the rmp.
    b. This phase requires the opening of voltage-gated K^+ channels.
    c. There is an overshoot, producing a hyperpolarization for a brief time.
    d. This phase requires the action of the Na^+/K^+ (ATPase) pumps.
11. The stronger the depolarization stimulus,
    a. the greater the frequency of action potential production.
    b. the greater the amplitude of the action potential.
    c. the greater the speed at which action potentials travel along the axon.
    d. the greater the duration of each action potential.
12. Which of the following is not a characteristic of action potentials?
    a. They are conducted without decrement.
    b. They can summate with each other.
    c. They are all-or-none.
    d. They have a refractory period.
13. Which of the following statements regarding the conduction of action potentials is false?
    a. An action potential at one site serves as the depolarization stimulus for an action potential at the next site.
    b. In an unmyelinated axon, an action potential is produced all along the axon.
    c. In a myelinated axon, an action potential is produced only at the nodes of Ranvier.
    d. Unmyelinated axons conduct action potentials faster than do myelinated axons.
14. Which of the following statements regarding neurotransmitters is true?
    a. The opening of voltage-gated Na^+ channels in the axon triggers the release of neurotransmitters from the axon terminals.
    b. Neurotransmitters leave the axon terminals by active transport across the plasma membrane.
    c. Neurotransmitters cause the opening of chemically regulated channels in the postsynaptic membrane.
    d. Neurotransmitters open voltage-gated channels in the dendrites.
15. Which of the following is not a characteristic of synaptic potentials?
    a. They can summate.
    b. They are all-or-none.
    c. They decrease in amplitude with distance.
    d. They don't have refractory periods.
16. Inhibitory postsynaptic potentials (IPSPs)
    a. are hyperpolarizations.
    b. can be produced by GABA.
    c. produce postsynaptic inhibition.
    d. all of these.

17. Which of the following types of receptor proteins (1) is also an ion channel and (2) always produces a depolarization (EPSP)?
    a. Nicotinic ACh receptors
    b. Muscarinic ACh receptors
    c. Receptors for GABA
    d. Receptors for monoamines

18. Which of the following types of receptor proteins is also an ion channel for Cl^-?
    a. Nicotinic ACh receptors
    b. Muscarinic ACh receptors
    c. Receptors for GABA
    d. Receptors for monoamines

19. A drug that inhibits acetylcholinesterase (AChE) will have all of the following effects except:
    a. Relaxation of skeletal muscles
    b. Increased amounts of ACh in the synaptic gap
    c. Increased production of EPSPs at cholinergic synapses
    d. Increased production of action potentials at cholinergic synapses

20. Which of the following neurotransmitters is not inactivated by transport out of the synaptic gap back into the presynaptic axon terminals?
    a. Serotonin
    b. Norepinephrine
    c. Dopamine
    d. Acetylcholine

21. A drug that inactivates monoamine oxidase (MAO) will increase the activity of all of the following neurotransmitters except:
    a. Serotonin
    b. Norepinephrine
    c. Dopamine
    d. Acetylcholine

22. The action of G-proteins is needed for all of the following except:
    a. Muscarinic ACh receptors
    b. Nicotinic ACh receptors
    c. Receptors for dopamine
    d. Receptors for norepinephrine

Essay Questions 1: *Test Your Understanding*

1. What is the blood-brain barrier, and what role do astrocytes play in its development?
2. Describe how the myelin sheath is formed in the PNS. How are the nodes of Ranvier formed? What is the relationship between the myelin sheath and the neurilemma?
3. Distinguish between the parts of a myelinated neuron that have a myelin sheath and the parts that do not have myelin. How does this relate to the structures contained in the gray matter and white matter of the CNS?
4. Define the terms *depolarization* and *hyperpolarization*, and describe how they appear in an oscilloscope screen.
5. Explain the role of positive feedback in the depolarization phase of an action potential.
6. Describe the cycle of changes that occur in a voltage-gated ion channel once it has been adequately stimulated, distinguishing between the different states of the channel.
7. Step-by-step, describe the sequential changes in the voltage-gated Na^+ and K^+ channels, and the net diffusion of these ions that results, during an action potential. Relate these events to the changes that occur in the membrane potential during an action potential.
8. Describe the all-or-none characteristic of action potentials; also, describe the refractory period. Explain how the strength of a stimulus is coded by action potentials.
9. Explain how action potentials are conducted by unmyelinated axons. Compare this with how myelinated axons conduct action potentials, and explain why myelinated axons have faster conduction.
10. Describe how neurotransmitters are stored and released by the presynaptic axon terminals. How does the frequency of action potentials influence the amount of neurotransmitter molecules released?
11. Construct a table that compares the characteristics of an EPSP with the corresponding characteristics of action potentials, so that the similarities and differences can be clearly seen.
12. Explain spatial summation of EPSPs and why such summation may be necessary to excite the postsynaptic neuron.
13. Describe an IPSP and how postsynaptic inhibition can occur.
14. Describe the chemical structure of nicotinic ACh receptors and explain how ACh binding to these receptors leads to the production of EPSPs. Where in the body are nicotinic ACh receptors located?
15. Explain how GABA causes hyperpolarization, and the significance of this action.
16. What are the G-proteins? What happens when a neurotransmitter binds to a G-protein-coupled receptor?
17. Using just the information in this chapter, explain how the binding of ACh to its muscarinic ACh receptors in the heart can cause slowing of the heart rate.
18. What are the monoamine neurotransmitters? How are these neurotransmitters inactivated after they're released into the synaptic gap?
19. Explain how a drug that interferes with the reuptake of monoamines into presynaptic axon terminals, and a drug that inactivates monoamine oxidase (MAO), affect the nervous system.
20. In a step-by-step manner, explain how the binding of norepinephrine to its receptor leads to production of cAMP in the postsynaptic cell. What is the function of cAMP, and what is its most direct action?

Essay Questions 2: *Test Your Analytical Ability*

1. The testes have a "germinal epithelium" forming sperm, which is protected by cells (Sertoli cells) that form a "blood-testis barrier." Explain how this compares with the blood-brain barrier, listing the analogous structures.
2. Sensory neurons are pseudounipolar. How might this structure, as opposed to a multipolar neuron structure, be beneficial for the function of sensory neurons?
3. Gated ion channels may be voltage-regulated or chemically regulated. Where is each type located in a neuron? How great a distance separates them? What places a limit on how far apart they can be?
4. What properties of the voltage-gated Na^+ and K^+ channels are responsible for the action potential having the appearance of a spike? That is, explain why the membrane potential changes so fast, going all the way to +30 mV and then stopping to suddenly reverse direction.
5. The action potential is produced by passive transport, but it can't be produced unless the neuron performs active transport. Explain this statement.
6. Where would you expect the voltage-regulated Na^+ and K^+ channels to be located in a myelinated neuron? Explain.
7. Is the neural code conducted by axons analogous to AM or FM radio? Explain. Also: Do action potentials form a digital or analog code? Explain.

8. Synaptic potentials are described as local events, whereas action potentials are conducted without decrement. Explain this distinction. What are the important implications of synaptic potentials being local events?
9. Most axons conducting sensations of pain are unmyelinated, whereas sensory axons conveying information regarding the degree of muscle stretch are myelinated. What are the implications of this difference, and what advantages might this difference confer?
10. If a person smokes, synapses located in which body regions would be stimulated by the nicotine? Given that sympathetic nerves cause blood vessels in the skin to constrict, by what mechanism might smoking result in reduced blood flow to the skin?
11. The receptors in the nose that interact with odorant molecules are G-coupled-receptors, and there can be hundreds of $\alpha\beta\gamma$ G-proteins associated with each receptor. How could such an association help provide great odor sensitivity?
12. What is the meaning of the terms *spastic* and *flaccid paralysis*? Using these terms, explain the mechanisms by which botulinum toxin, neostigmine, and curare influence skeletal muscles. Explain how these drugs may be used medically.

Web Activities

ARIS™ For additional study tools, go to: www.aris.mhhe.com. Click on your course topic and then this textbook's author/title. Register once for a semester's worth of interactive activities and practice quizzing to help boost your grade!

CHAPTER 5

Central Nervous System

Homeostasis

The central nervous system contains integrating centers for a great many negative feedback loops that are needed to maintain homeostasis. For example, the hypothalamus is a brain region that coordinates the responses of the body to changes in temperature; the medulla oblongata is a brain region that orchestrates the responses of the cardiovascular system to changes in blood pressure, and regulates breathing in response to changes in blood oxygen and carbon dioxide concentrations. More generally, the central nervous system governs how a person behaves when dealing with both the external and the internal environments, and so it is truly central to the maintenance of homeostasis.

But if homeostasis is not maintained . . .

Clinical Investigation

Paul was a 50-year-old custodian who visited the college clinic and flirted with a nursing student named Emma on a regular basis. Judging from some of Paul's symptoms, Emma thought he suffered from Parkinson's disease. When asked about this, Paul assured her that those symptoms were caused by a medicine he'd been taking called Haldol. Emma looked up the medicine on the Internet and was alarmed to learn that Haldol (haloperidol) blocks the action of dopamine and is often used to treat schizophrenia. Paul didn't return to see her for a couple of weeks, and when Emma asked about him, she was told that he suffered a stroke. When she saw him a few months later, he was partially paralyzed on his right side and couldn't speak, although he apparently understood everything Emma said.

What are some of the symptoms of Parkinson's disease, and what is its cause? How might a drug used to treat schizophrenia produce these symptoms? What particular brain areas were likely damaged by Paul's stroke, and how would this damage relate to his speech problems and partial paralysis?

Chapter Outline

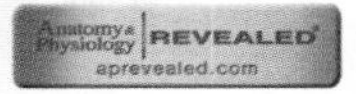

A virtual cadaver dissection experience

There Are Three Major Regions of the Brain

The brain and spinal cord develop from the embryonic neural tube. As the brain develops, it forms 3 regions: forebrain, midbrain, and hindbrain. The forebrain and hindbrain then each subdivide into 2 regions, forming 5 brain regions.

The brain is arguably the most important organ; after all, you can get other organs amputated or transplanted and still be *you*, but a person's death is defined by the lack of brain activity. And yet the brain is the least well-understood organ in the body, and is likely to remain so for the foreseeable future despite intense research. In order to begin your own study of brain physiology, you must first be familiar with the anatomical regions of the brain. You may have learned this in a previous anatomy course, or it may be new information; in either case, this anatomical base will help you build a good foundation for your pyramid of physiological knowledge.

The central nervous system (CNS; the brain and spinal cord) of an embryo begins as a *neural tube*. This tube is formed from the surface layer of embryonic tissue (*ectoderm*), the same embryonic layer that will form the epidermis of the skin. The surface ectoderm on the dorsal (backside) surface of the embryo furrows inward to produce a neural groove, which then fuses together to form the neural tube as a separate structure under the surface ectoderm (fig. 5.1). In figure 5.1, you can also see the formation of the embryonic *neural crest*, which will ultimately form the ganglia of the peripheral nervous system, among other structures.

By the middle of the fourth week after conception, there are three bulges on the anterior (front) end of the neural tube. These will become the 3 major brain regions: the **forebrain**, **midbrain**, and **hindbrain**. During the fifth week of development, the forebrain and hindbrain divide into 2 regions each, forming a total of 5 brain regions (fig. 5.2). The regions and the parts of the brain they include (fig. 5.3) are:

1. The forebrain and its two divisions: the **telencephalon** (including the *cerebrum*) and the **diencephalon** (including the *thalamus* and *hypothalamus*)

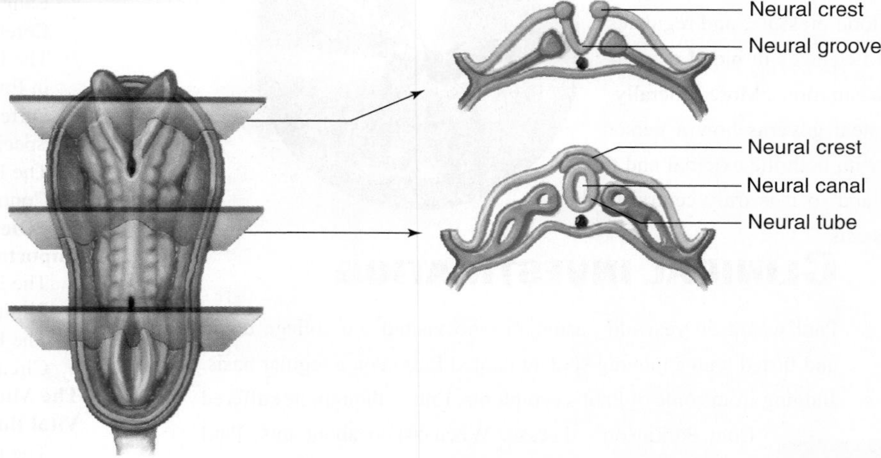

FIGURE 5.1 Embryonic development of the CNS. This is a dorsal view of a 22-day-old human embryo. The neural groove becomes the neural tube, which forms the brain and spinal cord. The lumen of the neural canal becomes the ventricles of the brain and the central canal of the spinal cord. The neural crest becomes the peripheral nervous system, among other structures.

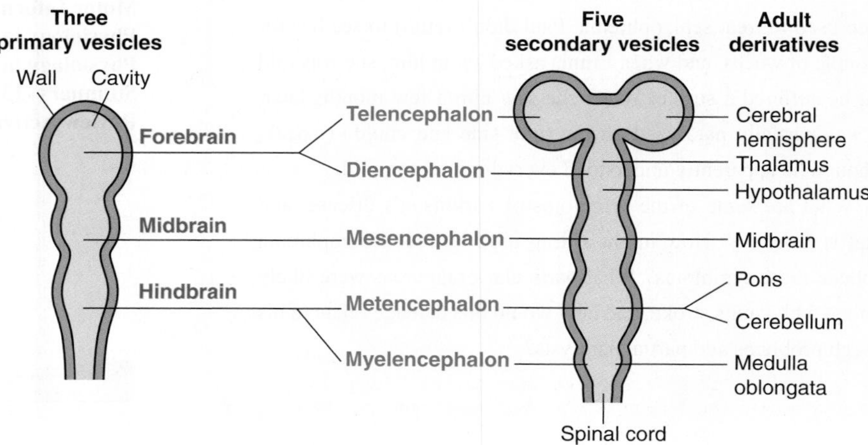

FIGURE 5.2 The developmental sequence of the brain. During the fourth week, 3 principal regions of the brain are formed. During the fifth week, a 5-regioned brain develops and specific structures begin to form.

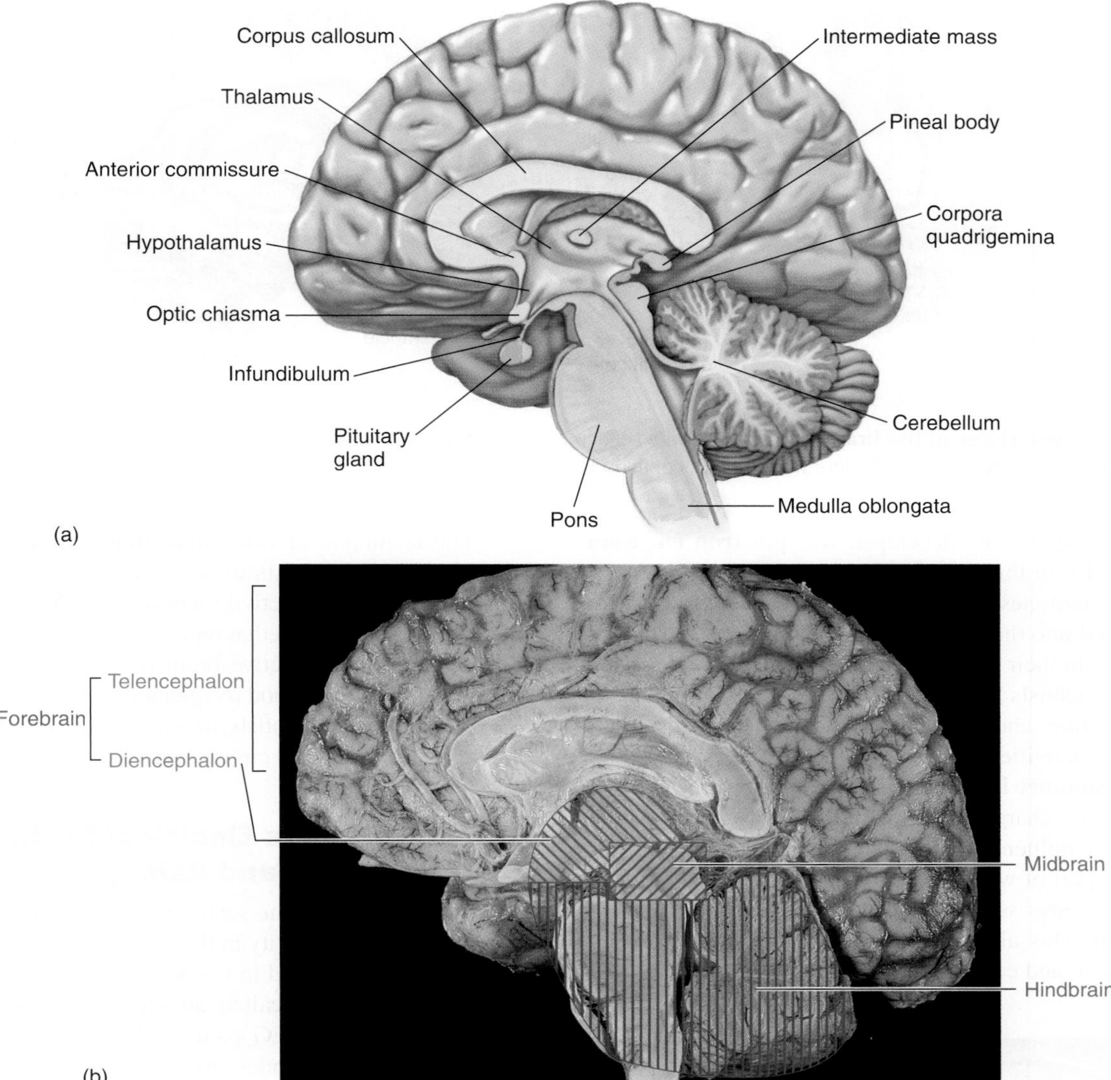

FIGURE 5.3 The adult brain seen in midsagittal section. The structures are labeled in the diagram shown in (*a*), and the brain regions are indicated in the photograph in (*b*). The diencephalon (shaded red) and telencephalon (unshaded area) make up the forebrain; the midbrain is shaded blue and the hindbrain is shaded green.

2. The midbrain (or **mesencephalon**), including the *corpora quadrigemina*
3. The hindbrain and its two divisions: the **metencephalon** (including the *pons* and *cerebellum*) and the **myelencephalon** (including the *medulla oblongata*)

Notice that the CNS begins in the embryo as a hollow tube. As the neural tube becomes the brain and spinal cord, its cavity becomes the **ventricles** of the brain[1] (fig. 5.4) and the **central canal** of the spinal cord. The ventricles and central canal contain *cerebrospinal fluid* (*CSF*), which also surrounds the brain and spinal cord in a cavity within the connective tissue **meninges** that cover the CNS. The CSF surrounding the brain and spinal cord serve as a shock-absorbing medium between these delicate neural organs and the surrounding bone of the skull and vertebrae.

FYI [1]Generations of physiology students were taught that an adult loses brain cells but never gets new ones. However, in recent years scientists have discovered that there are neural stem cells located in the "subventricular zone." These stem cells divide to replenish themselves and produce new neurons and glial cells, which can migrate to locations in the forebrain. Perhaps the lesson here is that you should study hard, but don't hold on to "facts" too tightly; you may have to modify some in the future.

Techniques Allow the Living Brain to Be Seen

The brains of living people can be seen using a few techniques, the first of which to be developed was **x-ray computed tomography** (**CT**). In this technique, the absorption of x-rays by tissues of different densities is analyzed by a computer and displayed as an image that can show sections (slices) of the brain at different depths.

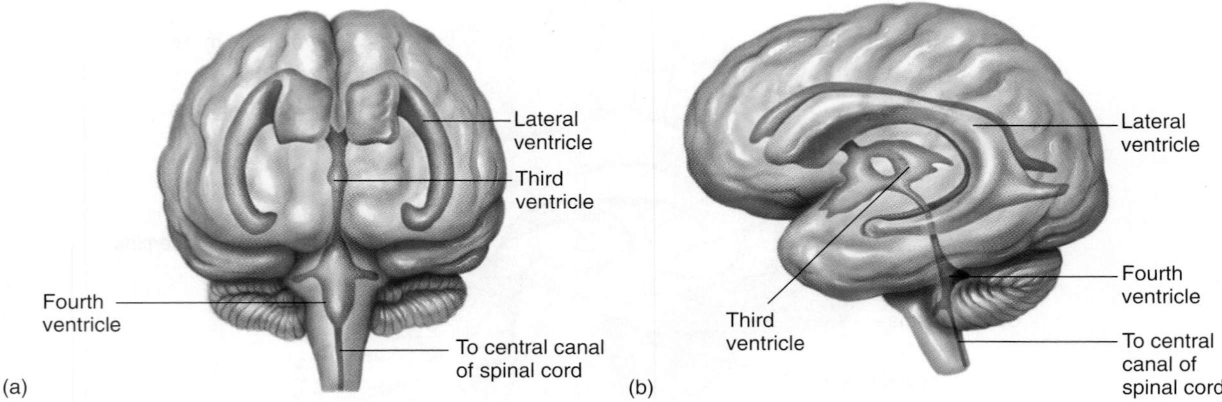

FIGURE 5.4 The ventricles of the brain. The ventricles are seen in (*a*) an anterior view, and (*b*) a lateral view. The ventricles are cavities in the brain, filled with cerebrospinal fluid (CSF). The fourth ventricle is continuous with the central canal of the spinal cord.

The next technique to be developed was **positron emission tomography (PET)**. In this technique, molecules that emit positrons (elementary particles like electrons, but carrying a positive charge) are injected into the blood. The collision of a positron and an electron results in their mutual annihilation, with the emission of gamma rays. Scientists have used PET to study brain metabolism, drug distribution, and the changes in blood flow that result from various brain activities.

Magnetic resonance imaging (MRI) is based on the concept that protons (H^+) are charged and spinning, and so are like little magnets that can be influenced by an external magnetic field. Most of the protons are part of water molecules, and the different water composition of various tissues provides the basis for distinguishing between them. This allows a clear distinction between gray matter, white matter, and cerebrospinal fluid (fig. 5.5).

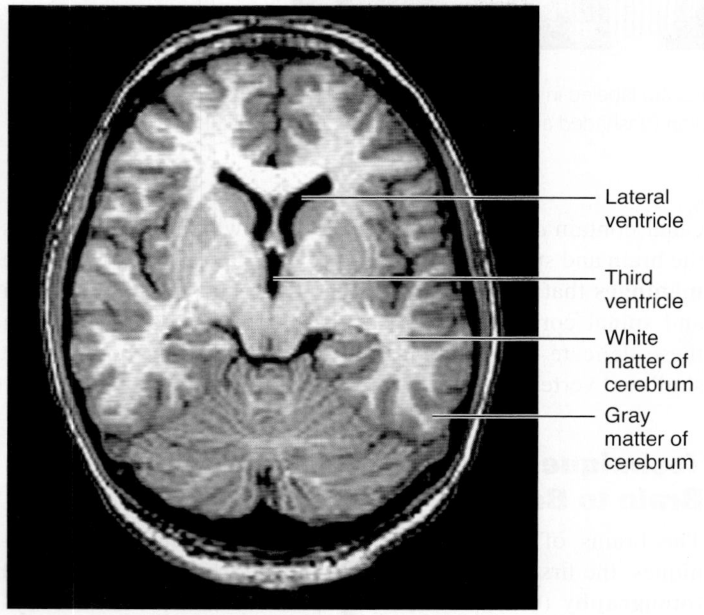

FIGURE 5.5 An MRI scan of the brain. Gray and white matter are easily distinguished, as are the ventricles containing cerebrospinal fluid.

The technique of **functional magnetic resonance imaging (fMRI)** allows scientists to study the function of a living brain. The increased neural activity within an active brain region causes increased blood flow to that region. As a result, there is more oxygen delivered to the active brain region. Because of this *BOLD response* (short for "blood oxygenation level dependent contrast"), the fMRI allows scientists to see which brain regions become active during particular circumstances.

The EEG Detects Electrical Brain Activity Awake and Asleep

Electrodes placed on the scalp can detect electrical currents produced by synaptic activity in the cerebral cortex (the outer layer of the cerebrum, discussed in the next section). A recording of these electrical currents is called an **electroencephalogram (EEG)**. There are different EEG patterns ("brain waves") that are produced by using electrodes over particular brain regions when a person is in different states of mental activity. The EEG patterns can be used by neurologists to help diagnose certain conditions, and by research scientists to learn more about brain activity.

For example, the EEG patterns change during sleep. Sleep can be divided into two major categories: **REM (rapid eye movement) sleep** and **non-REM sleep**. Dreams generally occur during REM sleep. During REM sleep, the EEG patterns are desynchronized, as during wakefulness. During non-REM sleep, the EEG shows high-amplitude, low-frequency waves. When people first fall asleep, they enter non-REM sleep of four stages, and then go backward through these stages to REM sleep. After REM sleep, they again descend through the different stages of non-REM sleep and back up to REM sleep. Each of these cycles lasts approximately 90 minutes, and a person typically goes through about five REM to non-REM cycles a night. When people are allowed to wake up naturally, they usually awaken from REM sleep.

One hypothesis regarding the importance of non-REM sleep is that it aids learning. For example, subjects who were allowed to have non-REM sleep after a learning trial had improved performance compared to those who were not allowed to have non-

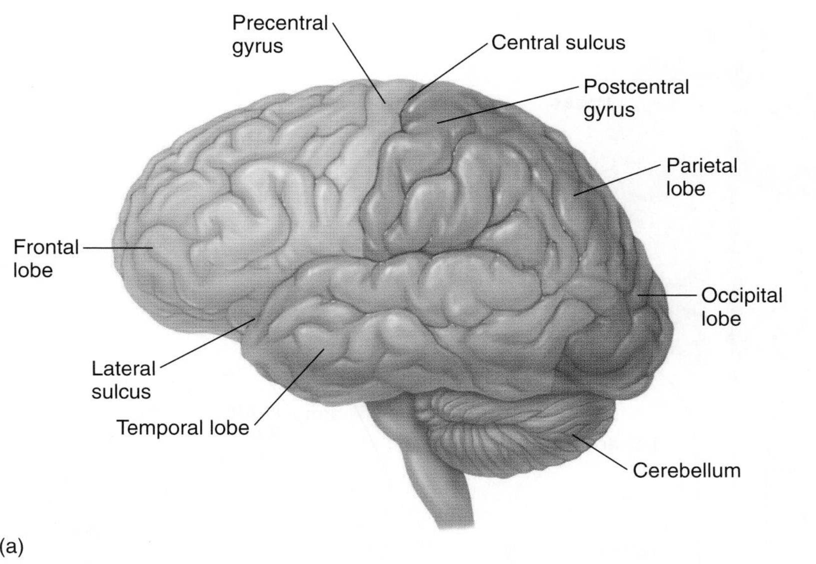

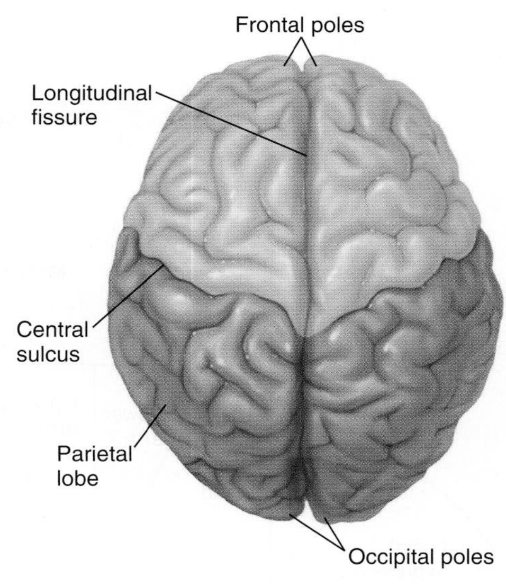

FIGURE 5.6 The cerebral cortex. (*a*) A lateral view and (*b*) a superior view.

REM sleep. In another study, slow-wave activity in an EEG (which indicates non-REM sleep) increased in trained subjects. These studies suggest that non-REM (slow-wave) sleep may aid the consolidation of memories. Although the relationship between sleep and memory consolidation is still not completely proven, these studies and others suggest that students who stay up all night for an exam would be better off if they began studying earlier and got some sleep the night before!

CHECK POINT

1. What are the three major brain divisions, and the 5 brain regions formed from them? Name some of the anatomical structures contained in each of the 5 brain regions.
2. What distinguishes CT, PET, and MRI techniques for seeing the living brain? What is the benefit of using fMRI in brain research?
3. How are EEG patterns obtained, and how has the EEG been used in sleep research?
4. Distinguish between REM and non-REM sleep.

The Cerebrum Performs Higher Brain Functions

The cerebrum is divided into 2 cerebral hemispheres, which have specialized functions. Each cerebral hemisphere is composed of surface gray matter, the cerebral cortex, and underlying white matter containing nuclei of gray matter that are involved in motor control. The functions of the cerebral cortex include sensory perception, motor control, language, emotion, and memory.

The **cerebrum** is the largest portion of the brain, accounting for 80% of its mass. It consists of left and right **cerebral hemispheres**, which are interconnected by a tract of axons that form the **corpus callosum** (see fig. 5.3). Each hemisphere has 2 to 4 mm of gray matter at its surface, containing neuron cell bodies and dendrites, and white matter underneath that contains myelinated axons (producing the white appearance). The outer gray layer of the cerebrum is the **cerebral cortex**.

The cerebral cortex has numerous folds and grooves, called *convolutions*. The elevated folds of the convolutions are called *gyri*, and the depressed grooves are the *sulci*. Deep sulci, or fissures, subdivide each cerebral hemisphere into five lobes (fig. 5.6). These are the **frontal lobe**, **parietal lobe**, **temporal lobe**, **occipital lobe**, and the **insula** (which cannot be seen in the surface view shown in fig. 5.6).

The frontal lobe and parietal lobe are separated by the *central sulcus*. Just anterior to the central sulcus, in the frontal lobe, is the **precentral gyrus**. Association neurons located in the precentral gyrus are called "upper motor neurons," because they help control the motor neurons of the spinal cord. For this reason, the precentral gyrus is sometimes referred to as the *motor cortex*. Just posterior to the central sulcus, in the parietal lobe, is the **postcentral gyrus**. This is the primary area of the cerebral cortex responsible for *somatesthetic sensations*—those arising from skin, muscles, tendons, and joints. These sensory and motor control areas are shown in figure 5.7.[2]

FYI [2]The map of the motor areas of the precentral gyrus looks like a distorted, upside-down man. Wilder Penfield (1891–1976), a neurosurgeon in Montreal, Canada, created this image while studying epilepsy in the 1950s. He exposed the brains of conscious patients (under only local anesthesia) and electrically stimulated different locations of the brain, in an effort to locate the source of their epileptic seizures. The patients had to be conscious to report their experiences as Dr. Penfield stimulated different brain locations. In the process, he mapped the motor and sensory cortex, and also discovered that electrical stimulation in the temporal lobe could evoke complete, vivid memories, even those that had been consciously forgotten.

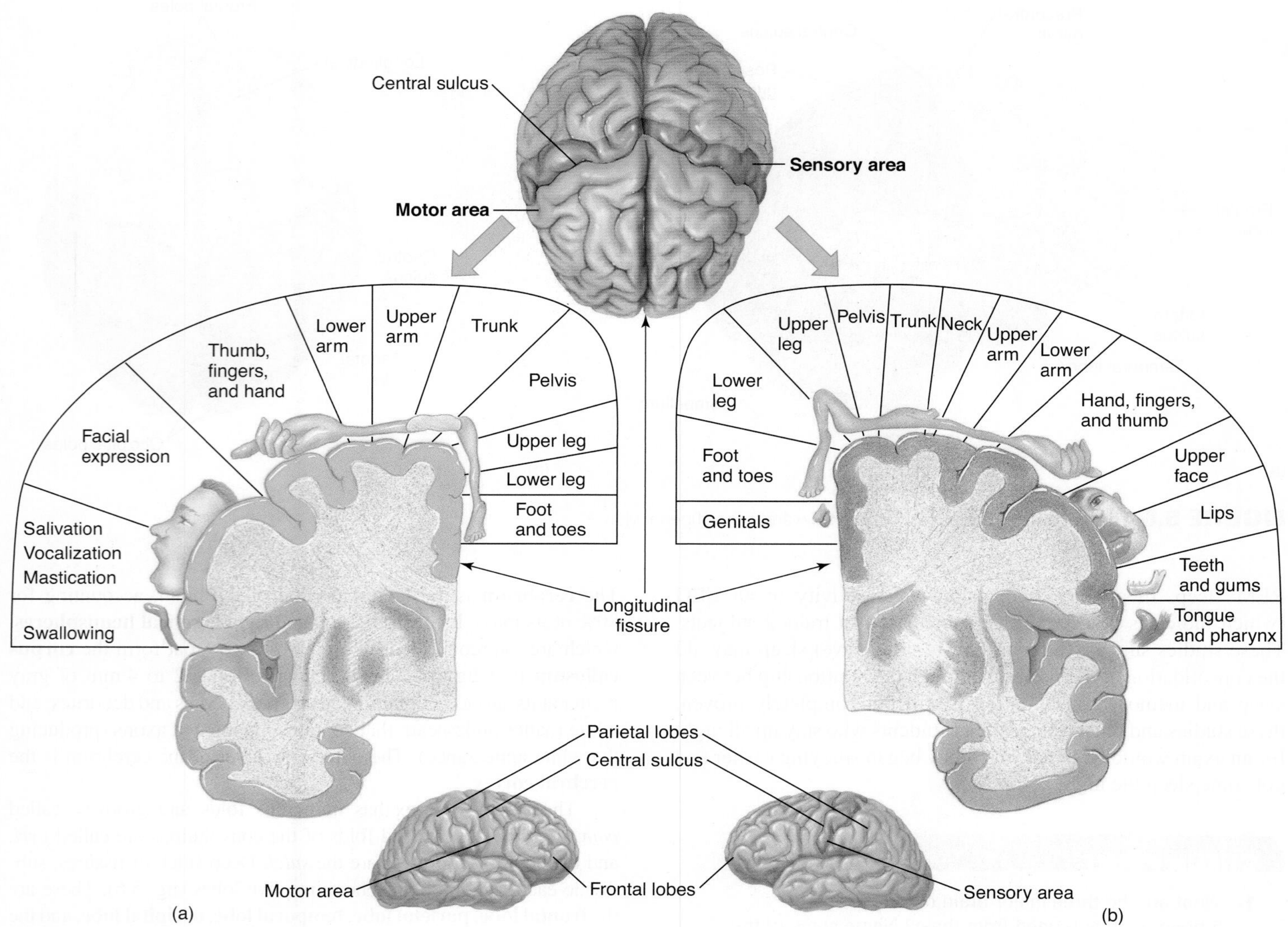

FIGURE 5.7 The motor and sensory cortex. (*a*) The motor cortex is the precentral gyrus, and (*b*) the somatesthetic sensory cortex (for senses coming from the skin, muscles, tendons, and joints) is the postcentral gyrus. The cartoons of distorted men represent the maps of these regions—that is, the parts of the body that these brain regions serve.

The areas of cerebral cortex devoted to sensory reception from, and motor control of, the face and hands are larger than for other parts of the body. This is because we need fine sensation and motor control of the muscles of facial expression, of the mouth, lips, and pharynx for speech, and of the hands and fingers for manual dexterity.

The **temporal lobe** contains auditory centers for receiving and analyzing the sense of hearing, and the **occipital lobe** is the primary area responsible for vision (fig. 5.8). The ear and the eye simply send action potentials to these two lobes of the cerebrum; the perceptions of sound and sight are produced in the brain. These lobes of the cerebrum have other very important functions in addition to sensory perception, some of which will be discussed later in this chapter. The **insula** appears to be involved in interpreting olfactory (smell) information and integrating sensations of pain with visceral responses.

The Right and Left Cerebral Hemispheres Are Specialized

Each cerebral hemisphere controls movement by means of the precentral gyrus, or motor cortex, of the *contralateral* (opposite) side of the body. This occurs as a result of the *decussation* (crossing over from one side to the other) of the axons from neurons in these cerebral hemispheres. Similarly, sensation from the skin, muscles, tendons, and joints in the right side of the body *project* (go) to the left hemisphere, and vice versa, because of the decussation of sensory axons.

However, each cerebral hemisphere receives information relating to both sides of the body, because the two hemispheres communicate with each other by means of the *corpus callosum*, a tract of about 200 million axons running from one hemisphere to the other. The left hemisphere (controlling the right side of the

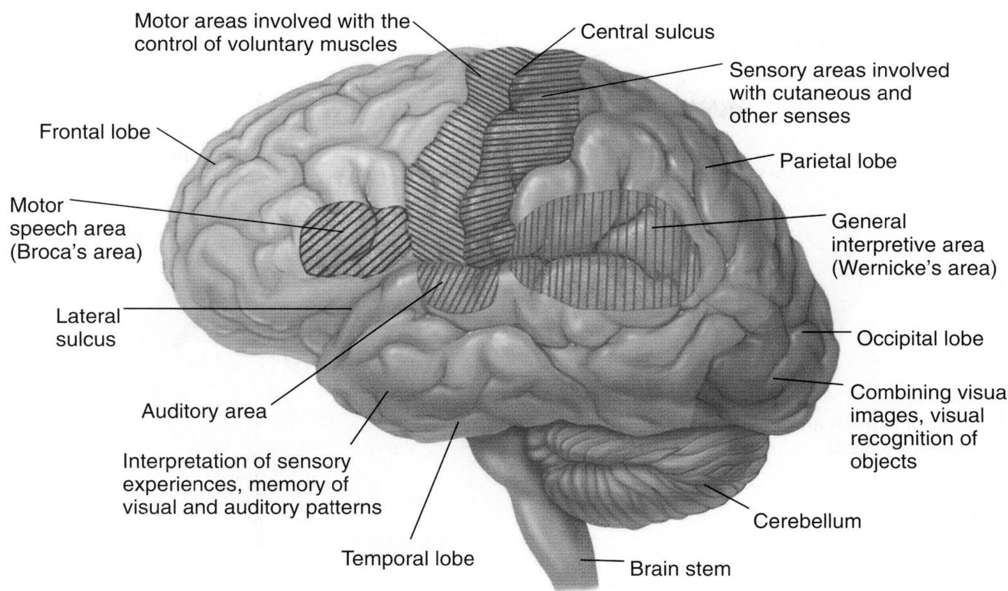

FIGURE 5.8 The lobes of the left cerebral hemisphere. This diagram shows the principal motor and sensory areas of the cerebral cortex.

body) is usually dominant, and research demonstrates that it's specialized for language and analytical ability. In contrast, the right hemisphere is specialized for spatial comprehension such as reading facial expressions and maps. This information was learned by experiments in humans in whom the corpus callosum had been cut to treat severe epilepsy.[3] The different specializations of the left and right cerebral hemispheres are called **cerebral lateralization**.

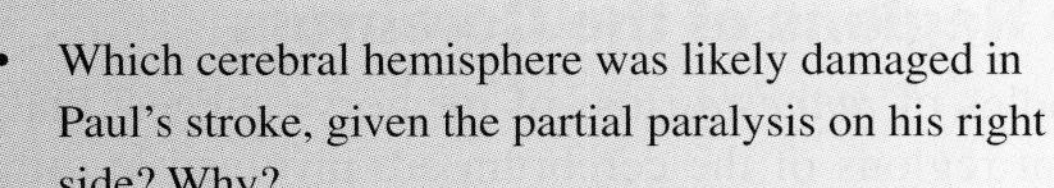

CLINICAL INVESTIGATION CLUES

Remember that Paul had great difficulty speaking, and was partially paralyzed on his right side, after his stroke.

- Which cerebral hemisphere was likely damaged in Paul's stroke, given the partial paralysis on his right side? Why?
- Which cerebral hemisphere was likely damaged in Paul's stroke, given his difficulty in speaking? Why?

Language Ability Resides in the Cerebral Cortex

Scientists have learned about the brain regions involved in language ability mostly through the study of *aphasias*—speech and language disorders. Aphasias are often produced by damage to the brain through head injury or stroke. *Broca's aphasia* is the result of damage to **Broca's area**, located in the left inferior frontal gyrus and surrounding areas (fig. 5.8). Because the left cerebral hemisphere controls muscles on the right side of the body, Broca's aphasia is often accompanied by weakness in the right arm and right side of the face. Although voluntary control over the muscles of speech is retained, when a person with Broca's aphasia tries to speak, the speech is slow and poorly articulated. People with Broca's aphasia understand what is spoken to them but have difficulty responding.

Wernicke's aphasia is caused by damage to **Wernicke's area**, usually located in the superior temporal gyrus of the left hemisphere. In this type of aphasia, language comprehension is destroyed. People with Wernicke's aphasia speak easily but use both real and made-up words tossed together like a mixed salad. People with Wernicke's aphasia can't understand either spoken or written language. From this and other information, it appears that oral and written language (involving the senses of hearing and vision) must project to Wernicke's area to be understood. Information from Wernicke's area must next project to Broca's area so that a person can respond. In order to speak or write, Broca's area must then project to the motor cortex of the precentral gyrus (fig. 5.9).

CLINICAL INVESTIGATION CLUES

Remember that Paul could understand Emma but couldn't speak to her.

- Paul's aphasia was likely caused by damage to which brain region of which cerebral hemisphere?
- If Paul couldn't comprehend what Emma said, which other brain region would likely have been affected by his stroke?

FYI [3]Roger Sperry (1913–1994) performed experiments on these "split-brain" patients that ingeniously revealed the nonverbal functions of the right hemisphere, demonstrating that it was conscious in its own artistic and creative way. Sperry won the Nobel Prize for this research in 1981, and upon receiving it in Stockholm, Sweden, said, "The whole world of inner experience (the world of the humanities), long rejected by twentieth-century scientific materialism, thus becomes recognized and included within the domain of science."

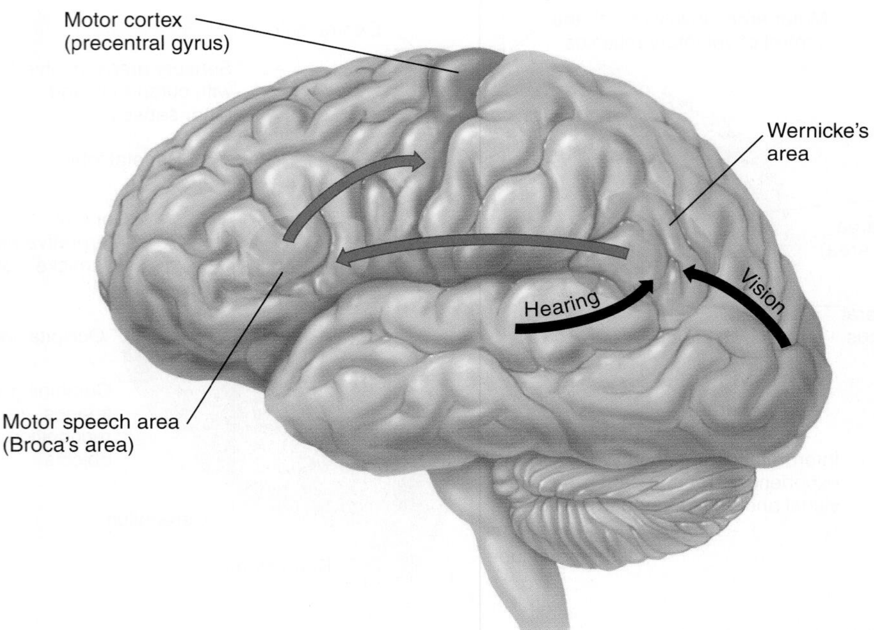

FIGURE 5.9 Brain areas involved in the control of speech. Damage to these areas produces speech deficits, known as aphasias. Wernicke's area, required for language comprehension, receives information from many areas of the brain, including the auditory cortex (for heard words), the visual cortex (for read words), and other brain areas. In order for a person to be able to speak intelligibly, Wernicke's area must send messages to Broca's area, which controls the motor aspects of speech by way of its input to the motor cortex.

The Limbic System Is Involved in Emotions

The **limbic system** consists of a group of cerebral nuclei (collections of neuron cell bodies) and axon tracts that form a ring around the brain stem (*limbus* = ring). The limbic system includes the *cingulate gyrus*, the *amygdaloid nucleus* (or *amygdala*), the *hippocampus*, and the *septal nuclei* (fig. 5.10). The hypothalamus, which is a part of the diencephalon (discussed later) rather than cerebrum, is also included with the limbic system because of its role in emotions.

The following are some of the functions of the limbic system and hypothalamus that relate to emotions:

1. **Aggression**. Stimulation of certain areas of the amygdala produces rage and aggression. Stimulation of particular areas of the hypothalamus can produce similar effects.
2. **Fear**. Fear can be produced by electrical stimulation of the amygdala and hypothalamus. People with damage to their amygdala have difficulty recognizing fear and anger in other people, implying that amygdala is required for learning and responding to fearful signals.
3. **Sex**. The hypothalamus and limbic system help regulate the sexual drive and sexual behavior. Other areas of the cerebrum are also involved in lower animals and even more so in humans.
4. **Goal-directed behavior (reward and punishment system)**. Experimental animals find shocks delivered by implanted electrodes in particular locations of the limbic system to be more rewarding than food or sex. Electric current that stimulates the same brain locations in humans provoke feelings of relaxation and relief from tension, but not ecstasy. Electrodes placed in slightly different brain locations in experimental animals stimulate a punishment system, so that the animals stop their behavior when those locations are stimulated.

Different Forms of Memory Involve Specific Regions of the Cerebrum

Clinical studies of *amnesia* (loss of memory) suggest that several different regions of the cerebrum are involved in different forms of memory. For example, people with head trauma may lose their memory of recent events but retain their older memories. This and other observations led scientists to distinguish between **short-term memory** and **long-term memory**. Long-term memory is divided into **nondeclarative** and **declarative memory**. Nondeclarative memory refers to the memory of simple skills and conditioning, such as how to tie shoelaces. Declarative memory refers to the memory of facts (such as names associated with the skeletal system) and events (such as a lab practical on the skeletal system you may have taken in an anatomy course).

People with amnesia have impaired declarative memory. Scientists have discovered that the **medial temporal lobes**, particularly the *hippocampus* and *amygdala* (fig. 5.10), are needed to convert short-term into long-term declarative memories. Although the hippocampus is needed for maintaining recent memories, it is no longer required once the memory has been consolidated into a

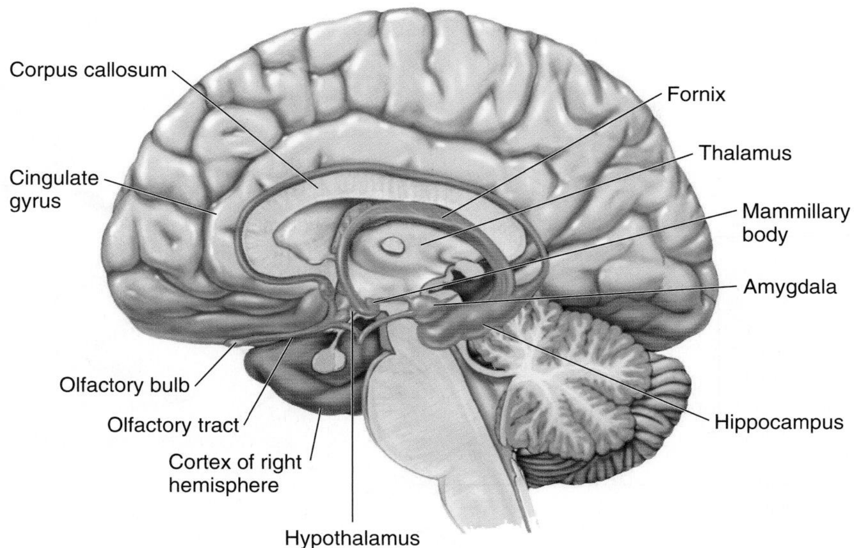

FIGURE 5.10 The limbic system. The left temporal lobe has been removed in this figure to show the structures of the limbic system (green). The limbic system consists of particular nuclei (aggregations of neuron cell bodies) and axon tracts of the cerebrum that cooperate in the generation of emotions. The hypothalamus, though part of the diencephalon rather than the cerebrum (telencephalon), participates with the limbic system in emotions.

more stable, long-term form.[4] The **inferior temporal lobes** appear to store long-term visual memories, and the left inferior temporal lobe is important for performing mathematical calculations. The **prefrontal cortex**, the most anterior portion of the frontal lobes, sends signals to the inferior temporal lobes and is needed for complex planning and problem-solving activities.

According to current thinking, particular aspects of a memory—visual, auditory, olfactory, spatial, and so on—are stored in different locations, and the cooperation of all of these areas is required to retrieve the complete memory. But how is the memory stored? Long-term memory involves long-term improvements in synaptic transmission in particular neural pathways. Scientists have shown that neurons require the synthesis of new proteins for long-term memory storage, and they have extensively studied synapses that use the neurotransmitter glutamate. Glutamate is the major transmitter in the hippocampus, and experiments demonstrate that the function of these synapses improves when they're stimulated in a particular way.

The Basal Nuclei (Basal Ganglia) Help Control Body Movements

The **basal nuclei** (commonly referred to as **basal ganglia**) are masses of gray matter (neuron cell bodies and dendrites) located deep within the cerebrum (fig. 5.11). The most prominent structure of the basal nuclei is the *corpus striatum*. The corpus striatum is actually a collection of gray-matter nuclei, as indicated in figure 5.11. The areas of the cerebral cortex that control movement (primarily the precentral gyrus, or motor cortex) send axons to the basal nuclei. In turn, neurons in the basal nuclei send axons to the thalamus (part of the diencephalon, discussed in the next section), which sends axons to the precentral gyrus, thereby completing a *motor circuit*.

The axons that extend from neurons in the basal nuclei release GABA, an inhibitory neurotransmitter (one that produces hyperpolarization, or IPSPs). This inhibition is important for the neural control of muscles and thus body movements. The activity of the GABA-releasing neurons of the basal nuclei is partly regulated by axons in the basal nuclei that extend from the *substantia nigra* (a part of the midbrain; see fig. 5.13). These neurons release dopamine as a neurotransmitter, and when these neurons degenerate, the resulting deficiency of dopamine in the basal ganglia produces Parkinson's disease.

CLINICAL APPLICATIONS

Alzheimer's disease is the most common neurodegenerative disease. People with Alzheimer's disease have (1) a loss of neurons in the hippocampus and cerebral cortex; (2) an accumulation of intracellular proteins forming *neurofibrillar tangles*; and (3) an accumulation of extracellular protein (mostly *amyloid β-peptide*, or *Aβ*) in deposits called *senile plaques*. The majority of Alzheimer's disease cases don't run in families and probably result from an interaction between many genes and the environment. Clinical observations support the notion that an intellectually rich and physically active lifestyle may offer some protection against Alzheimer's disease, perhaps by building up a "cognitive reserve." There is also evidence that a diet restricted in calories and saturated fats, and enriched in vitamins C, E, and folate, may afford some protection.

FYI [4]Two patients illustrate the importance of being able to consolidate short-term into long-term memories. An amnesiac patient ("E.P."), with bilateral damage to his medial temporal lobes, was able to remember the neighborhood he left 50 years earlier but not his current neighborhood. Another patient ("H.M.") had both right and left medial temporal lobes removed to treat his epilepsy. He could be introduced to a person and carry on a conversation, but would have no memory of the event a few minutes later. Every person that he met after his surgery, every event that he experienced, was as if for the first time.

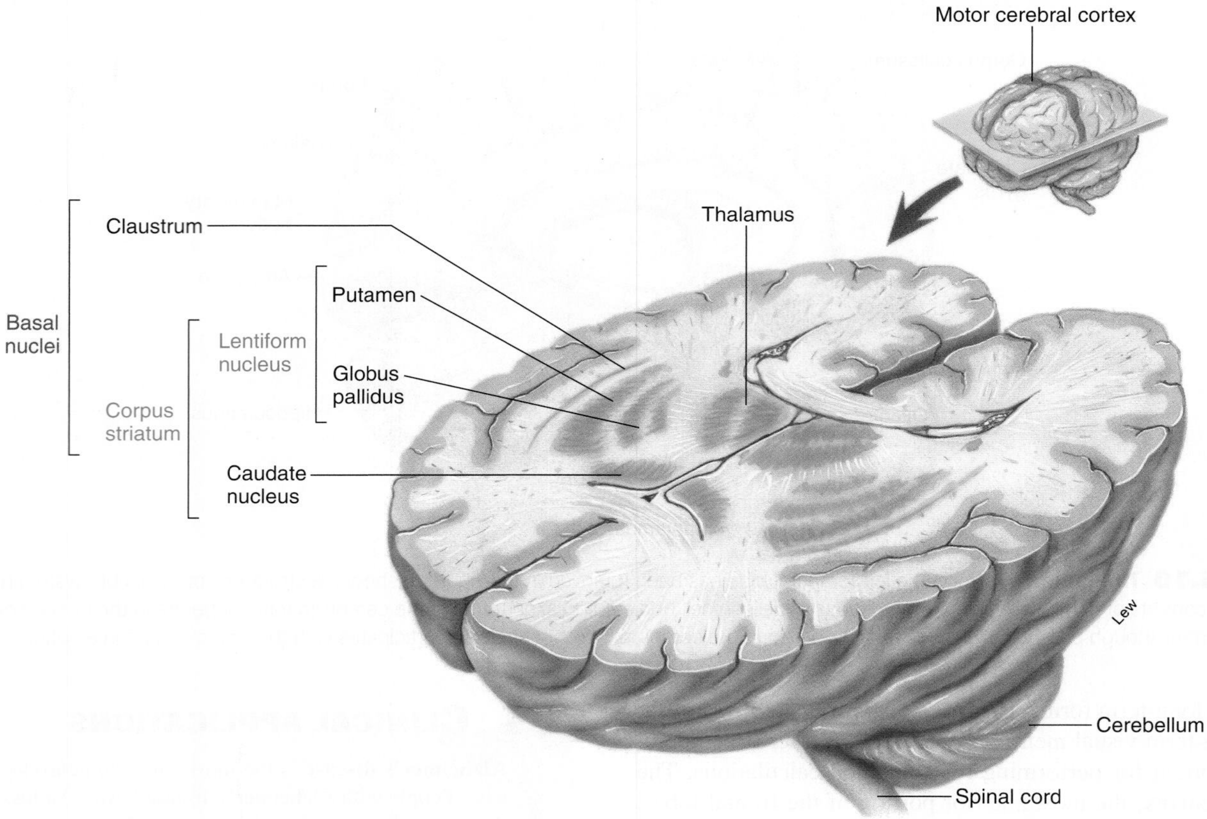

FIGURE 5.11 The basal nuclei. Also called the basal ganglia, the basal nuclei consist of several nuclei (groupings of neuron cell bodies), which form areas of gray matter deep in the cerebrum. The basal nuclei participate in the brain's control of body movement.

Clinical Applications

Degeneration of the neurons in the *caudate nucleus*, one the basal nuclei and part of the corpus striatum, occurs in **Huntington's disease.** This produces *chorea*—a disorder characterized by uncontrolled, jerky movements. Degeneration of dopamine-releasing neurons that go from the substantia nigra to the caudate nucleus produces the symptoms of **Parkinson's disease**. The symptoms of Parkinson's disease include muscular rigidity, resting tremor, and difficulty initiating voluntary movements.

Clinical Investigation Clues

Remember that Emma thought Paul had Parkinson's disease, but Paul claimed his symptoms were caused by a medicine he was taking, haloperidol.

- What are some of the symptoms of Parkinson's disease, and what causes this disease?
- Given that haloperidol blocks certain receptors for dopamine, how might this drug account for Parkinson's-diseaselike symptoms?

Check Point

1. Describe the structure of the cerebrum and some of the functions of its lobes.
2. Explain the specializations of the right and left cerebral hemispheres and the role of the corpus callosum.
3. What are the roles of Broca's and Wernicke's areas in language, and what are the types of aphasias produced by damage to each area?
4. What is the meaning of the upside-down people indicated as maps of the precentral and postcentral gyri? What does the term *decussation* mean, and what is its significance when discussing the motor cortex?
5. Describe the location and composition of the limbic system, and indicate some of its physiological functions.
6. Distinguish between short-term and long-term memory, and between nondeclarative and declarative memory. Indicate some of the brain regions involved.
7. Identify the basal nuclei (basal ganglia), and describe their physiological and medical importance.

The Diencephalon Is Small but Important

The diencephalon of the forebrain includes the thalamus, the hypothalamus, and part of the pituitary gland. The thalamus serves as a relay center for sensory information. The hypothalamus has many nuclei that perform a great variety of important functions, including controlling the pituitary gland, regulating circadian rhythms, the responses of thirst and hunger, and participating with the limbic system in emotions.

The hypothalamus plays an important role in the physiology of the nervous and endocrine systems. It also affects other body systems through its regulation of the pituitary gland, its control of circadian rhythms (those that repeat on a daily basis), its centers for hunger, thirst, and body temperature, and its contributions to emotions. This is our first discussion of the hypothalamus, but it won't be the last; we'll see it pop up frequently in future chapters as we discuss other aspects of physiology. So, get ready to add another important block to the pyramid of physiological knowledge you're constructing.

The *diencephalon* is part of the forebrain, and yet this small area is completely surrounded by the cerebral hemispheres (see fig. 5.3). Don't let its small size fool you; the diencephalon performs astonishingly diverse functions. The **thalamus** comprises four-fifths of the area of the diencephalon and functions primarily as a relay center. All sensory information coming into the CNS must ultimately be brought to the cerebral hemispheres for analysis and sensory perception, and all except smell are sent first to the thalamus. (Olfactory information goes directly from the nose to the cerebral cortex, bypassing the thalamus.) After synapsing in the thalamus, neurons send sensory information to the appropriate part of the cerebral cortex. For example, the *lateral geniculate nuclei* of the thalamus relay visual information to the occipital lobe, and the *medial geniculate nuclei* of the thalamus relay auditory information to the temporal lobe.

The **hypothalamus**, as its name implies, is located just below the thalamus (and just above the pituitary gland). Although quite small, it's packed with collections of neuron cell bodies, or nuclei (fig. 5.12). These nuclei have a great many functions. For example, electrical stimulation of the lateral hypothalamus can make an animal eat, whereas stimulation of the medial hypothalamus inhibits eating. There are centers in the hypothalamus that produce the sensation of thirst, causing drinking and other responses; other centers serve as a thermostat, causing shivering, sweating, panting, and other activities that regulate body temperature. In addition, centers in the hypothalamus contribute to the regulation of sleep and wakefulness, sexual arousal, and emotions of anger, fear, pain, and pleasure. In its regulation of emotions, the hypothalamus works together with the limbic system, as previously discussed.

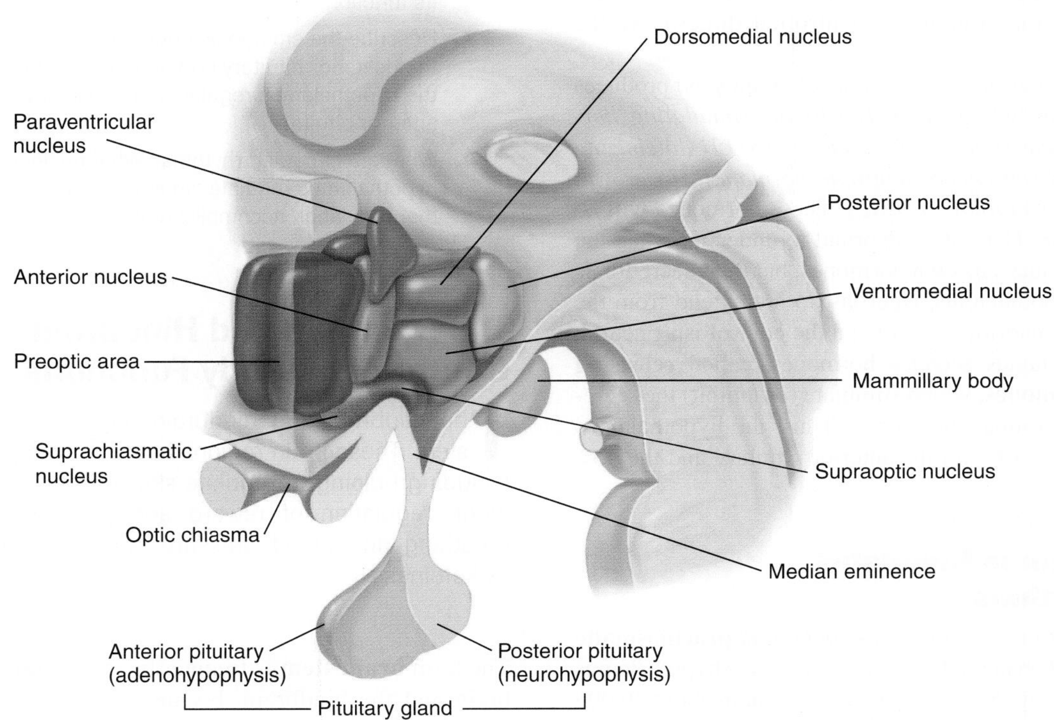

FIGURE 5.12 Nuclei within the hypothalamus. The hypothalamus is just a small area above the pituitary gland, but it contains many nuclei that perform various functions. For example, the suprachiasmatic nucleus (SCN) is the master clock for the body's circadian rhythms, and the supraoptic and paraventricular nuclei produce the hormones that the posterior pituitary secretes.

The Hypothalamus Controls the Pituitary Gland

The hypothalamus and **pituitary gland** (or **hypophysis**) have a close anatomical and physiological relationship. In the embryo, neural tissue of the diencephalon migrates downward to form a stalk connecting the hypothalamus to the **posterior pituitary**. Because the posterior pituitary is derived from neural tissue, it's also called the **neurohypophysis**. The **anterior pituitary** is not derived from neural tissue; instead, it's derived from epithelial tissue from the roof of the embryo's mouth, which detaches and migrates upward. As described in chapter 2, glands are classified as epithelial tissue, and so the anterior pituitary, or **adenohypophysis**, is truly glandular. In this text, the terms *anterior pituitary* and *posterior pituitary* will be used instead of *adenohypophysis* and *neurohypophysis*. Still, it's useful for you to know the latter pair of terms, so you can remember the embryonic origins and tissue compositions of the two.

The posterior pituitary secretes two hormones: (1) *ADH* (*antidiuretic hormone*), also known as *vasopressin*; and (2) *oxytocin*. The functions of these hormones will be discussed in chapter 8. However, here you should understand that these two hormones are *not* produced in the posterior pituitary. Instead, they are produced in two nuclei of the hypothalamus (the *supraoptic* and *paraventricular* nuclei) and are then transported within axons that run from the hypothalamus to the posterior pituitary. The posterior pituitary stores these two hormones and secretes them in response to appropriate stimulation from neurons that originate in the hypothalamus. Thus, ADH and oxytocin are hormones produced by the hypothalamus, and their secretion from the posterior pituitary is controlled directly by the hypothalamus.

In contrast, the anterior pituitary (adenohypophysis) produces its own hormones, including (1) *TSH* (*thyroid-stimulating hormone*), (2) *FSH* (*follicle-stimulating hormone*), (3) *LH* (*luteinizing hormone*), (4) *ACTH* (*adrenocorticotropic hormone*), (5) *growth hormone* (*GH*), and (6) *prolactin* (*PRL*). These hormones will be discussed in chapter 8. Here, it's important to understand that the anterior pituitary produces its own hormones, but their secretion is controlled by the hypothalamus. Neurons do not extend from the hypothalamus to the anterior pituitary, so the control isn't neural. Instead, the hypothalamus secretes hormones, called **releasing** (and **inhibiting**) **hormones**, which stimulate (or inhibit) the secretion of the anterior pituitary hormones. Thus, the hypothalamus regulates the secretions of both the anterior pituitary and the posterior pituitary.

The Hypothalamus Regulates Circadian Rhythms

The anterior region of the hypothalamus contains **suprachiasmatic nuclei (SCN)**, located on the right and left sides of the hypothalamus (only one is shown in fig. 5.12). These nuclei contain about 20,000 neurons that function as "clock cells," with electrical activity that automatically repeats itself every 24 hours. The SCN is believed to be the major brain region that regulates the body's **circadian rhythms** (*circa* = about; *dia* = day). You are probably aware of some of your circadian rhythms, such as the time of day you feel most alert and productive, and the time of day you feel drowsy and listless (hopefully, this doesn't coincide with the time you're in your physiology class). There are a great many physiological activities that similarly wax and wane on a 24-hour cycle, affecting most of your body functions.

In order for the 24-hour cycle set by the SCN to relate to the 24-hour cycle of day and night, information regarding light and dark must be relayed to the SCN. This information is received by the SCN from the retinas of the eyes by way of a tract of axons. The SCN, synchronized to the light-dark cycle by information from the eyes, must then be able to influence other organs in the body. This is accomplished through neural connections to other brain regions, and through anterior pituitary hormone secretion (because the pituitary gland is controlled by the hypothalamus, as described in the last section). The anterior pituitary then regulates the secretion of some other endocrine glands (the thyroid, adrenal cortex, and gonads). The SCN also controls the secretion of the hormone *melatonin* from the *pineal gland*, located within the brain. Melatonin is a major regulator of circadian rhythms (chapter 8; see fig. 8.28).

CHECK POINT

1. Name two pairs of nuclei within the thalamus and describe their function. What sensory modality does not go through the thalamus to the cerebral cortex?
2. Describe the location of the hypothalamus and several of its functions.
3. Describe the embryonic origin and tissue composition of the posterior pituitary and anterior pituitary. Explain how the hypothalamus regulates the posterior pituitary and anterior pituitary.
4. What are circadian rhythms? What nuclei in the hypothalamus regulate circadian rhythms, and how is that regulation accomplished?

The Midbrain and Hindbrain Control Vital Body Functions

The midbrain and hindbrain together compose the brain stem. These brain regions have many important functions—including helping to regulate skeletal muscles for body movement, regulation of reward and punishment, regulation of breathing and blood pressure, and regulation of sleep and wakefulness.

The term **brain stem** is frequently used to refer to both the midbrain and the hindbrain, because these structures appear as a stem holding up the larger cerebrum. The midbrain is located between the diencephalon and the pons (see fig. 5.3). It contains a number of nuclei with diverse functions. For example, the **corpora quadrigemina** are four rounded elevations consisting of two *superior colliculi*, which control visual reflexes, and two *inferior colliculi*, which control auditory reflexes. The midbrain also contains dopamine-releasing neurons in two axon tracts. One of these goes from the **substantia nigra** (a nucleus of the midbrain; fig. 5.13) to the basal nuclei of the cerebrum. This *nigrostriatal dopamine system* ("striatal" refers to the corpus striatum of the basal nuclei; see fig. 5.11) is involved in the control of body movements, and loss of these neurons produces Parkinson's disease.

The other dopamine-releasing tract of axons that originates in the midbrain goes to the limbic system of the forebrain. As previously discussed, the limbic system is a group of brain regions involved in emotions. This *mesolimbic dopamine system* (fig. 5.13) participates in emotional reward. This explains why the neurotransmitter dopamine is involved in goal-directed behavior, as seen in experimental animals when their dopamine system is blocked, and in people who abuse drugs. Abused drugs promote the release of dopamine in the **nucleus accumbens**, buried deep within the frontal lobe. However, other neurotransmitters, including serotonin and norepinephrine, can also be released in response to abused drugs.

CLINICAL INVESTIGATION CLUES

Remember that Paul claimed that he displayed Parkinson's-like symptoms because he was taking haloperidol, a drug used to treat schizophrenia.

- Given that haloperidol blocks the ability of dopamine to bind to particular dopamine receptors, which brain system is likely responsible for the ability of this drug to help treat schizophrenia?
- Given the dopamine-blocking action of haloperidol, which brain system is affected when the drug produces Parkinson's-like symptoms?

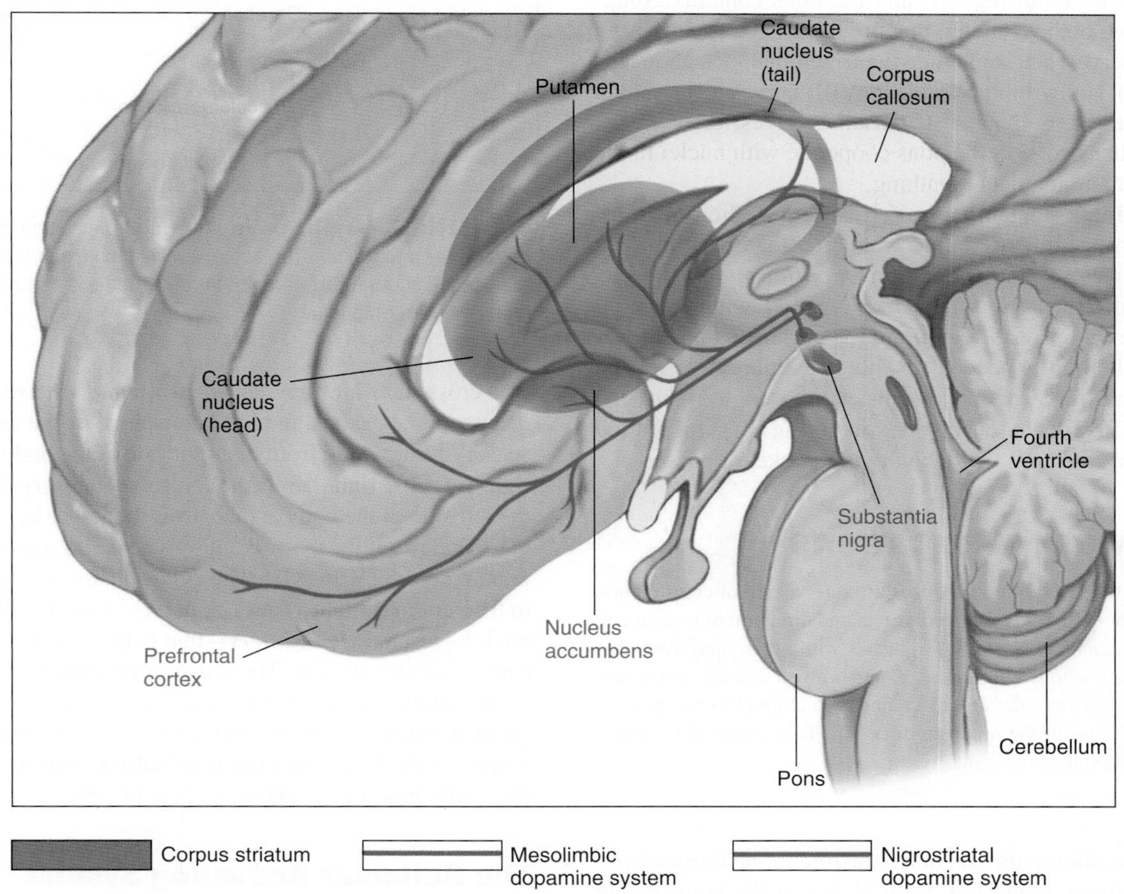

FIGURE 5.13 Dopamine-releasing neurons from the midbrain. There are two systems of axon tracts that originate in the midbrain and release dopamine at their synapses. The nigrostriatal dopamine system originates in the substantia nigra and synapses in the corpus striatum of the basal nuclei; it's used for the control of body movements. The mesolimbic dopamine system originates in the midbrain and synapses in the nucleus accumbens and prefrontal cortex; it's required for emotional reward.

Clinical Applications

Abused drugs may produce positive reinforcement through the activation of the mesolimbic dopamine system, causing dopamine to be released in the nucleus accumbens. *Nicotine* from tobacco stimulates dopamine-releasing neurons by means of its stimulation of nicotinic ACh receptors (chapter 4). *Heroin* and *morphine* activate the dopamine-releasing neurons by means of their interaction with *opioid receptors* in these neurons. *Cocaine* and *amphetamines* act in the nucleus accumbens to inhibit the reuptake of dopamine into presynaptic axon terminals (chapter 4). Ironically, drug abuse can desensitize neurons to dopamine, and so lessen the rewarding effect of dopamine release. This can lead to drug tolerance, so that higher amounts of the drug are needed to get a reward.

The Hindbrain Includes the Pons, Cerebellum, and Medulla Oblongata

The pons and the cerebellum together compose the part of the hindbrain called the *metencephalon*. The **pons** is seen as a rounded bulge on the underside of the brain (fig. 5.14). Axons interconnect the pons with other brain regions, and the pons contains groupings of neuron cell bodies (nuclei) that provide motor axons within several cranial nerves. These include the trigeminal (V), abducens (VI), facial (VII), and vestibulocochlear (VIII) nerves. (The roman numerals indicate the number of the cranial nerves; there are 12 all together). Other nuclei in the pons cooperate with nuclei in the medulla oblongata to regulate breathing.

The **cerebellum**, containing more than 100 billion neurons, is the second-largest structure in the brain after the cerebrum. The cerebellum receives sensory input from *proprioceptors* (muscle, tendon, and joint receptors) and works together with the basal nuclei and motor cortex of the cerebrum to coordinate body movements. The cerebellum is needed for motor learning, and for the proper timing and force of movements. For example, you need your cerebellum in order to touch your nose, bring a forkful of food to your mouth, or find keys by touch in your pocket or purse.

Clinical Applications

Damage to the cerebellum produces **ataxia**—a lack of coordination resulting from errors in the speed, force, and direction of movement. The movements and speech of a person with ataxia may resemble those of someone who is intoxicated. A person with damage to the cerebellum may reach and miss an object, and then attempt to compensate by moving the hand in the opposite direction. This can produce back-and-forth oscillations of the arm.

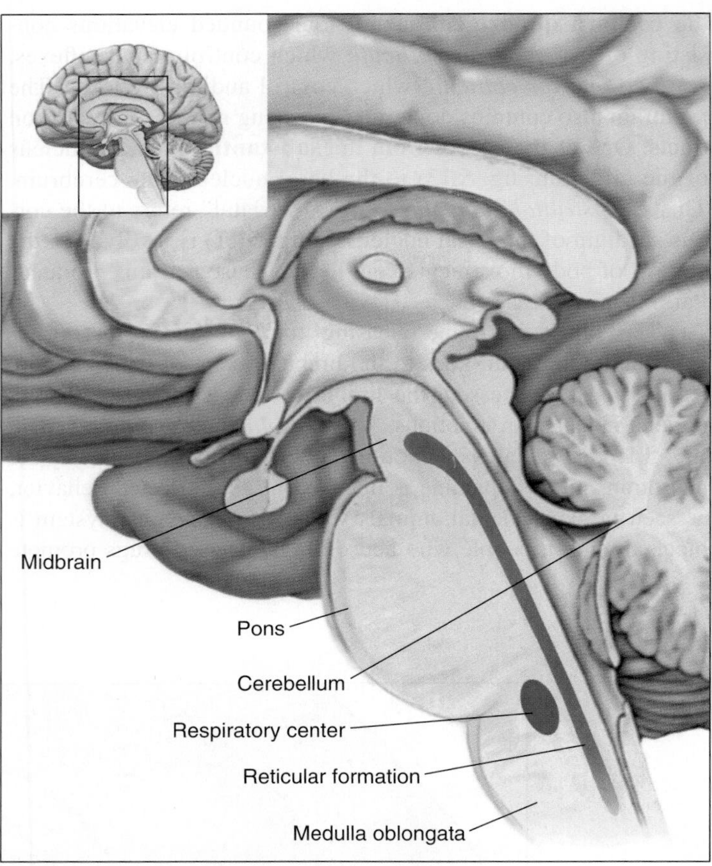

FIGURE 5.14 The brain stem. The brain stem consists of the midbrain and hindbrain. The reticular formation extends through both. The area containing the respiratory control center is indicated as a colored oval in the medulla oblongata.

The **medulla oblongata**, the only structure in the *myelencephalon* of the hindbrain, is the last 3 cm (1 in.) of the brain before reaching the spinal cord. All of the descending motor and ascending sensory tracts running between the spinal cord and the brain pass through the medulla. Many of these fiber tracts decussate (cross to the opposite side) in the medulla, forming elevated structures called the *pyramids*. (Motor axons descending from the precentral gyrus that cross here are thus called *pyramidal tracts*.) Because of the decussation of axons, the left hemisphere of the cerebrum receives sensory information from the right side of the body and controls the right side; the right hemisphere receives sensory information from the left side of the body and controls the left side.

Nuclei within the medulla give rise to axons within several cranial nerves (VIII–XII). The *vagus nuclei*, for example, give rise to the very important vagus (X) nerves. The medulla contains other nuclei—termed *vital centers*—that help regulate breathing and the cardiovascular system. The **vasomotor center** controls the activity of autonomic nerves that innervate blood vessels; the **cardiac control center** regulates the activity of autonomic nerves that innervate the heart; and the **respiratory center** directly controls the spinal motor neurons needed for breathing.

The Reticular Activating System Regulates Sleep and Wakefulness

To fall asleep, we must be able to "tune out" sensory stimulation that ascends to the cerebral cortex. Conversely, we quickly awaken when the cerebral cortex is alerted to incoming sensory information. These abilities, and the normal pattern of sleep and wakefulness that results, depend upon the **reticular formation** (fig. 5.14). The reticular formation is an interconnected group of neurons (*rete* = net) that go from the pons through the midbrain. The activity of these brain-stem neurons constitutes an arousal system known as the **reticular activating system (RAS)**.

The RAS (fig. 5.15) includes neurons that release ACh in the thalamus; neurons that release the monoamine neurotransmitters (dopamine, norepinephrine, histamine, and serotonin) in different locations in the cerebral cortex; and neurons that release a polypeptide neurotransmitter, called by some *hypocretin-1*. Scientists demonstrated that *narcolepsy* (a genetic disorder in which a person falls asleep at inappropriate times, despite having adequate amounts of sleep) is caused by degeneration of those polypeptide-releasing neurons.

The RAS is inhibited by another group of neurons in the hypothalamus that release GABA, an inhibitory neurotransmitter (chapter 4). The inhibitory effects of GABA and the arousal effects of the monoamine neurotransmitters are believed to function as a switch that controls when we fall asleep and wake up.

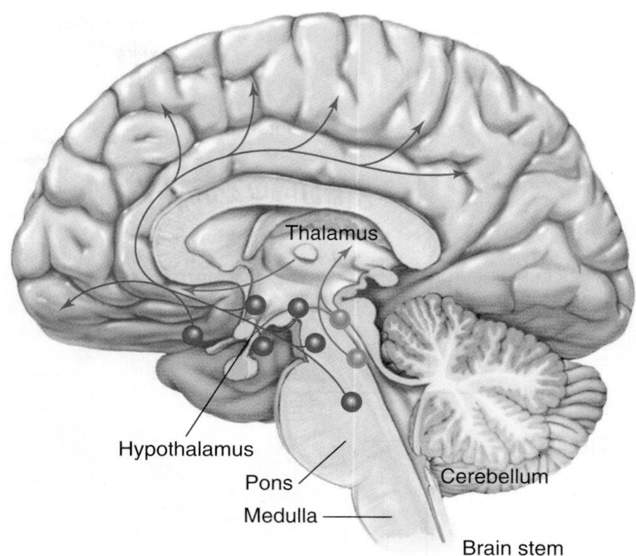

FIGURE 5.15 The reticular activating system (RAS). The groups of neurons shown in orange project to the thalamus, where they enhance the arousal of the cerebral cortex to sensory information relayed from the thalamus. The groups of neurons shown in red project to various locations in the cerebral cortex and more directly arouse the cerebral cortex to ascending sensory information. Activity of the RAS promotes wakefulness, and inhibition of the RAS promotes sleep.

Clinical Applications

The effectiveness of the monoamine-releasing neurons of the RAS is increased by *amphetamines*, which inhibit the dopamine reuptake transporter. This increases the amount of dopamine in the synaptic gap, promoting arousal. The antihistamine *Benadryl* can cross the blood-brain barrier and inhibit the effect of the histamine-releasing neurons of the RAS, causing drowsiness. (Antihistamines that don't cause drowsiness, such as *Claritin*, cannot cross the blood-brain barrier.) Drowsiness caused by *benzodiazepines* (such as *Valium*), *barbiturates*, *alcohol*, and most *anesthetic gases* is due to the ability of these chemicals to enhance the activity of GABA receptors. This increases the ability of GABA to inhibit the RAS, reducing arousal and promoting sleepiness.

Check Point

1. Describe the functions of the dopamine pathways from the midbrain, and some of the medical implications of these pathways.
2. What are some important functions of the cerebellum, and how is a person affected if the cerebellum is damaged?
3. Which cranial nerves originate in the pons and medulla oblongata? What vital centers are contained in the medulla oblongata?
4. Describe the RAS, identify some of the neurotransmitters involved, and explain the physiological significance of the RAS.

The Spinal Cord Transmits Sensory and Motor Information

The spinal cord contains a central area of gray matter surrounded by white matter. The white matter is composed of ascending sensory tracts and descending motor tracts. The gray matter contains the cell bodies of motor and association neurons.

The hindbrain ends and the spinal cord begins at the foramen magnum of the skull, and the spinal cord ends at the first lumbar vertebra. Like the brain, the spinal cord is part of the CNS and consists of gray and white matter. However, the arrangement of these two are reversed in the spinal cord compared to the brain; the white matter is on the outside of the spinal cord, and the gray matter is on the inside, shaped like a butterfly (fig. 5.16). The gray matter contains the neuron cell bodies of spinal motor and association neurons. If you examine figure 5.16, you'll see that the cell bodies of sensory neurons are packaged in dorsal root ganglia outside the spinal cord. (Remember, ganglia are collections of cell bodies in the PNS.) The PNS is covered in chapter 6.

The white matter contains axon tracts. Those that are sensory are *ascending tracts* (going to the brain); those that are motor are *descending tracts* (going from the brain down the spinal cord). The names of the ascending tracts start with the prefix *spino-* and end with the brain region where the axons synapse, such as the *spinothalamic tracts* (fig. 5.17). The names of the descending tracts begin with a prefix naming the brain region where the tract originates and ending with the suffix *-spinal*, such as the *corticospinal tracts* (fig. 5.18).

The descending tracts consist of two groups: the **corticospinal tracts** (also called **pyramidal tracts**) and the **extrapyramidal tracts**. The corticospinal tracts originate in the *precentral gyrus*, which is called the *motor cortex* (described earlier with the cerebrum). About 80% to 90% of these fibers decussate (cross) in the pyramids of the medulla oblongata, which is why these tracts are also called pyramidal tracts. Because of the decussation, the right

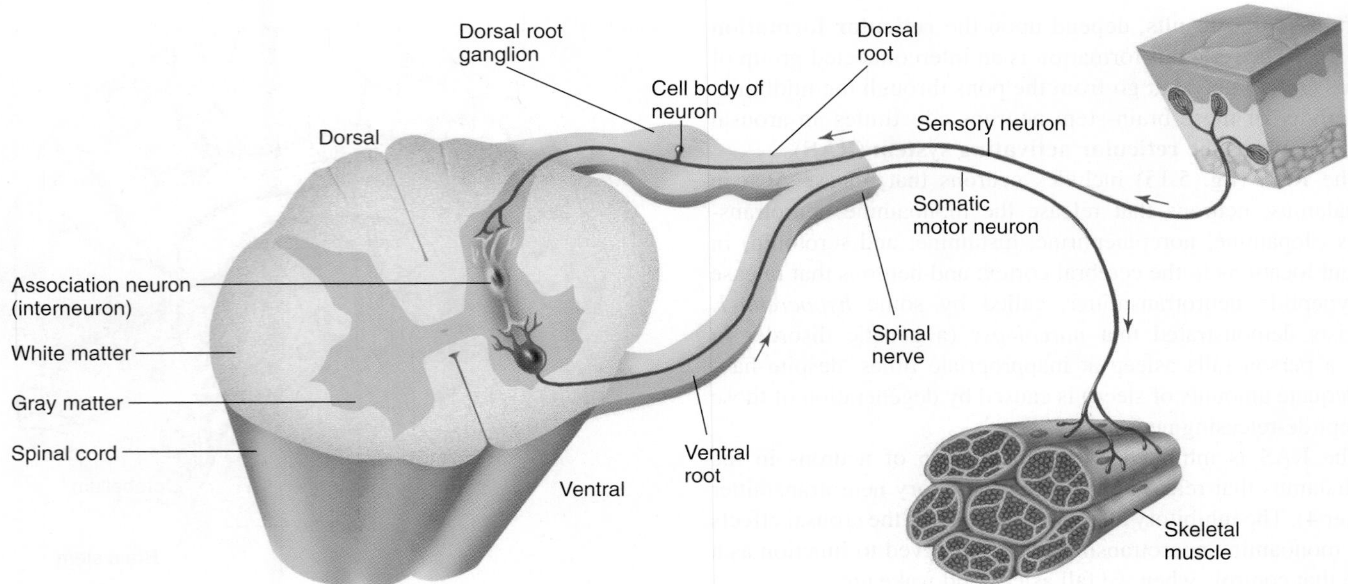

FIGURE 5.16 Cross section of the spinal cord. The gray matter is inside, in a butterfly shape, and the white matter is to the outside of the spinal cord. The gray matter contains neuron cell bodies of motor neurons (shown in purple) and association neurons (yellow). The cell bodies of sensory neurons (green) are outside the spinal cord, within dorsal root ganglia. The dorsal and ventral roots of a spinal nerve are also illustrated, and the "wiring diagram" presents a reflex arc.

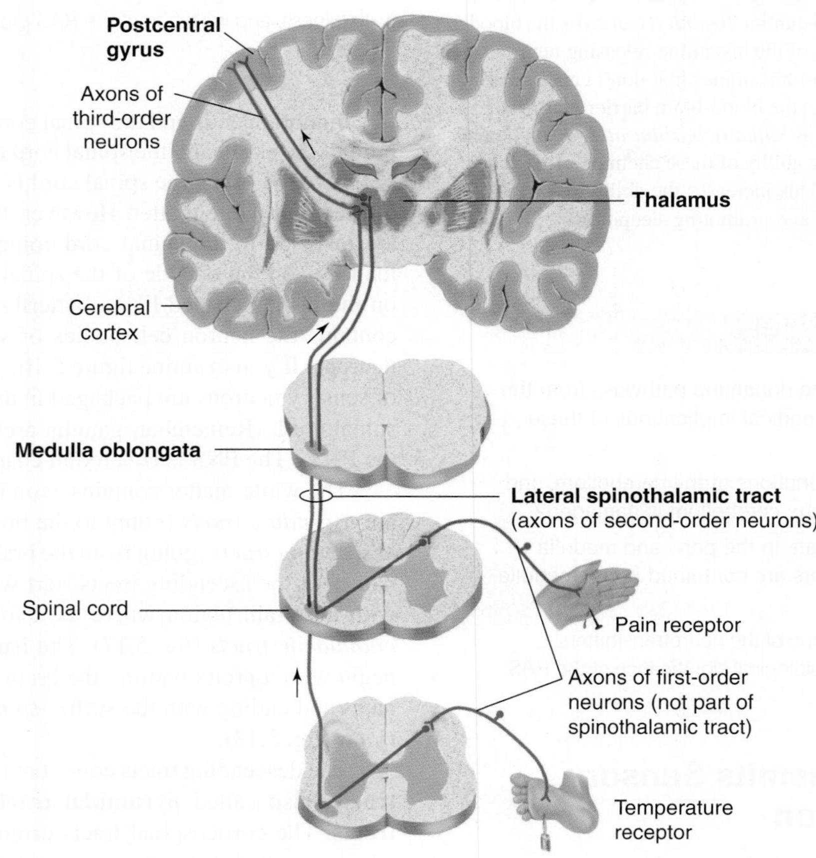

FIGURE 5.17 Ascending sensory neuron tracts. Sensory neurons of spinal nerves (first-order sensory neurons) synapse with second-order association neurons, which decussate (cross over to the opposite side) and project to the thalamus. These then synapse in the thalamus with third-order neurons, which project to the postcentral gyrus. This is the somatesthetic sensory cortex previously shown in figure 5.7.

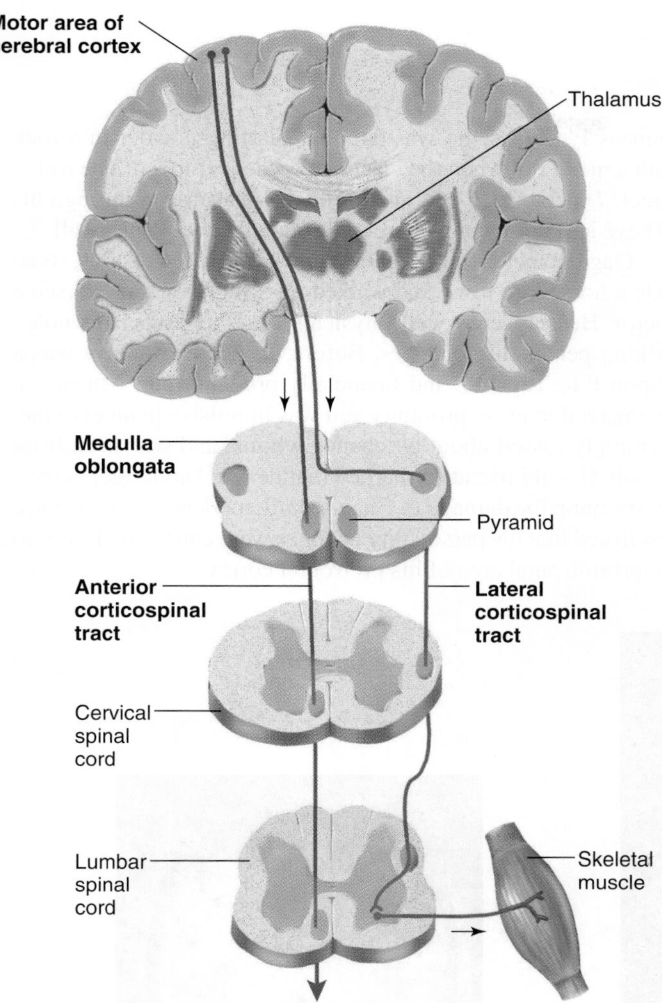

FIGURE 5.18 Descending corticospinal (pyramidal) motor tracts. These tracts contain axons that pass from the precentral gyrus of the cerebral cortex down the spinal cord to make synapses with spinal interneurons and lower motor neurons.

cerebral hemisphere controls the muscles on the left side of the body, and the left cerebral hemisphere controls the muscles on the right side of the body. The extrapyramidal motor tracts originate in various brain locations, including different parts of the basal nuclei (basal ganglia) of the cerebrum, the thalamus of the diencephalon, the substantia nigra of the midbrain, and the cerebellum of the hindbrain.

Because of the decussation of descending motor tracts, people with damage to the right cerebral hemisphere, particularly in the parietal lobe, have motor difficulties on the left side of their body. However, patients with damage in the left hemisphere often have impaired skilled motor activity of the right and left hands. These and other observations suggest that the left hemisphere is specialized for skilled motor control of both hands. But the left side of the body is controlled by the right cerebral hemisphere; how can the left hemisphere influence the left hand? This can result because of axons that project from the left to the right hemisphere through the corpus callosum. The right hemisphere directly controls the left hand, but the left hemisphere can influence this control by way of axons that pass through the corpus callosum.

Clinical Investigation Clues

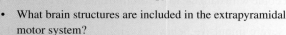

Remember that Paul took the drug Haldol (haloperidol) to treat his schizophrenia, and it caused Parkinson's-like symptoms. The medical literature describes these as "extrapyramidal effects."

- What brain structures are included in the extrapyramidal motor system?
- How does the extrapyramidal system relate to Parkinson's disease?
- How does the pyramidal motor system relate to Paul's paralysis on his right side as a result of his stroke?

Clinical Applications

The corticospinal tracts appear to be particularly important for voluntary, complex movements. For example, speech is impaired if there is damage to the corticospinal tracts in the thoracic (chest) region of the spinal cord, whereas involuntary breathing continues. Damage to the corticospinal tracts can be medically tested by the presence of the **Babinski reflex**. In this test, the sole of the foot is stimulated in a particular way that causes normal adults to produce a downward flexion, or curling, of the toes. When normal infants or adults with damage to their corticospinal tracts are stimulated in this way, they produce the Babinski reflex: their toes fan and their great toe extends upward.

Check Point

1. Explain how the right hemisphere receives sensory information from the left side of the body, and controls the left side. Describe the motor control exerted by the left hemisphere.
2. Distinguish between the pyramidal and the extrapyramidal motor systems.

PHYSIOLOGY IN HEALTH AND DISEASE

The prefrontal cortex is involved in higher brain functions, including memory, planning, and judgment. It's also required for normal motivation and social behavior. People with damage to the *lateral frontal area* of the prefrontal cortex show lack of motivation and sexual desire, and have defects in cognitive (thinking) ability. The *orbitofrontal area* of the prefrontal cortex (fig. 5.19) is required for pleasure and reward. It receives input from all the senses and has connection to the limbic system. Connections between the orbitofrontal cortex, the amygdala, and the cingulate gyrus (fig. 5.19) are needed for emotional reward and goal-directed behavior.

People with damage to the orbitofrontal area generally have normal cognitive function and memory. However, they experience severe impulsive behavior, verging on the sociopathic. A famous example of this was a railroad foreman in 1848 named Phineas P. Gage, who was tamping blasting powder in a rock with a metal rod when the blasting powder exploded. The rod—3 feet, 7 inches long and ¼-inch thick—was driven through his left eye and brain, and emerged through the top of his skull.

Gage experienced a few minutes of convulsions, then rode a horse into town and walked up a flight of stairs to see a doctor. He recovered well physically, but his associates noted striking personality changes. Before the accident, Gage was a responsible, capable, and financially prudent man. Afterward, he engaged in gross profanity and was impulsive to an extreme, seemingly tossed about by chance whims, and was fired from his job. His old friends remarked that he was "no longer Gage." By studying the damage to Gage's skull, modern scientists have confirmed that his personality changes were caused by injury to the orbitofrontal area of his prefrontal cortex.

(a)

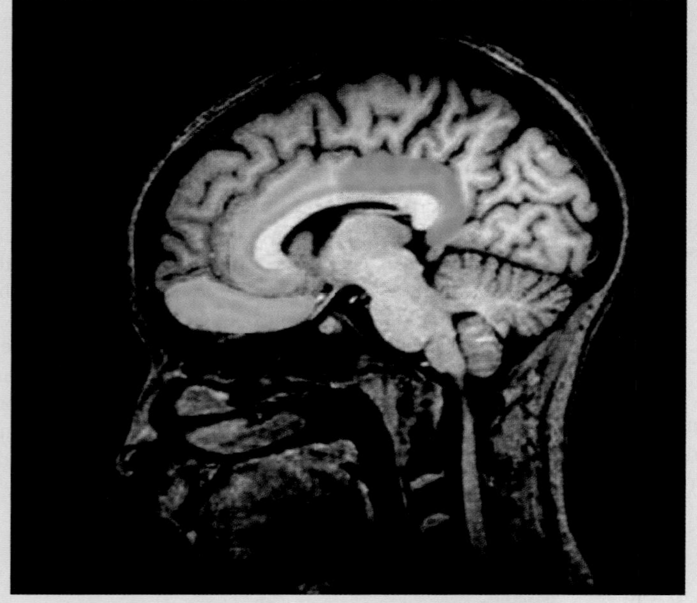

(b)

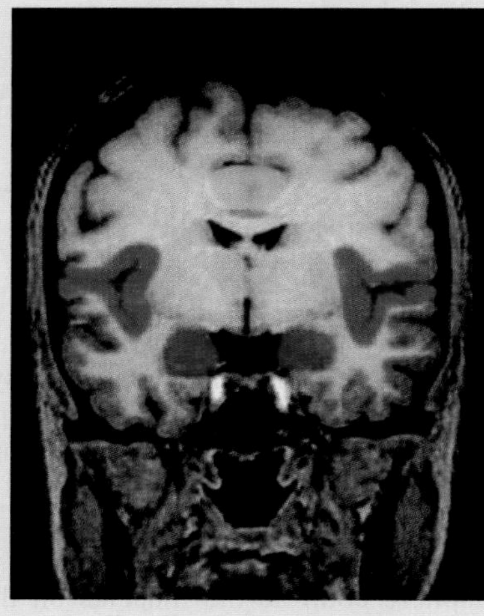

FIGURE 5.19 Some brain areas involved in emotion. (*a*) The orbitofrontal area of the prefrontal cortex is shown in yellow, and the cingulate gyrus of the limbic system is shown in blue-green (anterior portion) and green (posterior portion). (*b*) The insula of the cortex is shown in purple, the anterior cingulate gyrus of the limbic system in blue-green, and the amygdala in red.

Physiology in Balance

Central Nervous System

Working together with . . .

Endocrine System

The hypothalamus regulates the pituitary gland, which regulates other glands . . . p. 186

Respiratory System

The medulla oblongata has a center that controls breathing . . . p. 297

Muscular System

Spinal motor neurons stimulate muscle fibers to contract . . . p. 215

Cardiovascular System

The medulla oblongata regulates the heart rate and blood vessel constriction or dilation . . . p. 240

Digestive System

The brain stimulates secretions and contractions of the stomach . . . p. 348

Summary

There Are Three Major Regions of the Brain 114

- The 3 major brain regions are the forebrain, midbrain, and hindbrain.
- The forebrain contains the telencephalon (cerebrum) and the diencephalon; the hindbrain contains the metencephalon and myelencephalon.
- The brain contains hollow cavities called ventricles; these, and the spinal canal, are filled with cerebrospinal fluid.
- Various techniques allow the brains of living people to be seen, and the EEG is a recording of the brain's electrical activity.
- REM (rapid eye movement) sleep is when dreams occur; non-REM (slow-wave) sleep may be a period when memories are consolidated.

The Cerebrum Performs Higher Brain Functions 117

- The precentral gyrus, or motor cortex, controls muscles on the opposite side of the body.
- The postcentral gyrus is the somatesthetic sensory cortex, receiving sensory information from the skin, muscles, tendons, and joints on the opposite side of the body.
- The perception of hearing is primarily based in the temporal lobe, and the perception of vision is primarily based in the occipital lobe.
- The left hemisphere is dominant in most people, and that's where language and analytical ability mostly reside; the right hemisphere is more specialized for spatial comprehension.
- Speaking ability requires Broca's area in the left hemisphere, whereas language comprehension resides in Wernicke's area.
- The limbic system is a number of cerebral nuclei and axon tracts that are involved in emotions.
- The hippocampus and amygdala are needed to convert short-term into long-term memories, and other brain regions are involved in long-term storage of different aspects of memory.
- The basal nuclei, or basal ganglia, are a group of cerebral nuclei that includes the corpus striatum; these are involved in the control of body movements.

The Diencephalon Is Small but Important 123

- The thalamus is primarily a relay center; all sensory information except smell is sent to the thalamus, and from there it is relayed to the appropriate location in the cerebral cortex.
- The hypothalamus contains many nuclei that serve as centers for the control of hunger, thirst, body temperature, the pituitary gland, and some emotions.
- The posterior pituitary, or neurohypophysis, secretes two hormones that are produced by neurons in the hypothalamus; secretion is regulated by axons that extend from the hypothalamus into the posterior pituitary.
- The anterior pituitary, or adenohypophysis, produces its own hormones, but its secretion is regulated by the hypothalamus through releasing and inhibiting hormones secreted by the hypothalamus.
- The suprachiasmatic nucleus (SCN) of the hypothalamus is the major center that controls the body's circadian rhythms (those that repeat on a 24-hour basis).

The Midbrain and Hindbrain Control Vital Body Functions 124

- The corpora quadrigemina of the midbrain are centers for visual and auditory reflexes.
- An area of the midbrain, the substantia nigra, contains dopamine-releasing neurons that project to the corpus striatum of the basal nuclei; degeneration of these neurons of the nigrostriatal system produces Parkinson's disease.
- Other dopamine-releasing neurons from the midbrain project to the structures of the limbic system; these neurons of the mesolimbic system are involved in emotional reward and goal-directed behavior.
- The cerebellum, part of the metencephalon of the hindbrain, is needed for motor learning and the proper timing and force of movements.
- The medulla oblongata, which is the only structure of the myelencephalon of the hindbrain, contains vital centers that regulate breathing, heart rate, blood vessel constriction and dilation, and blood pressure.
- The reticular activating system (RAS) is an association of neurons in the brain stem that alerts the cerebral cortex to incoming sensory information; the RAS must be suppressed by the inhibitory neurotransmitter GABA in order to fall asleep.

The Spinal Cord Transmits Sensory and Motor Information 127

- Ascending fiber tracts are sensory; they decussate (cross over to the opposite side) and synapse in the thalamus, before being relayed to the cerebral cortex.
- The corticospinal tracts are descending motor tracts that originate in the motor cortex (precentral gyrus); most decussate in the pyramids of the medulla, and so these are also known as pyramidal motor tracts.
- Extrapyramidal motor tracts have many synaptic connections between various brain structures, including the basal nuclei (basal ganglia), cerebellum, thalamus, and cerebral cortex.

Review Activities

Objective Questions: Test Your Knowledge

1. Which of the following structures is not included in the forebrain?
   a. Cerebrum
   b. Thalamus
   c. Hypothalamus
   d. Cerebellum

2. Which of the following techniques allows scientists to visualize active brain areas by the increased amount of oxygen going to the active areas?
   a. CT
   b. fMRI
   c. EEG
   d. PET

3. Which of the following statements regarding sleep is true?
   a. People awaken naturally from non-REM sleep.
   b. People dream during non-REM sleep.
   c. Non-REM sleep is also known as slow (EEG) wave sleep.
   d. Memories are erased during non-REM sleep.

4. Which of the following brain areas contains the "motor cortex"?
   a. Precentral gyrus
   b. Postcentral gyrus
   c. Broca's area
   d. Occipital lobe
5. Sensations arising from the skin on the left side of the body are interpreted by the
   a. right precentral gyrus.
   b. right postcentral gyrus.
   c. left precentral gyrus.
   d. left postcentral gyrus.
6. The area of the cerebral cortex devoted to the sense of touch is greatest for which area of the body?
   a. Torso
   b. Legs
   c. Fingers
   d. Arms
7. The ability to speak requires which region of the brain?
   a. Wernicke's area in the right cerebral hemisphere
   b. Wernicke's area in the left cerebral hemisphere
   c. Broca's area in the right cerebral hemisphere
   d. Broca's area in the left cerebral hemisphere
8. Which of the following statements regarding the amygdala (amygdaloid nucleus) is true?
   a. It is in the forebrain.
   b. It is part of the limbic system.
   c. It is involved in fear and recognition of fearful stimuli.
   d. All of these statements are true.
9. Which of the following statements regarding the hippocampus is true?
   a. It is in the diencephalon.
   b. It is part of the basal nuclei.
   c. It is involved in consolidating short-term to long-term memory.
   d. All of these statements are true.
10. The corpus striatum is
    a. a group of nuclei located in the midbrain.
    b. a group of nuclei located in the diencephalon.
    c. a group of nuclei involved in emotions.
    d. a group of nuclei involved in control of body movement.
11. Dopamine-releasing neurons that go from the substantia nigra to the caudate nucleus
    a. degenerate in Parkinson's disease.
    b. are part of the pyramidal motor pathway.
    c. are needed for emotional reward and goal-directed behavior.
    d. are stimulated when people take abused drugs.
12. Which of the following statements about the suprachiasmatic nucleus is false?
    a. It is located in the hypothalamus.
    b. It is the master clock for the body's circadian rhythms.
    c. It is part of the limbic system.
    d. It influences the pineal gland.
13. Which of the following types of sensory information is not relayed through the thalamus?
    a. Taste
    b. Smell
    c. Touch
    d. Vision
14. Which of the following statements regarding the hormones ADH and oxytocin is false?
    a. They are produced in the posterior pituitary gland.
    b. They are transported along axons from the hypothalamus.
    c. They are secreted by the posterior pituitary gland.
    d. They are stored in the posterior pituitary gland.
15. Centers that regulate breathing and the cardiovascular system are located in the
    a. hypothalamus.
    b. thalamus.
    c. pons.
    d. medulla oblongata.
16. Which of the following regulates the timing and force of movements, is part of the extrapyramidal motor system, and is the second-largest brain structure?
    a. Cerebrum
    b. Cerebellum
    c. Diencephalon
    d. Myelencephalon
17. Which of the following statements regarding the reticular activating system (RAS) is false?
    a. It is located in the brain stem.
    b. It causes sleepiness by depressing the cerebrum.
    c. GABA inhibits the RAS so that we can fall asleep.
    d. Monoamine neurotransmitters in the RAS promote alertness.
18. Corticospinal tracts
    a. are ascending tracts.
    b. are sensory tracts.
    c. are extrapyramidal tracts.
    d. are motor tracts.

Essay Questions 1: *Test Your Understanding*

1. Describe the development of the brain, and explain the embryological reason why the brain has cavities.
2. How can REM and non-REM sleep be distinguished by the use of an EEG? What is the significance of these stages of sleep?
3. Describe the structure of the cerebrum and some of the functions of its lobes.
4. Explain the physiological significance of the maps of the precentral and postcentral gyri.
5. How do the right and left cerebral hemispheres differ in their specializations? How do they communicate with each other?
6. Describe the location and explain the significance of Broca's and Wernicke's areas.
7. List the brain areas that are part of the limbic system, and explain the physiological significance of the limbic system.
8. Identify the parts of the medial temporal lobes that are involved in memory, and explain their roles in memory.
9. What are the components of the basal nuclei (basal ganglia)? Describe the physiological significance of the basal nuclei.
10. List the functions of the hypothalamus, and explain how the hypothalamus regulates both the posterior and the anterior pituitary gland.
11. Define circadian rhythms, describe the location of the suprachiasmatic nucleus (SCN), and explain the role of the SCN in circadian rhythms.
12. Describe the structures involved in the nigrostriatal dopamine system, and explain this system's physiological and medical significance.

13. Describe the structures involved in the mesolimbic dopamine system, and explain this system's physiological and medical significance.
14. Where is the cerebellum located, and what is its physiological significance?
15. Where is the medulla oblongata located, and what "vital centers" are located in the medulla?
16. Describe the significance of the reticular activating system (RAS) and identify some of the neurotransmitters involved. How is the RAS turned off when we go to sleep?
17. Trace a sensory pathway from the toe to the cerebral cortex, and explain how the cerebral hemispheres receive information regarding the contralateral (opposite) sides of the body.
18. Distinguish between the pyramidal and extrapyramidal motor systems. Explain how the cerebral hemispheres control motor function on the contralateral (opposite) sides of the body.

Essay Questions 2: *Test Your Analytical Ability*

1. Compared to the human brain, a frog's brain has a proportionally larger midbrain and smaller forebrain. What might this imply regarding how frogs process visual sensory information compared to humans?
2. If you were running an experiment, how could you stop your subjects from experiencing specifically REM sleep or non-REM sleep? What effects might this sleep deprivation have?
3. Explain how the mapping of the somatesthetic sensory cortex and motor cortex was achieved. Why do you think the patients had to be awake? What does this information tell you about the density of touch receptors in the skin of the thumbs versus the skin of the back? Explain.
4. Which hemisphere contains the language centers? Which hemisphere do you think is used to appreciate a painting, a sculpture, and a symphony? Is it possible for someone to have aphasia and yet to be able to sing? Explain your answers.
5. Suppose some people have their corpus callosum cut to treat epilepsy (have a "split brain"). If they touch a set of keys (without seeing them) with the right hand, where in the cerebrum does this information go? Will they be able to name what they're touching? Suppose they do the same with their left hand—where does this information go, and will they be able to name what they're touching? How does this compare with people who have an intact corpus callosum?
6. Some of the structures of the limbic system are also involved in processing olfactory information. A smell can effectively evoke an emotionally charged memory. Explain the relationship between these observations.
7. Suppose a person with memory impairment due to bilateral damage to the medial temporal lobes goes through a day working in car sales. How could that person manage? What aids and help would be needed? Explain.
8. What specifically is impaired in a person with Parkinson's disease, and how is this condition aided by taking L-dopa, the precursor of dopamine? What would be required of regenerative medicine (stem cell medical therapies) to be able to correct this condition?
9. The anterior pituitary has sometimes been referred to as the "master gland" because some of its hormones (ACTH; TSH; FSH and LH) regulate other endocrine glands (the adrenal cortex, thyroid, and gonads, respectively). Evaluate the "master gland" concept, based only on what you've learned in this chapter.
10. What happens to the suprachiasmatic nucleus when a person's work requires a change from the day shift to the night shift? How does this influence the body? Explain.
11. Suppose a person has a tumor in the vicinity of the pons. What are the possible immediate consequences of this, and what possible future dangers could this present? Explain.
12. Given that drugs used to treat schizophrenia block a type of dopamine receptor, explain the "dopamine hypothesis" of schizophrenia. Identifying specific pathways, explain how such drugs cause symptoms that resemble Parkinson's disease.
13. Naming specific neural pathways and neurotransmitters, explain how abused drugs (including tobacco) can become addicting, and how drug withdrawal could cause anxiety and depression.
14. With the information given in this chapter on the suprachiasmatic nucleus (SCN) and the reticular activating system (RAS), propose an explanation of what might make you feel drowsy and listless at a particular time of day, and alert at another time of day. Which neurotransmitters would be involved?

Web Activities

ARIS™ For additional study tools, go to: www.aris.mhhe.com. Click on your course topic and then this textbook's author/title. Register once for a semester's worth of interactive activities and practice quizzing to help boost your grade!

Peripheral Nervous System

CHAPTER 6

Chapter Outline

Homeostasis

Normal functioning of the sensory and motor components of cranial and spinal nerves is required for homeostasis to be maintained. Sensory information from all body organs travels via sensory neurons into the CNS, and activation of motor neurons by the CNS regulates the responses of effector cells (muscles and glands) to correct deviations from homeostasis. Somatic motor neurons stimulate skeletal muscle contraction. Autonomic motor neurons maintain homeostasis by regulating the heart, smooth muscles within blood vessels and visceral organs, and glands.

Clinical Investigation

But if homeostasis is not maintained . . .

Joyce was a landscape architect who sustained a spinal cord injury at the fifth thoracic level by a construction site accident. Now paraplegic, she went to the physician complaining of a pounding headache and unusually severe spasms of her leg muscles. The physician noticed that Joyce was flushed and sweating profusely from her head and neck, although her lower body was cool. The physician discovered that Joyce's blood pressure was much higher than her usual (which, like most paraplegics, was lower than the average), although her heart rate was slower than usual. Further examination revealed that Joyce's indwelling Foley catheter (which drains her urinary bladder) was kinked, which the physician corrected. He stated that an overly full bladder was the most likely cause of Joyce's symptoms. Laboratory tests revealed that Joyce also had a urinary tract infection, and the physician prescribed an antibiotic and a drug called baclofen to reduce the muscle spasms.

Why was Joyce flushed and sweating in her head and neck, and yet cool in her lower body? What might have caused her high blood pressure, lower heart rate, and pounding headache? Why did she have severe leg muscle spasms? What is the possible relationship between the kinked catheter, urinary tract infection, and Joyce's symptoms, and how would the medication prescribed by the physician help alleviate these symptoms?

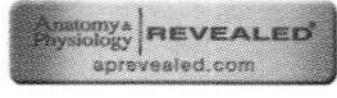

A virtual cadaver dissection experience

The PNS Includes Cranial and Spinal Nerves

The peripheral nervous system (PNS) includes the 12 pairs of cranial nerves that arise from the brain, and the 31 pairs of spinal nerves that arise from the spinal cord. Certain cranial nerves contain only sensory axons; however, most cranial nerves, and all spinal nerves, are mixed nerves containing both sensory and motor axons. The motor axons can be activated by association neurons in the CNS, and through reflex arcs involving sensory neurons.

You constructed a knowledge base when you learned about neurons and the CNS in chapters 4 and 5, and now you'll be building on that foundation. However, the information in this chapter is fundamental to an understanding of the physiology of skeletal muscles, the sensory system, and indeed all of the other body systems. Sensory neurons transmit information from sensory organs and sensory receptors in all body systems; motor neurons regulate all three types of muscles and glands. So, although the foundational layers are still under construction, be assured that you're well on your way to building a rock-solid pyramid of physiological knowledge.

Twelve Pairs of Cranial Nerves Arise from the Brain

There are 12 pairs of cranial nerves; 2 pairs arise from neuron cell bodies in the forebrain, and the other 10 pairs arise from the midbrain and hindbrain. These nerves are known by both their Roman numerals (numbered front to back) and names that indicate either the nerve's destination (such as the facial nerve) or its function (such as the oculomotor nerve). Nerves I, II, and VIII are strictly sensory; the others are *mixed nerves*, containing both sensory and motor axons (table 6.1).

Clinical Applications

Trigeminal neuralgia, or **tic douloureux**, is a relatively common condition among middle-aged and elderly people. This appears to be caused by compression of the sensory axons (usually by a blood vessel) in cranial nerve V, the trigeminal, just before it enters the pons. The compression causes action potentials to be produced in these sensory axons, resulting in severe pain in the lips, gums, cheeks, or chin. The pain, though intense, is short-lived; it thus causes the person to wince, producing a "tic." Another cranial nerve problem is **Bell's palsy**, the most common cause of facial paralysis. This involves cranial nerve VII, the facial, which innervates the muscles of facial expression. Both trigeminal neuralgia and Bell's palsy usually remit (go away) by themselves in a few weeks or months.

TABLE 6.1 The Cranial Nerves

Nerve Number and Name		Composition	Some Functions
I	Olfactory	Sensory only	Olfaction (smell)
II	Optic	Sensory only	Vision
III	Oculomotor	Motor and sensory	Serves muscles of the eye
IV	Trochlear	Motor and sensory	Serves the superior oblique eye muscle
V	Trigeminal	Motor and sensory	Sensory from face and mouth; motor to muscles of mastication (chewing)
VI	Abducens	Motor and sensory	Serves the lateral rectus eye muscle
VII	Facial	Motor and sensory	Serves the muscles of facial expression, lacrimal glands, and salivary glands
VIII	Vestibulocochlear	Sensory only	Equilibrium and hearing
IX	Glossopharyngeal	Motor and sensory	Serves the pharynx (throat) for swallowing, posterior third of tongue, parotid salivary gland
X	Vagus	Motor and sensory	Sensations from visceral (internal) organs, and parasympathetic motor regulation of visceral organs
XI	Accessory	Motor and sensory	Serves muscles that move head, neck, and shoulders
XII	Hypoglossal	Motor and sensory	Serves muscles of the tongue

Thirty-One Pairs of Nerves Arise from the Spinal Cord

There are 31 pairs of spinal nerves, designated by where they arise in the spinal cord: 8 cervical, 12 thoracic, 5 lumbar, 5 sacral, and 1 coccygeal (fig. 6.1). In some regions, axons from different levels of the spinal cord can come together in *plexuses*, which give rise to nerves that can be quite large because they contain axons from different spinal cord levels. The *sciatic nerve* (fig. 6.1) is a good example.

Spinal Nerves Provide Sensory and Motor Functions

Each spinal nerve is a mixed nerve composed of sensory and motor axons. These two kinds of axons separate near the root of the nerve in the spinal cord. The sensory axons enter the spinal cord on the dorsal (back) side; the motor axons exit on the ventral (belly) side. The sensory neurons form the **dorsal root** of the spinal nerve; the motor neurons form the **ventral root**. Think of the spinal nerve

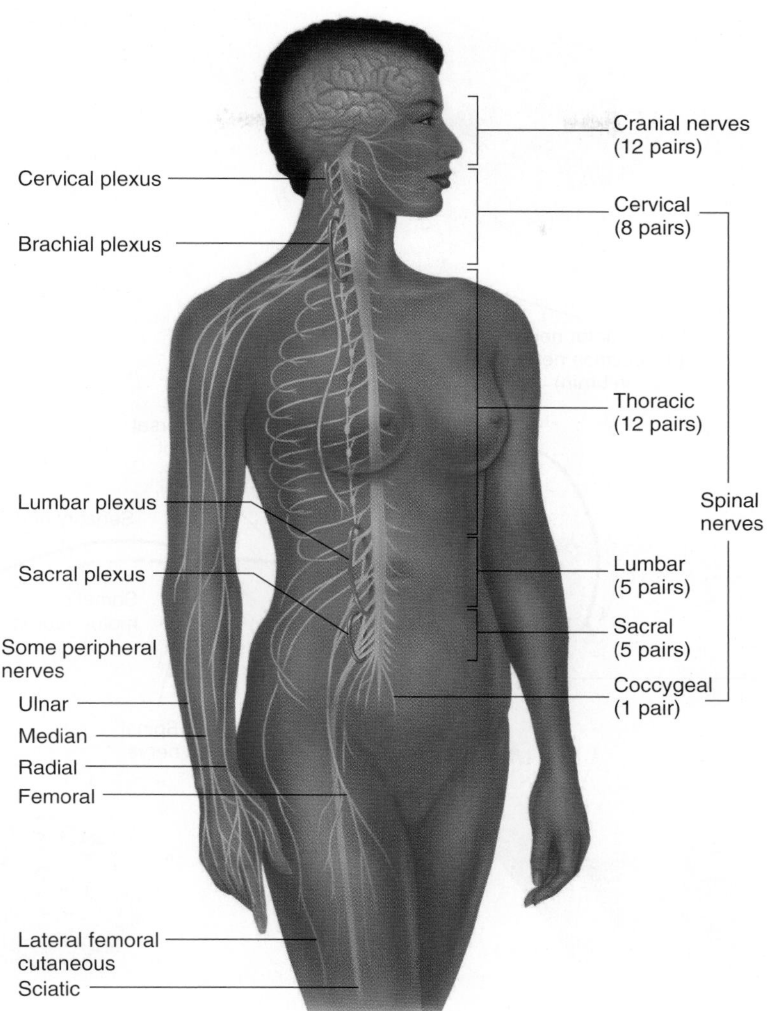

FIGURE 6.1 Nerves of the PNS. There are 12 pairs of cranial nerves and 31 pairs of spinal nerves. Plexuses are regions where axons from a number of different spinal cord levels come together to form larger nerves.

as a cable containing red and green wires, and the spinal cord as a junction box. The wires are packaged together in the cable (nerve), but the green and red wires separate as they enter the junction box (spinal cord), forming the two roots of the cable (spinal nerve).

The sensory axons synapse with motor and association neurons in the spinal cord. The motor neurons may be stimulated by either sensory or association neurons (fig. 6.2). Stimulation of spinal motor neurons by association neurons allows for motor control by **upper motor neurons** (association neurons in the motor areas of the brain) and for reflex activity. The **somatic motor neurons** (spinal motor neurons that stimulate skeletal muscles) are also called **lower motor neurons**. Autonomic motor neurons will be discussed later in this chapter.

The neuron cell bodies of the sensory neurons in a spinal nerve are grouped together, forming a bulge in the dorsal roots of each spinal nerve. These are the **dorsal root ganglia**. Remember that sensory neurons are pseudounipolar (chapter 4, fig. 4.4), so that a single long process can conduct action potentials along an uninterrupted axon from the sensory receptor into the spinal cord. In contrast, somatic motor neurons have their cell bodies in the gray matter of the spinal cord, not in ganglia.

Somatic Motor Neurons Participate in Spinal Reflexes

The simplest reflex is the **monosynaptic muscle stretch reflex** (fig. 6.3). It's called "monosynaptic" because only 1 synapse is crossed in the CNS: the sensory neuron synapses directly with the somatic motor neuron. This produces a **reflex arc**: action potentials "arc" into the spinal cord in the sensory neuron and out of the spinal cord in the motor neuron. The best-known example of a monosynaptic muscle stretch reflex is the knee-jerk reflex (fig. 6.3), in which the patellar ligament is struck with a rubber mallet. This produces a quick stretch of the quadriceps femoris muscles.

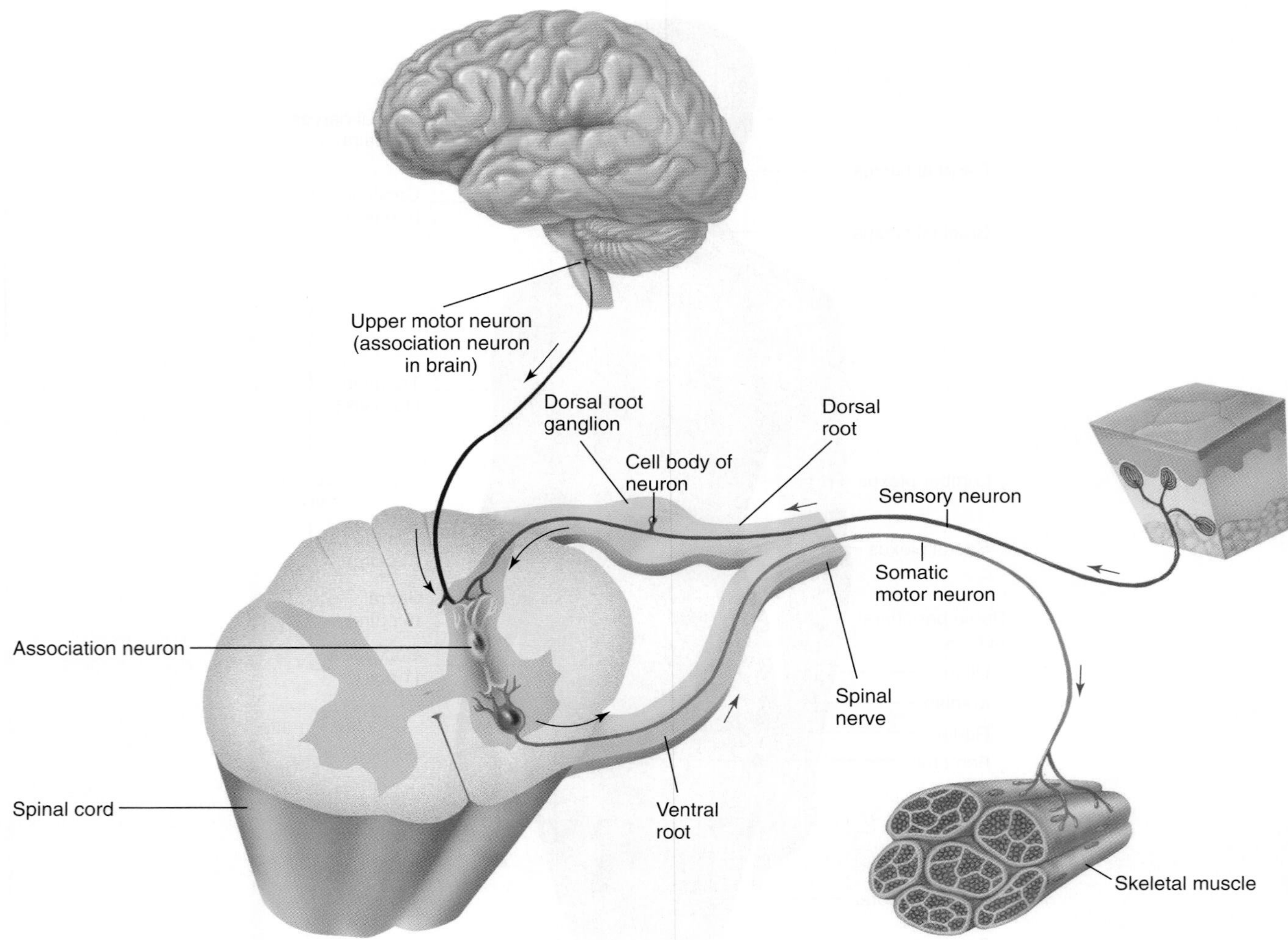

FIGURE 6.2 Activation of somatic motor neurons. Somatic motor neurons may be stimulated by spinal association neurons, as shown here, or directly by sensory neurons (see fig. 6.3), in a reflex arc that doesn't involve the brain. The spinal association neurons and motor neurons can also be stimulated by association neurons (called upper motor neurons) in the motor areas of the brain. This affords voluntary control of skeletal muscles.

The quick stretch of a skeletal muscle serves as the stimulus for the *muscle stretch receptor*, known as the **muscle spindle**, which is buried within the muscle. Each spindle contains several thin muscle fibers (called *intrafusal fibers* because they're inside the spindle), packaged together and attached to the muscle's tendons. When the muscle is stretched, the spindle is stretched. Stretching of the spindle stimulates sensory dendrites around the intrafusal fibers, producing action potentials that are conducted by the sensory neuron into the spinal cord. The sensory neuron synapses with the somatic motor neuron, which stimulates the muscle to contract and the knee to kick. The particular somatic motor neuron that stimulates contractions of muscle fibers outside the spindle (the *extrafusal fibers*) is known as an **alpha (α-) motoneuron**. There is another kind of somatic motor neuron (not shown in fig. 6.3) called a **gamma (γ-) motoneuron**, which stimulates only the few muscle fibers inside the spindle (the intrafusal fibers). This can't make the muscle shorten, but it does make the spindle tighter and thus more sensitive to stretch. In this way, the muscle spindle remains tight and can continue to monitor muscle length as the muscle shortens.

Clinical Applications

Spinal cord lesions (damage) first produce *spinal shock*, which may last several weeks. During this time, there is a temporary loss or reduction in spinal cord reflexes below the level of the lesion. After spinal shock wears off, muscle *spasticity* occurs below the level of the lesion. This is produced by exaggerated cutaneous and muscle spindle stretch reflexes. Muscle stretching, or any painful stimulus below the level of the lesion, can elicit these reflexes, and a urinary tract infection can make matters worse. In a person with an intact spinal cord, inhibitory motor tracts (releasing GABA, an inhibitory neurotransmitter) descending from the brain normally suppress inappropriate stretch reflexes; however, the spinal cord damage blocks these inhibitory tracts from going below the lesion. A medically useful drug, *baclofen*, is a GABA derivative that can help reduce muscle spasms in chronic conditions such as spinal cord lesions and multiple sclerosis. Since it works like the inhibitory neurotransmitter GABA, it helps suppress muscle contraction induced by stretch reflexes, thereby reducing muscle tone. Some spasticity may always be present, but the condition is improved by performing range-of-motion exercises.

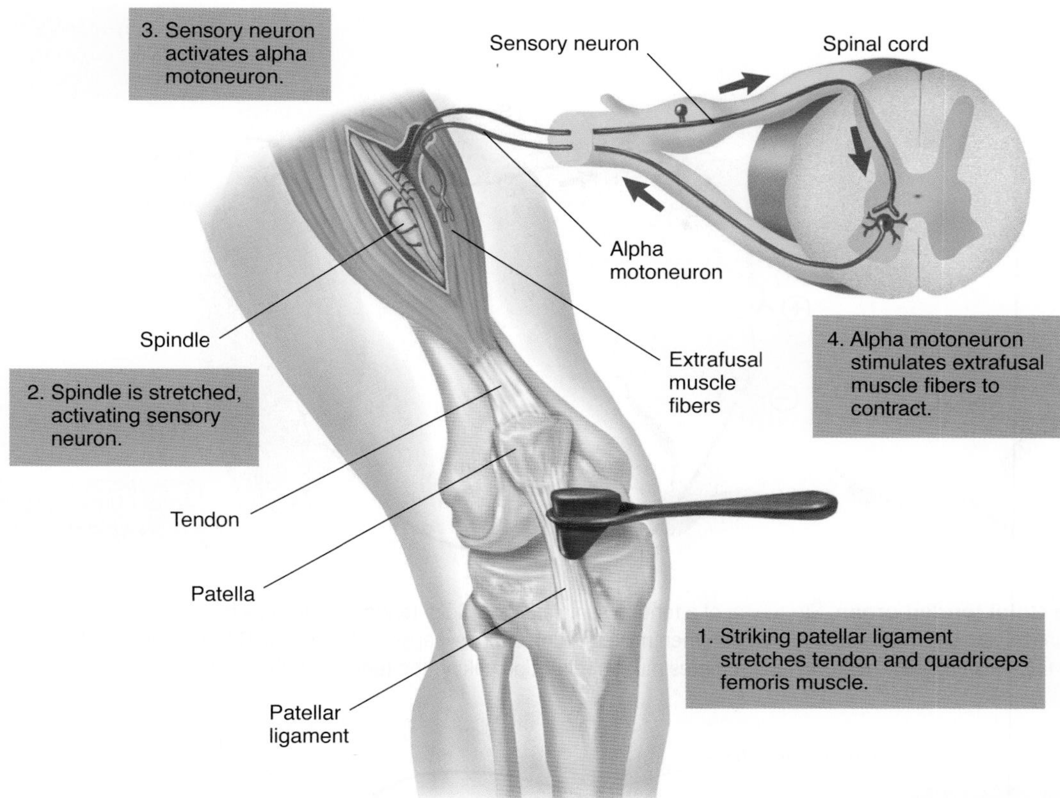
3. Sensory neuron activates alpha motoneuron.
Sensory neuron
Spinal cord
Alpha motoneuron
Spindle
2. Spindle is stretched, activating sensory neuron.
Extrafusal muscle fibers
4. Alpha motoneuron stimulates extrafusal muscle fibers to contract.
Tendon
Patella
1. Striking patellar ligament stretches tendon and quadriceps femoris muscle.
Patellar ligament

FIGURE 6.3 The knee-jerk reflex. This is an example of a monosynaptic reflex, because only 1 synapse is crossed in the CNS; the sensory neuron makes a direct synapse with a somatic motor neuron. The sensory organ of muscle stretch is the muscle spindle. The somatic motor neuron is called an alpha-motoneuron, because it stimulates the muscle fibers lying outside the spindle (the extrafusal muscle fibers). This stimulates the muscle to contract and shorten, making the leg kick.

When a muscle is stretched more slowly, the stretch reflex is activated to a lesser degree. Also, slower stretch allows another reflex to operate, one that inhibits contraction. This inhibitory reflex begins when stretching a tendon activates a receptor called the **Golgi tendon organ** (fig. 6.4). The Golgi tendon organ stimulates a sensory neuron, which synapses with an association neuron in the spinal cord. This association neuron then makes an inhibitory synapse (by releasing an inhibitory neurotransmitter—GABA or glycine) with a spinal motor neuron. In this way, the Golgi tendon organ reflex inhibits contraction of a muscle when its tendons are under increased tension. Slower stretching of a muscle allows this inhibitory reflex to prevent a painfully excessive muscle contraction (a spasm), which may occur if the muscle is stretched too rapidly.

Muscles that produce opposing movements of bones at a joint are called *antagonistic muscles*. For example, the biceps brachii flexes the arm at the elbow joint, whereas the triceps brachii extends the arm at the elbow joint. If you flex and extend your arm, notice that the antagonist muscle is stretched when you do a particular movement; for example, your triceps is stretched when your biceps contracts to flex your arm. This stimulates the stretch reflex of your triceps to some degree whenever you flex your arm. Thus, in order to produce a desired movement, the muscle that produces that movement (the *agonist muscle*) must be stimulated to contract while the stretch reflex of the antagonist muscle is inhibited. This inhibition is produced by association neurons, which release an inhibitory neurotransmitter (GABA or glycine) at their synapses with the motor neurons that innervate the antagonist muscle. Stimulation of the motor neurons to the agonist muscle and inhibition of the motor neurons to the antagonist muscle is known as **reciprocal innervation** (fig. 6.5).

Clinical Investigation Clues

Remember that Joyce had severe leg muscle spasms.

- What reflex is activated to produce the muscle spasms?
- Why are the muscle spasms produced below the level of the spinal cord injury but not above it?
- Given that a urinary tract infection exacerbates muscle spasms in people with spinal cord injury, how would both the medications prescribed by the physician help relieve Joyce's muscle spasms?

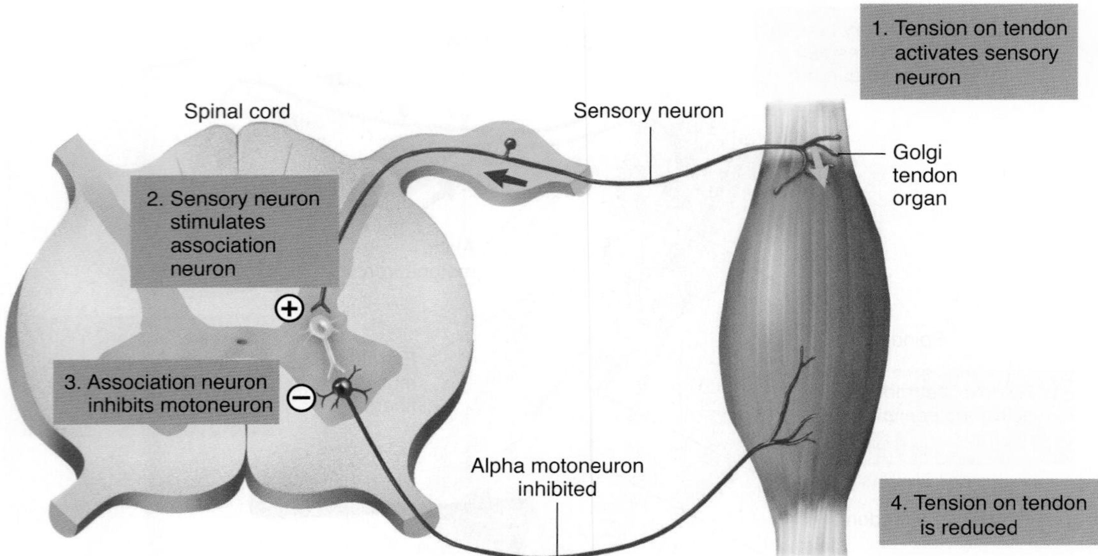

FIGURE 6.4 The Golgi tendon organ. Stretching of a tendon stimulates the Golgi tendon organ, which carries action potentials into the spinal cord and activates an association neuron. This association neuron then makes an inhibitory synapse (releasing GABA or glycine) with a somatic motor neuron. The reduced activity of the somatic motor neuron then promotes muscle relaxation, taking the tension off the tendon.

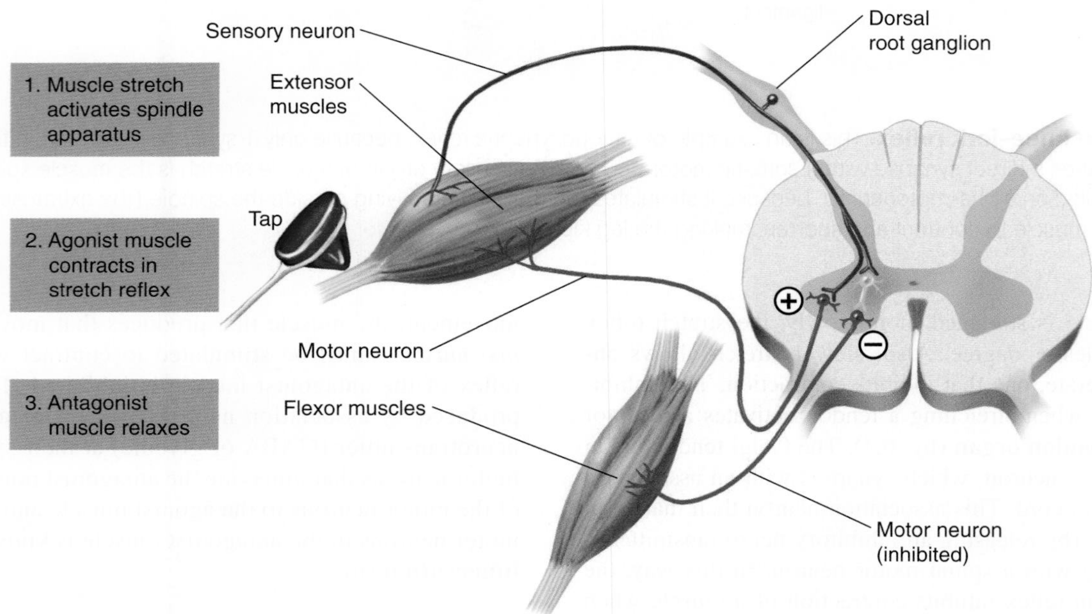

FIGURE 6.5 Reciprocal innervation. When the somatic motor neuron stimulates a particular muscle (the agonist) to contract, its antagonistic muscle is inhibited from contracting. This occurs because association neurons release inhibitory neurotransmitters (GABA or glycine) at synapses with the somatic motor neurons innervating the antagonistic muscles. Such inhibition prevents stretch reflexes from causing the contraction of antagonistic muscles when the agonist muscles contract.

Somatic Motor Neurons Form Motor Units

Each skeletal muscle fiber (cell) is stimulated by 1 somatic motor neuron, which makes a single synapse with the muscle fiber. The synapse between a somatic motor neuron and a skeletal muscle fiber is called a **neuromuscular junction**, or **myoneural junction**. As mentioned in chapter 4, somatic motor neurons release acetylcholine (ACh) as their neurotransmitter, and the ACh binds to its receptor proteins in the plasma membrane of the postsynaptic cell, in this case a skeletal muscle fiber. Receptors for ACh can be either of two types: nicotinic or muscarinic; skeletal muscles have nicotinic ACh receptors. For this type of ACh receptor, a chemically gated ion channel through the receptor protein opens when the receptor binds to ACh (chapter 4). This results in a depolarization (the EPSP, called an *end plate potential* in skeletal muscles), which stimulates the production of action potentials in the skeletal muscle fiber. These action potentials, in turn, stimulate the skeletal muscle to contract (chapter 9).

The axon of a somatic motor neuron produces a number of *collateral axon branches*, and each of these makes a synapse with a different muscle fiber (fig. 6.6). When the somatic motor neuron produces action potentials, these go along each axon collateral to stimulate the release of ACh from each axon terminal. As a result, all of the muscle fibers innervated by a single somatic motor neuron are stimulated to contract at the same time. A somatic motor neuron and the muscle fibers it innervates form a **motor unit**.

Motor unit size varies within a particular muscle; in the quadriceps femoris muscles of legs, for example, there can be up to a thousand muscle fibers per motor neuron. When you want to produce a stronger muscle contraction, you activate more (and larger) motor units, a process known as **recruitment**. Motor units in the extrinsic eye muscles (the muscles that move the eyeball) may contain as few as 3 or 4 muscle fibers. Although larger motor units produce greater contraction strength, smaller motor units afford a finer degree of neural control over the muscle's contraction. This is beneficial when we must make tiny adjustments in the position of our eyeballs, or tiny movements of our fingers when we write or thread a needle. Muscle contraction and its regulation are discussed more thoroughly in chapter 9.

Clinical Applications

Neuromuscular blocking drugs are often given intravenously after general anesthesia to relax muscles for surgery. These are chemicals based on *curare*, the drug used for poison arrows that blocks the ability of ACh to activate its nicotinic ACh receptors. Examples of medically used drugs that work in a way analogous to curare include *atracurium*, *mivacurium*, and *pancuronium*. *Succinylcholine* works a little differently, in that it at first causes some muscle depolarization (resulting in twitching) before the ACh receptor is blocked, producing muscle relaxation for a short time.

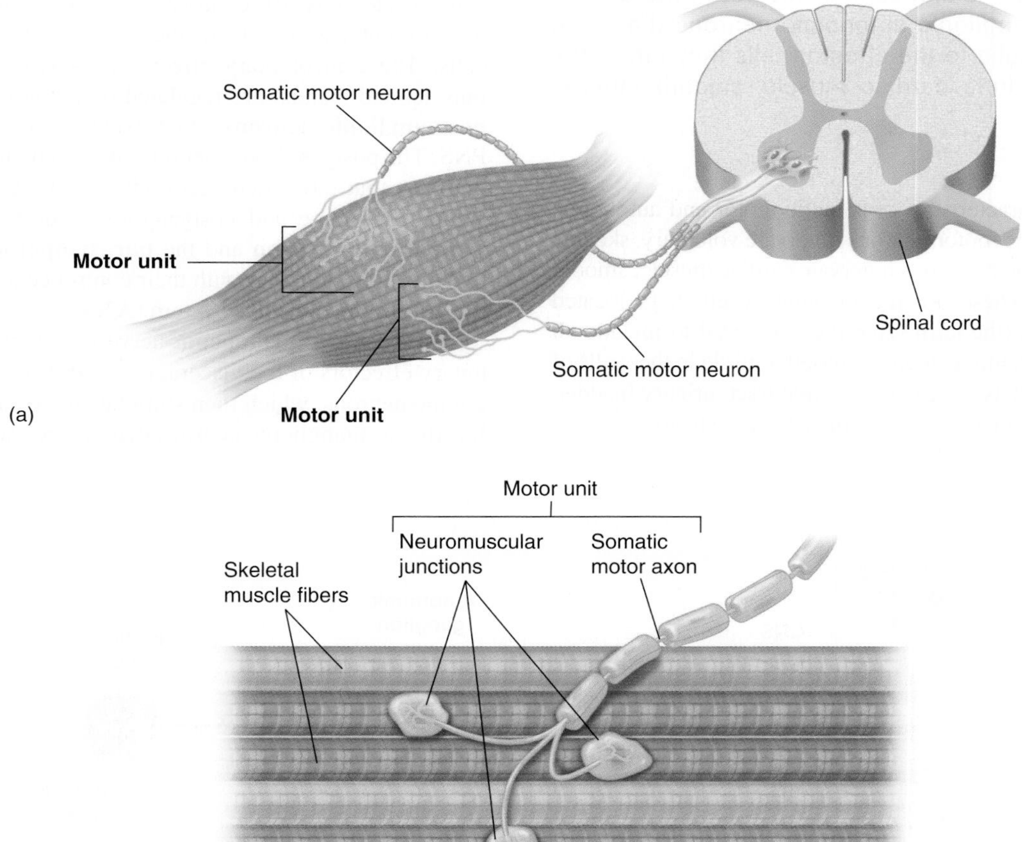

FIGURE 6.6 Motor units. A motor unit consists of a somatic motor neuron and the muscle fibers it innervates. (*a*) Illustration of a muscle containing 2 motor units. In reality, a muscle would contain many hundreds of motor units, and each motor unit would contain many more muscle fibers than are shown here. (*b*) A single motor unit consisting of a branched motor axon and the 3 muscle fibers it innervates (the fibers that are highlighted) is depicted. The other muscle fibers would be part of different motor units and would be innervated by different neurons (not shown).

CHECK POINT

1. Identify the composition of the ventral and dorsal roots of spinal nerves and of dorsal root ganglia.
2. Explain how striking the patellar ligament with a mallet causes the knee-jerk reflex.
3. Describe the role of GABA and glycine in the Golgi tendon organ reflex.
4. Describe reciprocal innervation, and explain its physiological significance.
5. Define a motor unit, and explain its physiological significance.

The Autonomic Nervous System Regulates Involuntary Effectors

The autonomic nervous system is composed of sympathetic and parasympathetic motor neurons and the brain regions that regulate their activities. There are preganglionic autonomic neurons that send axons out of the CNS to ganglia, and postganglionic autonomic neurons that send axons from the ganglia to the effector cells they innervate. These effectors include cardiac muscle, smooth muscle, and glands.

Let's compare the characteristics of somatic motor and autonomic motor neurons. Somatic motor neurons innervate voluntary, skeletal muscle; autonomic motor neurons innervate cardiac muscle, smooth muscle, and glands. These are the involuntary effectors located within visceral organs (the term *viscera* can be used to include all internal organs). For example, there is smooth muscle in the walls of all hollow organs, such as the gastrointestinal tract, urinary bladder, and blood vessels, and there is cardiac muscle in the heart.

If somatic motor axons to a skeletal muscle are cut, the skeletal muscle will not be stimulated to contract. Further, it will lose its muscle tone and eventually atrophy (lose structure). If autonomic nerves are cut, a smooth muscle will retain much of its tone, because it's less dependent on its innervation. Indeed, cardiac muscle and some smooth muscles (in the wall of the intestine, for example) contract and relax by themselves; they have automatic activity. In those cases, autonomic nerves regulate only the rate and strength of the contractions; they don't cause the contractions.

Autonomic Neurons Are Preganglionic or Postganglionic Neurons

Continuing with our comparison of somatic motor and autonomic motor neurons, there is a basic difference in the anatomy. Recall that motor neurons, which conduct action potentials out of the CNS, can be called *efferent* neurons. A somatic motor neuron has its cell body in the CNS, and its axon extends all the way to the skeletal muscle fibers in its motor unit (fig. 6.6). Thus, there is only 1 neuron in the efferent pathway from the CNS to the skeletal muscle cells. By contrast, there are *2* autonomic neurons in the efferent pathway from the CNS to the involuntary effector cells. These involuntary effector cells (cardiac muscle, smooth muscle, and glands) are regulated by autonomic neurons—called **postganglionic neurons**—that extend axons from ganglia in the PNS. The postganglionic neurons are stimulated by **preganglionic neurons** that have their cell bodies in the CNS (fig. 6.7). There are preganglionic and postganglionic autonomic neurons in the **sympathetic division** and the **parasympathetic division**. These two divisions, together with their control centers in the CNS, form the **autonomic nervous system** (**ANS**).

In order for the autonomic nervous system to regulate the involuntary effectors of the viscera, the CNS first activates the preganglionic neurons, which then stimulate the postganglionic neurons. It is the postganglionic neurons that synapse with the effector cells

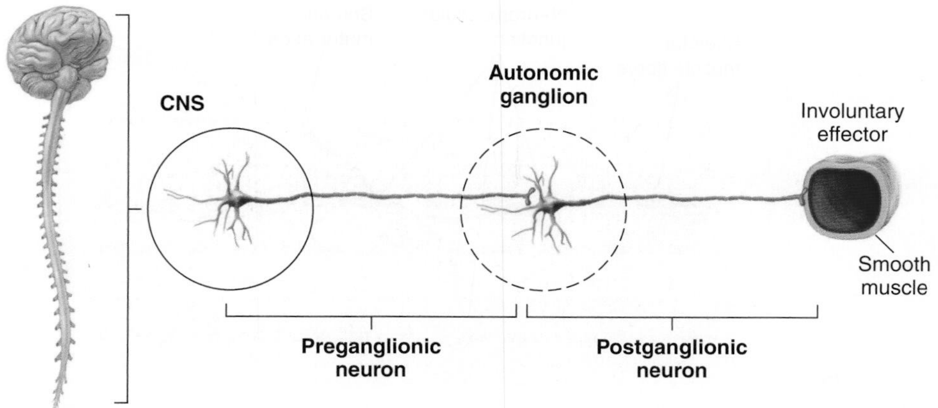

FIGURE 6.7 The ANS has preganglionic and postganglionic neurons. The preganglionic neurons of the ANS have cell bodies in the CNS, whereas the postganglionic neurons have cell bodies within autonomic ganglia. The sympathetic and parasympathetic divisions of the ANS differ in the particular locations of their preganglionic neuron cell bodies within the CNS, and in the location of their ganglia.

TABLE 6.2 Examples of Sympathetic and Parasympathetic Effects of the Autonomic Nervous System

Organ or Function Affected	Sympathetic Effects	Parasympathetic Effects
Heart rate	Increased	Decreased
Blood pressure	Increased	Slightly decreased
Urinary bladder	Increased sphincter tone	Decreased sphincter tone (for urinating)
Intestinal contractions	Decreased	Increased
Lungs	Dilation of bronchioles	Constriction of bronchioles
Pupils	Dilation	Constriction
Sexual function	Ejaculation and orgasm	Erection
Sweat glands	Sweating	No effect
Lacrimal glands	No effect	Tearing
Parotid glands	No effect	Salivation

Modified from Phillip A. Low and John W. Engstrom, "Disorders of the Autonomic System", in *Harrison's Principles of Internal Medicine* (New York: McGraw-Hill, 2005), p. 2428, table 354-1.

and regulate their activities. The sympathetic and parasympathetic divisions of the autonomic nervous system differ anatomically in the location of their preganglionic neurons and their ganglia. They also differ functionally; the physiology of the sympathetic and parasympathetic divisions will be considered in later sections, but you can see an overview of their actions in table 6.2.

The Sympathetic Division of the ANS Has Chains of Ganglia

The sympathetic division can also be called the *thoracolumbar division*, because its preganglionic neurons have cell bodies in each level of the spinal cord from the first thoracic to the second lumbar levels. Because of this arrangement, their axons exit the spinal cord in a row on each side of the spinal cord. Most of the preganglionic axons are short, synapsing with postganglionic neurons located in ganglia on each side of the spinal cord. These ganglia form a double chain on either side of the spinal cord, known as the **sympathetic chain of ganglia** (fig. 6.8). The preganglionic axons at each spinal level exit in the ventral roots of the spinal nerve, but then leave the nerve as they take a short detour to the sympathetic ganglion.

After the preganglionic axons synapse with postganglionic neurons in the sympathetic ganglion, the postganglionic axons leave the ganglion and rejoin the spinal nerve. (The short routes to and from the sympathetic ganglion are the *rami communicantes*; fig. 6.8). As a result, spinal nerves not only carry somatic motor neurons to innervate skeletal muscles, but they also carry postganglionic sympathetic axons to innervate blood vessels in the muscles and skin, as well as the sweat glands and arrector pili muscles of the skin.

When preganglionic axons enter the sympathetic ganglia, the axons branch to form collaterals that travel up and down in the sympathetic chain of ganglia, synapsing with postganglionic neurons at different levels (fig. 6.9). In the abdomen, some preganglionic axons enter the chain of ganglia but don't synapse there. Instead, they pass through the sympathetic chain to synapse with three *collateral sympathetic ganglia* associated with the digestive tract. These are exceptions to the rule that

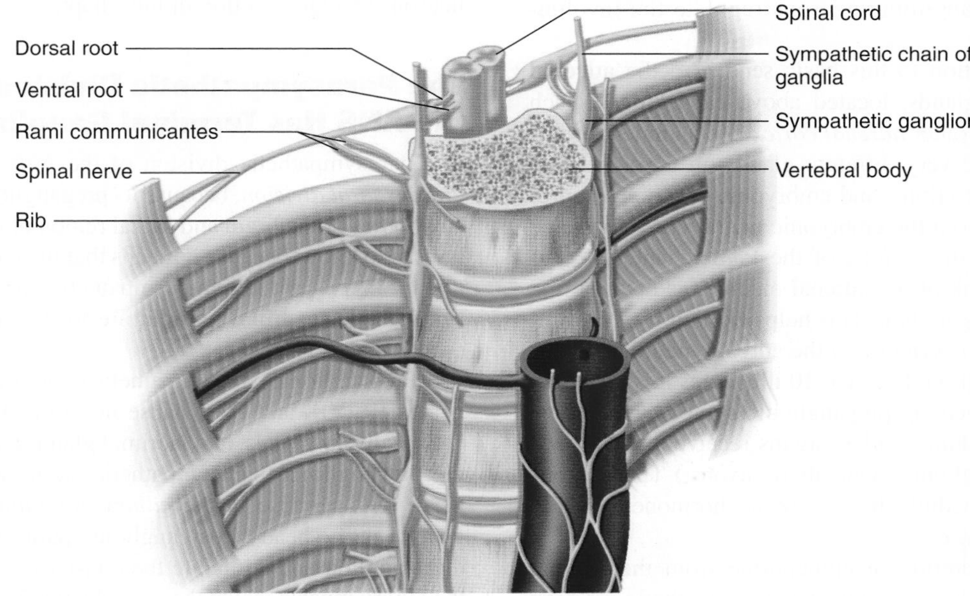

FIGURE 6.8 The sympathetic chain of ganglia. This diagram shows the anatomical relationship between the sympathetic ganglia and the vertebral column and spinal cord.

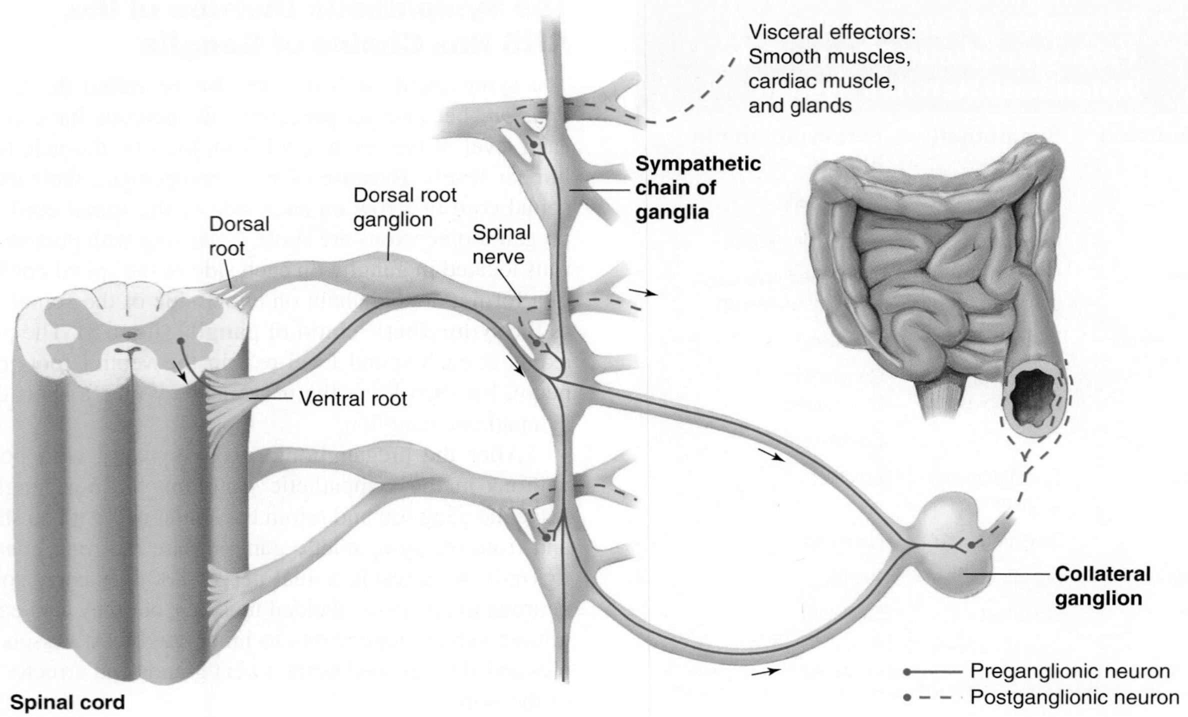

FIGURE 6.9 Sympathetic ganglia. Most preganglionic neurons of the sympathetic division synapse with postganglionic neurons in the sympathetic chain of ganglia. The preganglionic axon collaterals travel up and down the chain to synapse at different levels. In the abdomen, some preganglionic axons pass through the chain without synapsing. Instead, most of these synapse with postganglionic neurons in collateral ganglia, located near the digestive tract.

the postganglionic neurons are located in the sympathetic chain of ganglia. However, the basic sequence of events is still the same: the preganglionic neurons synapse with postganglionic neurons, and the postganglionic neurons regulate the involuntary effectors.

There is one exception to this basic sequence: the **adrenal medulla**. The adrenal glands, located above each kidney, each consists of two parts: an outer *adrenal cortex* and an inner adrenal medulla. These parts are very different glands, in terms of their regulation, hormonal secretions, and embryonic origin. The adrenal medulla is derived from the embryonic neural crest, the same tissue that gives rise to the ganglia of the PNS (chapter 5). For that reason, you can think of the adrenal medulla as being a first cousin to a sympathetic ganglion. This helps explain why preganglionic sympathetic axons synapse in the adrenal medulla, as if it were a sympathetic ganglion. Figure 6.10 illustrates this by showing a solid red line (indicating preganglionic sympathetic axons) entering the adrenal medulla. (Other organs receive a dashed red line, indicating postganglionic sympathetic axons.) These axons stimulate the adrenal medulla to secrete its hormones, mostly **epinephrine**, or *adrenaline*.

As a result, the secretion of epinephrine from the adrenal medulla accompanies the activation of the sympathetic division of the ANS. Because the sympathetic division of the ANS and the adrenal medulla function together, the two can be called the **sympathoadrenal system**. The logic of this grouping will become more apparent when we discuss the actions of the hormone epinephrine and of the neurotransmitters released by sympathetic neurons in a later section of this chapter.

The Parasympathetic Division of the ANS Has Terminal Ganglia

The parasympathetic division of the ANS is also known as the *craniosacral division*, because its preganglionic neurons have their cell bodies in the brain and sacral region of the spinal cord. This is helpful to know, because it means that the parasympathetic ganglia are not in a chain, like the sympathetic ganglia. Instead, the parasympathetic **terminal ganglia** are located next to, or within, the organs they serve.

Preganglionic parasympathetic axons that travel within cranial nerves III, VII, and IX synapse in terminal ganglia near the eyeballs, salivary glands, and lacrimal glands (fig. 6.10). For example, the preganglionic parasympathetic axons within the oculomotor nerve (III) synapse in the *ciliary ganglion* (one of the terminal ganglia in the head). Postganglionic parasympathetic axons leave that ganglion and stimulate the constrictor muscle of the iris and the ciliary muscle inside the eye (chapter 7). The *iris* is the colored part of the eye, surrounding an aperture called the *pupil*. It consists of two antagonistic smooth muscle layers. When the circular layer

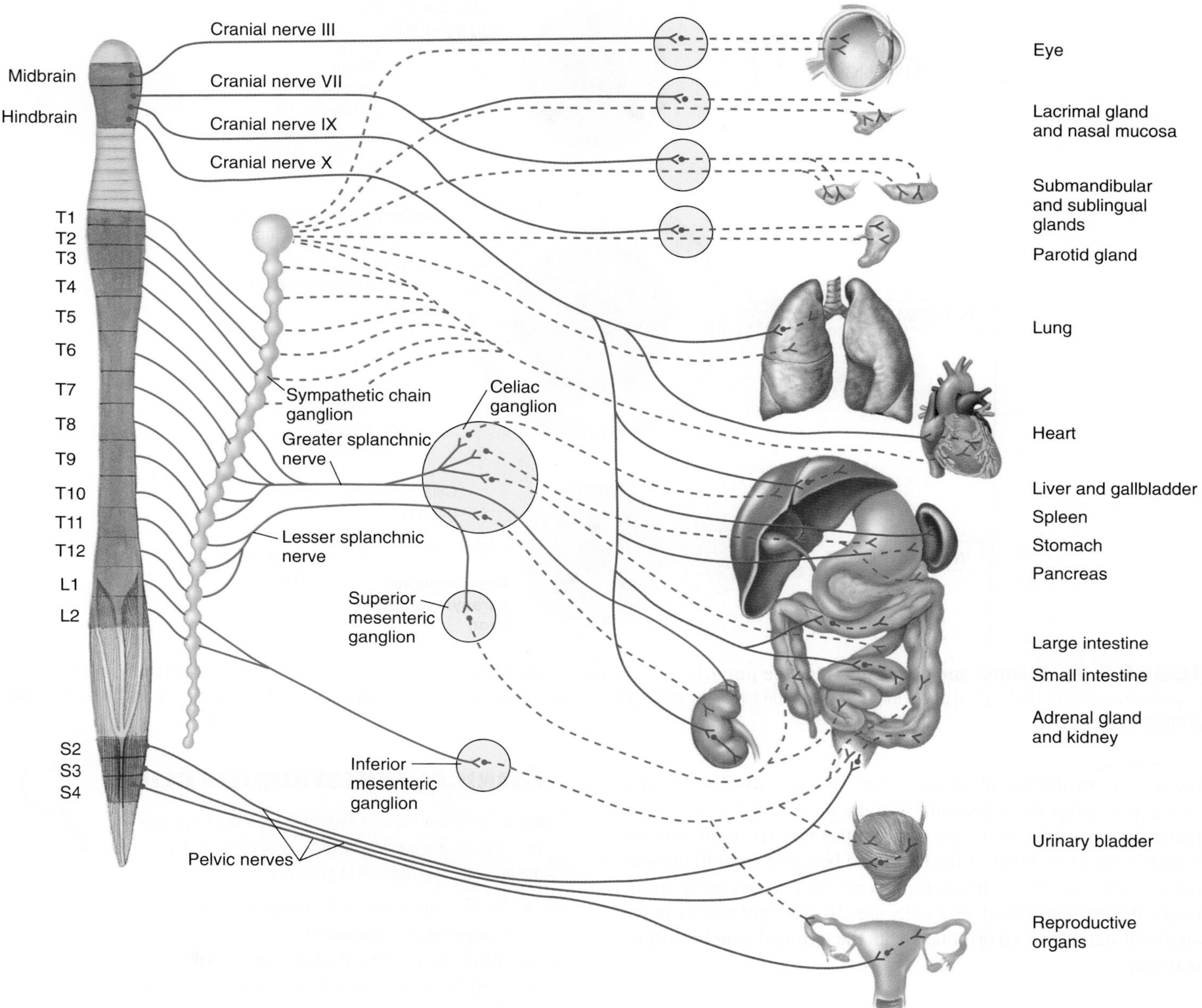

FIGURE 6.10 The autonomic nervous system. The sympathetic division is shown in red; the parasympathetic in blue. The solid lines indicate preganglionic fibers, and the dashed lines indicate postganglionic fibers. The celiac, superior mesenteric, and inferior mesenteric ganglia are sympathetic collateral ganglia. These ganglia receive preganglionic axons of the splanchnic nerves (as does the adrenal medulla).

gets parasympathetic stimulation, it constricts the pupil; when the radial layer gets sympathetic stimulation, it dilates the pupil (fig. 6.11). Remember that cranial nerve III also contains somatic motor axons, which stimulate skeletal muscles that attach to the eyeball and cause it to move.

The major parasympathetic nerves are the very long, paired **vagus (X) nerves**. The vagus nerves travel through the neck to enter the thoracic (chest) cavity, and then go through the diaphragm to extend into the abdominal cavity. The preganglionic neurons of the vagus have their cell bodies in the medulla oblongata, and their axons enter the organs of the thoracic and abdominal cavities. Thus, the preganglionic parasympathetic axons of the vagus nerves are quite long. This can be seen in figure 6.10 by the way that the vagus nerve is illustrated with solid blue lines (indicating preganglionic axons) that enter the organs. The vagus nerves also contain sensory axons that arise from receptors in the viscera, such as those that detect changes in blood pressure.

Examining figure 6.10 more closely reveals that the preganglionic parasympathetic axons (solid blue) of the vagus synapse with postganglionic neurons (dashed blue) inside the visceral organs. The same is true for the parasympathetic axons in the pelvic nerves

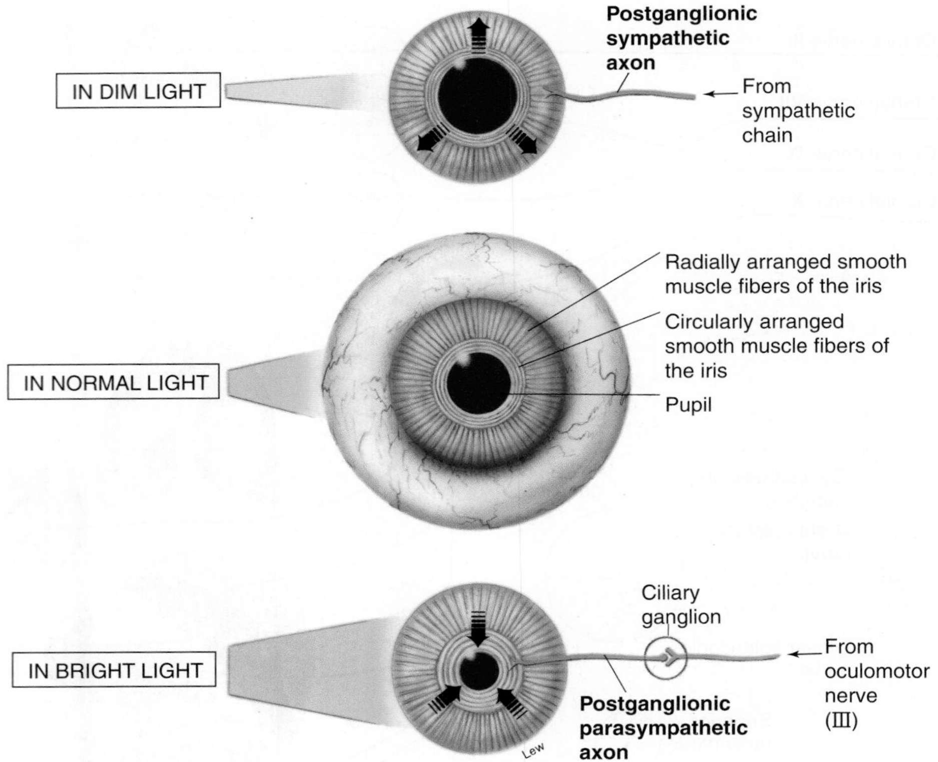

FIGURE 6.11 Dilation and constriction of the pupil. In dim light, the radially arranged smooth muscle fibers are stimulated to contract by sympathetic neurons, dilating the pupil. In bright light, the circularly arranged smooth muscle fibers are stimulated to contract by parasympathetic neurons, constricting the pupil.

that arise from the sacral levels of the spinal cord. This means that the parasympathetic terminal ganglia are located *inside* these organs. For example, there are parasympathetic terminal ganglia located in the outer layer of the heart and between smooth muscle layers of the intestine. Although preganglionic axons enter these organs, the involuntary effector cells are always regulated by postganglionic axons, which originate from the terminal ganglia within the organs.

Clinical Applications

People with lesions (damage) to the spinal cord at or above the sixth thoracic level may experience **autonomic dysreflexia**. In this condition, an unpleasant stimulus (often a full urinary bladder) can cause an excessive activation of the sympathetic division below the level of the lesion. This results in widespread constriction of blood vessels (vasoconstriction) in the viscera and the skin below the level of the lesion, making the skin pale and cold below the lesion. This vasoconstriction can produce a dangerous, even fatal, elevation in blood pressure. The elevated blood pressure can stimulate a reflex, acting through the medulla oblongata, that causes decreased sympathetic nerve and increased parasympathetic nerve activities above the level of the lesion. This results in vasodilation above the lesion (increasing blood flow), producing a flushed, sweaty skin above the lesion, painful headache, nasal congestion, and other symptoms.

Clinical Investigation Clues

Remember that Joyce's face and neck were flushed and sweaty, but her lower body was cool, and that she had an elevated blood pressure.

- What is the name of the condition in which these symptoms are produced?
- What produced the flushed, sweaty skin above the spinal cord lesion, and the cool skin below the lesion?
- What produced the elevated blood pressure?

Check Point

1. Compare the efferent pathways of somatic motor and autonomic motor neurons. Which autonomic motor neurons synapse with the effector cells?
2. List some of the actions of sympathetic and parasympathetic nerves on selected body organs.
3. Identify the origin of preganglionic sympathetic neurons. Where do these make synapses? What is the relationship between the sympathetic division of the ANS and the adrenal medulla?
4. Identify the origin of preganglionic parasympathetic neurons, and the location of the parasympathetic ganglia. Which parasympathetic neurons are the longest? Which are the shortest? Which synapse with the effector cells?

Regulation by the ANS Works Through Adrenergic and Cholinergic Effects

The preganglionic neurons of the ANS release ACh, which stimulates the postganglionic neurons. The postganglionic sympathetic axons release norepinephrine to produce adrenergic effects; the postganglionic parasympathetic axons release ACh to exert cholinergic effects. The adrenergic effects of the sympathoadrenal system prepare the body for "fight-or-flight." The cholinergic effects of the parasympathetic division promote effects that can be described as "rest and digest."

The autonomic nervous system (ANS) regulates many organs in different systems. Fortunately for students, the effects of ANS regulation can easily be predicted, because they tend to follow all-encompassing themes. For the sympathetic division of the ANS, the theme is described by the phrase **fight-or-flight**. For example: What should happen to your heart, the blood vessels in your muscles and skin, and your blood glucose concentration when you're scared or anxious? What changes should occur in your body when you're about to run an important race, go for an important interview, and so on? The changes produced by activation of the sympathetic division of the ANS—increased pumping by your heart, less blood flow to your skin and more to your muscles, a rise in blood glucose—can be logically predicted by the fight-or-flight theme.

Unlike the fight-or-flight phrase for the sympathetic division, the theme of the parasympathetic division isn't known by a universally recognized phrase. However, the phrase **rest-and-digest** captures the parasympathetic theme fairly well. In general, the effects of parasympathetic activation are opposite to the effects of sympathetic activation. Sympathetic nerves increase the heart rate, while parasympathetic nerves decrease the heart rate; sympathetic nerves dilate the pupils, while parasympathetic nerves constrict the pupils (fig. 6.11). Unlike sympathetic nerves, parasympathetic nerves promote the activity of the digestive system, as captured by the phrase *rest-and-digest*. You can get an overview of sympathetic and parasympathetic nerve effects by reviewing table 6.2.

The effects of a motor neuron are produced by the neurotransmitter it releases. As described earlier in this chapter, somatic motor neurons release ACh, which binds to the nicotinic type of ACh receptor in skeletal muscle cells. Since there are both preganglionic and postganglionic neurons of the ANS, and both sympathetic and parasympathetic divisions, the naming of neurotransmitters is a little more involved in the ANS (fig. 6.12).

1. All preganglionic neurons release ACh. These molecules bind to the nicotinic type of ACh receptors in the dendrites of the postganglionic neurons, thereby stimulating those neurons.
2. Most postganglionic axons of the parasympathetic division release ACh. These molecules bind to the muscarinic type of ACh receptor in the plasma membranes of cardiac muscle, smooth muscle, and glands.
3. Most postganglionic sympathetic axons release norepinephrine. These molecules bind to *adrenergic receptors* described in the next section.

Synaptic transmission that uses ACh as a neurotransmitter is described as **cholinergic**. Synaptic transmission that uses norepinephrine as a neurotransmitter is described as **adrenergic**. The regulation produced by postganglionic neurons of the ANS is due to their cholinergic or adrenergic effects on cardiac muscle, smooth muscle, and glands. When the sympathetic division is activated, the adrenal medulla is stimulated to secrete the hormone epinephrine, which also exerts adrenergic effects. Thus, when you think of the sympathoadrenal division and fight-or-flight, think of adrenergic effects (of epinephrine and norepinephrine). When you think of the parasympathetic division and rest-and-digest, think of cholinergic effects (of ACh on muscarinic ACh receptors).

Clinical Applications

Cocaine blocks the reuptake of dopamine and norepinephrine into presynaptic axon terminals (as described in chapter 4). As a result, excessive amounts of these neurotransmitters remain in the synaptic gap and stimulate the postsynaptic cells. Because norepinephrine produces sympathetic nerve effects, cocaine acts as a *sympathomimetic drug* (a drug that promotes sympathetic nerve effects). This can cause an increase in heart rate and constriction of coronary arteries, leading to heart damage. The combination of cocaine with alcohol (which depresses the CNS) is more deadly than either drug taken separately, and is a common cause of death from substance abuse.

The Sympathoadrenal System Activates Different Adrenergic Receptors

We are coming to information that is fundamental to later discussions of many body systems, particularly the cardiovascular and pulmonary systems. There are different types of adrenergic receptor proteins in different organs, causing different responses to the same neurotransmitter (norepinephrine) and hormone (epinephrine). Similarly, the muscarinic ACh receptors in many organs produce different responses to parasympathetic nerve activity. This information is important for an understanding of both normal physiology and many medical applications. Medically useful drugs stimulate or inhibit autonomic effects by working through the adrenergic and cholinergic receptors. So, be on the alert: the information in this section and the next will be fundamentally important to you in your later studies. Here is another opportunity to build a strong foundation for your knowledge pyramid in physiology.

When a postganglionic neuron of the sympathetic division is activated, it releases norepinephrine as a neurotransmitter. At the same time, the adrenal medulla is stimulated to secrete epinephrine (and lesser amounts of norepinephrine) into the blood. Both norepinephrine and epinephrine regulate their target cells by binding to *adrenergic receptors* in the plasma membrane. Remember the 2 types of ACh receptors, nicotinic

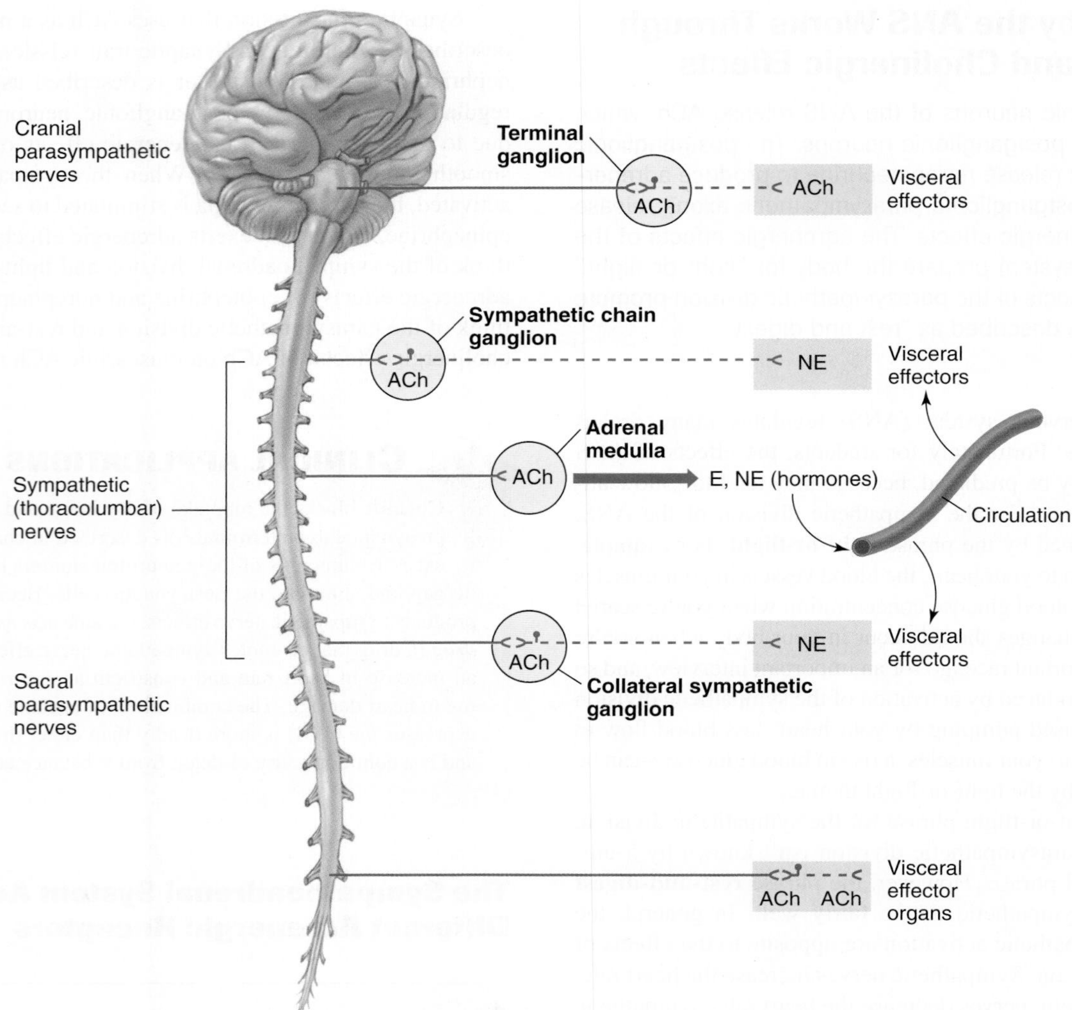

FIGURE 6.12 Neurotransmitters of the autonomic motor system. ACh = acetylcholine; NE = norepinephrine; E = epinephrine. Those nerves that release ACh are called cholinergic; those nerves that release NE are called adrenergic. The adrenal medulla secretes both epinephrine (85%) and norepinephrine (15%) as hormones into the blood.

and muscarinic (chapter 4)? Well, adrenergic receptors also come in two major types, the **alpha (α)** and **beta (β) adrenergic receptors**. Different tissues make one or the other, or sometimes both. Binding to norepinephrine and epinephrine activates both types of adrenergic receptors, but this binding produces different effects in different organs.

Just as there are different brands of cars, such as Toyotas and Fords, and different models of each, there are different models, or subtypes, of the α- and β-adrenergic receptors. These are designated α_1, α_2, β_1, and β_2. Organs contain effector cells that have more of one of these receptors than the others. Cardiac muscle cells, for example, have β_1-adrenergic receptors (which stimulate increased heart rate), whereas the smooth muscle cells of the pulmonary (lung) airways have β_2-adrenergic receptors (which promote relaxation of the smooth muscle and dilation of the airways). The α_1-adrenergic receptors are characteristic of vascular smooth muscle, stimulating their contraction and causing vasoconstriction.

In summary, sympathoadrenal stimulation of adrenergic receptors prepares the body for fight-or-flight. The changes produced by this stimulation include:

1. Dilation of the pupils.
2. Dilation of the airways (the bronchioles of the lungs; a β_2-adrenergic effect).
3. Increased rate of beat and strength of contraction of the heart (a β_1-adrenergic effect).
4. Breakdown of liver glycogen (glycogenolysis—chapter 2), so that the liver can secrete glucose into the blood, thereby increasing the blood glucose.
5. Vasoconstriction in the skin and visceral organs (an α_1-adrenergic effect); this reduces blood flow to these organs, so that more blood can be diverted to the heart and skeletal muscles for "fight-or-flight."
6. Vasodilation in the skeletal muscles, produced by stimulation of β_2-adrenergic receptors by the hormone epinephrine.

The α_2-adrenergic receptors are different from the others in that they are mostly found in presynaptic axon terminals. In the CNS, stimulation of the α_2-adrenergic receptors reduces the activity of the sympathoadrenal system. The distribution of some of these receptors is summarized in table 6.3, and some of their effects are illustrated in figure 6.13.

These differences in the type and subtype of adrenergic receptor between one organ and another are exploited for medical benefits. People with hypertension (high blood pressure), for example, may take a drug that blocks the β_1-adrenergic receptor, whereas people with asthma may use an inhaler that delivers a drug to stimulate the β_2-adrenergic receptors. Drugs that stimulate the heart activate β_1-adrenergic receptors; drugs that promote vasoconstriction stimulate the α_1-adrenergic receptor; and drugs that depress the sympathoadrenal system act on α_2-adrenergic receptors in the CNS. These are discussed in more detail in the "Physiology in Health and Disease" section at the end of this chapter.[1]

CLINICAL INVESTIGATION CLUES

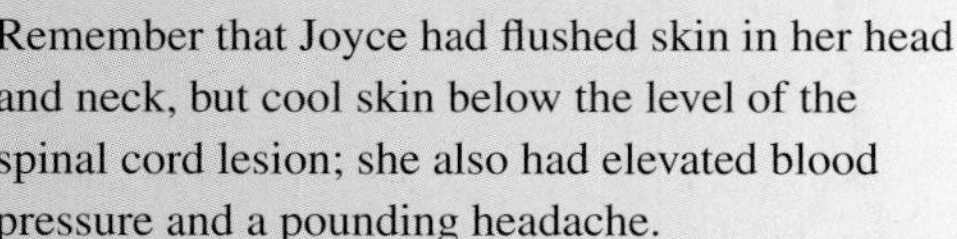

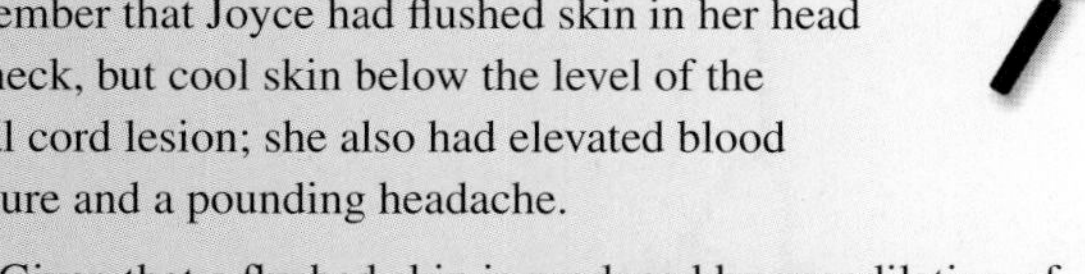

Remember that Joyce had flushed skin in her head and neck, but cool skin below the level of the spinal cord lesion; she also had elevated blood pressure and a pounding headache.

- Given that a flushed skin is produced by vasodilation of cutaneous blood vessels (which increases blood flow to the skin), and a cool skin is produced by constriction of these vessels, which type of adrenergic receptors were involved in these responses?
- How did the sympathetic division of the ANS produce these symptoms?

Parasympathetic Regulation Is Mostly Cholinergic

Synapses that use acetylcholine (ACh) as a neurotransmitter are described as *cholinergic synapses*. The ACh may bind to either of two types of receptors in the postsynaptic membrane, nicotinic or muscarinic. The ACh receptors in skeletal muscles, and in the dendrites of postganglionic neurons of the ANS, are of the nicotinic type. The ACh receptors in cardiac muscle, smooth muscles, and glands are muscarinic. These receive ACh released by postganglionic axons of the parasympathetic division (table 6.4). Thus, parasympathetic nerve effects on the organs of the viscera are muscarinic cholinergic effects.

Nicotinic ACh receptors are also ion channels that produce depolarizations (EPSPs); thus, they are always stimulatory (chapter 4). Muscarinic ACh receptors are separate from the ion channels and communicate with the ion channels by means of G-proteins (chapter 4). Because the ion channels are separate from the muscarinic receptors, when ACh binds to a muscarinic ACh receptor, it can cause either a depolarization and stimulation, or a hyperpolarization and inhibition (table 6.4). An example of a parasympathetic stimulatory effect is the promotion of contractions and secretions of the digestive system; an example of a parasympathetic inhibitory effect is the slowing of the heart rate. These and other parasympathetic effects can be inhibited by *atropine* and its analogues, which are drugs that specifically block the muscarinic ACh receptors.[2] The clinical uses of these drugs are described in the section "Physiology in Health and Disease" at the end of this chapter.

In summary, the effects of parasympathetic nerve stimulation promote the "rest-and-digest" theme through stimulation of muscarinic ACh receptors. These effects include:

1. Constriction of the pupils of the eyes (by stimulating the circular muscles of the iris)
2. Constriction of the pulmonary airways (bronchioles of the lungs)
3. Secretions of salivary glands and mucus glands
4. Slowing of the heart rate
5. Contractions of the gastrointestinal tract, and secretions of the organs of the digestive system
6. Vasodilation (thereby increasing the blood flow) in the genital organs, digestive system, and some other visceral organs.

TABLE 6.3 Selected Adrenergic Effects in Different Organs

Organ	Adrenergic Effects of Sympathoadrenal System	Adrenergic Receptor
Eye	Contraction of radial fibers of the iris dilates the pupils	α_1
Heart	Increase in heart rate and contraction strength	β_1 primarily
Skin and visceral vessels	Arterioles constrict due to smooth muscle contraction	α_1
Skeletal muscle vessels	Arterioles constrict due to sympathetic nerve activity	α_1
	Arterioles dilate due to hormone epinephrine	β_2
Lungs	Bronchioles (airways) dilate due to smooth muscle relaxation	β_2
Stomach and intestine	Contraction of sphincters slows passage of food	α_1
Liver	Glycogenolysis and secretion of glucose	α_1, β_2

Source: Simplified from table 6-1, pp. 110–111, of Goodman and Gilman's *The Pharmacological Basis of Therapeutics.* Ninth edition, J.E. Hardman et al., eds, 1996, McGraw-Hill.

FYI [1]Ephedrine and pseudoephedrine are molecules that occur naturally in plants of the genus *Ephedra*. Because ephedrine stimulates adrenergic receptors, in 2004 the FDA stopped the sale of ephedrine in dietary supplements. Pseudoephedrine stimulates α_1-adrenergic receptors, and can be used in nasal sprays. However, federal law now greatly restricts its sale because pseudoephedrine can be used illegally to produce methamphetamine, an abused, highly addictive, and dangerous drug.

FYI [2]Atropine is derived from the deadly nightshade plant, *Atropa belladonna*. During the Middle Ages, women dribbled extracts of this plant into their eyes to dilate their pupils (in Italian, *bella* = beautiful; *donna* = lady). This was done to enhance their beauty, and modern scientific studies have verified that, statistically, people find larger pupils more attractive in the opposite sex. Warning: while dilated pupils may have been romantic by candlelight in the Middle Ages, in today's bright lights it would lead to an unromantic headache.

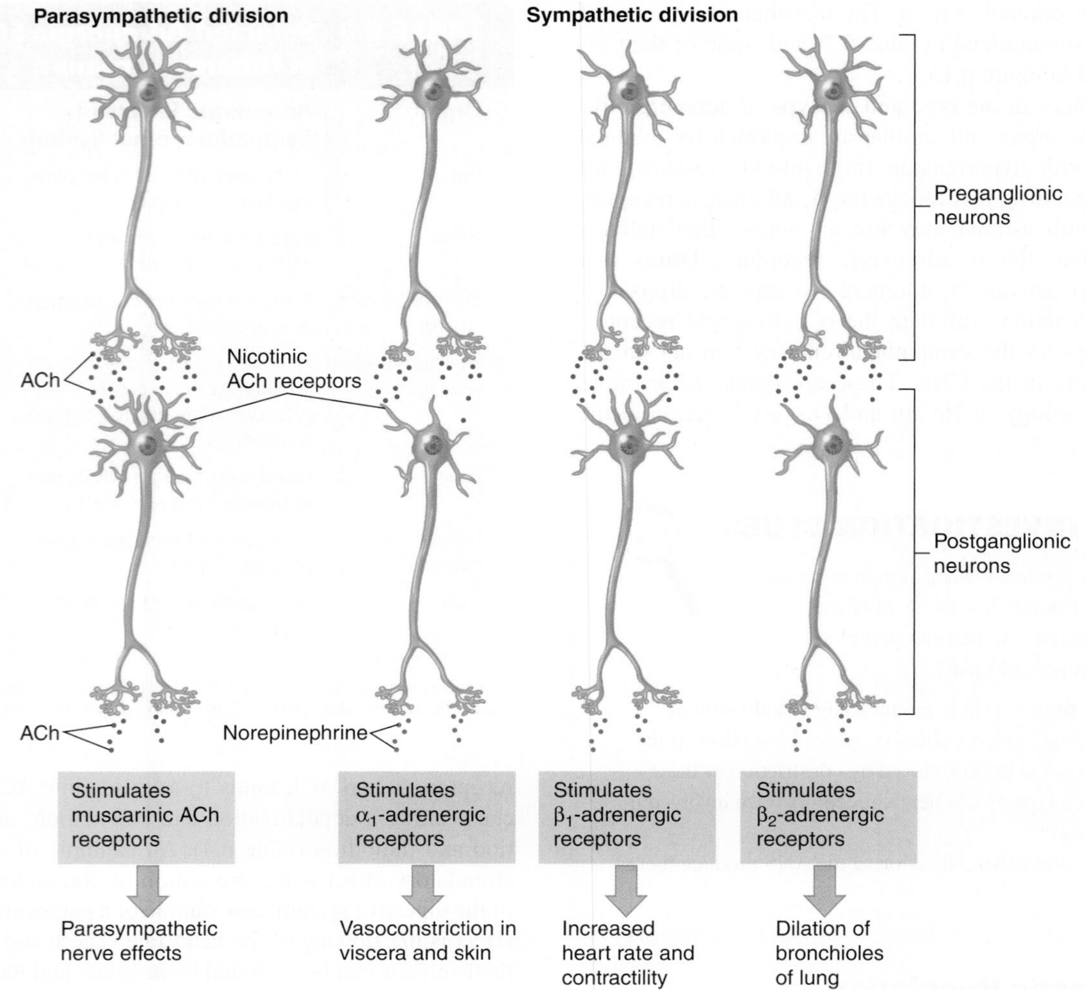

FIGURE 6.13 Receptors involved in autonomic regulation. Acetylcholine released by all preganglionic neurons stimulates the postganglionic neurons by means of nicotinic ACh receptors. Postganglionic parasympathetic axons regulate their target organs using muscarinic ACh receptors. Postganglionic sympathetic axons provide adrenergic regulation of their target organs by binding of norepinephrine to α_1, β_1, and β_2-adrenergic receptors.

TABLE 6.4 Effects of Acetylcholine (ACh) in the PNS

Neurons Releasing ACh	Location	Type of ACh Receptor	Response	Physiological Effect
Somatic motor	Skeletal muscles	Nicotinic	Depolarization, producing action potentials	Muscle contraction
Preganglionic neurons of ANS	Autonomic ganglia	Nicotinic	Depolarization, producing action potentials	Stimulates postganglionic neurons of the ANS
Postganglionic parasympathetic neurons	Smooth muscles, glands	Muscarinic	Depolarization, producing action potentials	Contraction of smooth muscles; secretion of glands
Postganglionic parasympathetic neurons	Heart	Muscarinic	Hyperpolarization, slowing the rate of automatic production of action potentials	Slowing of heart rate

Clinical Investigation Clues

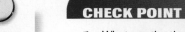

Remember that Joyce had a slower heart rate and a higher blood pressure.

- Which division of the ANS, and which nerve, was activated to produce the slower heart rate?
- Which neurotransmitter did the nerve release, and which type of receptor in the heart was activated by that neurotransmitter?
- Given information discussed in an earlier section, which division of the ANS produced the elevated blood pressure, and what caused the blood pressure to rise?
- Given that a slower heart rate is normally produced by an elevation in blood pressure, and that this reflex occurred in Joyce despite her spinal cord injury at T5, what does this imply about the location (in terms of body level) of the blood pressure receptors?

Check Point

1. What are the themes of the sympathoadrenal system and parasympathetic division? Give examples that illustrate these themes.
2. Name the neurotransmitters of all the neurons of the autonomic nervous system. Identify whether a cholinergic synapse is nicotinic or muscarinic.
3. List the types of adrenergic receptors and give one example of where in the body each type is found. What happens when each of the receptors in your examples is activated?
4. Which type of ACh receptor is activated by postganglionic parasympathetic axons? Describe some of the effects produced when this occurs.

Physiology in Health and Disease

There are a great many pharmaceutical drugs that treat different medical conditions by either stimulating or inhibiting different adrenergic receptors, thereby promoting or blocking sympathoadrenal effects. There are also drugs that promote or block the activation of muscarinic ACh receptors, and thereby stimulate or inhibit parasympathetic nerve effects.

For the acute treatment of **asthma**, people often use inhalers containing such compounds as *albuterol* and others that act as relatively specific agonists (stimulators) of the β_2-adrenergic receptors in the airways of the lungs. *Salmeterol* and others are longer-acting inhaled agonists of the β_2-adrenergic receptors. Selective β_1-adrenergic receptor agonists, such as *dobutamine*, may be given intravenously to increase cardiac output and blood pressure for people in **shock**. *Epinephrine*, which stimulates both α- and β-adrenergic receptors, is given either intravenously or directly into the heart to treat **cardiac arrest**. Epinephrine is also given subcutaneously or intramuscularly to treat **anaphylactic shock** (a rapid fall in blood pressure due to an excessive allergic reaction). On the other hand, drugs that block the β_1-adrenergic receptors, such as *atenolol* and *metoprolol*, can be useful in treating **hypertension** (high blood pressure).

Drugs that modify the activity of the α-adrenergic receptors are also medically useful. For example, *norepinephrine* (which acts primarily on the α-adrenergic receptors) may be used to help treat **hypotension** (low blood pressure) and shock. *Midodrine*, a long acting α_1-adrenergic receptor agonist, is used to help manage **orthostatic (postural) hypotension**. This is a relatively common condition where a person's blood pressure falls when they stand up too quickly. Similarly, *dihydroergotamine* stimulates α_1-adrenergic receptors to cause vasoconstriction of the carotid artery branches, helping to treat **vascular** and **migraine headaches**. This may be aided by *methysergide*, which inhibits the ability of serotonin to cause dilation of these vessels. The α_2-adrenergic receptors in the CNS are selectively stimulated by *clonidine*, and this effect reduces overall sympathoadrenal activity. Clonidine is thus sometimes useful in the treatment of hypertension and in the treatment of **drug withdrawal**.

Atropine and related drugs specifically block muscarinic receptors, and thereby inhibit parasympathetic nerve effects. Because parasympathetic nerves cause constriction of the pupils, atropine is dripped into the eye to **dilate pupils** for eye exams. Because parasympathetic nerves stimulate the gastrointestinal tract, antimuscarinic drugs such as atropine can help in the treatment of **peptic ulcers** and **irritable bowel syndrome**. For similar reasons, atropine is often used before surgery to inhibit salivation and excessive mucus secretions in the respiratory tract.

Perhaps the most dramatic use of atropine is for the treatment of **nerve gas** exposure. Nerve gases, such as tabun and sarin, are inhibitors of acetylcholinesterase (AChE; see chapter 4). As a result, these gases cause excessive ACh stimulation of all cholinergic synapses. This produces increased secretion from the nose, mouth, and pulmonary airways, spasms of the airways, abdominal cramping, pain, nausea, vomiting, and diarrhea. Atropine is given intramuscularly to help treat the immediate threat to the respiratory system. Since atropine has little effect on nicotinic ACh receptors, it can't treat the muscular spasms or sympathetic-nerve induced effects. However, the atropine is combined with another drug that helps restore the normal function of the AChE enzyme.

Physiology in Balance

Peripheral Nervous System

Working together with . . .

Endocrine System

Parasympathetic axons stimulate the pancreatic islets to secrete insulin . . . p. 186

Respiratory System

Somatic motor neurons stimulate the diaphragm and other respiratory muscles . . . p. 297

Cardiovascular System

Autonomic nerves regulate the heart and blood vessels, regulating blood pressure . . . p. 240

Sensory System

Cranial and spinal nerves carry action potentials from sensory organs into the CNS . . . p. 157

Digestive System

Autonomic nerves regulate contractions and secretions of the digestive tract . . . p. 348

Urinary System

Autonomic nerves control muscles for urination and the blood flow through the kidneys . . . p. 324

Homeostasis Revisited

Process	Set Point	Integrating Center	Sensors	Effectors
How the skeletal muscle stretch reflex and muscle spindle apparatus affect muscle length and tension when a muscle is stretched	Spindle length and tension	Spinal cord gray matter	Muscle spindles	Alpha motoneurons stimulate contraction of extrafusal muscle fibers, shortening the muscle and its spindles, and thereby reduce the tension on the spindles.
How the tension on tendons and the function of the Golgi tendon organs affect tendon tension when a tendon is stretched	Tension on a muscle's tendons	Spinal cord gray matter	Golgi tendon organs	Association neurons inhibit the alpha motoneurons that innervate a contracting muscle, reducing its contraction and thus the tension on its tendons.
How the sympathoadrenal system regulates blood glucose and liver glycogen when a person exercises (given that exercising skeletal muscles utilize blood glucose for energy)	Blood plasma glucose concentration	Brain regions regulating the autonomic nervous system	Brain regions regulating the autonomic nervous system	Norepinephrine and epinephrine, released by sympathetic axons and the adrenal medulla, stimulate the hydrolysis of liver glycogen and the secretion of glucose by the liver.
How parasympathetic nerves help maintain homeostasis when a hungry person, with low blood glucose, starts to eat (given that activation of the parasympathetic division occurs at this time, and that starch is digested into glucose)	Blood plasma glucose concentration	Brain regions regulating the autonomic nervous system	Brain regions regulating the autonomic nervous system	Parasympathetic axons, through the effects of acetylcholine, stimulate digestive functions and vasodilation in the digestive system, promoting the digestion of starch and the absorption of glucose into the blood.

Summary

The PNS Includes Cranial and Spinal Nerves 136

- There are 12 pairs of cranial nerves; some contain only sensory axons, but most are mixed, containing sensory and motor axons.
- Spinal nerves arise from each spinal cord segment, and are mixed nerves.
- Sensory neurons have their cell bodies in dorsal root ganglia and form the dorsal roots of spinal nerves; somatic motor neurons have their cell bodies in the spinal cord, and their axons form the ventral roots of spinal nerves.
- Stretching a skeletal muscle stimulates muscle spindles; this activates sensory neurons and initiates a monosynaptic stretch reflex.
- Alpha motoneurons are somatic motor neurons that innervate the muscle fibers outside the spindle (extrafusal fibers) and stimulate these fibers to contract; the knee-jerk reflex is produced in this way.
- Gamma motoneurons are somatic motor neurons that stimulate the muscle fibers inside the spindles (intrafusal fibers) to contract, tightening the spindle and making it more sensitive to stretch.
- Golgi tendon organs are sensory receptors that stimulate a reflex that inhibits muscle contraction through the release of GABA or glycine.
- Reciprocal innervation refers to the inhibitory synapses that association neurons make with somatic motor neurons that innervate antagonistic muscles; when an agonist muscle is stimulated, its antagonist muscle is inhibited.
- A motor unit is a somatic motor neuron and all of the muscle fibers its axon collaterals stimulate to contract when the neuron is activated.

The Autonomic Nervous System Regulates Involuntary Effectors 142

- Preganglionic neurons of the autonomic nervous system (ANS) have their cell bodies in the CNS; postganglionic neurons of the ANS have their cell bodies in ganglia.
- The postganglionic neurons of the ANS innervate the involuntary effectors: cardiac muscle, smooth muscles, and glands.
- The preganglionic neurons of the sympathetic division of the ANS exit the spinal cord at the thoracic and lumbar levels.
- Most sympathetic ganglia are located in two chains of ganglia parallel to the spinal cord; however, there are 3 collateral ganglia in the abdomen.
- Preganglionic axons of the sympathetic division synapse in sympathetic ganglia; the exception is the adrenal medulla, which receives preganglionic sympathetic axons.
- When the sympathetic division is activated, the adrenal medulla is stimulated to secrete the hormone epinephrine; thus, these are often together referred to as the sympathoadrenal system.

- Preganglionic neurons of the parasympathetic division have their cell bodies in the brain and sacral region of the spinal cord; those in the brain contribute axons to cranial nerves III, VII, IX, and X—cranial nerve X is the vagus nerve, the major parasympathetic nerve.
- Parasympathetic terminal ganglia are located next to or within the organs that they serve.

Regulation by the ANS Works Through Adrenergic and Cholinergic Effects 147

- The theme of the sympathoadrenal system is "fight-or-flight"; the theme of the parasympathetic division is "rest-and-digest."
- All preganglionic neurons of the ANS are cholinergic; they release ACh, which stimulates nicotinic ACh receptors in the postganglionic neurons.
- Postganglionic sympathetic axons release norepinephrine, and so provide adrenergic stimulation.
- Norepinephrine and epinephrine (from the adrenal medulla) stimulates α_1-, α_2-, β_1-, and β_2-adrenergic receptors; different organs make different types of these adrenergic receptors.
- Norepinephrine and epinephrine stimulate α_1-adrenergic receptors in blood vessels to cause constriction, β_1-adrenergic receptors in the heart to increase the rate and strength of its beat, and β_2-adrenergic receptors in the airways of the lungs to cause them to dilate.
- There are many pharmaceutical drugs that stimulate or inhibit sympathetic nerve effects by promoting or blocking specific types of adrenergic receptors.
- All of the adrenergic effects of the sympathoadrenal system occur when a person is frightened or anxious, and can be understood in terms of the fight-or-flight theme.
- Postganglionic parasympathetic axons are cholinergic; they release ACh, which stimulates the muscarinic type of ACh receptors in cardiac muscle, smooth muscle, and glands.
- The effects of parasympathetic stimulation include constriction of the pupils of the eyes, constriction of the airways, slowing of the heart rate, stimulation of salivary and mucus secretion, and stimulation of the contractions and secretions of the digestive tract.
- Parasympathetic nerve effects are inhibited by the drug atropine, which blocks the muscarinic ACh receptors.

Review Activities

Objective Questions: Test Your Knowledge

1. The cranial nerve that controls the muscles of facial expression is
   a. cranial nerve I.
   b. cranial nerve III.
   c. cranial nerve V.
   d. cranial nerve VII.

2. Sensory neurons have cell bodies in the
   a. gray matter of the spinal cord.
   b. white matter of the spinal cord.
   c. dorsal root ganglia.
   d. autonomic ganglia.

3. How many synapses are crossed in the spinal cord in a muscle stretch reflex?
   a. One
   b. Two
   c. Three
   d. Four

4. The neurons that stimulate extrafusal muscle fibers to contract, causing the muscle to shorten in response to stretch, are the
   a. preganglionic neurons.
   b. alpha motoneurons.
   c. postganglionic neurons.
   d. gamma motoneurons.

5. Reciprocal innervation refers to
   a. stimulation of the agonist and antagonist muscles.
   b. inhibition of the agonist and antagonist muscles.
   c. stimulation of the agonist and inhibition of the antagonist muscles.
   d. inhibition of the agonist and stimulation of the antagonist muscles.

6. Somatic motor neurons receive this neurotransmitter when the Golgi tendon organs of their muscle are stimulated:
   a. Acetylcholine
   b. GABA
   c. Norepinephrine
   d. Serotonin

7. A somatic motor neuron of a motor unit in the quadriceps femoris muscles of the leg may innervate as many as
   a. 3 muscle fibers.
   b. 100 muscle fibers.
   c. 1,000 muscle fibers.
   d. 1 million muscle fibers.

8. In the autonomic nervous system, the neurons that have their cell bodies in the thoracic or lumbar region of the spinal cord are
   a. preganglionic sympathetic neurons.
   b. postganglionic sympathetic neurons.
   c. preganglionic parasympathetic neurons.
   d. postganglionic parasympathetic neurons.

9. The adrenal medulla receives the axons of which neurons?
   a. Preganglionic sympathetic neurons
   b. Postganglionic sympathetic neurons
   c. Preganglionic parasympathetic neurons
   d. Postganglionic parasympathetic neurons

10. The vagus nerve contains the axons of which autonomic neurons?
    a. Preganglionic sympathetic neurons
    b. Postganglionic sympathetic neurons
    c. Preganglionic parasympathetic neurons
    d. Postganglionic parasympathetic neurons

11. All of the following neurons release ACh as their major neurotransmitter except
    a. preganglionic sympathetic neurons.
    b. postganglionic sympathetic neurons.
    c. preganglionic parasympathetic neurons.
    d. postganglionic parasympathetic neurons.

12. Which of the following contains muscarinic, but not nicotinic, ACh receptors?
    a. Skeletal muscles
    b. Autonomic ganglia
    c. Brain
    d. Smooth muscles

13. Which endocrine gland is derived from the same embryonic tissue as the autonomic ganglia, and secretes its hormones in response to sympathetic stimulation?
    a. Thyroid
    b. Adrenal cortex
    c. Adrenal medulla
    d. Pituitary

14. Autonomic axons that extend from the ciliary ganglion, a terminal ganglion near the eye, are
    a. preganglionic sympathetic axons.
    b. postganglionic sympathetic axons.
    c. preganglionic parasympathetic axons.
    d. postganglionic parasympathetic axons.

15. Adrenergic effects
    a. are produced by postganglionic sympathetic neurons.
    b. are produced by preganglionic sympathetic neurons.
    c. produce the effects described by the phrase *rest-and-digest.*
    d. produce the effects of parasympathetic nerve activation.

16. Muscarinic cholinergic effects are produced by the activation of
    a. preganglionic sympathetic axons.
    b. postganglionic sympathetic axons.
    c. preganglionic parasympathetic axons.
    d. postganglionic parasympathetic axons.

17. The drug atropine
    a. blocks sympathoadrenal effects.
    b. blocks parasympathetic effects.
    c. stimulates sympathoadrenal effects.
    d. stimulates parasympathetic effects.

Match the following effects with the responsible adrenergic receptors:

18. Heart beats faster and stronger. (a) α_1
19. Blood vessels in skin and viscera constrict. (b) α_2
20. Bronchioles (airways of lungs) dilate. (c) β_1
21. CNS reduces overall activity of sympathetic system. (d) β_2

Essay Questions 1: *Test Your Understanding*

1. Identify the cranial nerves by number and name, and say which are strictly sensory and which are mixed nerves.
2. Describe the composition of a spinal nerve, and the composition of the dorsal and ventral roots of a spinal nerve.
3. What is a muscle spindle, and how is it activated? Use this information to explain how a knee-jerk reflex is evoked when the patellar ligament is struck with a rubber mallet. Provide a circuit diagram of this reflex.
4. Distinguish between alpha and gamma motoneurons, and describe their functions.
5. Describe the Golgi tendon organ reflex and its function.
6. Explain reciprocal innervation and its physiological significance.
7. What is a motor unit? Identify muscles that have very small motor units and muscles that have very large motor units. What benefits are derived from this difference in motor unit size?
8. Compare the general efferent pathways to skeletal muscles with the efferent pathways to the involuntary effectors, naming the neurons involved and the location of their cell bodies.
9. Describe the location of the cell bodies of the preganglionic neurons of the sympathetic division of the ANS, and the location of the sympathetic ganglia.
10. Describe the location of the neuron cell bodies of the preganglionic neurons of the parasympathetic division of the ANS, and the location of the parasympathetic ganglia.
11. Explain the structural and functional relationship between the sympathetic division of the ANS and the adrenal medulla.
12. Which cranial nerves contain parasympathetic axons? Are they preganglionic or postganglionic? Which cranial nerve carries parasympathetic axons throughout the body? Where do these axons terminate to make synapses? Explain.
13. What is the theme of the sympathoadrenal system? Describe some of the body changes that follow this theme.
14. What is the theme of the parasympathetic division of the ANS? Describe some of the body changes that follow this theme.
15. List the neurotransmitters released by the preganglionic and postganglionic neurons of the ANS. Identify the types of receptors that these neurotransmitters stimulate.
16. Identify the different types of adrenergic receptor proteins in different organs, and describe some of the effects produced when they bind to norepinephrine or epinephrine.
17. Describe the effects produced in different organs when the parasympathetic division of the ANS is stimulated.
18. Identify the action of the drug atropine, and explain some of its effects.

Essay Questions 2: *Test Your Analytical Abillity*

1. Cranial nerve III (oculomotor) contains both somatic motor and autonomic motor axons. Identify the specific autonomic neurons, and describe the effects of both the somatic motor and autonomic motor neurons. Explain why there are no sympathetic axons in this nerve.
2. The blood vessels of the skin receive postganglionic sympathetic axons, but not postganglionic parasympathetic axons. Describe the differences in anatomy between these two divisions to explain why this is so.
3. Given that the muscle spindles are stimulated more by a quick than a slow stretch, explain why we kick our leg in the knee-jerk reflex in response to a slight stretch, yet can more greatly stretch our leg muscles before exercise without them going into spasm. Also explain how the Golgi tendon organ reflex helps us stretch our muscles.
4. Higher motor neurons inhibit inappropriate stretch reflexes, and so when there is a spinal cord lesion (damage) that blocks this inhibitory influence, there can be high muscle tone and spasticity below the level of the lesion. This also reduces reciprocal innervation. Explain how, in a person with a spinal cord lesion, the foot can flap (going up and down, called dorsiflexion and plantar flexion) when it is pushed up and then released.
5. Distinguish between alpha and gamma motoneurons, and describe their functions. What would happen if the alpha motoneurons to a muscle were cut? What do you think would happen if the gamma motoneurons were cut? Explain.
6. Given the difference in the size of the motor units between the eyes and legs, how would motor unit size compare between the fingers and the arms, and between the tongue and the torso? Explain.
7. How does your answer to question 6 relate to the size of the areas of cerebral cortex in the precentral gyrus devoted to these body areas (see chapter 5)? Explain the reasons for this relationship.
8. Which size motor units of your calf muscle (gastrocnemius) do you use in walking, running, and jumping? Given this, which motor units do you use most often, and which most rarely? Given that difference, which motor units do you suppose would have the most aerobically

adapted muscle fibers (thus most resistant to fatigue), and which the least? Explain.

9. Describe the body changes that occur during activation of the sympathoadrenal system, and explain how these relate to the "fight-or-flight" theme. How do you think these changes would affect the ability of the digestive system to process a meal? Relate this to the adage: "Don't eat before swimming."
10. A drug known as propranolol is a general β-adrenergic receptor blocker that can be used for treating high blood pressure, but which shouldn't be used if a person also has asthma. Speculate as to how this drug would help in the treatment of hypertension but would be contraindicated for a person with asthma.
11. Different drugs that are α-adrenergic receptor agonists (stimulators) can be used to either raise or lower the blood pressure. Explain how this is possible.
12. People waking up after general anesthesia for surgery may feel that their muscles are excessively relaxed and that their mouths feel very dry. Identify the two types of drugs, sometimes given to surgery patients, that can be responsible for those feelings, and explain why those drugs are used.

Web Activities

ARIS™ For additional study tools, go to www.aris.mhhe.com. Click on your course topic and then this textbook's author/title. Register once for a semester's worth of interactive activities and practice quizzing to help boost your grade!

CHAPTER 7

Sensory System

Homeostasis

In order for homeostasis to be maintained, there must be sensors to detect changes in the internal environment. As described in chapter 1, changes from a set point stimulate sensors, which relay information to an integrating center. The integrating center then activates effectors that complete a negative feedback loop to counter the change. In this way, a state of dynamic constancy of the internal environment—homeostasis—is maintained. The sensory system provides the sensors required for all negative feedback loops. Further, many sensory receptors (including the eyes, ears, taste, smell, and touch) allow us to function in the external world. If we were unable to interact effectively with our external environment, homeostasis of our internal environment would not be possible.

Clinical Investigation

But if homeostasis is not maintained . . .

"Your perfume reminds me of my mother!" Terri exclaimed, as the physician began to examine her. Terri wanted her hearing and vision checked because both had been giving her problems. She told the physician that she had lost equilibrium a number of times, with accompanying nausea, and had a continuous ringing in her ears. Her vision had been good all her life, Terri said, but now she had difficulty reading. This concerned her because her father and brother had retinitis pigmentosa, and she was worried that her symptoms might relate to this. The physician checked Terri's eyes with an ophthalmoscope and tested her peripheral vision. Afterward, the physician assured Terri that her vision was normal for a 50-year-old, and her difficulty reading could be treated with reading glasses. Terri asked if LASIK surgery would help, and the physician said that she didn't think it would. The physician told Terri that her hearing and equilibrium issues suggest that a particular drug, and perhaps other treatments, may be required.

How did the physician's perfume remind Terri of her mother? What disease might Terri have that accounts for her hearing and equilibrium problems? What's retinitis pigmentosa, and how would a check of peripheral vision relate to that condition? What caused Terri's near vision to deteriorate, and why wouldn't LASIK surgery help?

Chapter Outline

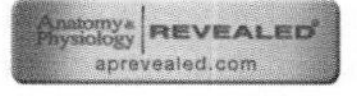

A virtual cadaver dissection experience

Sensory Receptors Stimulate Sensory Neurons

Sensory receptors respond to particular stimuli by causing action potentials to be produced in sensory neurons. The sensory receptors are classified according to their usual stimuli as being chemoreceptors, photoreceptors, thermoreceptors, or mechanoreceptors. Other categories are also recognized; for example, proprioceptors are muscle, joint and tendon receptors; and cutaneous receptors are the receptors in the skin. The senses of taste and smell are evoked by chemoreceptors that interact with molecules in the external environment.

You can't begin to study sensory physiology until you've had a previous background in the functions of neurons and the nervous system. So, you're already starting to reap some of the rewards of your previous efforts. When you examine the physiology of particular sensory organs, such as the eyes, you're studying highly specialized organs. However, the concept of sensory function is a fundamental one. All organs contain sensory receptors, which are needed for the body to respond to changes and thereby maintain homeostasis. We will encounter sensory receptors in later chapters when we discuss other systems, and so this information also serves as a foundation for your future advancement in physiological knowledge.

Our perceptions of the world—its textures, colors, and sounds, its warmth, smells, and tastes—are created in our brain from action potentials sent to it from sensory receptors. Because the sensory receptors *transduce* (change) many different forms of energy into the common language of action potentials, how do we perceive sight as being different from sound, or taste as different from touch? These different perceptions must be created in the brain, and so depend on the different neural pathways and synaptic connections made by these different *modalities* (types) of sensations. The specific neural pathways depend on the characteristics of the sensory receptor: light activates the photoreceptors in the eyes, not the taste buds on the tongue, for example. So, to begin a study of sensory physiology, we should start with the different categories of sensory receptors.

Sensory Receptors Are Categorized by Function

Sensory receptors are categorized according to the modality of the stimulus energy they transduce. On this basis, we have:[1]

1. **Chemoreceptors**, which respond to chemicals in the external environment (producing sensations of taste and smell) and internal environment (producing reflex responses to blood carbon dioxide, pH, and oxygen, for example).
2. **Thermoreceptors**, which respond to heat and cold (there are separate receptors for each).
3. **Mechanoreceptors**, which respond to stimuli that deform the plasma membrane of the receptor cell. These produce sensations of touch and pressure in the skin, and are also present in the inner ear, where they're responsible for the senses of equilibrium and hearing.
4. **Photoreceptors**, which respond to light. These are found in the neural layer of the eye, called the retina.
5. **Nociceptors**, which are pain receptors. There are separate receptors for pain, but pain can also be produced when other receptors (such as those for heat) are stimulated too intensely.
6. **Proprioceptors**, which are receptors in our muscles, joints, tendons, and ligaments. The muscle stretch receptors, called muscle spindles, and the Golgi tendon organs were discussed in chapter 6. Because of sensations arising from our proprioceptors, we can deftly pick up a pen from a desk without more than a casual glance, and hold up two fingers behind our back.
7. **Cutaneous receptors**, which are the sensory receptors in our skin. These include sensory receptors for touch, pressure, heat, cold, and pain.

Some receptors cause action potentials to be produced continuously from the time the stimulus is applied to the time the stimulus is withdrawn. Receptors that respond continuously are known as *tonic receptors*. Other receptors respond by producing a burst of action potentials when the stimulus is first applied, but then producing a gradually decreasing frequency of action potentials as the stimulus is maintained. These are called *phasic receptors* (fig. 7.1). Phasic receptors are partly responsible for the fact that

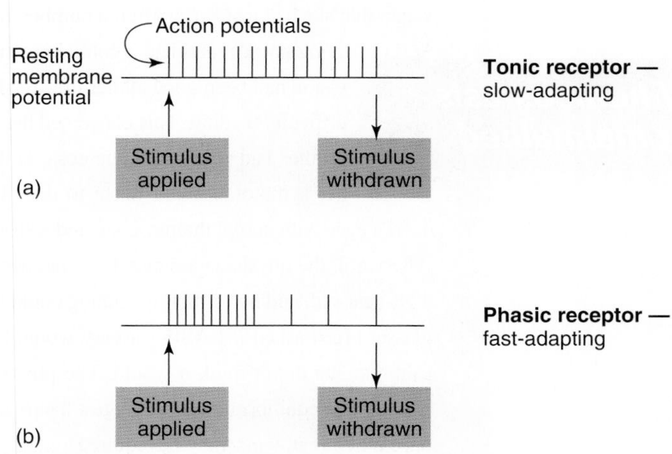

FIGURE 7.1 A comparison of tonic and phasic receptors. Tonic receptors (*a*) continue to fire at a relatively constant rate as long as the stimulus is maintained. These produce slow-adapting sensations. Phasic receptors (*b*) respond with a burst of action potentials when the stimulus is first applied, but then quickly reduce their rate of firing if the stimulus is maintained. This produces fast-adapting sensations.

FYI [1] Sharks and some other aquatic animals have a sense we lack: *electroreception*. This is because water (but not air) is a good conductor of electric currents. Electric currents are produced physically in the aquatic environment, helping some fishes navigate; aquatic animals also produce electric current, particularly when they swim and contract muscles. The electric currents emanating from fish allow them to be detected by predators. Sharks and rays have specialized electroreceptors (the *ampullae of Lorenzini*) that are especially sensitive, allowing the sharks to perceive their prey long before the prey can see the shark. But we shouldn't feel too deprived; actually, the duck-billed platypus is the only mammal that has electroreception.

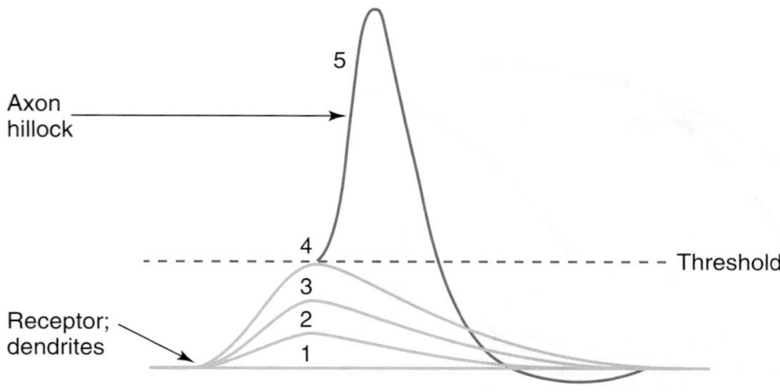

FIGURE 7.2 The receptor (generator) potential. This is a depolarization that must reach a threshold level in order to stimulate the production of action potentials.

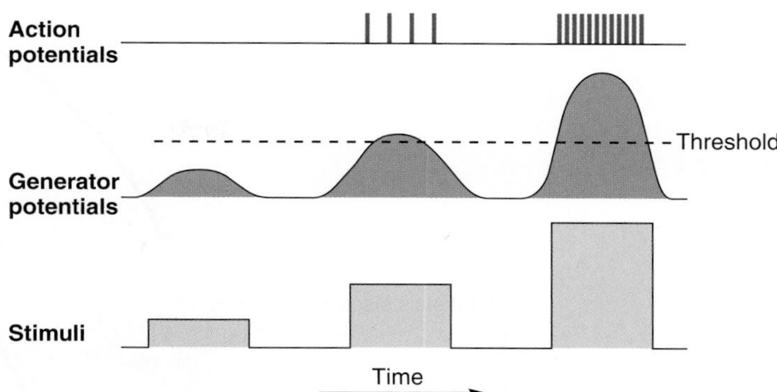

FIGURE 7.3 Effects of stronger stimuli. Stronger stimuli produce generator potentials of greater amplitude. The greater is the amplitude above threshold of the generator potential, the greater will be the frequency of action potentials produced in the sensory neuron.

we can stop paying attention to constant stimuli, a process known as **sensory adaptation**. Hot water feels hotter, for example, when we first get into the tub or shower, and the kitchen smells are more intense when we first enter the room.

Because stimulation of a particular receptor activates a neural pathway to a specific part of the brain, we experience only one characteristic sensation when that receptor is stimulated. This is the **law of specific nerve energies**. The normal, usual stimulus for that receptor is known as its *adequate stimulus*. For example, the adequate stimulus for the eyes is light, and the sensation it produces is sight. When the optic nerve produces action potentials in response to a different stimulus—say by a punch to the eye—the sensation it evokes is still sight ("seeing stars").

When a sensory receptor responds to its adequate stimulus (or when an unusual stimulus is strong enough), it causes the dendrites of sensory neurons to depolarize. This depolarization is called a **receptor potential**. Because the receptor potential generates action potentials in sensory axons, it's also called a **generator potential**. The receptor (generator) potential is equivalent to an EPSP (discussed in chapter 4); it must reach threshold (fig. 7.2) to stimulate an action potential, and its strength above threshold determines the frequency of the action potentials that will be produced (fig. 7.3). As discussed in chapter 4, stimulus strength is coded in the nervous system by action potential frequency, not by the amplitude of action potentials (which is all-or-none).

The Skin Contains Different Sensory Receptors

The **cutaneous sensations** are touch, pressure, heat, cold, and pain in the skin. The receptors for heat, cold, and pain are simply the naked dendrites of specialized sensory neurons. The sensations for touch are produced by dendrites surrounding hair follicles, and by expanded, specialized dendrites called *Ruffini endings* and *Merkel's discs* (fig. 7.4). Sensations of pressure are produced by *Meissner's corpuscles* and *pacinian* (*lamellated*) *corpuscles*, which are structures containing sensory dendrites. For example, a Pacinian corpuscle is a dendrite encased by 30 to 50 onion-like layers of connective tissues (fig. 7.4).

Nociceptors are free sensory dendrites; when these neurons are activated by mechanical stimuli that damage cells, their axons convey a sense of pain. The pain of a pinprick is transmitted by rapidly conducting myelinated axons, whereas a dull, persistent ache is transmitted by slower-conducting unmyelinated axons. Hot temperatures can produce sensations of pain through the action of a particular membrane protein in the sensory dendrites. This protein, called a *capsaicin receptor*, serves as both an ion channel and a receptor for the molecule capsaicin, which causes chili peppers to taste hot.

The sensations from cutaneous receptors and proprioceptors are together referred to as the **somatesthetic senses**. As described in chapter 5, these ascend on axon tracts that cross to the contralateral side of the CNS and project to the somatosensory cortex, or postcentral gyrus. As described in chapter 5, the different regions of the body that project to the postcentral gyrus form an upside-down map of the body (see fig. 5.7). However, sometimes the signals get crossed. This occurs when pain felt in one body location is actually caused by damage in another (such as in an internal organ). The pain is "referred" to a different body location, and is called a **referred pain**. The most famous example of a referred pain is *angina pectoris*—pain in the left pectoral region and left arm produced by damage in the heart (chapter 10).

CLINICAL APPLICATIONS

The term **phantom limb** refers to the pain felt by an amputee to originate in the missing limb. Many amputees experience complete sensations in the phantom missing limb (which is useful in fitting prosthetic devices), but pain in the phantom can be severe and persistent. Since phantom pain is experienced by 70% of amputees, this is a medically important phenomenon. One explanation for the referred pain of the phantom limb is that the remaining stump contains nerve endings that can generate action potentials. When these reach the location of the brain devoted to the missing limb, the person perceives the pain as originating in the phantom. Another explanation is that the sensations perceived as originating from the phantom result from brain reorganization produced by the absence of sensory information from the missing limb. This reorganization has been demonstrated to occur in both the thalamus and the postcentral gyrus.

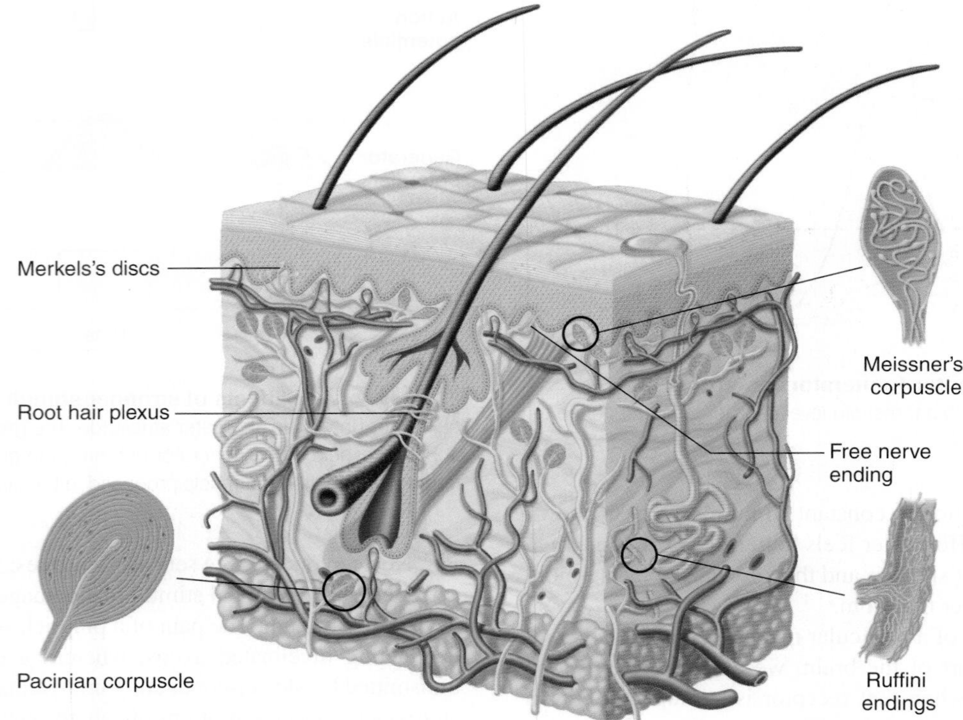

FIGURE 7.4 The cutaneous sensory receptors. Each of these structures is associated with a sensory (afferent) neuron. Free nerve endings are naked, dendritic branches that serve a variety of cutaneous sensations, including that of heat. Some cutaneous receptors are dendritic branches encapsulated within associated structures. Examples of this type include the Pacinian corpuscles, which provide a sense of deep pressure, and the Meissner's corpuscles, which provide cutaneous information related to changes in texture.

Taste and Smell Depend on Chemoreceptors

Taste and smell are chemical senses that respond to specific molecules in the external environment. For that reason, receptors for taste and smell are classified as *exteroceptors* to distinguish them from the chemoreceptors that respond to molecules within the body, called *interoceptors*. The senses of taste and smell complement each other, as can easily be verified by eating an onion (or almost anything) with and without the nostrils pinched together.

There Are Five Modalities of Taste

The *taste cells* are specialized epithelial cells within barrel-shaped **taste buds** (fig. 7.5), located on the tongue. The taste cells have long microvilli, which extend out into the external environment and are bathed in saliva. Although these taste cells are not neurons, they respond to particular molecules by depolarizing, and they release chemical transmitters that stimulate associated sensory neurons. Sensations of taste are carried from the anterior two-thirds of the tongue by cranial nerve VII, and from the posterior third of the tongue by cranial nerve IX.

Five categories of taste are now recognized: **salty**, **sour**, **sweet**, **bitter**, and **umami**. Umami is a Japanese term for savory, related to a meaty flavor and stimulated by the amino acid glutamate (and the flavor enhancer monosodium glutamate).

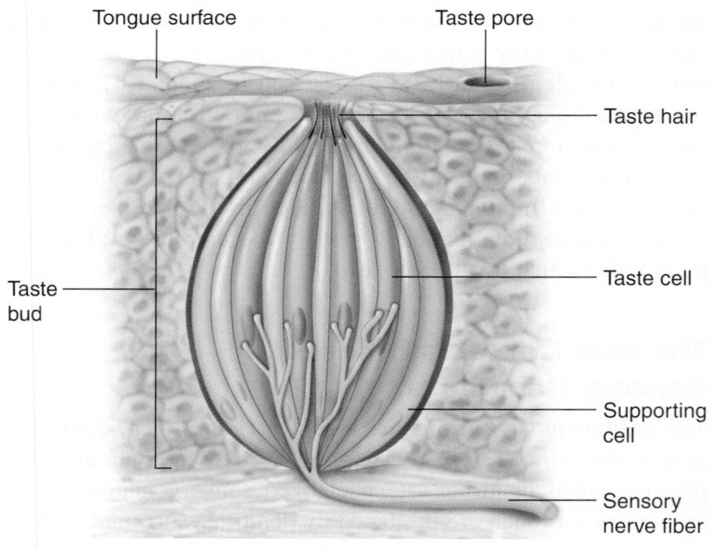

FIGURE 7.5 A taste bud. Chemicals dissolved in the fluid at the pore bind to receptor proteins in the microvilli of the sensory cells. This ultimately leads to the release of neurotransmitter, which activates the associated sensory neuron.

Scientists now believe that each taste bud contains taste cells sensitive to all five tastes, and that a given sensory neuron may have branches that go to different taste buds. However, a given sensory neuron is devoted to only one taste; for example, only taste cells that respond to sweet may stimulate it. Complex tastes activate different types of taste cells to various degrees, producing different patterns of activity in the associated sensory neurons. In this way, we get to experience tastes far more nuanced than simply the five taste modalities.

The salty taste of foods is due to the presence of sodium ions (Na^+), which pass into the sensitive taste cells through channels in their plasma membranes. In a similar manner, sour taste is due to hydrogen ions (H^+), which pass through channels in sensitive taste cells; all acids, therefore, taste sour. In contrast to salty and sour tastes, the other three tastes—sweet, bitter, and umami—are produced by the binding of molecules to specific receptor proteins in the plasma membranes of the sensitive taste cells. Sugars and some other organic molecules evoke a sweet taste when they bind to their receptor proteins; quinine and seemingly unrelated molecules evoke the bitter taste. It is the taste most associated with toxic molecules, although not all toxins taste bitter.

The Sense of Smell Involves Hundreds of Receptor Proteins

The receptors for **olfaction**, the sense of smell, are bipolar neurons in the olfactory epithelium (fig. 7.6). Each bipolar sensory neuron has one dendrite that projects into the nasal cavity and terminates in a knob containing cilia. It is the plasma membrane covering the cilia that contains the receptor proteins for smell. A family of about 1,000 genes codes for the receptor proteins, although it is believed that only about 300 different olfactory receptor proteins actually function in humans.

The olfactory receptor proteins are associated with G-proteins (chapter 3), which dissociate when the receptor protein binds to its odorant molecule (fig. 7.7). The G-protein subunits then activate the enzyme adenylate cyclase. This enzyme, built into the plasma membrane, now causes the second-messenger molecule, cyclic AMP (cAMP; see chapter 3), to be formed. The cAMP then opens ion channels in the plasma membrane, producing a depolarization that stimulates the sensory neuron. Because up to 50 G-protein complexes are associated with each receptor protein, many different ion channels may be opened when one odorant molecule binds to one receptor protein. This helps to increase the sensitivity to odors: the human nose can detect a billionth of an ounce of perfume in air.

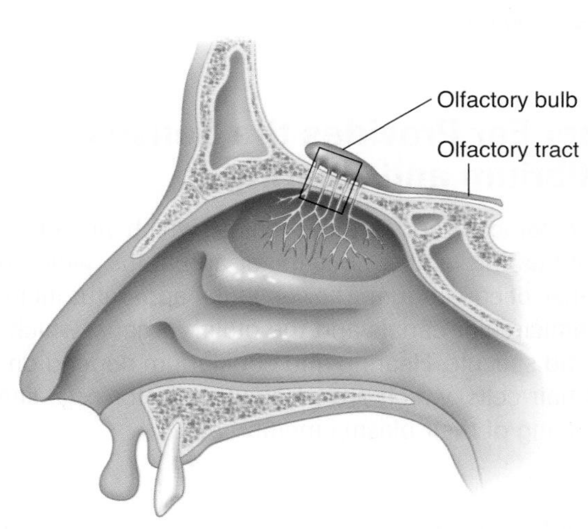

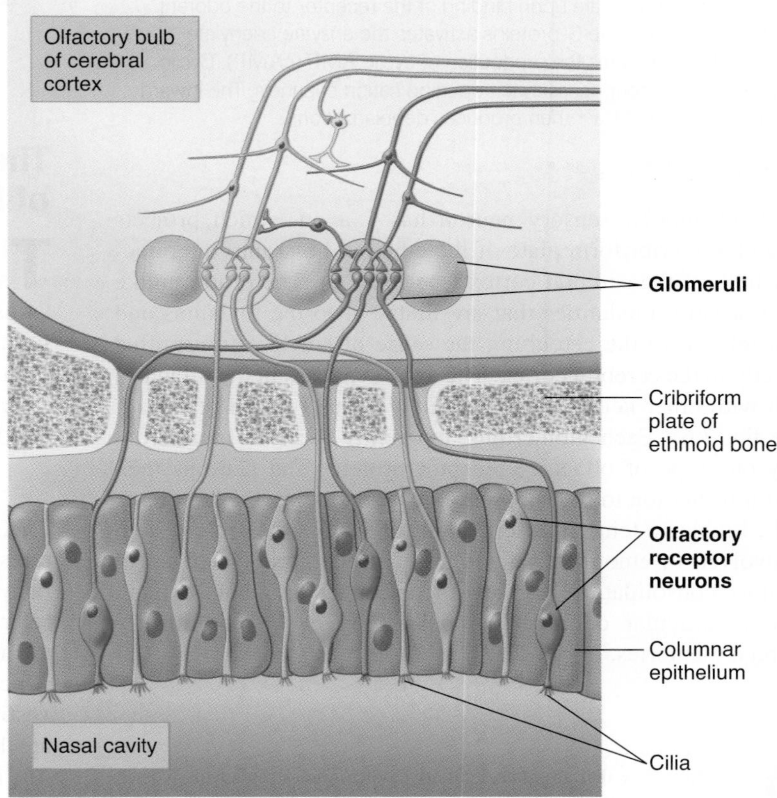

FIGURE 7.6 The neural pathway for olfaction. The olfactory epithelium of the nose contains receptor neurons that synapse with neurons in the olfactory bulb of the cerebral cortex. These synapses occur in rounded structures called glomeruli. Each glomerulus receives input from only one type of olfactory receptor. Neurons from the glomeruli send axons to other brain areas, including the medial temporal lobes.

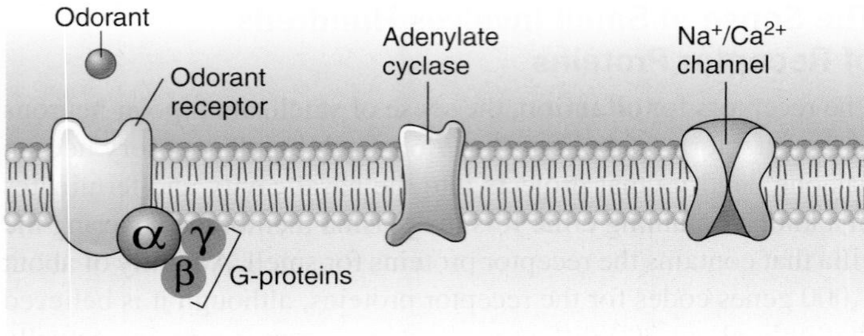

(a)

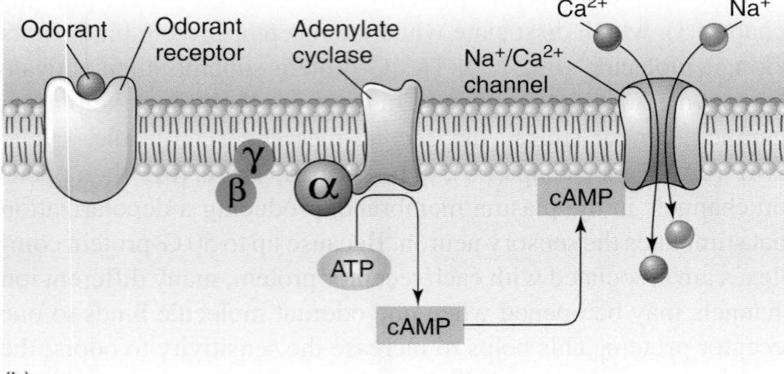

(b)

FIGURE 7.7 How an odorant molecule depolarizes an olfactory neuron. *(a)* The olfactory receptor is coupled to many G-proteins, which dissociate upon binding of the receptor to the odorant. *(b)* The α subunit of the G-proteins activates the enzyme adenylate cyclase, which catalyzes the production of cyclic AMP (cAMP). Cyclic AMP acts as a second messenger, opening cation channels. The inward diffusion of Na^+ and Ca^{2+} then produces depolarization.

Each bipolar sensory neuron has 1 axon, which projects through the cribriform plate of the ethmoid bone into the olfactory bulb of the cerebral cortex (see fig. 7.6). Therefore, unlike other sensory modalities that are first sent to the thalamus and then relayed to the cerebrum, the sense of smell is transmitted directly to the cerebral cortex. The axon synapses in the olfactory bulb with other neurons, grouped into clusters called *glomeruli* (see fig. 7.6). Each glomerulus apparently receives input from only one type of olfactory receptor protein. The neurons then send information to other brain regions, including the structures of the limbic system. As described in chapter 5, the limbic system is involved in emotions, and some of its parts are involved in memory consolidation. Perhaps these connections help explain why a particular odor can so powerfully evoke emotionally charged memories.[2]

FYI [2]Studies show that an infant finds its mother's breast by smell. Women have greater olfactory ability than men, and this ability is increased at about the time of ovulation. Olfactory sensitivity is great in humans, but the range of odors detectable by humans is less than that of many other mammals (although a trained person in the perfume industry can detect 10,000 odors). The relationship between the olfactory sense and the limbic system associates odors with emotions, and odors associated with pleasant experiences improve our mood.

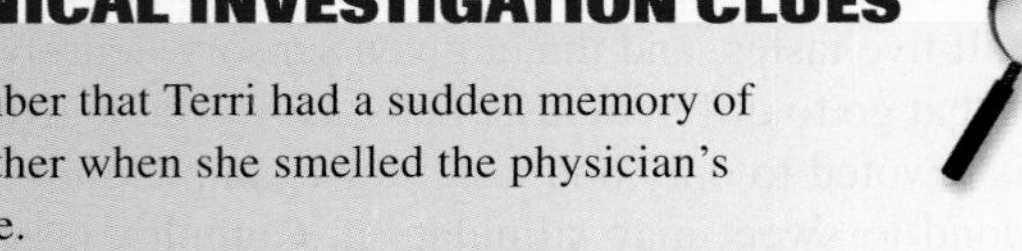

CLINICAL INVESTIGATION CLUES

Remember that Terri had a sudden memory of her mother when she smelled the physician's perfume.

- Why might the physician's perfume have reminded Terri of her mother?
- What brain structures could have been activated to trigger Terri's emotional response?
- What brain structures may have been involved to retrieve Terri's memory? (See chapter 5.)

CHECK POINT

1. List different functional categories of sensory receptors. What are some other categories of sensory receptors?
2. Describe the law of specific nerve energies and define the adequate stimulus. Give an example.
3. What are the cutaneous senses, and what types of sensory receptors produce them? What produces the hot taste of chili?
4. List the categories of taste and describe how these are produced.
5. Describe how we can discriminate so many different odors, and how our sense of smell can detect extremely low concentrations of an odorant.

The Inner Ear Provides the Senses of Equilibrium and Hearing

The inner ear consists of the cochlea, which provides the sense of hearing, and the vestibular apparatus, which provides the sense of equilibrium. The vestibular apparatus includes the three semicircular canals and the otolith organs, which are the utricle and saccule. Hearing and equilibrium depend on the function of hair cells, which are mechanoreceptor cells stimulated by bending of their plasma membranes.

The senses of hearing and equilibrium are provided by the structures of the **inner ear**. The inner ear is a small, complexly shaped bony structure called the **bony labyrinth** (the word *labyrinth* refers to intricate and confusing interconnected passageways, like a maze). The bony labyrinth consists of the snail-shaped **cochlea**, which is the organ of hearing, and structures that compose the **vestibular apparatus**. The structures of the vestibular apparatus are the 3 *semicircular canals*, the *utricle*, and the *saccule* (fig. 7.8). The inner ear is encased within the petrous portion of the temporal bone, the hardest bone in the skull.

Imagine a skilled balloon artist at a party, making intricately shaped giraffes and horses out of balloons. Now, imagine that person blowing a balloon into the shape of the inner-ear bony labyrinth. This balloon is analogous to the **membranous**

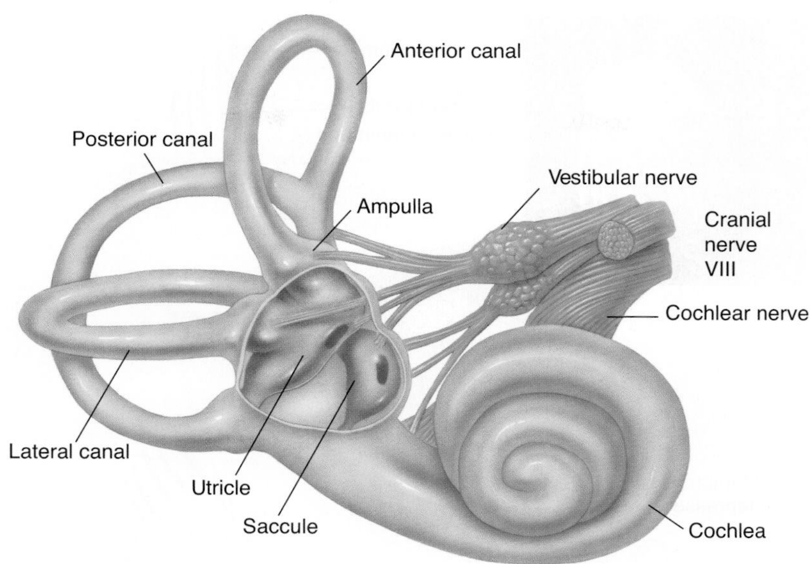

FIGURE 7.8 The cochlea and vestibular apparatus of the inner ear. The vestibular apparatus consists of the utricle and saccule (together called the otolith organs) and the 3 semicircular canals. The base of each semicircular canal is expanded into an ampulla that contains sensory hair cells.

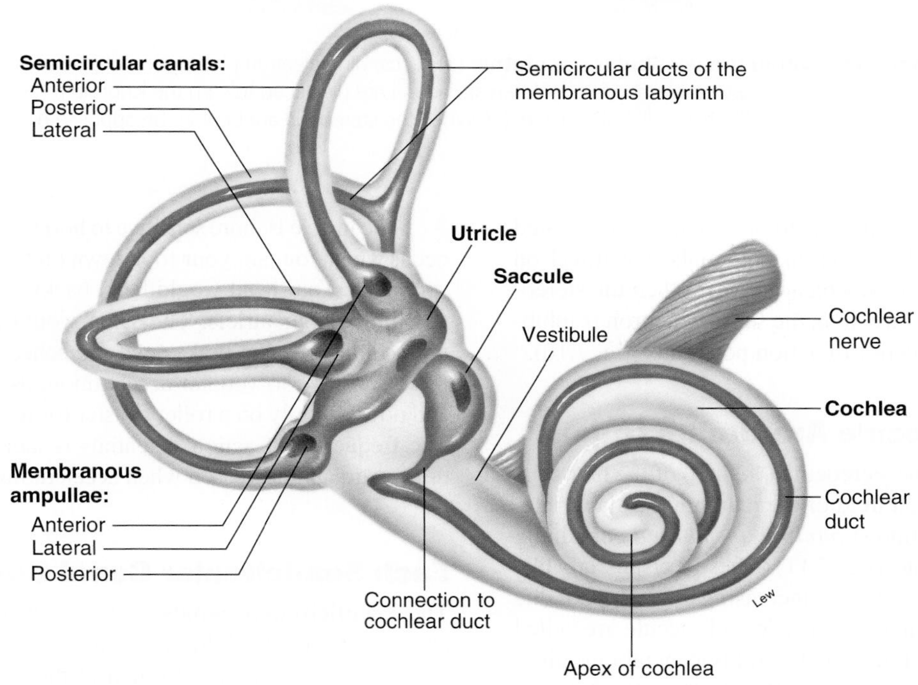

FIGURE 7.9 The labyrinths of the inner ear. The membranous labyrinth (darker blue) is contained within the bony labyrinth.

labyrinth, which is located within the bony labyrinth. The bony labyrinth is filled with a fluid called *perilymph* (colored light blue in fig. 7.9) that protects the delicate membranous labyrinth within it. The membranous labyrinth is actually more like a water balloon, filled with a fluid called *endolymph* (colored darker blue in fig. 7.9). It is the membranous labyrinth that contains the sensory **hair cells**.

Hair cells are mechanoreceptor cells responsible for the senses of hearing and equilibrium. Each hair cell has hairlike extensions known as *stereocilia*, which are processes containing protein filaments surrounded by part of the plasma membrane. One larger extension has the structure of a true cilium (chapter 2) and is known as a *kinocilium* (fig. 7.10). When the stereocilia are bent in the direction of the kinocilium, the plasma membrane becomes

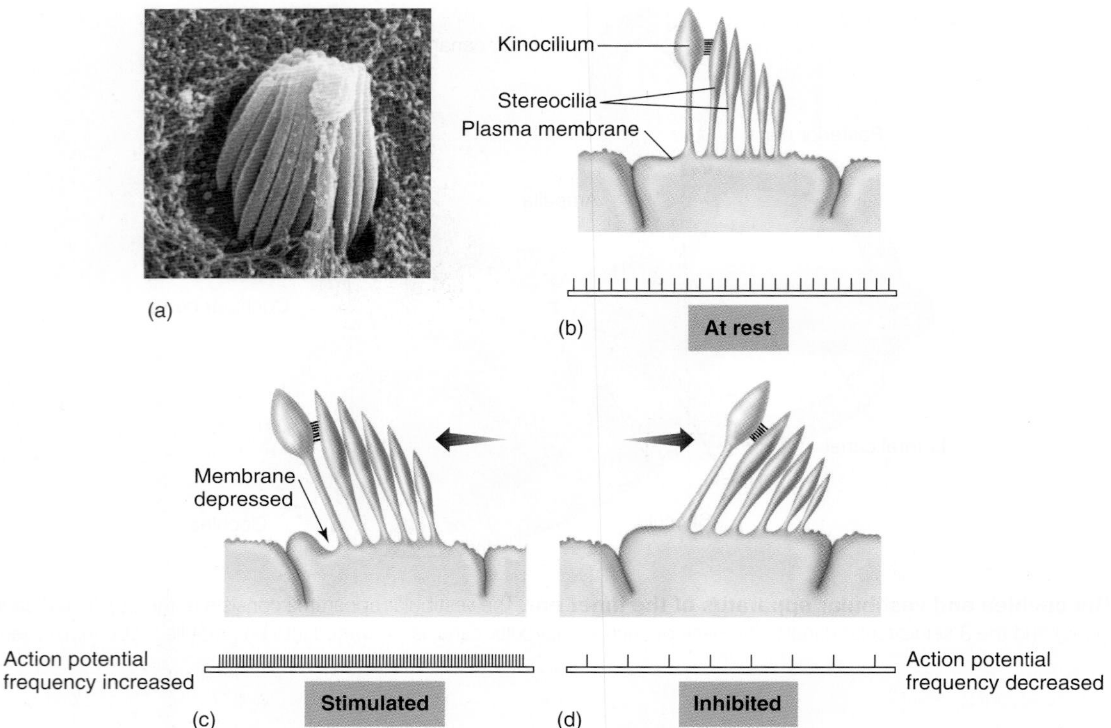

FIGURE 7.10 Sensory hair cells within the vestibular apparatus. (*a*) A scanning electron photograph of a kinocilium and stereocilia. (*b*) Each sensory hair cell contains a single kinocilium and several stereocilia. (*c*) When stereocilia are displaced toward the kinocilium (*arrow*), the cell membrane is depressed and the sensory neuron innervating the hair cell is stimulated. (*d*) When the stereocilia are bent in the opposite direction, away from the kinocilium, the sensory neuron is inhibited.

depolarized. This causes the hair cell to stimulate an associated sensory neuron, which produces action potentials that travel on cranial nerve VIII (the vestibulocochlear nerve). When the stereocilia are bent in the opposite direction, the sensory neuron is inhibited, producing a lower frequency of action potentials (fig. 7.10).

The Utricle and Saccule Are Otolith Organs

In the **utricle** and **saccule**, the stereocilia ("hairs") of the hair cells project up into the endolymph fluid of the membranous labyrinth. Here, they stick into a gelatinous *otolithic membrane* containing crystals of calcium carbonate (fig. 7.11). These crystals are like "stones" (*oto* = ear; *lith* = stone), increasing the mass of the membrane, and because of them the utricle and saccule are called the **otolith organs**. The higher mass of the otolithic membrane gives it a different inertia than the hair cells, so that the hairs bend during acceleration.[3] (*Inertia* refers to the tendency of an object to remain in its existing state of rest or linear motion; objects with more mass have more inertia, or resistance to change.)

The utricle is more sensitive to horizontal acceleration, as would occur when you jam your foot down on the accelerator of your new sports car. Your head would bend backward, and so would the stereocilia of your utricle. The same would happen in reverse when you stomped on the brakes and screeched to a halt. The saccule is more sensitive to vertical acceleration, as occurs when you go up and down rapidly on a roller coaster (or more gently in an elevator). The frequency of action potentials is increased when accelerating downward and decreased when accelerating upward.

Each Semicircular Canal Contains a Cupula

The 3 **semicircular canals**—*anterior*, *lateral*, and *posterior*—are arranged at nearly right angles to each other, like the sides of a cube. The hair cells are located at the base of each semicircular canal, in a swelling (the *ampulla*). The stereocilia of the hair cells project up into the endolymph of the membranous labyrinth here, embedded in a gelatinous membrane called the **cupula** (fig. 7.12). The endolymph can push the cupula the way wind pushes a sail; when the cupula bends, its embedded stereocilia bend as well, stimulating the production of action potentials in the associated sensory neurons of cranial nerve VIII.

The bending of the cupula and its embedded stereocilia occurs because the cupula has a different mass than the surrounding endolymph, giving it a different inertia when you turn your head along the plane of a semicircular canal. For example, if you turn your

FYI [3]Fish have a *lateral line organ* along their head and sides that contains hair cells used to detect pressure gradients and vibrations. This helps them gauge distances and water currents. Although fish lack visible ears, they have inner ears and can hear; they don't need external ears because sound travels from water through the fish's body to their inner ears. The organ of hearing in fish contains otoliths, analogous to those in our utricle and saccule, but unlike our own cochlea. However, in both fish and people, the sense of hearing results from the bending of the stereocilia on hair cells.

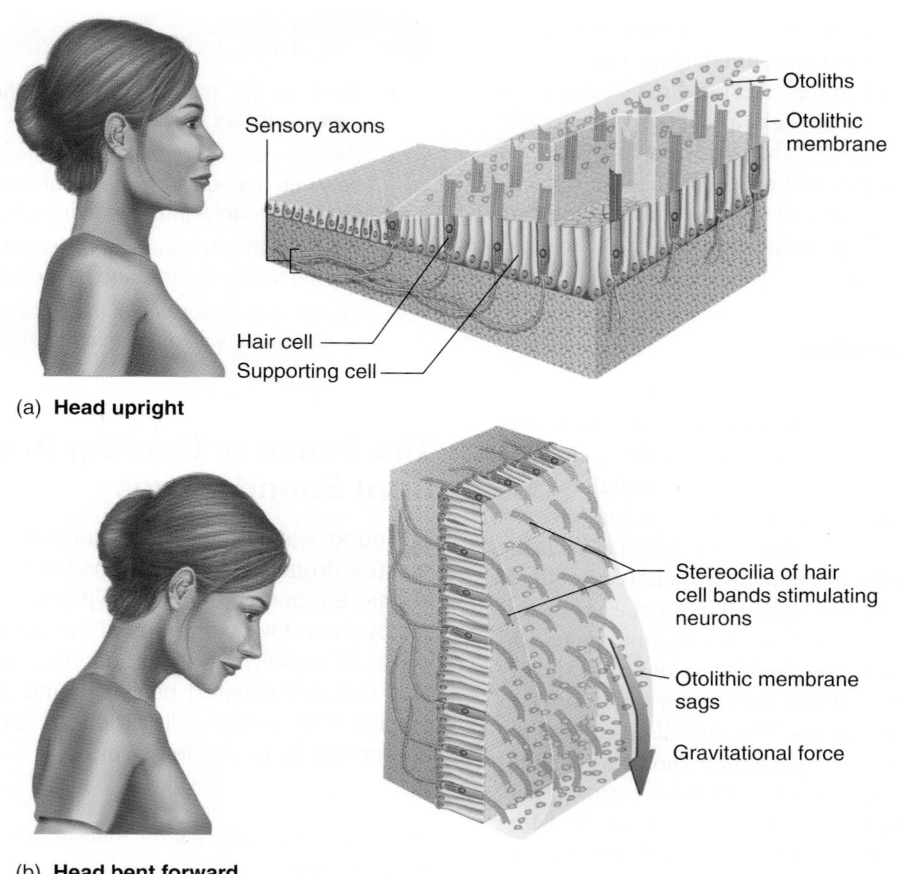

FIGURE 7.11 The otolith organ. (*a*) When the head is in an upright position, the weight of the otoliths applies direct pressure to the sensitive cytoplasmic extensions of the hair cells. (*b*) As the head is tilted forward, the extensions of the hair cells bend in response to gravitational force and cause the sensory nerve fibers to be stimulated.

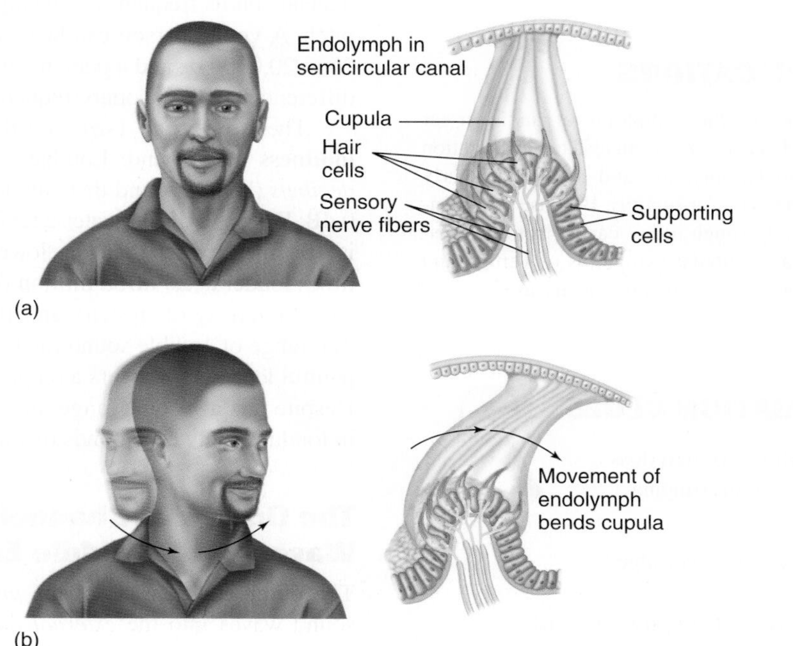

FIGURE 7.12 The cupula and hair cells within the semicircular canals. (*a*) Shown here, the structures are at rest or at a constant velocity. (*b*) Here, movement of the endolymph during rotation causes the cupula to bend, thus stimulating the hair cells.

head to the right, the endolymph in the lateral semicircular canal causes the cupula to bend to the left and stimulate the hair cells. In a similar manner, the hair cells in the anterior semicircular canal are stimulated when you do a somersault, and those in the posterior semicircular canal are stimulated when you do a cartwheel. Of course, you don't have to be a gymnast to stimulate these semicircular canals, and a head movement will often stimulate more than one, although generally it will stimulate one semicircular canal more than its stimulates the others.

The Vestibular Apparatus Can Cause Vertigo

Action potentials traveling on sensory axons from the vestibular apparatus go to the cerebellum and to the medulla oblongata (brain regions discussed in chapter 5). Axons leaving the medulla oblongata indirectly influence eye movements, so that you can track the visual field as you move your head. This is most dramatic during a spin, allowing you to maintain a stable visual field (especially if you're a trained gymnast, ice skater, or ballet dancer). However, if someone abruptly stops your spin, your eyes will continue to drift in the direction of your former spin, then be jerked back rapidly to the midline position. This produces involuntary oscillations of the eyes called **vestibular nystagmus**. You may also feel that the room is spinning, causing a loss of equilibrium known as **vertigo**.

Vertigo caused by spinning is normal. However, vertigo may have pathological (abnormal) causes; vertigo results when the action potential frequency coming from one inner ear is different from the frequency coming from the other. This is often caused by a viral infection and may be accompanied by other symptoms that result from activation of the autonomic nervous system—producing dizziness, pallor, sweating, nausea, and vomiting. These symptoms will sound familiar if you've ever experienced seasickness.

CLINICAL APPLICATIONS

Vestibular nystagmus is one of the symptoms of an inner-ear condition called **Ménière's disease**, which can lead to degeneration of the hair cells in the vestibular apparatus and cochlea. An early symptom is often "ringing in the ears," or *tinnitus*. Because the membranous labyrinth is continuous (through a tiny canal) from the vestibular apparatus to the cochlea, vestibular symptoms of vertigo and nystagmus often accompany hearing problems in this disease.

CLINICAL INVESTIGATION CLUES

Remember that Terri had episodes of vertigo (loss of equilibrium) and continuous tinnitus (ringing in her ears).

- Which anatomical structures are responsible for her tinnitus and vertigo?
- What is the relationship between these structures that might explain the association of vertigo and tinnitus?
- Which disease might be responsible for these symptoms?

CHECK POINT

1. What are the "hairs" of hair cells, and how do these cells have a sensory function?
2. Describe the structures of the inner ear, distinguishing between the bony and membranous labyrinths, and between endolymph and perilymph.
3. What are the structures of the vestibular apparatus? Which are the otolith organs, and how do they function?
4. Explain how stimulation of the semicircular canals leads to the production of action potentials in the vestibulocochlear nerve.

The Sense of Hearing Results from Sound Waves

Sound waves cause vibrations of the tympanic membrane (eardrum), which are transmitted via the movements of 3 middle-ear bones to produce vibrations of a smaller membrane, the oval window. Vibrations of the oval window cause pressure waves of endolymph in the cochlea, which produce vibrations of a hair-cell-covered basilar membrane. Bending of the stereocilia of these hair cells produces action potentials, which are interpreted by the brain as sound.

Sound waves are alternating zones of high and low pressure, where molecules are compacted or more spread out, traveling in a medium such as air or water. (There can't be sound in space, because there aren't molecules to produce the sound waves.) The **frequency** of the sound waves is measured in *cycles per second* (*cps*), which is now more commonly called *hertz* (*Hz*). The **pitch** of a sound depends on its frequency—the higher the frequency, the higher the pitch. A young person can hear sound frequencies in the range of 20 to 20,000 Hz; and a person who is musically trained can tell the difference between sound frequencies that differ by only 0.3%.

The **amplitude** (size) of the sound waves determines the **loudness** of a sound. Loudness is measured in units known as *decibels* (*dB*). A sound that can barely be heard has an intensity of 0 dB. Every 10 dB indicates a tenfold increase in loudness: a sound is 10 times louder than the lowest audible sound at 10 dB, 100 times louder at 20 dB, a million times louder at 60 dB (the typical sound intensity of speech), and 10 billion times louder at 100 dB. The range of audible sound intensities—from the barely audible to painful loudness—covers a range of 12 orders of magnitude (10^{12}). Despite this awesome range, the human ear can tell the difference in loudness between sounds that differ only in 0.1 to 0.5 dB.

The Outer Ear Channels Sound Waves to the Middle Ear

The **outer ear** consists of the *pinna*, or *auricle*, which transmits sound waves into the *external auditory meatus* (fig. 7.13). These sound waves then cause vibrations of the *tympanic membrane* (the eardrum). The **middle ear** is the cavity between the tympanic membrane and the cochlea (fig. 7.14). Within this cavity are the

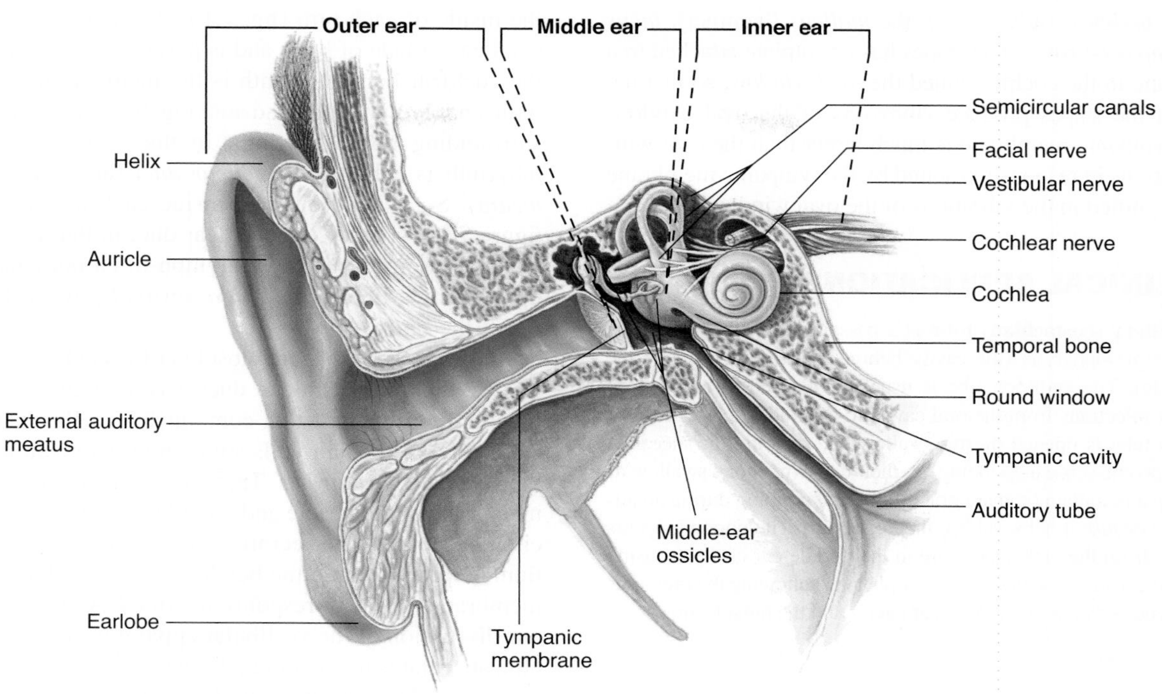

FIGURE 7.13 The ear. The structures of the outer, middle, and inner ear are shown.

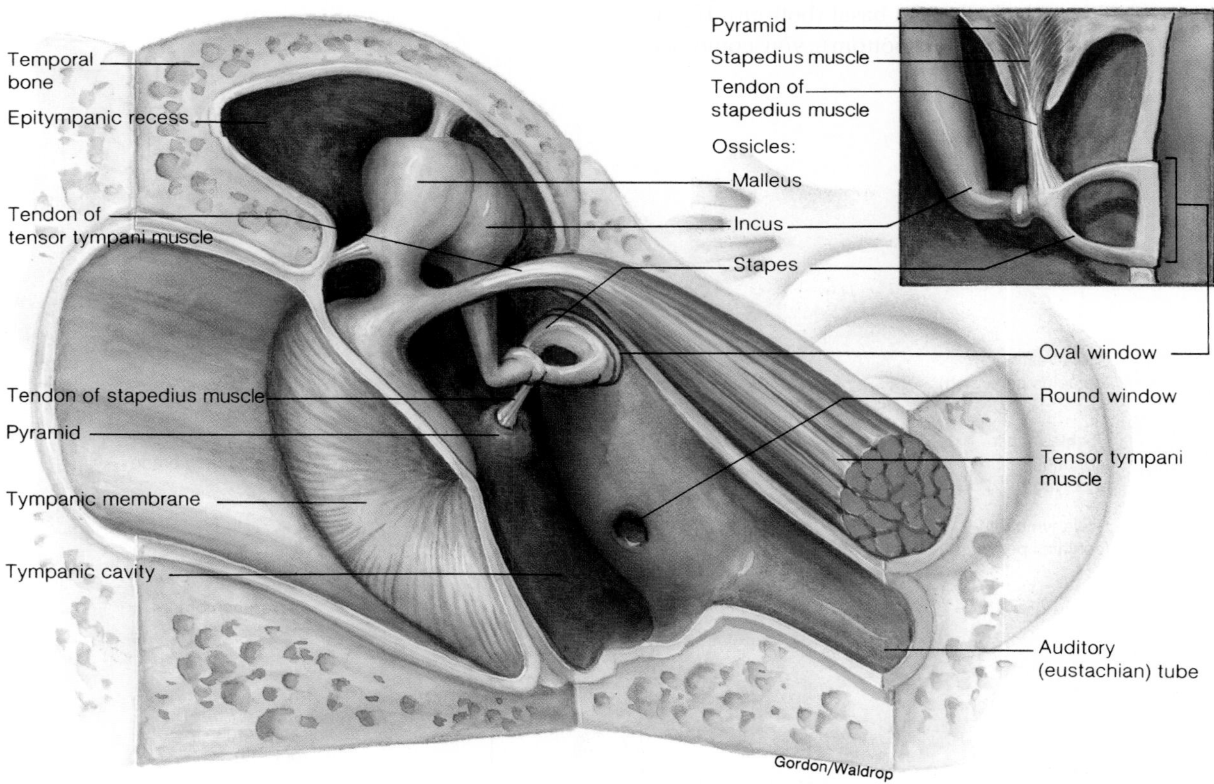

FIGURE 7.14 The middle ear. The relationship between the middle-ear ossicles and the tympanic membrane is shown, and the magnified inset depicts the footplate of the stapes against the oval window of the cochlea. The round window and the eustachian tube are depicted, and the muscles associated with the middle-ear ossicles are also indicated.

3 **middle-ear ossicles** (little bones): the *malleus* (hammer), *incus* (anvil), and *stapes* (stirrup). The stapes has a footplate attached to a small membrane in the cochlea called the *oval window*, so that the movements of the stapes produce vibrations of the oval window. Because the tympanic membrane is much larger than the oval window, the vibrations in response to sound by the tympanic membrane are greatly magnified in the vibrations of the oval window.

Clinical Applications

The **auditory (eustachian) tube** is a passageway from the middle ear to the nasopharynx (the cavity behind the nose and above the soft palate). The auditory tube is usually collapsed to prevent the spreading of infections from the oral cavity to the middle ear. However, the auditory tube is opened during swallowing, yawning, or sneezing. That's why people sense a "popping" in their ears when they swallow as they drive up a mountain (and are grateful for that popping during an airplane flight, because it helps relieve their ear pain). The popping occurs as air moves from the higher pressure in the middle-ear cavity (causing the pain) to the lower pressure in the nasopharynx, allowing the pressures to be equalized between the middle-ear cavity and the outside air.

The Cochlea Contains Hair Cells That Change Sound into Action Potentials

The **cochlea** is a part of the inner ear, as previously discussed. This structure, about the size of a pea, looks like a snail shell with three turns: apical (top), middle, and basal (bottom). If you were to slice it sagittally (from top to bottom), you could see the inside of each turn (fig. 7.15). Remember, the "shell" of the cochlea is made of bone and is part of the bony labyrinth. Inside the cochlear bony labyrinth is the membranous labyrinth, filled with endolymph fluid and making the same three turns as the surrounding bony labyrinth. In the cochlea, the membranous labyrinth is called the *cochlear duct* (also known as the *scala media*). So, when you look inside each turn of the cochlea in figure 7.15, you see the cochlear duct in the middle of the bony labyrinth, separating the perilymph-filled bony labyrinth into an upper compartment (the *scala vestibuli*) and a lower compartment (the *scala tympani*).

Now, let's look more closely at the cochlear duct (fig. 7.16). Notice that the cochlear duct has a bottom membrane—the **basilar membrane**—that contains hair cells on its surface. The stereocilia of the hair cells stick up into a gelatinous membrane, the *tectorial membrane*. This may remind you of the otolithic membrane of the utricle and saccule, or of the cupula of the semicircular canals. The tectorial membrane serves a similar function, helping to cause the bending of stereocilia when the basilar membrane moves in response to sound. As we've seen before in our discussion of the vestibular apparatus, the bending of stereocilia stimulates the associated sensory neurons to produce action potentials. The structures in the cochlea that convert sound into action potentials—including the basilar membrane, hair cells, sensory neurons, and tectorial membrane—together form the **spiral organ** (or **organ of Corti**). This is shown in the boxed portion of figure 7.16.

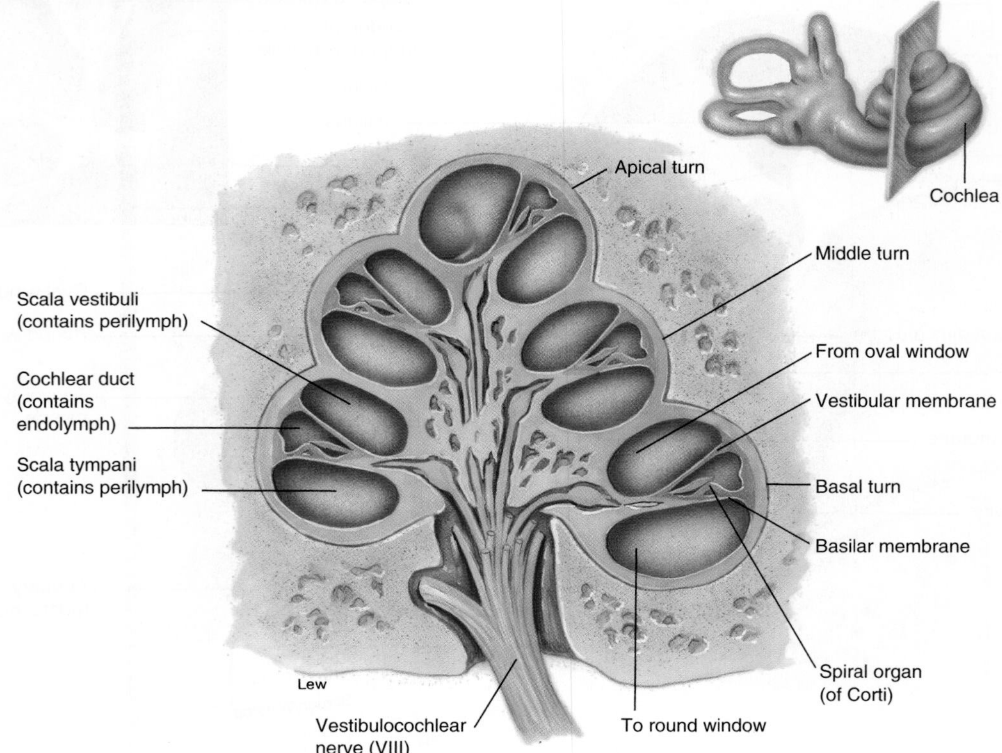

FIGURE 7.15 A section of the cochlea. The membranous labyrinth is the cochlear duct, containing endolymph. The cochlear duct separates the bony labyrinth (containing perilymph) into an upper scala vestibuli and lower scala tympani. The organ of Corti is the sensory apparatus of the cochlea.

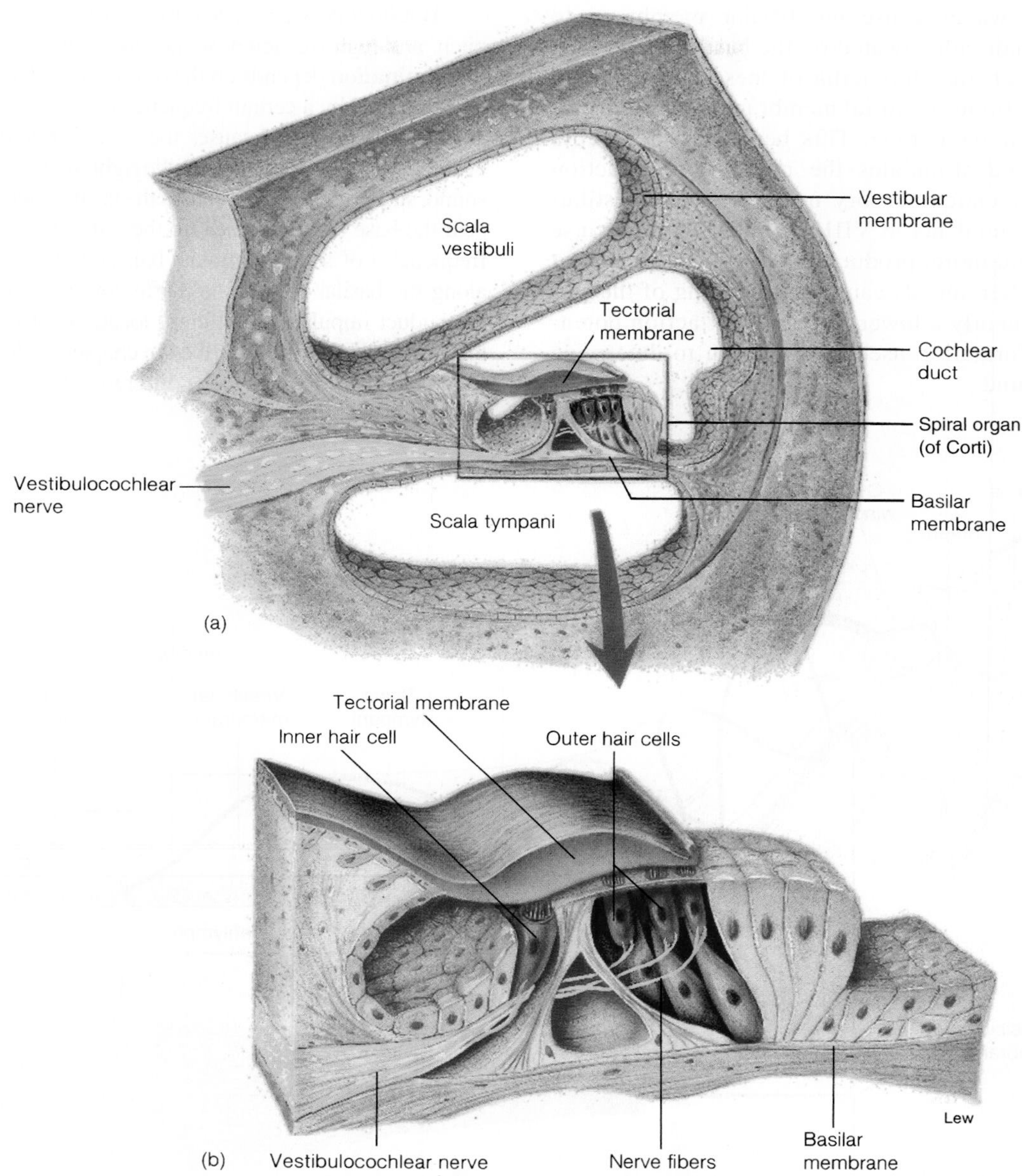

FIGURE 7.16 The spiral organ (organ of Corti). This functional unit of hearing is depicted (*a*) within the cochlear duct and (*b*) isolated to show greater detail.

But that's getting ahead of our story. We must first see how sound waves can affect the basilar membrane. Remember that the vibrations of the tympanic membrane cause movements of the middle-ear ossicles. Movements of the stapes, in turn, cause vibrations of a small, flexible membrane—the oval window—that is part of the cochlea (see fig. 7.14). The oval window looks into the perilymph within the upper scala vestibuli of the cochlea, so when the stapes presses the oval window inward, a pressure wave is transmitted by perilymph into the cochlea. Because fluid can't be compressed, some part of the membranous labyrinth must bulge out when the oval window presses in (think of squeezing a water balloon). This function is served by the round window, which can bulge out into the middle-ear cavity (see fig. 7.14).

Pitch Depends on the Location of the Stimulated Hair Cells

Imagine the cochlea untwisted, so it's straight rather than a spiral (fig. 7.17). The part to the left in figure 7.17, closest to the middle ear, is the part that would be in the basal turn; the part farthest to the right in the figure represents the part of the cochlea that would be in its apical turn. Movements of the stapes against the oval window produce pressure waves of perilymph in the upper scala vestibuli. However, the round window is located at the base of cochlea in the lower scala tympani (fig. 7.17). So, the pressure waves originating in the upper chamber must be transmitted through the cochlear duct to the lower chamber.

These pressure waves cause the basilar membrane to vibrate, so that the hair cells located on the basilar membrane also move. Remember, the stereocilia of these hair cells are embedded in the gelatinous tectorial membrane, which doesn't move like the basilar membrane. This bends the stereocilia of the hair cells, and stimulates the production of action potentials in the associated sensory neurons of the vestibulocochlear nerve (cranial nerve VIII). Louder sounds cause the stereocilia to bend more, producing a higher frequency of action potentials; softer sounds cause less bending of the stereocilia, and consequently a lower frequency of action potentials. This is the information used by the brain to determine the loudness of a sound.

But how do we perceive low-frequency sounds as being of low pitch, and high-frequency sounds as being of high pitch? This **pitch discrimination** depends on the region of the basilar membrane that is most affected by a certain frequency of sound. A low frequency, such as 500 Hz (fig. 7.17), causes the basilar membrane to vibrate most near the apex of the cochlea (to the right in fig. 7.17); a high-frequency sound, such as 20,000 Hz, causes the basilar membrane to vibrate most near the base of the cochlea (to the left in the figure). Thus, different frequencies of sound stimulate hair cells located on different places along the basilar membrane. Different axons are thereby stimulated to conduct impulses to different locations of the auditory cortex (in the temporal lobes, as described in chapter 5). As a result, we perceive different sound frequencies as different pitches.

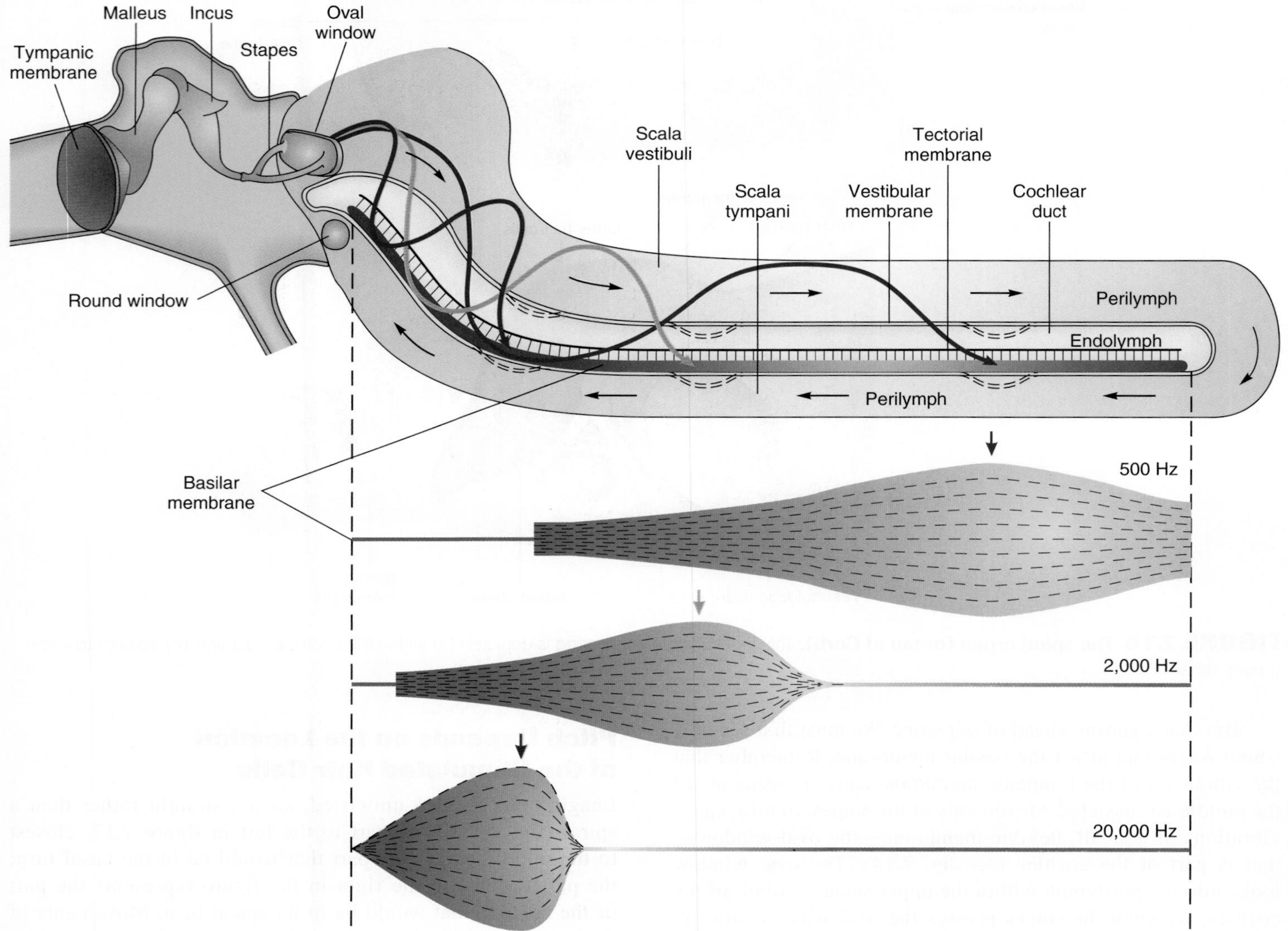

FIGURE 7.17 Effects of different sound pitches on the basilar membrane. Sounds of different pitch cause peak vibrations of the basilar membrane in different regions. Low-frequency (pitch) sounds, such as at 500 Hz, cause peak vibrations of the basilar membrane more toward the apex of the cochlea (to the right in this figure). High frequencies, such as 20,000 Hz, cause peak vibrations more toward the base of the cochlea (toward the left in the figure).

Clinical Applications

Damage to the tympanic membrane or middle-ear ossicles produces **conduction deafness**. Conduction deafness may be caused by a variety of problems, impairs hearing at all sound frequencies, and is often helped by the use of *hearing aids*. Hearing aids amplify sounds and conduct them through bone (bypassing the middle ear) to the inner ear. **Sensorineural**, or **perceptive**, **deafness** occurs when the transmission of impulses anywhere from the cochlea to the auditory cortex is impaired. Sensorineural deafness may result from a variety of pathological processes and from extremely loud sounds, as from gunshots or rock concerts. Age-related hearing impairments are common, particularly in men, where the ability to hear higher frequencies diminishes with age. Unfortunately, the hair cells of our inner ears can't regenerate. Some people with sensorineural deafness choose to have *cochlear implants*. These devices work because some sensory dendrites survive and can be stimulated electrically to produce action potentials, restoring some perception of speech in affected people.

CHECK POINT

1. Identify the structures of the outer and middle ears, and explain how these work to transmit vibrations to the cochlea.
2. Describe the structure of the cochlea and identify its three compartments. Which of these is the membranous labyrinth, and which the bony labyrinth?
3. Describe the location and structure of the organ of Corti, and explain how it functions to transduce sound into action potentials on sensory neurons.
4. Explain how the cochlea and nervous system respond to sounds of different loudness and frequency. How do we tell the difference between sounds of different pitches?

The Eyes Focus Images on the Retina

Each eye contains an opening, the pupil, through which light enters. The light bends as it passes through a clear layer over the pupil, the cornea, and bends again as it passes through a lens within the eye. This bending of light, or refraction, focuses images on the neural layer in the back of the eye, the retina. Focusing is achieved by changing the curvature of the lens through contraction and relaxation of the ciliary muscle, which encircles the lens and is joined to it by a suspensory ligament.

The outermost layer of the eye is a tough coat of connective tissue called the **sclera** (fig. 7.18), which can be seen as the whites of the eyes. The sclera is continuous with a transparent **cornea** at the front of the eye. Light passes through the cornea, and then through an opening called the **pupil** within a pigmented muscle, the **iris**. It is the iris that gives the eye its blue, green, brown, or black color, and that regulates the diameter of the pupil. As discussed in chapter 6, contraction of the radial muscle layer of the iris is stimulated by sympathetic axons and dilates the pupil; contraction of the circular muscle layer of the iris is stimulated by parasympathetic axons and constricts the pupil (see fig. 6.11). Varying the diameter of the pupil is like adjusting the f-stop of a camera.

The **lens** is suspended from a muscular process called the **ciliary body**. This is a circular structure that surrounds the lens and is connected to it through the **suspensory ligament**. The space between the cornea and iris is the **anterior chamber**, and the space between the iris and the ciliary body and lens is the **posterior chamber**. The anterior and posterior chambers are both filled with a fluid called **aqueous humor**. This fluid is secreted by the ciliary body into the posterior chamber and passes through

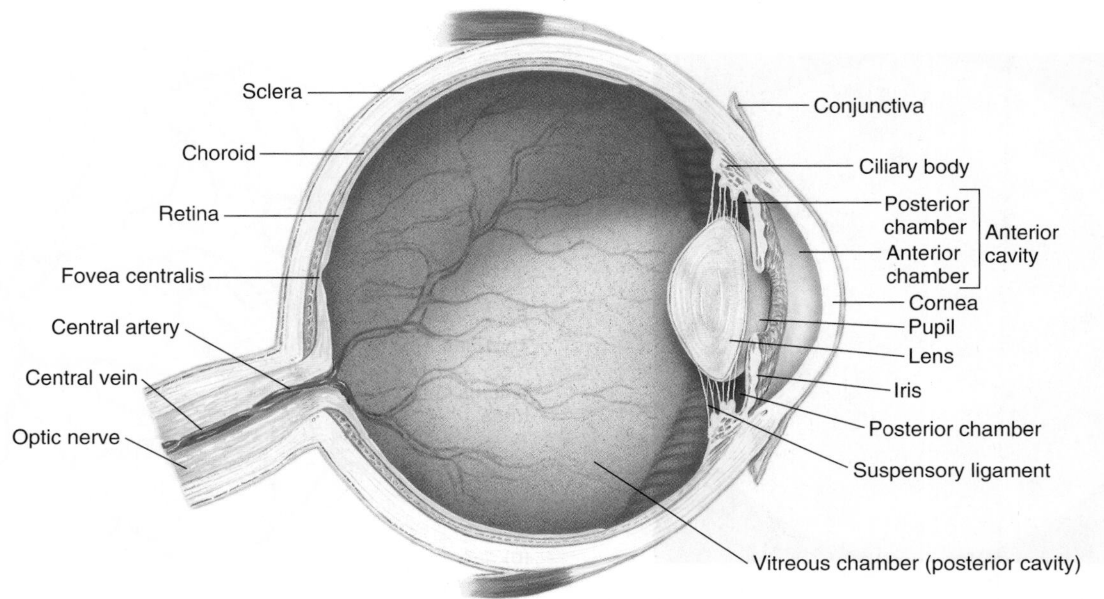

FIGURE 7.18 The internal anatomy of the eyeball. Light enters the eye from the right side of this figure and is focused on the retina.

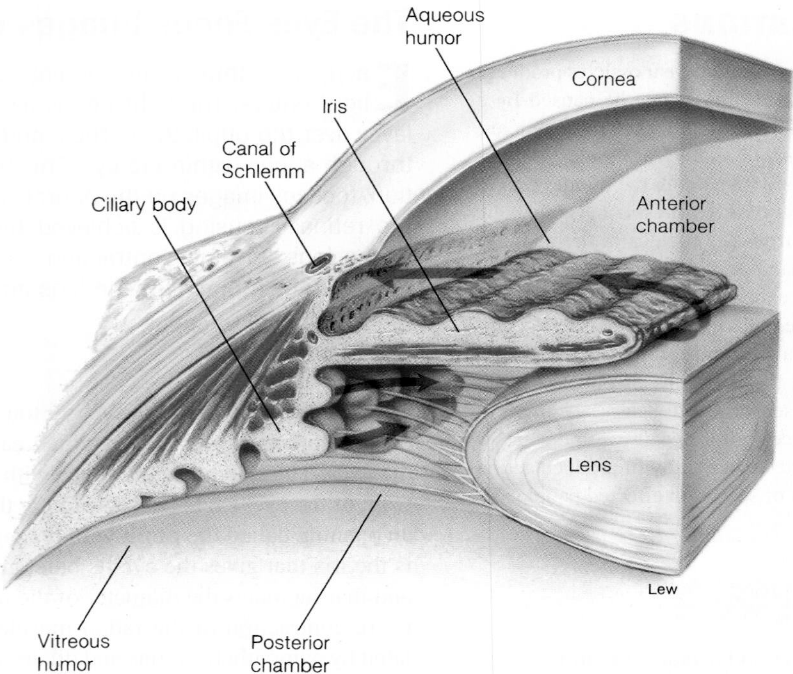

FIGURE 7.19 The production and drainage of aqueous humor. Aqueous humor maintains the intraocular pressure within the anterior and posterior chambers. It is secreted by the ciliary body into the posterior chamber, flows through the pupil into the anterior chamber, and drains from the eyeball through the canal of Schlemm.

the pupil into the anterior chamber (fig. 7.19). It is then drained from the anterior chamber by the *canal of Schlemm*, which returns it to the venous blood. Inadequate drainage can lead to an increase in intraocular pressure, a condition called *glaucoma*.

The portion of the eye located behind the lens is filled with a thick, viscous substance called the **vitreous body**, or **vitreous humor**. Light passing through the lens next passes through the vitreous body to enter the neural layer at the back of the eye. This neural layer, containing photoreceptors, is the **retina**. Neurons in the retina contribute axons that gather together at a region of the retina called the **optic disc** (fig. 7.20), where they exit the retina to form the optic nerve (cranial nerve II). Because the optic disc lacks photoreceptors, it is also known as the **blind spot**.

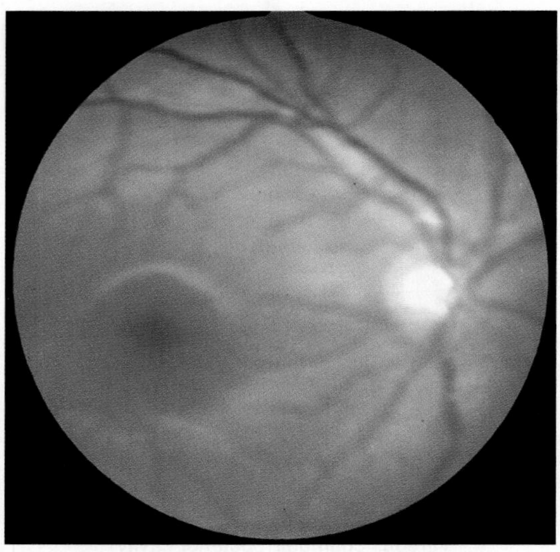

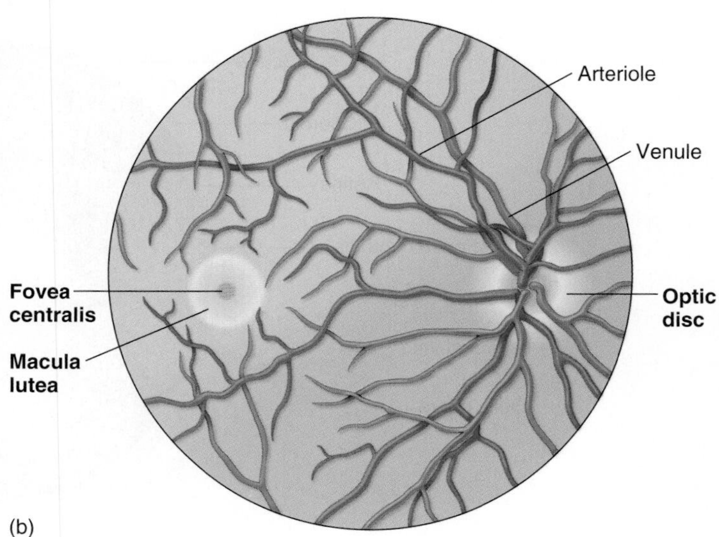

FIGURE 7.20 A view of the retina as seen with an ophthalmoscope. (*a*) A photograph and (*b*) an illustration of the optic fundus (back of the eye). Optic nerve fibers leave the eyeball at the optic disc to form the optic nerve. (Note the blood vessels that can be seen entering the eyeball at the optic disc.)

The Cornea and Lens Refract Light

Refraction refers to the bending of light when the light passes from a medium of one density to a medium of a different density. For example, when you stick a tree branch partway into a pond, it appears bent because the light is refracted when it passes from air to water. Similarly, light is refracted as it passes from the air into the cornea, from the cornea into the aqueous humor, from the aqueous humor into the lens, and from the lens into the vitreous body (fig. 7.21).

How light is refracted by a structure also depends on the shape of the structure. The cornea doesn't change its shape, so the degree of refraction by the cornea is constant. The shape and thickness of the lens, however, are adjustable. The lens is formed from about 1,000 layers of cells that are aligned in parallel and tightly joined together, so that gaps don't form as the shape of the lens is changed. The lens is transparent because (1) it's avascular (lacks blood vessels); (2) its cell organelles have been destroyed in a controlled process that stops before the cells die; and (3) the cell cytoplasm is filled with proteins called *crystallin*.

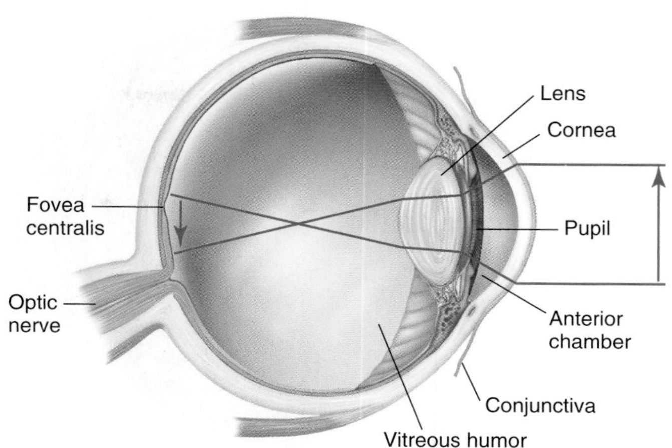

FIGURE 7.21 The cornea and lens refract light. Refraction, or bending, of light occurs at the cornea and lens. This refraction brings the image, which is inverted, to a focus on the retina. The curvature of the lens can be adjusted to keep the focus on the retina when the distance to the viewed object is changed.

Clinical Applications

Because of the unique structure of the lens, light passing through any of its regions is normally refracted to the same degree. However, damage from ultraviolet light, dehydration, or oxidation may cause the crystallin proteins to change shape and aggregate to produce cloudy patches in a person's visual field. These are known as **cataracts**. Cataracts interfere with vision in more than 50% of people over the age of 65. Cataracts are generally treated by surgically replacing the lens with an artificial lens.

The Lens Changes Shape to Keep the Image Focused on the Retina

As illustrated in figure 7.21, the refraction of light by the cornea and lens focuses an image on the retina. The focused image is upside-down, and reversed right-to-left; our brain rights the image we perceive. If this image is focused on the retina when the object is, say, 20 feet away, what would happen when we walk toward the object, so that now it's only 10 feet away? The image would be blurred, as anyone can tell by doing this with a camera requiring manual focusing. We would need to refocus our camera, and we do this by turning a knob that changes the distance between the lens and the film (or the digital sensor).

Accommodation is the ability of the eyes to keep the image focused on the retina as the distance between the eyes and the object changes. Accommodation is achieved by changing the shape and thickness of the lens, which occurs by changing the tension exerted on the lens by the ciliary muscle. This circular muscle surrounds the lens (fig. 7.22), much like a trampoline frame surrounds a trampoline mat. If the frame is widened, the mat is tightened; if the frame is narrowed, the mat is under less tension and sags.

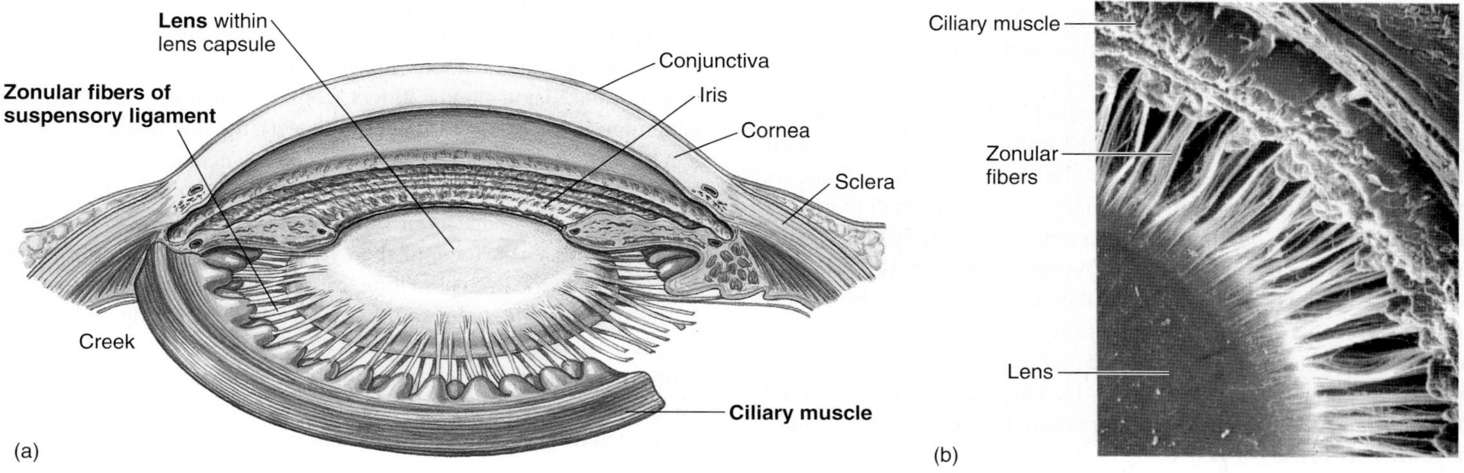

FIGURE 7.22 The relationship between the ciliary muscle and the lens. (*a*) A diagram, and (*b*) a scanning electron micrograph (from the eye of a 17-year-old boy) showing the relationship between the lens, zonular fibers, and ciliary muscle of the eye.

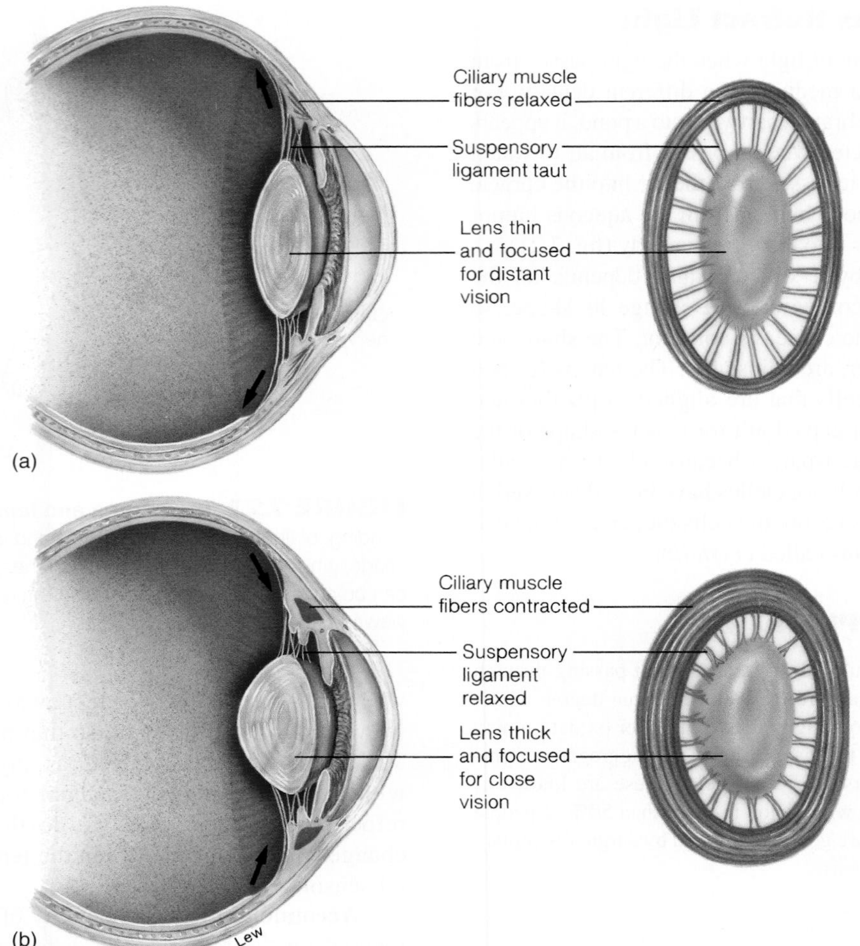

FIGURE 7.23 Changes in the shape of the lens permit accommodation. (*a*) The lens is flattened for distant vision when the ciliary muscle fibers are relaxed and the suspensory ligament is taut. (*b*) The lens is more spherical for close-up vision when the ciliary muscle fibers are contracted and the suspensory ligament is relaxed.

Similarly, when the ciliary muscle relaxes, its circle widens and puts more tension on the lens. The lens, as a result, gets stretched into a thinner shape. Conversely, when the ciliary muscle contracts, the circle narrows and puts less tension on the lens. This allows the lens to assume a thicker, rounder shape (fig. 7.23).

So, when we view an object located 20 feet or more from our eyes, our ciliary muscles are relaxed and our lenses are in their thinnest form, allowing the object to be focused on the retinas. As we walk closer to the object, our ciliary muscles contract to greater degrees, allowing our lenses to become increasingly thick. This maintains the focus on our retinas. As the object gets very close to our eyes, our lenses are as thick as they can get. This is the closest that we can focus and is known as the *near point of vision.* This distance becomes greater as we age and impairs the ability to read of almost everyone over the age of 45. The impaired ability of the lens to thicken, and consequent increase in the near point of vision, is known as **presbyopia**.

Myopia, Hyperopia, and Astigmatism Are Refractive Problems

Visual acuity refers to the sharpness of vision and is related to the ability of the eyes to focus images on the retina. When a person with normal vision stands 20 feet from a *Snellen eye chart* (so that accommodation doesn't influence the test), the line marked "20/20" can be read. If a person has **myopia** (*nearsightedness*), this line will be blurry because the image isn't focused on the retina. In myopia, the focus of the image is in front of the retina; in other words, it would have been in focus if the eye were shorter. The problem in myopia is that the eye is too long. Myopia is corrected with concave lenses that push the focus farther back from the lens (fig. 7.24).

If a person has **hyperopia** (*farsightedness*), the image is blurred because the eye is too short; the focus is behind the retina, so that the image would be clear if the eye were longer. Hyperopia is corrected with convex lenses that bring the focus up closer to

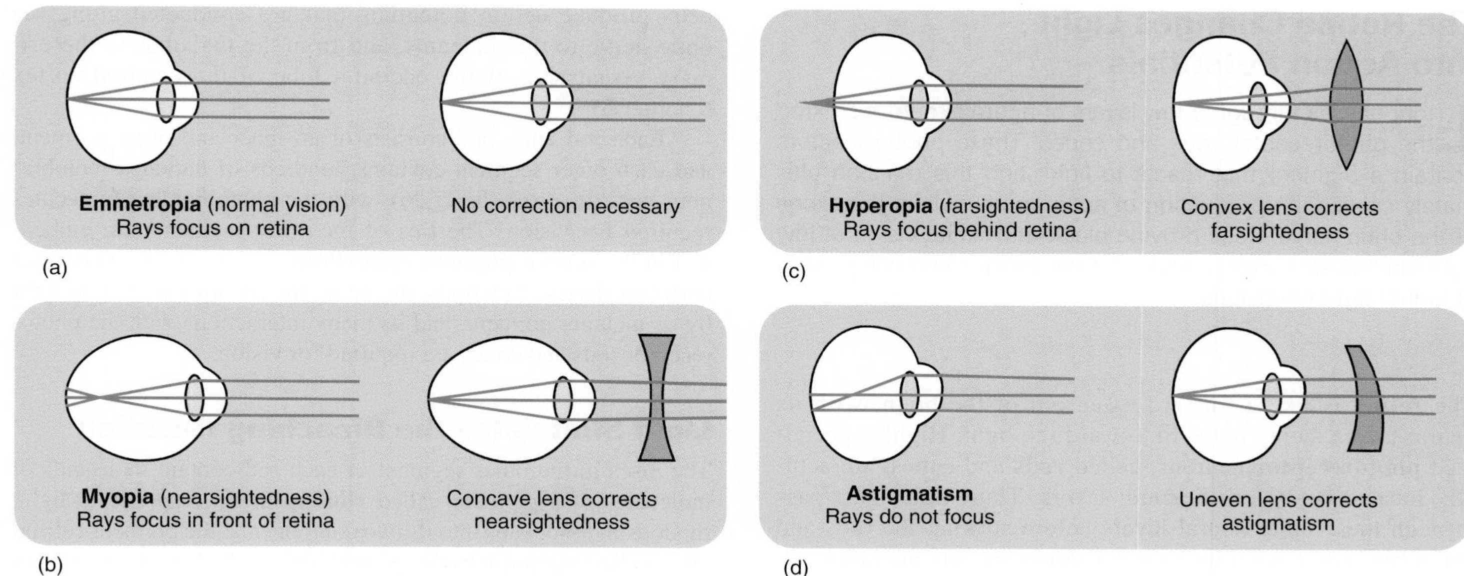

FIGURE 7.24 Problems of refraction and how they are corrected. In a normal eye (*a*), parallel rays of light are brought to a focus on the retina by refraction in the cornea and lens. If the eye is too long, as in myopia (*b*), the focus is in front of the retina. This can be corrected by a concave lens. If the eye is too short, as in hyperopia (*c*), the focus is behind the retina. This is corrected by a convex lens. In astigmatism (*d*), light refraction is uneven because of irregularities in the shape of the cornea or lens.

the lens (fig. 7.24). If a person has **astigmatism**, the cornea and/or lens is not curved symmetrically, so that different parts refract light to different degrees. If a person with astigmatism sees a circle of lines like the spokes of a wheel, the spokes will not appear as clear and dark in all 360 degrees. The parts of the circle that are blurry can be used to map the astigmatism. Astigmatism is corrected with circular lenses that compensate for the asymmetry (fig. 7.24).

Clinical Applications

Many people with refractive eye problems choose to have a surgical procedure known as **LASIK** (*laser-assisted in situ keratomileusis*). The surgeon first cuts a flap in the cornea, which is folded backward. Then a computer-guided laser burns corneal tissue as it reshapes the cornea, a procedure that takes a minute or so. The laser reduces the curve of the cornea for myopia, reducing refraction to move the focus back to the retina; it makes the cornea more steeply curved for hyperopia, increasing refraction to move the focus up to the retina. The laser makes the cornea more spherical to correct for astigmatism, so that the refraction is even all around the circle. Then the flap is put back and heals itself. LASIK can't correct for the problems of accommodation in *presbyopia*, so a person with this condition will need reading glasses if each eye is surgically corrected to 20/20. Alternatively, if the person also has myopia, one eye can be deliberately undercorrected for reading, while the other (dominant) eye is corrected for distance vision. This is called *monovision*, and is tolerated by some people better than others.

Clinical Investigation Clues

Remember that Terri had normal vision up until recently, when she noticed a difficulty in reading. Also, upon examination the physician told her that she could solve the problem with reading glasses but probably not with LASIK.

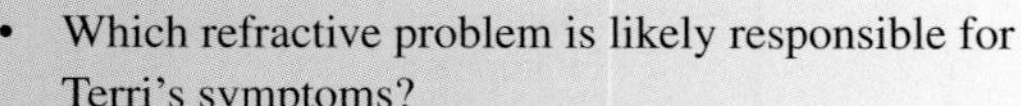

- Which refractive problem is likely responsible for Terri's symptoms?
- Why can't LASIK correct for this problem in a person who doesn't have myopia?

Check Point

1. Trace the path of light through the eye to retina, and indicate the parts of the eye that focus light onto the retina.
2. Define refraction, and describe the composition of the lens.
3. Define accommodation, and explain how it is achieved. What is presbyopia?
4. Explain the nature of myopia, hyperopia, and astigmatism.

The Retina Changes Light into Action Potentials

Light passes through a few layers of neurons before striking the photoreceptor rods and cones. These photoreceptors contain a pigment that reacts to light, and this reaction ultimately causes the production of action potentials in the axons of the optic nerve. Rods provide black-and-white vision at low light intensities; cones provide sharper images and color vision at higher light intensities.

The **retina** is like a forward extension of the brain, with its neural layers facing outward, toward the light. Highly specialized **photoreceptor** neurons, called **rods** and **cones**, are actually located below other neural layers. Thus, light must pass through these other neural layers before striking the rods and cones (fig. 7.25). After the rods and cones are activated by light, they stimulate neurons called *bipolar cells*, which in turn stimulate neurons called *ganglion cells*. The axons of the ganglion cells produce action potentials that are conducted along the optic nerve to the thalamus, and from the thalamus to the primary visual area of the occipital lobe of the cerebral cortex (chapter 5).

Each rod and cone consists of an inner and outer segment, and each outer segment contains hundreds of flattened membranous sacs, or *discs* (fig. 7.26), with the photopigment molecules required for vision. The tips of the outer segments are embedded in the **retinal pigment epithelium** (see fig. 7.25). This layer performs diverse functions, including the absorption of stray light by its melanin pigment, and its many interactions with the photoreceptor rods and cones are required for vision.

Light Stimulates the Bleaching Reaction

The discs in the outer segment of each rod contain thousands of molecules of a pigment called **rhodopsin**. In response to light, rhodopsin dissociates into its two components: the pigment **retinal** (also called **retinaldehyde**, or **retinene**), which is derived from vitamin A, and a protein called **opsin**. This dissociation reaction is known as the **bleaching reaction**.

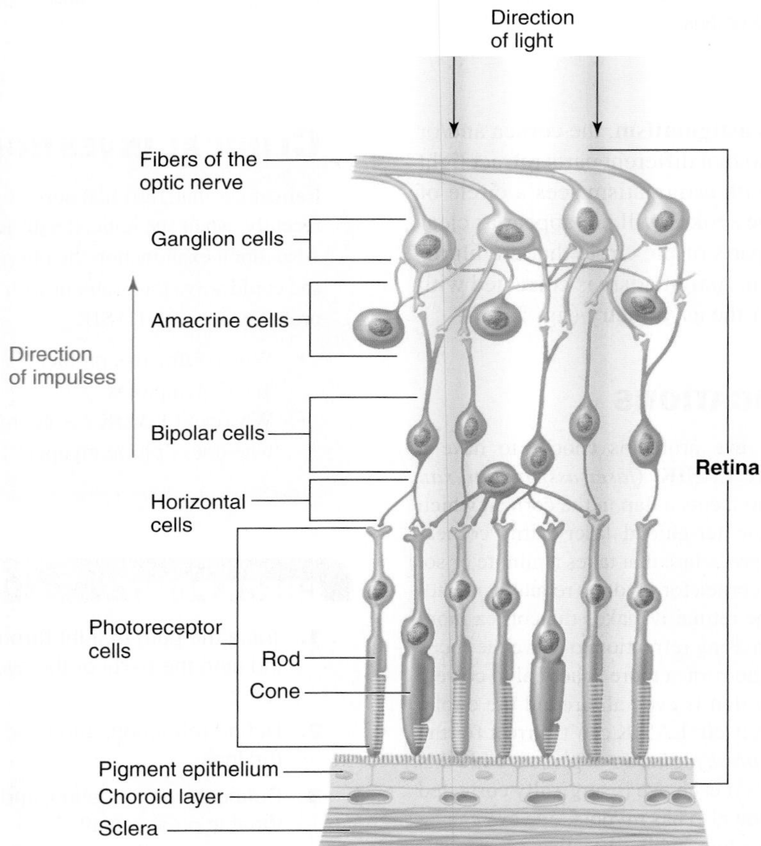

FIGURE 7.25 Layers of the retina. Light must pass through the ganglion cell layer and the bipolar cell layers before reaching the rods and cones. (Amacrine cells and horizontal cells are neurons that interconnect laterally in the retina.) Once the rods and cones are stimulated, impulses travel in the reverse direction, to the bipolar cells and then to the ganglion cells. Axons of the ganglion cells gather together as the optic nerve.

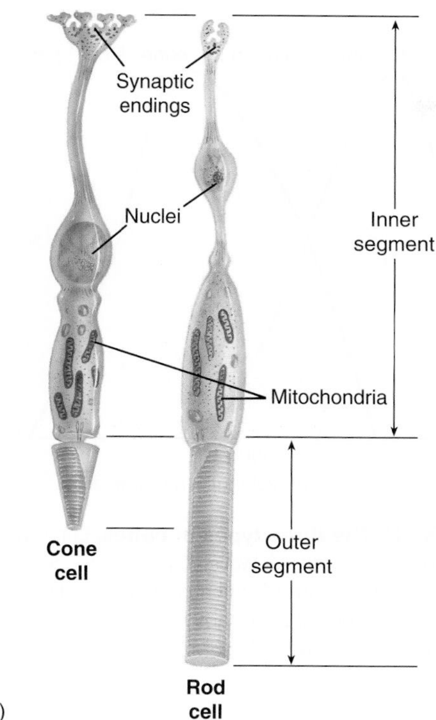

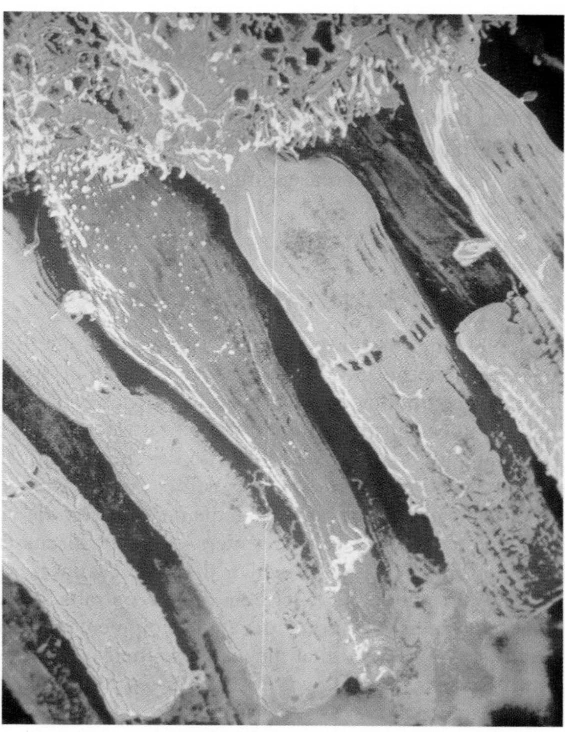

FIGURE 7.26 Rods and cones. (*a*) A diagram showing the structure of a rod and a cone. (*b*) A scanning electron micrograph of rods and cones. Note that each photoreceptor contains an outer and inner segment.

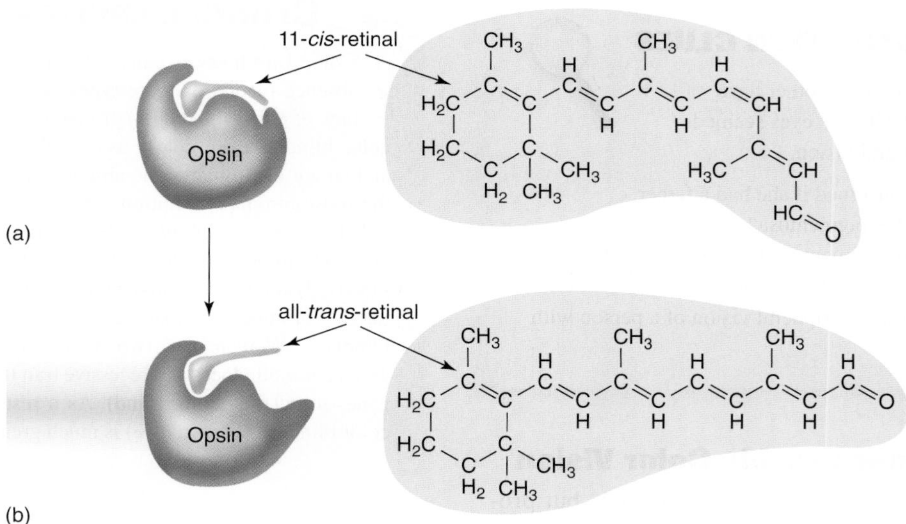

FIGURE 7.27 The bleaching reaction of rhodopsin. Rhodopsin consists of the protein opsin bound to the pigment retinal, which is in a shape called the 11-*cis* form. When light strikes the rhodopsin, a photodissociation reaction occurs, called the bleaching reaction. Retinal is converted into a different shape, known as the all-*trans* form. In this conformation (shape), the retinal dissociates from the opsin. This reaction stimulates the photoreceptor cells.

Retinal can exist in two forms—one called the all-*trans* form and one called the 11-*cis* form—that have different shapes. Only the 11-*cis* form can be attached to opsin. In response to absorbed light energy, the 11-*cis* form is converted into the all-*trans* form, causing it to dissociate from the opsin protein (fig. 7.27). This bleaching reaction stimulates the rods by changing their permeability to ions, and ultimately results in the production of action potentials in the axons of the ganglion cells.

The bleaching reaction lowers the amount of rhodopsin in the eyes, so that you have difficulty seeing when you go from daylight into a darkened room. A gradual increase in photoreceptor sensitivity, called **dark adaptation**, then occurs, reaching a maximum sensitivity in about 20 minutes. The increased sensitivity is due, at least in part, to the increased amounts of rhodopsin that can be reformed in the dark.

CLINICAL APPLICATIONS

People with **dominant retinitis pigmentosa** inherit a mutated gene for the opsin protein. A single change in the DNA base sequence, from cytosine to adenine, causes one amino acid in the opsin to be substituted for another (as discussed in chapter 2). This seemingly small change leads to degeneration of the photoreceptors, with rods degenerating before cones. This produces a loss of vision that progresses from the periphery of the visual field (involving mostly rods) to the center. As will be discussed shortly, the region of the retina on which the object's image falls when you look directly at the object, the fovea centralis, contains only cones. As a result, people with retinitis pigmentosa need to look directly at an object so that they can use their cones to see it. By contrast, people with *macular degeneration* lose vision from the center to the periphery of the visual field, and so must try to see from the corners of their eyes. Macular degeneration is described in the section "Physiology in Health and Disease" at the end of this chapter.

CLINICAL INVESTIGATION CLUES

Remember that Terri's father and brother have retinitis pigmentosa, but that Terri's eyes seemed normal, with normal peripheral vision.

- Why would Terri be concerned if she had a father and brother with retinitis pigmentosa?
- What occurs in retinitis pigmentosa to cause a loss of vision, and how does this affect the photoreceptors?
- What would happen to the peripheral vision of a person with retinitis pigmentosa?

Three Types of Cones Provide Color Vision

The cones require a greater light intensity than rods, but provide higher visual acuity and color vision during the day. Humans and other primates have **trichromatic color vision** (are *trichromats*). This means that our ability to perceive colors is produced by the stimulation of 3 types of cones, which differ in the nature of their opsin proteins. These are **blue**, or short-wavelength (**S**), **cones**; **green**, or medium-wavelength (**M**), **cones**; and **red**, or long-wavelength (**L**), **cones**. The colors and wavelengths refer to the color (wavelength) of light at which the cone pigment absorbs the most light (fig. 7.28). The ability to see all the colors we do from stimulation of these 3 types of cones is exploited by television and computer screens, which display only red, green, and blue pixels. Interestingly, many mammals with color vision have only two types of cones (they're *dichromats*).[4]

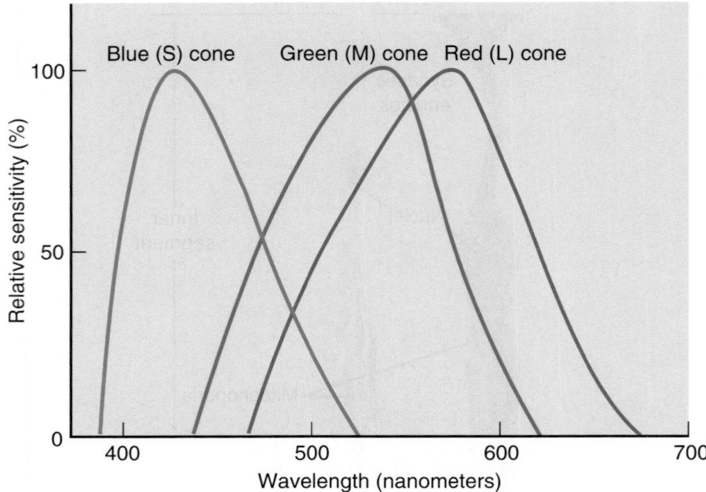

FIGURE 7.28 The three types of cones. Each type of cone absorbs light best at a particular wavelength, which corresponds to a particular color. These graphs of light absorption at different wavelengths reveal that the three types of cones absorb light best at wavelengths perceived as blue, green, and red.

CLINICAL APPLICATIONS

Color blindness is caused by a genetic defect that results in the absence of one or more types of cones. This usually involves the lack of either the red (L) or green (M) cones. People with red-green color blindness see colors using only two functioning cone systems, and so are dichromats. The absence of functioning green (M) cones is the most common condition, known as *deuteranopia*. The absence of red (L) cones (*protanopia*) is less common, and the absence of blue (S) cones (*tritanopia*) is the least common. The genes for the red (L) and green (M) cones are located on the X chromosome. Since men have only one X chromosome, a single defective gene produces red-green color blindness. Women have two X chromosomes, and so can carry red-green color blindness as a recessive trait (they must inherit two defective genes to become color blind). As a result, the incidence of red-green color blindness in men (8%) is much greater than in women (0.5%).

FYI [4]Dogs have two types of cones, and so their color vision is somewhat similar to that of people who are red-green color blind. Honeybees, like humans, have three types of cones, but they have a cone type that is sensitive to ultraviolet light (which we can't see), and they lack a cone sensitive to the red end of the spectrum. A bee's world thus looks very different from our own. Many species of birds, turtles, and fish have four types of cones, and some bird species can see ultraviolet light. Additionally, birds have differently colored oil droplets, which act as color filters, at the base of the outer segments of their cones. This may give the color vision of birds a much greater range than that provided by even their four types of cones. Thus, the world seen by birds may be far more colorful than the world that you and I inhabit.

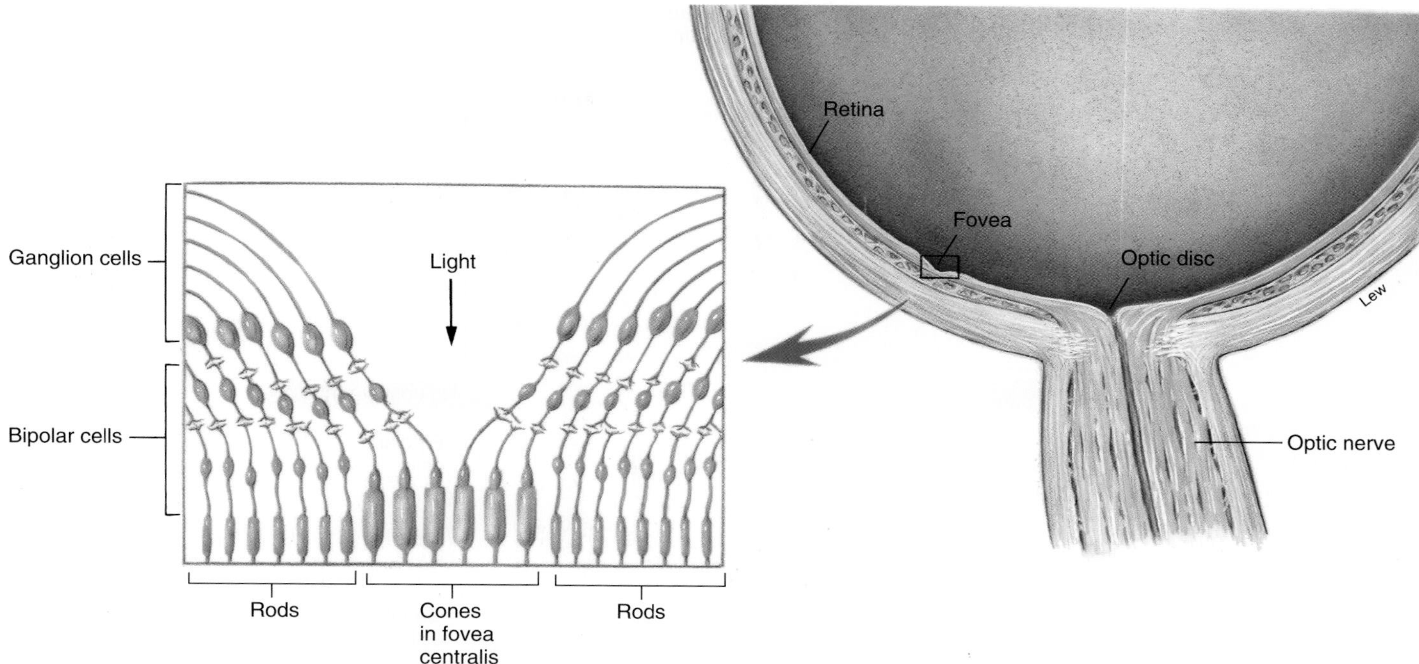

FIGURE 7.29 The fovea centralis. When the eyes "track" an object, the image is cast upon the fovea centralis of the retina. The fovea is literally a "pit" formed by parting of the neural layers. In this region, light thus falls directly on the photoreceptors (cones).

Cones Provide Visual Acuity, Rods Provide Sensitivity in Low Light

When we direct our visual attention at an object, we automatically adjust our eyes so that the image of that object falls on a tiny area of the retina called the **fovea centralis** (see fig. 7.20). As you read this sentence, your eyes move to keep each successive word focused on your fovea. The fovea is a pinhead-sized pit (*fovea* = pit) within a larger area of the retina called the *macula lutea*. The fovea contains only cones, and the pit is formed by the pushing aside of the neural layers above these cones, so that light falling on the fovea strikes the cones first (fig. 7.29).

The retina has about 6 million cones, which are most concentrated in the fovea, and about 120 million rods. In the fovea, one cone makes a synapse with one bipolar cell, which in turn makes a synapse with one ganglion cell (fig. 7.30). Remember, it's the axons of the ganglion cells that make up the optic nerve that carries visual information to the brain. So, each cone in the fovea has a direct line to the brain. There are about 4,000 cones in the fovea, transmitting information along 4,000 different lines to the brain. This provides the sharpest image possible, which is why we try to focus images on our fovea when we look directly at objects. However, cones require a relatively bright light for their activation, and so they aren't used when the light becomes too dim.

Rods, located away from the fovea, can be activated by lower light intensities. A number of different rods synapse with each bipolar cell. This *convergence* allows the stimulation by different rods to summate on the bipolar cell. Then, a number of bipolar cells synapse on a single ganglion cell (fig. 7.30). This convergence allows for summation to occur on the ganglion cells, so that a ganglion cell can be activated even when the light is very dim and only weakly stimulates a particular rod. Rods thus afford us the ability to see at low light intensities, as at night. This is why you can see a dim star better out of the "corner of the eyes" (so that the light stimulates rods) than by looking at it directly (so that the dim light falls on the fovea and fails to stimulate cones). This technique is known as *averted vision*, and it's also useful if you must watch your footing to hike down a mountain at night.

However, the costs of being able to see at night using vision produced by the rods are that (1) the images are in black and white (because we have only one type of rod), and (2) the images are less clear. The convergence of rods on bipolar cells, followed by the convergence of bipolar cells on ganglion cells, means that light striking a number of different rods will activate the same ganglion cell. This produces an image that is less sharp (has a lower resolution) than one that would have been produced if stronger light had stimulated the cones in the fovea.

CHECK POINT

1. Describe the composition of rhodopsin, and the bleaching reaction in response to light.
2. List the sequence of retinal layers through which light passes, and then list the cell sequence of neural impulses passing through the retina to the optic nerve.
3. Identify the types of cones that provide color vision. What is the most common cause of color blindness?
4. Explain the structure and significance of the fovea centralis.
5. Explain why cones provide greater visual acuity than rods, and why rods provide greater visual sensitivity than cones.

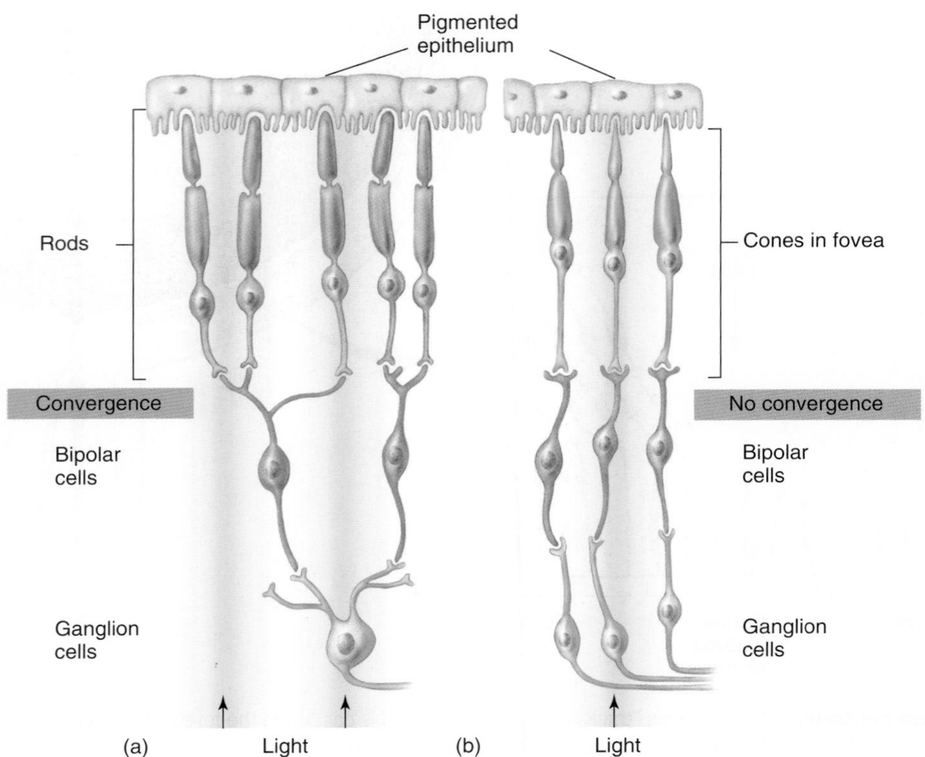

FIGURE 7.30 Convergence in the retina and light sensitivity. Since bipolar cells receive input from the convergence of many rods (*a*), and since a number of such bipolar cells converge on a single ganglion cell, rods maximize sensitivity to low levels of light at the expense of visual acuity. By contrast, the 1:1:1 ratio of cones to bipolar cells to ganglion cells in the fovea (*b*) provides high visual acuity, but sensitivity to light is reduced.

Physiology in Health and Disease

The only part of the visual field that is seen clearly is that tiny part that falls on the fovea centralis. We usually aren't aware of this because rapid eye movements automatically shift different parts of the visual field onto the fovea. **Macular degeneration**—degeneration of the macula lutea and its central fovea—is the most common cause of untreatable blindness in the Western world. About one in three people will have this condition to different degrees by the age of 75 years; its incidence increases from 0.05% before the age of 50 to 11.8% after the age of 80. Because of the increased incidence with age, this condition is often referred to as *age-related macular degeneration*. People with this condition lose the vision provided by their fovea and 30% of their vision in the central area of their visual field. As a result, they try to see out of the corners of their eyes and their vision is blurry.

When physicians look into patients' eyes with an ophthalmoscope, they can see differently colored dots, known as *drusen*. Macular degeneration appears to be caused by damage to the retinal pigment epithelium, which performs many functions essential to the rods and cones. The pigment epithelial cells may become damaged and die with age as a result of oxidative stress from molecules produced within them in response to light and from smoking (a clear risk factor). Certain genetic mutations have also been associated with an increased risk of macular degeneration. You can't do anything about the genes you inherit, but you can stop smoking, wear protective sunglasses when necessary, and take antioxidants such as zinc, beta carotene, and vitamins C and E.

The deterioration of the retina in people with macular degeneration sometimes causes blood vessels to grow into the retina from the underlying layer called the choroid. The growth of new blood vessels to a tissue is known as **neovascularization** (growth of blood vessels in general is called *angiogenesis*). Although it occurs in only about 10% of people with macular degeneration, this process is responsible for most of the cases of legal blindness caused by this disease. Neovascularization also occurs in cancer and provides increased blood flow to growing tumors. Knowing this, scientists have developed drugs to block the growth of new blood vessels as a cancer treatment. These drugs are antibodies that block the action of *vascular endothelial growth factor* (*VEGF*), a paracrine regulator (chapter 3) that stimulates the growth of blood vessels to the tumor. These antibodies against VEGF, developed as a treatment for cancer, have now also been successfully used to treat the neovascularization that contributes to blindness in macular degeneration.

Physiology in Balance

Sensory System

Working together with . . .

Endocrine System

Receptors in the hypothalamus sensitive to blood concentration regulate ADH secretion . . . p. 186

Respiratory System

Receptors sensitive to pH regulate the respiratory centers and control breathing . . . p. 297

Cardiovascular System

Receptors sensitive to blood pressure signal the brain to regulate heart rate and blood pressure . . . p. 240

Nervous System

Sensory receptors produce a depolarization that stimulates sensory neurons . . . p. 86

Digestive System

Receptors sensitive to the protein content of the stomach stimulate the secretion of gastric juice . . . p. 348

Reproductive System

Receptors in a nursing mother's nipple trigger the milk-ejection reflex . . . p. 373

Summary

Sensory Receptors Stimulate Sensory Neurons 158

- There are several functional categories of sensory receptors: chemoreceptors, thermoreceptors, mechanoreceptors, photoreceptors, nociceptors, proprioceptors, and cutaneous receptors.
- Sensory adaptation occurs when our awareness of a stimulus decreases as the stimulus is maintained.
- The adequate stimulus is the stimulus that normally excites a receptor; according to the law of specific nerve energies, we experience the sensation associated with the adequate stimulus regardless of how a sensory receptor is stimulated.
- Receptors cause the production of a receptor, or generator, potential, which is a depolarization that produces action potentials in sensory neurons.
- The skin contains a number of different cutaneous receptors, such as naked dendrites for hot, cold, touch, and pain, and pacinian corpuscles and others for pressure.
- A referred pain is a pain that is felt in one body location but caused by damage in a different location, such as angina pectoris caused by damage in the heart.
- There are five modalities of taste: salty (produced by Na^+), sour (produced by H^+), sweet (produced by sugars and some other organic molecules), bitter (produced by quinine and seemingly unrelated molecules), and umami (produced by glutamic acid, an amino acid).
- The sense of smell is very sensitive because there are many G-proteins associated with each receptor, so that one odorant molecule can cause the opening of many ion channels.
- We have hundreds of different receptors for odors; when the receptors are activated, information is sent directly to the cerebral cortex, including the limbic system for emotion and memory.

The Inner Ear Provides the Senses of Equilibrium and Hearing 162

- The inner ear consists of the bony labyrinth, filled with perilymph, and a membranous labyrinth, containing endolymph.
- The membranous labyrinth includes the organ of hearing, the cochlea, and the structures of the vestibular apparatus for equilibrium.
- The vestibular apparatus includes the 3 semicircular canals and the otolith organs: the utricle and saccule.
- The membranous labyrinth contains mechanoreceptor hair cells; the "hairs" are structures known as stereocilia, which, when bent, cause sensory neurons to become depolarized.
- In the utricle and saccule, the stereocilia of hair cells are embedded in an otolithic membrane; bending of the stereocilia produces sensations of acceleration in the horizontal and vertical directions.
- In the semicircular canals, the stereocilia are embedded in a gelatinous cupula, which is bent by the surrounding endolymph during acceleration; the semicircular canals provide sensations of movement in three planes, like the sides of a cube.
- Spinning, or abnormal function of the vestibular apparatus, can cause a loss of equilibrium known as vertigo.

The Sense of Hearing Results from Sound Waves 166

- The pinna and external auditory meatus channel sound waves to the tympanic membrane; vibrations of the tympanic membrane then cause movements of three middle-ear ossicles: the malleus, incus, and stapes.
- Movements of the stapes produce movements of an attached flexible membrane, the oval window, which is a part of the cochlea of the inner ear.
- The membranous labyrinth of the cochlea is the cochlear duct; it divides the bony labyrinth of the cochlea into an upper scala vestibuli and a lower scala tympani.
- The basilar membrane, the bottom membrane of the cochlear duct, vibrates in response to pressure waves in the cochlea; the basilar membrane contains the sensory hair cells.
- The stereocilia of the hair cells on the basilar membrane are embedded in a gelatinous tectorial membrane, so that the stereocilia can bend when the basilar membrane vibrates in response to sound.
- The organ of Corti, the sensory structure of hearing, includes the basilar membrane, hair cells, tectorial membrane, and sensory neurons.
- Different frequencies of sound produce sounds of different pitches because they cause peak vibration at different regions of the basilar membrane, thereby stimulating different hair cells and sensory neurons.

The Eyes Focus Images on the Retina 171

- Light enters the cornea and passes through the pupil, an opening in the iris, to reach the lens, which is suspended by a ligament from the ciliary body.
- Light is refracted (bent) when it passes into and through the cornea and lens, so that it comes to a focus on the retina, the neural photoreceptive layer of the retina.
- The optic disc, or blind spot, is the part of the retina that lacks photoreceptors because axons gather here to form the optic nerve.
- Accommodation is the ability of the lens to change its shape (and thus its refractive power), and thereby keep the image focused on the retina when the distance to the viewed object changes.
- When the image is closer to the eyes, the ciliary muscle contracts to take tension off the lens, so that the lens becomes thicker and more convex; when the image is farther from the eyes, the ciliary muscle relaxes to put more tension on the lens, which thereby becomes thinner.
- When a person has difficulty after the age of about 40 in accommodating to near distances, as in reading fine print, the condition is called presbyopia.
- Myopia (nearsightedness) occurs when the eye is too long; hyperopia (farsightedness) occurs when the eye is too short; astigmatism occurs when the degree of refraction of the cornea and/or lens is not the same in all 360 degrees.

The Retina Changes Light into Action Potentials 176

- The photoreceptor rods and cones synapse with bipolar cells, which in turn synapse with ganglion cells; light entering the retina first passes the ganglion cells, then the bipolar cells, before striking the photoreceptors.
- Rods and cones have inner and outer segments; the outer segments are in contact with the pigment epithelium, which interacts with the photoreceptors in several ways required for vision.
- In the bleaching reaction, light causes rhodopsin to dissociate into the pigment retinal and the protein opsin; the bleaching reaction is needed for stimulation of photoreceptors and the ultimate production of action potentials in the axons of ganglion cells.

- Dark adaption is partly due to the ability to re-form rhodopsin in the dark.
- We have three types of cones, named according to the color (wavelength) of light that they absorb best: blue (short-wavelength, or S) cones; green (medium-wavelength, or M) cones; and red (long-wavelength, or L) cones.
- We have highest visual acuity (clearest vision) in the fovea centralis, because here one cone synapses with one bipolar cell, which synapses with one ganglion cell.
- We have highest visual sensitivity in low light intensities when light strikes the retina away from the fovea centralis, and thereby stimulates the rods.
- A number of rods converge on one bipolar cell, and a number of bipolar cells converge on one ganglion cell; this increases visual sensitivity at the expense of acuity.

Review Activities

Objective Questions: Test Your Knowledge

1. Receptors that are stimulated by deformation of the plasma membrane are
   a. chemoreceptors.
   b. mechanoreceptors.
   c. nociceptors.
   d. proprioceptors.
2. Muscle spindles and Golgi tendon organs are
   a. chemoreceptors.
   b. mechanoreceptors.
   c. nociceptors.
   d. proprioceptors.
3. Phasic receptors
   a. do not respond to their adequate stimulus.
   b. respond continuously to a stimulus.
   c. are partly responsible for sensory adaptation.
   d. All of these are true.
4. A generator potential
   a. is a graded depolarization.
   b. must reach threshold to stimulate action potentials.
   c. is produced in response to a sensory stimulus.
   d. All of these are true.
5. Capsaicin receptors
   a. are stimulated by chili peppers.
   b. are located in sensory axons.
   c. are found within lamellated corpuscles.
   d. produce a sense of cold.
6. How many types of taste receptors do we have?
   a. Two
   b. Four
   c. Five
   d. Hundreds
7. The umami taste is produced by
   a. H^+.
   b. Na^+.
   c. glutamate.
   d. quinine.
8. Which of the following statements regarding the sense of smell is false?
   a. There are hundreds of olfactory receptors.
   b. The sensory receptors are bipolar neurons.
   c. The olfactory receptors send information to the thalamus, then the cerebrum.
   d. The olfactory receptor proteins are coupled to G-proteins.
9. Which sensation is transmitted directly to the cerebral cortex, and is noteworthy for its stimulation of the limbic system?
   a. Olfaction
   b. Taste
   c. Hearing
   d. Vision
10. Hair cells are located within all of the following except the
    a. membranous labyrinth of the semicircular canals.
    b. bony labyrinth of the otolith organs.
    c. membranous labyrinth of the cochlea.
    d. All of these.
11. The stereocilia of hair cells are bent by otoliths during horizontal acceleration (as in a car) in the
    a. utricle.
    b. saccule.
    c. semicircular canals.
    d. cochlea.
12. Which of the following structures have stereocilia embedded in a cupula?
    a. Utricle
    b. Saccule
    c. Semicircular canals
    d. Cochlea
13. Involuntary oscillation of the eyes when a person stops spinning is called
    a. vertigo.
    b. nystagmus.
    c. Ménière's disease.
    d. otosclerosis.
14. Which of the following statements about the oval window is false?
    a. It's vibrated by the footplate of the stapes.
    b. Its vibrations are coupled to the vibrations of the tympanic membrane.
    c. It's smaller than the tympanic membrane.
    d. It looks into the scala tympani of the cochlea.
15. The cochlear duct
    a. is the membranous labyrinth of the cochlea.
    b. contains endolymph.
    c. contains the sensory hair cells.
    d. All of these are true.
16. The organ of Corti (spiral organ) contains all of the following except the
    a. basilar membrane.
    b. sensory hair cells.
    c. round window.
    d. tectorial membrane.
17. Peak vibration of the basilar membrane in response to low-pitched sound occurs
    a. throughout the basilar membrane.
    b. closer to the base of the cochlea.
    c. close to the apex of the cochlea.
    d. near the middle turn of the cochlea.

18. The part of the eye that secretes aqueous humor is the
    a. cornea.
    b. lens.
    c. ciliary body.
    d. iris.

19. The blind spot is the
    a. optic disc.
    b. fovea centralis.
    c. place in the retina where the lens cannot focus an image.
    d. part of the visual field beyond the reach of peripheral vision.

20. When an image is brought closer to the eyes, the
    a. ciliary muscle relaxes to make the lens thinner.
    b. ciliary muscle contracts to make the lens thinner.
    c. ciliary muscle contracts to make the lens thicker.
    d. ciliary muscle relaxes to make the lens thicker.

Match the condition with its cause:

21. Hyperopia	a. The eye is too long.
22. Myopia	b. The eye is too short.
23. Astigmatism	c. The lens cannot accommodate well.
24. Presbyopia	d. The lens or cornea is not symmetrically refractive.

25. In the bleaching reaction in the retina, light causes
    a. the retinal to bind to the opsin.
    b. the opsin to change shape.
    c. 11-*cis* retinal to change to all-*trans* retinal.
    d. photodissociation in the pigment epithelium.

26. When you look directly at an object, you focus the image of that object on the
    a. cones in the fovea centralis.
    b. rods in the fovea centralis.
    c. cones in the optic disc.
    d. rods in the optic disc.

27. How many different types of cones are present in a person who isn't color-blind?
    a. Two
    b. Three
    c. Four
    d. Hundreds

28. Cones that absorb long-wavelength light best are also called
    a. blue cones.
    b. green cones.
    c. red cones.
    d. yellow cones.

29. Visual acuity is best, and sensitivity to low light levels is worst, when
    a. there is great convergence of photoreceptors on bipolar cells.
    b. there is a 1:1:1 ratio of photoreceptors to bipolar cells to ganglion cells.
    c. you see out of the "corners of the eyes."
    d. the image is focused on the optic disc.

Essay Questions 1: *Test Your Understanding*

1. List the functional categories of sensory receptors, and give examples.
2. Describe the receptor potential, and explain its significance.
3. What is the "law of specific nerve energies"? Use this law to explain how a person can see a flash of light when punched in the eye.
4. What are the somatesthetic senses, and to where in the cerebral cortex do they project?
5. What is a referred pain? Give an example.
6. What are the modalities of taste sensation? What stimulates the sour and salty taste? What is umami, and what molecule stimulates this taste?
7. Compare the number of olfactory receptors to the number of taste receptors, and explain how these two sensations are interrelated.
8. Explain how an odorant molecule causes depolarization of the sensory neuron.
9. Once an odorant has stimulated an olfactory receptor, where does the sensory information travel to? How does an odor trigger an emotion and a memory?
10. What structures compose the membranous labyrinth, and how does the membranous labyrinth relate to the bony labyrinth?
11. Explain how a hair cell in the vestibular apparatus becomes depolarized and hyperpolarized.
12. Which are the otolith organs, and how are they stimulated?
13. Describe the location and structure of the sensory apparatus in the semicircular canals, and explain how it is stimulated.
14. After a person spins to the right and is suddenly stopped, the eyes keep drifting to the right and are quickly pulled back to the left, producing oscillations of the eyes. What is this called, and how is it produced?
15. Giving a cause-and-effect sequence, explain how sound waves cause vibrations of the oval window of the cochlea.
16. Giving a cause-and-effect sequence, explain how vibrations of the oval window produce vibrations of the basilar membrane.
17. Giving a cause-and-effect sequence, explain how vibrations of the basilar membrane cause the production of action potentials in cranial nerve VIII.
18. Explain how different pitches of sound are detected by the ear and brain.
19. Define refraction, and describe the structures in the eye that use refraction to focus light on the retina.
20. Describe how the ciliary muscle and lens respond to keep the image focused on the retina as the distance to a viewed object changes; that is, explain accommodation.
21. Describe why an image appears blurred in myopia, hyperopia, and astigmatism.
22. Describe the structure of rods and cones, and the location and structure of rhodopsin. Explain what happens to rhodopsin when it is exposed to light.
23. Distinguish between the different types of cones. What happens to vision when one cone type is missing or defective?
24. Identify the macula lutea and fovea centralis, and describe the composition of the fovea. Where on the retina is the image focused when we look directly at it, and where is it focused when we look "out of the corners of our eyes"?
25. Explain why vision provided by cones is sharper, but vision provided by rods is more sensitive to low levels of light. Use this information to explain the benefit of "averted vision."

Essay Questions 2: *Test Your Analytical Ability*

1. In the movie *The Matrix*, the life that a person experienced was an illusion created by robots that stimulated the person's brain. Using the law of specific nerve energies, explain how that's possible.
2. "Hot" isn't one of the taste modalities, and yet chili peppers taste hot. Explain how.

3. A person with a gallstone may develop a pain below the right scapula (shoulder blade) when oily food is eaten. What kind of pain is this? How does this relate to the pain felt by an amputee in the missing limb?
4. Explain how a food can taste somewhat salty, somewhat sour, and somewhat sweet at the same time. How might it taste if the nose is plugged?
5. Most (but not all) toxic molecules taste bitter, and acids taste sour. What advantages do these tastes confer? Explain.
6. The olfactory receptors are coupled to many G-protein complexes. What benefit does this confer? Explain.
7. Suppose you blindfolded people and had them smell freshly barbecued chicken. Then, suppose you rubbed the chicken on the grass of a hill to create a scent trail, and asked the people to crawl through the grass and follow the scent trail. How might they do at this task compared to a dog? Explain. (Something similar has actually been done, with some surprising results!).
8. Explain how hair cells function, and how they are used in several different sensory systems.
9. Which sensory organs are stimulated when a person goes on the "tea cup" ride at an amusement park? Explain.
10. Define *vertigo* and *nystagmus*, and explain how these may appear in a person who just finished the ride described in the previous question.
11. Explain the concept of inertia, and how it relates to the function of the vestibular apparatus.
12. Explain the significance of the differences in size between the tympanic membrane and the oval window, and function of the round window of the middle ear.
13. Compare the structure of the organ of Corti with that of the otolith organs. What function do the tectorial membrane and otolithic membrane serve?
14. Does a sound of a particular frequency stimulate only one small group of hair cells? If not, what must the nervous system do for us to hear sound of only one pitch in response to sound of a particular frequency?
15. Describe the structure and significance of the optic disc. Why do you think we don't see a hole in each side of our visual field due to the optic disc in each eye?
16. Why do you think getting monovision through LASIK surgery can help a person with myopia who later develops presbyopia, but not help a person with normal vision who develops presbyopia?
17. "We see with our brain, not our eyes." Explain why that statement is true, using trichromatic color vision as an example.
18. Suppose you were observing a person with retinitis pigmentosa and another with age-related macular degeneration, both trying to watch a TV screen. How could you tell which person had which condition? Explain.

Web Activities

ARIS™ **For additional study tools, go to www.aris.mhhe.com. Click on your course topic and then this textbook's author/title. Register once for a semester's worth of interactive activities and practice quizzing to help boost your grade!**

CHAPTER 8

Endocrine System

Chapter Outline

HOMEOSTASIS

The functions of the body systems must be coordinated to maintain homeostasis, and this coordination requires the action of the nervous and endocrine systems. The endocrine system secretes hormones that coordinate the metabolism of such organs as skeletal muscles, adipose tissue, and the liver, so that metabolic balance is maintained. For example, homeostasis of body weight requires the proper action of hormones that promote anabolism (the synthesis of large, energy storage molecules) and others that promote catabolism (the breakdown of larger molecules into smaller ones that can be used for cell respiration). Further, homeostasis of blood glucose, which is required for the metabolism of the CNS, is dependent upon the action of insulin and several other hormones. The endocrine system is also essential in the regulation of the reproductive system in both sexes.

But if homeostasis is not maintained . . .

CLINICAL INVESTIGATION

Tony had his thyroid gland surgically removed when he was in his late twenties; now in his sixties, he takes Synthroid (thyroxine) pills daily and periodically has his blood levels checked for T_4 and TSH. Tony had recently also been taking prednisolone (a glucocorticoid) pills for a severe flare-up of an inflammation caused by herpes virus. When he visited his physician, the physician said that the prednisolone doses would be tapered off. She also cautioned Tony to lose weight and exercise, because his father had diabetes and, since Tony was overweight, he was also in danger of becoming diabetic.

Why could Tony take thyroxine orally, without requiring injections? What are T_4 and TSH, and what could be learned from laboratory tests of their levels in the blood? What is a glucocorticoid, and why would Tony take this drug for a severe inflammation? Why would he need to taper off this drug? Why would the physician think that Tony was in danger of becoming diabetic, and why would she recommend diet and exercise?

Endocrine Glands Secrete Chemically Different Hormones

Hormones travel in the blood from the endocrine glands to their target organs. Steroid and thyroid hormones are nonpolar; these are lipid-soluble, and so can pass through plasma membranes to enter their target cells. Most of the other hormones are polar molecules. These are water-soluble, so they can dissolve in the blood plasma, but they can't get through the plasma membrane to enter their target cells. Therefore, the polar hormones must regulate their target cells through the action of second messengers.

Working together with the nervous system, the endocrine system regulates the other systems of the body. Therefore, we need a foundation in endocrine physiology to properly understand the function of the cardiovascular, digestive, reproductive, and other body systems. Because the regulation by hormones involves molecular events occurring inside of cells, we need to study how these mechanisms operate in order to understand the actions of hormones and other regulatory molecules. Further, exogenously administered hormones, and drugs that promote or block hormone action, have wide medical application. So be advised: by studying the endocrine system in this chapter, you are laying an important foundation for your future success in physiology and its health applications.

Endocrine glands lack the ducts present in exocrine glands (chapter 2). **Hormones** are regulatory molecules that endocrine glands secrete into the blood. Because hormones circulate in the blood, they are delivered to every cell in every organ of the body.

However, only certain organs—called the **target organs**—can respond to each hormone. For example, the uterus and the big toe both receive the hormone estradiol, secreted by the ovaries, but the big toe (unlike the uterus) isn't a target organ for estradiol and so can't respond to this hormone. This is because the big toe lacks receptor proteins (chapter 3) for estradiol.

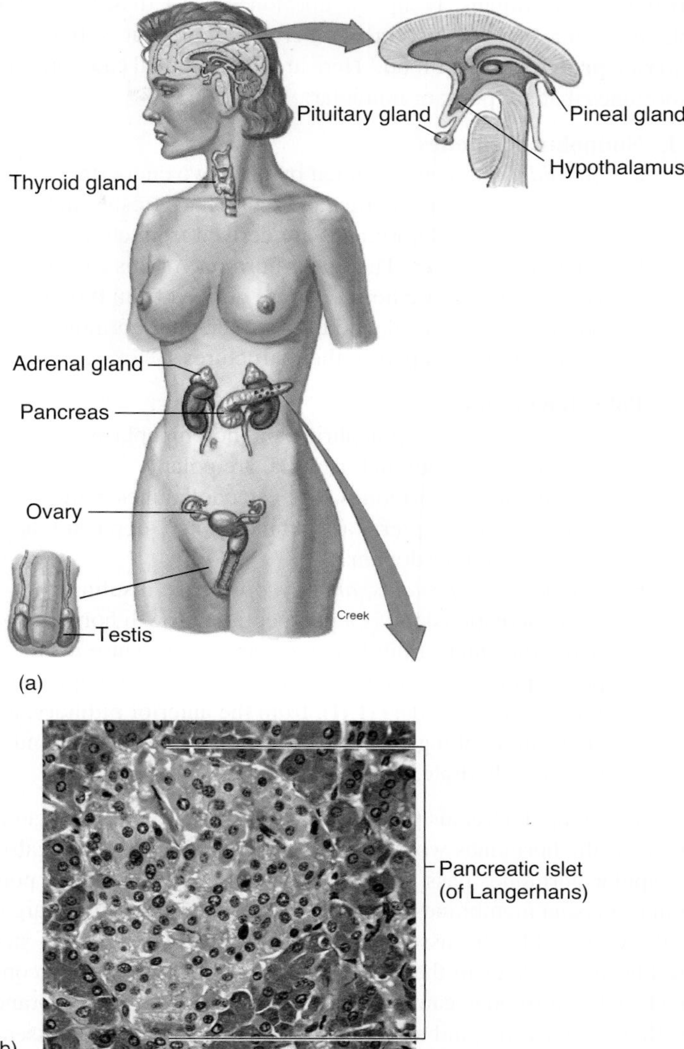

FIGURE 8.1 The major endocrine glands. (*a*) The anatomical location of some of the endocrine glands. (*b*) A photomicrograph of a pancreatic islet (of Langerhans) within the pancreas.

The Endocrine System Includes All Organs That Secrete Hormones

The **endocrine system** includes organs that secrete hormones as their primary function, such as the pituitary, thyroid, and adrenal glands (fig. 8.1). The pancreas is shown as an insert in figure 8.1 because the pancreas, though mostly exocrine (as will be discussed with the digestive system in chapter 14), also has endocrine units called the pancreatic islets (or islets of Langerhans). However, the endocrine system includes many more organs than are shown in figure 8.1, because organs that have other important functions also secrete hormones, and so should be considered part of the endocrine system. For example, the heart secretes a hormone that stimulates salt and water excretion by the kidneys; the brain secretes a number of hormones that relate to the pituitary gland (as will be described later); the stomach and intestines secrete hormones that regulate the digestive system; the adipose tissue secretes several hormones that regulate hunger and metabolism; and the kidneys secrete a hormone that regulates red blood cell production by the bone marrow. These hormones will be discussed in later chapters together with the systems they regulate.

Chemically Different Hormones May Be Polar or Nonpolar

As explained in chapter 1, polar molecules have plus and minus poles and so are soluble in water, a polar solvent. By contrast, nonpolar molecules are insoluble in water but are soluble in other nonpolar molecules, including the different lipids. Because polar hormones are soluble in water, they can dissolve in the blood plasma (which is mostly water) and be carried in a dissolved state to their target organs. Nonpolar hormones can't dissolve in the blood plasma, and so must travel on cruise ships; that is, bound to **carrier proteins** in the blood.[1] Here are the chemical categories of hormones that are polar or nonpolar:

1. **Nonpolar hormones**
   a. *Steroids*. These are secreted by only two endocrine glands: the adrenal cortex and the gonads (testes and ovaries). Steroid hormones are derived from cholesterol.
   b. *Thyroid hormones*. The major hormone in this category is *thyroxine*. These hormones are derived from tyrosine, a nonpolar amino acid, and they are the only hormones that include iodine as part of their structures.
2. **Polar hormones**
   a. *Catecholamines*. Epinephrine and norepinephrine, secreted by the adrenal medulla, are polar catecholamine hormones. The catecholamines were mentioned in chapter 6 as also including L-dopa and the neurotransmitter dopamine.
   b. *Polypeptides, proteins, and glycoproteins*. Insulin (from the pancreatic islets) is a polypeptide; growth hormone (from the anterior pituitary) is a protein, because it's a polypeptide more than 100 amino acids long (chapter 1). Luteinizing hormone (LH), from the anterior pituitary, is an example of a glycoprotein hormone (a protein bound to a carbohydrate, see chapter 1).

The nonpolar versus polar distinction is useful in understanding how the hormones work and how they can be used medically. Nonpolar hormones pass through the nonpolar phospholipid portion of plasma membranes (chapter 3) and so can enter their target cells. This will be discussed in conjunction with the action of steroid hormones later in the section on the adrenal cortex. By contrast, polar hormones cannot pass through the plasma membrane of their target cells, and so require the action of second messengers (such as cyclic AMP) to produce their effects. This will be discussed later in conjunction with the action of epinephrine and norepinephrine in the section on the adrenal medulla.

Steroid hormones are produced from cholesterol by a stepwise pathway (fig. 8.2). Examples of steroid hormones and the endocrine glands provided in figure 8.2 include progesterone, secreted from the corpus luteum of the ovaries (chapter 15); cortisol, secreted by the adrenal cortex; testosterone, secreted by the Leydig cells of the testes (chapter 15); and estradiol, secreted by the ovarian follicles. Don't worry about memorizing the hormone structures or the steps of the metabolic pathway; just notice that (1) starting from cholesterol, one steroid is made from another (estradiol is made from testosterone, for example); and (2) small differences in chemical structures can make major differences in function. Although the molecular structures may look the same at first glance, these hormones have different target organs and produce very different physiological effects. You need only consider the secondary sex characteristics produced by testosterone in males and estradiol in females to appreciate this.

Because steroid hormones are nonpolar, they can be taken orally (by mouth, as pills) and be absorbed through the intestinal lining into the blood. Birth control pills, containing derivatives of estradiol and progesterone, are an example. By contrast, polar hormones (such as insulin) can't be taken orally because they would be digested before they could be absorbed. As will be described in a later section, diabetics who need insulin must thus take insulin by injection.

CLINICAL APPLICATIONS

Anabolic steroids are synthetic androgens (male hormones) that promote protein synthesis in muscles and other organs. Use of anabolic steroids is prohibited by most athletic organizations and has a variety of possibly dangerous side effects. Since the liver and adipose tissue can change androgens into estrogens (see the last conversion step in fig. 8.2), male athletes who take exogenous androgens may develop *gynecomastia*—the growth of female-like mammary tissue. High levels of androgens also inhibit the pituitary's secretion of FSH and LH (discussed shortly); this produces atrophy of the testes and erectile dysfunction. Exogenous androgens promote acne, aggressive behavior, male-pattern baldness, and premature cessation of growth in adolescents. Female users of anabolic steroids display masculinization and antisocial behavior. In both sexes, anabolic steroids raise the levels of LDL cholesterol (the "bad cholesterol") and triglycerides, while lowering the level of HDL cholesterol (the "good cholesterol"), leading to increased risk of heart disease and stroke.

Although thyroid hormones are not steroids, they are also nonpolar (fig. 8.3; notice the alternating double bonds in the ring structures—this is known as an "aromatic ring" and indicates a nonpolar structure). The major thyroid hormone is **thyroxine**. Because it contains 4 iodine atoms, thyroxine can also be called *tetraiodothyronine,* commonly abbreviated **T_4**. However, the thyroid also secretes smaller amounts of another hormone with the same basic structure but with only 3 iodine atoms (fig. 8.3). This hormone is commonly abbreviated **T_3**, which stands for **triiodothyronine**. Because T_4 and T_3 are nonpolar hormones, they—like the steroid hormones—can be taken as pills. People who are hypothyroid (described in a later section) commonly take thyroid hormone pills.

CLINICAL INVESTIGATION CLUES

Remember that Tony was taking Synthroid (thyroxine) pills to treat his hypothyroidism.

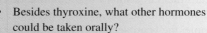

- Besides thyroxine, what other hormones could be taken orally?
- Why is it that thyroxine can be taken as a pill, but insulin must be injected?

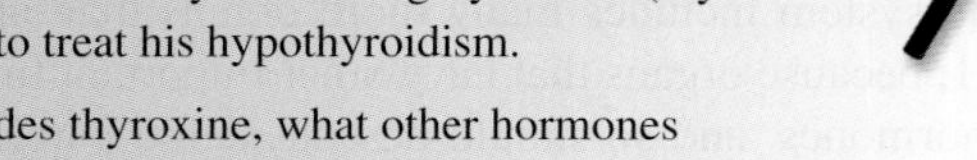

FYI [1]Carrier proteins also transport other nonpolar molecules, including cholesterol and triglycerides. Cholesterol travels in the blood bound to carrier proteins made famous by their association with cardiovascular health. The so-called "bad cholesterol" is cholesterol bound to carriers called *LDL* (low-density lipoproteins); the "good cholesterol" is cholesterol bound to *HDL* (high-density lipoproteins). It's not the cholesterol that's good or bad but the carrier proteins, because they determine if the cholesterol is going to an artery (LDL; bad) or the liver (HDL; good). This is discussed in chapter 10.

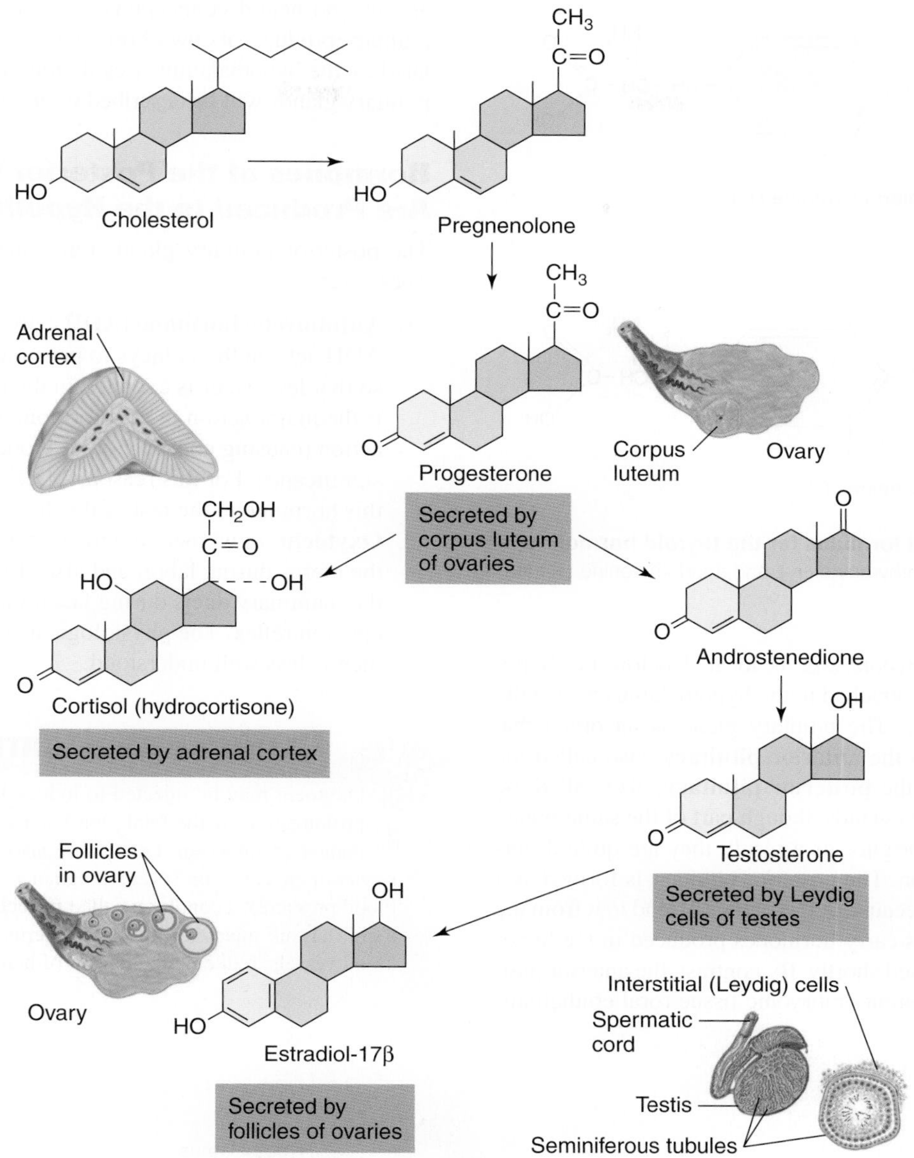

Cholesterol → Pregnenolone → Progesterone (Secreted by corpus luteum of ovaries) → Cortisol (hydrocortisone) (Secreted by adrenal cortex); Progesterone → Androstenedione → Testosterone (Secreted by Leydig cells of testes) → Estradiol-17β (Secreted by follicles of ovaries)

Adrenal cortex

Corpus luteum

Ovary

Follicles in ovary

Ovary

Interstitial (Leydig) cells

Spermatic cord

Testis

Seminiferous tubules

FIGURE 8.2 Simplified biosynthetic pathways for steroid hormones. Notice that progesterone (a hormone secreted by the corpus luteum of the ovaries) is a common precursor of all other steroid hormones and that testosterone (the major androgen secreted by the Leydig cells of the testes) is a precursor of estradiol-17β, the major estrogen secreted by the follicles of the ovaries.

CHECK POINT

1. How can the kidneys and heart be considered endocrine glands? What are the characteristics of a hormone?
2. Which are the nonpolar hormones? What solubility properties do they have that are different from the polar hormones?
3. How does the polar and nonpolar nature of hormones influence how they are transported in blood, and how they act on their target cells?
4. Why can people take birth control pills and thyroid pills, but not insulin pills?

The Pituitary Gland and Hypothalamus Work Together

The pituitary gland is actually two glands in one. The anterior pituitary secretes a variety of hormones, some of which regulate other endocrine glands. The anterior pituitary is itself controlled by hormones, which are secreted from the hypothalamus. The posterior pituitary stores and secretes two hormones that are produced by neuron cell bodies in the hypothalamus and transported into the posterior pituitary by an axon tract.

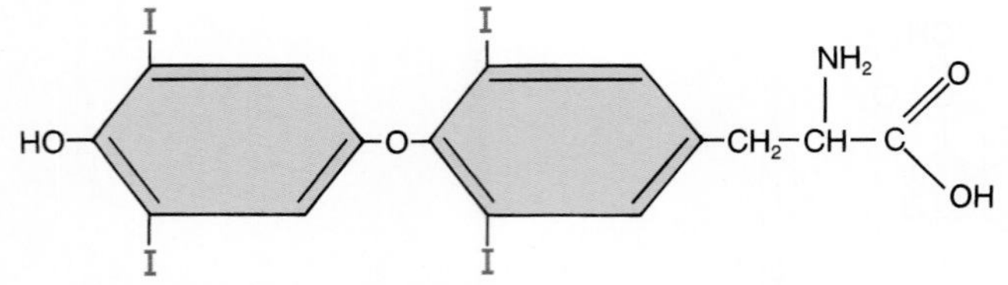

Thyroxine, or **tetraiodothyronine (T_4)**

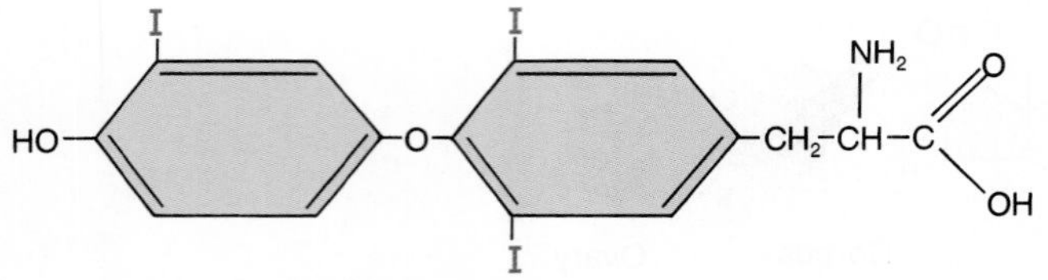

Triiodothyronine (T_3)

FIGURE 8.3 Structural formulas for the thyroid hormones. Thyroxine, also called tetraiodothyronine (T_4), and triiodothyronine (T_3) are secreted in a ratio of 9 to 1.

The **pituitary gland**, or *hypophysis,* is located below the hypothalamus (chapter 5) and is attached to the hypothalamus by a stalk, the *infundibulum* (fig. 8.4). The pituitary gland is an organ that consists of two glands: (1) the **anterior pituitary**, also called the *adenohypophysis*; and (2) the **posterior pituitary**, also called the *neurohypophysis*. These two glands, though part of the same organ, have different embryonic origins; as a result, they are quite different in structure and function. The posterior pituitary is formed as a downgrowth of the brain; because of this, axons extend to it from the hypothalamus. These axons carry hormones produced in the hypothalamus, as will be discussed shortly. By contrast, the anterior pituitary is formed from a different embryonic tissue (oral epithelium) and has no neural connections to the hypothalamus. The anterior pituitary produces its own hormones, but its secretion is still regulated by the hypothalamus. Regulation of the anterior and posterior pituitary glands will be described in more detail shortly.

Hormones of the Posterior Pituitary Are Produced in the Hypothalamus

The posterior pituitary gland stores and secretes two hormones. These are:

1. **Antidiuretic hormone (ADH)**, also known as **vasopressin**. ADH acts on the kidneys to promote the retention of water, so that less water is excreted in the urine (chapter 10). This is the major action of this hormone in humans; vasopressin action (causing constriction of blood vessels) is of secondary significance. For this reason, ADH will be the name used for this hormone in the rest of this text.
2. **Oxytocin**. In women, oxytocin stimulates contractions of the uterus during labor, and also stimulates contractions of the mammary ducts during lactation, resulting in the milk-ejection reflex. The physiological significance of oxytocin in men is less well understood.

CLINICAL APPLICATIONS

Oxytocin may be injected to induce labor if the pregnancy is too prolonged, or if the fetal membranes have ruptured and there is a danger of infection. Labor may also be induced by injections of oxytocin in cases of severe pregnancy-induced hypertension (high blood pressure), a condition called **preeclampsia.** After delivery, oxytocin may be injected to help the uterus regress in size and squeeze blood vessels, reducing the danger of hemorrhage.

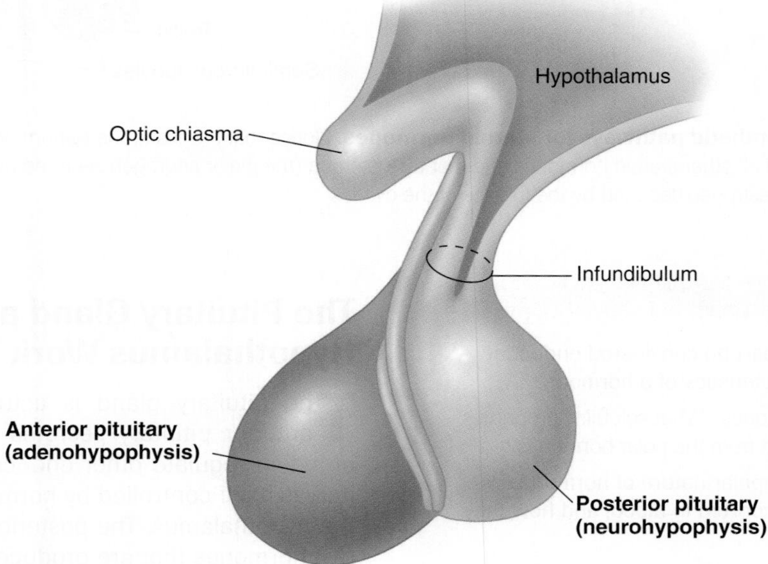

FIGURE 8.4 The structure of the pituitary gland. The anterior lobe is composed of glandular tissue, whereas the posterior lobe is composed largely of neuroglia and nerve fibers.

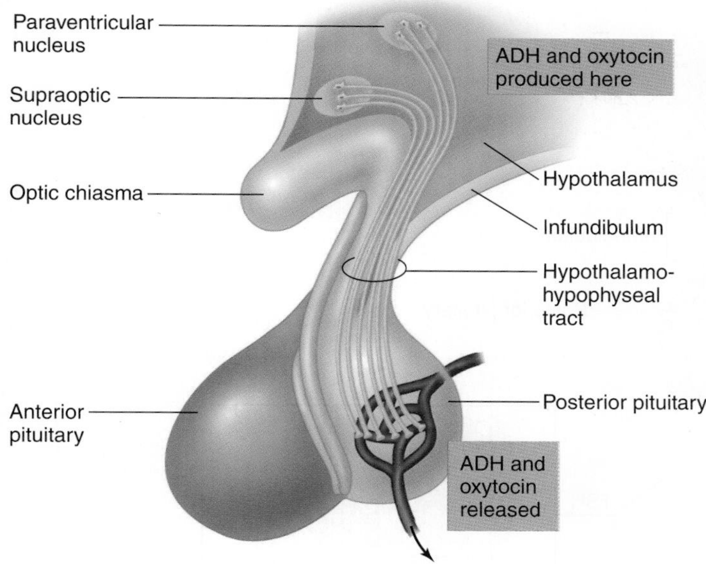

FIGURE 8.5 Hypothalamic control of the posterior pituitary. The posterior pituitary, or neurohypophysis, stores and releases hormones—ADH and oxytocin—that are actually produced in neurons within the supraoptic and paraventricular nuclei of the hypothalamus. These hormones are transported to the posterior pituitary by axons in the hypothalamo-hypophyseal tract.

Although the posterior pituitary secretes ADH and oxytocin, it doesn't produce these hormones. These two hormones are produced by neuron cell bodies in two nuclei (the supraoptic and paraventricular nuclei) in the hypothalamus, and then transported to the posterior pituitary in axons of the **hypothalamo-hypophyseal tract** (fig. 8.5). The posterior pituitary gland stores ADH and oxytocin until it is stimulated to secrete each hormone by the activation of appropriate neurons in the hypothalamus. For example, ADH will be secreted when osmoreceptor neurons in the hypothalamus are stimulated (chapter 13), and oxytocin will be secreted for the milk-ejection reflex in response to sensory stimuli of the baby suckling (chapter 15). **Neuroendocrine reflexes** thus control the secretion of ADH and oxytocin.

The Anterior Pituitary Is Regulated by Hypothalamic Hormones

The anterior pituitary gland produces and secretes its own hormones (fig. 8.6). These are generically known as *trophic hormones* (*trophic* = feed), a term used because higher than normal amounts of these hormones cause their target organs to grow, and lower amounts cause their target organs to shrink. When used as a suffix in a name for one of these hormones, the "trophic" is shortened to *tropic,* meaning "turning." The following hormones are produced and secreted by the anterior pituitary gland:

1. **Growth hormone** (**GH**, also called *somatotropin*). GH stimulates protein synthesis and promotes tissue and organ growth. Many of these growth-promoting effects are produced indirectly: GH stimulates the liver to produce molecules called *somatomedins,* which stimulate growth.
2. **Thyroid-stimulating hormone** (**TSH**, or *thyrotropin*). TSH stimulates the thyroid to secrete its hormones, thyroxine (T_4) and triiodothyronine (T_3).
3. **Adrenocorticotropic hormone** (**ACTH**, or *corticotropin*). ACTH stimulates the adrenal cortex to secrete its steroid hormones (the corticosteroids), notably cortisol (hydrocortisone).
4. **Follicle-stimulating hormone** (**FSH**, or *folliculotropin*). FSH stimulates the growth of ovarian follicles in women and the production of sperm in the testes of men.
5. **Luteinizing hormone** (**LH**, or *luteotropin*). LH and FSH are collectively termed the **gonadotropic hormones**, or **gonadotropins**. In women, LH stimulates ovulation and the formation of a corpus luteum in the ovaries. In men, LH stimulates the interstitial Leydig cells of the testes to secrete testosterone. For that reason, LH in men is sometimes called *interstitial cell-stimulating hormone* (*ICSH*). However, LH is the more common term in both sexes, because it can be confusing to use two different names for the same hormone.
6. **Prolactin**. In women, this hormone stimulates milk production by the mammary glands after a baby is born. Prolactin also has a number of other supporting roles in the reproductive system and kidneys in both men and women.

CLINICAL APPLICATIONS

Inadequate growth hormone (GH) secretion during childhood causes **pituitary dwarfism**. Inadequate GH secretion in adults causes a rare condition known as **pituitary cachexia** (**Simmonds' disease**). One of the symptoms of this disease is premature aging. Oversecretion of GH during childhood causes **gigantism**. In an adult, excessive GH can't cause continued growth of long bones (because the growth plates, or epiphyseal discs, of cartilage have ossified); it instead causes **acromegaly**. In acromegaly, a person's appearance gradually changes as a result of the thickening of bones and the growth of soft tissues, particularly in the face, hands, and feet.

Hypothalamic Releasing Hormones Stimulate the Anterior Pituitary

Notice that the anterior pituitary secretes some hormones that regulate other endocrine glands. TSH stimulates the thyroid; ACTH stimulates the adrenal cortex; and the gonadotropic hormones (FSH and LH) regulate the gonads. For this reason, some people refer to the anterior pituitary as a "master gland." This term is misleading on three counts: (1) the anterior pituitary regulates only the thyroid, adrenal cortex, and gonads, not the other endocrine glands; (2) the secretion of anterior pituitary hormones is itself regulated by the hypothalamus; (3) not even the hypothalamus is a master gland, because it's controlled by higher brain areas and by feedback effects from other endocrine glands (discussed shortly).

There are no axons that extend from the hypothalamus to the anterior pituitary, and so the hypothalamus can't control the anterior pituitary through neural means, as it does the posterior pituitary. Instead, there is an endocrine link between the hypothalamus

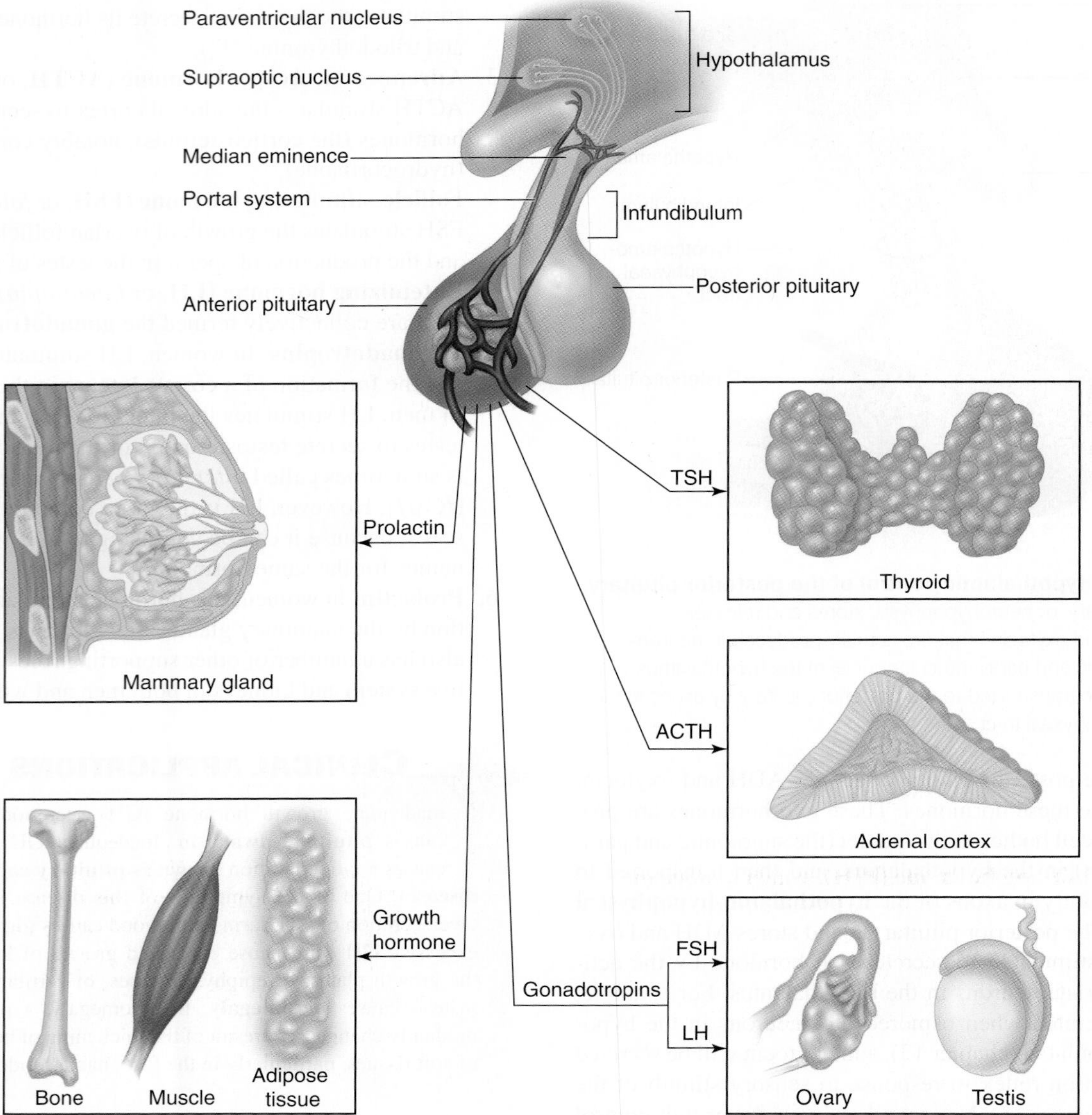

FIGURE 8.6 Hormones secreted by the anterior pituitary gland. The hormones shown on the right stimulate other endocrine glands: the thyroid, adrenal cortex, and gonads (testes and ovaries). Two hormones, FSH and LH, are called gonadotropins because they both stimulate the gonads. The two hormones shown on the left, prolactin and growth hormone, are secreted by the anterior pituitary but have functions other than the regulation of specific glands.

and the anterior pituitary gland. A system of blood capillaries at the base of the hypothalamus (in the *median eminence*) is drained into venules that carry blood to a second bed of capillaries in the anterior pituitary. This system of blood vessels is the **hypothalamo-hypophyseal portal system**. (In a portal system, blood from one organ is delivered downstream to another organ; we see this also in the blood that goes from the intestine to the liver in the hepatic portal vein.)

The hypothalamus regulates the anterior pituitary by means of regulatory hormones it secretes into the hypothalamo-hypophyseal portal system (fig. 8.7). These hypothalamic regulatory hormones include:

1. **Corticotropin-releasing hormone (CRH)**, which stimulates the anterior pituitary to secrete adrenocorticotropic hormone (ACTH).
2. **Thyrotropin-releasing hormone (TRH)**, which stimulates the anterior pituitary to secrete thyroid-stimulating hormone (TSH).
3. **Gonadotropin-releasing hormone (GnRH)**, which stimulates the anterior pituitary to secrete both gonadotropic hormones (FSH and LH).
4. **Growth-hormone-releasing hormone (GHRH)**, which stimulates the anterior pituitary to secrete growth hormone (GH).
5. **Somatostatin**, which inhibits the anterior pituitary from secreting growth hormone.

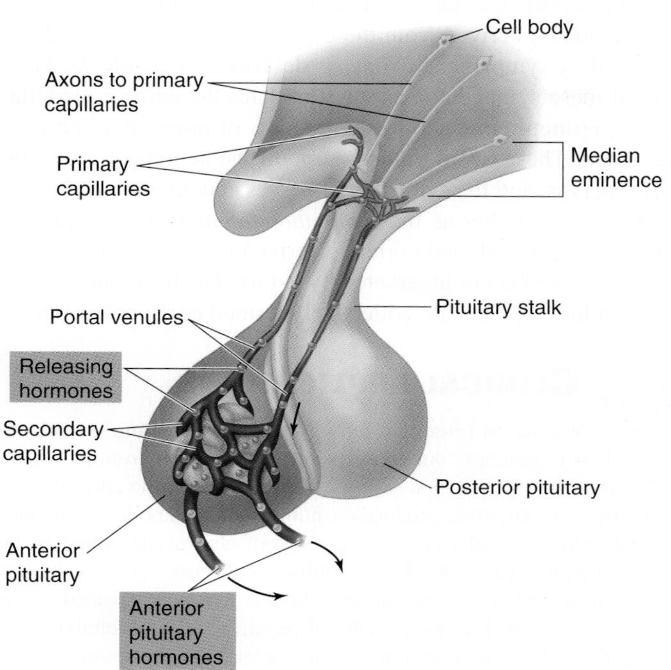

FIGURE 8.7 Hypothalamic control of the anterior pituitary. Neurons in the hypothalamus secrete releasing hormones (shown as green spheres) into the blood vessels of the hypothalamo-hypophyseal portal system. These releasing hormones stimulate the anterior pituitary to secrete its hormones (shown as pink spheres) into the general circulation.

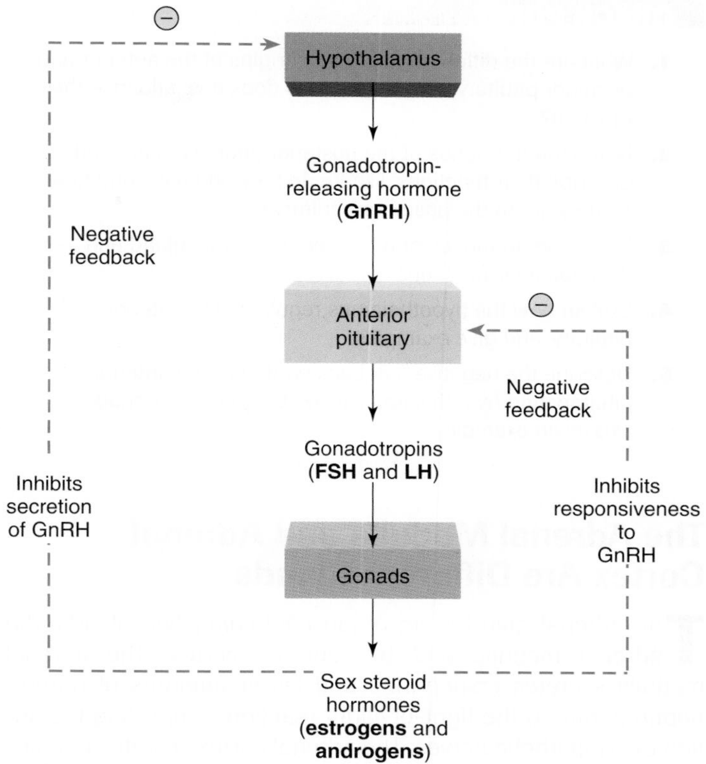

FIGURE 8.8 The hypothalamus-pituitary-gonad axis (control system). The hypothalamus secretes GnRH, which stimulates the anterior pituitary to secrete the gonadotropins (FSH and LH). These, in turn, stimulate the gonads to secrete the sex steroids. The secretions of the hypothalamus and anterior pituitary are themselves regulated by negative feedback inhibition (blue arrows) from the sex steroids.

6. **Prolactin-inhibiting hormone (PIH)**, which inhibits the anterior pituitary from secreting prolactin.

Notice that the anterior pituitary hormones that control other endocrine glands (that is, TSH, ACTH, FSH, and LH) are themselves regulated by releasing hormones—TRH, CRH, and GnRH. It seems that growth hormone and prolactin—the anterior pituitary hormones that don't specifically stimulate other endocrine glands—are the only ones controlled by hypothalamic inhibitory hormones, although both a releasing and an inhibiting hormone control GH.

The Hypothalamus and Anterior Pituitary Are Regulated by Negative Feedback

The anterior pituitary controls the thyroid, adrenal cortex, and gonads, and the hypothalamus controls the anterior pituitary. So, does that make the hypothalamus a "master gland"? Not really, because the hypothalamus and anterior pituitary are themselves regulated by the thyroid, adrenal cortex, and gonads. The control system is circular because of negative feedback loops.

The hormones secreted by the thyroid, adrenal cortex, and gonads exert **negative feedback inhibition** on the anterior pituitary and hypothalamus. We can use the hypothalamus–anterior pituitary–gonad control system (the *pituitary-gonad axis*, for short) as an example. The hypothalamus secretes gonadotropin-releasing hormone (GnRH) into the hypothalamo-hypophyseal portal system, which carries it to the anterior pituitary. In response to GnRH, the anterior pituitary secretes its gonadotropic hormones (FSH and LH). These enter the blood circulation and travel to the gonads. In response, the gonads secrete the different sex steroid hormones (estradiol and progesterone from the ovaries, testosterone from the testes). These enter the blood and travel to the anterior pituitary, *inhibiting* the secretion of FSH and LH, and to the hypothalamus, *inhibiting* the secretion of GnRH (fig. 8.8).

Negative feedback inhibition from target gland hormones prevents the hypothalamus and anterior pituitary from "driving" the target glands (thyroid, adrenal cortex, and gonads) excessively. As described in chapter 1, negative feedback is the means by which homeostasis is maintained. In the example given in figure 8.8, it also keeps the hormonal secretion of the testes relatively constant. In women, the negative feedback inhibition produced by estradiol and progesterone on the secretion of FSH and LH prevents ovulation at a certain time of the cycle. This is also how the birth control pills work; the pills contain estradiol and progesterone, which exert negative feedback inhibition of FSH and LH secretion and thereby prevent ovulation. However, there is also a unique positive feedback effect (stimulating ovulation) that operates at a certain time of the menstrual cycle, as will be described in chapter 15.

CHECK POINT

1. What are the different embryonic origins of the anterior and posterior pituitary glands, and how does this influence their function?
2. Name the hormones of the posterior pituitary gland and describe their functions. Where do they originate, and how do they get to the posterior pituitary?
3. Name the hormones of the anterior pituitary gland and describe their functions.
4. Explain how the hypothalamus regulates the anterior pituitary, and give examples.
5. Describe the negative feedback control of the anterior pituitary and hypothalamus, using the pituitary-gonad axis as an example.

The Adrenal Medulla and Adrenal Cortex Are Different Glands

The adrenal gland is an organ containing two glands: the adrenal medulla and the adrenal cortex. The adrenal medulla secretes epinephrine and lesser amounts of norepinephrine during the fight-or-flight reaction, supporting the action of sympathetic nerves. The adrenal cortex secretes corticosteroid hormones that regulate metabolism and electrolyte (Na^+ and K^+) homeostasis. Hormones, including ACTH, regulate the secretions of the adrenal cortex.

The **adrenal gland** is really two glands in one (fig. 8.9): the **adrenal cortex** (the outer part) and the **adrenal medulla** (the inner part). Like the anterior and posterior pituitary glands, the adrenal cortex and medulla have different embryonic origins, secrete different hormones, and are regulated in different ways. The adrenal medulla is derived from the embryonic neural crest and is innervated by sympathetic axons, as described in chapter 6. Activation of these sympathetic axons stimulates the adrenal medulla to secrete **epinephrine**, and lesser amounts of **norepinephrine**, into the blood. These two hormones support the actions of the sympathetic nerves, and the sympathetic nerves and adrenal medulla can be grouped together as the *sympathoadrenal system* (chapter 6). By contrast, the adrenal cortex is derived from a different embryonic tissue and is not innervated by axons. The hormones secreted by the adrenal cortex are collectively termed **corticosteroids**.

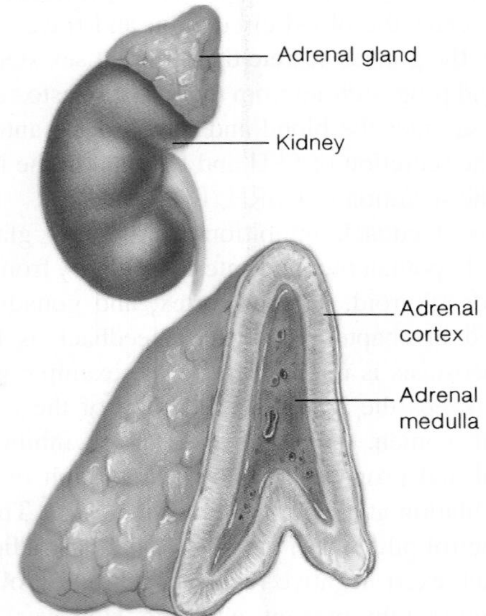

FIGURE 8.9 The adrenal gland. Each of the two adrenal glands is located on top of the kidney. A section through the adrenal gland reveals that it is composed of an outer adrenal cortex and an inner adrenal medulla.

CLINICAL APPLICATIONS

Oversecretion of corticosteroids results in **Cushing's syndrome**. This is generally due to oversecretion of ACTH from the anterior pituitary, but it can also result from a tumor of the adrenal cortex. Cushing's syndrome produces a number of metabolic disturbances, and results in a puffy appearance and body changes described as "buffalo hump" and "moon face." **Addison's disease** is caused by inadequate secretion of corticosteroids (and is thus accompanied by high secretion of ACTH, due to reduced negative feedback inhibition; see fig. 8.12). Low corticosteroids produce a variety of symptoms, and this condition is fatal if untreated because of electrolyte (ion) imbalances and dehydration. Fortunately, Addison's disease can be treated easily because corticosteroids can be taken orally.[2]

Epinephrine Exerts Its Effect Through Cyclic AMP

Epinephrine and lesser amounts of norepinephrine are secreted from the adrenal medulla as a result of sympathetic nerve stimulation during the fight-or-flight reaction (chapter 6). Their various effects include stimulation of heart rate and strength of contraction, and dilation of the bronchioles (pulmonary airways). Epinephrine also has metabolic effects; for example, it stimulates the hydrolysis (breakdown; see chapter 1) of glycogen in the liver. Hydrolysis of liver glycogen releases free glucose into the blood during emergencies, when the need for glucose would be raised by increased muscle metabolism. But how does epinephrine produce these effects, when it's a polar hormone that can't pass through the plasma membrane of its target cells?

We can use epinephrine (and norepinephrine) action as an example of how polar hormones work in general. Because they can't pass through the phospholipid layers of plasma membranes, the receptor proteins for polar hormones are built into the plasma membranes and bind to their hormones in the extracellular fluid. This activates second-messenger systems within the target cells,

FYI [2]President John F. Kennedy had Addison's disease, but few knew of it because it was well controlled by corticosteroids. If untreated, there is very high ACTH secretion because of reduced negative feedback from corticosteroids. High ACTH has an interesting effect: the skin color changes, producing colors that range from smoky, to amber, to chestnut brown (in the terms used by Dr. Addison in describing this disease). The discoloration is due to ACTH stimulation of melanocytes in the skin.

which actually carry out the effects of the hormones (chapter 3). There are different second-messenger systems used by different hormones; here, we will consider only the *cyclic AMP (cAMP)* system used by epinephrine and norepinephrine.

Epinephrine and norepinephrine (or other hormones that also use cAMP as a second messenger) operate in the following manner (fig. 8.10):

1. The polar hormone binds to its receptor in the plasma membrane.
2. This causes the G-protein complex (chapter 3) that is bound to the receptor to dissociate into its subunits and detach from the receptor.
3. The α subunit moves through the plasma membrane to an enzyme protein, adenylate cyclase, activating this enzyme (which was inactive before this occurs).
4. The active adenylate cyclase catalyzes (chapter 1) a chemical reaction in which ATP is converted into cyclic AMP (cAMP).
5. cAMP binds to an inactive protein kinase enzyme in the cytoplasm, breaking off an inhibitory subunit from that enzyme.
6. The active protein kinase enzyme catalyzes the addition of phosphate groups to (the phosphorylation of) inactive enzyme proteins; phosphorylation converts these previously inactive enzymes to their active forms. By contrast, the phosphorylation of different enzymes can cause their inactivation (fig. 8.10).
7. Now that the activities of specific enzymes have been changed, the enzymes catalyze reactions that produce the effects of that hormone in the target cell.

In the case of epinephrine acting on a liver cell, the activated protein kinase phosphorylates and activates an enzyme that promotes the hydrolysis of glycogen to glucose; it also inactivates an enzyme that produces glycogen from glucose. This results in the liver's response to cAMP production when stimulated by epinephrine: the breakdown of glycogen and secretion of glucose into the blood.

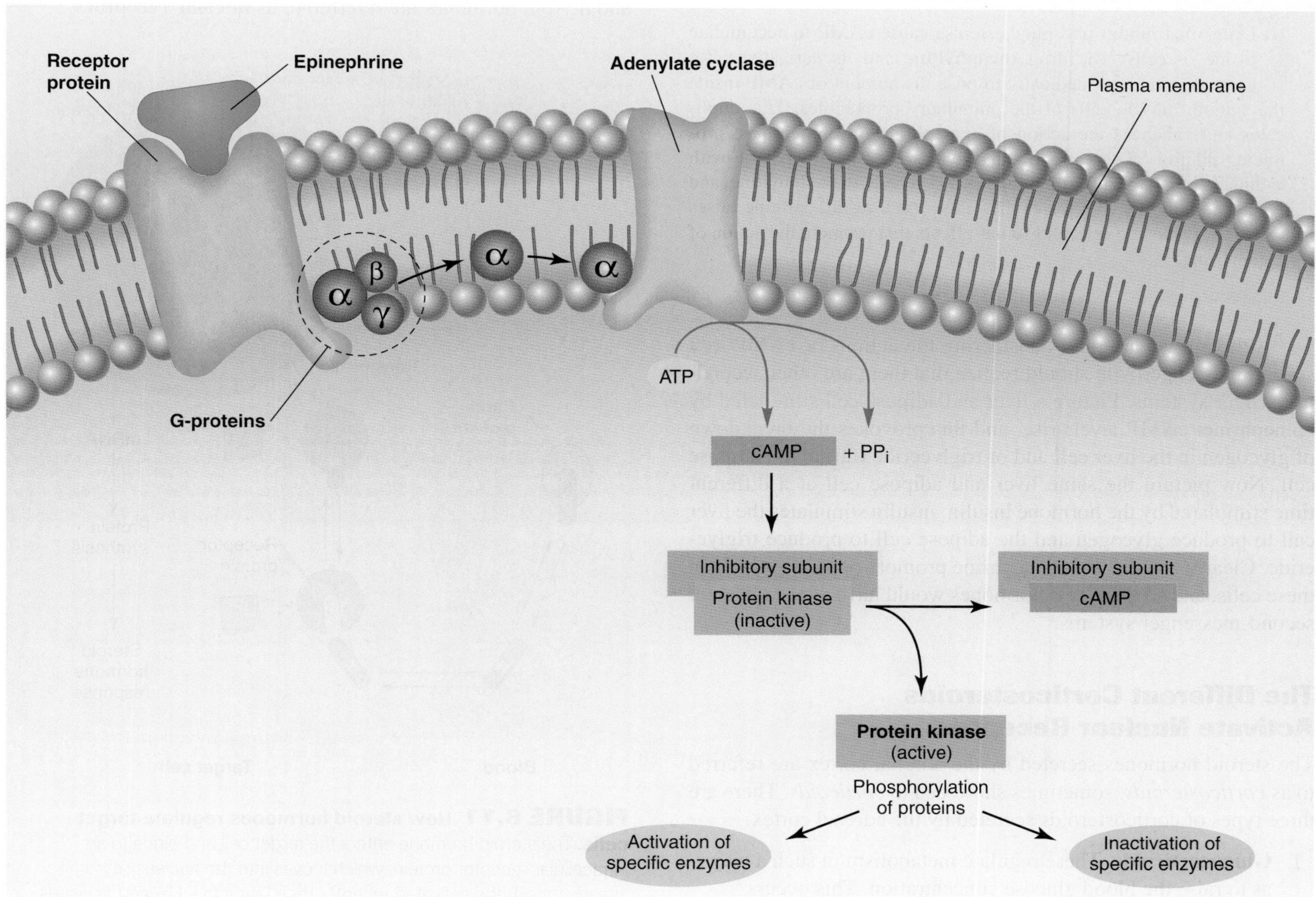

FIGURE 8.10 The cyclic AMP second-messenger system. The hormone causes the production of cAMP within the target cell cytoplasm, and cAMP activates protein kinase. The activated protein kinase then causes the activation or inactivation of a number of specific enzymes. These changes lead to the characteristic effects of epinephrine in its target cells.

In other target cells, cAMP still activates protein kinase, but the effects of this active enzyme are different because different genes are active, producing different enzymes. For example, when epinephrine acts on the smooth muscles of the bronchioles (pulmonary airways), the cAMP causes smooth muscle relaxation and dilation of the bronchioles.

So, can physicians give patients cyclic AMP instead of epinephrine? No, because cAMP, like all organic molecules with phosphate groups, is too polar to cross plasma membranes. Each target cell has to make its own cAMP when stimulated by epinephrine (or other hormones that act in a similar manner). Then, something has to stop the effects of cAMP when epinephrine is removed. This requires the breakdown of cAMP within the target cells, produced by the enzymatic action of **phosphodiesterase**. In this way, a hormone such as epinephrine must continue to be secreted in order to continue to stimulate its target cells. The "off switch" provided by phosphodiesterase is required for regulation.

Clinical Applications

Drugs that inhibit phosphodiesterase cause cAMP to accumulate inside of cells. The drug **theophylline** and its derivatives, for example, are used medically to raise the amount of cAMP inside the smooth muscle cells of the pulmonary bronchioles. This duplicates and enhances the action of epinephrine on bronchioles, producing dilation of the airways that makes it easier for people with asthma to breathe. **Caffeine** is chemically related to theophylline, and is also an inhibitor of phosphodiesterase. As a result, caffeine raises cAMP levels inside cells and so has effects that promote the action of epinephrine.

Although we are only considering the actions of cAMP as a second messenger, you should realize that there are other second-messenger systems. Picture a liver and adipose cell stimulated by epinephrine: cAMP levels rise, and this provokes the breakdown of glycogen in the liver cell and of triglyceride (fat) in the adipose cell. Now picture the same liver and adipose cell at a different time stimulated by the hormone insulin: insulin stimulates the liver cell to produce glycogen and the adipose cell to produce triglyceride. Clearly, insulin and epinephrine promote opposite effects in these cells, and so these two hormones would have to use different second-messenger systems.

The Different Corticosteroids Activate Nuclear Receptors

The steroid hormones secreted by the adrenal cortex are referred to as *corticosteroids*, sometimes shortened to *corticoids*. There are three types of corticosteroids secreted by the adrenal cortex:

1. **Glucocorticoids**. These regulate metabolism in such a way as to raise the blood glucose concentration. This occurs partly because of the breakdown of liver glycogen (*glycogenolysis*), but is mainly due to the conversion of non-carbohydrate molecules (such as amino acids) into glucose, a process called *gluconeogenesis*. The glucocorticoids also promote catabolism (breakdown) of proteins, increasing blood levels of amino acids. The major glucocorticoid in humans is **cortisol** (also called **hydrocortisone**).
2. **Mineralocorticoids**. These regulate mineral, or electrolyte (ion), concentrations in the blood, particularly of Na^+ and K^+. The major mineralocorticoid in humans is **aldosterone**.
3. **Weak androgens**. The functions of these in men are not well understood; in women, they are the only androgens (male hormones) present, and may contribute to sexual drive. The principal adrenal androgen is **androstenedione**.

The corticosteroids, like other steroid hormones and the thyroid hormones, are nonpolar. Thus, they can easily pass through the phospholipid layers of the plasma membranes of their target cells. The receptor proteins for nonpolar hormones are generally located inside their target cells. Once the nonpolar hormone binds to its receptor protein, the receptor becomes activated and moves into the nucleus (if it isn't there already), where it binds to the DNA (fig. 8.11). Because of this, the receptor proteins for nonpolar steroid and thyroid hormones are referred to as **nuclear receptors**.

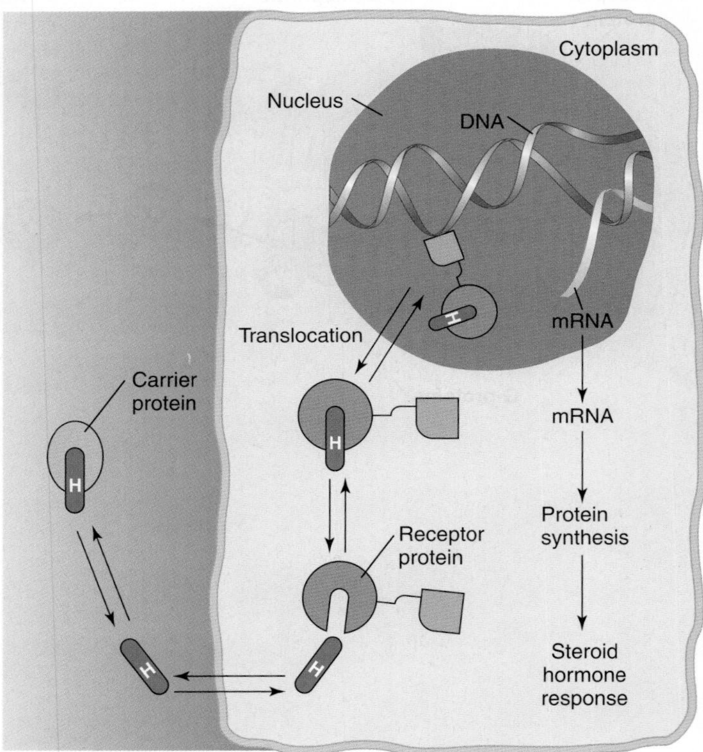

FIGURE 8.11 How steroid hormones regulate target cells. The steroid hormone enters the target cell and binds to an intracellular receptor protein, which moves into the nucleus (if it isn't already there); the activated receptor then binds to DNA and activates specific genes. The genes are transcribed into messenger RNA (mRNA), which is translated into specific proteins. The newly synthesized proteins then produce the response of the target cell to the steroid hormone.

The activated nuclear receptors bind to specific locations in the DNA and activate specific genes. These activated genes then produce specific mRNA (chapter 2), which codes for specific proteins. Thus, in response to the stimulation of a target cell by a nonpolar hormone, genes are activated and new proteins are produced. Some of these new proteins are enzyme proteins, which catalyze specific reactions in the cell. As a result, the metabolism of the target cells is changed in such a manner as to produce the characteristic response to the hormone (fig. 8.11).

Secretions of the Adrenal Cortex Are Regulated by Hormones

ACTH, secreted by the anterior pituitary, stimulates the adrenal cortex to secrete the glucocorticoid hormones (principally cortisol). The secretion of ACTH is stimulated by CRH, a releasing hormone secreted by the hypothalamus. To complete the circular control system, cortisol exerts a negative feedback inhibition of both ACTH and CRH secretion (fig. 8.12). This helps maintain homeostasis of the pituitary-adrenal axis.

However, higher brain areas can "drive" the secretion of CRH, and thus ACTH, and thereby increase the secretion of cortisol (fig. 8.12). This occurs in response to stress as part of the **general adaptation syndrome (GAS)**. The increased cortisol is a beneficial response to the stress of severe infections, burns, and surgery, where cortisol levels can rise to as high as six times the unstressed levels. The elevated cortisol causes increased blood levels of glucose for the CNS and of amino acids for the repair of injured tissues. However, in other stressful situations the rise in blood cortisol levels may be detrimental, because the glucocorticoids can suppress the immune system. A suppressed immune system increases the risk of illness, as many students during finals week can attest. High cortisol secretion due to prolonged stress can have other deleterious effects, including memory deficits, anxiety, and depression.

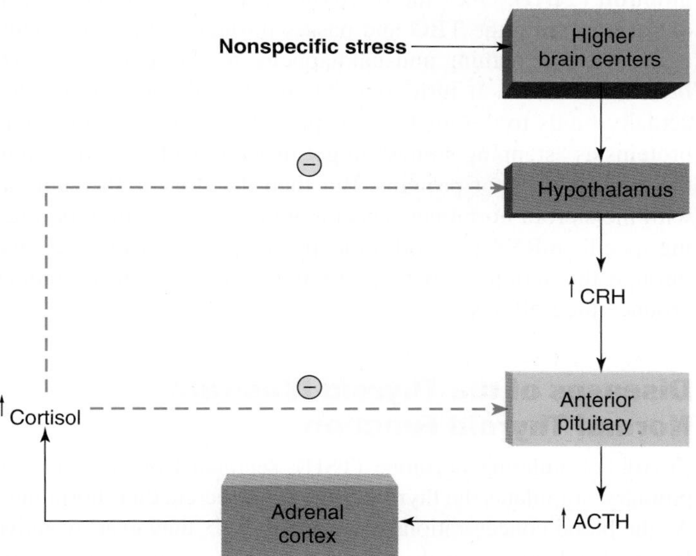

FIGURE 8.12 Activation of the pituitary-adrenal axis by nonspecific stress. Negative feedback control of the adrenal cortex (blue arrows) is also shown.

Clinical Applications

Since glucocorticoids such as cortisol (hydrocortisone) can inhibit the immune system and suppress inflammation, glucocorticoids such as **prednisolone** and **dexamethasone** are used medically. They are given as topical ointments, pills, or injections to treat various inflammatory conditions and to suppress the immune rejection of transplanted organs. As would be expected, raising blood glucocorticoid levels by giving these as exogenous drugs would suppress the secretion of ACTH (through negative feedback inhibition), suppressing the secretion of cortisol from the adrenal cortex. Because of this, medical treatment with glucocorticoid pills must be tapered off gradually rather than end abruptly.

Clinical Investigation Clues

Remember, Tony was taking glucocorticoid (prednisolone) pills to treat an inflammatory condition.

- What benefit would be derived from taking this drug?
- What effect would these pills have on the secretion of ACTH from Tony's anterior pituitary gland?
- What effect would this have on the secretion of cortisol from Tony's adrenal cortex?
- Why do the hormones in these pills need to be tapered off, rather than discontinued abruptly?

The adrenal cortex also secretes mineralocorticoid hormones, principally aldosterone. Aldosterone stimulates the kidneys to (1) retain more Na^+ and water in the blood and (2) excrete more K^+ in the urine. We need aldosterone to live, because without it we'd lose too much Na^+ and water in the urine, becoming dehydrated, and we'd develop hyperkalemia (high blood K^+), which could cause the heart to fibrillate (stop beating). Aldosterone secretion isn't stimulated by ACTH, but by other chemical regulators. If you eat a banana or other potassium-rich foods, the rise in blood K^+ stimulates the adrenal cortex to secrete aldosterone, which helps eliminate the extra K^+ in the urine. Also, a chemical regulator known as *angiotensin II*, produced in the blood when there is a fall in blood volume and pressure, stimulates the adrenal cortex to secrete aldosterone (chapter 10). Aldosterone then stimulates the kidneys to retain Na^+ and water in the blood, counteracting the changes and helping to maintain homeostasis.

CHECK POINT

1. Step-by-step, explain how epinephrine and norepinephrine stimulate their target cells.
2. List the different categories of corticosteroids and provide examples of each.
3. Step-by-step, explain how steroid hormones stimulate their target cells.
4. Draw the negative feedback loop that controls ACTH and cortisol secretion, and explain how this pituitary-adrenal axis is affected by stress.
5. What is the GAS? When is this beneficial? How can this reaction be detrimental?
6. Describe the actions of aldosterone and how its secretion is regulated.

The Thyroid and Parathyroid Glands Have Different Functions

The thyroid gland contains hollow structures called thyroid follicles that trap iodine and use it to produce the thyroid hormones, thyroxine (T_4) and triiodothyronine (T_3). These hormones stimulate cell respiration, helping to set the basal metabolic rate (BMR); they are also needed for growth and maturation of the CNS. The parathyroid glands secrete parathyroid hormone (PTH), which helps regulate calcium balance in the body.

The **thyroid gland** is located just below the larynx (fig. 8.13) and contains many microscopic-size **thyroid follicles** (fig. 8.14). The thyroid follicles are hollow structures, like water balloons. Instead of water, the follicles contain a protein-rich fluid called *colloid.*

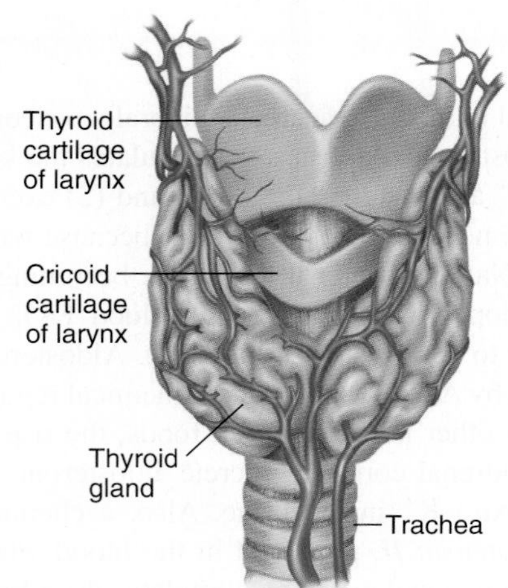

FIGURE 8.13 The thyroid gland. The thyroid gland is located on the anterior surface of the larynx (voice box), above the trachea (windpipe). Two of the cartilages of the larynx are also labeled.

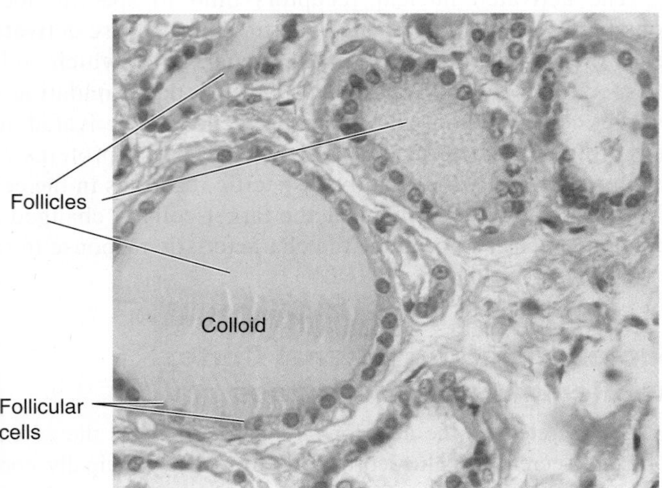

FIGURE 8.14 A photomicrograph (250×) of a thyroid gland. Numerous thyroid follicles are visible. Each follicle consists of follicular cells surrounding the fluid known as colloid, which contains thyroglobulin.

The follicles trap iodine, which is transported out of the blood and combined in the colloid with a protein called **thyroglobulin.** The cells surrounding the thyroid follicles then use the thyroglobulin to make the two thyroid hormones, *thyroxine* (T_4) and *triiodothyronine* (T_3).

The thyroid hormones are nonpolar molecules, as previously mentioned. Thyroxine is particularly nonpolar, and must be bound to a carrier protein in order to be transported by the blood. The carrier protein for thyroxine is **thyroxine-binding globulin (TBG).** Once the thyroxine arrives at its target cell, it dissociates from the TBG and passes through the plasma membrane. Then something unusual happens: *the target cell converts thyroxine into* T_3. It turns out that it's the T_3, not the T_4, that actually binds to the nuclear receptor. The T_3 uses cytoplasmic proteins as "stepping stones" to get into the nucleus and bind to the nuclear receptor proteins. But then the story is the same as with the steroid hormones: specific genes are activated, producing specific mRNA that codes for specific proteins (fig. 8.15). It's through the action of these proteins that the thyroid hormones produce their effects.

Diseases of the Thyroid Illustrate Normal Thyroid Function

Thyroid-stimulating hormone (TSH), secreted from the anterior pituitary, stimulates the thyroid follicles to secrete their hormones. As the blood concentrations of T_4 and T_3 rise, they exert negative feedback inhibition on the anterior pituitary gland, reducing its secretion of TSH. (There doesn't appear to be negative feedback inhibition of TRH secretion from the hypothalamus.) This maintains normal homeostasis of thyroxine and T_3 levels in the blood (fig. 8.16, upper right).

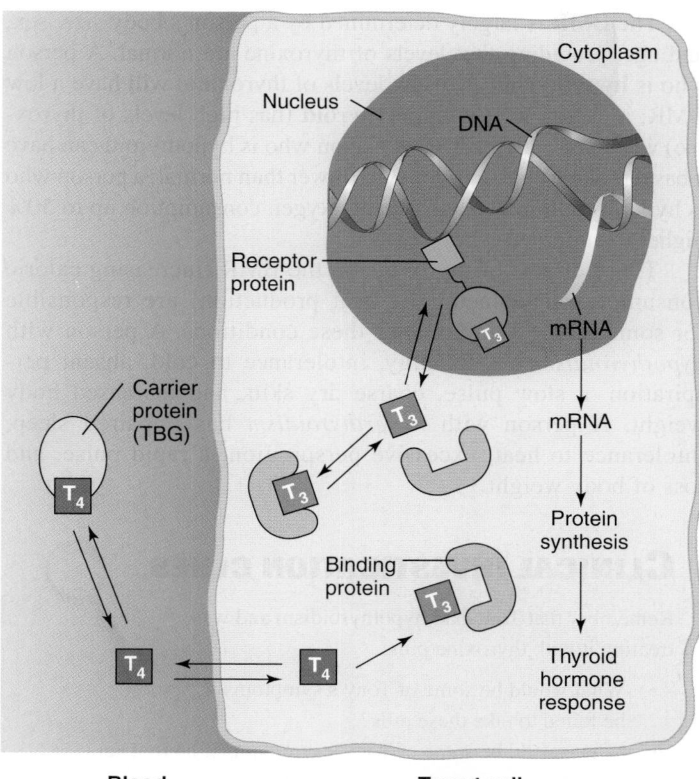

FIGURE 8.15 The mechanism of action of thyroid hormones on their target cells. T_4 is first converted into T_3 within the cytoplasm of the target cell. T_3 then enters the nucleus and binds to its nuclear receptor. The hormone-receptor complex can then bind to a specific area of DNA and activate specific genes.

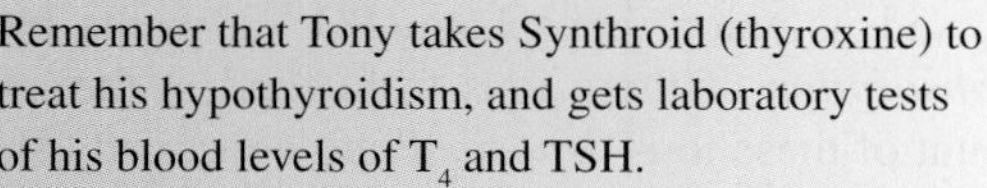

Clinical Investigation Clues

Remember that Tony takes Synthroid (thyroxine) to treat his hypothyroidism, and gets laboratory tests of his blood levels of T_4 and TSH.

- *T_4* and *TSH* are abbreviations for which terms?
- If the dose of his thyroxine pills is too high, what would happen to his blood levels of T_4 and TSH?
- By what mechanism does the thyroxine in his pills affect his blood levels of TSH?

Abnormal Growth of the Thyroid Is a Goiter

Suppose that the thyroid gland lacks sufficient iodine to make thyroxine and T_3. Blood levels of these hormones will necessarily decline, a condition called *hypothyroidism*. As the thyroid hormone levels go down, TSH must rise, because of the decline in negative feedback inhibition. Abnormally high TSH levels can't cause increased thyroxine secretion (and thereby return to homeostasis), because the thyroid can't make thyroxine without iodine. So TSH levels continue to rise. Remember, the anterior pituitary hormones have a "trophic" (feeding) effect on their target glands. Here's a good example: the high TSH stimulates the thyroid gland to grow

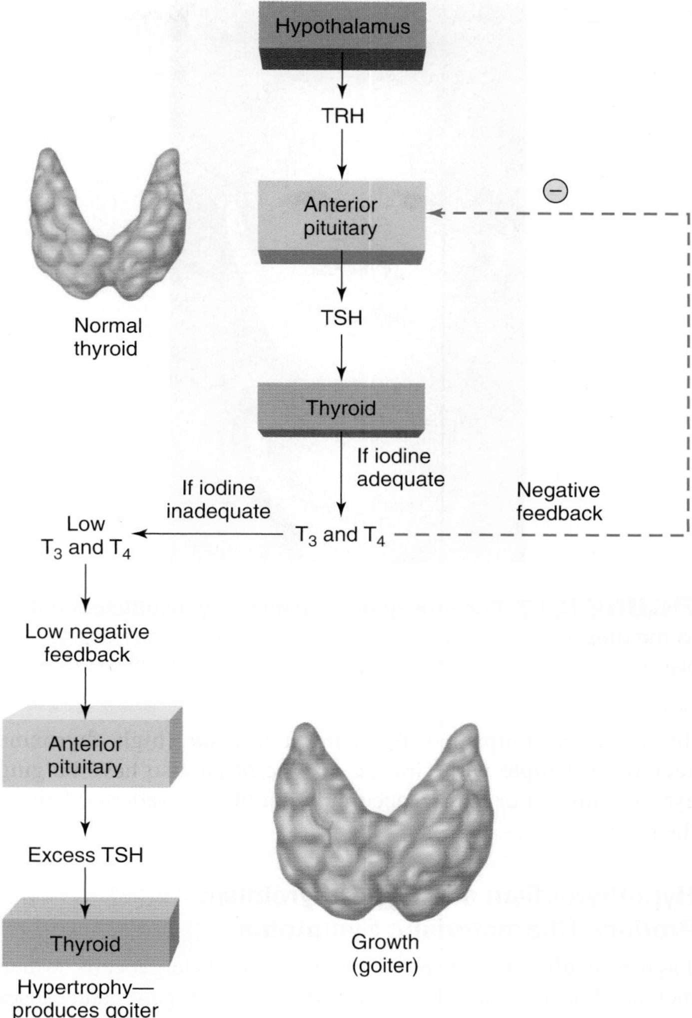

FIGURE 8.16 Negative feedback control of the pituitary-thyroid axis. The upper portion of this figure illustrates normal negative feedback inhibition of TSH secretion. The lower portion of this figure illustrates how an endemic goiter is produced when there is an iodine deficiency in the diet. Lack of negative feedback causes excessive TSH secretion, which stimulates abnormal growth of the thyroid.

excessively large (fig. 8.16, lower left). This is a **goiter** (fig. 8.17). Because it's caused by low dietary iodine, an endemic condition in some parts of the world, a goiter caused by iodine deficiency is known as an *endemic goiter*.

You should understand that not everyone with a goiter lacks dietary iodine. In **Graves' disease**, a goiter is caused by stimulation of the thyroid by "autoantibodies"—antibody proteins that bind to molecules that are a normal part of a person's own body. This can produce many different diseases, known as autoimmune diseases (chapter 11). The antibodies that cause Graves' disease stimulate the thyroid to grow (producing a goiter) while, at the same time, stimulating the thyroid to secrete excessive amounts of thyroxine (because the antibodies are not suppressed by negative feedback). Thus, unlike endemic goiter, the goiter of Graves'

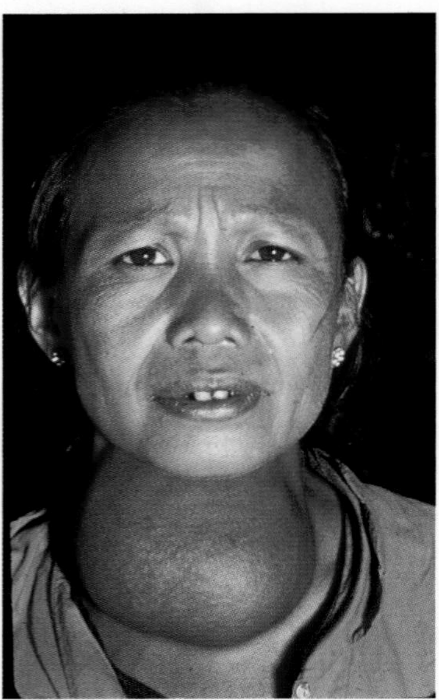

FIGURE 8.17 Endemic goiter is caused by insufficient iodine in the diet. A lack of iodine causes hypothyroidism, and the resulting elevation in TSH secretion stimulates the excessive growth of the thyroid.

disease is accompanied by *hyperthyroidism* (high thyroxine secretion). People with Graves' disease often also have bulging eyes because of excessive accumulation of fluid (edema) behind the eyes.

Hypothyroidism and Hyperthyroidism Produce Characteristic Symptoms

Thyroxine, after it's converted into T_3 within its target cells (which include almost all the cells of the body), activates genes and causes the production of enzyme proteins that have a number of important effects:

1. In babies and growing children, thyroxine is required for proper growth and development, particularly of the central nervous system (CNS). Thyroxine is needed to prevent the development of **cretinism**, a condition of severe mental retardation caused by inadequate thyroxine. Fortunately, babies less than 1 month old with cretinism can take thyroxine to prevent the mental retardation.
2. Thyroxine stimulates the rate of cell respiration (chapter 2) in the body. By doing so, it helps set the rate at which the body consumes energy (calories and oxygen) at rest, called the **basal metabolic rate (BMR)**.

The basal metabolic rate is more formally described as the energy expenditure of a relaxed, resting person who is in a neutral room temperature (about 28° C) and has not eaten in 8 to 12 hours. The BMR accounts for about 60% of the caloric expenditure of an average adult. The remaining calories are used in (1) *adaptive thermogenesis,* which refers to the energy spent adjusting to different ambient temperatures and the energy required to digest and absorb food; and (2) *physical activity*, which raises the metabolic rate of skeletal muscles.

The BMR is largely determined by a person's body size, sex, and age, providing that levels of thyroxine are normal. A person who is **hypothyroid** (has low levels of thyroxine) will have a low BMR; a person who is **hyperthyroid** (has high levels of thyroxine) will have a high BMR. A person who is hypothyroid can have a basal oxygen consumption 30% lower than normal; a person who is hyperthyroid may have a basal oxygen consumption up to 50% higher than normal.

These effects of thyroxine on the BMR (increasing calorie consumption and metabolic heat production) are responsible for some of the symptoms of these conditions. A person with *hypothyroidism* has lethargy, intolerance to cold, absent perspiration, a slow pulse, coarse dry skin, and increased body weight. A person with *hyperthyroidism* has impaired sleep, intolerance to heat, excessive perspiration, a rapid pulse, and loss of body weight.[3]

CLINICAL INVESTIGATION CLUES

Remember that Tony had hypothyroidism and was treating it with thyroxine pills.

- What would be some of Tony's symptoms if he failed to take these pills?
- What would be some of Tony's symptoms if he took an overdose of these pills over an extended period of time?

The Parathyroid Glands Regulate Calcium Balance

The four small, flattened **parathyroid glands** are embedded in the posterior surface of the thyroids gland (fig. 8.18). These glands secrete one hormone, **parathyroid hormone (PTH)**, which is the major hormone regulating the blood levels of calcium (Ca^{2+}). The amount of Ca^{2+}, and its anion, phosphate (PO_4^{3-}), in the blood reflects (1) the *absorption* of these ions from food in the intestine, (2) the amount of these ions *deposited* into bone, and (3) the amount of these ions *resorbed* (removed) from bone.

The skeleton, in addition to its other functions, serves to store calcium and phosphate in the form of calcium phosphate crystals, known as *hydroxyapatite*. The calcium and phosphate for these crystals come from the blood and are deposited in bone by the action of **osteoblasts**. The osteoblasts secrete an organic matrix composed of collagen proteins, which becomes hardened by deposits of hydroxyapatite. Bone resorption, the dissolution of the hydroxyapatite crystals to release Ca^{2+} and PO_4^{3-} into the blood, is produced by the action of **osteoclasts**.

FYI [3]Some people want to take thyroxine pills to lose weight. There are problems with this. Raising thyroxine levels causes negative feedback inhibition of TSH, resulting in less stimulation of the thyroid gland. If people with normal thyroid function take thyroxine pills, the reduced TSH will cause their own thyroid glands to atrophy and underproduce thyroxine. As a result, they will put the weight back on. In the meantime, high thyroxine levels increase the pulse rate and blood pressure, endangering their health; it can also cause osteoporosis to progress at a faster rate.

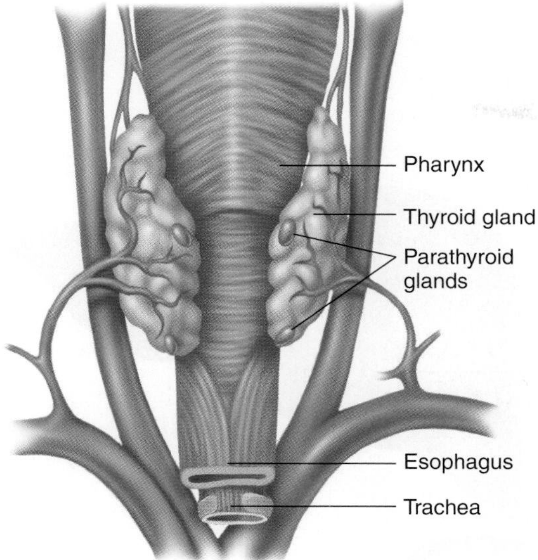

FIGURE 8.18 A posterior view of the parathyroid glands. The parathyroids are embedded in the tissue of the thyroid gland.

Whenever the blood Ca^{2+} level begins to fall, the parathyroid glands are stimulated to secrete PTH. PTH then stimulates osteoclast activity, causing bone resorption. This raises the blood Ca^{2+} level and thus provides for the completion of a negative feedback loop, maintaining homeostasis of the blood Ca^{2+} levels (fig. 8.19). Notice that the kidneys are also stimulated to retain more Ca^{2+}, so that less is excreted in the urine. Now, let's get personal: these

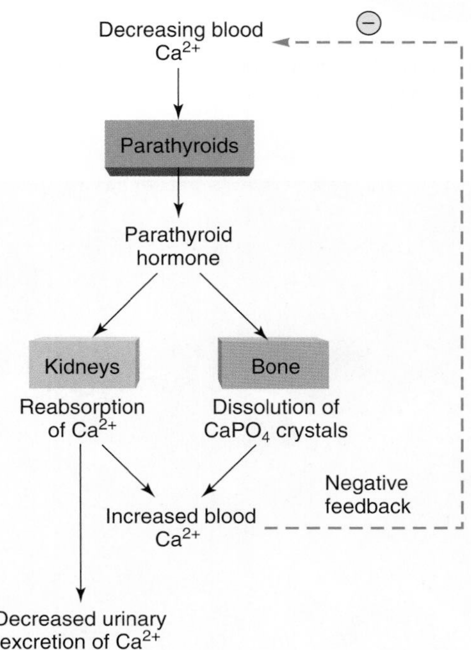

FIGURE 8.19 The actions of parathyroid hormone and the control of its secretion. An increased level of parathyroid hormone causes the bones to release calcium and the kidneys to conserve calcium that would otherwise be lost through the urine. A rise in blood Ca^{+} can then exert negative feedback inhibition on parathyroid hormone secretion.

mechanisms will maintain homeostasis of blood Ca^{2+} *at the expense of your bones*. If you aren't getting enough calcium in your diet and your blood Ca^{2+} starts to fall, PTH will stimulate bone resorption. If you want to maintain strong bones, you must take in adequate amounts of calcium in your food and drink, or take supplements.

Another hormone important in the regulation of blood Ca^{2+} levels is **1,25-dihydroxyvitamin D_3**. This is made from vitamin D produced in the skin in response to sunlight (fig. 8.20). The vitamin D is a **prehormone**; it must be converted into the active hormone by enzymes in the liver and kidneys. Then 1,25-dihydroxyvitamin D stimulates the small intestine to absorb Ca^{2+} and PO_4^{3-} into the blood, so they can be deposited in bone. Because our skin doesn't make sufficient vitamin D, we must also get it from our diet (that's why it's a vitamin). Children who don't get sufficient vitamin D in their diets develop soft bones, a condition called **rickets**.

CLINICAL APPLICATIONS

The most common bone disorder in elderly people is **osteoporosis,** which is produced by parallel losses of both organic matrix and hydroxyapatite deposits (fig. 8.21). Osteoporosis reduces bone mass and density, increasing the risk of fractures. This is more common in women than in men, suggesting that the fall in estrogen secretion at menopause is a contributing factor. Premenopausal women with a low percent body fat and amenorrhea (cessation of menstruation) can also develop osteoporosis. It's important to take sufficient dietary calcium in food or supplements, and exercise also helps to strengthen bones. The most commonly used drugs for treating osteoporosis are the *bisphosphonates,* which promote the death of osteoclasts. A newer drug approved for the treatment of osteoporosis is *raloxifene,* a drug that stimulates estrogen receptors in some organs (bone) but not others (uterus).

CHECK POINT

1. Describe how the hypothalamus and pituitary regulate the thyroid, and describe the negative feedback inhibition of TSH. Use this information to explain how an endemic goiter is produced.
2. Describe the mechanism by which thyroxine stimulates its target cells.
3. What are the physiological effects of thyroxine? Describe how these relate to the symptoms of hypothyroidism and hyperthyroidism.
4. What causes PTH to be secreted, and how does PTH help maintain homeostasis of blood Ca^{2+}?
5. Describe the formation and action of 1,25-dihydroxyvitamin D_3.

The Pancreatic Islets Secrete Insulin and Glucagon

The pancreatic islets (islets of Langerhans) contain alpha cells that secrete glucagon and beta cells that secrete insulin. Glucagon acts to raise the blood glucose, and insulin acts to lower it by promoting the entry of blood glucose into

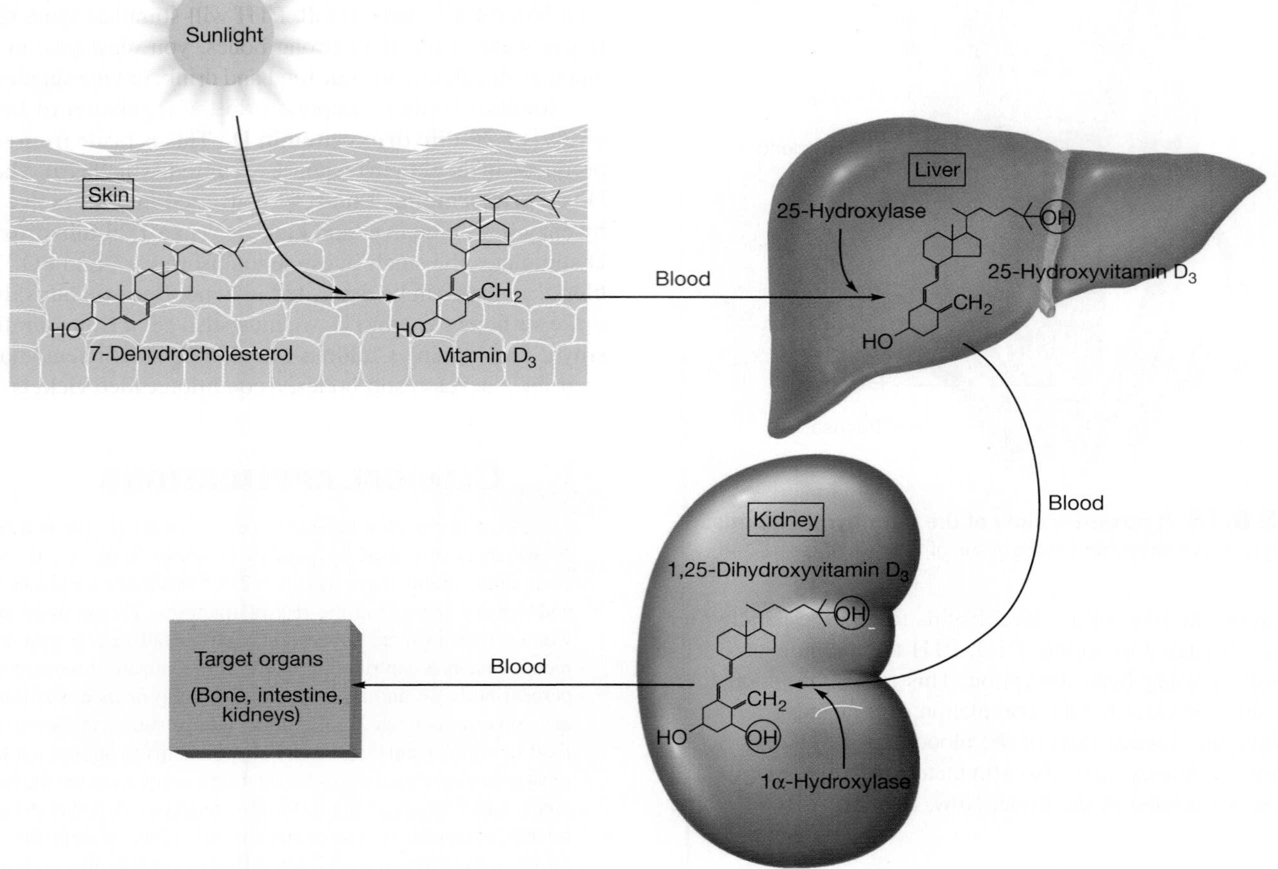

FIGURE 8.20 The production of 1,25-dihydroxyvitamin D_3. This hormone is produced in the kidneys from the inactive precursor 25-hydroxyvitamin D_3 (formed in the liver). The latter molecule is produced from vitamin D_3 secreted by the skin.

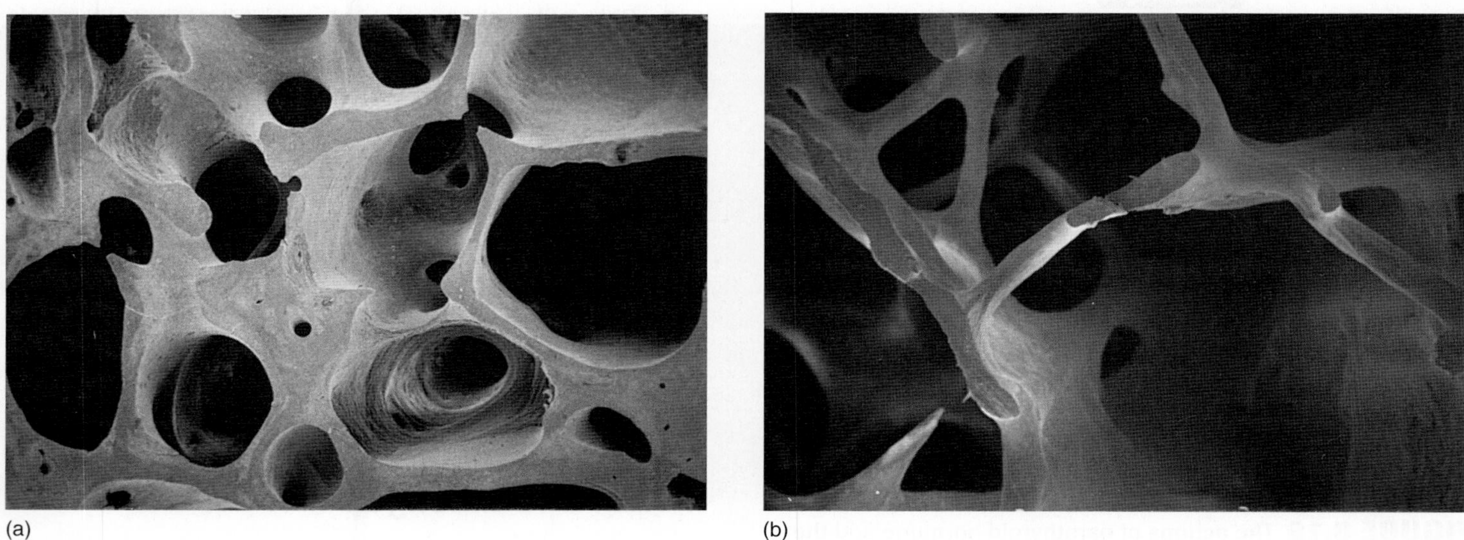

FIGURE 8.21 Scanning electron micrographs of bone. These biopsy specimens were taken from the iliac crest. Compare bone thickness in (*a*) a normal specimen and (*b*) a specimen from a person with osteoporosis. From L. G. Raisz, S. W. Dempster, et al., "Mechanisms of Disease" in *New England Journal of Medicine.* Vol. 218 (13):818,

tissue cells, particularly of skeletal muscles, liver, and adipose tissue. This promotes the synthesis of glycogen and fat, which are energy storage molecules. Diabetes mellitus is caused by inadequate secretion and/or action of insulin.

The **pancreatic islets (islets of Langerhans)** are islands of endocrine units within a sea of exocrine gland tissue (acini) in the pancreas (fig. 8.22). The two principal endocrine cells within the islets are the **alpha cells**, which secrete the hormone **glucagon**, and the **beta cells**, which secrete the hormone **insulin**. Insulin and glucagon are two of the most important hormones that regulate metabolism.

Metabolism refers to all of the chemical changes within the cells of the body. Metabolism can be divided into anabolism and catabolism. **Anabolism** refers to the conversion of smaller molecules into larger ones by dehydration synthesis reactions (chapter 1). Anabolic reactions convert smaller molecules of glucose into glycogen, fatty acids and glycerol into triglycerides (fats), and amino acids into proteins. Glycogen is stored in liver and skeletal muscle cells; triglyceride is stored in adipose cells; and protein isn't stored, but there are large amounts of usable proteins in the skeletal muscles. **Catabolism** refers to the hydrolysis (breakdown; chapter 1) of these molecules into their subunits (fig. 8.23) and the use of these subunits in cell respiration (chapter 2).

Different hormones stimulate either anabolism or catabolism, and so tilt metabolic balance in one direction or the other (fig. 8.23). Glucagon is a catabolic hormone, promoting the breakdown of large energy-storage molecules into their smaller subunits. Insulin is the chief anabolic hormone, tilting metabolic balance toward anabolism and energy storage.

Insulin and Glucagon Have Opposite Effects on Blood Glucose

Insulin and glucagon are antagonistic hormones; when the secretion of one is increased, the secretion of the other is reduced. After you eat a meal containing carbohydrates, your blood glucose starts to rise. This rising blood glucose concentration stimulates insulin secretion from the beta cells of the islets and inhibits the secretion of glucagon from the alpha cells. The increased insulin then promotes the movement of glucose out of the blood plasma and into insulin's target cells, chiefly skeletal muscles, liver, and adipose tissue. As a result, the blood glucose concentration falls, helping to correct the initial rise in blood glucose and maintain homeostasis (fig. 8.24).

Notice that insulin promotes the conversion of glucose into (a) glycogen in the liver and skeletal muscles, and (b) triglycerides in the adipose tissue. Thus, after a meal, glucose and other energy molecules derived from food in the intestine are stored as larger molecules inside cells for later use. This is done for the same reason we store money in a bank account—so it will be there when we need it.

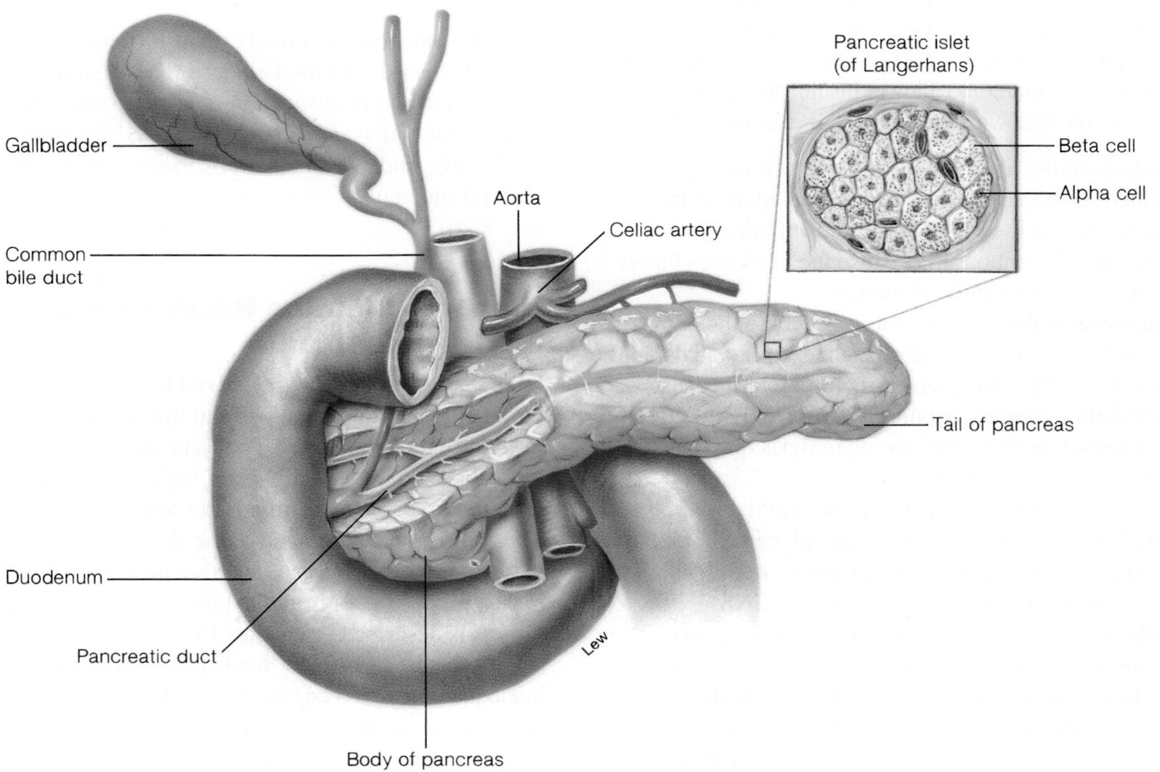

FIGURE 8.22 The pancreas and associated pancreatic islets (islets of Langerhans). Alpha cells secrete glucagon and beta cells secrete insulin. The pancreas is also exocrine, producing pancreatic juice for transport via the pancreatic duct to the duodenum of the small intestine.

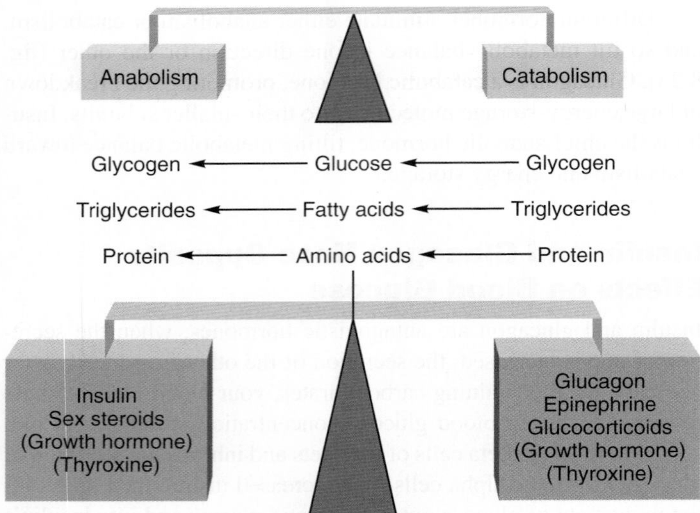

FIGURE 8.23 The regulation of metabolic balance. The balance of metabolism can be tilted toward anabolism (synthesis of energy reserves) or catabolism (utilization of energy reserves) by the combined actions of various hormones. Growth hormone and thyroxine (in parantheses) have both anabolic and catabolic effects. Growth hormone stimulates the synthesis of proteins (anabolism), but the breakdown of glycogen and fat (catabolism). Thyroxine stimulates protein synthesis (anabolism), but also cell respiration (catabolism).

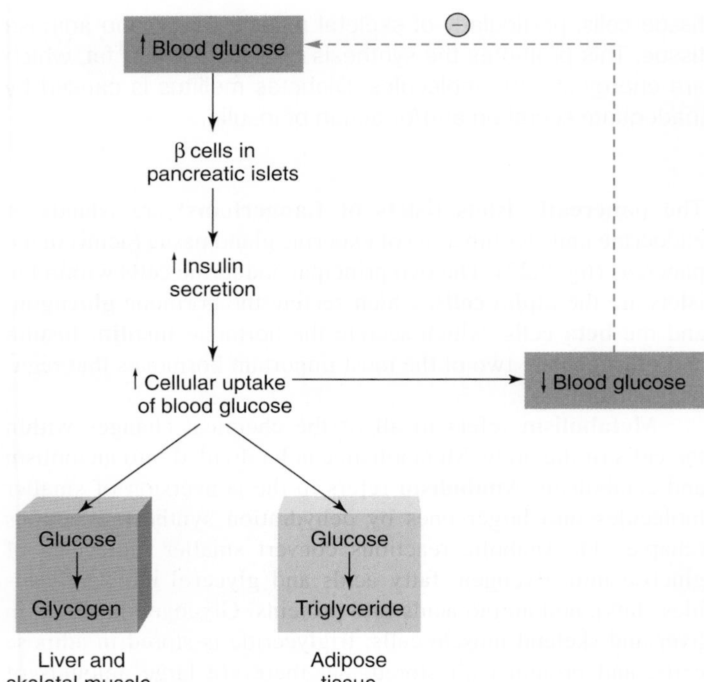

FIGURE 8.24 Homeostasis of blood glucose. A rise in blood glucose concentration stimulates insulin secretion. Insulin promotes a fall in blood glucose by stimulating the cellular uptake of glucose and the conversion of glucose to glycogen and fat.

Between meals, when the blood glucose concentration starts to fall, we need to withdraw money (energy molecules) from our bank account (deposits of glycogen and fat). The falling blood glucose causes the alpha cells to secrete more glucagon, and the beta cells to secrete less insulin. The conditions of lower insulin and higher glucagon levels stimulates the following catabolic processes (fig. 8.25):

1. **Glycogenolysis** in the liver. This is the hydrolysis of glycogen to glucose, so that the liver can secrete glucose into the blood and prevent the blood glucose from falling too low. This is critically important because the CNS uses blood glucose as its exclusive source of energy.
2. **Gluconeogenesis** in the liver. This term refers to the conversion of noncarbohydrate molecules, such as amino acids, into glucose (fig. 8.25). Gluconeogenesis is important because we can store only a limited amount of glycogen in the liver. Gluconeogenesis, together with glycogenolysis, prevents the blood glucose from falling too low.
3. **Lipolysis** in adipose tissue. This term refers to the hydrolysis of stored triglycerides into fatty acids and glycerol. Free fatty acids and glycerol provide alternate energy sources (other than glucose) for use by organs other than the CNS.
4. **Ketogenesis**. This is the formation of ketone bodies from fatty acids, by way of acetyl-Coenzyme A (fig. 8.25; see chapter 2). Ketone bodies are 4-carbon-long acids derived from fatty acids and used for energy by many organs. Since a number of ketone bodies are derived from one fatty acid, ketogenesis increases the number of acidic molecules in the blood. This isn't normally a problem, but it can become one in diabetes (discussed next).

The antagonistic effects of insulin and glucagon help maintain homeostasis of blood glucose. These antagonistic mechanisms operate during (1) *absorption* of a meal (after the food is digested into its subunits and absorbed across the intestinal lining into the blood); and (2) the *postabsorptive*, or *fasting*, period, when no food molecules are entering the blood from the digestive tract (fig. 8.26).

Diabetes Mellitus Results from Inadequate Insulin

Diabetes mellitus is characterized by fasting *hyperglycemia* (high blood glucose concentration) and often by the presence of glucose in the urine (*glycosuria*). The term *diabetes* comes from a Greek term for "siphon," relating to a high frequency of urination. The term *mellitus* comes from a Latin term for "honeyed" or "sweet." The urine is sweet because of the abnormal presence of glucose in the urine, caused by an excessive degree of hyperglycemia. The reason glucose goes from the blood into the urine under these conditions is described in chapter 13.

Diabetes is due to inadequate secretion of insulin and/or inadequate tissue responsiveness (sensitivity) to insulin. As a result, a rise in the blood glucose concentration can't be adequately lowered and corrected by the action of insulin. This produces hyperglycemia, which results in glycosuria when the blood glucose concentration reaches a particular level (chapter 13). Diabetes can thus be suspected in the presence of hyperglycemia or glycosuria.

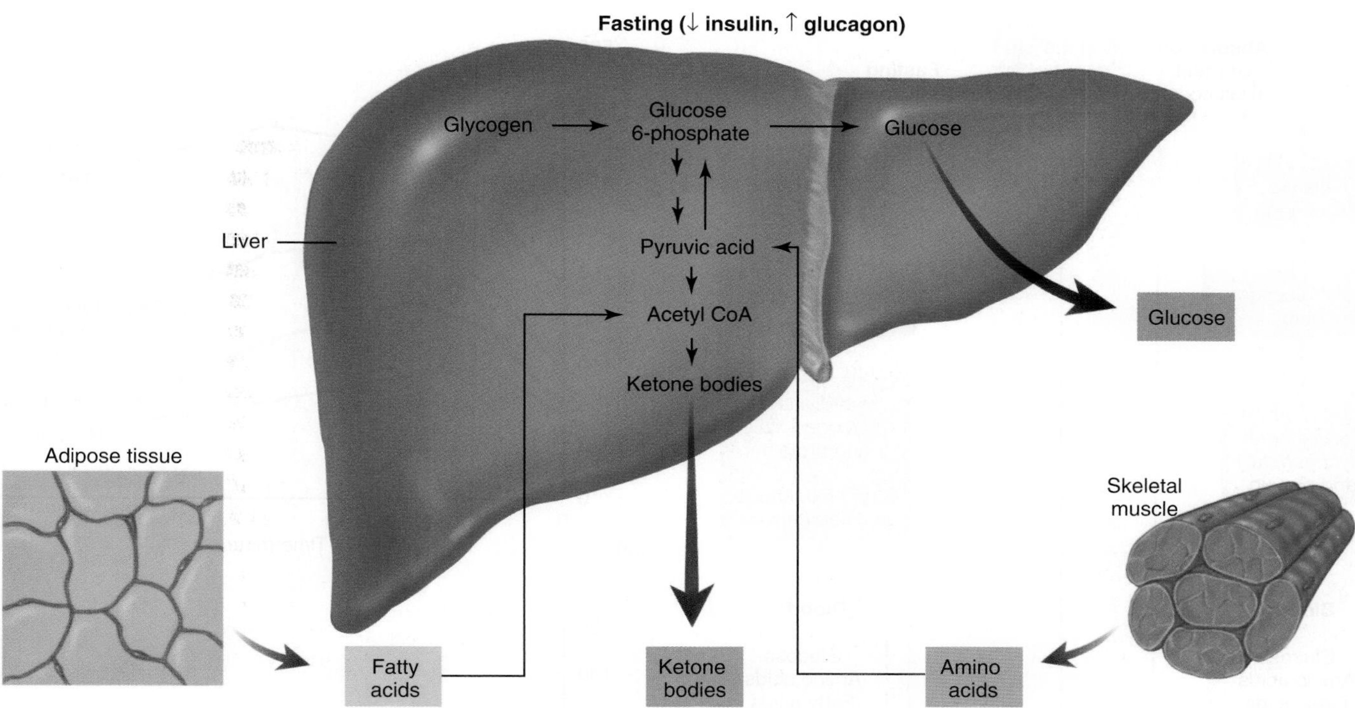

FIGURE 8.25 Catabolism during fasting. Increased glucagon secretion and decreased insulin secretion during fasting favors catabolism. These hormonal changes promote the release of glucose, fatty acids, ketone bodies, and amino acids into the blood. Notice that the liver secretes glucose that is derived both from the breakdown of liver glycogen and from the conversion of amino acids in gluconeogenesis.

However, the best laboratory test for diabetes mellitus is the **oral glucose tolerance test**. In this procedure, a person drinks a solution of glucose and the blood glucose levels are tested periodically over a few hours (fig. 8.27).

There are two major forms of diabetes mellitus. About 5% of diabetics have **type 1 diabetes mellitus**. In type 1 diabetes mellitus, the beta cells are destroyed by the person's own immune system—type 1 diabetes is an autoimmune disease. With little or no insulin, the secretion of glucagon is greatly elevated and contributes to the symptoms. People with type 1 diabetes must take insulin injections to prevent and treat the symptoms described in the next "Clinical Applications" box.

Clinical Applications

In **type 1 diabetes mellitus**, the absence of insulin and the elevated glucagon cause very high blood glucose levels, which "spill" into the urine; the glycosuria then causes frequent urination, promoting dehydration. There is also a rapid breakdown of fat, and as a result, a greatly elevated production of ketone bodies. The large amount of ketone bodies in the blood can cause the blood pH to fall, a condition called *ketoacidosis*. The ketone bodies also "spill" into the urine and promote frequent urination, contributing to dehydration. The combination of acidosis and dehydration can produce electrolyte (ion) imbalances. These very dangerous consequences of untreated type 1 diabetes are accompanied by a "fruity breath," because some of the ketone bodies become acetone (the solvent in nail polish remover), a volatile (gaseous) molecule in the breath.

About 95% of diabetics have **type 2 diabetes**. In people with type 2 diabetes mellitus, there is a reduced tissue responsiveness (sensitivity) to insulin; thus, more insulin is required to maintain homeostasis of the blood glucose. This results in abnormal patterns of insulin secretion (fig. 8.27); eventually the islets may not be able to produce sufficient insulin. Depending on severity, type 2 diabetes can be treated with diet and exercise, and with oral drugs that improve the sensitivity of the target organs to insulin. Type 2 diabetes is described in more detail in the "Physiology in Health and Disease" section at the end of this chapter.

Clinical Investigation Clues

Remember, Tony's father had diabetes, and Tony was cautioned by his physician that he might also get diabetes if he doesn't lose weight and exercise.

- The physician believes that Tony is at risk for which type of diabetes? (See the last section of this chapter, "Physiology in Health and Disease.")
- Given that heredity and obesity contribute to risk, how might Tony reduce his risk of getting diabetes?
- What oral drugs are used to treat this type of diabetes? (See the last section of this chapter, "Physiology in Health and Disease.")

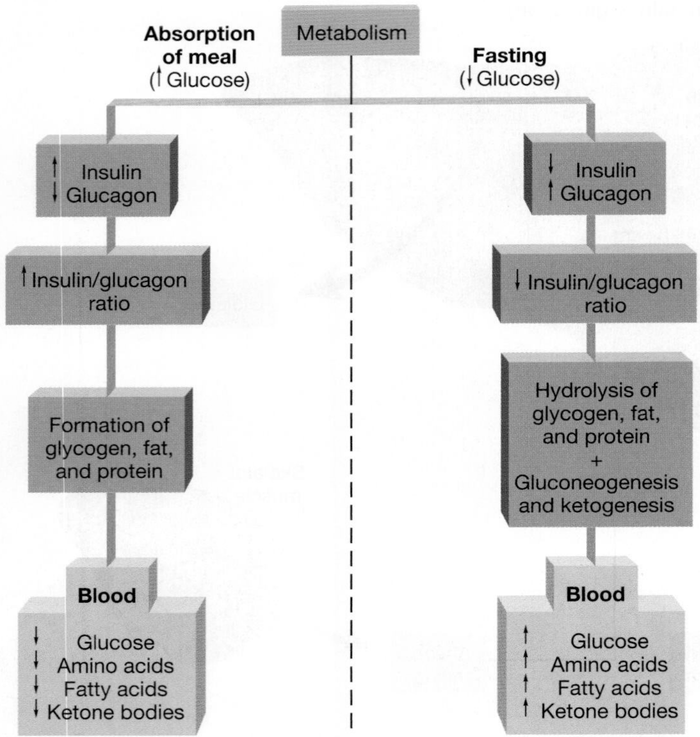

FIGURE 8.26 The effect of feeding and fasting on metabolism. Metabolic balance is tilted toward anabolism by feeding (absorption of a meal) and toward catabolism by fasting. This occurs because of an inverse relationship between insulin and glucagon secretion. Insulin secretion rises and glucagon secretion falls during food absorption, whereas the opposite occurs during fasting.

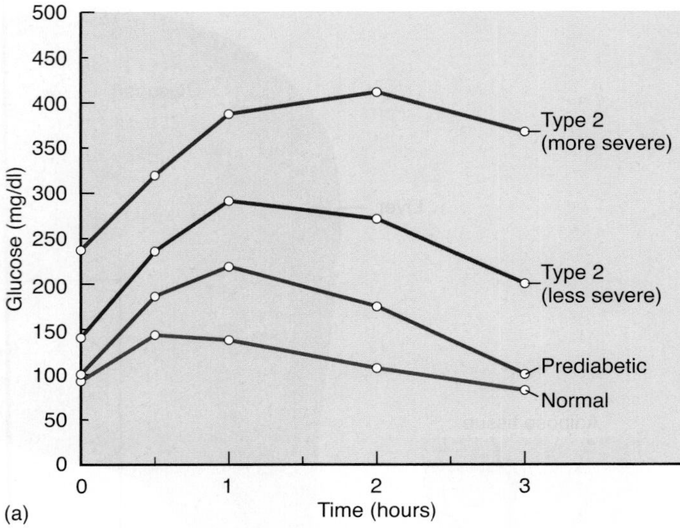

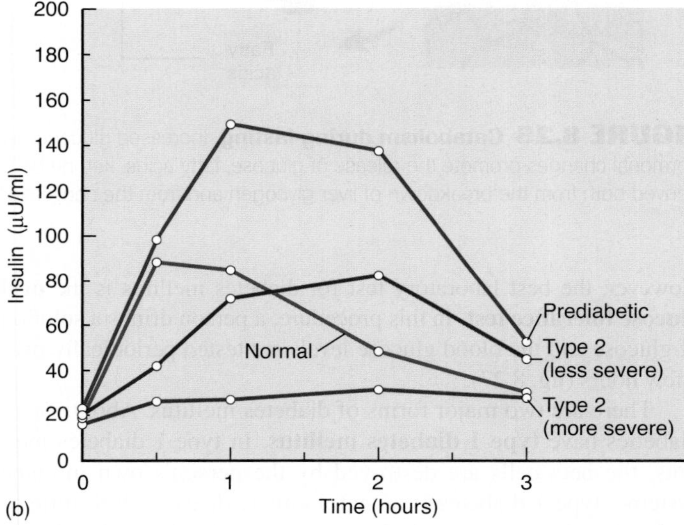

FIGURE 8.27 Oral glucose tolerance in prediabetes and type 2 diabetes. The oral glucose tolerance test showing (*a*) blood glucose concentrations and (*b*) insulin values following the ingestion of a glucose solution. Values are shown for people who are normal, prediabetic, and type 2 diabetic. Prediabetics (those who demonstrate "insulin resistance") often show impaired glucose tolerance without fasting hyperglycemia.

CHECK POINT

1. Compare anabolism and catabolism, and list hormones that promote each.
2. Describe how insulin and glucagon secretion change during the absorption of a carbohydrate meal, and how these hormones contribute to homeostasis.
3. Describe how insulin and glucagon secretion change during the postabsorptive (fasting) period. What metabolic changes are produced by glucagon during this time?
4. What are the characteristics of diabetes mellitus, and what are the primary defects in type 1 and type 2 diabetes?

The Pineal Gland and Other Endocrine Glands Regulate Body Function

The pineal gland works with the suprachiasmatic nucleus of the hypothalamus to regulate the circadian rhythms of the body. Many other endocrine glands help regulate the functions of all body systems.

The pineal gland is an organ that is exclusively an endocrine gland, but there are many organs in the body that perform different functions along with the secretion of hormones. The gonads (testes and ovaries) produce the sex cells (sperm and ova), for example, and the placenta secretes hormones while it performs functions similar to those of the lungs and liver. The stomach and intestines are digestive organs, but they also secrete hormones that regulate digestive functions and even the pancreatic islets.

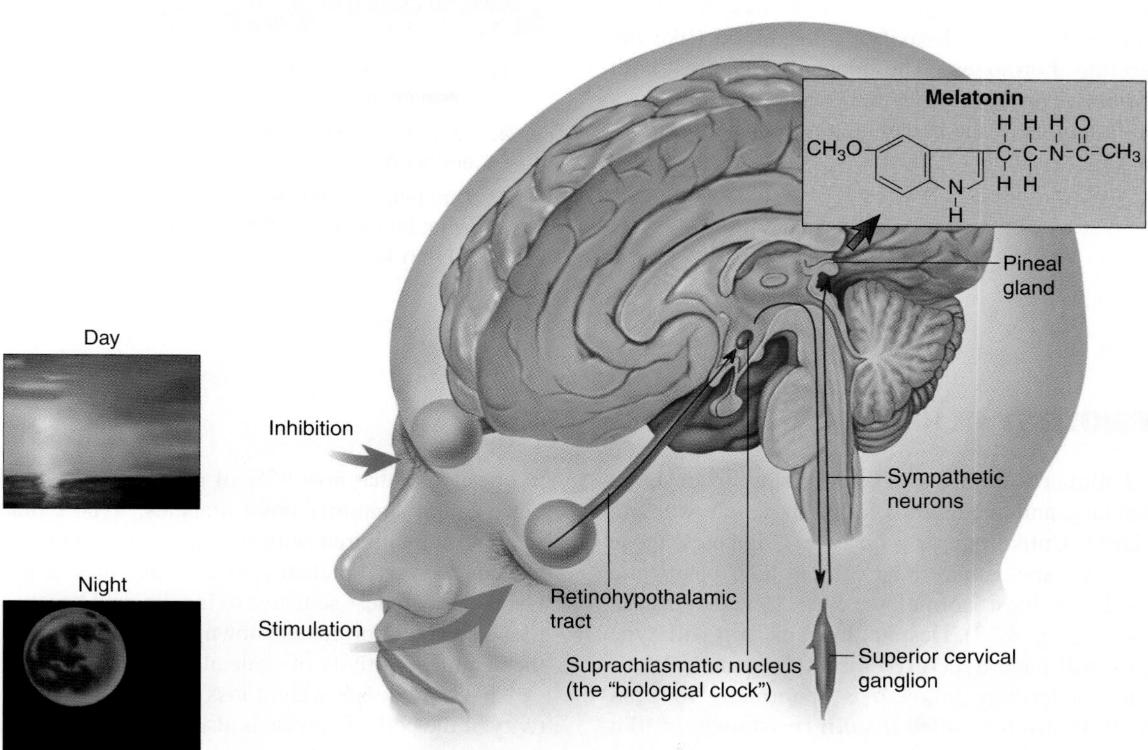

FIGURE 8.28 The secretion of melatonin. The secretion of melatonin by the pineal gland is stimulated by sympathetic axons originating in the superior cervical ganglion. Activity of these neurons is regulated by the cyclic activity of the suprachiasmatic nucleus of the hypothalamus, which sets a circadian rhythm. This rhythm is entrained to light/dark cycles by neurons in the retina.

The Pineal Gland Helps Regulate Circadian Rhythms

The **pineal gland** is located on the roof of the third ventricle, the brain cavity in the diencephalon (chapter 5). Its principal hormone is **melatonin** (fig. 8.28). Secretion of melatonin is stimulated by the **suprachiasmatic nucleus (SCN)** of the hypothalamus (chapter 5), the center identified as the "master clock" for the body's *circadian rhythms* (rhythms that follow a 24-hour pattern). The daily rhythm of the SCN is entrained (synchronized) to the light/dark cycle, increasing during darkness and reaching a peak by the middle of the night. During the day, neural pathways from the retinas of the eyes to the hypothalamus depress the activity of the SCN. Melatonin secretion from the pineal gland thereby increases at night and decreases during the day (fig. 8.28).

Some of the proposed physiological effects of melatonin in humans are controversial, including a proposed role in regulating the onset of puberty. It is known that melatonin pills decrease the time required to fall asleep and increase the duration of rapid eye movement (REM) sleep, so they can be useful in helping people to fall asleep. Exogenous melatonin makes most people sleepy 30 to 120 minutes after taking the pills, although this is highly variable. Melatonin pills may thus be beneficial for treating insomnia, especially in elderly people who have the lowest endogenous nighttime levels of melatonin secretion.

The Gastrointestinal Tract, Gonads, and Placenta Are Endocrine Glands

The **stomach** and **small intestine** secrete a number of hormones that act on the digestive tract itself and on the liver, gallbladder, and pancreas, which are accessory digestive organs (chapter 14). Indeed, the first hormone ever discovered (given the enviable name *secretin*) is secreted by the small intestine. Although most of these hormones regulate digestion, some may have other effects. For example, there are hormones secreted by the small intestine that stimulate the beta cells of the pancreatic islets to secrete insulin. Thus, insulin secretion can begin to rise as soon as food reaches the small intestine, in anticipation of a rise in the blood glucose concentration that hasn't yet occurred.

The gonads (**testes** and **ovaries**) secrete sex steroids: the testes secrete *testosterone*, the ovaries secrete *estradiol* and *progesterone*. Secretion of these hormones is stimulated by the gonadotropic hormones (FSH and LH) from the anterior pituitary gland, as was described earlier in this chapter and will be described in more detail in chapter 15. The sex steroids have numerous effects on the reproductive system and other body organs.

The **placenta** is present in a woman only when she is pregnant, and so secretes some hormones that are unique for pregnancy. Indeed, one of the placental hormones (*human chorionic gonadotropin, hCG*) is used to test for pregnancy. The placenta is a remarkable organ, providing gas exchange (like the lungs) between the fetus and mother, detoxifying the blood (like the liver), and secreting hormones. The placenta secretes both pituitary-like hormones (such as hCG), and ovary-like hormones (estrogens). The functions of the placenta are described in more detail in chapter 15.

CHECK POINT

1. What hormone does the pineal gland secrete, and how is this secretion regulated?
2. Why can the stomach and small intestine be considered endocrine glands? Give examples.
3. The gonads and placenta are endocrine glands and more. What hormones do they secrete, and what other functions do they perform?

PHYSIOLOGY IN HEALTH AND DISEASE

Type 2 diabetes mellitus is usually slow to develop, is hereditary, and occurs most often in people who are overweight. Unlike people with type 1 diabetes, those with type 2 diabetes have beta cells in their pancreatic islets and can have normal or even elevated insulin secretion (see fig. 8.27). Despite this, a person with type 2 diabetes will have hyperglycemia if untreated. That's because the underlying problem is abnormally low tissue sensitivity to insulin, called **insulin resistance**. Insulin resistance reduces the ability of insulin to stimulate the skeletal muscles and adipose tissue to take glucose out of the blood. Also, the liver is stimulated by glucagon to secrete glucose (derived from stored glycogen or amino acids) into the blood, because of a reduced ability of insulin to block this response. People who are **prediabetic** have **impaired glucose tolerance**, which is defined as a plasma glucose level of 140 to 200 mg/dl at 2 hours following the ingestion of glucose in an oral glucose tolerance test.

Although the tendency toward insulin resistance is inherited, the expression of this genetic tendency is increased by obesity. This is particularly true if the person has an "apple shape" obesity, where there is a high amount of *visceral fat* (fat in the greater omentum, a membrane in the abdominal cavity); the smaller adipocytes of subcutaneous fat (producing a "pear shape") contribute less to diabetes. The incidence of type 2 diabetes has tripled in the past 30 years because of the increase in obesity, so that now 95% of diabetics have type 2 diabetes. Once called "maturity-onset diabetes," type 2 diabetes is now seen among children with increasing frequency. Obesity contributes to type 2 diabetes because the large adipocytes of the visceral fat are less sensitive to insulin, and because the adipose tissue secretes hormones known as **adipokines**, which regulate the insulin sensitivity of skeletal muscles.

Even moderate weight loss can increase the insulin sensitivity of the body. Exercise is also beneficial, because it not only contributes to weight loss, it also independently makes the skeletal muscles more sensitive to insulin. If these lifestyle changes are insufficient, there are oral drugs available—*metformin, rosiglitazone,* and *pioglitazone*—that reduce the insulin resistance of body tissues.

The insulin resistance of type 2 diabetes may be accompanied by elevated blood triglycerides, a lowered HDL cholesterol (the "good cholesterol"), and hypertension (high blood pressure). A person who has two of these conditions combined with *central obesity* (defined by a waist circumference greater than a specific value, which differs by sex and ethnicity) has a condition known as the **metabolic syndrome**. The nature and dangers of the metabolic syndrome have been recognized only relatively recently, as the incidence of this condition and type 2 diabetes have increased in the population. One study estimated that people with the metabolic syndrome have their risk of coronary heart disease and stroke increased by a factor of three.

Physiology in Balance

Endocrine System

Working together with . . .

Nervous System

Hormones act on the hypothalamus to regulate the secretion of GnRH . . . p. 113

Urinary System

ADH stimulates the kidneys to retain more water, excreting less urine . . . p. 324

Cardiovascular System

Aldosterone stimulates the kidneys to retain salt and water, raising the blood volume and thus the blood pressure . . . p. 240

Immune System

Glucocorticoids (such as cortisol) suppress the immune system and inflammation . . . p. 279

Digestive System

Gastrin from the stomach stimulates gastric secretion; secretin and cholecystokinin from the small intestine stimulate the pancreas, liver, and gallbladder . . . p. 348

Reproductive System

Estradiol from the ovaries stimulates growth of the endometrium of the uterus . . . p. 373

Homeostasis Revisited

Process	Set Point	Integrating Center	Sensors	Effectors
How secretion of ADH and water retention by the kidneys are affected by an increased concentration (osmolarity) of the plasma during dehydration	Plasma osmolarity (concentration)	Hypothalamus of brain	Hypothalamus of brain	Posterior pituitary secretes increased amount of ADH, which stimulates the kidneys to reabsorb (retain) water.
How the secretion of LH in men is affected by an increase in the blood concentration of testosterone	Plasma testosterone concentration	Hypothalamus of brain	Hypothalamus of brain	Hypothalamus secretes less GnRH, causing the anterior pituitary to secrete less LH, which causes the testes to secrete less testosterone.
How the secretion of ACTH is affected by an increase in the blood concentration of cortisol	Plasma cortisol concentration	Hypothalamus of brain	Hypothalamus of brain	Hypothalamus secretes less CRH, causing the anterior pituitary to secrete less ACTH, which causes the adrenal cortex to secrete less cortisol.
How the secretion of TSH is affected by a decrease in the blood concentration of thyroxine	Plasma thyroxine concentration	Hypothalamus of brain	Hypothalamus of brain	Hypothalamus secretes more TRH, causing the anterior pituitary to secrete more TSH, which causes the thyroid to secrete more thyroxine.
How the secretion of parathyroid hormone maintains plasma Ca^{2+} homeostasis when the plasma Ca^{2+} concentration decreases	Plasma Ca^{2+} concentration	Parathyroid glands	Parathyroid glands	Parathyroid hormone stimulates osteoclasts to resorb bone, so that Ca^{2+} from hydroxyapatite crystals enters the plasma.
How the secretion of insulin maintains plasma glucose homeostasis when a person eats a carbohydrate meal	Plasma glucose concentration	Pancreatic islets (of Langerhans)	Pancreatic islets (of Langerhans)	Beta cells in islets secrete more insulin, which stimulates the uptake of plasma glucose into skeletal muscle and other tissues, lowering the plasma glucose concentration.
How the secretion of glucagon maintains plasma glucose homeostasis when a person eats a carbohydrate meal	Plasma glucose concentration	Pancreatic islets (of Langerhans)	Pancreatic islets (of Langerhans)	Alpha cells secrete less glucagon, causing less hydrolysis of liver glycogen so that the liver secretes less glucose into the blood.

Summary

Endocrine Glands Secrete Chemically Different Hormones 187

- Steroid and thyroid hormones are nonpolar; catecholamine, polypeptide, protein, and glycoprotein hormones are polar.
- Polar hormones are soluble in water and in the blood plasma; they cannot pass through the plasma membranes of their target cells, and thus require a second messenger to exert their effects.
- Nonpolar hormones can pass through the plasma membranes and enter their target cells; these are the only hormones that can be taken orally (as pills).

The Pituitary Gland and Hypothalamus Work Together 189

- The posterior pituitary secretes two hormones—ADH (antidiuretic hormone) and oxytocin—that are produced by neurons in the hypothalamus and transported along axons into the posterior pituitary.
- The anterior pituitary produces and secretes GH (growth hormone), prolactin, TSH (thyroid-stimulating hormone), ACTH, (adreno-corticotropic hormone); and the two gonadotropic hormones: FSH (follicle-stimulating hormone) and LH (luteinizing hormone).
- The hypothalamus controls the anterior pituitary by means of releasing hormones (CRH, TRH, GnRH, and GHRH) secreted into the hypothalamo-hypophyseal portal system; there are also inhibiting hormones, PIH and somatostatin, for the control of prolactin and growth hormone, respectively.
- Both the secretion of the releasing hormones from the hypothalamus and the secretion of certain anterior pituitary hormones (TSH, ACTH, and the gonadotropins FSH and LH) are regulated by feedback from hormones secreted by the target endocrine glands (thyroid, adrenal cortex, and gonads, respectively).
- A rise in thyroxine inhibits TSH secretion, a rise in cortisol inhibits ACTH secretion, and a rise in sex steroid hormone secretion inhibits the secretion of the gonadotropins (FSH and LH).

The Adrenal Medulla and Adrenal Cortex Are Different Glands 194

- The adrenal medulla secretes epinephrine and lesser amounts of norepinephrine into the blood when stimulated by sympathetic nerves during the fight-or-flight reaction.
- The catecholamines (epinephrine and norepinephrine) bind to receptor proteins in the plasma membrane of their target cells and, through G-proteins, activate adenylate cyclase, which produces cyclic AMP (cAMP) inside the target cells.
- cAMP then activates protein kinase, a previously inactive enzyme, which phosphorylates other enzymes, activating or inactivating them and thereby producing the responses of the target cells to the hormone epinephrine.
- The adrenal cortex secretes different types of corticosteroid hormones: glucocorticoids (principally cortisol), mineralocorticoids (principally aldosterone), and weak androgens (principally androstenedione).
- Steroid hormones travel in the blood bound to carrier proteins; they enter their target cells and bind to nuclear receptor proteins, which then bind to DNA and activate specific genes.
- The activated genes produce messenger RNA (mRNA), which codes for the production of new proteins; these newly synthesized proteins then evoke the responses of the target cells to the steroid hormones.
- ACTH from the anterior pituitary stimulates the secretion of the glucocorticoids (cortisol); this pituitary-adrenal axis is activated by stress in the general adaptation syndrome (GAS).
- A rise in cortisol secretion during the GAS is beneficial in some cases because it raises the blood levels of glucose and amino acids, but detrimental in others because high cortisol levels can suppress the immune system.
- The mineralocorticoids (chiefly aldosterone) stimulate the kidneys to retain Na^+ and water, while excreting more K^+; aldosterone secretion is stimulated by a rise in blood K^+, helping to maintain homeostasis of the blood K^+ level; aldosterone secretion is also stimulated by angiotensin II, produced in the blood when there is a fall in blood volume and pressure.

The Thyroid and Parathyroid Glands Have Different Functions 198

- The thyroid is composed of follicles containing a fluid known as colloid; the follicles trap iodine and bind it to thyroglobulin protein in the colloid; the thyroglobulin then serves as the precursor for the thyroid hormones.
- Thyroxine travels in the blood bound to a carrier protein, TBG; when it reaches its target cells, thyroxine dissociates from the TBG and enters the cells.
- Once inside its target cells, thyroxine is converted into triiodothyronine (T_3), which binds to nuclear receptors and activates genes, similar to the mechanism of action of steroid hormones.
- Negative feedback inhibition by thyroxine on TSH secretion maintains homeostasis of thyroxine secretion.
- If there is insufficient dietary iodine, the blood thyroxine levels fall and TSH secretion rises (because of less negative feedback inhibition); high TSH stimulates the abnormal growth of the thyroid and the formation of an endemic goiter.
- Inadequate thyroxine secretion (hypothyroidism) causes cretinism if it occurs in very young children.
- Hypothyroidism causes a low BMR (basal metabolic rate) and other symptoms; hyperthyroidism causes a high BMR and other symptoms.
- Calcium (Ca^{2+}) and phosphate (PO_4^{3-}) ions are deposited as hydroxyapatite crystals in bone by osteoblasts; osteoclasts resorb bone and release these ions into the blood.
- Parathyroid hormone (PTH) stimulates osteoclasts to resorb bone, helping to raise blood calcium levels if they start to fall.
- 1,25-dihydroxyvitamin D_3 is formed from vitamin D produced in the skin or taken in the diet; this hormone stimulates the small intestine to absorb Ca^{2+} and PO_4^{3-} into the blood.

The Pancreatic Islets Secrete Insulin and Glucagon 201

- The pancreatic islets (Islets of Langerhans) contain alpha cells that secrete glucagon and beta cells that secrete insulin, which help regulate metabolism.
- Metabolism consists of anabolic reactions that build glycogen, fat, and protein, which are large energy-storage molecules; and catabolic reactions, that break down the large energy-storage molecules into their subunits to be used for cell respiration.
- Insulin promotes anabolism by stimulating the movement of blood glucose into the cells of the liver and skeletal muscles, where it is converted into glycogen; insulin also stimulates the movement of glucose into adipose tissue, where it is converted into triglycerides.
- Insulin secretion rises and glucagon secretion falls after eating a meal, promoting anabolism; insulin secretion falls and glucagon secretion rises during the post-absorptive, or fasting, state, promoting catabolism.
- During fasting, glucagon stimulates glycogenolysis, gluconeogenesis, lipolysis, and ketogenesis; this results in a rise in blood glucose, fatty acids. glycerol, and ketone bodies for the body cells to use in cell respiration.
- Diabetes occurs when there is fasting hyperglycemia (often leading to glycosuria) as a result of inadequate insulin; diabetes can be detected by an oral glucose tolerance test.
- In type 1 diabetes, the beta cells are destroyed and there is little or no insulin; in type 2 diabetes (the most common form), there are beta cells secreting insulin, but the tissues of the body have a reduced sensitivity to insulin.

The Pineal Gland and Other Endocrine Glands Regulate Body Function 206

- The pineal gland is stimulated by the suprachiasmatic nucleus of the hypothalamus, the "master clock" for circadian rhythms (those body rhythms that repeat in a 24-hour pattern).
- The suprachiasmatic nucleus and pineal gland are suppressed by daylight, so that the pineal is stimulated to secrete its hormone, melatonin, at night; this hormone appears to promote sleep.
- The digestive tract secretes a variety of hormones that regulate the digestive system; some hormones of the small intestine also stimulate the beta cells of the pancreatic islets to secrete insulin before there is a rise in blood glucose.
- The gonads (ovaries and testes) produce gametes (eggs and sperm) and also secrete sex steroid hormones—estradiol and progesterone from the ovaries, testosterone from the testes—that help regulate the reproductive system, among other effects.
- The placenta secretes both pituitary-like hormones (such as hCG, which is used to test for pregnancy) and steroid hormones, including estrogens; it also performs many other functions needed for the health of the fetus.

Review Activities

Objective Questions: Test Your Knowledge

1. Which of the following statements about nonpolar hormones is true?
   a. Steroid hormones are nonpolar.
   b. Polypeptide hormones are nonpolar.
   c. Nonpolar hormones cannot be taken orally and must be injected.
   d. Nonpolar hormones cannot enter their target cells.
2. Which of the following organs secretes hormones?
   a. Kidneys
   b. Pancreas
   c. Brain
   d. All of these secrete hormones.
3. Which of the following statements about steroid hormones is false?
   a. Steroid hormones are originally derived from cholesterol.
   b. Steroid hormones are secreted by the gonads.
   c. Steroid hormones are secreted by the adrenal medulla.
   d. Estradiol is derived from testosterone.
4. Which of the following statements about thyroid hormones is false?
   a. They are nonpolar hormones.
   b. They are derived from progesterone.
   c. They contain iodine.
   d. They can be taken orally, as pills.
5. The posterior pituitary gland
   a. is derived from embryonic oral epithelium.
   b. secretes growth hormone and prolactin.
   c. secretes its hormones in response to hypothalamic releasing hormones.
   d. does not produce the hormones it secretes.
6. Which of the following statements about the hypothalamo-hypophyseal tract is true?
   a. It goes from the hypothalamus to the posterior pituitary.
   b. It goes from the hypothalamus to the anterior pituitary.
   c. It carries releasing hormones from the hypothalamus.
   d. It carries anterior pituitary hormones.
7. Which statement regarding the gonadotropic hormones is false?
   a. They include FSH and LH.
   b. Their secretion is stimulated by sex steroid hormones.
   c. Their secretion is stimulated by GnRH.
   d. They are secreted by the anterior pituitary.
8. Which of the following pituitary hormones stimulates the liver to secrete somatomedins?
   a. Prolactin
   b. Oxytocin
   c. ACTH
   d. Growth hormone
9. Hormone (1) stimulates milk production by the mammary gland; hormone (2) stimulates the milk ejection reflex:
   a. Prolactin; oxytocin
   b. Oxytocin; prolactin
   c. Luteinizing hormone; follicle-stimulating hormone
   d. Follicle-stimulating hormone; luteinizing hormone
10. If a pet dog is "fixed" (castrated; has its gonads removed), which of the following statements would be true?
    a. Secretion of FSH and LH will become very low.
    b. Secretion of FSH and LH will become very high.
    c. Secretion of FSH will rise and LH will fall.
    d. Secretion of LH will rise and FSH will fall.
11. The hypothalamo-hypophyseal portal system carries
    a. blood from the posterior pituitary to the hypothalamus.
    b. blood from the anterior pituitary to the hypothalamus.
    c. hormones secreted by the anterior pituitary.
    d. hormones secreted by the hypothalamus.
12. Which of the following statements about TRH is true?
    a. It's secreted by the hypothalamus.
    b. It's carried to the anterior pituitary in the hypothalamo-hypophyseal tract.
    c. It stimulates the anterior pituitary to secrete ACTH.
    d. It's inhibited by a rising secretion of estradiol or testosterone.
13. Which of the following statements about cyclic AMP (cAMP) is false?
    a. It's produced by the action of adenylate cyclase.
    b. Its production is stimulated by epinephrine acting on its target cells.
    c. It activates a previously inactive enzyme, protein kinase.
    d. It's activated by the enzyme phosphodiesterase.
14. Which of the following hormones is a mineralocorticoid?
    a. Cortisol
    b. Androstenedione
    c. Aldosterone
    d. Hydrocortisone
15. Steroid hormones
    a. bind to receptor proteins in the plasma membrane.
    b. stimulate the production of cAMP in the target cells.
    c. travel in the blood dissolved in the plasma.
    d. activate genes, leading to the synthesis of new proteins.
16. In the general adaptation syndrome, which of the following occurs?
    a. CRH secretion is increased.
    b. ACTH secretion is increased.
    c. Cortisol secretion is increased.
    d. All of these occur.

17. If a person takes hydrocortisone pills, which of the following will occur?
    a. CRH secretion will rise.
    b. ACTH secretion will fall.
    c. Cortisol secretion from the adrenal cortex will rise.
    d. All of these will occur.

18. Which of the following statements about aldosterone is true?
    a. It stimulates the excretion of more Na^+ and water in the urine.
    b. It exerts negative feedback inhibition on ACTH secretion.
    c. It stimulates the excretion of more K^+ in the urine.
    d. Its secretion is stimulated by ACTH.

19. Thyroxine travels in the blood
    a. bound to TBG.
    b. bound to thyroglobulin.
    c. bound to colloid.
    d. dissolved in the blood plasma.

20. Which of the following statements about thyroxine is true?
    a. It binds to receptor proteins in the plasma membrane.
    b. It must first be converted into T_3 to become active in the target cell.
    c. It enters the nucleus and binds to nuclear receptor proteins.
    d. It requires cAMP to regulate its target cells.

21. If a person took a high dose of thyroxine pills over an extended period, which of the following statements would be true?
    a. Blood thyroxine levels would be high.
    b. TSH secretion would be low.
    c. Thyroxine secretion from the thyroid would be reduced.
    d. All of these are true.

22. In a person with endemic goiter
    a. the thyroid gland is enlarged.
    b. TSH secretion is suppressed.
    c. thyroxine secretion is elevated.
    d. All of these are true.

23. Hypothyroidism can cause all of the following except
    a. a low BMR.
    b. cretinism.
    c. a rapid pulse.
    d. intolerance to cold.

24. Parathyroid hormone stimulates
    a. osteoblasts to deposit calcium phosphate crystals in bone.
    c. osteoclasts to resorb calcium and phosphate from bone.
    c. osteoblasts to resorb calcium and phosphate from bone.
    d. osteoclasts to deposit calcium and phosphate crystals in bone.

25. If a person doesn't get sufficient vitamin D
    a. the blood calcium levels will rise.
    b. the activity of osteoclasts will be inhibited.
    c. the intestine will not be able to absorb calcium and phosphate.
    d. the secretion of parathyroid hormone will fall.

26. The formation of glycogen from glucose is an example of which type of metabolism, and is stimulated by which hormone?
    a. anabolism, stimulated by insulin
    b. anabolism, stimulated by glucagon
    c. catabolism, stimulated by insulin
    d. catabolism, stimulated by glucagon

27. Which of the following stimulates the secretion of insulin?
    a. A fall in glucagon secretion
    b. A rise in blood levels of thyroxine
    c. A rise in blood levels of glucose
    d. A fall in blood glucose levels

28. Which of the following statements accurately describes what occurs during fasting?
    a. Insulin secretion falls and glucagon secretion falls.
    b. Insulin secretion rises and glucagon secretion rises.
    c. Insulin secretion rises and glucagon secretion falls.
    d. Insulin secretion falls and glucagon secretion rises.

29. Glycogenolysis, gluconeogenesis, lipolysis, and ketogenesis occur in response to which hormone during which state?
    a. Insulin, during the absorption of a meal.
    b. Glucagon, during the fasting state.
    c. Insulin, during the fasting state.
    d. Glucagon, during the absorption of a meal.

30. Which of the following statements about type 1 diabetes mellitus is false?
    a. If untreated, there is fasting hyperglycemia.
    b. If untreated, there is glycosuria.
    c. The beta cells are destroyed by the person's immune system.
    d. About 95% of diabetics have this type of diabetes.

Essay Questions 1: *Test Your Understanding*

1. The brain, heart, kidneys, and adipose tissue can be considered endocrine glands. Explain why this statement is true.
2. List the different chemical types of hormones, give examples of each, and indicate if the hormone type is polar or nonpolar.
3. Which hormones are soluble in water? How are these carried in the blood?
4. Which hormones aren't soluble in water? How are these carried in the blood?
5. Which hormones can be taken as pills, and which need to be injected? Explain why this is so.
6. Describe the different embryonic origins of the anterior and posterior pituitary glands, and explain how this relates to the way they are regulated by the hypothalamus.
7. Which hormones are secreted by the posterior pituitary gland? Describe where they are produced and how the secretion of these hormones is regulated.
8. List the hormones secreted by the anterior pituitary gland. Indicate which hormones are considered to be gonadotropic hormones.
9. Describe the hypothalamo-hypophyseal portal system, list the hormones secreted into it by the hypothalamus, and indicate which anterior pituitary hormones are regulated by each of the hypothalamic hormones.
10. Given that FSH and LH are required for ovulation, explain how the birth control pills work.
11. Describe the role of adenylate cyclase, protein kinase, and phosphodiesterase enzymes in the mechanism of action of those hormones that use cAMP as a second messenger.
12. List the different types of corticosteroids and give examples of each. Which type is secreted in response to ACTH stimulation? Which hormone suppresses ACTH secretion by negative feedback inhibition? Explain.
13. Provide a step-by-step description of how steroid hormones stimulate their target cells.

14. Describe the general adaptation syndrome. Is this beneficial or detrimental? Explain.
15. Describe the action of aldosterone and how its secretion is regulated.
16. Describe the location and significance of thyroglobulin and thyroxine-binding globulin (TBG).
17. Describe the structure of thyroxine and T_3. Describe the relative amounts of these secreted by the thyroid, and the relationship between the two within the target cells.
18. Step-by-step, describe how thyroxine stimulates its target cells.
19. A person who has a dietary deficiency in iodine can develop a goiter. Explain how this occurs.
20. What is the basal metabolic rate (BMR)? How is it affected by the function of the thyroid gland?
21. Compare the oxygen consumption and other symptoms of a hypothyroid person with one who is hyperthyroid.
22. Describe the action of osteoblasts and osteoclasts. How is the balance of their actions affected by parathyroid hormone?
23. What stimulates the secretion of parathyroid hormone, and how does this hormone influence the blood levels of Ca^{2+}?
24. Describe the formation and physiological significance of 1,25-dihydroxyvitamin D_3.
25. Describe anabolism and catabolism, and indicate hormones that promote each.
26. Describe what happens to the secretion of insulin and glucagon after a meal, and how these hormones influence metabolism at that time.
27. Describe what happens to insulin and glucagon secretions during fasting, and the resulting metabolic changes that occur at that time.
28. Explain why a person with diabetes can develop fasting hyperglycemia.
29. Describe the oral glucose tolerance test, and how the different causes of type 1 and type 2 diabetes mellitus affect this test.
30. Describe the location and regulation of the pineal gland, and discuss the possible functions of its hormone.

Essay Questions 2: *Test Your Analytical Ability*

1. If you wanted a somewhat polar hormone to be better absorbed across the skin or intestinal lining, what kind of change would you want to make in its chemical nature? Explain.
2. People often think of cholesterol as an unhealthy molecule in the body. Explain the origin of this belief, and why it's misleading.
3. Describe the biosynthetic pathway for the production of estradiol. Using this pathway, explain how a male bodybuilder taking anabolic steroids could develop female-like breast tissue.
4. Use the regulation in the pituitary-gonad axis to explain how a man taking anabolic steroids could develop shrunken testes.
5. Explain why thyroxine, steroid hormones, cholesterol, and triglycerides cannot be dissolved in the blood plasma. How are these molecules transported in the blood?
6. Given that dehydration raises the concentration and osmotic pressure of the blood, explain how the posterior pituitary can help maintain homeostasis of blood concentration.
7. Compare how the hypothalamus regulates prolactin secretion with how it regulates TSH secretion. What might account for this difference?
8. Suppose an animal is castrated (has its gonads removed). What would happen to the secretion of its gonadotropins? Explain. Now, imagine that hormones from the gonads are added back (in pills or injections); how would that influence the secretion of its gonadotropins? Explain.
9. Suppose a blood test revealed very low levels of ACTH, and a hormone-secreting tumor was suspected. What possible location of the tumor could account for the low ACTH? Explain.
10. Explain how a phosphodiesterase inhibitor drug could promote dilation of the bronchioles (airways) as a treatment for asthma.
11. A different variety of phosphodiesterase breaks down a different, but related, second messenger: cyclic GMP (cGMP). It turns out that cGMP produces dilation of blood vessels in the penis, causing erection. Now, explain how Viagra, an inhibitor of this phosphodiesterase, produces its effects.
12. Cushing's syndrome may be caused by a tumor of the anterior pituitary or by a tumor of the adrenal cortex. Explain how these two possibilities can be distinguished by measuring blood levels of ACTH and of cortisol.
13. Hyperkalemia (high blood K^+) can cause death by producing fibrillation of the heart (a condition that causes it to stop beating). Despite this, a healthy person can eat all the potassium-rich bananas they want. Explain how the adrenal cortex makes this possible.
14. Triiodothyronine pills are faster acting than thyroxine pills. Partly, that's because a lower proportion of T_3 than T_4 is bound to TBG; and partly it's due to the events that occur inside the target cells. Explain these statements.
15. Explain why an overweight person who is euthyroid (has normal thyroid function) might be tempted to take thyroxine pills. Also, explain why this person would eventually put the weight back on. What other reasons are there for not using thyroxine pills to lose weight?
16. Explain the body's need for iodine in the diet. Can you raise your metabolic rate by eating extra amounts of iodine? Explain.
17. You must eat sufficient amounts of calcium in your diet or you will lose bone mass. Explain the physiological mechanisms involved.
18. Vitamin D is both a vitamin and a prehormone. Explain why this statement is true.
19. How is homeostasis of blood glucose maintained after eating a meal containing carbohydrates? Explain the hormonal mechanisms involved.
20. Explain the oral glucose tolerance test, and how it's affected by diabetes.
21. Describe the changes that occur in insulin and glucagon secretion during fasting, and the changes in metabolism that result.
22. Given your answer to the previous question, explain how type 1 diabetes mellitus is similar to an extreme fasting state.
23. A friend advises a person who is going to go overseas by air to take melatonin pills to avoid jet lag. What is melatonin, which gland secretes it, and how is its secretion regulated? Is there a scientific rationale for this advice? Explain.

Web Activities

For additional study tools, go to: www.aris.mhhe.com. Click on your course topic and then this textbook's author/title. Register once for a semester's worth of interactive activities and practice quizzing to help boost your grade!

Muscle Physiology

CHAPTER 9

Chapter Outline

Homeostasis

We need the contractions of skeletal muscles to pull on bones at their articulations (joints) in order to move our body. This includes lifting food and liquid to eat and drink, moving away from danger and toward things we want and need, and communicating both orally and in writing. These and many other activities are required for homeostasis, and we couldn't live unaided without skeletal muscle function. Cardiac muscle is responsible for the beating of the heart, and thus for the pumping action required for circulation of the blood. Likewise, smooth muscles are necessary for the functions of all of the organ systems—such as the movement of the food through the gastrointestinal tract and the contractions of the urinary and gallbladders—and homeostasis is thus also dependent on the functioning of cardiac and smooth muscles.

But if homeostasis is not maintained . . .

Clinical Investigation

Rachel, a 63-year-old woman, went to the gym and met her new trainer. Rachel reported that she had heart problems a few years earlier, and at that time lab tests found elevated levels of troponin and CPK in her blood. Her physician put her on a calcium channel blocker. The trainer said that Rachel would have to exercise very carefully because of this medical history. When Rachel was doing leg lifts (extensions of the knee joint) on a machine that provided resistance, the trainer mentioned that she was working her "quads" (quadriceps femoris muscles) both in the lift and in the return movement. Rachel complained that her muscles felt fatigued, and her trainer said that they were building up lactic acid. The trainer said that this exercise would help her "quads" to get larger and stronger, but that Rachel would have to use the treadmill or other "aerobic" machines to improve her endurance.

What are troponin and CPK, and what might an elevation of their blood levels indicate? What is a calcium channel blocker, and how might this drug help? How do the leg lifts and the return movement exercise Rachel's quadriceps muscles? Why would lactic acid be produced, and how would the exercise build up muscle size? How does this exercise differ from "aerobic" exercise in its effect on muscles?

Skeletal Muscles Are Specialized for Contraction

When skeletal muscles contract, they exert tension on their tendons. Contractions that result in shortening of the muscle produce movements of the skeleton. However, a muscle may contract and not shorten if the force it produces in the contraction is less than the resistance it must overcome. Muscle contractions are graded, because a muscle is composed of individual muscle fibers; each fiber contains myofibrils, and each myofibril contains myofilaments. Muscle twitches can be summated, and if the summation is sufficiently rapid, a smooth, sustained contraction called tetanus is produced.

As discussed in chapter 2, there are three types of muscle tissue: smooth muscle, cardiac muscle, and skeletal muscle. Cardiac and skeletal muscles are striated, for reasons discussed in this section. Because they are both striated, cardiac and skeletal muscles contract by means of a similar mechanism, whereas smooth muscle has a somewhat different mechanism of contraction. Most of this chapter is devoted to the physiology of skeletal muscles, although cardiac and smooth muscles are discussed near the end of this chapter as a gateway to understanding the systems covered in the rest of this textbook.

A skeletal muscle is composed of many individual **muscle fibers**, or **myofibers**. Each muscle fiber (cell) is surrounded by a plasma membrane, or **sarcolemma**, enveloped by a thin connective tissue membrane called an *endomysium* (fig. 9.1). Muscle fibers within a

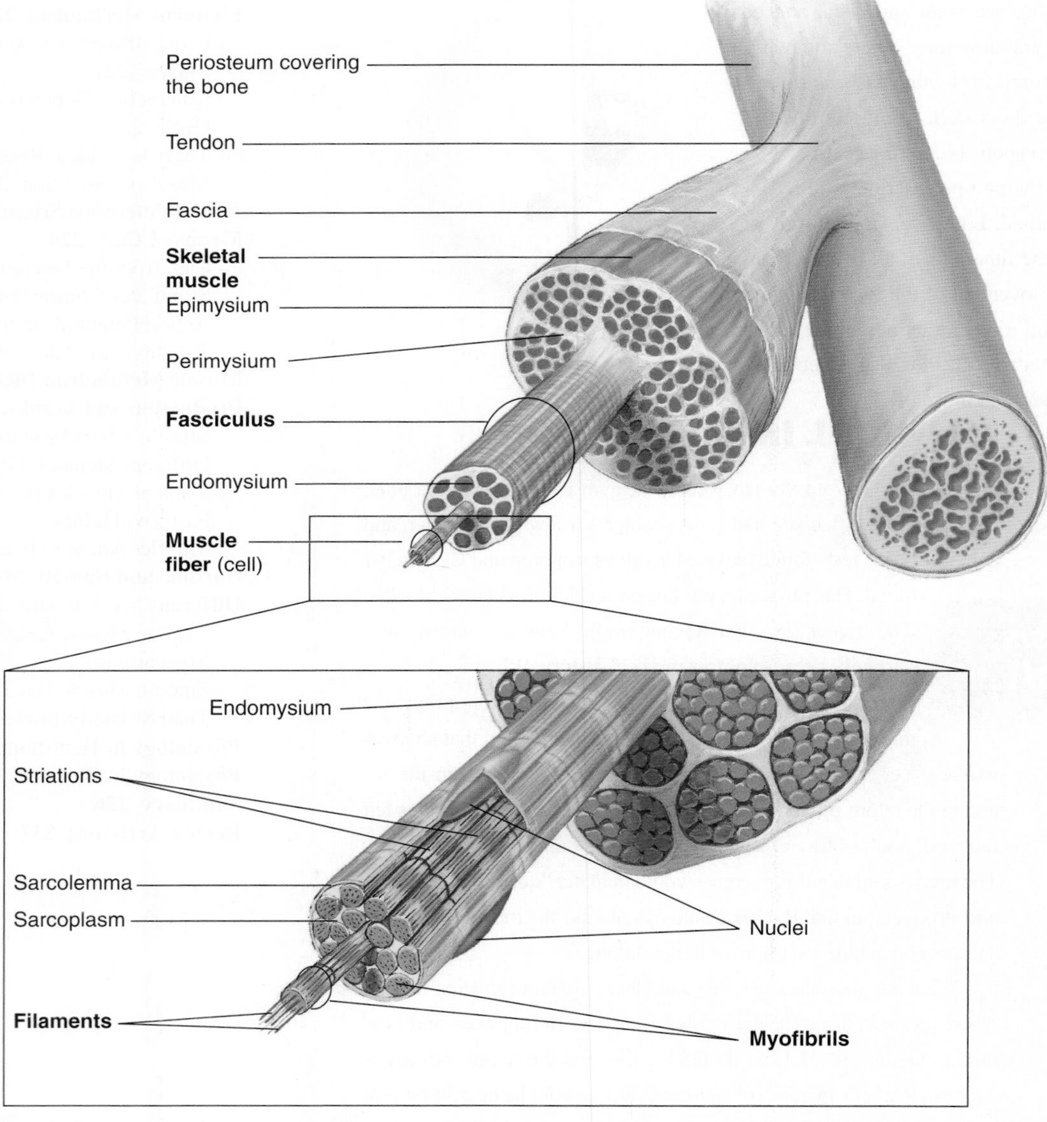

FIGURE 9.1 The structure of a skeletal muscle. The relationship between muscle fibers and the connective tissues of the tendon, epimysium, perimysium, and endomysium is depicted in the upper figure. Below is a close-up of a single muscle fiber.

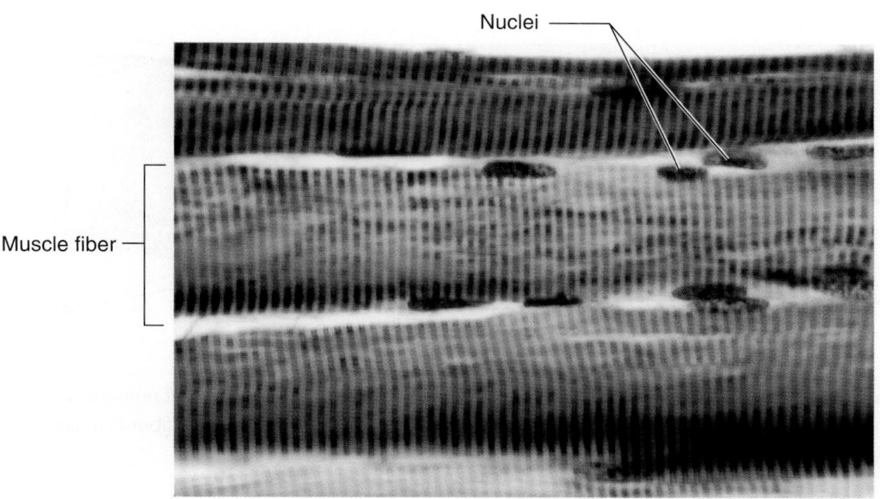

FIGURE 9.2 Skeletal muscle fibers seen through a microscope. Notice the dark A bands; the I bands are the lighter spaces between A bands. These bands give skeletal muscle fibers a striated appearance.

muscle are grouped into **muscle fascicles** by connective tissue known as a *perimysium.*[1] The entire muscle is surrounded by connective tissue called an *epimysium*. The connective tissue of the endomysium, perimysium, and epimysium is continuous with the connective tissue of the *tendons* that connect the **origin** and **insertion** of the muscle to the skeleton. The insertion is the more moveable attachment, so that contraction of the muscle produces movement of the insertion toward the origin. Groups of different muscles may be packaged together by connective tissue known as *fascia* (fig. 9.1).

Each muscle fiber has many nuclei (for reasons discussed in chapter 2), mitochondria, and other cellular organelles. However, the most distinctive microscopic feature of a muscle fiber is its striated (striped) appearance, produced by alternating dark and light bands (fig. 9.2). The dark bands are called **A bands**, and the light bands are called **I bands**. At very high magnification, thin dark lines can be seen in the middle of the I bands; these are *Z lines* (see fig. 9.4). The Z lines are actually three-dimensional, and so are more accurately called **Z discs**. The alternating A and I bands give skeletal muscle its striated appearance and are produced by structures within each muscle fiber.

A Muscle Fiber Contains Myofibrils and Myofilaments

Figure 9.3 illustrates a single muscle fiber. It contains a number of nuclei located just under the sarcolemma, and exhibits a striated appearance. We can see through the sarcolemma in this figure, and observe that the striated appearance of the muscle fiber is produced by striations of other structures within it. These run longitudinally, along the axis of the muscle fiber, and are also long and thin—that is, fibrous in appearance. We can't use the word *fiber* to describe these (because *fiber* refers to the whole muscle cell); these are smaller, subcellular structures called **myofibrils** (*fibril* = little fiber). You can think of the fiber as a cable containing many wires, the myofibrils.

Figure 9.3 shows one myofibril pulled partway out to the left of the fiber. This myofibril contains thinner structures that are also long and thin like fibers, but the terms *fiber* and *fibril* are already taken; these are known as **myofilaments**. There are two types of myofilaments: **thick filaments** and **thin filaments**. We can see how these are arranged by looking at a photograph of a myofibril taken with an electron microscope. Figure 9.4*a* shows an extremely thin section (slice) of a myofibril; notice the striations and the labeled A and I bands, as well as the Z discs. Figure 9.4*b* illustrates how the thick and thin filaments are arranged to produce the banding pattern.

Within each myofibril, the thick filaments are stacked one over the other to produce the dark A bands (fig. 9.4). The lighter I bands contain the thin filaments. However, you can see in figure 9.4*b* that the thin filaments also extend partway into the A band,

CLINICAL APPLICATIONS

Duchenne's muscular dystrophy is the most severe of the muscular dystrophies, affecting 1 out of every 3,500 boys each year. This disease is inherited on the X chromosome and produces progressive muscle wasting. The product of the defective gene is a protein called *dystrophin*, located just under the sarcolemma, which provides support for the muscle fiber by bridging its interior structures with the extracellular matrix (chapter 2). Using this information, scientists have developed laboratory tests that can detect this disease in fetal cells obtained by amniocentesis. They have also produced a strain of mice that develop the mouse equivalent of this disease, and have shown that inserting the "good gene" for dystrophin into these mice prevents the disease. However, scientists have not yet been able to insert sufficiently large numbers of the gene into mature muscle fibers to produce a cure, and other avenues of research are ongoing.

FYI [1]Have you ever eaten stringy meat and had the "strings" catch between your teeth? The strings between your teeth are the muscle fascicles. If you hold up a piece of stringy meat and tease apart one of these strings (fascicles), you can see the tearing of connective tissue; this is the perimysium.

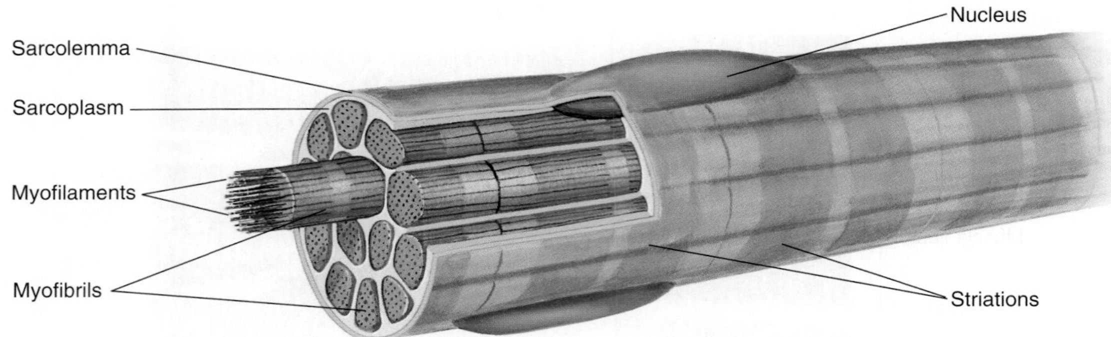

FIGURE 9.3 The components of a skeletal muscle fiber. A skeletal muscle fiber is composed of numerous myofibrils that contain myofilaments of actin and myosin. Overlapping of the myofilaments produces a striated appearance. Each skeletal muscle fiber is multinucleated.

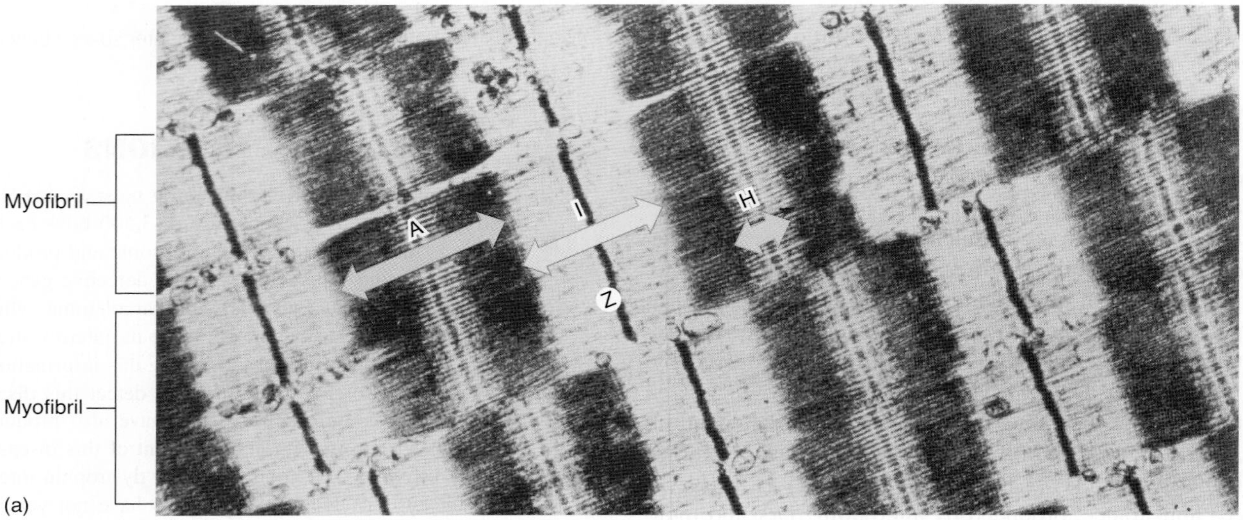

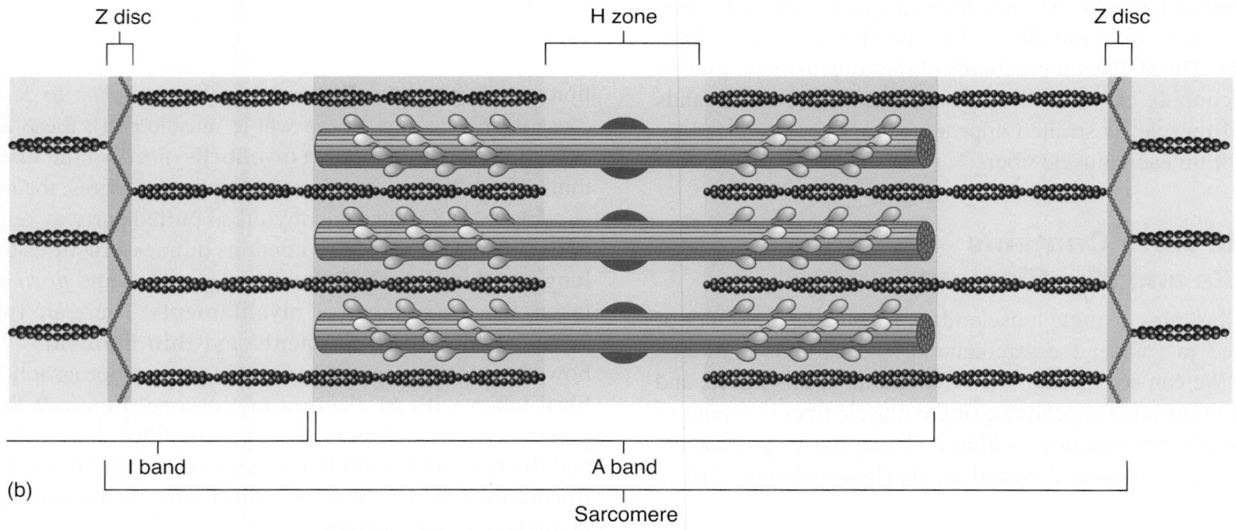

FIGURE 9.4 The striations of skeletal muscles are produced by thick and thin filaments. (*a*) Electron micrograph of a longitudinal section of myofibrils, showing the banding pattern characteristic of striated muscle. (*b*) Illustration of the arrangement of thick and thin filaments that produces the banding pattern. The colors used in (*a*) to depict different bands and structures correspond to the colors of (*b*).

overlapping with part of the thick filaments. The Z discs are in the middle of the I bands (fig. 9.4*a*), and serve to anchor the thin filaments (fig. 9.4*b*). Notice that the structure from one Z disc to the next is repeated along the myofibril, serving as a basic subunit of muscle structure. This structure, from one Z disc to the next ("from Z to shining Z"), is called a **sarcomere**. Figure 9.4 also shows an **H zone**: this is in the center of the A band, and is lighter because there isn't overlap of thick and thin filaments here. In other words, the H zone contains only thick filaments, whereas the rest of the A band contains both thick and thin filaments. The H zone is in the middle of each sarcomere.

The thick filaments are composed of hundreds of **myosin** proteins, and the thin filaments are composed primarily of the protein **actin**. The myosin and actin proteins provide interactions of the thick and thin filaments that lead to shortening of the sarcomeres. This produces shortening of the myofibrils, and thus shortening of the muscle fibers. The mechanism of muscle contraction will be described in the next major section.

There Are Different Types of Skeletal Muscle Contractions

Muscle contractions generally produce movements of bones at joints. In order for a muscle to shorten when it contracts, and thus to move its insertion toward its origin, the noncontractile parts of the muscle and its tendons must first be pulled tight; that is, slack must be removed. These structures have some elasticity—they resist distension (stretching), and when the distending force is released, they spring back to their resting lengths (like a rubber band). Thus, tendons provide what is called a **series elastic component**: they are somewhat elastic and in series (in line) with the force of the muscle contraction. When a muscle relaxes, the *elastic recoil* provided by the series elastic component helps the muscle return to its resting length.

Twitches Can Summate and Produce Tetanus

Contractions of isolated muscles (*in vitro*) in response to electric shocks mimic the contractions of muscles in the body (*in vivo*). When a muscle is stimulated with a single electric shock, it quickly contracts and relaxes. This is called a **twitch** (fig. 9.5). Increasing the stimulus voltage increases the strength of the twitch, up to a maximum, demonstrating that contraction strength is *graded*, or varied. This is because increasing the strength of the electric shock stimulates more of the muscle fibers within the muscle, producing a stronger contraction. This is similar to what happens in the body when more motor units are activated. As discussed in chapter 6, a *motor unit* is a somatic motor neuron and all of the muscle fibers it innervates. When more motor neurons are activated to stimulate more muscle fibers, a stronger contraction is produced.

If two shocks are given in rapid succession to a muscle, the muscle will twitch twice so rapidly that the second contraction begins before the first has finished. The first contraction takes some of the slack out of the series elastic component, and the second contraction begins before relaxation has completed. As a result, the second contraction is greater than the first, demonstrating the **summation** of twitches (fig. 9.5). If the electric shocks are delivered more and more rapidly, the relaxation time between successive twitches becomes

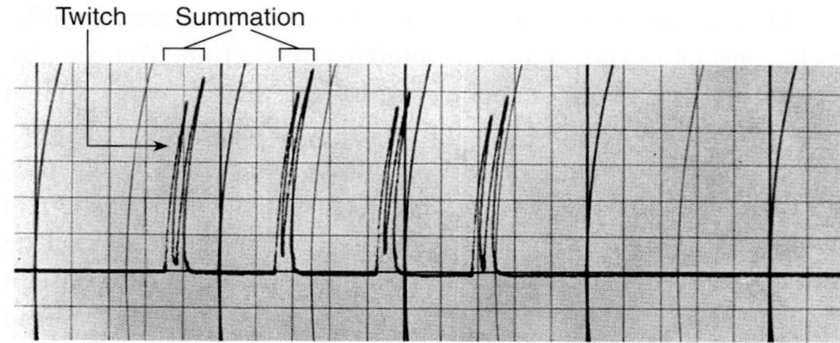

FIGURE 9.5 Recording muscle contractions. Recorder tracings demonstrating twitch and summation of an isolated frog gastrocnemius muscle.

shorter and shorter and the strength of contraction will increase, resulting in an **incomplete tetanus** (fig. 9.6). Finally, when the electric shocks occur so rapidly that there is no relaxation time between successive twitches, the contraction will appear smooth and sustained. In the body, the same effect is produced by the activation of many different motor units in quick succession. A smooth, sustained muscle contraction is called **complete tetanus**. Note that the term *tetanus* in muscle physiology has a different meaning than the disease of the same name, which is caused by a bacterial toxin that produces a painful state of muscle contracture, known as *tetany*.

Muscles in Vivo May or May Not Shorten when They Contract

Muscle contraction always generates a force causing the muscle to shorten. However, muscles in the body may or may not shorten when they contract. Imagine doing a "curl" using a 10-pound weight. This weight is a load on your biceps brachii muscle, providing a resistance to its ability to shorten when you flex your arm for the curl. Now imagine attempting to do a curl using a 100-pound weight. You may not be able to flex your arm and perform the curl, although your muscle is contracting and exerting tension on its insertion.

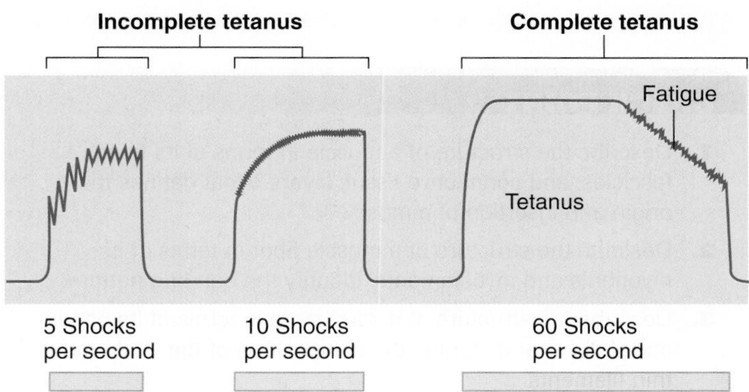

FIGURE 9.6 Incomplete and complete tetanus. When an isolated muscle is shocked repeatedly, the separate twitches summate to produce a sustained contraction. At a relatively slow rate of stimulation (5 or 10 per second), the separate muscle twitches can still be observed. This is incomplete tetanus. When the frequency of stimulation increases to 60 shocks per second, however, complete tetanus—a smooth, sustained contraction—is observed. If the stimulation is continued, the muscle will demonstrate fatigue.

Muscle contraction in which the muscle length remains relatively constant is known as an **isometric contraction**. You can also perform an isometric contraction by curling the 10-pound weight partway and then holding your forearm in a partially flexed position. Now if you increase the force of your muscle contraction sufficiently, you will complete the curl as your muscle shortens. Once the strength of the muscle's contraction has reached the level needed to shorten the muscle, the strength of the muscle's contraction is relatively constant throughout the contraction. Thus, a muscle contraction that results in a change in muscle length is called an **isotonic contraction**.

A muscle will shorten when the force of a muscle's contraction is greater than the load that resists the shortening; doing a curl with a 10-pound dumbbell is an example. This is a type of isotonic contraction known as a **concentric contraction**. You can also perform a different sort of isotonic contraction. For example, suppose that after you complete your curl, you lower the dumbbell slowly, so that it comes to a rest gently on its supports. To do this, you must contract your biceps to control the dumbbell's fall. In that case, your muscle is lengthening as it contracts; more accurately, it's lengthening *despite* its contraction. This is called an **eccentric contraction**. In an eccentric contraction, the muscle lengthens because the load forcing the muscle to lengthen is greater than the force of the muscle contraction.

Clinical Investigation Clues

Remember that Rachel exercised on a machine that worked her quadriceps femoris muscles (which extend the knee joint) by doing leg lifts, and that her trainer stated that she was exercising her muscles in both the lift and the return.

- What type of contraction did Rachel perform in the leg lift?
- What type of contraction did Rachel perform when she gently returned to the resting position?
- If Rachel lifted and then held her leg straight for a time, what type of contraction would she be performing?

Check Point

1. Describe the structure of a muscle in terms of its fibers, fascicles, and connective tissue layers. What defines the origin and insertion of a muscle?
2. Describe the structure of a muscle fiber in terms of its myofibrils and myofilaments. Identify the banding pattern.
3. Describe the structure of a sarcomere in terms of its bands and Z discs, and identify the composition of the thick and thin filaments.
4. What is the series elastic component of a muscle, and what is its significance?
5. Describe the nature and significance of muscle twitch, summation, and tetanus.
6. Distinguish between isotonic and isometric contractions, and between concentric and eccentric contractions.

Striated Muscles Contract by a Sliding Filament Mechanism

When striated muscles contract, the sarcomeres become shorter because the thin filaments slide over the thick filaments, producing greater amounts of overlap. However, the lengths of the thick and thin filaments remain the same. The heads of myosin proteins, which form cross bridges to actin proteins, produce this sliding of the myofilaments through their power strokes. The greatest strength of contraction occurs when the muscle is at its normal resting length before the contraction begins.

When a muscle fiber contracts and shortens, its myofibrils shorten as a result of the shortening of their sarcomeres. The A bands stay the same length while the I bands shorten. This is because the thin filaments slide deeper into the A bands (fig. 9.7)—the myofilaments do *not* shorten. Because sliding of the myofilaments causes

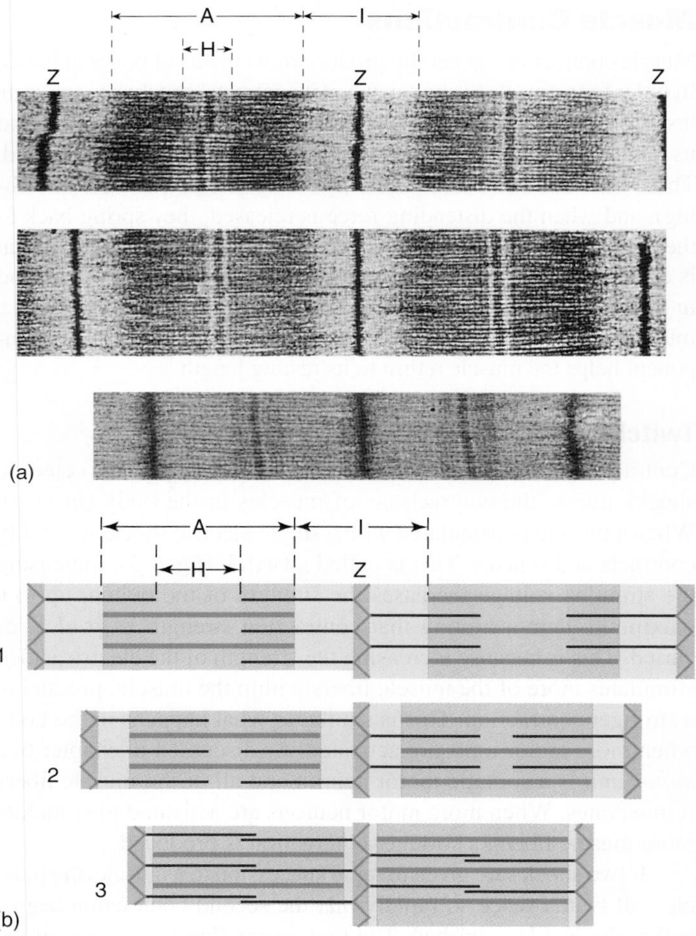

FIGURE 9.7 The sliding filament model of muscle contraction. (*a*) An electron micrograph and (*b*) a diagram of the sliding filament model of contraction. As the filaments slide, the Z lines are brought closer together and the sarcomeres get shorter. (*1*) Relaxed muscle; (*2*) partially contracted muscle; (*3*) fully contracted muscle.

increased overlap of thick and thin filaments in the A bands, the H zones become shorter as the muscle contracts and shortens. At a maximum shortening, the H zone disappears (fig. 9.7). The shortening of a muscle produced by sliding of the myofilaments is known as the **sliding filament theory of contraction**.

Cross Bridges Are Activated in the Resting Muscle

Sliding of the filaments is produced by the action of **cross bridges** that extend from the thick filaments toward the thin filaments. These cross bridges are part of the myosin proteins that compose the thick filaments. A portion of each myosin protein helps form the thick filament, and a portion of the myosin protrudes from the thick filament like a little arm, which consists of a *myosin tail* ending in a *myosin head* (fig. 9.8). It is the myosin heads that form the cross bridges to the actin in the thin filaments. Figure 9.8 also shows that the thin filaments are composed of a double row of globular actin proteins, twisted into a helix. The thin filaments also contain two other proteins. *Tropomyosin* is a filamentous protein bound to actin that lies in the groove created by the helical twisting of the globular actin proteins. *Troponin* is actually a complex of three proteins attached to the tropomyosin. Troponin and tropomyosin have regulatory functions, as will be discussed in the next major section.

Each myosin head has two active regions: (1) an *actin-binding site*, and (2) an *ATP-binding site*. The actin-binding site is specialized for attaching to a particular *binding site* in an actin protein of the thin filament, forming a cross bridge (shown on the right side of fig. 9.8). However, it is important to know that the myosin head does *not* bind to the actin when a muscle is at rest. The ATP-binding site is also an enzyme, known as an *ATPase*, that breaks down ATP into ADP and P_i (inorganic phosphate; see chapter 2). After this reaction, the resulting ADP and P_i remain bound to the myosin head (as shown to the right in fig. 9.8). Although this myosin is shown attached to the actin in figure 9.8, you should remember that the head does not bind to the actin if the muscle is relaxed.

Compare the angle of the myosin heads on the left and right in figure 9.8. Notice that after the ATPase has hydrolyzed ATP to ADP and P_i, the head is "cocked." This is analogous to cocking, or pulling back, the string of a bow. The pulled bowstring has the energy needed to shoot the arrow. Similarly, the myosin head, activated by the hydrolysis of ATP, can later undergo a power stroke and produce sliding of the filaments. However, the activation of the myosin heads by the hydrolysis of ATP occurs before this, in the resting muscle. By analogy, the bowstring is cocked and ready, but not yet fired. The myosin heads are not yet bound to actin in the resting muscle, and so do not yet produce contraction.

Contraction Depends on the Cross-Bridge Cycle

When a muscle is stimulated to contract, changes occur that allow the myosin heads to bind to actin, forming cross bridges. The changes that occur upon stimulation that allow cross-bridge formation and muscle contraction are discussed in the next section.

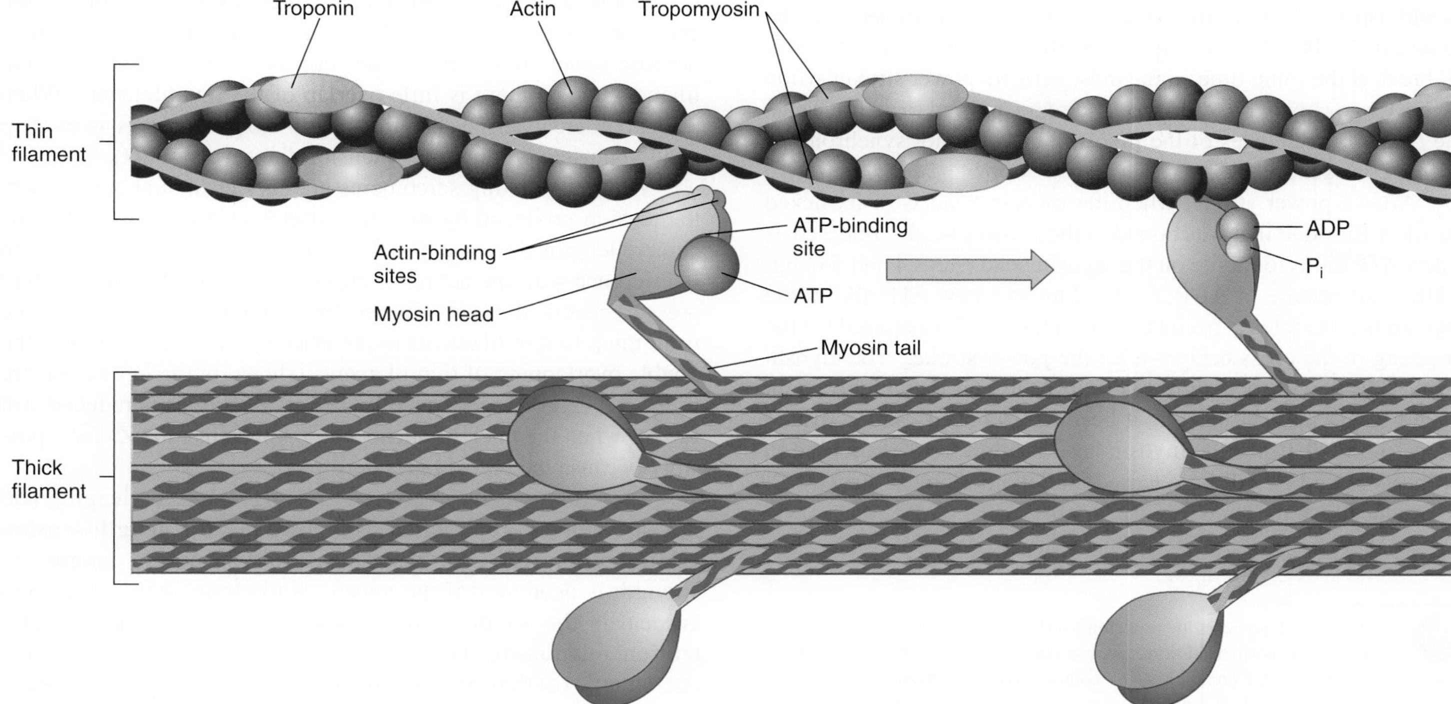

FIGURE 9.8 The structure of myosin, showing its binding sites for ATP and actin. Once the myosin head binds to ATP, it is hydrolyzed into ADP and inorganic phosphate (P_i). This activates the myosin head, "cocking it" to put it into position to bind to attachment sites in the actin molecules.

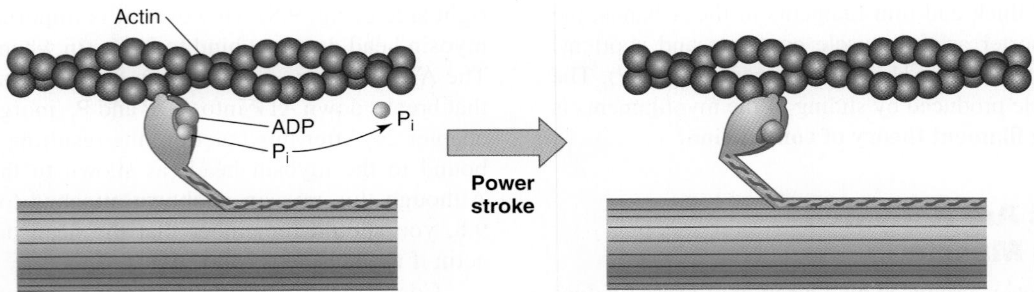

FIGURE 9.9 The power stroke of the cross bridge. After the myosin head binds to actin to form a cross bridge, inorganic phosphate (P_i) is released. This causes a conformational change in the myosin head, resulting in a power stroke that produces sliding of the thin filament over the thick filament.

When a myosin head is allowed to bind to an actin protein, the phosphate group (P_i) detaches from the myosin (fig. 9.9). This produces a conformational (shape) change in the myosin, causing it to rotate and produce a **power stroke**. It is the power stroke that slides the thin filament over the thick filament.

If you think of the cross bridges as little arms pulling on the thin filaments, you'll realize that just one power stroke can't pull the thin filaments very far; indeed, a single power stroke can shorten a muscle by only about 1% of its resting length. Here's an analogy: in order for people's arms to pull on a rope and compete in a tug-of-war, they must be able to let go after they pull, swing back around, regrasp the rope, and pull again; and they must repeat this cycle a number of times. Since muscles can shorten up to 60% of their resting lengths, the cross bridges must similarly have a **cross-bridge cycle**. Before describing this, there is one more benefit to be derived from the tug-of-war analogy. What would happen if all the arms of the people in one team let go at the same time? (They'd lose.) Similarly, the little cross bridges can't all break at the same time; some must perform power strokes while others let go, swing back around, and regrasp (bind to actin). Thus, the cross-bridge cycles of the myosin heads are not synchronized during contraction.

After a power stroke, when the myosin head is in its flexed position, the ADP that was bound to the myosin head is released as a new ATP binds to its site on the myosin head (steps 4 and 5 in fig. 9.10). This release of ADP and binding to a new ATP allows the myosin head to release from the actin. Thus, ATP is required for the breaking of the cross bridges after the power stroke.[2] The myosin head ATPase will then split ATP into ADP and P_i, which remain bound to the myosin head as it becomes "cocked" and activated (step 1 in fig. 9.10). If the myosin head is allowed to again bind to actin, its bound P_i will be released and another power stroke will be produced (steps 2 and 3 in fig. 9.10).

FYI [2]What would happen if there weren't ATP available in the muscle fiber to bind to the myosin heads after a power stroke? The cross bridges wouldn't release, and these "rigor complexes" of bound myosin heads would cause the muscles to stay contracted. This happens when muscle fibers die, and is responsible for *rigor mortis*—the stiffening of a body after death. A living muscle cell always has sufficient ATP to prevent muscles from forming these rigor complexes while we're alive.

There Is an Ideal Resting Length for Muscle Contraction

There is an ideal resting length for striated muscles. This is the length at which they can generate the maximum force when they contract. This force is measured as the *tension* produced by the contraction of the muscle. As illustrated in figure 9.11, this tension is greatest when the sarcomeres are at a length of 2.0 to 2.2 µm. As it turns out, this is the length of the sarcomeres when skeletal muscles are at their normal resting lengths. This normal resting length is maintained by muscle stretch reflexes, which result in contraction when stretching of the muscle activates stretch receptors called *muscle spindles* (chapter 6). These monosynaptic muscle stretch reflexes (chapter 6) help maintain muscle tone and keep the muscle at its ideal resting length.

When the muscle is stretched so that its sarcomeres are longer than about 2.2 µm, the tension produced by muscle contraction decreases as the sarcomere length increases (fig. 9.11). This is because fewer cross bridges can form between the thick and thin filaments when there is little overlap of the myofilaments. When the sarcomere reaches a length of about 3.6 µm, there is no overlap and the muscle produces zero tension (it can't contract).

When the resting sarcomeres are shorter than about 2.0 µm, the tension produced by muscle contraction decreases as the sarcomere lengths decrease (fig. 9.11). This is believed to result from interference with the action of cross bridges by a buildup of fluid pressure as the fiber gets shorter; from an increase in the distance from thick to thin filaments as the fiber gets thicker; and from the double overlapping of thin filaments (shown in the left sarcomere in fig. 9.11). Force generated by the cross bridges is reduced still further when the thick filaments bump up against the Z discs, possibly because the myosin gets deformed.

The relationship between resting muscle length and the strength of contraction is known as the **length-tension relationship**. This is a "three bears story": muscle contraction strength is decreased if the muscle is too long or too short, and is ideal in the middle range of sarcomere lengths. The length-tension relationship helps explain the function of the muscle stretch reflexes described in chapter 6. This relationship between sarcomere length and the strength of contraction is also important in understanding how the contraction strength of the heart is regulated (chapter 10).

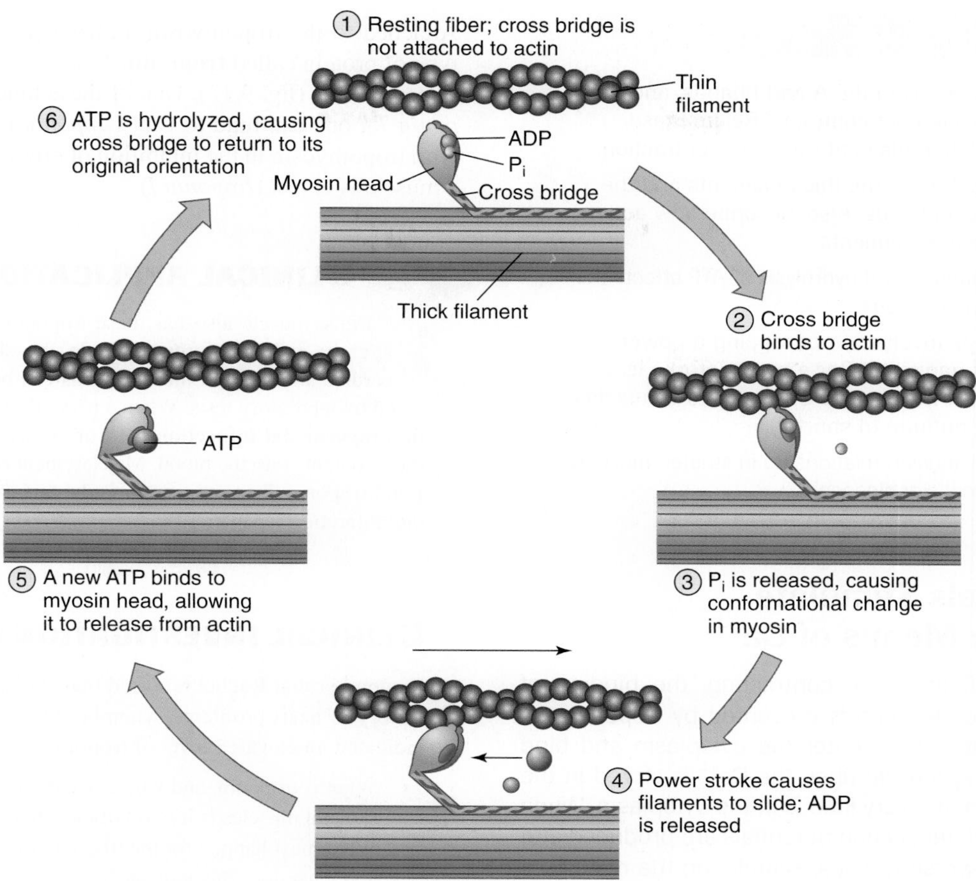

FIGURE 9.10 The cross-bridge cycle that causes sliding of the filaments and muscle contraction. Hydrolysis of ATP is required for activation of the cross bridge, and the binding of a new ATP is required for the cross bridge to release from the actin at the end of a cycle.

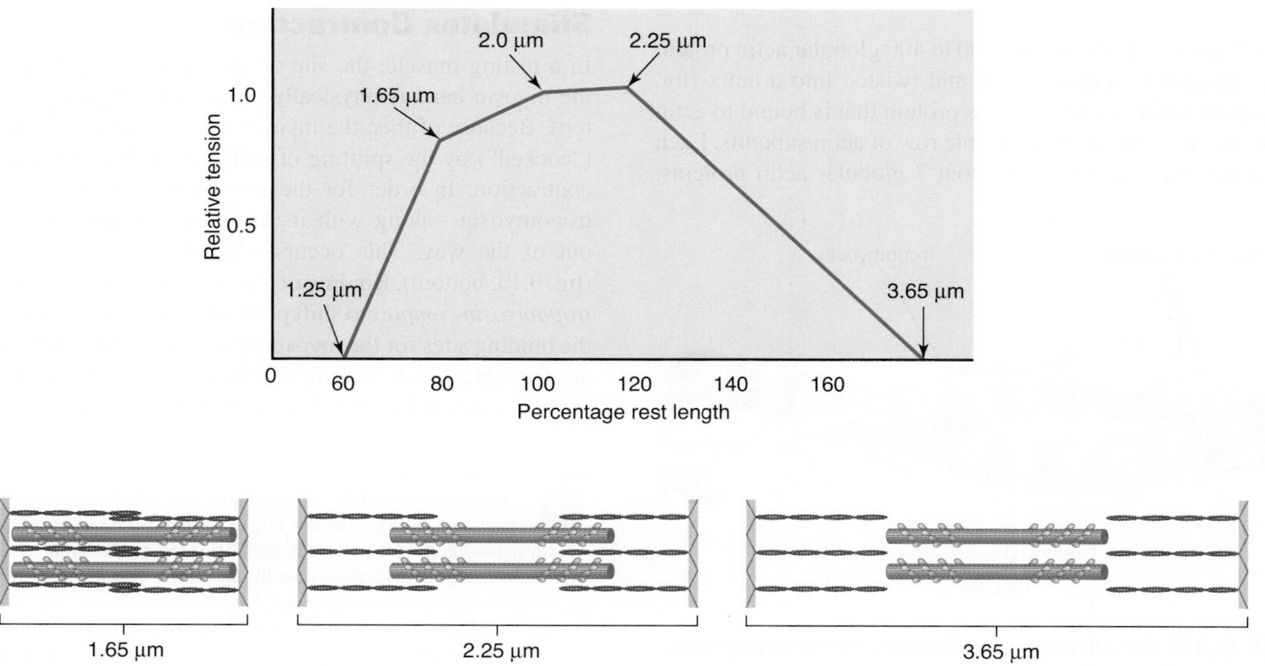

FIGURE 9.11 The length-tension relationship in skeletal muscles. Maximum relative tension (1.0 on the *y* axis) is achieved when the muscle is 100% to 120% of its resting length (sarcomere lengths from 2.0 to 2.25 μm). Increases or decreases in muscle (and sarcomere) lengths result in rapid decreases in tension.

CHECK POINT

1. Describe what happens to the A and I bands, and the H zones, as a muscle fiber contracts. Relate these changes to the sliding filament theory of contraction.
2. Describe the structure of the thick filaments and the nature of the myosin heads. Also, describe how actin proteins form the thin filaments.
3. Explain how the binding and hydrolysis of ATP affects the position of the myosin heads.
4. Describe the steps involved in producing a power stroke, and what happens after a power stroke is finished. Explain how the cross bridges allow muscles to contract and continue to shorten.
5. Explain the length-tension relationship in striated muscles, and its physiological significance.

Action Potentials Stimulate Contraction by Means of Ca^{2+}

When a muscle fiber is not contracting, the binding of the myosin heads to actin is prevented by tropomyosin. Contraction requires that Ca^{2+} enter the cytoplasm and bind to troponin. In a resting muscle fiber, the Ca^{2+} is stored in the sarcoplasmic reticulum and very little is in the cytoplasm. When a muscle fiber is stimulated, action potentials are produced and conducted along transverse tubules, stimulating the release of Ca^{2+} from the sarcoplasmic reticulum. This Ca^{2+} can then bind to troponin, causing the troponin-tropomyosin complexes to move and allowing the myosin heads to bind to actin.

Each thin filament is formed from 300 to 400 globular actin protein subunits, arranged in a double row and twisted into a helix (fig. 9.12). **Tropomyosin** is a filamentous protein that is bound to actin and lies within the groove of the double row of actin subunits. Each tropomyosin spans a distance of about 7 globular actin proteins.

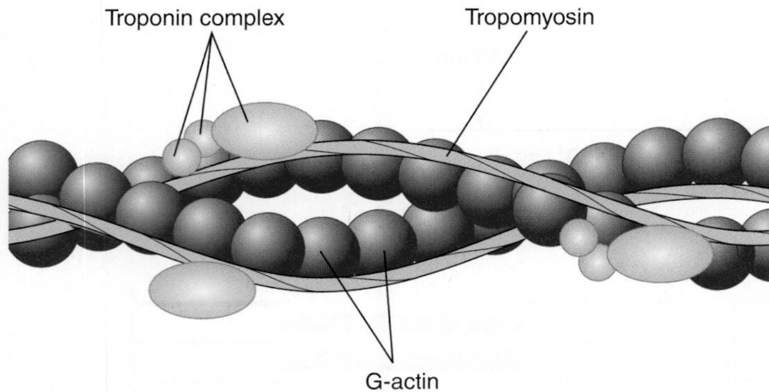

FIGURE 9.12 The structural relationship between troponin, tropomyosin, and actin. The tropomyosin is attached to actin, whereas the troponin complex of three subunits is attached to tropomyosin (not directly to actin).

Attached to the tropomyosin, rather than to the actin, is another type of protein called **troponin**. Troponin is actually a complex of three proteins (fig. 9.12). One of these binds to tropomyosin (*troponin T*); one can bind to Ca^{2+} (*troponin C*); and one participates with tropomyosin in the inhibition of cross-bridge formation when a muscle is at rest (*troponin I*).

CLINICAL APPLICATIONS

Cardiac muscle also has these troponin units; however, it makes slightly different forms than do the skeletal muscles, and the cardiac and skeletal muscle forms can be distinguished and measured by laboratory tests. When a person's cardiac muscle is damaged in a **myocardial infarction** (MI, or "heart attack"), the cells release their contents into the blood. Measurement of blood levels of troponin I and troponin T are now commonly performed to diagnose myocardial infarction (chapter 10).

CLINICAL INVESTIGATION CLUES

Remember that Rachel reported that she had a history of heart problems, when her blood work indicated an elevated level of troponin.

- What is troponin, and where in cardiac and skeletal muscles is it normally located?
- What must happen for the troponin to be found in the blood, and what does this indicate?

Ca^{2+} from the Sarcoplasmic Reticulum Stimulates Contraction

In a resting muscle, the site on the actin protein that can bind to the myosin head is physically blocked by tropomyosin (fig. 9.13, top). Because of this, the myosin head, although already activated ("cocked") by the splitting of ATP into ADP and P_i, cannot cause contraction. In order for the myosin head to bind to actin, the tropomyosin—along with its attached troponin—must be moved out of the way. This occurs when Ca^{2+} binds to the troponin (fig. 9.13, bottom). Binding of Ca^{2+} to troponin causes the *troponin-tropomyosin complex* to shift positions in the thin filament, exposing the binding sites for the myosin heads. As a result, the cross bridges are formed, P_i is released, and the power stroke occurs as previously described. In short, *Ca^{2+} stimulates contraction.*[3]

FYI [3]Sidney Ringer first documented the requirement of muscles for Ca^{2+} in a paper published in 1883. He was able to get isolated rat hearts to beat when they were in an isotonic saline solution (chapter 3) that was made with water piped in from a river. However, the hearts beat abnormally and soon stopped beating when they were in an isotonic saline solution made by dissolving NaCl in distilled water. Through a series of brilliant and meticulous experiments, he demonstrated that Ca^{2+}, present in the river water but absent from distilled water, is required for contraction. He also demonstrated that proper proportions of potassium and other ions are needed, and that this applies to skeletal muscle as well as the heart. Today, "Ringer's solutions," containing the needed concentration of ions, are commonly used in physiology laboratories and hospitals.

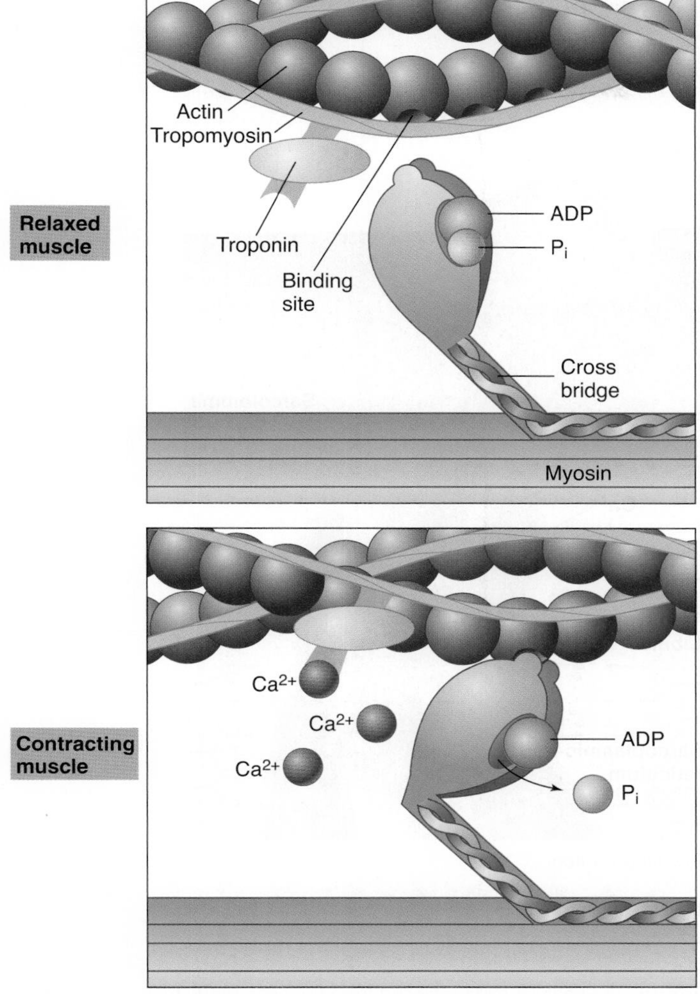

FIGURE 9.13 The role of Ca^{2+} in muscle contraction. The attachment of Ca^{2+} to troponin causes movement of the troponin-tropomyosin complex, which exposes binding sites on the actin. The myosin cross bridges can then attach to actin and undergo a power stroke.

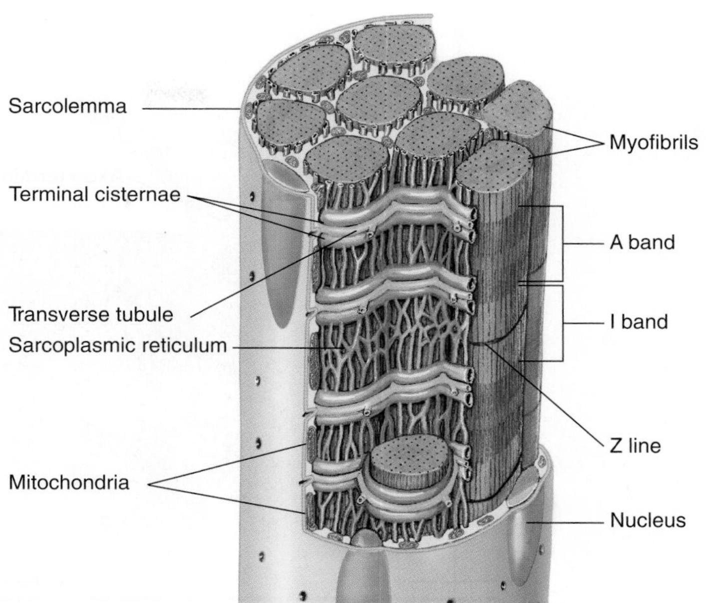

FIGURE 9.14 The sarcoplasmic reticulum. This figure depicts the relationship between myofibrils, the transverse tubules, and the sarcoplasmic reticulum. The sarcoplasmic reticulum (*green*) stores Ca^{2+} and is stimulated to release it by action potentials arriving in the transverse tubules.

In a resting muscle fiber, Ca^{2+} is essentially absent from the cytoplasm and so can't bind to the troponin. This is because, when a muscle fiber isn't contracting, the specialized endoplasmic reticulum (an intracellular system of membranous sacs; see chapter 2) of a muscle fiber accumulates the Ca^{2+} inside its enlarged *terminal cisternae* (cavities). The endoplasmic reticulum of muscle cells is called a **sarcoplasmic reticulum (SR)** (fig. 9.14). An active transport process (chapter 3) enables the sarcoplasmic reticulum to accumulate a high concentration of Ca^{2+} within its terminal cisternae, while also producing a very low concentration of Ca^{2+} in the cytoplasm.

Action Potentials in the Transverse Tubules Stimulate Ca^{2+} Release

Another structure visible in figure 9.14 is the **transverse tubules**. These are little inward extensions of the sarcolemma (plasma membrane) that tunnel deep into the muscle fiber, so that they run very close to the terminal cisternae of the sarcoplasmic reticulum. Each transverse tubule contains extracellular fluid and brings the sarcolemma to even the deepest part of the muscle fiber. Because of this, an action potential produced at the surface will be conducted along the sarcolemma and down the transverse tubules so that it's brought next to the terminal cisternae of the sarcoplasmic reticulum.

The stage is now set for a discussion of **excitation-contraction coupling**: how action potentials cause muscle contraction. In the body, the contraction of skeletal muscles is stimulated by somatic motor neurons. The somatic motor neurons release acetylcholine (ACh), which diffuses across the synaptic cleft and binds to its receptor proteins in the sarcolemma (fig. 9.15). This opens chemically gated ion channels that produce an excitatory postsynaptic potential (EPSP). As in the plasma membrane of neurons (chapter 4), the EPSPs stimulates the opening of voltage-gated Na^+ channels and results in the production of an action potential. However, in a muscle fiber the voltage-gated channels are immediately adjacent to the chemically gated channels, which are located in the middle of the muscle fiber at the synapse.

Action potentials are conducted along the sarcolemma in the same way that action potentials are conducted by unmyelinated axons. However, because the synapse is in the middle of the muscle fiber, the action potentials spread out from there throughout the muscle fiber. As the action potential is conducted along the sarcolemma, it follows the transverse tubules deep into the muscle fiber.

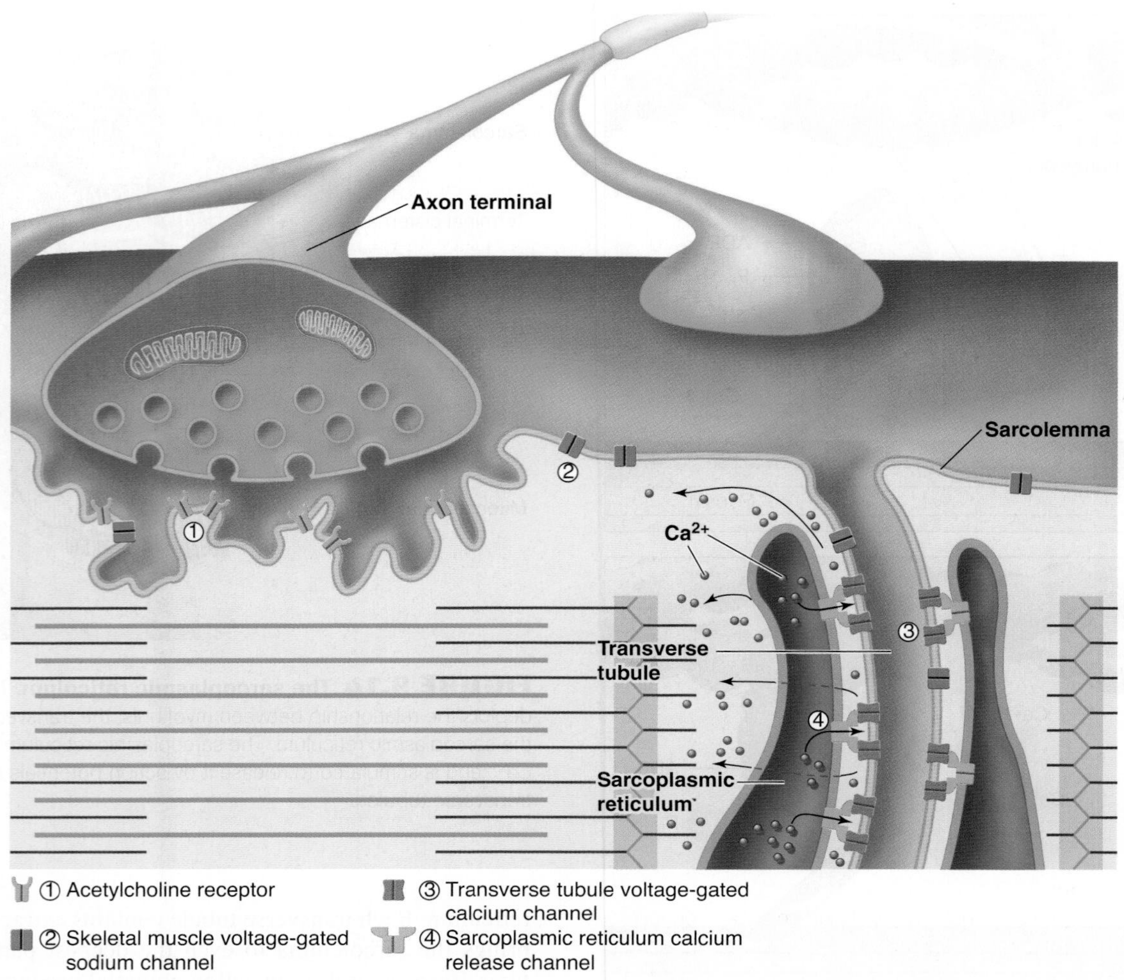

FIGURE 9.15 The structures involved in excitation-contraction coupling. The acetylcholine released from the axon binds to its receptors in the motor end plate (the postsynaptic membrane). This stimulates the production of a depolarization, which causes the opening of voltage-gated Na^+ channels and the resulting production of action potentials along the sarcolemma. The spread of action potentials into the transverse tubules stimulates the opening of their voltage-gated Ca^{2+} channels, which (directly or indirectly) causes the opening of voltage-gated Ca^{2+} channels in the sarcoplasmic reticulum. Calcium diffuses out of the sarcoplasmic reticulum, binds to troponin, and stimulates contraction.

The membrane of the sarcolemma and transverse tubules contains **voltage-gated Ca^{2+} channels**. These channels are stimulated to open by the depolarization of an action potential, allowing Ca^{2+} to diffuse from the extracellular fluid (within the transverse tubules) into the cytoplasm. In skeletal muscles, the voltage-gated Ca^{2+} channels in the transverse tubules are mechanically coupled to **Ca^{2+} release channels** in the SR (sarcoplasmic reticulum). Because of this mechanical coupling, opening of the voltage-gated Ca^{2+} channels in the transverse tubules directly causes the opening of the Ca^{2+} release channels in the SR (fig. 9.15). In addition, there are other Ca^{2+} release channels in the SR that are opened indirectly; these open in response to Ca^{2+} diffusing from the extracellular fluid into the cytoplasm through the voltage-gated Ca^{2+} channels in the transverse tubules. This mechanism is called *Ca^{2+}-induced Ca^{2+} release*. The end result is that a great amount of Ca^{2+} diffuses out of the sarcoplasmic reticulum in response to action potentials.

Once Ca^{2+} enters the cytoplasm, it can bind to troponin, causing the troponin-tropomyosin complex to move out of the way of the actin-binding sites. Now the myosin heads can bind to actin and the cross-bridge cycle can proceed, causing muscle contraction. These events are summarized in figure 9.16.

But how can a muscle relax? First, action potentials in the somatic motor neuron must stop, causing action potentials in the muscle fibers to stop. When this happens, the voltage-gated Ca^{2+} channels in the transverse tubules close, and likewise the Ca^{2+} release channels in the sarcoplasmic reticulum close. Now, there are active transport **Ca^{2+} pumps** in the sarcoplasmic reticulum that can do their job unhindered, actively accumulating Ca^{2+} from the cytoplasm and storing it in the terminal cisternae. Notice something important and perhaps surprising: because of active transport, relaxation of a muscle, like contraction, requires ATP! No wonder we burn so many additional calories in exercise.

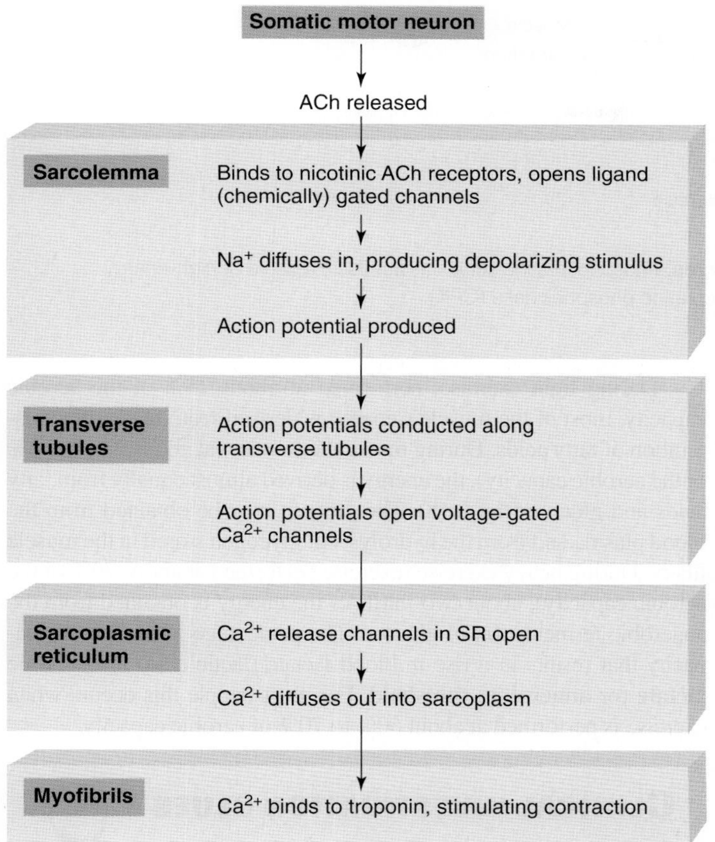
Somatic motor neuron
ACh released
Sarcolemma
Binds to nicotinic ACh receptors, opens ligand (chemically) gated channels
Na+ diffuses in, producing depolarizing stimulus
Action potential produced
Transverse tubules
Action potentials conducted along transverse tubules
Action potentials open voltage-gated Ca2+ channels
Sarcoplasmic reticulum
Ca2+ release channels in SR open
Ca2+ diffuses out into sarcoplasm
Myofibrils
Ca2+ binds to troponin, stimulating contraction

FIGURE 9.16 Summary of excitation-contraction coupling. Electrical excitation of the muscle fiber—that is, action potentials conducted along the sarcolemma and down the transverse tubules—triggers the release of Ca^{2+} from the sarcoplasmic reticulum. Since Ca^{2+} binding to troponin leads to contraction, the Ca^{2+} can be said to couple excitation to contraction.

CHECK POINT

1. Describe the location and functions of troponin and tropomyosin. How do these compare in a relaxed and in a contracting muscle?
2. Identify the transverse tubules and sarcoplasmic reticulum, and describe their functions.
3. How are action potentials produced in a muscle fiber, and where are they conducted to?
4. How do action potentials stimulate the entry of Ca^{2+} into the cytoplasm?
5. Step-by-step, explain how Ca^{2+} stimulates muscle contraction.

Muscle Metabolism Includes Lactic Acid Production and Aerobic Respiration

Muscles require ATP for both contraction and relaxation. The ATP in the muscle fibers is ultimately produced by cell respiration, but it can be made quickly using the phosphate group of phosphocreatine, which stores "high-energy phosphate" in the cell. Lactic acid fermentation generates ATP but is less efficient than aerobic cell respiration. People with a higher aerobic capacity can consequently perform more exercise without fatigue. Slow-twitch fibers are adapted for aerobic respiration, whereas fast-twitch fibers are adapted for lactic acid fermentation.

As explained in the last section, muscle contraction requires ATP to (1) activate the myosin head cross bridges prior to contraction, and (2) bind to the myosin heads after the power stroke, so that the cross bridges can detach from the actin and again become activated for another cycle. Muscle relaxation is also dependent on ATP, because ATP is required to power the active transport Ca^{2+} pumps in the membrane of the sarcoplasmic reticulum. These pumps transport the Ca^{2+} into the terminal cisternae of the SR, so that there is very little left in the cytoplasm to bind to troponin. As a result, the tropomyosin returns to its inhibitory position, preventing the binding of the myosin heads to actin.

Muscle Fibers Produce ATP Using Different Metabolic Pathways

Cells can't build up a large store of extra ATP for later use, because ATP exerts a negative feedback inhibition on its own production by cell respiration. This might present a problem for muscles, because during intense exercise muscles can use ATP at a faster rate than it can be produced through cell respiration. Fortunately, muscles at rest can use the ATP they make through cell respiration to build up a store of "high-energy phosphate." The ATP in a resting muscle fiber can donate its phosphate (and thus be converted into ADP) to a molecule called **creatine**, forming **phosphocreatine** (fig. 9.17). Large amounts of phosphocreatine can be stored in the resting muscle fiber for later use. During intense exercise, as ATP is being used for muscle contraction and relaxation, the ATP can be quickly regenerated using phosphocreatine.[4] In this case, phosphocreatine is enzymatically converted back into creatine, and the phosphate group is donated to ADP, forming ATP (fig. 9.17).

CLINICAL APPLICATIONS

The enzyme that transfers phosphate between creatine and ATP is called **creatine kinase** (**CK**), or **creatine phosphokinase** (**CPK**). Skeletal muscle and cardiac muscle have two different forms of this enzyme, or isoenzymes. The isoenzyme of CPK found in skeletal muscles is elevated in the blood of people with *muscular dystrophy*. The isoenzyme of CPK found in cardiac muscle is elevated in the blood of people with a *myocardial infarction* ("heart attack"), and laboratory tests for this are used to diagnose heart disease.

FYI [4]People wishing to improve their muscle performance sometimes take creatine supplements. Studies show that these supplements increase the amount of phosphocreatine in muscle fibers and appear to improve performance by a few percentage points in high-intensity exercise, such as in sprints and weight lifting. Short-term studies have so far failed to prove that taking exogenous creatine (creatine supplements) has deleterious side effects. However, (1) exogenous creatine appears to suppress the ability of muscles to produce their own creatine; and (2) the kidneys excrete a great amount of extra creatine in the urine, and the long-term effects of this on kidney health are presently unknown but of concern. So, if you're contemplating using creatine supplements, get the facts and weigh the potential risks against whatever benefits you might realistically expect to derive.

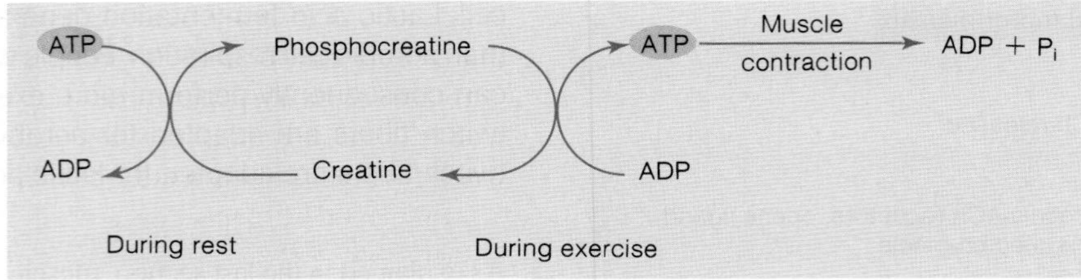

FIGURE 9.17 The production and use of phosphocreatine in muscles. Phosphocreatine serves as a muscle reserve of high-energy phosphate, used for the rapid formation of ATP. These reactions are catalyzed by creatine phosphokinase (CPK).

Clinical Investigation Clues

Remember that Rachel had a history of heart problems, and at that time her blood work indicated an elevated level of CPK.

- What is CPK, and what is its function?
- Where is CPK normally located, and what could an elevated blood level indicate?

Muscle Metabolism Depends on Exercise Intensity

When you do moderate to heavy exercise, the first 45 to 90 seconds of the exercise is mostly performed using *lactic acid fermentation* (also called *anaerobic respiration*; see chapter 2) to generate ATP. This means that the muscles first convert glucose into pyruvic acid (a metabolic pathway called *glycolysis*; see chapter 2), and then convert the pyruvic acid into lactic acid. This produces 2 ATP per glucose, which is substantially less than would have been produced aerobically. Because of this inefficiency, more glucose molecules must be consumed to produce the amount of ATP that would have been produced aerobically. Also, the lactic acid produced by this metabolic pathway may contribute to muscle fatigue. However, there isn't a choice: the muscles are forced to get their ATP anaerobically at the beginning of strenuous exercise because there hasn't yet been sufficient time to make the physiological adjustments necessary to increase the oxygen delivery to meet the increased metabolic demand. If exercise is moderate, aerobic cell respiration will contribute most of the energy needed by the muscles following the first 2 minutes of exercise.

Whereas glucose is the only molecule that can be used for lactic acid fermentation, cells can use fatty acids as well as glucose for aerobic respiration. Indeed, resting skeletal muscles, and those performing mild exercise, use the aerobic respiration of fatty acids in the blood plasma (the fluid portion of the blood) as their primary source of energy (fig. 9.18). Whether exercise is light, moderate, or heavy for a given person depends on that person's **maximal oxygen uptake**, or **aerobic capacity**. This is the maximum rate of a person's oxygen consumption during intense exercise. It's 15% to 20% higher for males than for females, and higher at age 20 for both sexes. Some world-class athletes have aerobic capacities that are twice the average for their age and sex. This depends mostly on genetics, but aerobic training can increase aerobic capacity by about 20%.

During light exercise, performed at about 25% of the aerobic capacity, most of the muscle's energy is derived from the aerobic respiration of fatty acids. During moderate exercise (at about 50% to 70% of the aerobic capacity), the energy is derived almost equally from fatty acids and glucose (fig. 9.18). The glucose may be obtained from the blood plasma and from the hydrolysis of glycogen stored in the muscle fibers. During heavy exercise (exercise performed at above 70% of the aerobic capacity), about two-thirds of the energy is obtained from the anaerobic fermentation of glucose. The percentage of the aerobic capacity that results in a rise in blood lactate (lactic acid) levels is the **lactate** (or **anaerobic**) **threshold**. For most people this occurs when exercise is performed at about 50% to 70% of aerobic capacity.

Clinical Investigation Clues

Remember that Rachel complained of muscle fatigue after working on the leg-lift machine, and the trainer responded that her muscles were producing lactic acid.

- Which metabolic pathway, starting with which molecule, produces lactic acid?
- Why did Rachel's quadriceps femoris muscles use this pathway for this exercise?
- At what percentage range of Rachel's aerobic capacity was this exercise performed?
- If Rachel used a much lighter resistance (weight) and could continue this exercise for a few more minutes without muscle fatigue, which metabolic pathway and which energy molecules would her muscles use to make ATP?

The Oxygen Debt Must Be Repaid After Exercise

When you stop exercising, you continue to breathe harder for a period of time. During this time following exercise, your body continues to consume oxygen (in aerobic cell respiration) at a faster rate than at rest. This extra oxygen is used to repay the **oxygen debt** incurred during exercise. There are several reasons for this oxygen debt:

1. Oxygen must be repaid that was withdrawn from the blood (where it is bound to hemoglobin in red blood cells) and from the muscle fibers (where it is bound to a molecule related to hemoglobin called *myoglobin*). The storage of oxygen bound to myoglobin in muscles occurs more in some fibers than in others, as described in the next section.

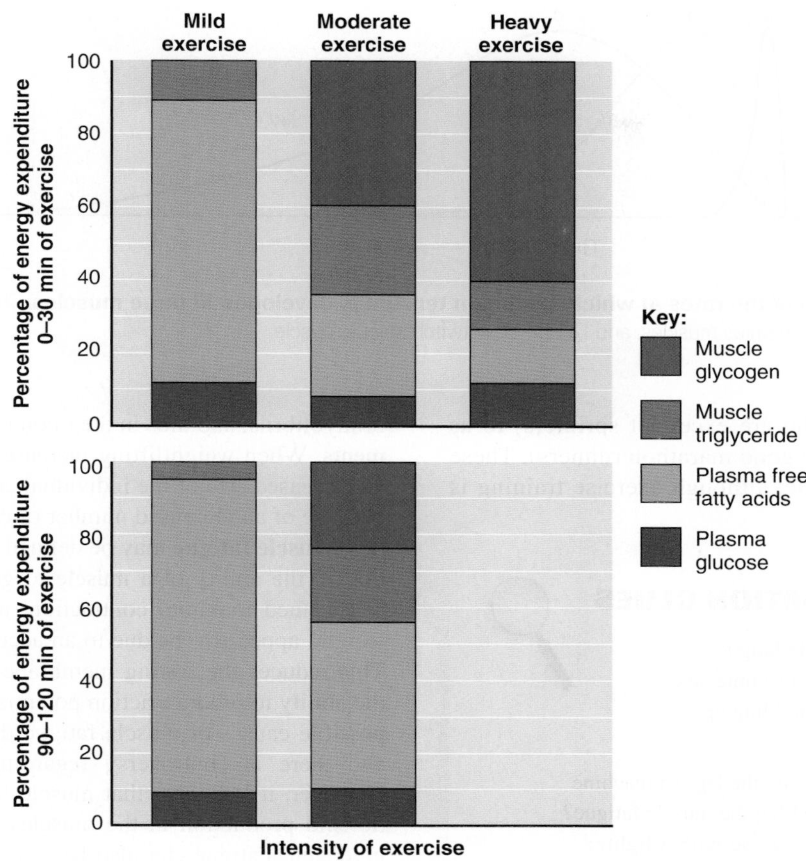

FIGURE 9.18 Muscle metabolism during exercise. The relative contributions of plasma glucose, plasma free fatty acid, muscle glycogen, and muscle triglycerides (fats) are indicated at different levels of exercise. Exercise is divided into mild exercise (at 25% of aerobic capacity); moderate exercise (at 65% of aerobic capacity); and heavy exercise (at 85% of aerobic capacity). Data for heavy exercise at 90 to 120 minutes were not available.

2. The muscles consume extra oxygen after exercise because they are still warm, with a faster metabolic rate than at rest.
3. Extra oxygen is needed to metabolize the lactic acid produced by the exercising skeletal muscles. Lactic acid can be used in aerobic cell respiration, where it is metabolized to carbon dioxide and water and releases energy for ATP formation.

Skeletal Muscles Have Slow- and Fast-Twitch Fibers

Skeletal muscle fibers can be divided on the basis of their contraction speed into **slow-twitch**, or **type I**, **fibers**, and **fast-twitch**, or **type II**, **fibers**. The extraocular muscles that move the eyes, for example, have a high proportion of fast-twitch fibers; the soleus muscle in the leg, by contrast, has a high proportion of slow-twitch fibers. The gastrocnemius (calf) muscle has a mixture of these two types, and its contraction speed is midway between that of the other two muscles (fig. 9.19). Most muscles of the body contain mixtures of fast- and slow-twitch fibers.

Slow-twitch (type I) fibers are adapted for aerobic respiration, and because of this they are also referred to as *slow oxidative fibers*. Their adaptations for aerobic respiration include a rich blood supply, numerous mitochondria (chapter 2), a high concentration of the enzymes needed for aerobic respiration, and a large amount of the protein **myoglobin**. Myoglobin is an iron-containing red pigment, similar to the hemoglobin in red blood cells, which can bond to oxygen. Because slow-twitch fibers have high myoglobin content, they are also called *red fibers*.

Fast-twitch (type II) fibers are thicker and have a less extensive blood supply, fewer mitochondria, and a lower amount of myoglobin than the slow-twitch fibers. Because of their reduced myoglobin content, fast-twitch fibers have also been called *white fibers*. The fast-twitch fibers, adapted for lactic acid production by having a large store of glycogen and a high content of the enzymes needed for glycolysis, are known as *fast glycolytic fibers*. Other fast-twitch fibers have a greater capacity for aerobic respiration, and are thus called *fast oxidative fibers*. People vary tremendously in their proportion of slow- and fast-twitch fibers. For example, the percentage of slow-twitch fibers in the quadriceps femoris muscles of the leg can vary

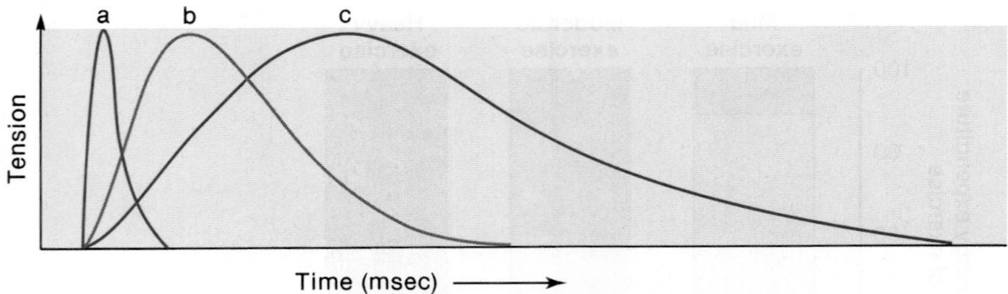

FIGURE 9.19 A comparison of the rates at which maximum tension is developed in three muscles. These are (*a*) the relatively fast-twitch extraocular and (*b*) gastrocnemius muscles, and (*c*) the slow-twitch soleus muscle.

from under 20% (in people who are excellent sprinters) to as high as 95% (in people who are good marathon runners). These differences are primarily genetic, although exercise training is also important.

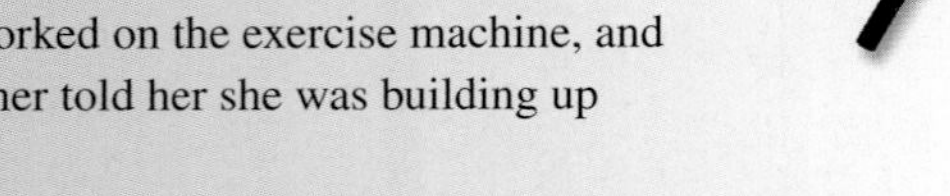

CLINICAL INVESTIGATION CLUES

Remember that Rachel felt muscle fatigue when she worked on the exercise machine, and that her trainer told her she was building up lactic acid.

- Which type of muscle fibers did the leg-lift machine mostly exercise when Rachel felt the muscle fatigue?
- If Rachel performed this exercise with a lighter resistance (weight) that didn't cause muscle fatigue, which muscle fiber types would be mostly involved in this exercise?

Muscles Adapt to Exercise Training

All muscle fiber types adapt to endurance training by an increase in mitochondria and other changes that can increase the aerobic capacity by as much as 20%. The higher aerobic capacity of trained endurance athletes (such as swimmers and long-distance runners) increases the amount of exercise they can perform before reaching their lactate threshold. For example, the lactate threshold of an untrained person may be 60% of their lower aerobic capacity, whereas the lactate threshold of a trained athlete may be 70% or more of their higher aerobic capacity. This increases the amount of exercise that the athlete can perform aerobically, reducing fatigue. Also, endurance training increases the proportion of muscle energy derived from the aerobic respiration of fatty acids. This decreases the amount of energy that must be derived from glucose, allowing the muscle's store of glycogen to last longer.

Endurance training doesn't increase the size of muscles. Muscle enlargement is produced only by frequent periods of high-intensity exercise in which muscles work against a high resistance, as in weightlifting. This increases the thickness of fast-twitch (type II) muscle fibers because of an increased size and number of myofibrils, and thus in their content of sarcomeres and myofilaments. When weightlifting increases muscle size, it's because of an increased size of the individual cells[5] (*hypertrophy*) rather than because of an increased number of cells (*hyperplasia*).

Muscle fatigue may be defined as an exercise-induced reduction in the ability of a muscle to generate force. Fatigue during a sustained maximal contraction, as when lifting a very heavy weight, appears to be due to an accumulation of extracellular K^+. This reduces the resting membrane potential and interferes with the ability to produce action potentials. There appear to be several possible causes of muscle fatigue during most types of exercise, and there is controversy regarding their relative importance. However, it is known that muscle fatigue is associated with lactic acid production in the muscles and depletion of their stored glycogen. Fatigue can also be produced by changes in the CNS, rather than in the muscles, so that it sets in before the muscles themselves fatigue. The causes of this *central fatigue* are not presently well understood.

CLINICAL INVESTIGATION CLUES

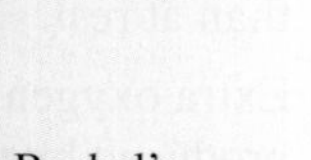

Remember that Rachel felt muscle fatigue when she exercised her quadriceps femoris muscles on the leg-lift machine, and that her trainer said she was producing lactic acid.

- How would you expect this exercise to affect Rachel's endurance and the size of her quadriceps femoris muscles?
- If her muscles get larger from this exercise, how is this enlargement produced?
- If Rachel were to perform this exercise using a lighter resistance (weight) for a longer time, how would this exercise affect her endurance and muscle size?

FYI [5]Muscle atrophy (reduction in size) occurs when astronauts experience *microgravity* (weightlessness) for extended periods of time, as in *Skylab 4*, which lasted 84 days. Adjustments in diet and an exercise program were able to reduce this effect. Here on earth, weight-bearing muscles can be similarly unloaded in people who are bedridden or in a cast, promoting muscle atrophy.

CHECK POINT

1. Describe the reactions that interchange the phosphate group between ATP and creatine during rest and exercise, and explain the significance of phosphocreatine.
2. When do muscles perform anaerobic respiration? What are the benefits and costs of this metabolic pathway?
3. How do muscles obtain energy at rest and following 2 minutes of moderate exercise? Which molecules are used for this process?
4. Define the oxygen debt and explain the reasons it exists.
5. Distinguish between the different types of skeletal muscle fibers, and explain how these fibers are affected by exercise training.

Cardiac and Smooth Muscles Contract by Different Mechanisms

Cardiac muscle is striated, and so contains sarcomeres and contracts by the same sliding filament mechanism as skeletal muscle. However, the action potentials start spontaneously and are conducted from one cell to the next in cardiac muscle. Smooth muscle contains actin and myosin, but these aren't arranged in sarcomeres and the mechanism of contraction is somewhat different than in striated muscles. Also, smooth and striated muscles have different mechanisms of excitation-contraction coupling.

The description of cardiac muscle in this section will serve as a basis for a later discussion of the nature of the heartbeat and the electrocardiogram, and for the regulation of heart function in the next chapter. Similarly, the description of smooth muscles in this section will provide a basis for aspects of the physiology of the digestive, urinary, and reproductive systems that will be discussed in later chapters. Thus, this section on cardiac and smooth muscle will provide part of a sturdy foundation for your knowledge pyramid in the physiology of the body systems.

Cardiac and smooth muscles are similar in that they're found in the organs of the internal environment and are innervated by the autonomic nervous system (chapter 6). This is different from skeletal muscles, which are innervated by somatic motor neurons (chapter 6). Because of this difference in innervation, skeletal muscle is *voluntary muscle*, whereas cardiac and smooth muscles are *involuntary muscles* (chapter 2). However, cardiac and smooth muscles differ in a number of important respects, as discussed in this section.

Cardiac Muscle Cells Form a Functioning Myocardium

Cardiac muscle cells, or **myocardial cells**, are striated; they contain thick filaments composed of myosin and thin filaments composed of actin, and these myofilaments are organized into sarcomeres. The thin filaments contain troponin and tropomyosin, and contraction is stimulated when Ca^{2+} binds to troponin. As in skeletal muscles, the sarcoplasmic reticulum of myocardial cells stores Ca^{2+} and releases it in response to action potentials that are conducted by the sarcolemma and transverse tubules.

However, the organization of skeletal muscle fibers in a skeletal muscle is different from the organization of myocardial cells in the heart. The skeletal muscle fibers are structurally and functionally separate. Remember, each fiber is surrounded by its own connective tissue endomysium (see fig. 9.1), and each runs from the tendon of origin to the tendon of insertion. This permits variations in the numbers of muscle fibers stimulated, and thus provides gradations in the strength of contraction. By contrast, the myocardial cells are short, branched, tubular cells that are physically interconnected at the *intercalated discs* (fig. 9.20; also see chapter 2).

Myocardial cells in the heart aren't simply physically interconnected; they are also electrically joined together by **gap junctions** that are concentrated in the intercalated discs (fig. 9.21). As mentioned in chapter 4, gap junctions are electrical synapses; they are regions where the plasma membranes of adjacent cells fuse together, with special proteins that form channels through both membranes. In this way, ions (Na^+, K^+, and Ca^{2+}) can diffuse from the cytoplasm of one cell to the next, as if they were both one cell. As a result, an action potential can be conducted from one cell to the next. The physically and electrically interconnected mass of myocardial cells, known as a **myocardium**, behaves as a single functional unit. The myocardium conducts an action potential as a unit (as if it were one cell), and contracts as a unit. Thus, unlike the graded contraction of skeletal muscles, a myocardial contraction involves all of its component cells; the myocardium puts its whole heart into the contraction.

Actually, there are two myocardia in the heart (the two atria form one myocardium, and the two ventricles form another myocardium; see chapter 10). The heart has cells in a region (the right atrium) that function as a *pacemaker*. The pacemaker region

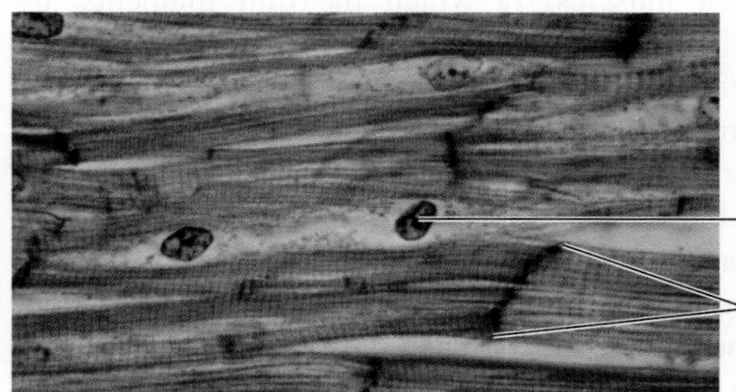

FIGURE 9.20 Cardiac muscle. Notice that the cells are short, branched, and striated and that they are interconnected by intercalated discs.

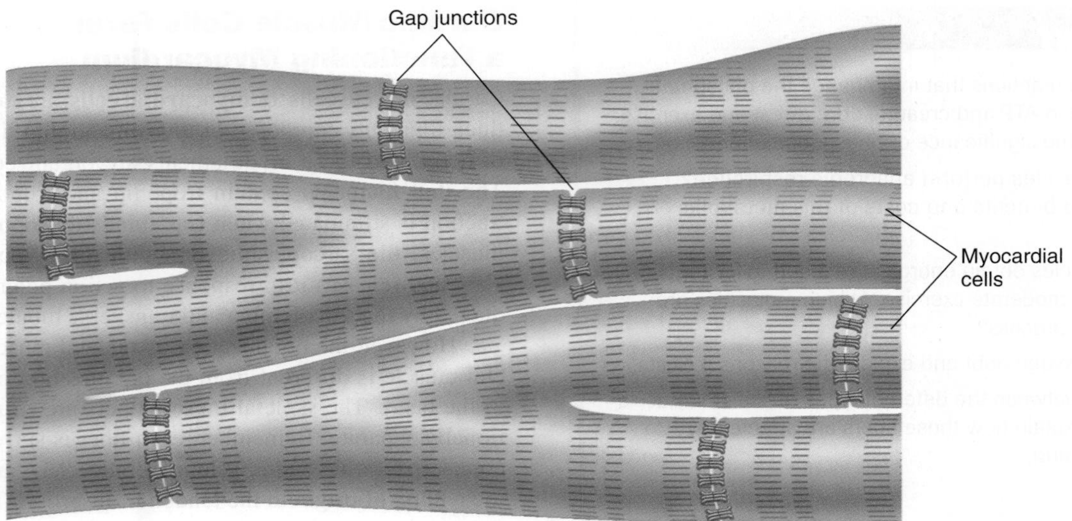

FIGURE 9.21 Myocardial cells are interconnected by gap junctions. The gap junctions are fluid-filled channels through the plasma membrane of adjacent cells that permit the conduction of impulses from one cell to the next. The gap junctions are concentrated at the ends of each myocardial cell, in the location of the intercalated discs.

undergoes a *spontaneous, automatic depolarization*, generating action potentials at a certain rate. Autonomic nerves only speed up or slow this rate; they don't cause the depolarization or action potential to occur. This contrasts with skeletal muscles, which depolarize and produce action potentials only in response to ACh released from somatic motor neurons. The electrical activity of the heart is discussed in chapter 10.

Smooth Muscle Has Different Properties Than Striated Muscles

Smooth muscle is arranged in circular layers in the walls of hollow visceral organs, such as in the gastrointestinal tract and the bronchioles (airways). There are also longitudinal layers of smooth muscles in the gastrointestinal tract and in the uterine tubes (which transport eggs), for example. The combination of circular and longitudinal smooth muscles permits **peristaltic contraction (peristalsis)**, a wavelike contraction that propels the contents of these tubes in one direction.

Unlike striated muscles, in which the thin filaments are relatively short, the thin filaments in smooth muscles are quite long. In the smooth muscle cell, the thin filaments attach to regions of the plasma membrane or to cytoplasmic structures called *dense bodies*, which are analogous to the Z discs of striated muscles. Unlike in striated muscles, the myosin proteins are arranged vertically to the long axis of the thick filaments (fig. 9.22). Its unique arrangement of myosin proteins and thin filaments gives smooth muscles a major advantage over striated muscles: *smooth muscles can contract even when greatly stretched*.

The smooth muscles in the urinary bladder must be able to contract strongly even though they are often stretched up to two and a half times their resting length. However, the most striking example of smooth muscle stretching is the uterus: by the end of pregnancy, the uterine smooth muscle cells may be stretched up to eight times their original length! What would happen if the uterus were instead composed of striated muscle? A glance back at figure 9.11 will reveal the answer—a sarcomere stretched to that degree simply couldn't contract. In contrast, stretched smooth muscles contract strongly, because their myosin heads can still bind to the actin proteins in the thin filaments.

Like in striated muscle, the contraction of smooth muscle is triggered by a sharp rise in the Ca^{2+} concentration within the cytoplasm. Unlike in striated muscles, where the Ca^{2+} for contraction comes primarily from the SR, in smooth muscle the Ca^{2+} comes in through voltage-gated Ca^{2+} channels in the plasma membrane. This Ca^{2+} entering the cytoplasm stimulates smooth muscle contraction by a different mechanism than in striated muscles. Although more complex than the mechanism responsible for excitation-contraction coupling in striated muscles, the mechanism by which Ca^{2+} stimulates contraction in smooth muscle permits the smooth muscles to contract in a slow, sustained, and energy-efficient manner.

CLINICAL APPLICATIONS

Drugs such as *nifedipine* (*Procardia*) and newer compounds are **calcium channel blockers**. These drugs block the voltage-gated Ca^{2+} channels in the plasma membrane of smooth muscle cells located in the walls of blood vessels, causing the smooth muscles to relax. This dilates the blood vessels, a response that can be useful in treating *hypertension* (high blood pressure). Calcium channel blockers are also used when spasm of the coronary arteries produces *angina pectoris*, a referred pain (chapter 7) caused by insufficient blood flow to the myocardium of the heart. By relaxing the smooth muscles of arteries and causing their dilation, calcium channel blockers can improve the blood flow to the heart and reduce the work the heart performs in ejecting blood into the arterial system.

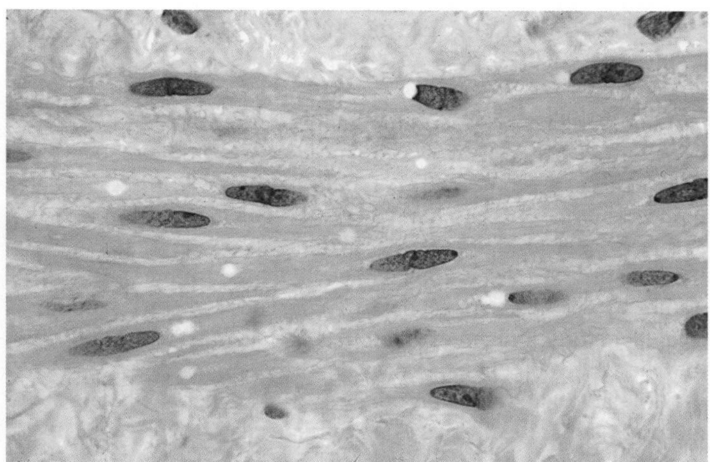

(a)

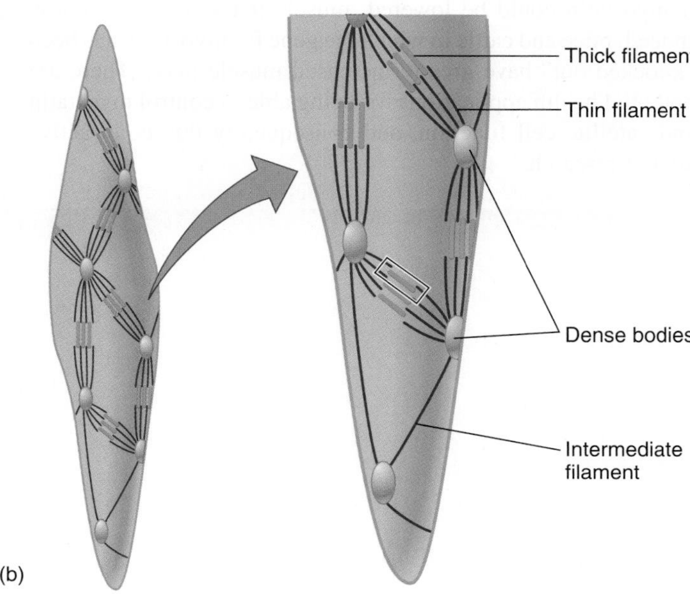

(b)

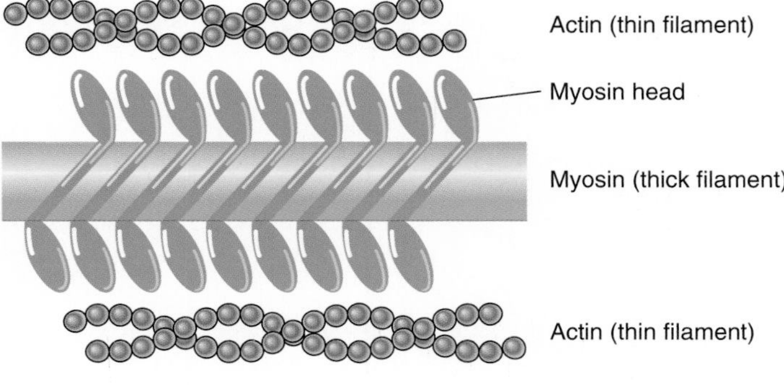

(c)

FIGURE 9.22 Smooth muscle and its contractile apparatus. (*a*) A photomicrograph of smooth muscle cells in the wall of a blood vessel. (*b*) Arrangement of thick and thin filaments in smooth muscles. Note that dense bodies are also interconnected by intermediate fibers. (*c*) The myosin proteins are stacked in a different arrangement in smooth muscles than in striated muscles.

Clinical Investigation Clues

Remember that Rachel reported that she had heart problems and that she was taking calcium channel blockers.

- What cellular process does calcium channel blockers inhibit?
- What effect does the action of these drugs have on the body systems?
- How would calcium channel blockers help the heart?

The smooth muscles in different organs can have different properties. Most smooth muscles—including those in the gastrointestinal tract and uterus—are **single-unit smooth muscles**. Single-unit smooth muscles have numerous gap junctions between adjacent cells that weld them together electrically. This is similar to cardiac muscle. Another similarity is that single-unit smooth muscles display *pacemaker* activity, in which certain cells stimulate others in the unit. However, unlike cardiac muscle, some single-unit smooth muscles can generate action potentials and contract in response to stretching (by increased intestinal contents, for example). Autonomic nerves influence the rate and strength of the contractions of single-unit smooth muscles, but the contractions themselves result from the intrinsic activities of the smooth muscle.

Some smooth muscles—including the ciliary muscle that adjusts the lens of the eye—are **multiunit smooth muscles**. Unlike the single-unit smooth muscles, the multiunit smooth muscles don't contract without stimulation from autonomic axons. Multiunit smooth muscle generally lacks gap junctions between its cells, and so each smooth muscle cell must be stimulated by an autonomic axon, similar to the way that skeletal muscle is regulated. In the case of the ciliary muscle, this allows the nervous system to finely control its degree of contraction, and thus finely control the shape of the lens to maintain a focus on the retina (chapter 7).

Check Point

1. How is cardiac muscle similar to skeletal muscle, and how is it different? Also, how does neural regulation compare in skeletal and cardiac muscles?
2. Describe how the arrangement of thick and thin filaments, and myosin proteins in the thick filaments, compare in striated and smooth muscles.
3. From where in a smooth muscle cell does Ca^{2+} enter to stimulate contraction? How does this differ from striated muscles?
4. Distinguish between single-unit and multiunit smooth muscles.

Physiology in Health and Disease

There are commonly recognized muscle changes that occur with age. Older people experience a loss of muscle fibers, a decreased size of fast-twitch muscle fibers, a reduced amount of glycogen stored in the muscle fibers, and changes that decrease their aerobic capacity. Resistance training can help the surviving muscle fibers to hypertrophy and partially compensate for the decline in the number of muscle fibers. Endurance training can increase the supply of blood to the muscle fibers, improving the ability of the fibers to respire aerobically. Also, endurance training can increase the muscle's store of glycogen, although it can't raise it to the levels present in youth. These physiological changes that occur with age justify the grouping of contestants into age categories for athletic competitions

Destruction of striated muscle fibers is particularly damaging because the remaining healthy fibers can't divide to replace the damaged ones. However, skeletal muscles have stem cells (chapter 2), known as **satellite cells**, located outside the muscle fibers. The satellite cells can proliferate (divide) after injury, and undergo changes that allow them to fuse together to form new muscle fibers. However, the number of satellite cells and their ability to proliferate after injury declines with age. The reduced number of satellite cells in the elderly also limits the ability of muscle fibers to grow by hypertrophy, because satellite cells provide the additional nuclei required for muscle fibers as they grow thicker in response to resistance (weight) training. The declining numbers of satellite cells may also contribute to the muscle atrophy that can occur in the elderly.

A paracrine regulator (chapter 3) known as **myostatin** inhibits satellite cell function and muscle growth. Consistent with this, studies have found that elderly people with reduced muscle mass have elevated levels of myostatin. Thus, if myostatin could be lowered, muscle mass should increase. Indeed, mice and cattle in which the gene for myostatin has been "knocked out" have greatly increased muscle mass. There are potential health applications for being able to control myostatin and satellite cell function, and consequently this is an active area of research.

Physiology in Balance

Muscle Physiology

Working together with . . .

Nervous System

Motor neurons stimulate the contraction of all muscle types . . . p. 135

Urinary System

Micturition (urination) involves smooth muscle contraction . . . p. 324

Cardiovascular System

The contractions of the heart serve as a pump to move blood through the circulation to the organs . . . p. 240

Sensory System

Smooth muscle in the ciliary muscle contracts and relaxes to adjust the refraction of the lens . . . p. 157

Digestive System

Smooth muscles in the stomach and intestine contract to move food by peristalsis from one region of the digestive tract to another . . . p. 348

Summary

Skeletal Muscles Are Specialized for Contraction 216

- Skeletal muscles are composed of muscle fibers, organized into muscle fascicles; each muscle is attached by means of a tendon to its origin and insertion.
- The muscles fibers are striated, with dark A bands and light I bands, and with Z discs in the middle of the I bands.
- Each muscle fiber contains numerous myofibrils, and each myofibril contains many myofilaments that are arranged to produce the A and I bands.
- A sarcomere is the region from Z disc to Z disc; the A band is in the middle and is produced by the stacking of thick filaments.
- The thin filaments are anchored by the Z discs and extend into the I bands and partway into the A bands; the H zones are the middle regions of the A bands that contain only the thick filaments.
- Thick filaments are composed of the protein myosin; thin filaments are composed mostly of the protein actin.
- Skeletal muscles can produce twitches, and the twitches can summate to produce a smooth sustained contraction known as tetanus.
- Contractions in which the muscle stays the same length are isometric contractions, whereas contractions in which the muscle changes length are isotonic contractions.
- Concentric contractions occur when the muscle shortens, because the contraction strength is greater than the load; eccentric contractions are those in which a force causes a muscle to lengthen despite its contraction, because the load is greater than the contraction strength.

Striated Muscles Contract by a Sliding Filament Mechanism 220

- The myofilaments do not shorten in muscle contraction; instead, the thin filaments slide over the thick filaments due to movements of cross bridges between them; this causes the sarcomeres to become shorter.
- The myosin head has an actin-binding site, for forming cross bridges with the actin in the thin filaments; it also has an ATPase site that hydrolyzes ATP into ADP and P_i, which stay bound to the myosin head.
- The hydrolysis of ATP activates ("cocks") the myosin head, changing its orientation so that it can undergo a power stroke and produce sliding of the filaments; however, the myosin heads do not bind to actin when a muscle is at rest.
- When the myosin head is allowed to bind to actin, the P_i group is released and the power stroke occurs; after that, the myosin head releases the ADP and binds a new ATP.
- Binding to a new ATP allows the myosin head to release from the actin; then, the ATPase breaks down that ATP to again activate the myosin head so that another power stroke can occur and the muscle can shorten to a greater extent.
- Contraction strength is greatest when the resting muscle length allows the ideal amount of interaction between the thick and thin filaments; if the resting sarcomere is too long or too short, contraction strength is decreased.

Action Potentials Stimulate Contraction by Means of Ca^{2+} 224

- The globular actin proteins in the thin filaments form a double row containing tropomyosin; a group of 3 troponin proteins is bound to the tropomyosin.
- In a resting muscle, tropomyosin physically blocks the binding of the myosin heads to actin, inhibiting contraction.
- In a resting muscle, the Ca^{2+} is stored in the terminal cisternae of the sarcoplasmic reticulum, with very little in the cytoplasm; this is an active transport process.
- The sarcolemma, which is the plasma membrane of muscle fibers, tunnels inward to form the transverse tubules.
- When a somatic motor neuron releases ACh, it stimulates an EPSP in the postsynaptic sarcolemma membrane; this causes adjacent voltage-gated channels in the sarcolemma to produce and conduct an action potential.
- The transverse tubules conduct action potentials deep within the muscle fiber, and the action potentials cause the opening of voltage-gated Ca^{2+} channels in the transverse tubules.
- The opening of voltage-gated Ca^{2+} channels in the transverse tubules causes the opening of larger Ca^{2+} release channels in the sarcoplasmic reticulum (SR), permitting the diffusion of Ca^{2+} out of the SR.
- When Ca^{2+} enters the cytoplasm, it binds to troponin and causes the troponin-tropomyosin complex to shift position in the thin filaments, exposing the myosin-binding sites on the actin.
- The myosin heads then bind to the actin, forming cross bridges that undergo power strokes; thus, Ca^{2+} links electrical excitation to contraction in a process known as excitation-contraction coupling.
- After the action potentials have stopped, the Ca^{2+} is moved by active transport Ca^{2+} pumps in the sarcoplasmic reticulum out of the cytoplasm and back into the terminal cisternae; muscle relaxation thus requires ATP.

Muscle Metabolism Includes Lactic Acid Production and Aerobic Respiration 227

- Muscle fibers can produce ATP quickly during intense exercise by donating the "high-energy phosphate" from phosphocreatine to ADP; during rest, the fiber can build a store of phosphocreatine by donating the phosphate from ATP to creatine.
- The first 45 to 90 seconds of moderate to heavy exercise is performed by muscles using energy obtained through the anaerobic process of lactic acid fermentation.
- Adjustments in the ability to obtain oxygen from the blood permit the muscles to obtain energy through the aerobic respiration of fatty acids and glucose following 2 minutes of exercise.
- Whether exercise is light, moderate, or heavy for a person depends on that person's maximal oxygen uptake, or aerobic capacity.
- Lactic acid production and the blood levels of lactic acid rise when most people exercise at about 50% to 70% of their aerobic capacity; this is termed their lactate (or anaerobic) threshold.
- Extra oxygen is consumed after exercise has ended; this is the oxygen debt, used to repay oxygen withdrawn from hemoglobin and myoglobin during exercise; to supply the muscles still warm after exercise; and to metabolize lactic acid.
- Muscles contain fast-twitch and slow-twitch fibers; some muscles (such as the soleus) are mostly slow-twitch; others (such as the extraocular muscles) are mostly fast-twitch; most muscles (such as the gastrocnemius) contain a mixture of both types and have an intermediate speed of contraction.
- Slow-twitch fibers are adapted for aerobic respiration; they have much myoglobin (making them red fibers), a high number of mitochondria, and a rich blood supply.
- Some fast-twitch fibers are also adapted for aerobic respiration (fast oxidative fibers), while others are adapted for lactic acid fermentation (fast glycolytic fibers).
- Endurance training increases the aerobic ability of all fiber types and can raise a person's aerobic capacity by as much as 20%; high-resistance (weight) training causes fast-twitch muscle fibers to hypertrophy; that is, they get thicker as their content of myofibrils and sarcomeres increases.

- Muscle fatigue has a number of causes and is generally associated with lactic acid production and the depletion of stored glycogen.

Cardiac and Smooth Muscles Contract by Different Mechanisms 231

- Myocardial cells are short, branched, and interconnected by intercalated discs; there are also gap junctions (electrical synapses) concentrated at the regions of the intercalated discs.
- The interconnected myocardial cells form a functioning unit known as a myocardium; in the myocardium, action potentials can travel from one cell to the next, as if the myocardium were a single cell.
- Cardiac muscle has a pacemaker region that automatically and spontaneously depolarizes to stimulate the production of action potentials at a certain rate; autonomic nerves modify this rate, they don't cause the heartbeat.
- Cardiac muscle is striated, so it has sarcomeres like skeletal muscle; also, like in skeletal muscle, contraction is stimulated by a rise in the concentration of Ca^{2+} in the cytoplasm and the binding of Ca^{2+} to troponin.
- Smooth muscle cells have actin and myosin, and thick and thin filaments, but the thin filaments are long and connected either to the plasma membrane or to dense bodies, which are analogous to the Z discs of striated muscles.
- The myosin proteins are organized differently in smooth muscles, allowing the smooth muscles to contract even when they are greatly stretched.
- Ca^{2+} entering the cytoplasm through voltage-gated Ca^{2+} channels in the plasma membrane stimulates contraction of smooth muscles; however, the mechanism of excitation-contraction coupling is different in smooth muscles than in striated muscles.
- Single-unit smooth muscles (as in the gastrointestinal tract) have cells joined by gap junctions and have pacemaker activity; autonomic nerves only modify the contractions of these smooth muscle.
- Multiunit smooth muscles (as in the ciliary muscle of the eye) generally lack gap junctions; each of their cells must be separately stimulated by an autonomic neuron in order to contract.

Review Activities

Objective Questions: Test Your Knowledge

1. Which of the following is in correct order of size, from largest to smallest?
   a. Fiber, myofibril, fascicle, myofilament
   b. Fiber, myofibril, myofilament, fascicle
   c. Fascicle, fiber, myofibril, myofilament
   d. Myofilament, myofibril, fiber, fascicle

2. Which of the following statements about a sarcomere is false?
   a. A sarcomere extends from one Z disc to the next.
   b. The A bands contain only thick filaments.
   c. The I bands contain only thin filaments.
   d. The thick filaments are composed of myosin.

3. Which of the following statements about the series elastic component is true?
   a. It must first be stretched tight in order for a muscle to shorten.
   b. It has elastic properties.
   c. When a muscle relaxes, it exhibits elastic recoil.
   d. All of these are true.

4. A muscle contraction in which there is no significant change in the muscle's length is a(n)
   a. isometric contraction.
   b. isotonic contraction.
   c. concentric contraction.
   d. eccentric contraction.

5. Which of the following illustrates a concentric contraction?
   a. A gymnast holding his legs in an "L" position
   b. A person lowering the weight back to the chest after doing a "bench press"
   c. A gymnast lifting one leg up laterally until her foot's above her head
   d. All of these.

6. Which of the following statements about a smooth, sustained contraction is false?
   a. It's called tetany.
   b. It's called tetanus.
   c. It's produced by rapid fusion of muscle twitches.
   d. It's produced *in vivo* by the activation of motor units in rapid succession.

7. When a muscle contracts and shortens, which of the following does not occur?
   a. The sarcomeres get shorter.
   b. The myofilaments get shorter.
   c. The myofibrils get shorter.
   d. The I bands get shorter.

8. The cross bridges are
   a. part of the actin proteins.
   b. composed of tropomyosin.
   c. extensions from the thin filaments.
   d. extensions from the thick filaments.

9. In order for a cross bridge to get activated, which of the following must immediately occur?
   a. It must bind to actin.
   b. It must release a P_i.
   c. It must release ADP and bind to a new ATP.
   d. It must hydrolyze ATP into ADP and P_i.

10. In order for a cross bridge to undergo a power stroke, which of the following must immediately occur?
    a. It must bind to actin.
    b. It must release a P_i.
    c. It must release ADP and bind a new ATP.
    d. It must hydrolyze ATP into ADP and P_i.

11. In order for a cross bridge to release from the thin filament after a power stroke is completed, which of the following must immediately occur?
    a. It must bind to actin.
    b. It must release a P_i.
    c. It must release ADP and bind to a new ATP.
    d. It must hydrolyze ATP into ADP and P_i.

12. The greatest strength of contraction is produced when a resting muscle has
    a. just some overlap of thin and thick filaments in the A bands.
    b. the greatest amount of overlap between the thick and thin filaments in the A bands.
    c. no overlap between thick and thin filaments in the A bands.
    d. the thick filaments of the A bands abutting the Z discs.

13. Which of the following is a true statement about tropomyosin?
    a. There are three different tropomyosins.
    b. It is located in the thick filaments.
    c. It lies in the groove of a helical double row of globular actin proteins.
    d. It binds to Ca^{2+}.

14. In a resting muscle fiber, the Ca^{2+} is located
    a. in the sarcoplasmic reticulum.
    b. in the cytoplasm.
    c. in the transverse tubules.
    d. in the sarcolemma.

15. Which of the following statements about Ca^{2+} release channels is false?
    a. They are located in the sarcoplasmic reticulum.
    b. They are opened by voltage-gated Ca^{2+} channels.
    c. They permit diffusion of Ca^{2+} into the terminal cisternae.
    d. They are 10 times larger than the voltage-gated Ca^{2+} channels.

16. Which of the following statements about action potentials in a skeletal muscle fiber is false?
    a. They are produced by the sarcolemma.
    b. They are conducted along the transverse tubules.
    c. They are produced in response to EPSPs.
    d. They originate spontaneously in the skeletal muscle fiber.

17. Which of the following processes does not directly require ATP?
    a. Binding of the myosin head to actin
    b. Activation of the myosin head
    c. Release of the cross bridge from actin after a power stroke
    d. Action of Ca^{2+} pumps in the sarcoplasmic reticulum

18. Energy obtained during the first 15 seconds or so of moderate to heavy exercise is obtained from
    a. the lactic acid fermentation of fatty acids.
    b. the lactic acid fermentation of glucose.
    c. the aerobic respiration of fatty acids.
    d. the aerobic respiration of glucose and fatty acids.

19. On the average, the blood levels of lactic acid begin to rise when a person exercises at about
    a. 15% to 25% of the aerobic capacity.
    b. 25% to 50% of the aerobic capacity.
    c. 50% to 70% of the aerobic capacity.
    d. 80% to 95% of the aerobic capacity.

20. The extra oxygen required by the body at the end of exercise is the
    a. aerobic capacity.
    b. oxygen debt.
    c. anaerobic threshold.
    d. maximal oxygen uptake.

21. Which of the following muscle fibers has high amounts of myoglobin, a rich blood supply, and a slow rate of twitch?
    a. Type I fibers
    b. Fast glycolytic fibers
    c. Fast oxidative fibers
    d. Type II fibers

22. Which of the following statements regarding endurance training is false?
    a. It increases aerobic capacity by about 20%.
    b. It increases the proportion of energy obtained from fatty acids.
    c. It allows muscle glycogen to become depleted at a slower rate.
    d. It promotes hypertrophy of type II fibers.

23. Which of the following statements regarding cardiac muscle is false?
    a. It has numerous gap junctions between its myocardial cells.
    b. It spontaneously depolarizes to automatically produce action potentials.
    c. It has dense bodies in the myocardial cells.
    d. Its cells are physically interconnected at the intercalated discs.

24. The type of muscle in which most of the Ca^{2+} entering the cytoplasm during contraction goes through voltage-gated Ca^{2+} channels in the plasma membrane is
    a. cardiac muscle.
    b. smooth muscle.
    c. skeletal muscle.
    d. All of these.

25. The types of muscle in which there are numerous gap junctions and pacemaker activity are
    a. multiunit smooth muscle and skeletal muscle.
    b. multiunit smooth muscle and cardiac muscle.
    c. single-unit smooth muscle and cardiac muscle.
    d. single-unit smooth muscle and skeletal muscle.

Essay Questions 1: *Test Your Understanding*

1. Describe the structure of a muscle, including its connective tissue layers, down to the level of the muscle fibers.
2. Describe the structure of a muscle fiber, down to the level of the sarcomeres.
3. Explain how the dark and light banding pattern in a striated muscle is produced, and indicate the type of myofilaments in each band.
4. Explain how summation of muscle twitches, incomplete tetanus, and complete tetanus are produced *in vitro* and *in vivo*.
5. Explain how graded skeletal muscle contractions can be produced *in vitro* and how they are normally produced *in vivo*.
6. Distinguish between an isometric and an isotonic contraction.
7. Distinguish between a concentric and an eccentric contraction.
8. Draw a sarcomere in a resting muscle, one in a partially contracted muscle, and one in a fully contracted muscle, labeling the A and I bands, Z discs, and H zones.
9. Explain the sliding filament theory of contraction, and explain the role of cross bridges in this mechanism.
10. Describe the structure and function of the myosin heads, and how they form cross bridges with the thin filaments.
11. Step-by-step, describe the cross-bridge cycle that occurs when the myosin heads are allowed to bind to actin.
12. Explain the length-tension relationship in striated muscles. How is the ideal resting muscle length for skeletal muscles maintained?
13. Identify the location, and describe the structure and functions, of troponin and tropomyosin.
14. What is the sarcoplasmic reticulum, and what is its function in muscle contraction and relaxation?
15. Identify the location of the voltage-gated Ca^{2+} channels and the Ca^{2+} release channels, and explain their functional relationship.
16. Step-by-step, explain how a somatic motor neuron stimulates a skeletal muscle fiber to contract; that is, explain excitation-contraction coupling, beginning with the release of ACh and ending at the binding of the myosin heads to actin.
17. Explain all of the ways that ATP is required in muscle contraction and relaxation.
18. Describe two ways that the opening of voltage-gated Ca^{2+} channels results in the opening of the Ca^{2+} release channels.

19. Describe how muscles use aerobic respiration and lactic acid fermentation for energy, and indicate under which conditions each is used.
20. Explain the physiological significance of creatine and phosphocreatine.
21. Define the oxygen debt and explain why it occurs.
22. Distinguish between the different muscle fiber types, and describe the effect of endurance and weight training on the muscle fibers.
23. Describe the structure of cardiac muscle, and explain how a myocardium can be said to function as a unit.
24. Explain how cardiac and skeletal muscles are similar, and how they are different.
25. Distinguish between single-unit and multiunit smooth muscles, and explain how autonomic nerves influence each.

Essay Questions 2: *Test Your Analytical Ability*

1. Compare what happens to the H zones during an isometric and an isotonic concentric contraction. Describe the action of the cross bridges in the two types of contractions.
2. Suppose a person is doing a curl (flexing the biceps brachii) with a heavy weight, and midway into the curl the arm starts shaking. Then, arm still shaking, the person lowers the weight as gently as possible back down to the support. Describe the muscle contractions occurring at each stage of this process, including the shaking.
3. Suppose a gymnast does a dismount from the horizontal bar, "sticking" the dismount without taking steps. However, he does crouch partway to absorb the shock, and then stands up straight. Describe the types of contractions of the quadriceps femoris muscles in this dismount, and the forces acting on the muscles.
4. What is an ATPase? There are two locations of an ATPase given in the story of skeletal muscle contraction: where are they, and what are their functions?
5. ATP is needed in muscle contraction, but it doesn't cause the contraction to occur at a particular time. Explain this statement.
6. What are rigor complexes, and what stops our muscles from getting them while we're alive? What role could phosphocreatine play in this?
7. Using the length-tension relationship, explain the effect of muscle spindle reflexes and muscle tone on the ability of muscles to contract.
8. Which is a better analogy with the cross-bridge cycles: a team of competitive rowers in a boat, or a team engaged in a tug-of-war? Explain.
9. Tropomyosin is sometimes compared to a "safety" catch on a gun. Explain this analogy.
10. Activation of the myosin heads by the hydrolysis of ATP may be likened to pulling back the string of a bow. Explain this analogy.
11. What would happen to the contraction and relaxation of an isolated skeletal muscle or heart if it were bathed in a solution containing an abnormally high concentration of Ca^{2+}? Explain.
12. What would happen to the ability of an isolated skeletal muscle or heart to contract if it were bathed in a solution containing an abnormally high concentration of K^+? Explain. (Hint: Review the effects of hyperkalemia on the resting membrane potential in chapter 4.)
13. Ca^{2+} can be considered a second messenger in muscle contraction, analogous to the second messengers used in hormone action (chapter 8). In this analogy, what would be the first messenger, and how does Ca^{2+} mediate its response?
14. A runner may experience a "second wind" after about 2 minutes of exercise. What might be responsible for this effect? Explain.
15. A sprinter, or a gymnast specializing in floor exercise, often has large quadriceps femoris muscles, whereas those muscles in a marathon runner are generally thinner. Explain this observation in terms of the muscle fiber types and the particular exercise.
16. Long periods of walking for exercise may be more effective at reducing fat stores than short periods of high-intensity exercise, even though the latter consumes more calories per minute. Explain this statement, in terms of the metabolic pathways involved.
17. "An athlete is made, not born." Evaluate this statement in terms of the types of muscle fibers and the effects of physical training.
18. If myocardial cells were organized like skeletal muscle fibers, and were innervated like skeletal muscles, could the heart function effectively as a pump? Explain.
19. There are regions of the heart other than the pacemaker region that could (under abnormal conditions) produce action potentials automatically, but those regions would do so at a slower rate than the normal pacemaker. Given refractory periods (chapter 4) and the way action potentials are conducted by the myocardium, explain why those regions instead follow the rhythm of the normal pacemaker.
20. During the course of a pregnancy, the smooth muscle cells of the uterus gain increasing numbers of gap junctions. Explain the physiological consequences of this change.
21. The smooth muscle of the gastrointestinal tract resembles cardiac muscle in its function and innervation, whereas the smooth muscle of the ciliary body (in the eye) resembles skeletal muscle in its function and innervation. Explain this statement.

Web Activities

ARIS™ **For additional study tools, go to: www.aris.mhhe.com. Click on your course topic and then this textbook's author/title. Register once for a semester's worth of interactive activities and practice quizzing to help boost your grade!**

CHAPTER 10

Blood and Circulation

Chapter Outline

A virtual cadaver dissection experience

Homeostasis

The circulation is the major avenue of communication between the different organs and systems of the body. All organs release carbon dioxide and metabolic wastes into the blood; these are carried to the lungs, liver, and kidneys for elimination. The lungs oxygenate the blood, which then carries the oxygen to all body cells. Food molecules needed for metabolism, as well as vitamins and essential amino acids and fatty acids, enter the blood from the digestive tract and are delivered by the blood to all cells. Hormones circulate in the blood to their target organs, and the cells and molecules of the immune system likewise travel through the blood and lymph to protect the body from disease. These and other functions of the circulation place it front and center in the body's ability to maintain homeostasis.

But if homeostasis is not maintained . . .

Clinical Investigation

Ron, a supermarket checker, complained to his physician about his bulging veins and swollen feet. The physician advised Ron to rock up on his toes periodically and walk when he could. Ron also complained that he became very dizzy when he first stood up after lying down, and the physician stated that he would change Ron's blood pressure medication from a β_1-adrenergic receptor blocker to an ARB to see if that would help. Ron also was concerned because he felt a thump in his chest when, as a member of the "Polar Bear Club," he last dived into ice-cold water. The physician said that this was normal, but that—even though Ron's ECG didn't show myocardial ischemia—it would be safer to enter the water more slowly. However, after checking Ron's heart sounds with a stethoscope and confirming that he had a murmur produced by mitral valve prolapse, the physician strongly advised Ron not to engage in that kind of activity at all.

What made Ron's veins bulge and his feet swell? What could have accounted for Ron's dizziness upon standing? How do β_1-adrenergic receptor blockers work to lower blood pressure, and how could these drugs have contributed to Ron's dizziness? What are ARBs, and how do they work to lower blood pressure? What caused the thump in Ron's chest when he dived into the ice-cold water? What are mitral value prolapse and myocardial ischemia, and why would it be dangerous for someone with those condition to dive into ice-cold water?

Blood Consists of Plasma and Formed Elements

The plasma is the liquid portion of the blood, and contains proteins and dissolved organic molecules, as well as many ions. The formed elements of the blood include red blood cells, white blood cells, and platelets. The blood type refers to ABO and Rh antigens on the surface of the red blood cells. The ability of the blood to clot relates to the function of platelets and of clotting factors in the blood plasma.

The different organs and systems communicate with each other through the blood. Hormones secreted by endocrine glands travel through the blood to their target organs; oxygen obtained in the lungs is transported by the blood to all body cells, and carbon dioxide travels in the blood from the metabolizing tissues to the lungs, where it can be exhaled. Waste products of metabolism are carried in the blood to the kidneys for excretion in the urine, and to the liver for elimination in the bile. Food molecules that are absorbed by the intestine into the blood travel to all organs. And the list goes on. A study of the blood and circulation is thus central to all aspects of human physiology, and so will serve as a major foundation for later study of other systems.

The **blood plasma**, the liquid portion of the blood, can be separated from the **formed elements**, the solid portion, by centrifugation (fig. 10.1). This process reveals the proportion of the packed red blood cells to the total volume of the blood sample. This measurement, the *hematocrit*, is commonly performed in medical laboratories (table 10.1). Plasma contains soluble molecules called *clotting factors* that are able to form a blood clot (discussed shortly), and when a clot is formed in a glass tube, those clotting factors become the insoluble fibers of a blood clot. **Serum** is the liquid portion from clotted blood, and so lacks the clotting factors.

TABLE 10.1 Some Normal Blood Values

Measurement	Normal Range
Blood volume	80–85 ml/kg body weight
Blood osmolarity	285–295 mOsM
Blood pH	7.38–7.44
Hematocrit	Female: 36%–46%; Male: 41%–53%
Hemoglobin	Female: 13–16 g/100 ml; Male: 13.5–17.5 g/100 ml
Red blood cell count	4.50–5.90 million/mm^3
White blood cell count	4.500–11,000/mm^3

Source: Excepted from material appearing in The New England Journal of Medicine, "Case Records of the Massachusetts General Hospital", 302: 37–38; 314: 39–49; 351: 1548–1563. 1980; 1986; 2004.

The plasma also contains dissolved ions and organic molecules, such as glucose and urea. Not quite dissolved, but rather present as a "colloidal suspension" in the plasma, are the plasma proteins. **Albumen** accounts for most (60% to 80%) of the plasma proteins, and the rest are **alpha globulins**, **beta globulins**, and **gamma globulins**. As the major type of plasma protein, albumen contributes most to the osmotic pressure (chapter 3) of the plasma, an important force helping to maintain blood volume (discussed later). The alpha and beta globulins include carrier proteins that transport such nonpolar molecules as cholesterol, triglycerides, and nonpolar hormones (steroids and thyroxine). The gamma globulins include antibody proteins, produced and secreted by lymphocytes (a type of white blood cell). This is discussed together with the immune system in chapter 11.

Red Blood Cells (Erythrocytes) Contain Hemoglobin

Red blood cells (**RBCs**), or **erythrocytes**, are biconcave discs (fig. 10.2), a shape that aids their function of transporting oxygen from the lungs throughout the body. RBCs lack nuclei and mitochondria, and as a result have a life span of about 120 days in

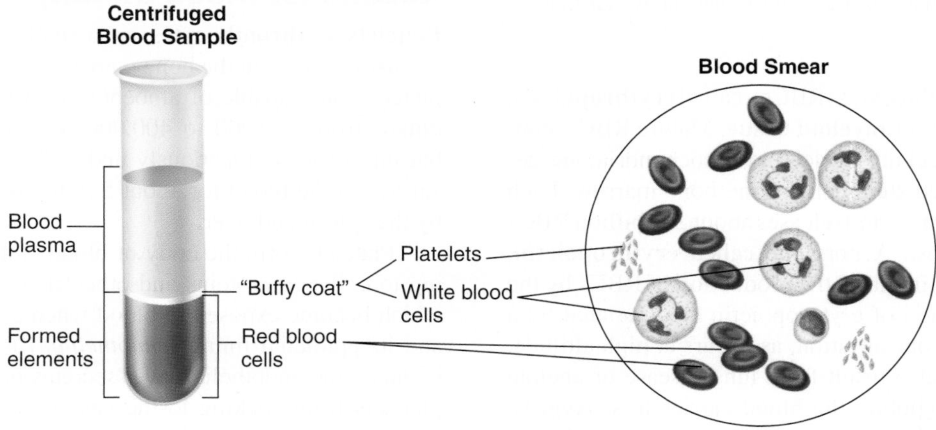

FIGURE 10.1 The constituents of blood. Blood cells become packed at the bottom of the test tube when whole blood is centrifuged, leaving the fluid plasma at the top of the tube. Red blood cells are the most abundant of the blood cells—white blood cells and platelets form only a thin, light-colored "buffy coat" at the interface between the packed red blood cells and the plasma.

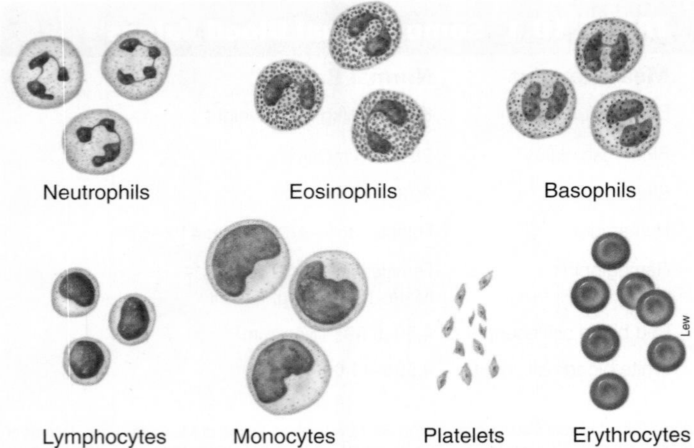

FIGURE 10.2 The blood cells and platelets. The white blood cells depicted above are granular leukocytes; the lymphocytes and monocytes are nongranular leukocytes.

the blood. Phagocytic cells in the spleen, bone marrow, and liver remove the older RBCs. RBCs are able to transport oxygen because they contain **hemoglobin**, a red, iron-containing pigment that binds oxygen in the lungs and releases some of it as the blood goes through the different organs (chapter 12); each RBC contains about 280 million hemoglobin molecules. *Anemia* refers to an abnormally low RBC count and/or hemoglobin concentration and hematocrit.

CLINICAL APPLICATIONS

The most common type of anemia is **iron-deficiency anemia**. Because the iron from destroyed RBCs is mostly recycled back to the bone marrow, the dietary requirement is low, and iron-deficiency anemia is caused more by blood loss (usually through menstruation) than by inadequate iron in the diet. In **pernicious anemia**, there is an inadequate amount of vitamin B_{12}, which is needed for red blood cell production. This is usually due to atrophy of the glandular mucosa of the stomach, which normally secretes a protein called *intrinsic factor*. In the absence of intrinsic factor, the intestine can't absorb the vitamin B_{12} from food. **Aplastic anemia** is due to destruction of the bone marrow, which may be caused by chemicals (including benzene and arsenic) or by radiation.

The production of erythrocytes (RBCs), called **erythropoiesis**, occurs in the bone marrow, or **myeloid tissue**. Mature RBCs, containing hemoglobin but lacking nuclei and mitochondria, are derived from less-specialized stem cells in the bone marrow. Each day the bone marrow produces and releases about 200 billion RBCs to replace the number lost. A hormone called **erythropoietin**, secreted by the kidneys, stimulates the production of RBCs by the bone marrow. The secretion of erythropoietin is stimulated by a fall in the blood oxygen concentration, as occurs at high altitude. Low blood oxygen may also result from lung disease or anemia (because, with less hemoglobin, the blood carries less oxygen). The increased secretion of erythropoietin under these conditions stimulates an increased rate of red blood cell production, helping to compensate for these changes and maintain homeostasis of the blood oxygen concentration.

White Blood Cells (Leukocytes) Function in Immunity

White blood cells (**WBCs**), or **leukocytes**, can be divided into those that are *granular* and those that are *nongranular* (or *agranular*). The granular leukocytes include the **neutrophils**, **eosinophils**, and **basophils**; the agranular leukocytes include the **lymphocytes** and **monocytes** (fig. 10.2). All of the WBCs (like the RBCs) are produced in the bone marrow. However, the **lymphoid tissue**—including lymph nodes, tonsils, spleen, and thymus—also produces lymphocytes, which are derived from cells that had seeded the lymphoid tissue from the bone marrow.

The different types of WBCs function in different aspects of the immune system (chapter 11). For example, neutrophils (the most common type of WBC) are phagocytic cells, which can leave the blood and enter the connective tissues. Monocytes, which also are phagocytic cells, can enter the connective tissues and become transformed into *macrophages* (literally, "big eaters"). Eosinophils contain *histamine* and other chemicals that are important in inflammation; basophils, the rarest type of WBC, contain the anticoagulant *heparin*. Lymphocytes (the second most common type of WBC, after neutrophils) are involved in specific immunity—that is, immune attack directed at specific molecules known as *antigens*.

CLINICAL APPLICATIONS

An elevated WBC count is called **leukocytosis** and is often associated with infections. Large numbers of immature WBCs in a blood sample are diagnostic of **leukemia**. A low WBC count is called **leukopenia**. This may be caused by a variety of factors; low numbers of lymphocytes, for example, may result from poor nutrition or from whole-body radiation treatment for cancer. Low numbers of eosinophils (*eosinopenia*) may result from elevated levels of glucocorticoids (such as cortisol; see chapter 8), which can be produced by stress or by taking exogenous glucocorticoid drugs.

Platelets (Thrombocytes) Are Needed for Blood Clotting

Platelets, or **thrombocytes**, are actually fragments of cells, called *megakaryocytes*, in the bone marrow. Although not complete cells, platelets are capable of amoeboid movement. The platelet count ranges from 130,000 to 400,000 per cubic millimeter of blood, but this varies tremendously under different conditions. Platelets survive in the blood for about 5 to 9 days before being destroyed by the spleen and liver.

Platelets form the body of blood clots. They do this by sticking to collagen proteins and other elements of connective tissue, which become exposed to blood when a blood vessel is damaged and its epithelial lining (*endothelium*) is broken. When a vessel is intact, the endothelial layer secretes molecules that prevent the platelets from sticking to the vessel wall (fig. 10.3*a*). When the endothelium is damaged, platelets stick to the exposed connective tissue and release molecules that attract other platelets (fig. 10.3*b*), causing many platelets to aggregate, stick together, and form a **platelet plug** (fig. 10.3*c*).

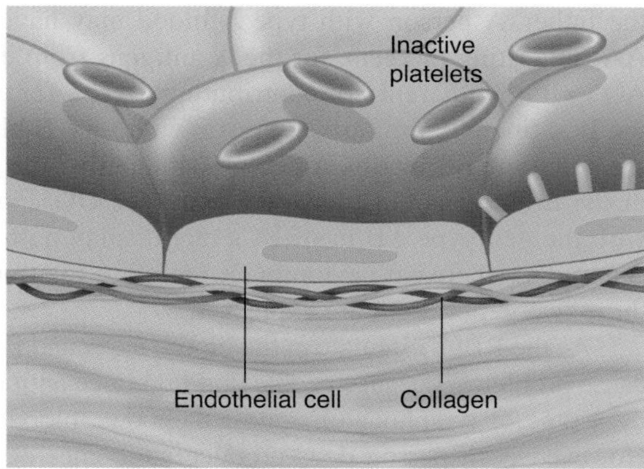

(a)

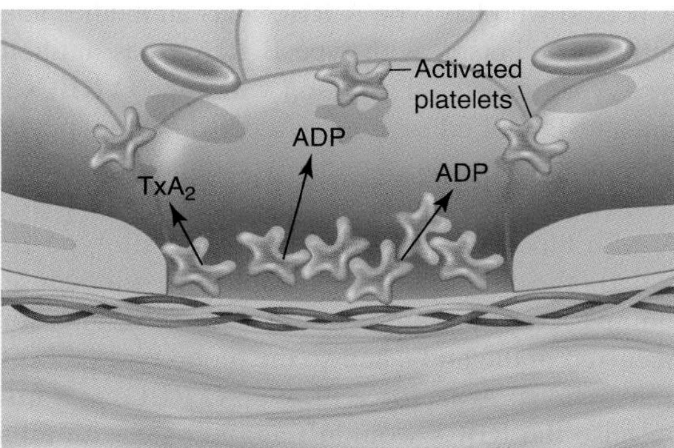

(b)

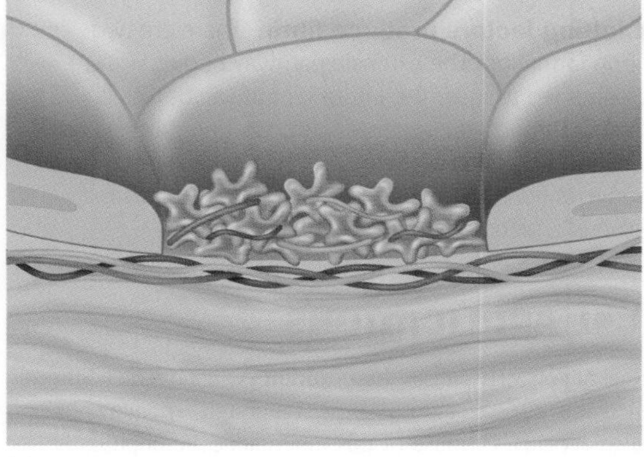

(c)

FIGURE 10.3 Platelet aggregation to form a blood clot. (*a*) In the intact vessel, platelets do not stick to each other or to the vessel wall. (*b*) When the endothelium is damaged, platelets stick to the exposed collagen proteins and release chemicals, including ADP and thromboxane A_2 (TxA_2), a prostaglandin. These promote platelet stickiness. (*c*) Platelets aggregate and stick together to form a platelet plug, which will form the body of the blood clot and will later be reinforced with fibrin proteins.

Clinical Applications

Prostaglandins are a family of regulatory fatty acids that perform many functions in different organs. One type of prostaglandin, *prostacyclin*, is secreted by an intact endothelium and helps prevent platelets from aggregating. Platelets release a different prostaglandin, called *thromboxane* A_2, which enables platelets to stick together in a blood clot. The enzyme that forms prostaglandins is known as **cyclo-oxygenase (COX)**. **Aspirin** inhibits COX and, by preventing prostaglandin formation, inhibits platelet aggregation. This effect prolongs bleeding time, and it lasts for the life of the platelets (5 to 9 days).

The platelet plug must be reinforced by strong fibers, formed from threads of an insoluble protein called **fibrin** (fig. 10.4). Fibrin is derived from a soluble protein in the plasma, *fibrinogen* (*gen* = origin; when you see this as a suffix, you are alerted that the named substance will become something else). The conversion of fibrinogen to fibrin is regulated very carefully to prevent the dangerous formation of inappropriate clots. Because of this, there are a number of **clotting factors** in the plasma that must be activated in sequence in order for fibrin to be formed.

Fibrinogen is converted into fibrin by the action of an enzyme known as **thrombin**. Thrombin formation must therefore be a carefully controlled process. Thrombin is formed from its inactive precursor, *prothrombin*, by activated clotting factors; Ca^{2+} is also required (fig. 10.5).

There are two pathways that lead to the activation of clotting factors and thus the formation of thrombin from prothrombin. One pathway may be activated by the contact of blood with a glass tube; since no chemical is added from outside the blood, this is called the *intrinsic pathway*. The other pathway occurs when damaged tissue releases an activator molecule; this is the *extrinsic pathway*. Both pathways eventually activate factor X, which works together with other activated factors and Ca^{2+} to cause the formation of thrombin, and thus of fibrin (fig. 10.5).

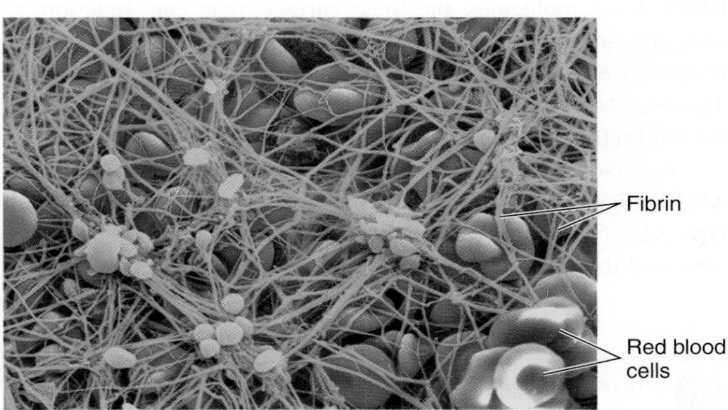

FIGURE 10.4 Colorized scanning electron micrograph of a blood clot. The threads of fibrin have been colored green, the erythrocytes are shown red, and the platelets have been colored purple.

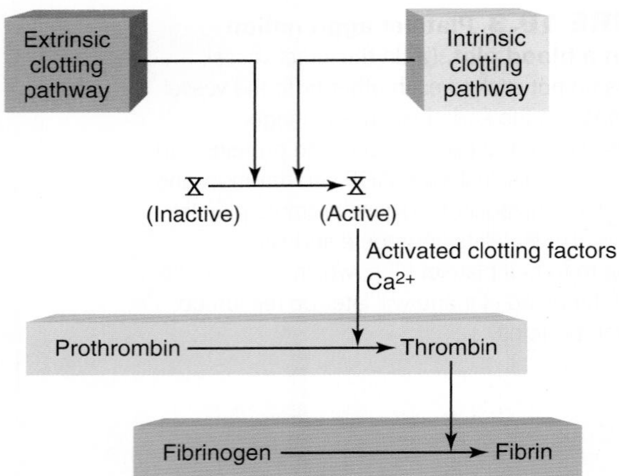

FIGURE 10.5 Clotting factors produce fibrin. There are two pathways leading to fibrin formation. The extrinsic pathway begins with activators released by damaged tissues. The intrinsic pathway begins by activation of clotting factors through their adhesion to collagen proteins or glass. Both pathways lead to activation of factor X, which causes thrombin formation. Thrombin converts fibrinogen to fibrin, which form threads that reinforce the platelet plug in a blood clot.

Clinical Applications

Defective clotting factor VIII causes **hemophilia A**, a rare bleeding disorder inherited on the X chromosome (although spontaneous mutations can also appear). Because of this inheritance, hemophilia is much more common in males (who have only one X chromosome) than in females (who have two X chromosomes, and can thus carry this as a recessive trait).[1] A variety of chemical **anticoagulants** can inhibit blood clotting. *Citrate* inhibits clotting by removing Ca^{2+}; *heparin* inhibits clotting by promoting the inactivation of thrombin. The *coumarin* drugs (such as *warfarin*) inhibit clotting by interfering with the cellular activation of vitamin K, which is required for the functioning of several clotting factors.

Blood Type Indicates Antigens on the RBC Surface

RBCs have molecules on their surface that can function as *antigens*—molecules that can be recognized and attacked by lymphocytes as part of the specific immune response (chapter 11). There are two major types of RBC antigens: the **ABO system**, and the **Rh factor** (antigen). Let's first consider the ABO system.

A person may have RBCs with A antigens only (**blood type A**); B antigens only (**blood type B**); both A and B antigens (**blood type AB**), or neither A nor B antigens (**blood type O**). These are inherited through two genes: one from the mother and one from the father. A person with type A blood may have the genotype AA (inheriting the gene for the A antigens from each parent) or AO (with one parent not contributing a gene for either the A or the B antigens). Similarly, a person with type B blood may have the genotype BB or BO. A person with neither the gene for the A or the B antigens must have the genotype OO. Finally, a person with the genotype AB has RBCs that display both the A and the B antigens (there isn't a dominant-recessive relationship between these two genes; both are expressed).

As the RBCs circulate in the blood, the A and B antigens don't cause problems, because people don't produce antibodies (secreted by a type of lymphocyte) against their own blood type antigens. However, each person does produce antibodies against the other blood type antigens, and these antibodies circulate in the plasma. A person who has type A RBCs has antibodies against the type B antigens; a person with type B RBCs has antibodies against the type A antigens; and a person with type O RBCs has both anti-A and anti-B antibodies. Finally, a person with type AB RBCs doesn't have antibodies against either type of antigens.

This makes sense, because if you mix RBCs that are type A with blood plasma from a person with type B blood (or vice versa), the antibodies in the plasma will cause the RBCs to clump together, or **agglutinate** (fig. 10.6). This could be fatal if it occurred in the body, and so the blood types must be matched for transfusions. A person with type O RBCs is a **universal donor**, because the RBCs wouldn't agglutinate with either antibody. A person with type AB RBCs is a **universal recipient**, because this person doesn't have antibodies in the plasma against either the A or the B antigens. The concepts of universal donor and recipient hold only for emergencies and when only small amounts of blood need to be transfused.

The *Rh factor* is another antigen that can be on the RBCs. People with that antigen are **Rh positive**; people without it are **Rh negative**. For example, a person with blood type "A positive" has both the type A antigens and the Rh antigen on the RBCs, and both must be matched for transfusions. The Rh factor is particularly significant when a woman is Rh negative, because if she becomes pregnant and the father is Rh positive, there is a chance that the fetus could also be Rh positive. If so, the mother could become exposed to the Rh antigen when the baby is born. Since the antigen is foreign to her, her immune system could produce antibodies against the Rh antigen.

Clinical Applications

Antibodies against the Rh factor don't harm the Rh negative mother, because her RBCs lack the Rh antigen. However, if an Rh negative mother, exposed to the Rh antigen in a previous pregnancy, becomes pregnant again, there is a potential danger. If the next fetus is also Rh positive, the antibodies could cross the placenta and cause *hemolysis* (rupture of red blood cells) in the fetus. The baby could be born anemic, with **erythroblastosis fetalis**, or **hemolytic disease of the newborn**. This can be prevented by injecting the Rh negative mother with an antibody preparation—known as *RhoGAM*—against the Rh factor.

FYI [1]Hemophilia A has been called the "royal disease" because it appeared in the descendants of Queen Victoria of Great Britain. For example, a granddaughter of Queen Victoria, Alexandra, was a carrier, and married Czar Nikolas II of Russia. Their four daughters didn't inherit the disease, but their one son, Alexis, was hemophilic. The desperate parents hired the "mad monk" Rasputin to care for Alexis, and it is possible that the parents' distraction, and the wide dislike for Rasputin, contributed to the Russian revolution.

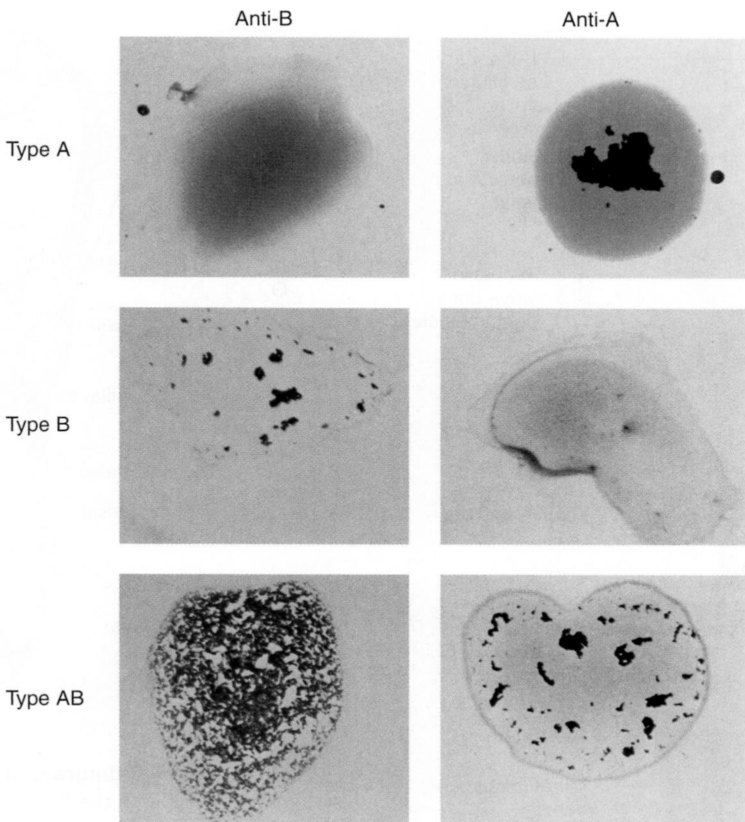

FIGURE 10.6 Blood typing. Agglutination (clumping) of red blood cells occurs when cells with A-type antigens are mixed with anti-A antibodies and when cells with B-type antigens are mixed with anti-B antibodies. No agglutination would occur with type O blood (not shown).

CHECK POINT

1. What are the plasma proteins, and what are their functions?
2. Describe the structure and function of the red blood cells.
3. List the different types of white blood cells and describe some of their functions.
4. Discuss the role of blood platelets in blood clotting, and how the clotting factors contribute to the formation of a blood clot.
5. Discuss the blood types, and explain why blood types should be matched for a blood transfusion.

The Heart Pumps Blood Through the Circulation

The heart has two atria that receive blood from veins, and two ventricles that pump blood into arteries. The right ventricle pumps blood into the pulmonary circulation, and the left ventricle pumps blood into the systemic circulation. The cardiac cycle consists of systole (contraction) and diastole (relaxation) of the atria and ventricles, and this cycle in the ventricles is associated with the opening and closing of the heart valves. Closing of the AV and semilunar valves produce the heart sounds.

Arteries are vessels that carry blood away from the heart; **veins** are vessels that carry blood to the heart. The two chambers of the heart that receive blood from veins are the **atria** (singular, *atrium*). Blood from the atria next enters the two **ventricles**, which are the heart chambers that pump blood into the arteries. The right atrium and ventricle (sometimes called the *right pump*) are separated from the left atrium and ventricle (the *left pump*) by a muscular wall, or *septum*. The two atria form a single functioning myocardium (chapter 9) and contract together. Similarly, the two ventricles form a single myocardium and contract together.

The Ventricles Pump Blood into the Pulmonary and Systemic Circulation

The heart has two pairs of valves. One pair, the **atrioventricular valves (AV valves)**, separates the two atria from the two ventricles. These are one-way valves that allow the blood to go from each atrium into each ventricle, but not the other way. The AV valve separating the right atrium from the right ventricle is the *tricuspid valve*; the one separating the left atrium from the left ventricle is the *bicuspid*, or *mitral*, *valve*. These valves, and other features of the heart's internal structure, are illustrated in figure 10.7. The second pair of valves is the **semilunar valves**, located at the opening where blood leaves each ventricle to enter the arterial system. The semilunar valve at the

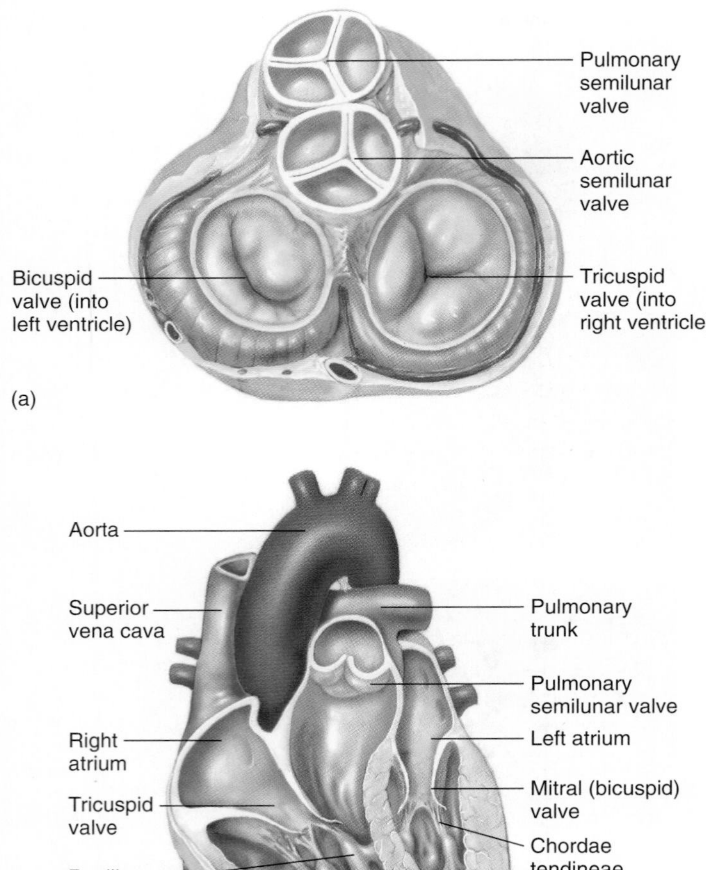

FIGURE 10.7 The heart valves. (*a*) A superior view of the heart valves. (*b*) A sagittal section through the heart, showing both AV valves and the pulmonary semilunar valve (the aortic semilunar valve is not visible in this view).

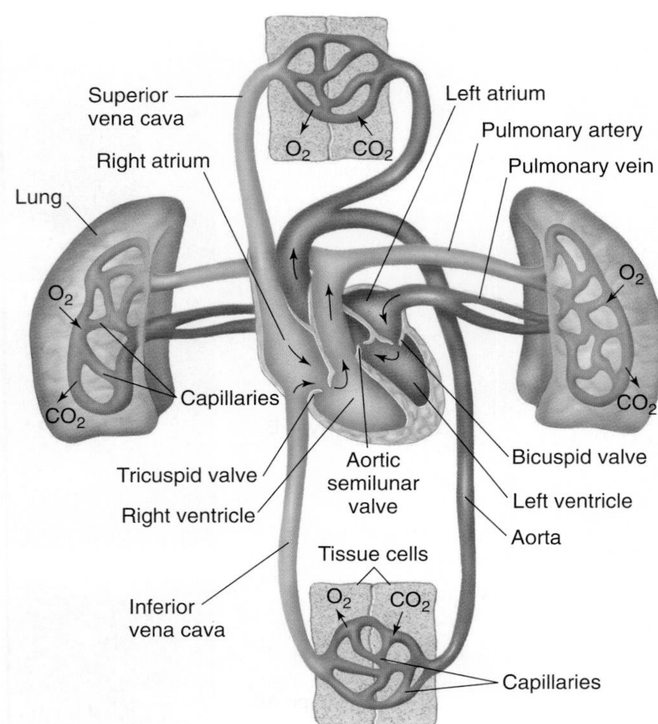

FIGURE 10.8 A diagram of the circulatory system. The systemic circulation includes the aorta and venae cavae; the pulmonary circulation includes the pulmonary arteries and pulmonary veins.

exit of the right ventricle is the *pulmonic valve*; the semilunar valve at the exit of the left ventricle is the *aortic valve*. These are also one-way valves, allowing blood to exit each ventricle but preventing blood from backing up into the ventricles from the arteries.

The right ventricle pumps blood into the short pulmonary trunk, which quickly divides into the right and left *pulmonary arteries*. These arteries carry blood to the lungs, where gas exchange occurs. During gas exchange in the lungs, the blood gains oxygen and loses carbon dioxide by the diffusion of these gases down their concentration gradients (chapter 3). This oxygenated blood leaves the lungs in four *pulmonary veins* (two from each lung) that return the blood to the right atrium. This circuit—from the heart (right ventricle) to the lungs and back to the heart (left atrium)—is the **pulmonary circulation** (fig. 10.8).

Oxygenated blood from the left atrium then goes through the mitral valve into the left ventricle. The left ventricle pumps blood into the largest artery in the body—the *aorta*. The aorta goes slightly above the heart, makes an arch, and then descends through the thoracic and abdominal cavities. All along the way, arteries branch from the aorta and supply oxygenated blood to every organ of the body. Then, blood that's partially depleted of oxygen leaves the organs in veins. Smaller veins converge into larger veins, eventually forming two large veins, the *superior* and *inferior venae cavae*, that return the blood to the right atrium. This circuit—from the heart (left ventricle) to all of the body systems and back to the heart (right atrium)—is the **systemic circulation** (fig. 10.8). Blood from the right atrium then goes through the tricuspid valve to again enter the right ventricle.

Notice in figure 10.8 that blood entering and leaving the left pump (left atrium and ventricle) is shown in red. This is done to indicate oxygenated blood that came from the lungs, because this blood has a bright, tomato-juice-red color. The blood entering and leaving the right pump (right atrium and ventricle) is shown in blue. This is done to indicate blood that is partially depleted of oxygen, due to the removal of oxygen from the blood by the metabolizing tissues. This blood isn't really blue; you've seen its color when you've had blood drawn for medical tests. The blood is more a darker shade of red, but it gives the tissues around it a bluish color. This is why, by convention, the blood is color-coded blue in illustrations.

One more point regarding figure 10.8: notice that an artery (the aorta) carries oxygen-rich (red) blood, but another artery (a pulmonary artery) carries blood partially depleted of oxygen (blue). Similarly, the venae cavae carry oxygen-depleted blood (blue) from the body organs, but the pulmonary veins carry oxygen-rich blood (red) from the lungs. Thus, we can't define arteries or veins by the type of blood they carry, but only by the direction of blood flow: away from the heart for arteries, to the heart for veins.

The Cardiac Cycle Is Accompanied by Heart Sounds

The **cardiac cycle** is the repeating pattern of contraction and relaxation of the heart's chambers. The phase of contraction is **systole**; the phase of relaxation is **diastole**. The two atria contract together at atrial systole and relax together at atrial diastole. Similarly, the two ventricles undergo systole and diastole at the same time. When you see the terms *systole* and *diastole* used without reference to atria or ventricles, you can assume that the terms refer to ventricular systole and diastole.

The two atria contract together, followed about 0.2 second later by contraction of the two ventricles. When the atria and ventricles are relaxed, blood fills the atria and then goes through the AV valves into the ventricles; this occurs because the pressure in the atria is greater (due to their filling with blood) than the pressure in the ventricles. The ventricles become about 80% filled even *before* the atria contract. Even if the atria were to stop pumping (as they do in *atrial fibrillation*), the ventricles would still receive and eject blood. This is why people can live with atrial fibrillation, but not with *ventricular fibrillation* (discussed later).

Contraction of the atria adds to the total amount of blood that will be in the ventricles at the end of their diastole; this is called the **end-diastolic volume**. When the ventricles contract, the amount of blood they eject is the **stroke volume**. The stroke volume of a resting person is about two-thirds of the end-diastolic volume. In other words, there is some blood in reserve within the ventricles that isn't ejected (the *end-systolic volume*), which could be ejected if the ventricles were to contract more strongly (as they do during exercise). The events of the cardiac cycle, assuming a **cardiac rate** (heart rate) of 75 beats per minute, are shown in figure 10.9.

Figure 10.10 (top left) shows pressure changes during the cardiac cycle in the left ventricle. The numbers refer to the following events, illustrated in figure 10.10, right:

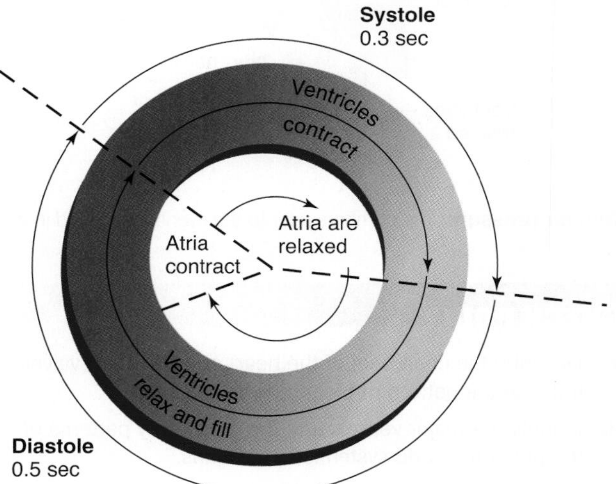

FIGURE 10.9 The cardiac cycle of ventricular systole and diastole. Contraction of the atria occurs in the last 0.1 second of ventricular diastole. Relaxation of the atria occurs during ventricular systole. The durations given for systole and diastole relate to a cardiac rate of 75 beats per minute.

1. **Isovolumetric contraction**. As the ventricles begin contracting at systole, the pressure within them rises sharply, becoming greater than the pressure in the atria; this snaps the AV valves shut. This is called the phase of *isovolumetric contraction*, because blood is not yet ejected. Notice that the intraventricular pressure of the left ventricle rises to about 120 mmHg (mercury).
2. **Ejection**. When the pressure in the ventricles becomes greater than the pressure in the arteries, the semilunar valves open and the phase of *ejection* begins. The pressure within the left ventricle starts to fall as blood leaves it, but the pressure in the aorta will rise to about 120 mmHg as a result of systole.
3. **Isovolumetric relaxation**. As the pressure in the ventricles falls below the pressure in the arteries, the pressure difference causes the semilunar valves to snap shut, preventing backflow. The phase of *isovolumetric relaxation* occurs when the AV valves and semilunar valves are closed, so that no blood can flow into the ventricles. This is the beginning of diastole.
4. **Rapid filling**. When the pressure in the ventricles falls below the pressure in the atria, the AV valves open and a phase of *rapid filling* of the ventricles occurs. This is still diastole, and is important because this is when the ventricles fill with blood.
5. **Atrial contraction**. Atrial contraction empties the final amount of blood into the ventricles to complete the end-diastolic volume, just before the next ventricular contraction (systole).

Figure 10.10 (bottom left) indicates the **heart sounds**. There is a **first heart sound** (**lub**), followed by a **second heart sound** (**dub**). The first heart sound is produced by closing of the AV valves at the beginning of systole (step 1 of the previous list). The second heart sound is produced by closing of the semilunar valves at the beginning of diastole (step 3 in the previous list). Notice that it is the closing, not the opening, of the heart valves that produces the sounds. The heart sounds can be heard using a stethoscope placed in certain positions on the chest (fig. 10.11), and this procedure may detect **heart murmurs**. Murmurs are abnormal heart sounds usually produced by defective valves or other structural defects that result in blood flowing in an abnormal way in the heart.

CLINICAL APPLICATIONS

There are many defects that can result in heart murmurs. In **mitral stenosis**, the mitral valve becomes thickened and calcified. **Mitral valve prolapse** is a relatively common condition in which the flaps of the mitral valve "prolapse," or protrude up into the left atrium because of the rising pressure in the left ventricle during systole. You can visualize this as similar to the way a sail billows in the wind. The prolapsed mitral valve allows some blood to leak back up into the left atrium. This can be caused by congenital defects in the mitral valve, but it can also be caused by damage to the chordae tendineae supporting the valves (see fig. 10.7), or other conditions. Similarly, defects of the other heart valves (the tricuspid, pulmonic, and aortic) sometimes produce heart murmurs. Also, **septal defects**, involving the septum between the atria or ventricles, can produce heart murmurs.

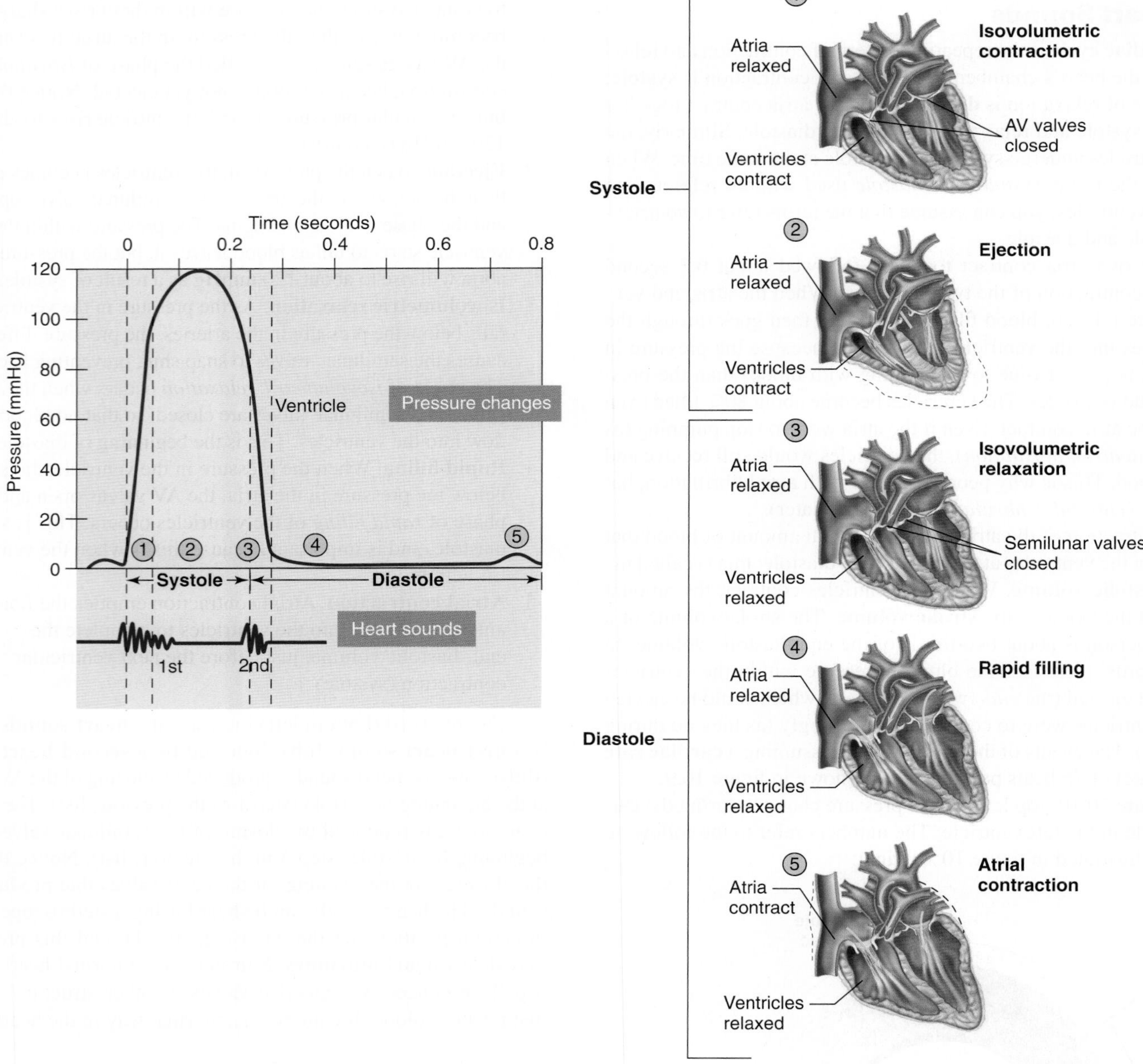

FIGURE 10.10 The relationship between heart sounds and the intraventricular pressure. The numbers refer to the events described in the text.

Clinical Investigation Clues

Remember that Ron's physician detected a heart murmur that was produced by mitral valve prolapse.

- What is the mitral valve, what is prolapse, and how does that produce a heart murmur?
- Given that an extra-strong contraction of the left ventricle could produce the "thump" that Ron felt in his chest, why would the physician be concerned about this happening to a patient with mitral valve prolapse?

Check Point

1. Describe the chambers of the heart and the heart valves. Indicate the pattern of blood flow in the heart.
2. Identify the major vessels and the circulatory patterns of the pulmonary and systemic circulations.
3. Describe the cardiac cycle of systole and diastole, indicating the pressure changes and the phases in which the heart valves open and close.
4. Explain how and when the heart sounds are produced, and how murmurs occur.

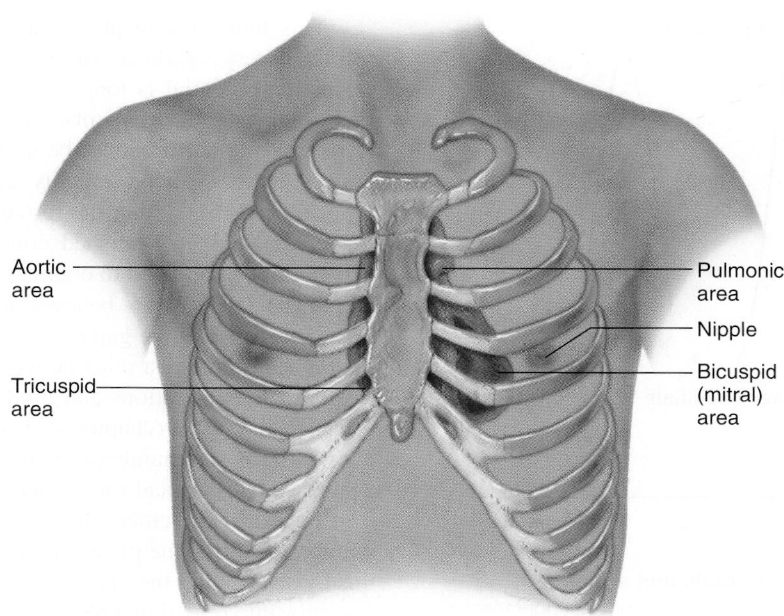

FIGURE 10.11 Routine stethoscope positions for listening to the heart sounds. The first heart sound is caused by closing of the AV valves; the second by closing of the semilunar valves.

Action Potentials in the Heart Stimulate Contraction and Produce the ECG

The SA node is the pacemaker region of the heart, generating action potentials due to an automatic depolarization. Action potentials spread through gap junctions in the atria, and then are conducted by specialized tissues into the ventricles. Depolarization produces contraction (systole); repolarization produces relaxation (diastole). The repeating pattern of electrical activity in the heart generates electrical currents at the body surface that are recorded as an electrocardiogram.

The normal pacemaker region of the heart is called the **sinoatrial (SA) node**, located in the right atrium near the opening of the superior vena cava (fig. 10.12). The action potentials spread through gap junctions from the SA node to both the right and the left atria. This conduction is very rapid, so that both atria are stimulated to contract at the same time. However, the ventricles are a different, thicker myocardium than the atria, and they require specialized conducting tissue to deliver the action potential. This specialized conducting tissue includes the **AV node**, **AV bundle** (**bundle of His**), and **Purkinje fibers** (fig. 10.12). Notice that the bundle of His divides into a right and left branch as it travels along the septum between the ventricles. These conducting tissues are specialized myocardial cells, not neural tissue.

Action Potentials Stimulate Contraction of the Myocardium

The SA node undergoes an automatic, spontaneous depolarization; in other words, it depolarizes itself, without neural stimulation. This automatic depolarization occurs during diastole, and can be called the **pacemaker potential** (fig. 10.13). When the depolarization reaches a threshold level, it stimulates the production of an action potential. Almost immediately after the repolarization phase of the action potential, the membrane of the pacemaker cells of the SA node begins to again spontaneously depolarize, in preparation for the next cycle. The sympathoadrenal system (through the action of epinephrine and norepinephrine) makes the automatic depolarization faster, increasing the cardiac rate; the parasympathetic nerves to the SA node (through the action of acetylcholine) make the automatic depolarization slower, decreasing the cardiac rate.

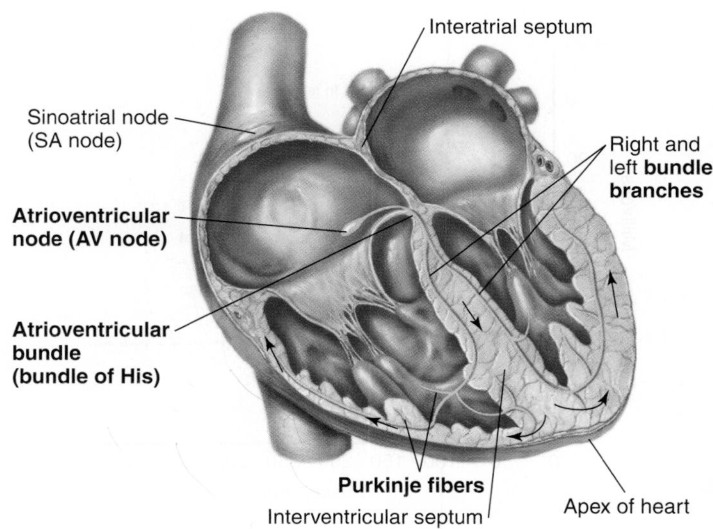

FIGURE 10.12 The conduction system of the heart. The conduction system consists of specialized myocardial cells that rapidly conduct the impulses from the atria into the ventricles.

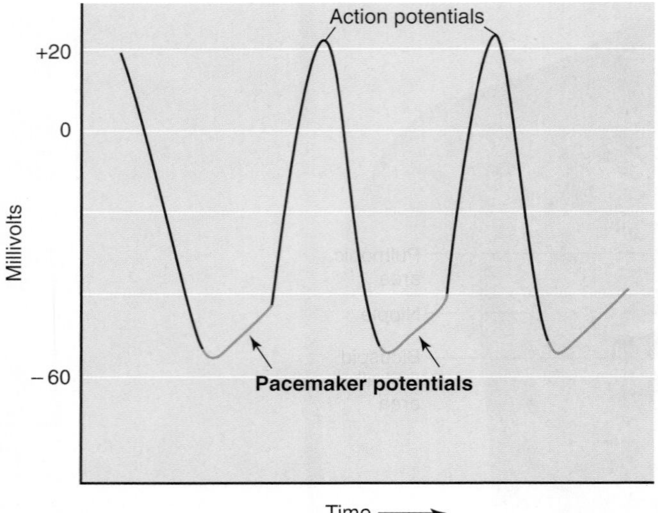

FIGURE 10.13 Pacemaker potentials and action potentials in the SA node. The pacemaker potentials are spontaneous depolarizations. When they reach threshold, they trigger action potentials.

Once a myocardial cell in the ventricles has been stimulated by an action potential that originated in the SA node, it produces its own action potential. This action potential looks quite different from the "spike" action potentials produced by axons and by skeletal muscle fibers (fig. 10.14). The myocardial action potential begins with the opening of voltage-gated Na^+ channels, producing a rapid depolarization phase in which the membrane polarity is reversed. However, unlike other action potentials, there is a very long *plateau phase*, lasting 200–300 milliseconds before repolarization. This plateau phase results from an inward diffusion of Ca^{2+} through *slow Ca^{2+} channels* that balances a slow outward diffusion of K^+. Repolarization is then produced by the closing of the slow Ca^{2+} channels and the opening of voltage-gated K^+ channels, allowing a rapid diffusion of K^+ out of the membrane.

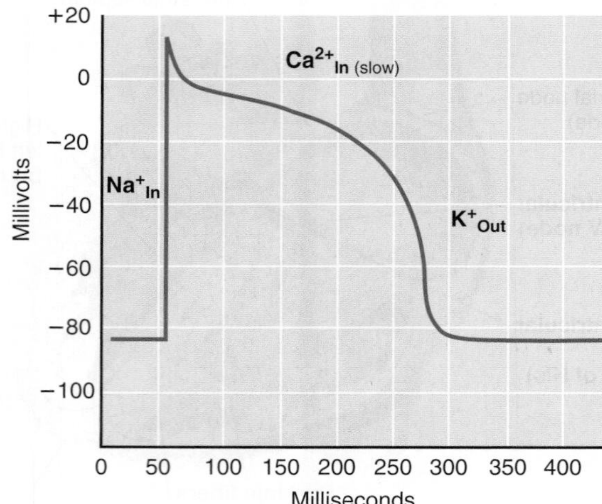

FIGURE 10.14 An action potential in a myocardial cell from the ventricles. The plateau phase of the action potential is maintained by a slow inward diffusion of Ca^{2+}. The cardiac action potential, as a result, is about 100 times longer in duration than the "spike-like" action potential in an axon.

This long plateau phase, and the long action potential that results, is very significant for the functioning of the heart. Because the action potential is long, the refractory period is correspondingly long as well. Remember, while a plasma membrane is in a refractory period, it cannot be stimulated again (chapter 4). The long refractory period roughly corresponds to the contraction time (fig. 10.15), so that the refractory period isn't over until the myocardial cell has finished contracting. Thus, the myocardial cell can't be stimulated to contract again until it's had a chance to relax. The myocardium behaves as if it were a single cell in this regard (because of the gap junction, as described in chapter 9), so the myocardium must relax between beats—it can't be stimulated to summate contractions and produce a sustained contraction, as can skeletal muscle (chapter 9). The relaxation of the myocardium allows the heart chambers to fill with blood before contracting again, which is critical for the heart to work as a pump.

The Ca^{2+} that enters the myocardial cell through the plasma membrane during the plateau phase stimulates contraction. It does this by stimulating the opening of the Ca^{2+} release channels in the sarcoplasmic reticulum (SR), which allows the Ca^{2+} from the SR to diffuse into the cytoplasm and bind to troponin (chapter 9). This process is known as **Ca^{2+}-stimulated Ca^{2+} release**. An important cardiac drug, *digitalis*, indirectly acts to raise the intracellular concentration of Ca^{2+}, thereby increasing the strength of ventricular contraction in people suffering from *congestive heart failure*.

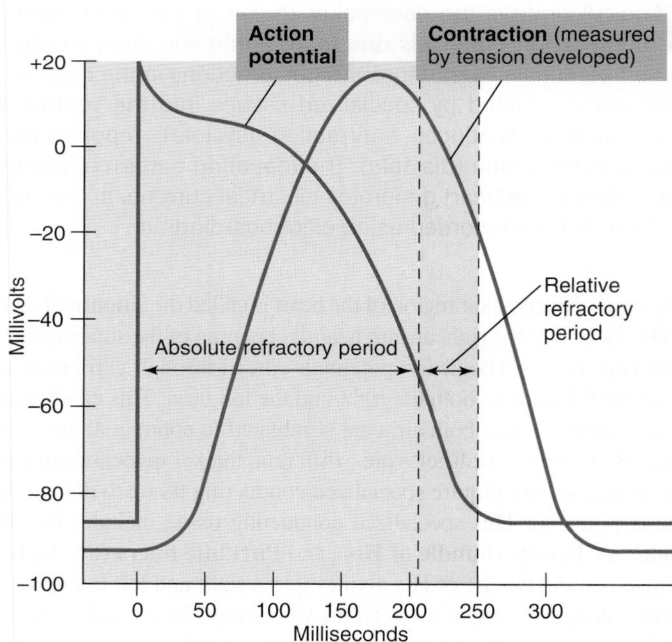

FIGURE 10.15 Correlation of the myocardial action potential with myocardial contraction. The time course for the myocardial action potential is compared with the duration of contraction. Notice that the long action potential results in a correspondingly long absolute refractory period and relative refractory period. These refractory periods last almost as long as the contraction, so that the myocardial cells cannot be stimulated a second time until they have completed their contraction from the first stimulus.

A very important concept emerges from the preceding discussion. Depolarization causes, and occurs during, myocardial contraction; repolarization causes, and occurs during, the early phase of myocardial relaxation. In other words, when the atria or ventricles are producing an action potential and are depolarized, they are in systole. When the atria or ventricles are repolarizing, they are beginning diastole. You can think of *depolarization/contraction/systole* and of *repolarization/relaxation/beginning of diastole* as associated terms.

Clinical Applications

Abnormal patters of electrical conduction in the heart are called **arrhythmias**. These can seriously interfere with the normal cardiac cycle and compromise the function of the heart. Arrhythmias may be treated with drugs that inhibit aspects of the myocardial action potentials and their abnormal conduction. These include drugs that (1) block the Na^+ channel (*quinidine*, *procainamide*, and *lidocaine*); (2) block the slow Ca^{2+} channel (*verapamil*); and (3) block β-adrenergic receptors for epinephrine and norepinephrine (*atenolol*).

The ECG Correlates with Events in the Cardiac Cycle

Electrodes placed in different positions on the surface of the body can produce a recording of the heart's electrical activity as it goes through its cycle (fig. 10.16). This recording is known as an **electrocardiogram (ECG** or **EKG**). These surface electrodes can't record action potentials, but they do record electric currents that result from the action potentials produced and conducted by the heart. The relationship between the myocardial action potential and the ECG waves is illustrated in figure 10.17, and the different components of the ECG are shown in figure 10.18.

The electrical events in the heart begin when the SA node in the right atrium becomes depolarized and the action potential spreads through both atria. This is indicated by the **P wave** of an ECG. Once the right and left atria are completely depolarized, the ECG wave returns to the baseline. Then, when the conducting tissue (AV node, bundle of His, and Purkinje fibers) carries action potentials into the ventricles, the ventricles become depolarized. This produces the **QRS complex**. The atria become repolarized during this time, but this isn't seen in the ECG pattern. Finally, the ventricles become repolarized, producing the **T wave**.

We can now correlate the ECG waves with the heart sounds and cardiac cycle (fig. 10.19). The P wave occurs when the atria depolarize and contract, emptying their content of blood into the ventricles. This occurs at the end of ventricular diastole, completing the final end-diastolic volume of the ventricles. No heart sounds are produced, because the heart valves don't close when the atria contract. Now the ventricles depolarize and produce the QRS complex. This depolarization causes contraction, so the QRS complex must occur at the very beginning of systole when the AV valves snap shut to produce the first heart sound ("lub"). Then the ventricles repolarize, producing the T wave as they relax. Thus, the T wave must occur at the beginning of diastole, when the semilunar valves snap shut to produce the second heart sound ("dub").

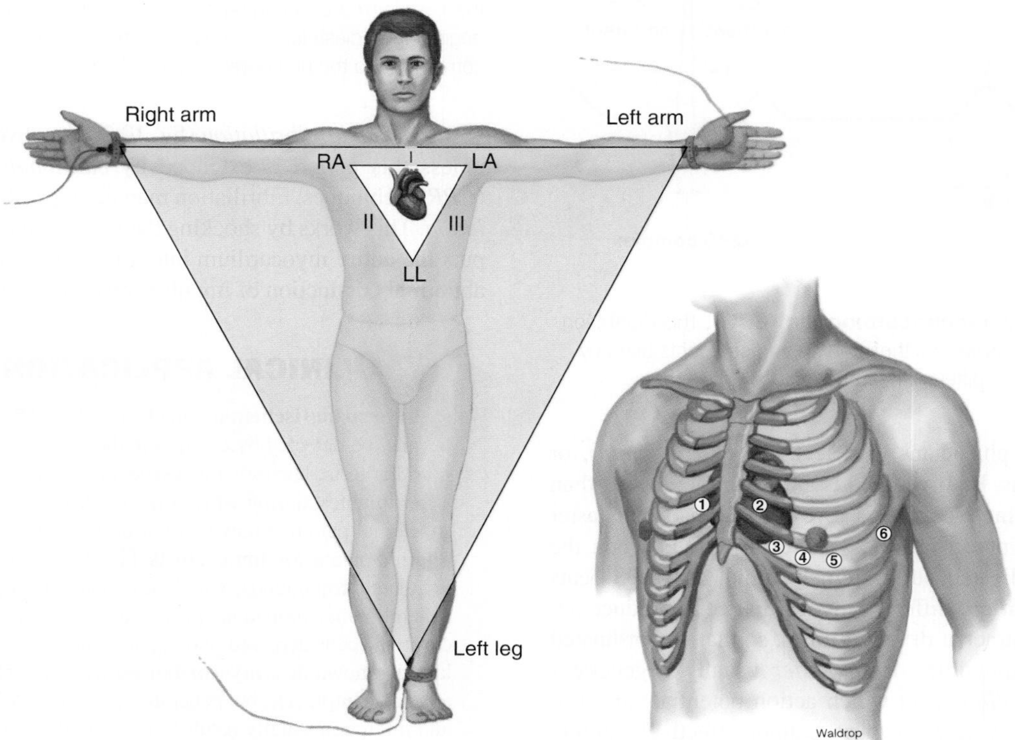

FIGURE 10.16 The electrocardiograph leads. The placement of the bipolar limb leads and the exploratory electrode for the unipolar chest leads in an electrocardiogram (ECG). The circled numbers correspond to standard placement positions for chest ECG electrodes.

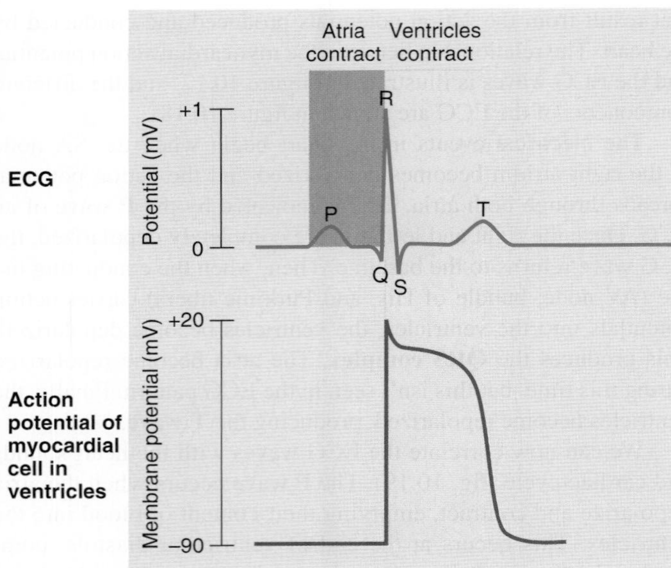

FIGURE 10.17 Correlation of action potential and ECG. The action potential of a typical myocardial cell in the ventricles (bottom) is correlated with an ECG (top). Notice that the QRS wave of the ECG occurs during depolarization and the T wave of the ECG occurs during repolarization of the ventricles.

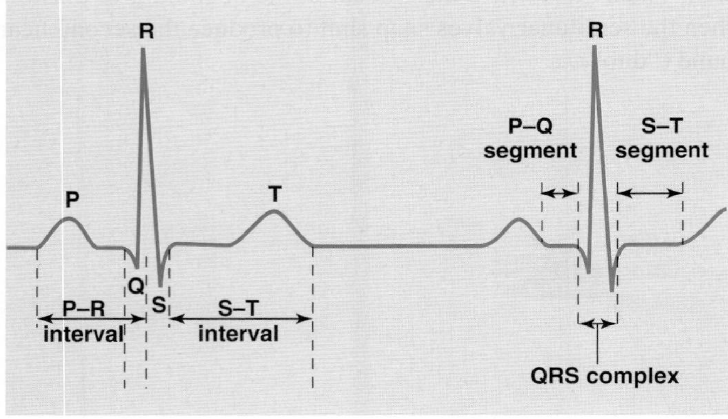

FIGURE 10.18 The electrocardiogram (ECG). This illustration indicates the ECG waves, as well as the intervals and segments used by physicians to help diagnose different heart conditions.

The ECG allows physicians to detect many *arrhythmias*, or abnormal heart rhythms. **Bradycardia** is a cardiac rate slower than 60 beats per minute; **tachycardia** is a resting cardiac rate faster than 100 beats per minute (fig. 10.20). **Flutter** occurs when the contractions are coordinated but extremely rapid (200–300 beats per minute). In **fibrillation**, different myocardial cells produce action potentials and contract at different times, so that a coordinated pumping action is impossible. *Atrial flutter* usually degenerates quickly into *atrial fibrillation*, in which action potentials are produced extremely rapidly and the atria cannot effectively pump. However, remember that the ventricles are normally 80% filled even before the atria contract; because of this, the ineffective contractions of the atria are not immediately life threatening. By contrast, people

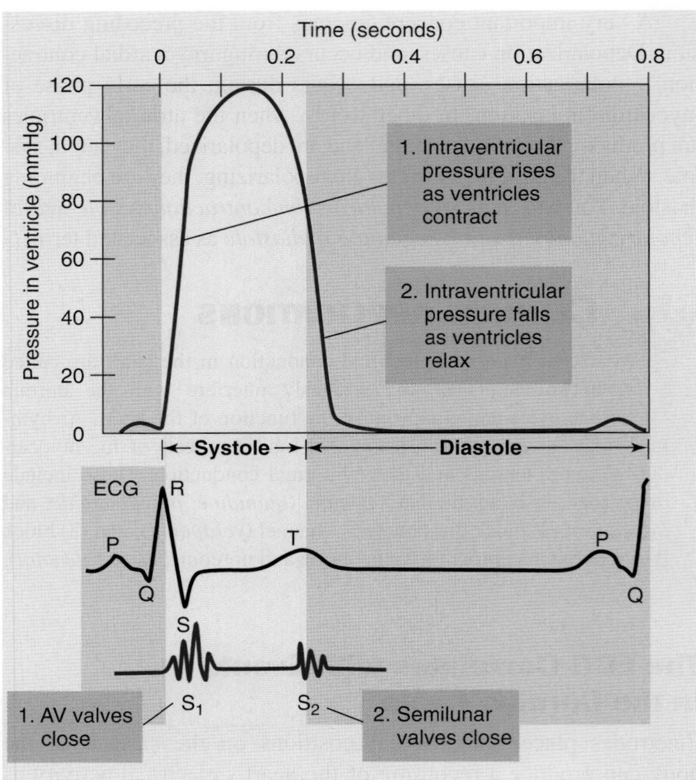

FIGURE 10.19 The relationship between changes in intraventricular pressure and the ECG. The QRS wave (representing depolarization of the ventricles) occurs at the beginning of systole, whereas the T wave (representing repolarization of the ventricles) occurs at the beginning of diastole. The numbered steps at the bottom of the figure correspond to the numbered steps at the top.

with *ventricular fibrillation* (fig. 10.20) can live only a few minutes, unless this period is extended by *cardiopulmonary resuscitation (CPR)* techniques. Fibrillation may be stopped by *electrical defibrillation*. This works by shocking the heart with an electric current that puts the entire myocardium into a refractory period, abolishing the abnormal conduction of impulses responsible for the fibrillation.

CLINICAL APPLICATIONS

Myocardial ischemia—inadequate blood flow to the myocardium—may be detected by changes in the S-T segment of the ECG and by other tests. The ischemic tissue may obtain energy anaerobically for a few minutes, during which time it releases lactic acid and produces a substernal pain that may be referred to the left shoulder and arm. This referred pain is **angina pectoris**. People with angina take nitroglycerin or related compounds, which stimulate vasodilation. If ischemia and anaerobic metabolism are prolonged, *necrosis* (cellular death) may occur in the cells deprived of oxygen. A sudden, irreversible injury of this kind is known as a **myocardial infarction** (**MI**). The lay term "heart attack" is imprecise, but generally refers to an MI. Myocardial ischemia and infarction usually result from the buildup of plaque in the walls of the coronary arteries, a condition called *atherosclerosis* (described in the "Physiology in Health and Disease" section of this chapter).

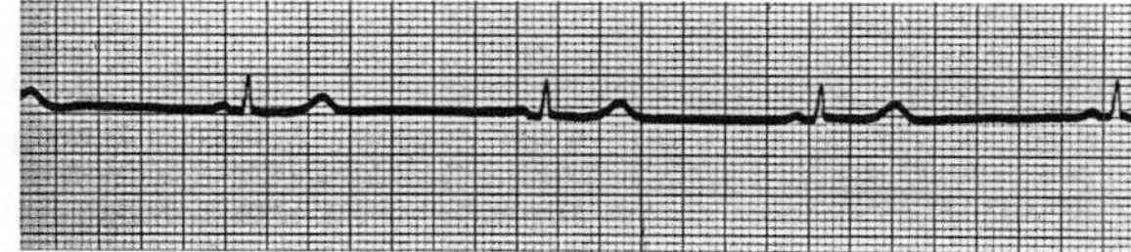

Sinus bradycardia

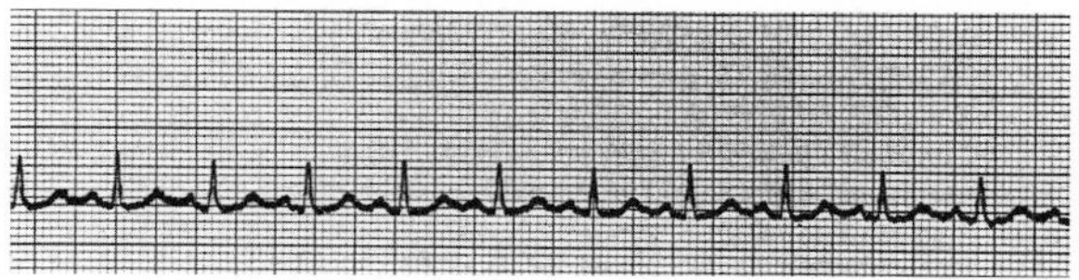

Sinus tachycardia

(a)

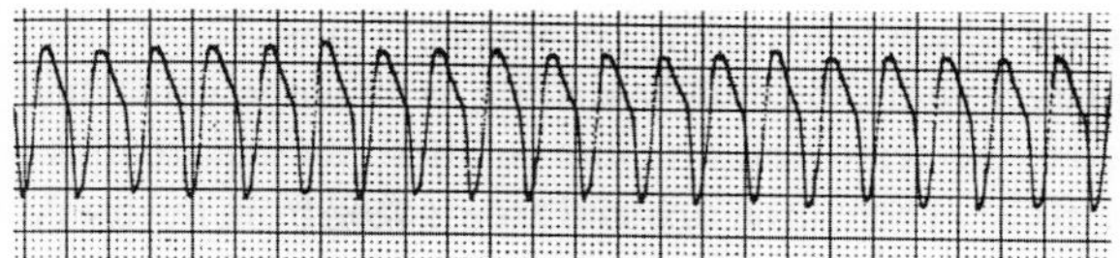

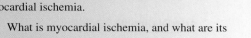

Ventricular tachycardia

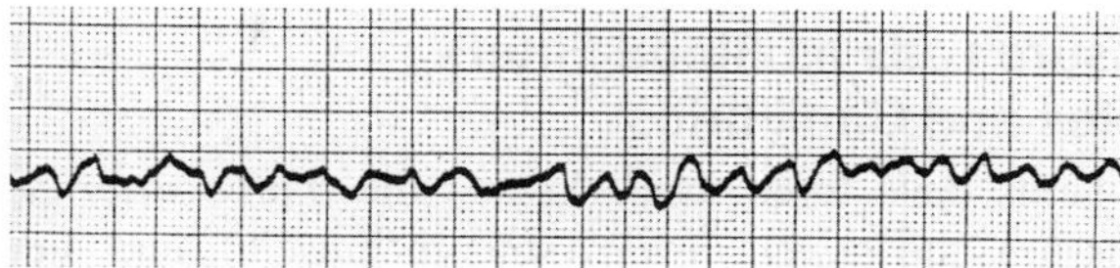

Ventricular fibrillation

(b)

FIGURE 10.20 Some arrhythmias detected by the ECG. In (*a*) the heartbeat is paced by the normal pacemaker—the SA node (hence the name *sinus rhythm*). This can be abnormally slow (bradycardia—42 beats per minute in this example) or fast (tachycardia—125 beats per minute in this example). Compare the pattern of tachycardia in (*a*) with the tachycardia in (*b*). Ventricular tachycardia is produced by an ectopic pacemaker in the ventricles. This dangerous condition can quickly lead to ventricular fibrillation, also shown in (*b*).

CLINICAL INVESTIGATION CLUES

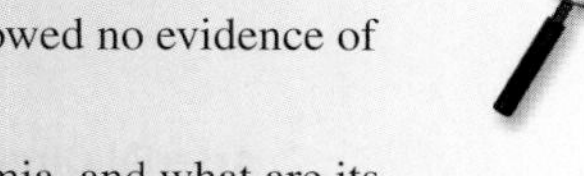

Remember that Ron's ECG showed no evidence of myocardial ischemia.

- What is myocardial ischemia, and what are its dangers?
- Why would it be especially dangerous for a person with myocardial ischemia to force the heart to work harder than usual (producing the "thump" in Ron's chest)?

CHECK POINT

1. Identify the normal pacemaker of the heart, and describe how action potentials are conducted through the atria and into the ventricles.
2. Describe the myocardial action potential and explain its correlation with the refractory period and the myocardial contraction.
3. Identify the waves of the ECG and explain how each is produced.
4. Correlate the ECG waves with the events of the cardiac cycle and the heart sounds.

Cardiac Output Indicates the Heart's Pumping Ability

The cardiac output is the volume of blood pumped per minute by each ventricle. This is regulated by adjustments in the cardiac rate and stroke volume. The stroke volume depends on the end-diastolic volume and on the contraction strength of the ventricle. Contraction strength, in turn, depends on the end-diastolic volume; it is also increased by sympathoadrenal stimulation. Arteries, capillaries, and veins differ in their properties, allowing the blood vessels to perform different functions as they transport the cardiac output through the circulation.

The **cardiac output** measures the pumping ability of the ventricles, and is equal to the stroke volume (in ml/beat) multiplied by the cardiac rate (in beats/min). Thus, the cardiac output will be expressed in units of milliliters (or liters) per minute. For example, if the stroke volume is 80 ml per beat, and the cardiac rate is 70 beats per minute, the cardiac output is 5,600 ml/min (or 5.6 L/min). The cardiac output of adults averages about 5.5 L/min, although this varies depending on body size and other factors.

This is the rate of blood flow out of one ventricle. Which one? The right ventricle pumps blood through the pulmonary circulation, so the entire output of the right ventricle goes through only one organ, the lungs. The output of the left ventricle is delivered to all the organs of the body. If you reasoned that, because of this difference, the output of one ventricle is greater than the other, you'd be wrong. The reason will be clear if you follow this supposedly higher output through the circulation; if the right ventricle pumped more blood per minute than the left (or vice versa), where would the extra blood go? Since the cardiovascular system is a closed circulation, it would fail if the output of one ventricle became higher than the other. Thus, right and left ventricles normally pump the same amount of blood per minute through both pulmonary and systemic circulations.

The pulmonary circulation offers less resistance to blood flow (discussed shortly) than the systemic circulation, so the right ventricle generates less pressure than the left ventricle when it contracts, resulting in the same rate of blood flow through the two circulations. That contraction of the left ventricle produces a higher pressure may be inferred from its thicker and stronger walls compared to the right ventricle.

The Arterial Tree Transports Blood from the Ventricles

Blood ejected from each ventricle enters large **elastic arteries** and then passes into smaller **muscular arteries**. From these, the blood enters microscopic-size arteries, called **arterioles**, which deliver blood into the smallest and most numerous of the blood vessels, the **capillaries**. Blood from capillaries is drained into microscopic-size veins, or **venules**, which converge to form small veins. These converge to form larger veins, eventually leading to the large veins that return the blood to the heart. The structures of these vessels are illustrated in figure 10.21. There is an **arterial tree**, going from the trunk (the aorta, for the systemic circulation) to the branches, and a **venous tree**, going from the branches to the trunk (the venae cavae).

Blood Pressure Falls as Blood Enters the Arterioles and Capillaries

Arteries and veins have three tissue layers: the *tunica externa* (connective tissue), the *tunica media* (smooth muscle), and the *tunica interna* (including the lining endothelium, a simple squamous epithelium). Notice in figure 10.21 that an artery has a much thicker layer of smooth muscle than does a vein of similar size. This is needed because the artery is subject to a much higher blood pressure, created by the pumping action of the heart. Figure 10.22 illustrates the blood pressures in the vessels of the systemic circulation. Notice that the pressure drops as the blood enters the small arteries and arterioles, and then falls even more as the blood enters the capillaries.

The elastic arteries, subject to the highest blood pressure, expand at systole and then recoil at diastole, giving the blood an added push. The muscular arteries hold their caliber more as the blood pressure fluctuates within them. The blood pressure falls as the blood flows through the arterioles, because although each arteriole is narrow, there are so many of them that their total cross-sectional area is very large. This is even truer of capillaries, which are by far the most numerous vessels with the greatest cross-sectional area (fig. 10.22). After all, there is only one aorta, and all of its blood must enter the capillaries of the systemic circulation. There are so many capillaries that every cell in the body is less than 0.1 mm from a capillary; each cell has "riverfront property."

Smaller Arteries Have Greater Resistance to Blood Flow

Arteries offer resistance to blood flow, caused by friction of the blood against their walls. The total resistance to blood flow that the left ventricle must overcome in order to push blood through the systemic circulation is known as the **total peripheral resistance**. Narrower arteries offer more resistance than wider arteries. Said

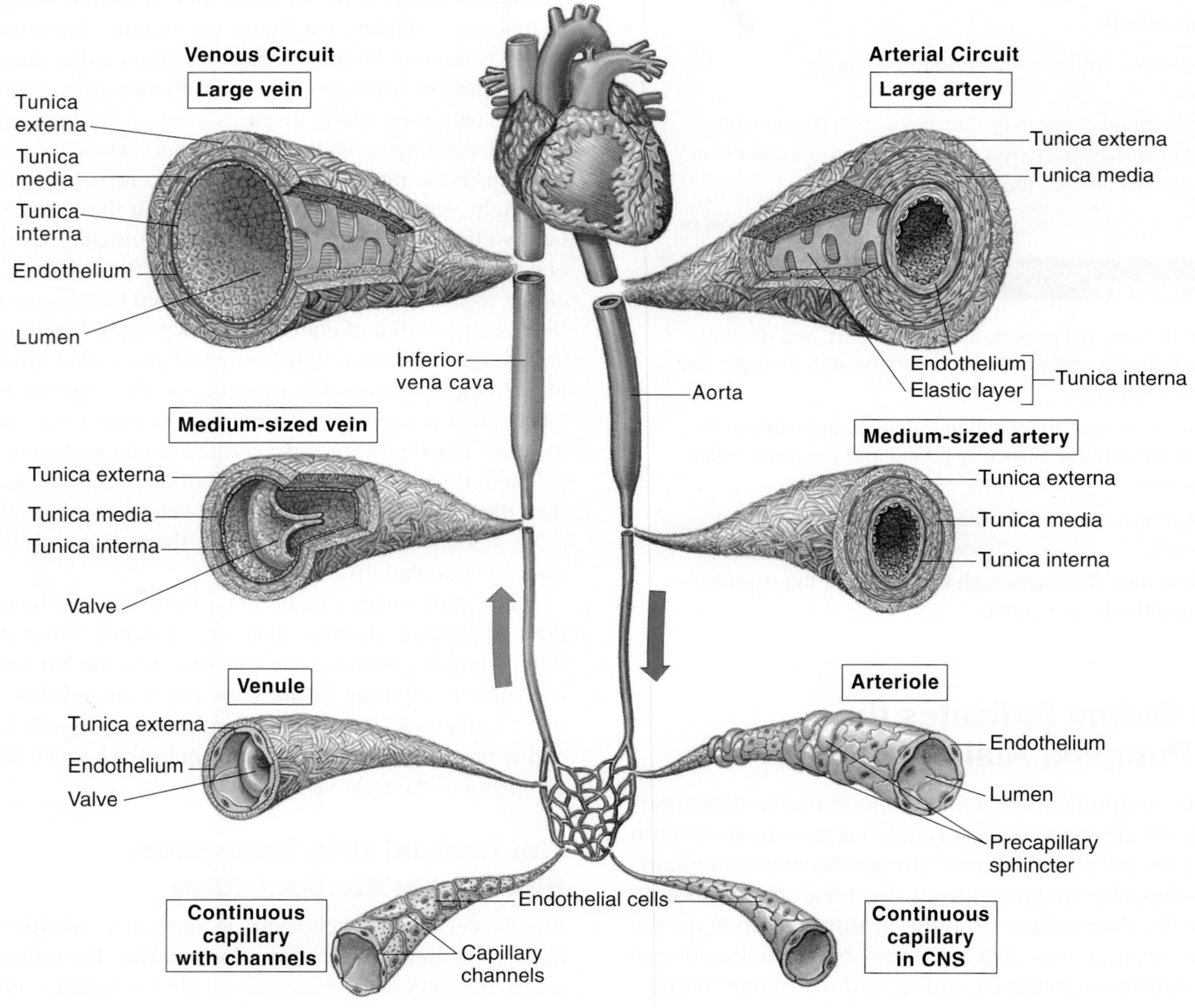

FIGURE 10.21 The structure of blood vessels. Notice the relative thickness and composition of the tunicas (layers) in comparable arteries and veins.

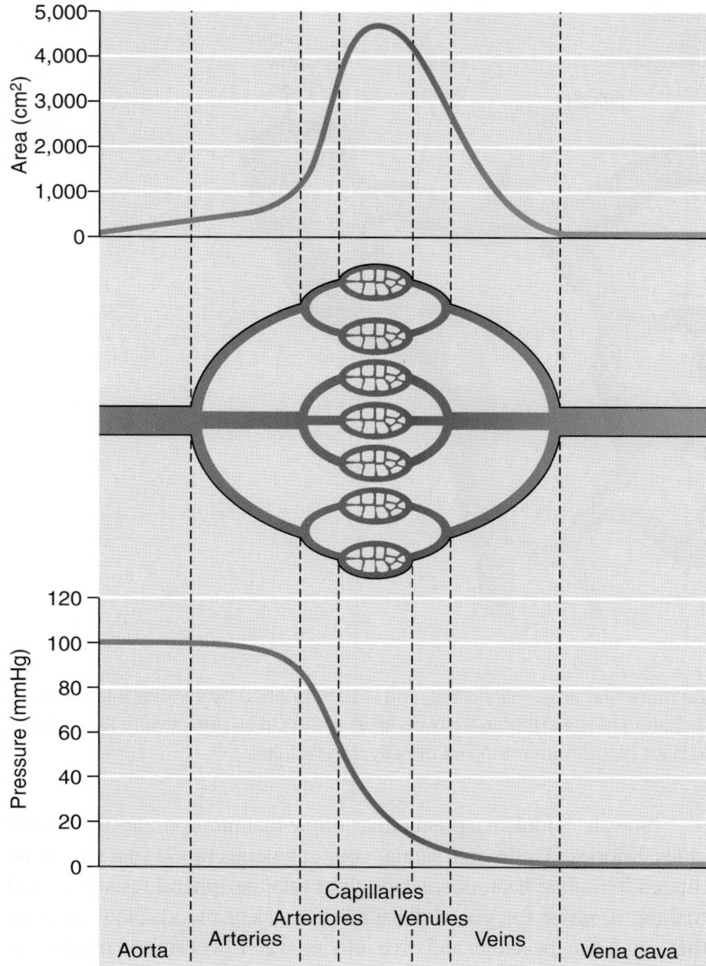

FIGURE 10.22 The relationship between blood pressure and the cross-sectional area of vessels. As blood passes from the aorta to the smaller arteries, arterioles, and capillaries, the cross-sectional area increases as the pressure decreases.

another way, as the radius gets smaller, the resistance becomes greater; the resistance is inversely proportional to the radius of the vessel. When this is stated as an equation, the radius is shown in the denominator to indicate inverse proportionality:

$$\text{Resistance} \propto \frac{\text{vessel length} \times \text{blood viscosity}}{\text{radius}^4}$$

Notice that vessel length and blood viscosity are in the numerator; this indicates that resistance is directly proportional to these factors. Increasing vessel length, or increasing blood viscosity (the "thickness" of blood; molasses has a high viscosity, for example), increases the resistance to flow. Radius is in the denominator, indicating inverse proportionality; the smaller the radius, the greater the resistance. The fourth power is a magnification effect: small changes in radius have a greatly magnified effect on resistance. For example, if the radius were to decrease by one half, the resistance will increase by 16 ($2^4 = 2 \times 2 \times 2 \times 2 = 16$). The inverse relationship between radius and resistance is common experience: for example, you know intuitively that it's easier to use a wider straw (lower resistance) than a thinner straw (higher resistance) to suck a milkshake (which has a high viscosity).

This inverse relationship has great physiological significance:

1. The part of the arterial tree with the smallest radius, the small arteries and arterioles, provide the greatest resistance to blood flow. The *arterioles are the major resistance vessels*; the ventricles must generate enough pressure to overcome their resistance, so that blood can flow through them into the capillaries.
2. If an arteriole were to constrict (by contraction of the smooth muscle in the tunica media) to a smaller radius, its resistance will increase greatly (because of the fourth-power effect). Conversely, if the arteriole were to dilate (by relaxation of the smooth muscle in the tunica media), its resistance will decrease greatly.

The constriction and dilation of small arteries and arterioles provides a way to regulate blood flow to different organs in the body; blood can be diverted away from the organs with a higher resistance and into the organs with a lower resistance. This will be described in a later section describing regulation of blood flow. Note: the fourth-power effect of radius on resistance also holds for airflow through tubes, so constriction of the bronchioles (narrow airways, analogous to arterioles) in the lungs greatly increases the difficulty of breathing for people with asthma (chapter 12).

Capillaries Provide Exchanges Between the Blood and Tissues

Capillaries are where the business of the circulation occurs: that is, the exchanges of fluid and molecules between the blood and the *interstitial fluid* (*tissue fluid*; chapter 3) occur across the walls of the capillaries. This is because the capillary wall is composed only of a simple squamous endothelium. Blood flows into capillaries because of the pressure drop between the arterioles and the capillaries, and the rate of this blood flow is regulated by vasoconstriction and vasodilation of the arterioles. However, blood flow into a particular capillary bed may be diverted through an *arteriovenous shunt* (*shunt* = diversion), which offers a path of low resistance, or may be cut off by the contraction of *precapillary sphincters* (fig. 10.23).

Anatomically, capillaries may be classified in various ways. **Continuous capillaries** are those in which the adjacent endothelial cells are closely joined together. In most organs with continuous capillaries, there are narrow channels between the endothelial cells that permit the passage of dissolved molecules in the plasma (but not plasma proteins) to exit the capillary blood and enter the interstitial fluid. However, the continuous capillaries in the CNS lack these channels; as a result, molecules that pass from the blood into the CNS must go through the cytoplasm of the endothelial cells. This is the basis of the selective *blood-brain barrier* (chapter 5). **Discontinuous capillaries**, with wide distances between endothelial cells, are found in the liver, bone marrow, and spleen.

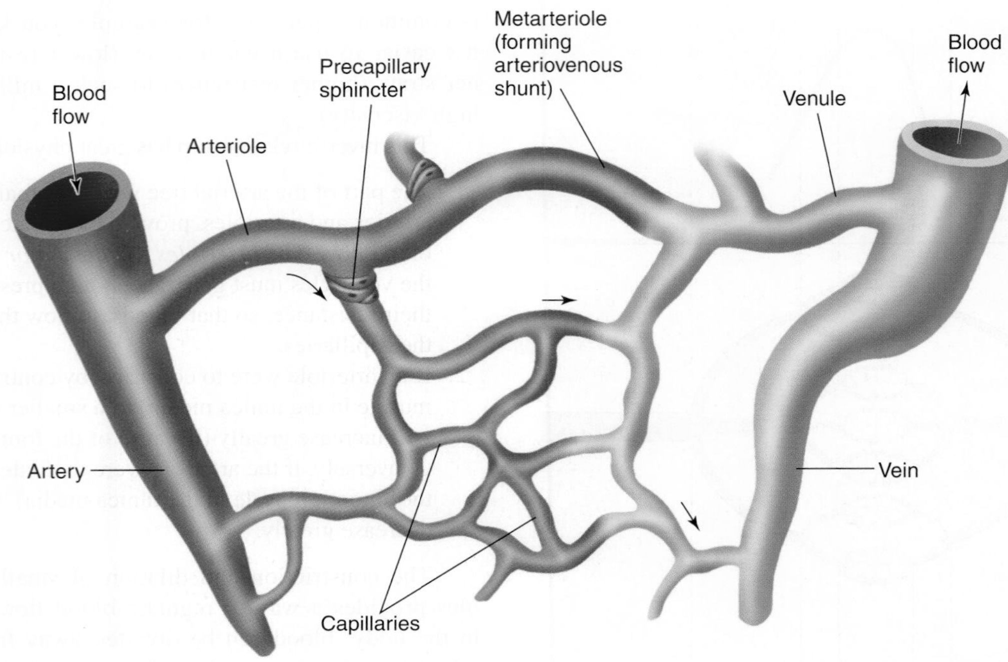

FIGURE 10.23 The microcirculation. Blood from arterioles enters into the more numerous capillaries, and is then drained by venules into veins. Blood flow to a bed of capillaries is regulated primarily by vasoconstriction and vasodilation of the arterioles. However, in some cases precapillary sphincters can also regulate blood flow to capillaries, and, in certain organs, metarterioles can divert blood from arterioles directly into venules.

Stroke Volume Is Regulated Intrinsically and Extrinsically

Stroke volume depends on three factors:

1. The **end-diastolic volume**, which is the amount of blood in the ventricles at the end of diastole, immediately prior to contraction
2. The **total peripheral resistance**, which is the frictional resistance to blood flow in the arteries (primarily in the arterioles) that provides an impedance to the ejection of blood from the ventricles
3. The **contractility**, which is the strength of the ventricle's contraction

How much blood a ventricle ejects per beat (the stroke volume) is related to how much blood is available to be ejected. As a silly (but useful) example, if there's no blood in the ventricle at the end of diastole, no blood can be ejected. So the stroke volume must depend on the end-diastolic volume. With a given end-diastolic volume, the amount of blood ejected is related to the contraction strength (contractility). Again, a silly but useful example may be helpful: if the contraction strength were zero, the stroke volume would be zero. Finally, at a given end-diastolic volume and contractility, the amount of blood ejected depends on how hard it is for the ventricle to eject blood into the arteries. That is, the stroke volume is inversely related to the total peripheral resistance. If all other factors are equal, the stroke volume will decrease when the total peripheral resistance increases. If the total peripheral resistance were sufficiently high, a contraction of a given strength might not be able to overcome it and eject blood.

Now, let's look at a practical example that incorporates these variables. Suppose you're in a sauna; your cutaneous blood vessels will be dilated to radiate heat, decreasing your total peripheral resistance and making it easier for your left ventricle to eject blood. Now suppose that you leave the sauna and dive into ice water (a common practice in Scandinavian countries). Your cutaneous vessels will constrict, significantly increasing your total peripheral resistance. Your left ventricle contracts with the same strength as when you were in the sauna, and ejects less blood. If this were to continue, the cardiac output of the left ventricle would be less than the output of the right ventricle, and this mismatch could be fatal. However, a healthy heart can compensate for this suddenly increased total peripheral resistance with a stronger contraction, which you may feel as a thump in your chest.

As a result of ejecting less blood in the previous beat, more blood is left in the ventricle for the next cycle. The left ventricle again receives blood from the left atrium, but now it has a higher end-diastolic volume than it did before. This higher end-diastolic volume stretches the myocardium, so that the thin filaments overlap more advantageously with the thick filaments (fig. 10.24). As a result, the contraction strength is increased (producing the thump). This can be seen by comparing fig. 10.24*d*, which illustrates greater stretching at a higher end-diastolic volume, with the other parts of the figure that show less stretching at lower end-diastolic volumes. The direct relationship between end-diastolic volume, myocardial stretch, and the strength of myocardial contraction is known as the **Frank-Starling Law of the Heart**.

The Frank-Starling Law of the Heart is an *intrinsic* (built-in) adjustment of contraction strength that matches the output of the two ventricles, so that circulation can be maintained. Contractility (strength) can also be increased by an *extrinsic* (outside) source of regulation—the sympathoadrenal system. Increasing amounts

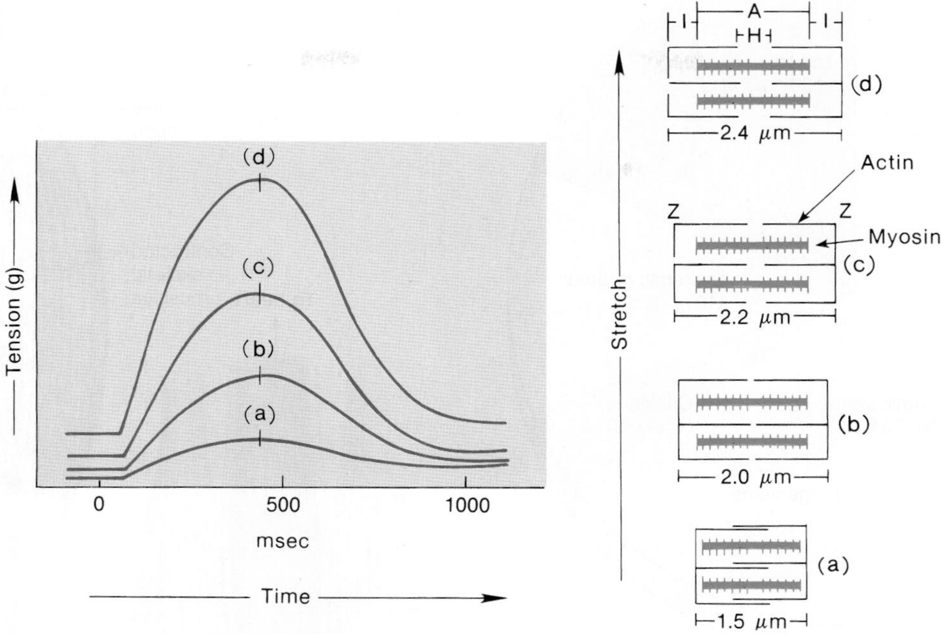

FIGURE 10.24 The Frank-Starling Law of the Heart. Figures (*a*) through (*d*) show the strength of contraction (measured by tension) increasing (left) as the degree of stretching increases (right). At (*d*), the interaction between thick and thin filaments provides the maximum strength of contraction. Stretching is increased by increases in the end-diastolic volume, so the ventricle contracts more strongly when there is more blood to be pumped.

of epinephrine and norepinephrine released during a fight-or-flight reaction (which would be aroused by bathing in ice water after a sauna) stimulates contractility. Thus, after some moments in the ice water, the increased epinephrine and norepinephrine make the ventricles contract more strongly. More accurately, the contraction is stronger with epinephrine (extrinsic effect) at a given degree of stretch (intrinsic effect). A summary of the regulation of cardiac output discussed thus far is provided in figure 10.25.

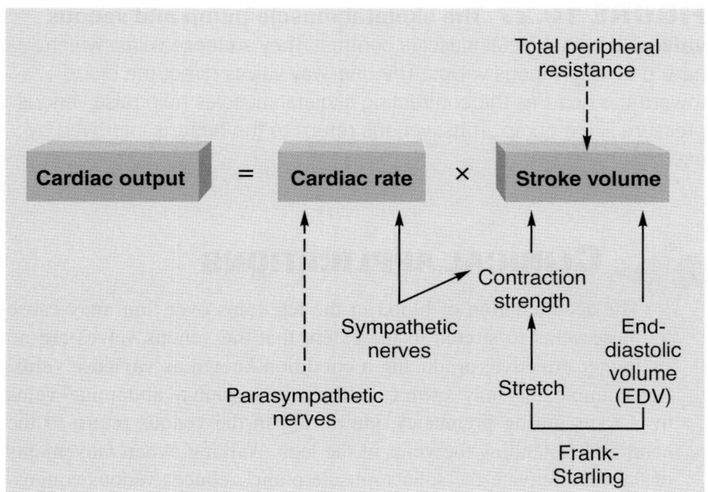

FIGURE 10.25 The regulation of cardiac output. Factors that stimulate cardiac output are shown as solid arrows; factors that inhibit cardiac output are shown as dashed arrows.

Clinical Investigation Clues

Remember that Ron felt a thump in his chest when he dove into ice-cold water; the physician said that this was normal, but was concerned because of Ron's heart murmur.

- What physiological mechanism produced a stronger contraction of the left ventricle, which Ron felt as a thump in his chest, when he dived into the cold water?
- Why might this be dangerous for a person with myocardial ischemia or a mitral valve prolapse?
- After that first thump, what physiological mechanism allows the ventricles to contract more strongly with every beat?

End-Diastolic Volume Depends on the Venous Return

End-diastolic volume is an important factor regulating stroke volume, and thus cardiac output. However, the end-diastolic volume is itself subject to regulation. The end-diastolic volume depends on the amount of blood during diastole flowing in veins back to the heart. This is called the **venous return**, and its regulation is based on the properties of veins. Remember, the walls of veins are thinner than those of arteries (see fig. 10.21), and so veins can expand to hold extra amounts of blood; they serve as reservoirs for blood, and contain most of the total blood volume (fig. 10.26). This storage function can be verified by looking at the veins in your feet after standing for hours; they bulge with extra blood.

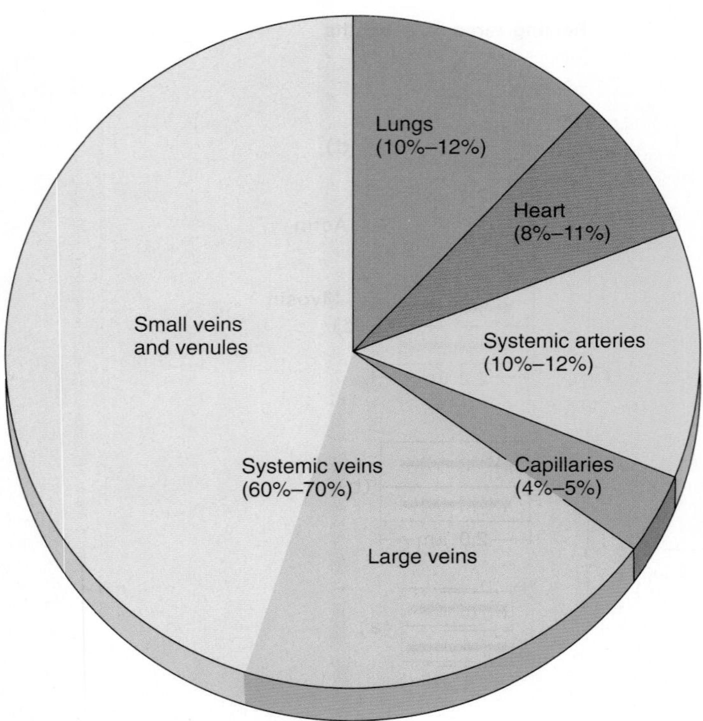

FIGURE 10.26 The distribution of blood within the circulatory system at rest. Notice that the venous system contains most of the blood; it functions as a reservoir from which more blood can be added to the circulation under appropriate conditions (such as exercise).

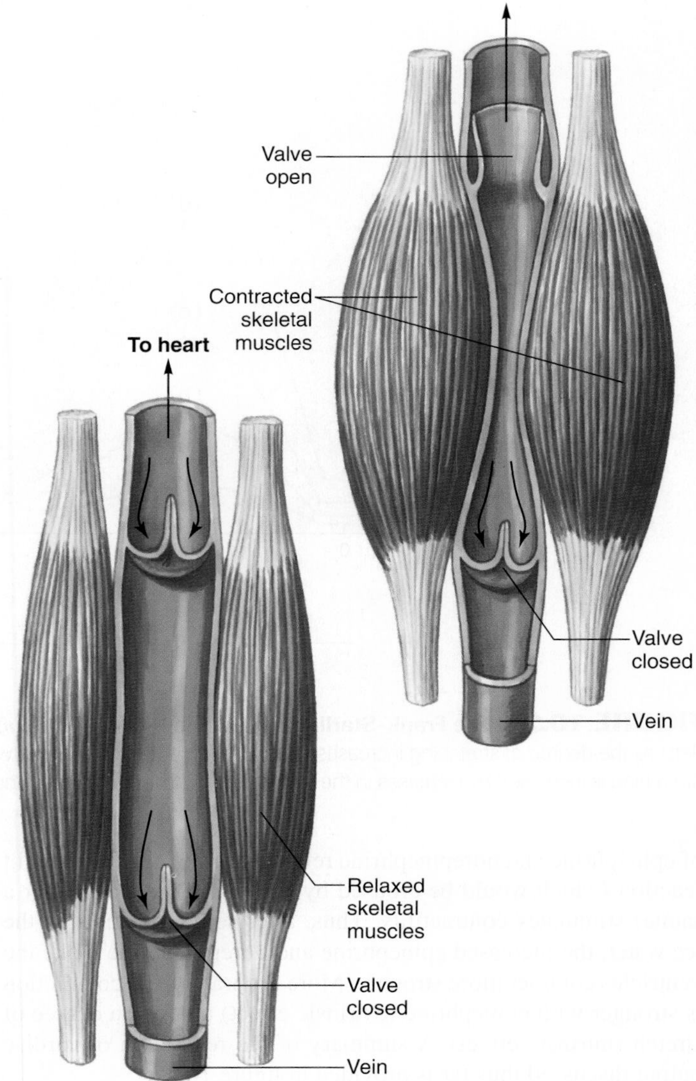

FIGURE 10.27 The skeletal muscle pump and venous valves. When skeletal muscles contract, they squeeze veins, which have one-way venous valves. The one-way valves direct the blood toward the heart as the contracting skeletal muscles help raise venous pressure like a pump. Thus, venous return to the heart is improved by exercise.

The mean arterial pressure is close to 100 mmHg, but the mean venous pressure is closer to 2 mmHg (see fig. 10.22). This pressure is too low to get blood back to the heart from the legs; it requires assistance. Contractions of skeletal muscles give the blood a push by squeezing veins located between the muscles. This action is known as the **skeletal muscle pump**. In order for this to move blood to the heart, veins have one-way **venous valves** (fig. 10.27). Think of squeezing a water-filled hose in the middle; without the one-way valves, the water would squirt from both ends of the hose. Unlike veins, arteries don't require one-way valves, because the higher pressure generated by contraction of the ventricles ensures that blood will flow in only one direction, away from the heart.

During exercise, the skeletal muscle pumps squeeze the veins, increasing the venous return and thus the stroke volume and cardiac output. Venous return is also aided by contractions of the diaphragm during breathing. The diaphragm is a muscular sheet separating the thoracic cavity (containing the lungs) from the abdominal cavity, and when it contracts during inhalation, the pressure in the thoracic cavity drops (causing air to enter the lungs). The higher pressure in the abdominal cavity helps blood move up the inferior vena cava back to the right atrium of the heart. This action of the diaphragm to increase venous return can be called the **breathing pump**. The more active the skeletal muscle and breathing pumps, the more blood flows through the veins back to the heart. This helps sustain the increased cardiac output during exercise, as will be described in a later section.

Clinical Applications

The accumulation of blood in the leg veins over time may cause these veins to stretch to the extent that the venous valves are no longer effective, producing a condition known as **varicose veins**. For example, this may result from the compression of abdominal veins by a fetus during pregnancy, interfering in the venous return to the heart and distending the veins in the legs. Walking, when movements of the foot activate the soleus muscle pump, reduces venous congestion in the legs. This effect can be produced in bedridden people by extending and flexing the ankle joints.

Clinical Investigation Clues

Remember that Ron, a supermarket checker, complained about the bulging veins in his swollen feet, and the physician advised Ron to rock up on his toes periodically and walk as frequently as possible.

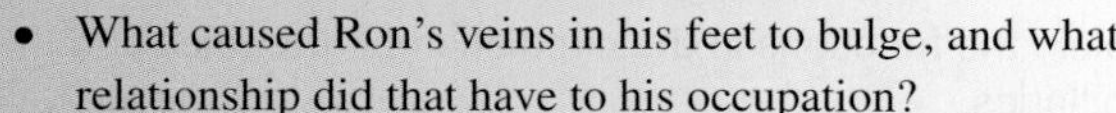

- What caused Ron's veins in his feet to bulge, and what relationship did that have to his occupation?
- How could rocking up on his toes periodically and frequent walking help alleviate the bulging veins?

Cardiac Rate Is Adjusted by the Baroreceptor Reflex

As you go through the day, your heart rate increases and decreases; you can tell this is true by measuring your pulse at different times, or by looking at an ECG strip recorded over a period of time. This is because your cardiac rate is adjusted to maintain homeostasis of blood pressure by means of a negative feedback loop. As discussed in chapter 1, negative feedback loops require sensors to detect changes in the internal environment. The sensors for arterial blood pressure are called **baroreceptors** and are located in the *aortic arch* and *carotid sinuses* (fig. 10.28). These relay information via sensory neurons to integrating centers: the *cardiac control center* and *vasomotor center* in the medulla oblongata.

If the stimulus for this negative feedback loop is a fall in blood pressure, these centers then increase the activity of sympathetic neurons, and reduce the activity of parasympathetic neurons, causing the heart's pacemaker (SA node) to increase the cardiac rate. The increased cardiac rate raises the blood pressure, helping to correct the deviation from homeostasis (fig. 10.29). At the same time, the vasomotor center of the medulla oblongata activates sympathetic neurons that cause vasoconstriction of arterioles in the viscera and skin, increasing the total peripheral resistance. This also helps raise the blood pressure and maintain homeostasis.

Thus, if a person has a fall in blood pressure for any reason—such as loss of blood volume through hemorrhage or dehydration, or through other mechanisms—you should expect this person to have a (1) rapid pulse, due to an increased cardiac rate; and (2) cold, clammy skin, due to cutaneous vasoconstriction. These effects are produced by

Clinical Applications

Many people feel dizzy and disoriented if they stand up too quickly. This is because there is a momentary fall in blood pressure upon standing, and the baroreceptor reflex may require a few seconds before it is fully effective. If the baroreceptor sensitivity is abnormally reduced, an uncompensated fall in blood pressure may occur upon standing. This is called **postural**, or **orthostatic, hypotension**. In severe cases it can make a person faint, because the low blood pressure results in an inadequate blood flow to the brain.

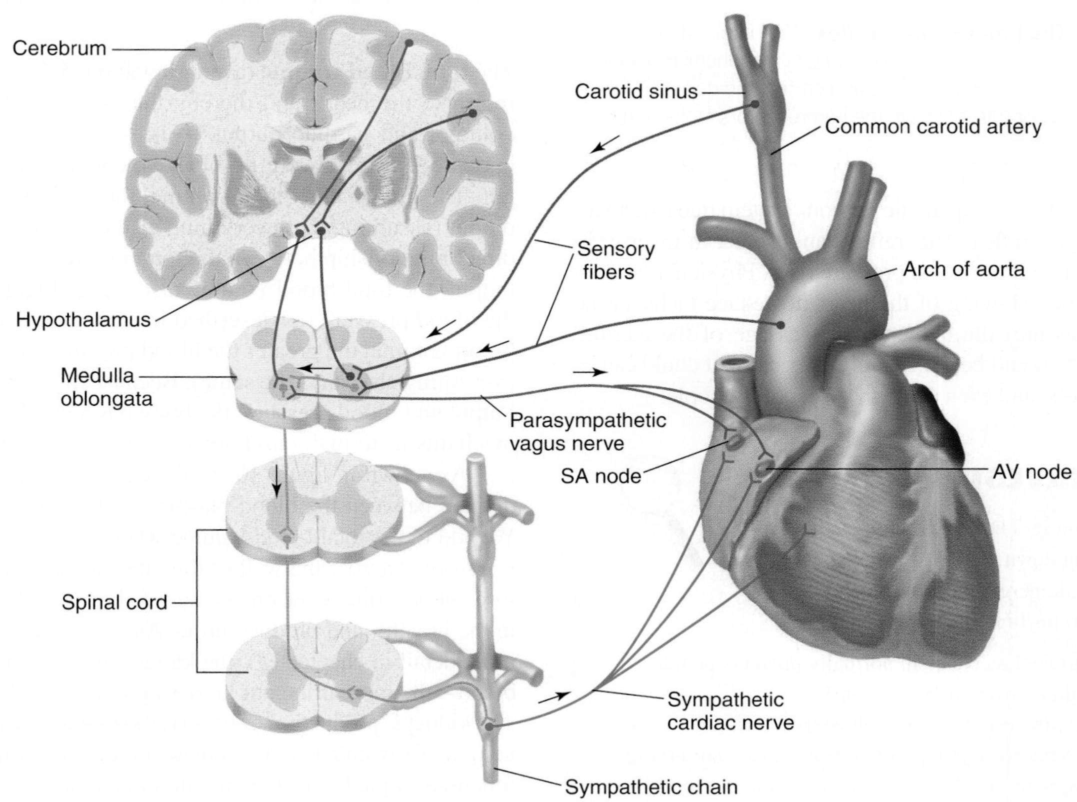

FIGURE 10.28 Structures involved in the baroreceptor reflex. Sensory stimuli from baroreceptors in the carotid sinus and the aortic arch, acting via control centers in the medulla oblongata, affect the activity of sympathetic and parasympathetic nerve fibers in the heart.

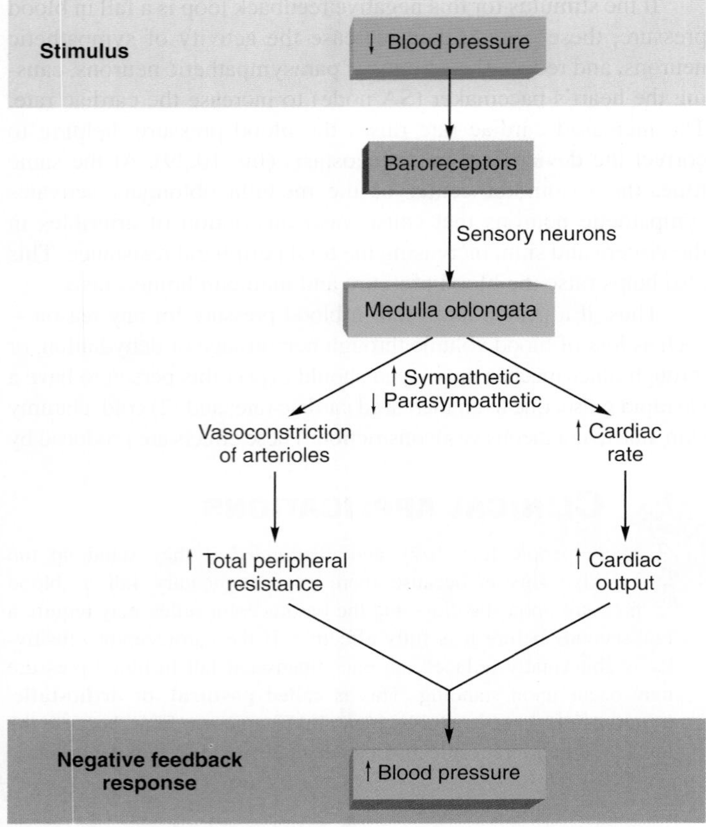

FIGURE 10.29 The baroreceptor reflex. A fall in blood pressure stimulates a reflex increase in the cardiac rate and total peripheral resistance. These changes help raise the blood pressure and correct the deviation from homeostasis. The sensors for this reflex are the baroreceptors in the aortic arch and carotid sinuses.

an increased activity of the sympathetic nervous system due to activation of the baroreceptor reflex. This reflex can also act in reverse: it can slow the cardiac rate if the blood pressure rises. Physicians sometimes induce this reflex slowing of the heart (to reduce tachycardia and lower blood pressure) through manual massage of the carotid sinuses, although this should be done carefully because it could cause a loss of consciousness, and even death in susceptible people.

Clinical Investigation Clues

Remember that Ron said he felt dizzy when he stood up after lying down, and the physician took him off of the β_1-adrenergic receptor blocker he'd been taking to treat his high blood pressure.

- What physiological mechanism normally prevents people from getting dizzy when they stand up?
- Given that β_1-adrenergic receptor blockers reduce sympathetic nerve stimulation of the heart, how would these drugs interfere with the physiological mechanism responsible for preventing dizziness upon standing?

Check Point

1. Define cardiac output, and explain how it relates to cardiac rate and stroke volume.
2. Describe the arterial tree, and explain the relationship between vessel radius and the resistance to flow.
3. Explain the functions in the circulation of arterioles and capillaries.
4. State the Frank-Starling Law of the Heart, and explain its significance.
5. Compare the structure and function of veins and arteries, and explain the significance of the skeletal muscle pump.
6. Describe the baroreceptor reflex, and explain how it helps maintain homeostasis of blood pressure.

Blood Volume Is Regulated to Maintain Homeostasis

The total blood volume influences cardiac output and blood pressure, and so must be carefully regulated. There is a dynamic relationship between capillary blood and interstitial fluid that influences blood volume and replenishes the extracellular environment of the tissues. The kidneys regulate blood volume by adjusting the amount of water excreted in the urine to maintain homeostasis, and this function is finely tuned by the action of ADH and aldosterone on the kidneys.

The **total blood volume** (averaging about 5.5 L) affects the venous return to the heart, and thereby the end-diastolic volume, stroke volume, and cardiac output. This is evident if we take it to an extreme—if there were no blood, there would be no cardiac output. The greater the blood volume, the greater will be the cardiac output (assuming that everything else is the same). Because of the direct relationship between the total blood volume and the cardiac output, the total blood volume also has a direct relationship with the blood pressure (as described in a later section): the greater the blood volume, the greater the blood pressure (again, assuming that everything else stays the same). Because of its influence on cardiac output and blood pressure, the body has a number of physiological mechanisms to maintain homeostasis of the total blood volume.

As shown in figure 10.30, the extracellular fluid of the body is divided between the blood plasma and the interstitial (tissue) fluid. We add to the total blood volume when we ingest liquids: the water is absorbed across the wall of the intestine and enters the blood. We lose blood volume when we sweat, breathe (because of moisture in the breath), and produce urine. As will be described later (and in more detail in chapter 13), the kidneys produce urine as a filtrate of blood—the more urine our kidneys produce, the less blood we have. The kidneys' production of urine is the major way that the total blood volume is regulated, and this regulation is one of the most important functions of the kidneys. Thus, although kidney function is discussed in chapter 13, we must examine aspects of kidney function in this chapter to understand how the cardiovascular system is regulated.

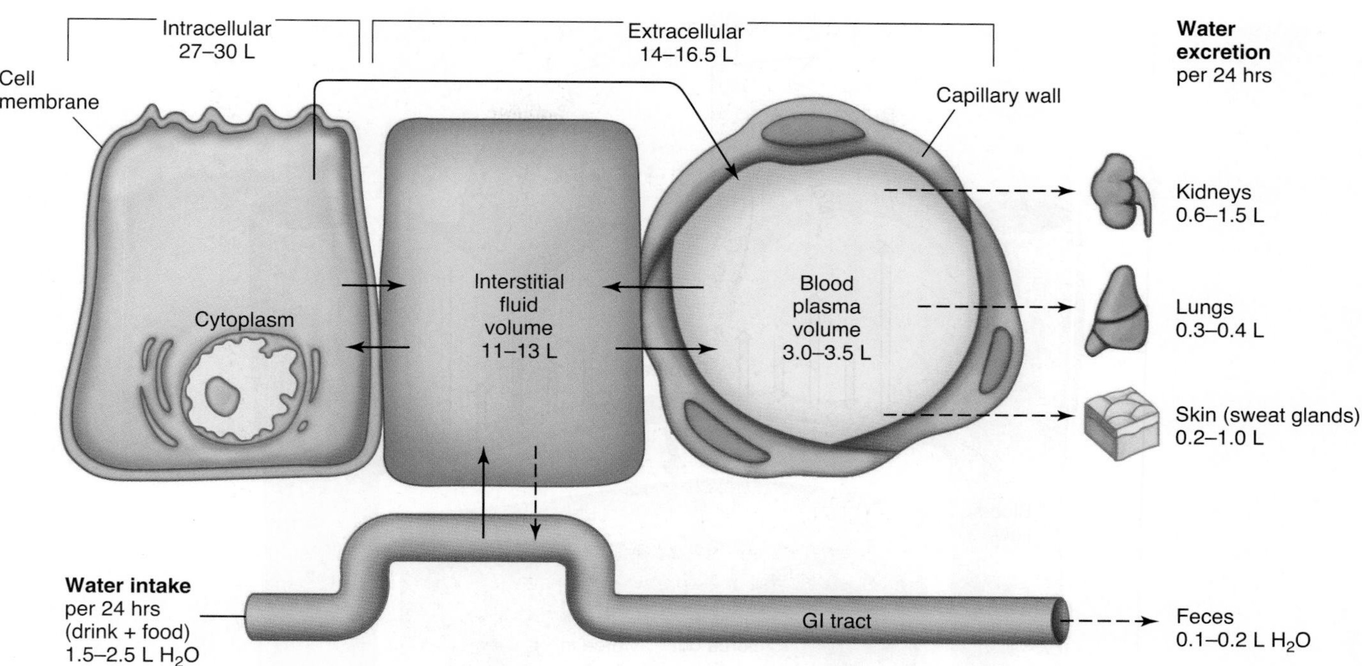

FIGURE 10.30 The distribution of body water between the intracellular and extracellular compartments. The extracellular compartment includes the blood plasma and the interstitial (tissue) fluid.

Capillary Blood and Interstitial Fluid Have a Dynamic Relationship

As mentioned previously, most capillaries in the body (other than those in the CNS) have channels between the adjacent endothelial cells. These channels permit a filtration process to occur. Under the force of the capillary blood pressure, plasma is forced out through these channels and enters the interstitial fluid of the surrounding tissues. The force of the blood pressure is a hydrostatic pressure exerted on the inside of the capillary wall. The hydrostatic pressure of the interstitial fluid opposes this force, but this countervailing force is normally so small it can be ignored. Thus, fluid is filtered out of the capillary wall due to the capillary blood pressure, which acts as a **filtration pressure** (fig. 10.31).

The blood volume and pressure inside a capillary decrease as fluid filters out of the capillary. This is shown by the smaller outward arrows as you go from the arterial end of the capillary (on the left) toward the venous end (on the right) in figure 10.31. The outward movement of fluid is opposed by a **colloid osmotic pressure** produced by plasma proteins. As discussed in chapter 3, the osmotic pressure of a solution is directly proportional to its solute concentration, and solutes that are osmotically active cannot pass through the membrane separating the solutions. Plasma proteins (in a colloidal suspension) are too large to exit through the channels between capillary endothelial cells. Because of this, the concentration of plasma proteins (normally 6–8 g/100 ml) is higher than the protein concentration of the interstitial fluid (2 g/100 ml). The greater protein concentration in the capillary blood provides a colloid osmotic pressure that draws water by osmosis from the interstitial fluid.

As blood leaves the capillary by filtration, the concentration of the blood left behind increases. This is why the inward arrows, indicating the colloid osmotic pressure, increase going from the arterial end (on the left) to the venous end (on the right) of the capillary in figure 10.31. As a result, fluid circulates from the arterial end of a capillary, into the interstitial fluid, and then backs into the venous end of the capillary. However, the amount of fluid returning to the venous end of a capillary is less than the amount leaving at the arterial end. This imbalance could result in the accumulation of interstitial fluid, but normally any excessive interstitial fluid is drained into lymphatic vessels, as discussed in the next section.

Clinical Investigation Clues

Remember that Ron—who, because of his occupation, had to stand for long periods—complained of swollen feet as well as bulging veins.

- How did standing for long periods influence the dynamic relationship between capillary blood and interstitial fluid in his feet?
- How did this cause his feet to swell?
- How could rocking up on his toes and walking change the relationship between capillary blood and interstitial fluid in his feet?

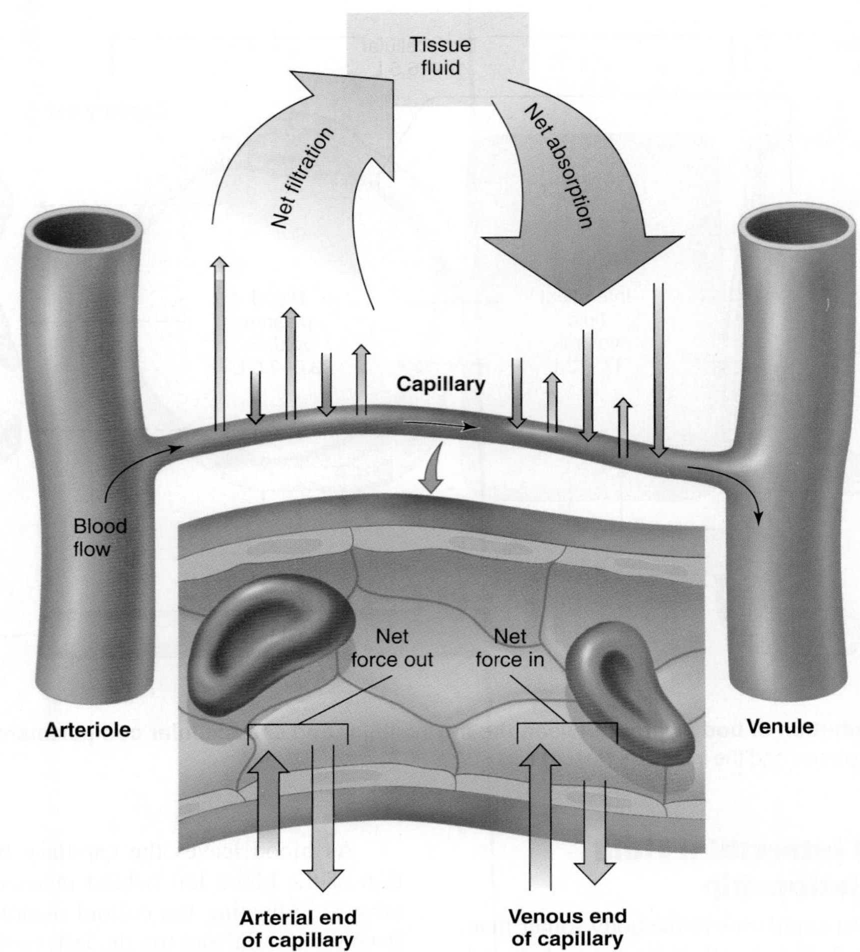

FIGURE 10.31 The distribution of fluid across the walls of a capillary. Tissue, or interstitial, fluid is formed by filtration (*yellow arrows*) as a result of blood pressures at the arteriolar ends of capillaries; it is returned to the venular ends of capillaries by the colloid osmotic pressure of plasma proteins (*orange arrows*).

Lymphatic Drainage Normally Prevents Edema

The **lymphatic system** has three functions: (1) it transports excess interstitial fluid back into the blood; (2) it transports absorbed fat from the small intestine into the blood (chapter 14); and (3) its cells—lymphocytes—help provide immunological defense against disease-causing agents (chapter 11). Here we will consider only its first function—the return of excess interstitial fluid to the vascular system.

The smallest vessels of the lymphatic system are the **lymphatic capillaries**. These are microscopic, closed-ended tubes that form a vast network in the intercellular spaces within most organs. Although they are closed-ended, lymphatic capillaries are so porous that interstitial fluid, proteins, bacteria and viruses, and even leukocytes can enter them (fig. 10.32). Once fluid enters the lymphatic capillaries, it is referred to as **lymph**. This lymph drains into ever-larger lymphatic vessels called **lymph ducts**, and is filtered through **lymph nodes** (chapter 11). Finally, the lymph is returned to the blood, as the *thoracic duct* and the *right lymphatic duct* drain into the left and right *subclavian veins* (located below the clavicles). In this way, the lymphatic fluid derived from blood plasma is returned to the blood plasma (fig. 10.33).

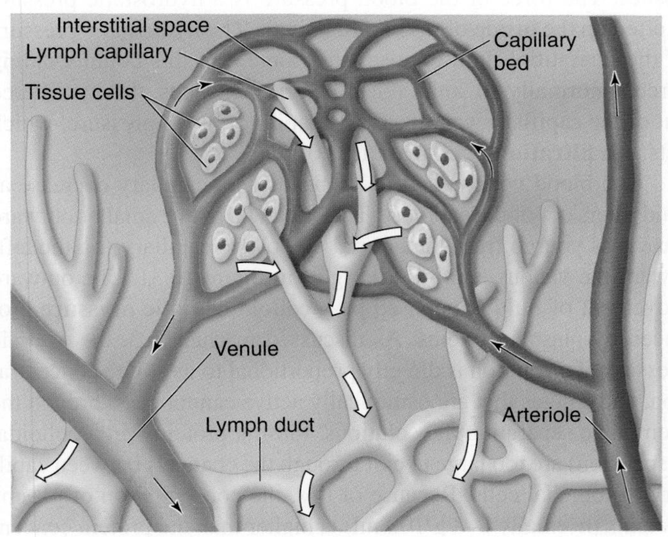

FIGURE 10.32 The relationship between blood capillaries and lymphatic capillaries. Notice that lymphatic capillaries are blind-ended. They are, however, highly permeable, so that excess fluid and protein within the interstitial space can drain into the lymphatic system.

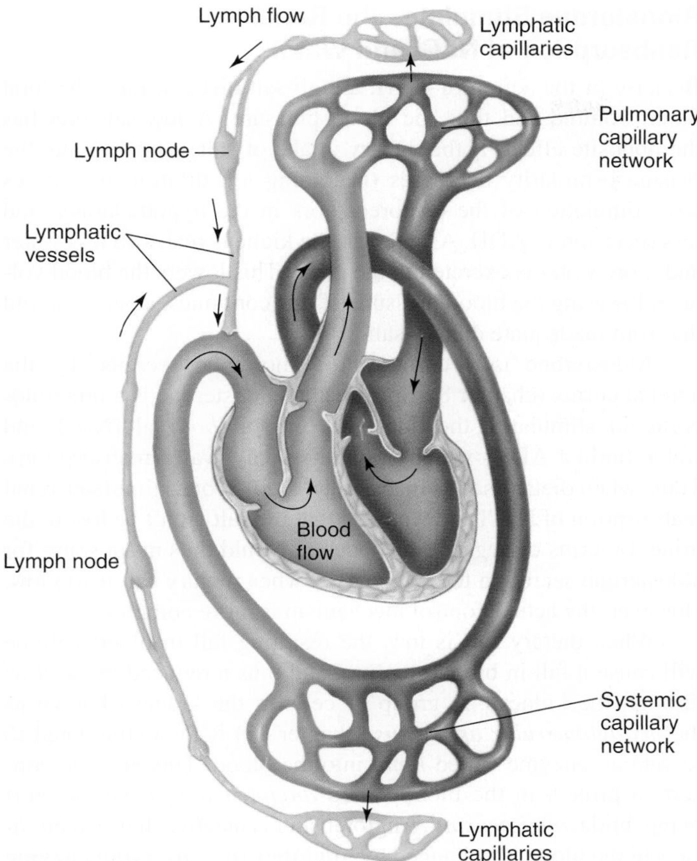

FIGURE 10.33 The relationship between the circulatory and lymphatic systems. This schematic illustrates that the lymphatic system transports fluid from the interstitial space back to the blood through a system of lymphatic vessels. Lymph is eventually returned to the vascular system at the subclavian veins.

Excessive accumulation of interstitial fluid is known as **edema**. Examination of figures 10.31 and 10.32 can reveal how edema may be produced. The possible causes include:

1. *High arterial blood pressure*, increasing the filtration pressure in the capillaries
2. *Venous obstruction* (as may be produced by a blood clot or compression of veins in pregnancy), which produces a congestive increase in capillary filtration pressure
3. *Leakage of plasma proteins into the interstitial fluid*, which increases the colloid osmotic pressure of the interstitial fluid and thereby reduces the osmotic return of water to the capillaries
4. *Decreased plasma protein concentration*, as a result of liver disease (because the liver makes most of the plasma proteins) or other causes, which reduces the colloid osmotic pressure of the plasma
5. *Obstruction of the lymphatic drainage*, which prevents the elimination of the excessive interstitial fluid in the lymph

CLINICAL APPLICATIONS

In the tropical disease *filariasis*, mosquitos transmit a nematode worm parasite to humans. The larvae of these worms invade lymphatic vessels and block the lymphatic drainage. The edema that results can be so severe that the tissues swell to produce an elephant-like appearance; this condition is thus named **elephantiasis** (fig. 10.34). There is a drug regimen effective against the parasite, and an international effort is under way to treat this disease.

ADH and Aldosterone Adjust the Renal Regulation of Blood Volume

The kidneys produce urine in the same way that capillaries produce interstitial fluid—by filtration through the channels in the capillary walls. Under the pressure of the blood in special capillaries called *glomeruli*, a filtrate is produced that enters a system of microscopic tubes, or *tubules*, that transform the plasma filtrate into urine (chapter 13). Both kidneys produce an astounding 180 L of filtrate per day. This is far greater than the total blood volume (of about 5.5 L), so it follows that most of the filtrate must re-enter the blood. This process is called **reabsorption**. The kidneys typically reabsorb more than 99% of the water in the filtrate; this leaves about 1% of the volume of filtrate to become urine (people excrete an average of about 1.5 L/day of urine). However, this *renal* (an adjective referring to the kidneys) reabsorption must be regulated to maintain homeostasis of the blood volume.

FIGURE 10.34 The severe edema of elephantiasis. Parasitic larvae that block lymphatic drainage produce tissue edema and the tremendous enlargement of the limbs and external genitalia in elephantiasis.

Antidiuretic Hormone (ADH) Stimulates the Renal Reabsorption of Water

Antidiuretic hormone (ADH) is secreted by the posterior pituitary gland when stimulated by axons descending from the hypothalamus (chapter 8). The secretion of ADH is influenced by the osmolarity (total solute concentration—see chapter 3) of the plasma. The plasma osmolarity can be raised by either (1) dehydration, which reduces the amount of solvent (water); or (2) eating salt (NaCl), which raises the solute concentration of the plasma (the major plasma solute is Na^+).

When there is an increased plasma osmolarity, **osmoreceptors** are stimulated in the hypothalamus. This produces a sense of thirst, and also stimulates the posterior pituitary to secrete more ADH. ADH stimulates the renal reabsorption of filtered water; that is, ADH stimulates the movement of water by osmosis from the tubules back into the blood. This effect, combined with increased water intake, dilutes the blood plasma and lowers its osmolarity back to the normal range. As a result, when we're dehydrated or eat salty food, we drink more and urinate less, thereby maintaining homeostasis of the plasma osmolarity (fig. 10.35).

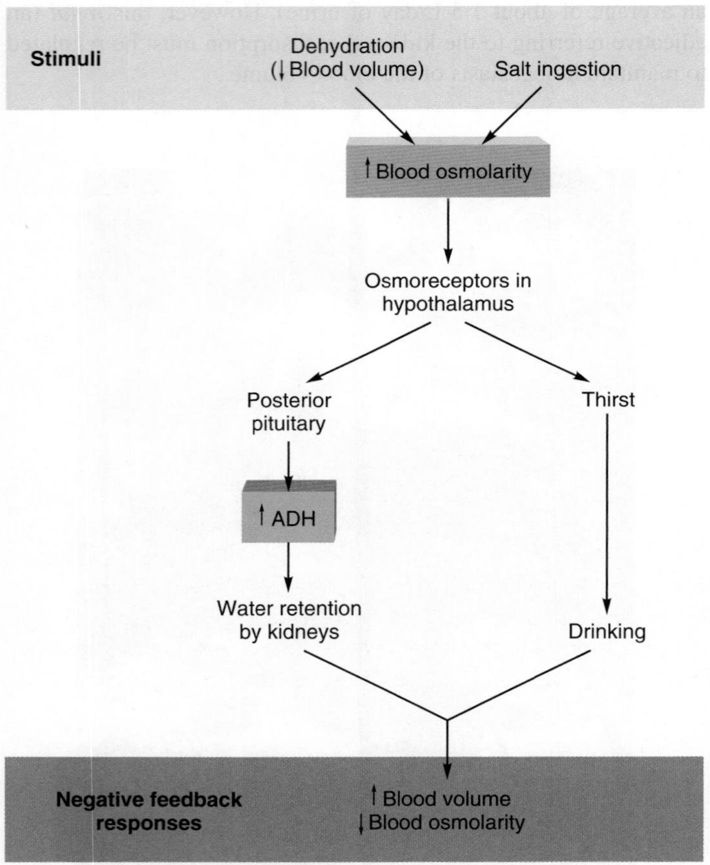

FIGURE 10.35 The ADH regulation of blood volume. Dehydration or salt ingestion increases the osmolarity (concentration) of the blood. This is corrected by a negative feedback mechanism: the increased blood osmolarity stimulates osmoreceptors in the hypothalamus. As a result, we get thirsty and drink. Also, the posterior pituitary secretes increased amounts of ADH, which stimulates the kidneys to retain more water. This increases the blood volume and maintains homeostasis of the blood osmolarity.

Aldosterone Stimulates the Renal Reabsorption of NaCl and Water

Because of the action of ADH, a high-salt diet can raise the total blood volume and thus the blood pressure. A low-salt diet has the opposite effect. If the dietary intake of salt is inadequate, the plasma osmolarity decreases (becoming too dilute); this causes less stimulation of the osmoreceptors in the hypothalamus, and less secretion of ADH. As a result, the kidneys reabsorb less water and more water is excreted in the urine. This lowers the blood volume, lowering the blood pressure. If this continues, a person could die from inadequate dietary salt.[2]

Aldosterone is a corticosteroid hormone secreted by the adrenal cortex (chapter 8). Specifically, aldosterone is a mineralocorticoid, stimulating the kidneys to reabsorb *both* salt (NaCl) and water (unlike ADH, which stimulates only water reabsorption). Thus, when dietary salt is inadequate, aldosterone stimulates renal reabsorption of NaCl and water so that the salt won't be lost in the urine. In terms of negative feedback, it would thus make sense for aldosterone secretion to be increased when dietary salt is too low. However, the actual control mechanism is more complex.

When dietary salt is low, the resulting fall in blood volume will cause a fall in blood pressure, and thus a reduced blood flow through the kidneys. A group of cells in the kidneys known as the *juxtaglomerular apparatus* (chapter 13) is then stimulated to secrete an enzyme called *renin* into the blood. This enzyme converts a protein in the blood called *angiotensinogen* into a short polypeptide, *angiotensin I*. Angiotensin I is inactive, but as it circulates in the blood it's changed by **angiotensin-converting enzyme (ACE)** into a slightly shorter polypeptide, **angiotensin II.** Angiotensin II is a very potent regulatory molecule. It has a number of effects, but its two major actions are: (1) it simulates the adrenal cortex to secrete aldosterone; and (2) it stimulates vasoconstriction of arterioles. Because of both actions of angiotensin II, blood pressure is raised as a negative feedback correction to the initial fall in blood pressure, and homeostasis is maintained (fig. 10.36). The relationship between renin, angiotensin, and aldosterone is known as the **renin-angiotensin-aldosterone system**.

CLINICAL APPLICATIONS

Drugs known as **angiotensin-converting enzyme inhibitors**, or **ACE inhibitors**, are often used to control **hypertension** (high blood pressure). These drugs block the formation of angiotensin II, and thereby lower pressure through two mechanisms. They (1) reduce the stimulation of aldosterone secretion, so that more salt and water are excreted in the urine, and thereby lower the blood volume; and (2) reduce the angiotensin II stimulation of vasoconstriction. Another class of drugs used to treat hypertension in a similar manner is the **angiotensin receptor blockers (ARBs)**. These drugs allow angiotensin II to be formed, but block its interaction with its receptor proteins.

FYI [2]Throughout most of human history, salt was in short supply and highly valued. Moorish merchants in the sixth century traded an ounce of salt for an ounce of gold, and salt cakes were used as money in Abyssinia. Part of a Roman soldier's pay was given in salt—which is the origin of the word *salary* (*sal* = salt). Because slaves were sometimes purchased with salt, we have the expression "worth his salt."

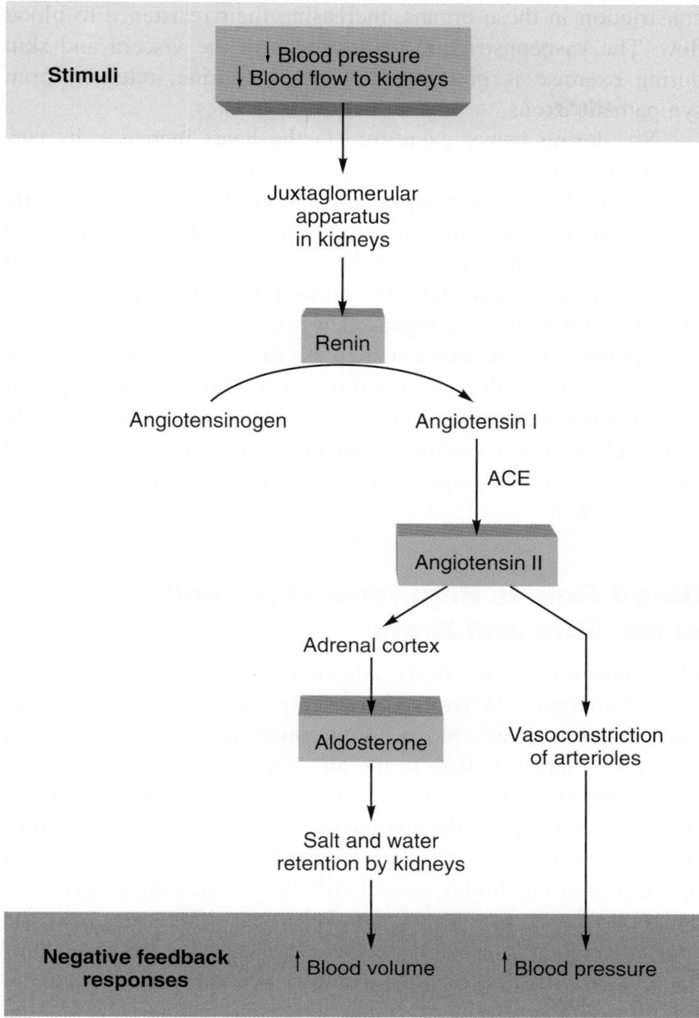

FIGURE 10.36 The renin-angiotensin-aldosterone system. If there is a fall in blood flow through the kidneys, produced by lowered blood pressure, a group of cells called the juxtaglomerular apparatus secretes the enzyme renin into the blood. This causes the production of angiotensin I, which is converted by ACE (angiotensin-converting enzyme) into angiotensin II. Angiotensin II stimulates vasoconstriction and the secretion of aldosterone from the adrenal cortex. Aldosterone then stimulates the kidneys to retain salt and water, helping to restore homeostasis.

CLINICAL INVESTIGATION CLUES

Remember that Ron complained of dizziness upon standing, and his physician changed his blood pressure medication from a β_1-adrenergic receptor blocker to an ARB to reduce that problem.

- What is an ARB, and how does it help reduce blood pressure?
- What is the name of the condition where a person has a fall in blood pressure and becomes dizzy upon standing?
- Why might this condition be aggravated by a β_1-adrenergic receptor blocker but not by an ARB?

CHECK POINT

1. Describe the movement of fluid across the wall of capillaries, referencing the filtration pressure and colloid osmotic pressure.
2. Describe the role of the lymphatic system in preventing the accumulation of excess tissue fluid, and list the different possible causes of edema.
3. Explain how the kidneys maintain homeostasis of blood volume and pressure when a person is dehydrated, referencing the role of ADH in this process.
4. Describe how a high-salt diet can raise the blood volume and pressure, and how a low-salt diet can lower blood volume and pressure.
5. Explain how blood volume affects the secretion of aldosterone, and the physiological significance of angiotensin II and aldosterone in the regulation of blood volume and pressure.

Blood Flow and Blood Pressure Are Related and Regulated

Arterial blood pressure serves as the driving force for blood flow through the arterial trees; the peripheral resistance opposes this flow. The resistance to flow through a particular organ can be varied by constriction or dilation of its arterioles. Resistance is decreased in the heart and skeletal muscles during exercise. Resistance and blood flow through the skin varies greatly, whereas the blood flow to the brain is maintained constant despite changes in blood pressure. Blood pressure is directly proportional to cardiac output and total peripheral resistance, and is measured by listening for the sounds of Korotkoff.

A fluid will flow through a tube from the regions of higher pressure to lower pressure. In the cardiovascular system, the highest pressure is produced by contractions of the ventricles, pushing blood into the arteries. The difference between the arterial blood pressure and the lower blood pressure downstream, in the capillaries and venous system, serves as the driving force for blood flow through an organ. This pressure difference (gradient), or pressure "head" driving the blood flow, can be symbolized as "ΔP" in the following expression:

$$\text{Blood flow} \propto \frac{\Delta P}{\text{resistance}}$$

This equation states that the blood flow is directly proportional to the pressure head and inversely proportional to the resistance to blood flow. The resistance to blood flow was discussed earlier in this chapter. Remember: the resistance is inversely proportional to the fourth power of the radius. If vasoconstriction of the arterioles in an organ were to decrease their radius to one-third of their previous radius, the resistance to blood flow through this organ would increase by 3^4 ($3 \times 3 \times 3 \times 3$); that is, the resistance would be 81 times greater. With the same pressure head, that would reduce the blood flow by this great a factor.

Vasodilation of the arterioles in an organ produces the opposite effect. If the arterioles dilated to three times their previous radius, their resistance to flow would decrease by a factor of 81, and so the blood flow through this organ would increase by a factor of 81 (assuming the same pressure head). In this way, blood can be shunted (diverted) from one organ to another, and even from one part of an organ to another part, by vasoconstriction and vasodilation of the resistance vessels, the small arteries and arterioles.

Blood Flow to the Heart and Muscles Is Regulated Intrinsically

The coronary arteries supply the myocardium with oxygen-rich arterial blood, and the unusually dense network of capillaries in the heart help provide enough oxygen to sustain aerobic respiration. Indeed, the capillary network is so extensive that myocardial cells are within 10 μm of a capillary, compared to an average distance in other organs of 70 μm. Myocardial cells also contain myoglobin (chapter 9) and have other adaptations for aerobic cell respiration.

The coronary arterioles can constrict in response to norepinephrine (an α-adrenergic effect; see chapter 6) and dilate in response to epinephrine (a β-adrenergic effect). However, the great vasodilation that occurs in the coronary circulation during exercise is mainly due to an *intrinsic* (built-in) mechanism. As the heart's metabolism increases, local chemical changes in the myocardium (increased CO_2, depletion of O_2, and other changes) act on the vascular smooth muscle to cause vasodilation. We can term this effect **intrinsic metabolic vasodilation**. This reduces the resistance to flow and thereby increases the blood flow through the heart. As a result, even during heavy exercise, the normal myocardium obtains sufficient blood flow to sustain aerobic cell respiration.

Like the heart, the arterioles in skeletal muscles can respond to regulation by the autonomic nervous system. This is important primarily at rest and at the anticipation of exercise. However, during exercise the primary regulation of blood flow is through *intrinsic metabolic vasodilation*, similar to the mechanism operating in the heart. That is, the increased metabolism of the exercising muscles creates local chemical changes within the muscles that promote vasodilation. The greater the exercise, the faster the muscle's metabolism, and thus the greater will be the degree of vasodilation of arterioles within the muscle. This reduces the resistance to blood flow through the exercising muscles and thereby increases their blood flow in proportion to their metabolic work. However, the great increase in blood flow through the skeletal muscles—from 0.75 L/min at rest to 20 L/min during heavy exercise (fig. 10.37)—cannot be accounted for solely by the reduced resistance in the skeletal muscles.

Figure 10.37 shows that cardiac output increases to 5 times the original output, from 5 L/min at rest to 25 L/min, during heavy exercise. If the resistances didn't change with exercise, this would increase blood flow to every organ by a factor of 5. However, the blood flow to skeletal muscles increases by more than 20 times. Where is this "extra" blood coming from? Examination of figure 10.37 shows that the blood flow through the viscera (the digestive organs shown at the left of the figure) and the skin decreases substantially during heavy exercise. This occurs because of vasoconstriction in these organs, increasing their resistance to blood flow. The vasoconstriction of arterioles in the viscera and skin during exercise is produced by norepinephrine, released from sympathetic axons.

So, during heavy exercise: (1) the heart increases its cardiac output, and thus the total blood flow, fivefold; (2) there is vasoconstriction in the viscera and skin produced by sympathetic nerve activity, reducing the blood flow to these organs; and (3) there is an intrinsic metabolic vasodilation in the heart and skeletal muscles, increasing the proportion of the cardiac output that flows through these organs. The values given in figure 10.37 are averages; elite athletes can increase their cardiac output during exercise by as much as 6 or 7 times. The high cardiac output of elite athletes during strenuous exercise is the major contributor to their high aerobic capacities (chapter 9). A summary of some of the cardiovascular changes that occur during exercise is provided in figure 10.38.

Blood Flow Is Regulated Differently in the Skin and Brain

Blood flow through the skin is adjusted to maintain homeostasis of body temperature. When the body temperature is high, cutaneous vasodilation increases blood flow through the skin, allowing heat from the blood to radiate to the air. When the body temperature falls, cutaneous vasoconstriction reduces blood flow through the skin, thereby reducing the amount of heat lost to the cold air. Thus, cutaneous blood flow can vary tremendously, from as high as 3 to 4 L/min at maximum vasodilation to less than 20 ml per minute at maximum vasoconstriction. When the skin is very cold, the cutaneous blood flow can be so low as to cause the tissue death of *frostbite*. These adjustments in cutaneous blood flow are made by variations in the activity of sympathetic neurons that innervate the cutaneous vessels.

If you again examine figure 10.37, you can see that the blood flow to the brain is the same at rest (15% of 5 L/min) as during heavy exercise (3% of 25 L/min). Exercise may "clear your brain," but that doesn't occur because of increased blood flow. This makes sense, when you remember that the brain is encased in the skull, and increased blood flow would cause brain tissue to expand and become compressed against the bone. Indeed, headaches are often caused by a relatively small increase in cerebral blood flow produced by cerebral vasodilation.

Normally the blood flow to the brain is maintained relatively constant despite changes in arterial blood pressure. A different type of intrinsic (built-in) mechanism maintains this constancy of blood flow: **intrinsic myogenic regulation**. A rise in arterial blood pressure acts on the smooth muscle cells in the arterial walls to cause constriction of the cerebral arteries (thereby preventing increased blood flow); a fall in arterial pressure causes the cerebral arteries to dilate (thereby preventing a reduction in blood flow). The ability of an organ to self-regulate its blood flow is termed **autoregulation**.

Although the total blood flow to the brain remains relatively constant, the distribution of that blood flow to different brain regions varies, depending on the activity of these regions. Active brain regions experience *hyperemia* (increased blood flow);

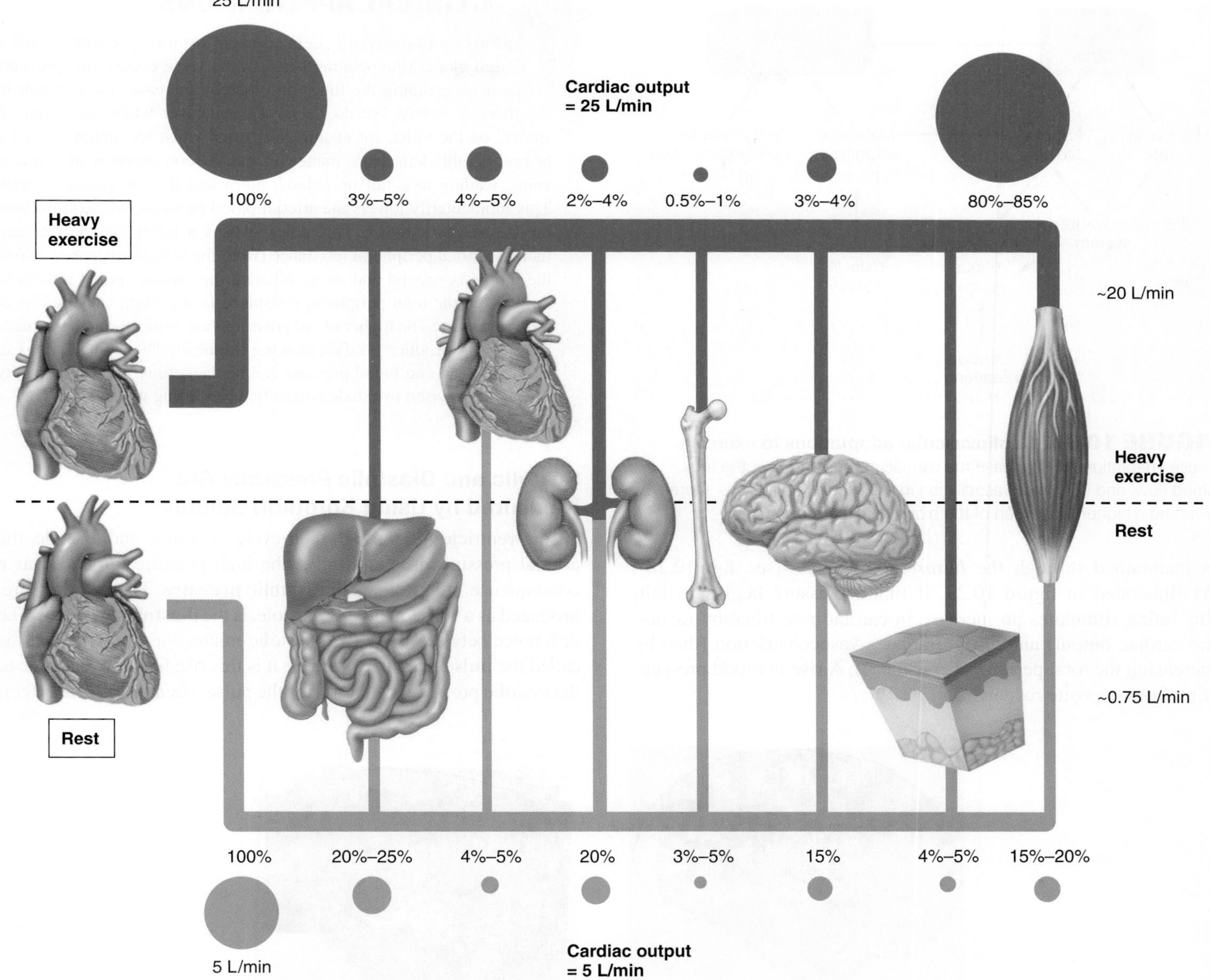

FIGURE 10.37 The distribution of blood flow (cardiac output) during rest and heavy exercise. At rest, the cardiac output is 5 L per minute (*bottom of figure*); during heavy exercise the cardiac output increases to 25 L per minute (*top of figure*). At rest, for example, the brain receives 15% of 5 L per minute (= 750 ml/min), whereas during exercise it receives 3% to 4% of 25 L per minute (0.03 25 = 750 ml/min). Flow to the skeletal muscles increases more than twenty-fold because the total cardiac output increases (from 5 L/min to 25 L/min) and because the percentage of the total received by the muscles increases from 15% to 80%.

indeed, the supply of oxygenated blood to the active region actually increases more than the increase in the region's metabolism (this serves as the basis for the fMRI brain scans; chapter 5). This shunting, or diversion, of blood to the active brain regions occurs as a result of intrinsic metabolic vasodilation. Visual and auditory stimuli, for example, increase the metabolism of the occipital and temporal lobes, respectively, increasing the blood flow to these regions (fig. 10.39).

Blood Pressure Depends on Cardiac Output and Total Peripheral Resistance

Arterial blood pressure is directly proportional to cardiac output and total peripheral resistance. This means that if there is an uncompensated increase in either the cardiac output or the total peripheral resistance, blood pressure will increase. Similarly, an uncompensated decrease in either variable will lower the blood pressure. As described earlier, homeostasis of the blood pressure

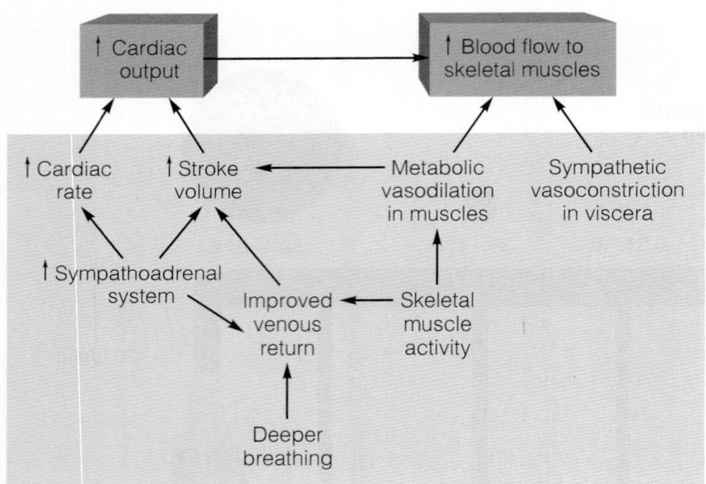

FIGURE 10.38 Cardiovascular adaptations to exercise. These adaptations (1) increase the cardiac output, and thus the total blood flow; and (2) cause vasodilation in the exercising muscles, thereby diverting a higher proportion of the blood flow to those muscles.

is maintained through the *baroreceptor reflex* (see fig. 10.28). As illustrated in figure 10.29, if blood pressure begins to fall, this reflex stimulates an increase in cardiac rate (thereby raising the cardiac output) and also stimulates vasoconstriction (thereby increasing the total peripheral resistance). A rise in blood pressure stimulates opposite responses.

Clinical Applications

Valsalva's maneuver is defined as an expiratory effort against a closed glottis (the opening between the vocal cords); this prevents air from escaping the lungs and thereby increases the pressure in the thoracic cavity. We do Valsalva's maneuver when we "strain at stools" on the toilet, for example, or often when we attempt to lift a heavy weight. Valsalva's maneuver causes compression of thoracic veins, leading to a fall in venous return and thus in cardiac output. This momentarily lowers the arterial blood pressure, which stimulates the baroreceptor reflex. This reflex evokes a faster cardiac rate and increased total peripheral resistance (from the vasoconstriction). When the glottis is opened and air is exhaled, the cardiac output returns to normal but the total peripheral resistance is still high, causing higher blood pressure. The higher blood pressure then works through the baroreceptors to produce a reflex slowing the heart. These fluctuations in cardiac output and blood pressure can be dangerous, and even healthy people are advised to exhale normally when lifting weights.

Systolic and Diastolic Pressures Are Measured by Using Korotkoff Sounds

As the ventricles go through their cycle of systole and diastole, the arterial pressure rises and falls. The high pressure, produced as a consequence of systole, is the **systolic pressure**. The low pressure, produced as a consequence of diastole, is the **diastolic pressure**. The difference between these two (systolic minus diastolic pressures) is called the **pulse pressure**, because it is this rise from the diastolic to the systolic pressure that produces the pulse when you *palpate* (feel)

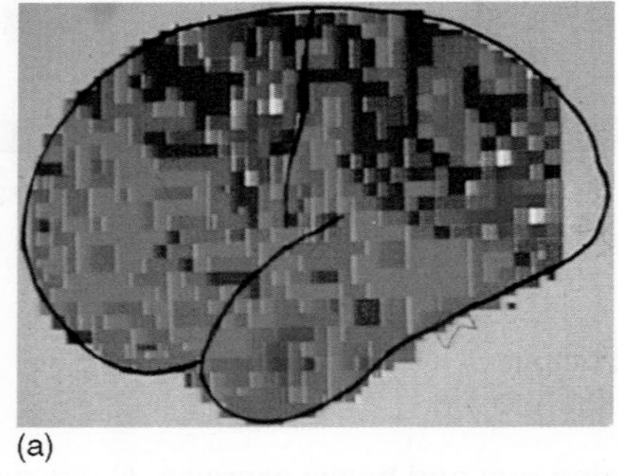
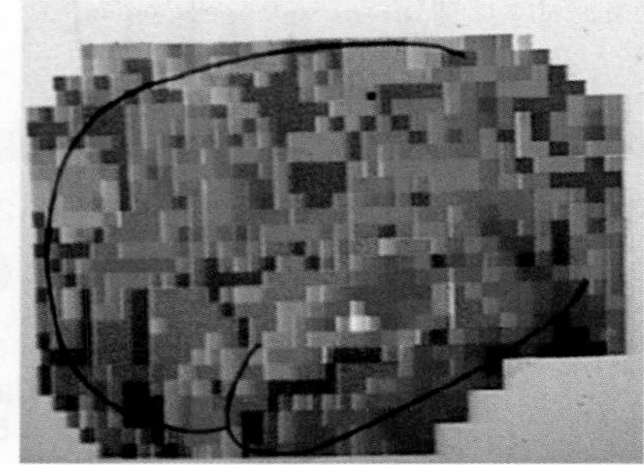

(a)

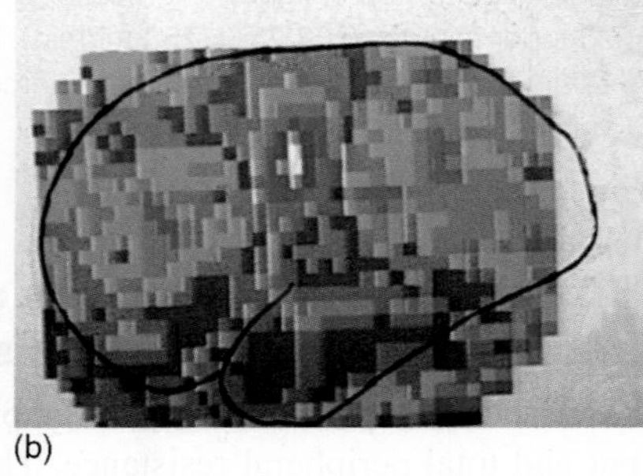
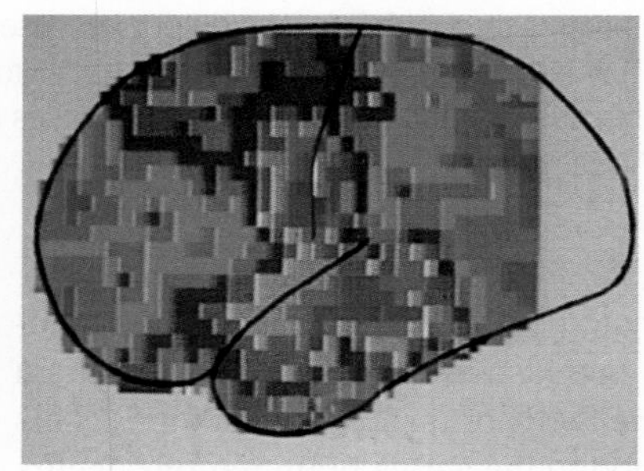

(b)

FIGURE 10.39 Changing patterns of blood flow in the brain. A computerized picture of blood-flow distribution in the brain after injecting the carotid artery with a radioactive isotope. In (*a*), on the left, the subject followed a moving object with his eyes. High activity is seen over the occipital lobe of the brain. In (*a*), on the right, the subject listened to spoken words. Notice that the high activity is seen over the temporal lobe (the auditory cortex). In (*b*), on the left, the subject moved his fingers on the side of the body opposite to the cerebral hemisphere being studied. In (*b*), on the right, the subject counted to 20. High activity is seen over the mouth area of the motor cortex, the supplementary motor area, and the auditory cortex.

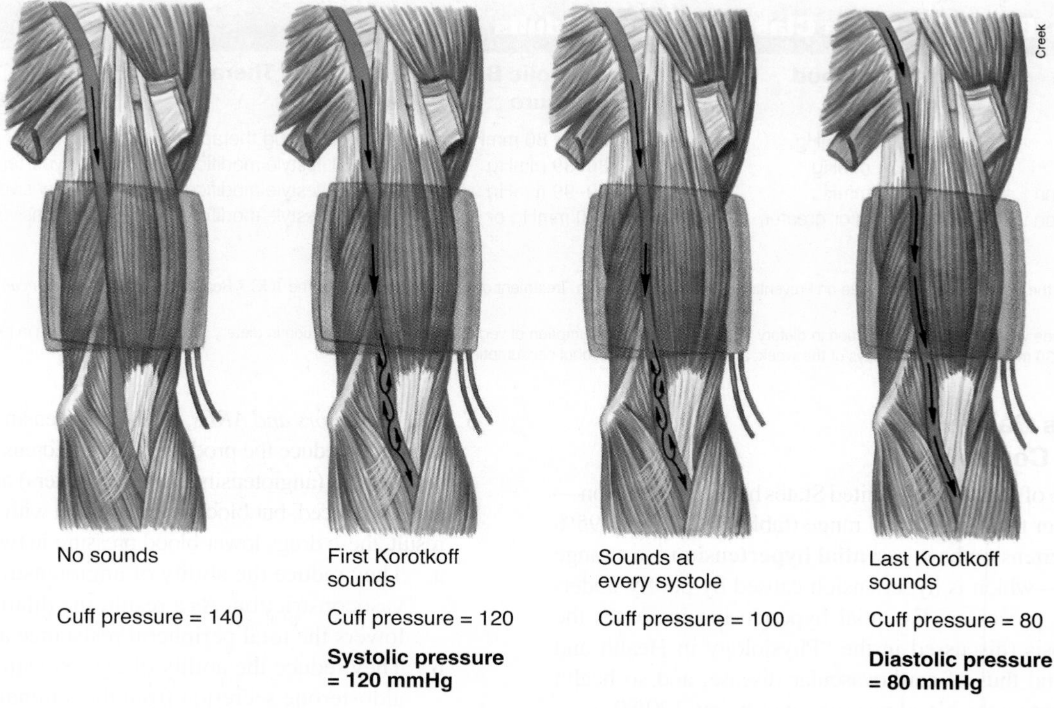

FIGURE 10.40 The blood flow and Korotkoff sounds during a blood pressure measurement. When the cuff pressure is above the systolic pressure, the artery is constricted. When the cuff pressure is below the diastolic pressure, the artery is open and flow is laminar. When the cuff pressure is between the diastolic and systolic pressure, blood flow is turbulent and the Korotkoff sounds are heard with each systole.

an artery. The **mean arterial pressure** is an average, but not a simple arithmetic average because the heart spends longer in diastole than in systole. Its value is significant because the difference between the mean arterial pressure and the pressure in the venous system drives the blood through the capillary beds.

Measurement of the systolic and diastolic blood pressures is a common medical procedure that uses a simple, **auscultatory** (listening) method. This technique utilizes a stethoscope, an inflatable cuff with a rubber bulb, and a manometer (device to measure pressure) called a **sphygmomanometer**, which measures the blood pressure in millimeters of mercury (mmHg). A person using this equipment listens for sounds that emanate from the artery compressed by the cuff; these sounds don't come from the heart valves. The sounds produced by the compressed artery when blood pressure measurements are taken, called the **sounds of Korotkoff**, are named after the originator of this technique.[3]

The artery is silent before the cuff is inflated because blood flow in arteries is mostly smooth and *laminar* (layered). When the cuff is inflated to a pressure greater than the highest pressure in the artery (that is, to a pressure above the systolic pressure), the artery is pinched shut, and thus silent because there is no blood flow. However, when the cuff pressure is lower than the systolic pressure, but greater than the diastolic pressure, Korotkoff sounds are heard with every heartbeat. This is because the artery begins to open at every systole. As it does, blood rushes through the partially constricted opening in a *turbulent flow*. This causes vibrations and sounds (fig. 10.40). Think of a kinked garden hose; it gurgles and vibrates. Korotkoff sounds are more a tapping sound than a gurgle, because the turbulent flow is very brief; at the next diastole, the artery is again pinched shut.

So the artery opens at each systole and closes at each diastole, producing sounds of Korotkoff at every beat due to turbulent blood flow through the partially constricted opening. This begins when the cuff pressure is first equal to the systolic pressure, so that the artery can start to open at systole. Thus, the pressure at which the *first Korotkoff sound* is heard is taken as the systolic pressure. When the sounds disappear, it must mean that even the lowest pressure in the artery (the diastolic pressure) is greater than the cuff pressure, so that laminar flow and silence can resume. Thus, the pressure at which the *last sound of Korotkoff* is heard is measured as the diastolic pressure (fig. 10.40).

When recording the blood pressure readings, it's customary to record the systolic pressure first, and indicate it over the diastolic pressure. For example, the average healthy adult blood pressure is about 120/80: a systolic pressure of 120 mmHg, and a diastolic pressure of 80 mmHg. A blood pressure in excess of certain limits is termed *hypertension* (table 10.2).

FYI [3]Nicolai Sergeievich Korotkoff first presented the auscultatory method of measuring blood pressure to the Imperial Military Academy in St. Petersburg, Russia, in 1905, and this method rapidly became (and still is) the worldwide standard method of measuring blood pressures. Dr. Korotkoff was a physician during the Russian-Japanese war and later in St. Petersburg, and was the first to use the cuff and sphygmomanometer described in 1896 by an Italian scientist, Riva-Rocci, to measure systolic and diastolic blood pressures. He also performed experiments with dogs demonstrating that the sounds were produced by the compressed artery and not by the heart valves.

TABLE 10.2 Blood Pressure Classification in Adults

Blood Pressure Classification	Systolic Blood Pressure		Diastolic Blood Pressure	Drug Therapy
Normal	Under 120 mmHg	and	Under 80 mmHg	No drug therapy
Prehypertension	120–139 mmHg	or	80–89 mmHg	Lifestyle modification;* no antihypertensive drug indicated
Stage 1 Hypertension	140–159 mmHg	or	90–99 mmHg	Lifestyle modification; antihypertensive drugs
Stage 2 Hypertension	160 mmHg or greater	or	100 mmHg or greater	Lifestyle modification; antihypertensive drugs

From the Seventh Report of the Joint National Committee on Prevention, Detection, Evaluation, Treatment of High Blood Pressure: The JNC 7 Report. *Journal of the American Medical Association*; 289 (2003): 2560–2572.

*Lifestyle modifications include weight reduction; reduction in dietary fat and increased consumption of vegetables and fruit; reduction in dietary sodium (salt); engaging in regular aerobic exercise, such as brisk walking for at least 30 minutes a day, most days of the week; and moderation of alcohol consumption.

Hypertension Is Dangerous and Should Be Controlled

Approximately 20% of adults in the United States have hypertension—blood pressure higher than the normal range (table 10.2). About 95% of people with hypertension have **essential hypertension**—a strange term for a disorder—which is hypertension caused by poorly understood and complex processes. Essential hypertension increases the risk of atherosclerosis (discussed in the "Physiology in Health and Disease" section) and thus of cardiovascular disease, and so health personnel try to maintain the blood pressure at or below 120/80.

Because blood pressure is proportional to cardiac output and total peripheral resistance, either or both of these must be elevated to produce hypertension. Although the causes of essential hypertension are controversial, it's generally agreed that a high-salt diet should be avoided. Increased activity of the sympathoadrenal system may also contribute to elevated blood pressure, and may interact with the Na^+ from salt to raise blood pressure.

If changes in diet, weight reduction, cessation of smoking, and moderation in alcohol intake are insufficient to lower the blood pressure to 120/80 or less, then medications that lower the cardiac rate, total peripheral resistance, and/or the total blood volume may be necessary. These medications include drugs that are:

1. *Diuretics*. These act on the kidney to reduce the amount of water reabsorbed (through mechanisms discussed in chapter 13), so that more water is excreted in the urine. This lowers the blood volume, thereby lowering the blood pressure.
2. *Beta-blockers*. These block the β_1-adrenergic receptors in the heart (chapter 6), so that epinephrine and norepinephrine have less stimulatory effect on the cardiac rate. By lowering the cardiac rate, these drugs lower the cardiac output and thereby reduce the blood pressure.
3. *Alpha-blockers and -stimulators*. Some of these stimulate the α_1-adrenergic receptors in the brain, which reduce the activation of the sympathoadrenal system (chapter 6). Others block the α_2-adrenergic receptors in vascular smooth muscle, thereby preventing sympathetic axons from stimulating vasoconstriction. The vasodilation that results lowers the total peripheral resistance, thereby lowering the blood pressure.
4. *Calcium channel blockers*. These block the Ca^{2+} channels in the plasma membrane of vascular smooth muscle cells, reducing their ability to contract. As a result, the arterioles dilate to lower the total peripheral resistance and thus blood pressure.
5. *ACE inhibitors and ARBs*. ACE (angiotensin-converting enzyme) inhibitors reduce the production of angiotensin II from angiotensin I. ARBs (angiotensin receptor blockers) allow angiotensin II to be produced, but block its interaction with its receptors. As a result, these drugs lower blood pressure in two ways:
   a. They reduce the ability of angiotensin II to stimulate vasoconstriction. As a result, the dilation of arterioles lowers the total peripheral resistance and blood pressure.
   b. They reduce the ability of angiotensin II to stimulate aldosterone secretion from the adrenal cortex. With less aldosterone, the kidneys reabsorb less salt and water, excreting more in the urine. This lowers the blood volume and thereby lowers blood pressure.

CLINICAL INVESTIGATION CLUES

Remember that Ron had hypertension, and the physician switched his medication from a β_1-adrenergic receptor blocker to an ARB to reduce Ron's dizziness upon standing.

- Besides the β_1-adrenergic receptor blockers, what other blood pressure medications could also exacerbate Ron's orthostatic hypotension?
- What other class of blood pressure medication is most similar to the ARBs in its action and effects on blood pressure?

CHECK POINT

1. Describe the relationship between blood flow, blood pressure, and total peripheral resistance.
2. Explain how vasoconstriction and vasodilation influence the peripheral resistance and blood flow.
3. Describe how blood flow to the heart and skeletal muscles increases, and how blood flow to the viscera and skin decreases, during exercise.
4. Explain the mechanisms that regulate blood flow to the brain, and blood flow within different regions of the brain.
5. Describe the origin of the sounds of Korotkoff, and explain how blood pressure is measured.
6. Describe the effects of cardiac output and total peripheral resistance on the arterial blood pressure, and explain how different medications act to help lower blood pressure.

PHYSIOLOGY IN HEALTH AND DISEASE

Atherosclerosis is the most common form of arteriosclerosis (hardening of the arteries). Through its contribution to heart disease and stroke, atherosclerosis is responsible for about 50% of the deaths in the developed world. Atherosclerosis is characterized by *plaques*, or *atheromas*, that protrude into the lumen of the coronary or cerebral arteries and thus reduce blood flow. Additionally, the plaques serve as sites for *thrombus* (blood clot) formation, which can further occlude the artery (fig. 10.41).

It's currently believed that plaque formation begins as a result of damage to the endothelium of the artery. The progression of plaque development is then promoted by the participation of lymphocytes, monocytes, macrophages, and platelets; atherosclerosis is now believed to be an inflammatory disease, to a significant degree. For example, monocytes enter into the subendothelial connective tissue of the artery and become transformed into macrophages, which engulf lipids and take on a *foamy cell* appearance. The endothelium normally protects against this progression, but the protective action of the endothelium can be damaged by (1) hypertension, (2) smoking, and (3) high blood cholesterol. The most effective action smokers can take to lower their risk of atherosclerosis is to quit smoking.

Cholesterol is carried to the arteries by proteins called **low-density lipoproteins (LDLs)**. People who eat a diet high in cholesterol and saturated fat, and people with an inherited tendency for this condition, have *hypercholesteremia* (high blood cholesterol). The endothelial cells of the artery engulf the LDL particles, and then oxidize them to a product called *oxidized LDL*, which contributes to endothelial cell injury. Antioxidants, such as vitamin C, vitamin E, and beta-carotene, may thus offer some protective effect. **High-density lipoproteins (HDLs)** protect against atherosclerosis because these particles take cholesterol from the foamy cells of the artery and carry it to the liver, which converts cholesterol into bile salts and then excretes the bile salts in the bile (chapter 14). So, to protect against atherosclerosis, we should attempt to lower our LDL cholesterol while raising our HDL cholesterol. The levels of HDL are largely determined by genetics and gender (women prior to menopause have higher levels than men), but exercise can also help raise the HDL level.

The liver produces bile salts from cholesterol, and obtains this cholesterol through (1) the synthesis of new cholesterol, and (2) uptake of cholesterol bound to HDL particles in the blood. This physiology is the basis for how drugs called **statins** (such as Lipitor) work to help people with dangerously high levels of LDL cholesterol. The statins act in the liver cells to inhibit an enzyme needed for the synthesis of cholesterol. By inhibiting the ability of the liver cells to produce new cholesterol, the statins force the liver to obtain more cholesterol from the blood. This more active traffic of cholesterol from the arteries to the liver lowers the cholesterol concentration of the blood.

In addition to stopping smoking (for smokers), controlling hypertension (if present), reducing the percentage of saturated fat and cholesterol in the diet, ingesting antioxidants, and exercising, is there any other lifestyle change that might offer some protection from atherosclerosis? Yes: eat fish once or twice a week. Fish, particularly oily fish, such as salmon, sardines, mackerel, and swordfish, have omega-3 fatty acids (this refers to double bonds between the third and fourth carbons). For reasons not fully understood, omega-3 fatty acids help protect the heart by lowering the risk of atherosclerosis.

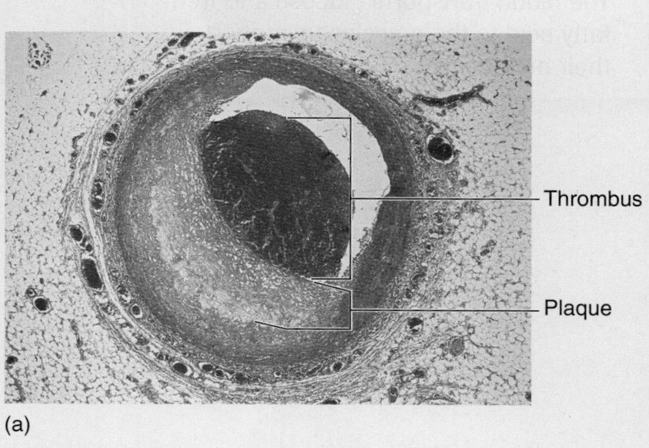

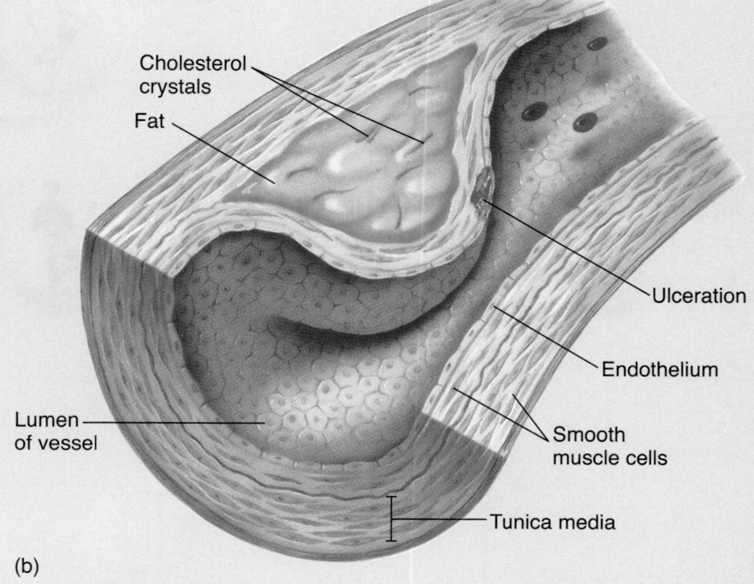

FIGURE 10.41 Atherosclerosis. *(a)* A photograph of the lumen (cavity) of a human coronary artery that is partially occluded by an atherosclerotic plaque and a thrombus. *(b)* A diagram of the structure of an atherosclerotic plaque.

Physiology in Balance

Blood and Circulation

Working together with . . .

Respiratory System

Oxygen is transported by the blood to all body cells . . . p. 297

Endocrine System

Hormones secreted by endocrine glands are transported by the blood to their target cells . . . p. 186

Digestive System

Food molecules absorbed through the intestine into the blood are transported by the blood to all body cells . . . p. 348

Muscular System

The blood transports glucose and free fatty acid to the muscles to support their metabolism . . . p. 215

Urinary System

Waste products in the blood get filtered out and excreted in urine . . . p. 324

Homeostasis Revisited

Topic	Set Point	Integrating Center	Sensors	Effectors
How the baroreceptor reflex maintains homeostasis of blood pressure through the regulation of the cardiac rate when blood pressure falls	Arterial blood pressure	Cardiac control center of the medulla oblongata	Baroreceptors in the aortic arch and carotid sinuses	Sympathetic nerves stimulate the SA node to increase the heart rate, raising the cardiac output and blood pressure.
How the baroreceptor reflex maintains homeostasis of blood pressure through regulation of the total peripheral resistance when blood pressure falls	Arterial blood pressure	Vasomotor center of the medulla oblongata	Baroreceptors in the aortic arch and carotid sinuses	Sympathetic nerves stimulate smooth muscle in arterial walls to cause vasoconstriction of arterioles, increasing the total peripheral resistance and blood pressure.
How ADH secretion maintains homeostasis of blood volume and pressure when a person is dehydrated	Plasma osmolarity (concentration)	Hypothalamus	Osmoreceptor neurons	Increased secretion of ADH from the posterior pituitary stimulates the kidneys to reabsorb more water, maintaining blood volume and pressure.
How aldosterone secretion is adjusted to help maintain homeostasis of blood volume and pressure when a person eats excessive amounts of salt	Blood flow through the kidneys	Juxtaglomerular apparatus of the kidneys	Juxtaglomerular apparatus of the kidneys	Secretion of renin from the juxtaglomerular apparatus is reduced; angiotensin II production is reduced; secretion of aldosterone, from the adrenal cortex is reduced, decreasing the reabsorption of salt and water and causing more to be excreted in the urine.
How renin secretion, by indirectly affecting the total peripheral resistance, helps maintain homeostasis of blood pressure when blood pressure falls	Blood flow through the kidneys	Juxtaglomerular apparatus of the kidneys	Juxtaglomerular apparatus of the kidneys	Increased secretion of renin from the juxtaglomerular apparatus results in increased production of angiotensin II, which causes vasoconstriction of arterioles, raising the total peripheral resistance and blood pressure.
How intrinsic metabolic vasodilation in exercising skeletal muscles helps support the aerobic respiration of the muscles	Tissue concentrations of oxygen, carbon dioxide, and other metabolites	Smooth muscle cells within arterioles in the skeletal muscles	Smooth muscle cells within arterioles in the skeletal muscles	Smooth muscle cells within the arterioles in exercising skeletal muscles relax, reducing the resistance and increasing the blood flow.
How intrinsic myogenic regulation maintains homeostasis of cerebral blood flow when arterial blood pressure rises	Cerebral blood flow	Smooth muscle cells within the walls of cerebral arteries	Mechanical distortion of the plasma membrane of the smooth muscle cells in cerebral arteries as the arteries stretch	Contraction of the smooth muscle layer of cerebral arteries causes constriction and prevents blood flow from increasing when the arterial pressure rises.

Summary

Blood Consists of Plasma and Formed Elements 241

- Blood plasma contains dissolved molecules and ions, as well as proteins (albumin and the globulins) that are in a colloidal suspension.
- Red blood cells (RBCs, or erythrocytes) lack nuclei and mitochondria, but contain hemoglobin molecules that bind to oxygen; because of this, RBCs function in oxygen transport.
- White blood cells (WBCs, or leukocytes) function in immunity; they include the granular leukocytes (neutrophils, eosinophils, and basophils) and the nongranular leukocytes (lymphocytes and monocytes).
- Platelets (thrombocytes) function in blood clotting by aggregating at the site of injury and sticking together; the platelet plug is reinforced by threads of fibrin, formed by clotting factors in the plasma that become sequentially activated.
- The ABO blood-typing system consists of antigens on the red blood cells that may be type A, B, AB, or O; people have antibodies in the plasma against the ABO antigens missing from their RBCs, so mixing blood of different types causes the red blood cells to clump together, or agglutinate.
- People who are Rh positive have the Rh factor (antigen) on their RBCs; those who are Rh negative lack the Rh antigen.

The Heart Pumps Blood Through the Circulation 245

- The right ventricle pumps blood into the pulmonary arteries, which carry blood low in oxygen to the lungs; blood rich in oxygen returns via pulmonary veins to the left atrium to complete the pulmonary circulation.
- The left ventricle pumps oxygen-rich blood into the aorta, which gives rise to branches that supply all organs with blood; blood partially depleted of oxygen then returns via the superior and inferior venae cavae to the right atrium, completing the systemic circulation.
- One-way atrioventricular (AV) valves, including the tricuspid on the right and the mitral on the left, separate the atria and ventricles; the exits of each ventricle are guarded by one-way semilunar valves (the pulmonic and aortic valves).
- Both atria contract and relax together, as do both ventricles; the phase of contraction is called systole, and the phase of relaxation is called diastole.
- When the ventricles contract at systole, the pressure within them rises to close the AV valves; this produces the first heart sound ("lub").
- When the ventricles relax at the beginning of diastole, the pressure within them falls below the pressure of the arteries, closing the semilunar valves; this produces the second heart sound ("dub").
- During diastole, the blood in the atria fills the ventricles about 80% even before the atria contract; when the atria contract, they add to the final end-diastolic volume.
- When the ventricles contract, they eject about two-thirds of their end-diastolic volume; the amount of blood they eject per beat is the stroke volume.

Action Potentials in the Heart Stimulate Contraction and Produce the ECG 249

- The normal pacemaker is the sinoatrial (SA) node; action potentials spread from here through gap junctions to stimulate both atria.
- Action potentials from the atria are conducted into the ventricles by specialized conducting tissue—the AV node, bundle of His, and Purkinje fibers.
- The pacemaker cells undergo an automatic, spontaneous depolarization called the pacemaker potential, which triggers action potentials.
- Epinephrine and norepinephrine from sympathoadrenal stimulation increase the rate of the pacemaker depolarization (and thus the cardiac rate), whereas ACh from parasympathetic axons decreases the rate of the pacemaker depolarization (and thus the cardiac rate).
- The action potential from myocardial cells in the ventricles has a long plateau phase, in which Ca^{2+} enters through the plasma membrane and stimulates the release of Ca^{2+} from the sarcoplasmic reticulum (SR).
- Ca^{2+} from the SR stimulates a contraction that corresponds in time to the long action potential; thus, the myocardium is refractory while it is depolarizing and contracting, and also while it is repolarizing and relaxing.
- Electrodes from the body surface record the ECG; although it is not a recording of action potentials, the ECG reflects the conduction of action potentials in the heart.
- The P wave is produced by depolarization of the atria; the QRS complex is produced by depolarization of the ventricles; and the T wave is produced by repolarization of the ventricles.
- The QRS complex is seen as the ventricles depolarize at the beginning of systole, and thus correlates with the first heart sound; the T wave occurs as the ventricles repolarize at the beginning of diastole, and thus correlates with the second heart sound.

Cardiac Output Indicates the Heart's Pumping Ability 253

- Cardiac output is the volume of blood ejected by each ventricle per minute, and is equal to the product of the stroke volume and the cardiac rate.
- The ventricles eject blood into elastic arteries, which transport the blood into muscular arteries and then arterioles; these provide a frictional resistance to blood flow called total peripheral resistance.
- Resistance is inversely proportional to the fourth power of the radius of the vessels, so that the narrowest arteries—the arterioles—provide the greatest resistance to blood flow; their resistance is increased by vasoconstriction, and decreased by vasodilation.
- Exchanges between blood and interstitial fluid occur across the walls of capillaries, which are the most numerous vessels with the greatest total cross-sectional area; because of this, there is a large drop in pressure as blood passes through capillaries from the arterial to the venous systems.
- As the end-diastolic volume is increased, there is more stretching of the myocardium, causing the ventricles to contract more strongly to eject the greater amount of blood; this intrinsic relationship between end-diastolic volume and contraction strength is known as the Frank-Starling Law of the Heart.
- Stroke volume is also increased when epinephrine and norepinephrine, through sympathoadrenal stimulation, cause increased contraction strength.
- Stroke volume is inversely related to total peripheral resistance: when total peripheral resistance rises, stroke volume falls, if other factors remain constant; a normal heart, however, compensates for this by an increased contraction strength.
- The end-diastolic volume depends on the venous return to the heart; this is aided by contractions of skeletal muscles (the skeletal muscle pump) and the action of one-way venous valves.
- A fall in blood pressure is detected by baroreceptors in the aortic arch and carotid sinuses; these initiate a baroreceptor reflex, which increases the cardiac rate and stimulates vasoconstriction in the viscera and skin.

Blood Volume Is Regulated to Maintain Homeostasis 260

- Blood plasma is filtered through capillary channels under the pressure of the blood, forming interstitial fluid; water is returned to the capillary by osmosis due to the colloid osmotic pressure produced by plasma proteins.
- The lymphatic system normally drains away excess interstitial fluid; an excessive accumulation of interstitial fluid is known as edema.
- The kidneys are the major organs regulating blood volume, because urine originates as a blood filtrate; the kidneys regulate blood volume by adjusting the amount of filtered water that they reabsorb back into the blood.
- Antidiuretic hormone (ADH) stimulates the kidneys to reabsorb more of the filtered water, so that more remains in the blood and less is excreted in the urine; ADH secretion is increased by a rise in the blood osmolarity (concentration), stimulating osmoreceptors in the hypothalamus.
- Aldosterone stimulates the kidneys to reabsorb more salt (NaCl) and water; aldosterone secretion is stimulated by angiotensin II, formed when the renin-angiotensin-aldosterone system is activated by a fall in blood pressure and blood flow through the kidneys.
- Angiotensin II is formed from angiotensin I by angiotensin-converting enzyme (ACE), and raises blood pressure by stimulating vasoconstriction as well as by stimulating aldosterone secretion.

Blood Flow and Blood Pressure Are Related and Regulated 265

- Blood flow through the arterial system is directly proportional to the pressure head, generated by contraction of the ventricles, and inversely proportional to total peripheral resistance.
- Blood flow increases to the heart and skeletal muscles during exercise by intrinsic metabolic vasodilation; when these organs work harder and their metabolism increases, local chemical changes within them cause the smooth muscle of their arterioles to relax, producing vasodilation and thus increased blood flow.
- During exercise, sympathetic nerves cause constriction of arterioles in the viscera and skin, so that blood is diverted away from these organs to the exercising skeletal muscles and heart; cardiac output (and thus total blood flow) also increases with exercise.
- Blood flow to the skin can vary greatly to help maintain homeostasis of deep body temperature; blood flow to the brain remains relatively constant because of intrinsic myogenic regulation, where changes in blood pressure act directly on the vascular smooth muscle to cause vasoconstriction or vasodilation (in response to a rise or a fall in blood pressure, respectively).
- Blood pressure is directly proportional to cardiac output and total peripheral resistance, so an uncompensated increase in either would cause a rise in blood pressure.
- Blood pressure measurements are taken using the auscultatory method, where Korotkoff sounds are produced by turbulent blood flow through an artery partially constricted by a cuff.
- The cuff pressure at which the first Korotkoff sound occurs is the systolic pressure; the cuff pressure at which the last Korotkoff sound occurs is the diastolic pressure.
- Hypertension (high blood pressure) can be treated by lifestyle changes or by drugs that lower the cardiac rate, total peripheral resistance, or total blood volume.

Review Activities

Objective Questions: Test Your Knowledge

1. Which of the following statements about erythropoietin is true?
   a. Erythropoietin is secreted by the kidneys.
   b. Erythropoietin stimulates the production of white blood cells.
   c. Erythropoietin secretion increases when the blood oxygen content increases.
   d. Erythropoietin stimulates the lymphoid tissue.
2. Which of the following cell types contains hemoglobin and functions in oxygen transport?
   a. Thrombocytes
   b. Leukocytes
   c. Erythrocytes
   d. Megakaryocytes
3. Which of the following cells is an agranular leukocyte?
   a. Neutrophil
   b. Eosinophil
   c. Basophil
   d. Lymphocyte
4. Which of the following statements regarding thrombin is true?
   a. It is released from damaged tissues to start clot formation.
   b. It changes into prothrombin during clot formation.
   c. It converts fibrinogen into fibrin.
   d. It makes thrombocytes sticky in clot formation.
5. Which of the following blood types is the universal donor?
   a. Type A
   b. Type B
   c. Type AB
   d. Type O
6. Which of the following statements about veins is true?
   a. They always transport blood away from the heart.
   b. They always carry blood partially depleted in oxygen.
   c. They always transport blood to the heart.
   d. They always carry oxygenated blood.
7. The pressure inside the ventricles (intraventricular pressure) rises when
   a. the ventricles contract at the beginning of diastole.
   b. the ventricles contract at the beginning of systole.
   c. the ventricles relax at the beginning of diastole.
   d. the ventricles relax at the beginning of systole.
8. The amount of blood in the ventricles just before they contract is called the
   a. stroke volume.
   b. end-systolic volume.
   c. end-diastolic volume.
   d. end-systolic volume.
9. The second heart sound occurs when
   a. the AV valves close at the beginning of diastole.
   b. the AV valves close at the beginning of systole.
   c. the semilunar valves close at the beginning of diastole.
   d. the semilunar valves close at the beginning of systole.
10. The normal pacemaker of the heart is the
    a. Purkinje fibers.
    b. bundle of His.
    c. AV node.
    d. SA node.

11. Which of the following is not a characteristic of a myocardial action potential?
   a. It has a long plateau phase.
   b. Ca^{2+} enters through slow Ca^{2+} channels to stimulate contraction.
   c. Its duration lasts about as long as the myocardial contraction.
   d. It has a short refractory period.

12. The T wave of the ECG occurs when the
   a. ventricles depolarize.
   b. ventricles repolarize.
   c. atria depolarize.
   d. atria repolarize.

13. In relation to the ECG, the first heart sound occurs when
   a. the P wave is produced and the atria contract.
   b. the P wave is produced and the atria relax.
   c. the QRS wave is produced and the ventricles contract.
   d. the QRS wave is produced and the ventricles relax.

14. Which statement regarding cardiac output is true?
   a. Cardiac output equals stroke volume divided by cardiac rate.
   b. Cardiac output of both ventricles is normally equal.
   c. Cardiac output is normally greater for the right ventricle.
   d. Cardiac output is normally greater for the left ventricle.

15. The vessels that provide the greatest resistance to the ejection of blood from the ventricles are the
   a. large elastic arteries.
   b. small muscular arteries.
   c. arterioles.
   d. capillaries.

16. The vessels with the greatest total cross-sectional area, where the blood pressure falls the most, are the
   a. large elastic arteries.
   b. small muscular arteries.
   c. arterioles.
   d. capillaries.

17. If an arteriole were to constrict to one-fourth its previous radius, its resistance to blood flow would
   a. increase by a factor of 256.
   b. decrease by a factor of 256.
   c. increase by a factor of 81.
   d. decrease by a factor of 81.

18. Cardiac output is inversely proportional to
   a. end-diastolic volume.
   b. contractility.
   c. total peripheral resistance.
   d. stroke volume.

19. According to the Frank-Starling Law of the Heart, the strength of the ventricle's contraction is increased as a result of the
   a. increase in end-diastolic volume.
   b. decrease in end-diastolic volume.
   c. increase in total peripheral resistance.
   d. decrease in total peripheral resistance.

20. The end-diastolic volume is increased by all of the following except
   a. increased activity of the skeletal muscle pumps.
   b. the action of one-way venous valves.
   c. increased activity of the breathing pump.
   d. increased total peripheral resistance.

21. Because of the baroreceptor reflex, a person with a fall in blood pressure would have all of the following except
   a. vasoconstriction in the viscera.
   b. vasoconstriction in the skin.
   c. vasoconstriction in the skeletal muscles.
   d. a faster pulse.

22. At the arterial end of a capillary, there is a net movement of fluid
   a. out of the capillary, because of the colloid osmotic pressure of the blood.
   b. out of the capillary, because of the capillary blood pressure.
   c. into the capillary, because of the colloid osmotic pressure of the blood.
   d. into the capillary, because of the capillary blood pressure.

23. An abnormally low plasma protein concentration
   a. lowers the colloid osmotic pressure of the blood.
   b. lowers the filtration pressure in the capillaries.
   c. lowers the amount of interstitial fluid.
   d. prevents edema from occurring.

24. When a person becomes dehydrated, all of the following occur except
   a. blood volume decreases.
   b. blood osmolarity decreases.
   c. ADH secretion increases.
   d. the kidneys reabsorb more water.

25. When blood volume and blood pressure decrease, all of the following occur except
   a. aldosterone stimulates more salt and water excretion in the urine.
   b. the juxtaglomerular apparatus secretes more renin.
   c. more angiotensin I is produced.
   d. more angiotensin II is produced.

26. Angiotensin II stimulates
   a. vasodilation.
   b. ADH secretion.
   c. ADH and aldosterone secretion.
   d. vasoconstriction and aldosterone secretion.

27. During exercise,
   a. there is intrinsic metabolic vasodilation in the exercising muscles.
   b. sympathetic axons cause vasoconstriction in the viscera and skin.
   c. cardiac output and total blood flow increase.
   d. all of these are true.

28. Which of the following statements is true?
   a. Blood flow to the brain is relatively constant.
   b. Blood flow to the skin is relatively constant.
   c. Blood flow to the brain and skin is relatively constant.
   d. Blood flow to the skin is regulated by intrinsic myogenic regulation.

29. The sounds of Korotkoff are produced by
   a. closing of the AV and semilunar valves.
   b. opening of the AV and semilunar valves.
   c. laminar flow of blood through an artery.
   d. turbulent blood flow through a constricted artery.

30. If a person has a blood pressure of 130/90,
   a. the diastolic pressure is 130 mmHg.
   b. the pulse pressure is 40 mmHg.
   c. the systolic pressure is 90 mmHg.
   d. the person does not have hypertension.

Essay Questions 1: *Test Your Understanding*

1. What is blood plasma, and what does it contain? If you drew blood into a test tube, what would you have to do to obtain serum?
2. Describe the function of red blood cells, and explain why anemia can be a problem.
3. Explain how the red blood cell count is affected in a person who lives at a high altitude.
4. Distinguish between the different types of white blood cells. What are their functions?
5. Describe the role of platelets in blood clotting. Also, explain how the clotting factors in plasma participate in the formation of a blood clot.
6. What does it mean if a person's blood type is type O positive, or type A negative? Explain why blood types are matched before blood is transfused from one person to another.
7. Under what circumstances is there reason for concern when an Rh-negative woman becomes pregnant? Explain the reasons for this concern.
8. Identify the "right pump" and "left pump," and explain how the blood within them differs.
9. Identify the heart valves, and describe their functions.
10. Describe the phases of the cardiac cycle, and correlate the changes in intraventricular pressure with the opening and closing of the heart valves. Identify when the first and second heart sounds are produced.
11. What is the normal pacemaker of the heart? How does the action potential it produces travel through the atria?
12. Identify the conducting tissues of the heart and their locations. What functions do these tissues serve?
13. Describe the characteristics of the pacemaker potential, and the way the autonomic nervous system influences cardiac rate.
14. Describe the characteristics of the action potential produced by a typical myocardial cell. Explain how depolarization and contraction, and repolarization and relaxation, are correlated.
15. Explain the significance of the fact that the action potential, and the refractory period, have a long duration.
16. Identify the ECG waves and their causes. In which part of the cardiac cycle is each ECG wave produced?
17. Which ECG waves are produced when you hear the first and second heart sounds? Explain the reasons for this association.
18. Define the following terms: *bradycardia*, *tachycardia*, *flutter*, and *fibrillation*.
19. How is cardiac output related to cardiac rate, stroke volume, and total peripheral resistance?
20. How do the cardiac outputs of the left and right ventricles compare? How do the thicknesses of the two ventricles compare? Explain the reasons for both of your answers, and how these two answers are related.
21. Describe the components of the arterial tree, and the characteristics of each type of artery.
22. Explain how resistance to blood flow is related to the radius of an artery, and the significance of this relationship when an artery constricts or dilates.
23. State the Frank-Starling Law of the Heart, and explain its significance.
24. How does the blood pressure in veins compare to the blood pressure in arteries? How does this difference relate to the anatomy of veins compared to arteries?
25. Describe how the venous return to the heart is influenced by the contractions of skeletal muscles and by breathing.
26. Describe the baroreceptor reflex, starting from a fall in blood pressure (such as when a person quickly goes from a lying to a standing position).
27. Describe how blood plasma and interstitial fluid are interchanged across the walls of capillaries, explaining the forces involved.
28. Explain the different ways edema can be produced, and how the lymphatic system normally prevents edema.
29. Step by step, describe the role of ADH in maintaining homeostasis when a person is dehydrated.
30. Step by step, describe the role of the renin-angiotensin-aldosterone system in maintaining homeostasis when a person has a fall in blood volume and pressure.
31. Explain how blood flow to an organ is related to blood pressure and peripheral resistance.
32. Describe how the blood flow to the skeletal muscles and heart is increased during exercise.
33. Compare how blood flow is regulated through the brain and skin.
34. Identify the variables that influence the arterial blood pressure.
35. What produces the sounds of Korotkoff? How are these sounds used to obtain a measurement of the systolic and diastolic blood pressure?

Essay Questions 2: *Test Your Analytical Ability*

1. Which of the plasma proteins is most responsible for the colloid osmotic pressure of the blood? Given the interaction between capillary plasma and interstitial fluid, explain how liver disease could result in edema.
2. Explain the similarity between a person with anemia and a person with a normal RBC count and hemoglobin concentration but newly arrived at a high altitude.
3. People with kidney disease often suffer from anemia. Explain the relationship between these conditions.
4. People who want to donate blood are asked if they've recently taken aspirin. Explain why this question is relevant.
5. Laboratory personnel often add citrate or EDTA to test tubes that will be used for blood collection. These compounds chelate (bind to) Ca^{2+}. How would that affect the blood? Explain.
6. A differential WBC count provides the percentage of each type of leukocyte. What sorts of medically useful information can a differential WBC count provide that isn't provided by a total WBC count?
7. Suppose a woman with blood type A negative gives birth to a child who is type O positive. She claims that a man who is type B positive is the father. Do the blood types indicate that this is possible? Explain.
8. Does the woman in the previous question need to be concerned about the health of the baby, and if so, what can be done? Explain.
9. How would you expect the pressure in each of the pulmonary arteries to compare with the pressure in the aorta? How would the blood flow in the two pulmonary arteries compare to the blood flow in the aorta? Explain your answers.
10. If the ventricles fibrillate and stop pumping, the person still has a few minutes' worth of blood oxygen in reserve (in the absence of CPR). How does that relate to the observation that blood in the systemic veins has about 20% less oxygen than the blood contained in systemic arteries?
11. The heart valves open and close in response to differences in pressure on each side of the valve flaps. Use this information to explain why the AV and semilunar valves open and close at their specific times during the cardiac cycle.
12. Explain how epinephrine causes an increase, and acetylcholine causes a decrease, in the cardiac rate.
13. When do the slow Ca^{2+} channels open in the myocardial plasma membrane? Why does this correspond to the time of myocardial contraction? Explain the mechanisms involved.

14. Explain why the repolarization phase of the myocardial action potential correlates with the beginning of diastole. Also, explain why the myocardium cannot be repeatedly stimulated to produce a continuous, tetanus-like contraction.
15. A person with a condition known as AV node block may skip ventricular contractions; for example, the ventricles might depolarize and contract only on every other cycle, although the atria beat normally. Explain how this would affect the ECG.
16. Given the condition described in the previous question, how would the stroke volume and cardiac output be affected? Explain.
17. Using the Frank-Starling Law of the Heart, explain how the strength of the ventricles' contractions would be affected by bradycardia.
18. Suppose an artery gives rise to two branches of equal size that enter different organs. Suppose that the arterioles in each organ are equal in number and size. Given a constant blood pressure, explain how blood flow would be affected if the arterioles in one organ dilate to twice their previous radius, while the arterioles in the other organ constrict to half their previous radius.
19. What force drives blood through the capillary beds? How is that affected by (a) hypertension (high arterial pressure); and (b) venous congestion (raising the venous blood pressure)? Explain.
20. Given your answers to the previous question, how would the amount of interstitial fluid be influenced by (a) or (b)? Explain.
21. What happens to total peripheral resistance when you go from a warm room to cold outside air? How does this influence cardiac output and blood pressure? What compensations may occur to maintain homeostasis?
22. Given that the liver produces plasma albumin, explain how a person with cirrhosis could get edema.
23. You shouldn't try on new shoes after standing for a long period. Use the physiology of veins and of interstitial fluid formation to explain this statement.
24. Suppose you come across a person crawling along the hot desert sand who had been lost for more than a day. When you feel her pulse, you find that it is weak but rapid, and her skin feels cold despite the heat. Use the baroreceptor reflex to explain these observations.
25. The person in the desert in the previous question produces a very small quantity of highly concentrated urine. How does this relate to the maintenance of blood volume and pressure? Explain the physiological mechanisms involved.
26. When we force ourselves to drink more water than we want to, we later excrete a large volume of dilute urine. How does this relate to homeostasis of blood volume and pressure? What mechanism allows us to excrete the excess water?
27. Suppose you lose a large quantity of salt by sweating—explain the mechanisms by which your blood volume and pressure will be affected. Also, explain the physiological mechanisms that will help counteract these changes to maintain homeostasis.
28. Describe how blood flow to the skeletal muscles and skin changes during exercise, explaining the regulatory mechanisms involved. Also, explain why a person's face feels hotter just after stopping the exercise than it did during the exercise.
29. How does eating salty food affect blood volume and pressure? Explain. Also, explain what must happen in order to excrete the extra Na^+ in the urine.
30. During blood pressure measurements of the brachial artery (in the arm), no sounds are heard in the stethoscope after the cuff is inflated to a pressure higher than the systolic pressure, or after the cuff pressure becomes lower than the diastolic pressure. Explain why this is so.

Web Activities

CHAPTER 11

Immune System

Homeostasis

We live in a sea of viruses and bacteria. Some of these are beneficial, most are harmless, but many are potentially pathogenic (have the ability to cause disease). These and other microorganisms can disturb the delicate homeostatic balance that our regulatory systems labor so hard to maintain. If this happens, we become ill. The immune system functions to combat these foreign invaders, and so indirectly helps the body to maintain homeostasis. However, the "standing army" of the immune system needs policing, because it occasionally commits domestic offenses, causing allergy and an array of autoimmune diseases. Illness may result when either the ability of the immune system to combat invading pathogens is deficient, or its ability to police itself proves inadequate.

But if homeostasis is not maintained . . .

Clinical Investigation

Helen sneezed just before getting her flu shot, saying, "There go my allergies again." She asked the doctor why she sneezed, and he answered that there was pollen in the air that time of year, and hoped it wasn't because the room was dusty. He gave her a prescription for an antihistamine. A month later, while walking her mountain bike up a steep trail in the hills, she was bitten by a rattlesnake. Fortunately, she was biking with a friend, who helped rush her to an emergency room. She was given antitoxin, and experienced a gradual but full recovery. Helen asked if the antitoxin would immunize her against any future rattlesnake bites, and was told that it wouldn't. If she ever got bitten again, she would again have to undergo the same treatment. Helen responded: "Oh, is it just like getting a new flu shot each year?" "No," the physician responded, "the reasons why antitoxin would be needed again are different."

What might cause Helen to sneeze in response to pollen or dust, and how would antihistamines help? How do vaccines help prevent disease, and why didn't last year's flu vaccine work against this year's flu? What is antitoxin, and how did it help Helen recover from the snakebite? Why would she need to get another antitoxin treatment if a rattlesnake bit her in the future?

Chapter Outline

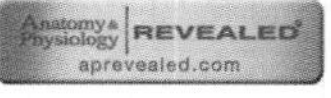

A virtual cadaver dissection experience

The Immune System Defends Against Pathogens

The immune system consists of innate mechanisms that confer a nonspecific defense against disease-causing agents (pathogens), and adaptive immune responses, which offer immune defenses against specific molecular targets, or antigens. Nonspecific defense mechanisms, such as phagocytosis by neutrophils and macrophages, are often supplemented by the adaptive immune responses during a local inflammation. Adaptive immunity against specific antigen-bearing pathogens is afforded by B and T lymphocytes, which are produced in the bone marrow and lymphoid organs.

Immunology, the scientific investigation of the immune system, is so medically important and complex that it's a separate discipline, which students can study in advanced courses. However, students learning the fundamentals of human physiology also become acquainted with the basic concepts of immune function. These concepts not only are important for an understanding of human physiology, they also support the study of microbiology. If you are like most students who take a human physiology course, you will also take a course in microbiology. If so, you may study aspects of the immune system in both your physiology and microbiology courses, because the subject is fundamental to both disciplines. So, get set to add to your knowledge pyramid in human physiology as you also contribute to your present or future study of microbiology.

The **immune system** includes all of the structures and processes that provide a defense against potential **pathogens** (disease-causing agents). These defenses can be divided into two categories: **innate** (or **nonspecific**) **immunity** and **adaptive** (or **specific**) **immunity**.

Innate immunity is provided by:

1. *Epithelial membranes*, which nonspecifically block the entrance of most pathogens into the body
2. *Stomach acidity*, which kills many microorganisms before they can damage the body
3. *Phagocytosis* by neutrophils, monocytes, macrophages, and other phagocytes in the body
4. *Complement proteins* in the plasma, which become activated to destroy pathogens by antibodies produced in an adaptive immune response
5. *Natural killer (NK) cells*—named because of their ability to kill tumor cells—which are part of innate immunity but work in a manner similar to the "killer T cells" of the adaptive immune response

Adaptive immunity is provided by lymphocytes. The terms *adaptive immunity* and *specific immunity* refer to the processes by which the immune system acquires the ability to defend against pathogens by exposure of lymphocytes to specific molecular targets, or **antigens**, which are part of those pathogens.

The difference between innate and adaptive immunity can be understood by examining the receptor proteins on the cell surface that provide each category of immunity. The innate immune system responds to molecules characteristic of invading microorganisms (such as bacteria) but not of human cells. At present, scientists have identified 10 different receptors on certain cells of the innate immune system that respond to these pathogen-specific molecules. By contrast, the adaptive immune response is mounted against very specific molecular targets, or antigens, which are usually foreign molecules but which can also be molecules produced by human cells. There are hundreds of thousands of different receptor proteins (each on different lymphocytes) specific for different antigens. Prior exposure to a specific antigen can produce active immunity to that antigen, and this acquired immunity is specific for that antigen.

Innate Immunity Includes Phagocytosis and Fever

There are three groups of phagocytic cells:

1. **Neutrophils**, the most abundant type of white blood cell.
2. **Mononuclear phagocyte system**, including *monocytes* from the blood and *macrophages* (derived from monocytes) in the connective tissues.
3. **Organ-specific phagocytes**, found in the liver, spleen, lymph nodes, lungs, and brain. These include the *microglia* of the brain and *fixed phagocytes* (phagocytic cells that can't move around in the organ). Fixed phagocytes are found in the lining of blood sinusoids (a type of capillary) in the liver and spleen, which help to eliminate pathogens from the blood, and in the sinusoids of lymph nodes, which provide a similar function for the lymph.

For example, if there is a break in the epithelial barrier provided by the epidermis of the skin, bacteria can enter the connective tissue dermis and produce a **local inflammation**. There, the bacteria will be gobbled up by resident macrophages or neutrophils (fig. 11.1). If there are more bacteria than the local population of phagocytic cells can handle, chemical signals will be sent that attract other leukocytes to the infected area. Neutrophils and monocytes can squeeze themselves through the small channels between the adjacent endothelial cells that form the walls of capillaries and the smallest venules (chapter 10). This process, known as **extravasation**, adds new recruits to the battle (fig. 11.2).

The neutrophils arrive at the scene early, and eliminate bacteria through phagocytosis and by releasing various chemicals into the extracellular matrix. Some of these are *proteases* (protein-digesting enzymes) that liquify the surrounding tissue, creating *pus*. The neutrophils die in the battle and are phagocytosed by macrophages, which are increased in number by the arrival of new monocytes that become transformed into macrophages. **Mast cells** in the connective tissues contain **histamine** and are stimulated to release their histamine and other pro-inflammatory **cytokines** (autocrine regulatory molecules; chapter 3). Cytokines are released

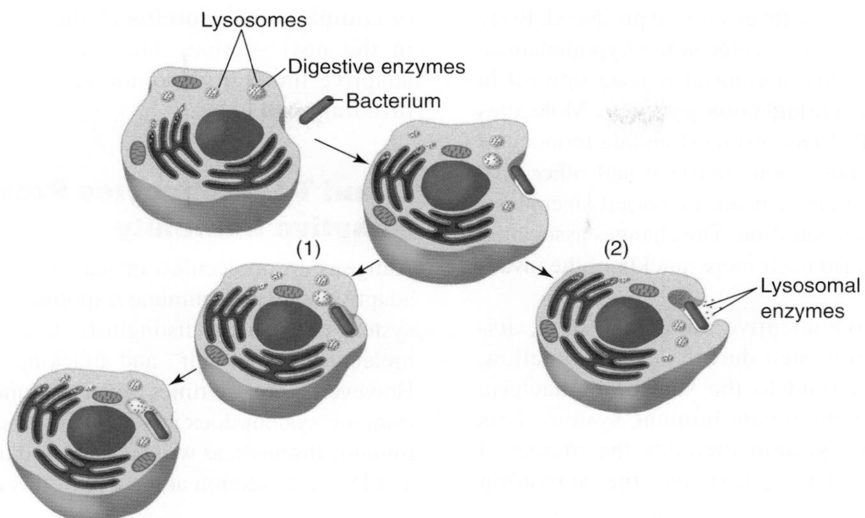

FIGURE 11.1 Phagocytosis by a neutrophil or macrophage. A phagocytic cell extends its pseudopods around the object to be engulfed (such as a bacterium). (Blue dots represent lysosomal enzymes.) (*1*) If the pseudopods fuse to form a complete food vacuole, lysosomal enzymes are restricted to the organelle formed by the lysosome and food vacuole. (*2*) If the lysosome fuses with the vacuole before fusion of the pseudopods is complete, lysosomal enzymes are released into the infected area of tissue.

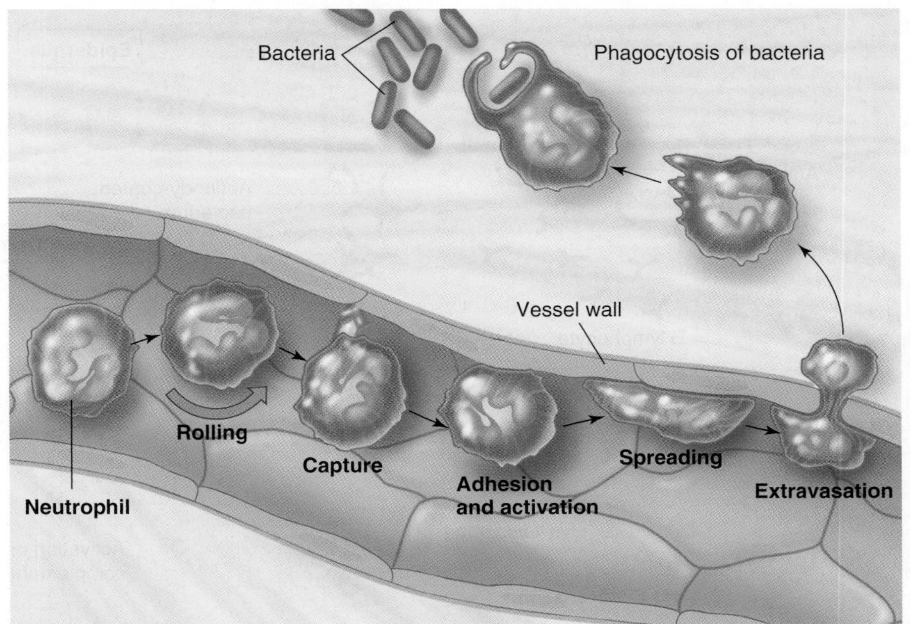

FIGURE 11.2 The migration of white blood cells from blood vessels into tissues. The figure depicts a neutrophil that goes through the stages of rolling, capture, adhesion and activation, and finally extravasation through the blood vessel wall.

by many cells of the immune system and serve to regulate different immune responses. Some of these, including histamine, promote the immune responses of an inflammation. In the lungs, histamine stimulates constriction of the bronchioles (small airways), contributing to asthma; however, histamine has the opposite effect on small blood vessels, producing vasodilation. This causes the inflamed area to become red, as blood flow increases (fig. 11.3). Histamine also increases capillary permeability to proteins, so that proteins leak into the interstitial fluid and cause a local edema. These effects produce the characteristic symptoms of a local inflammation: *redness* and *warmth* (due to vasodilation); *swelling* (edema); and *pain.*[1]

FYI [1]Celsus first described these symptoms around A.D. 40. It seems more poetic using his rhyming Latin terms: *rubor* (redness), *calor* (heat), *dolor* (pain), and *tumor* (swelling).

If the inflammation continues, a **fever** may be produced. Body temperature is controlled by a neural center in the hypothalamus, which acts like a thermostat. This thermostat is reset upward in response to a chemical called **endogenous pyrogen**. Molecules in bacteria (including one called *endotoxin*) stimulate monocytes and macrophages to release endogenous pyrogen and other pro-inflammatory cytokines, which cause fever, increased sleepiness, and a fall in the plasma iron concentration. The changes associated with a fever may help the body rid itself more quickly of the invading bacteria.

B lymphocytes (part of the adaptive immune system, discussed next) may also be stimulated during an inflammation. They secrete antibodies that bind to the surface of bacteria and target them for attack by the innate immune system. This attack by the innate immune system includes the increased activity of phagocytic cells (fig. 11.3) and the activation of **complement proteins** in the plasma, as will be described in the next section. Thus, in an inflammation, innate and adaptive immunity provide an integrated defense against the invading pathogens.

B and T Lymphocytes Provide Adaptive Immunity

Antigens are molecules, or parts of molecules, that stimulate the adaptive (specific) immune response by lymphocytes. The immune system can usually distinguish "*self*" from "*nonself*", tolerating molecules of the "self" and attacking foreign "nonself" antigens. However, there are times when tolerance mechanisms fail and the immune system does attack "self" antigens. This produces autoimmune diseases, as will be discussed in the "Physiology in Health and Disease" section at the end of this chapter.

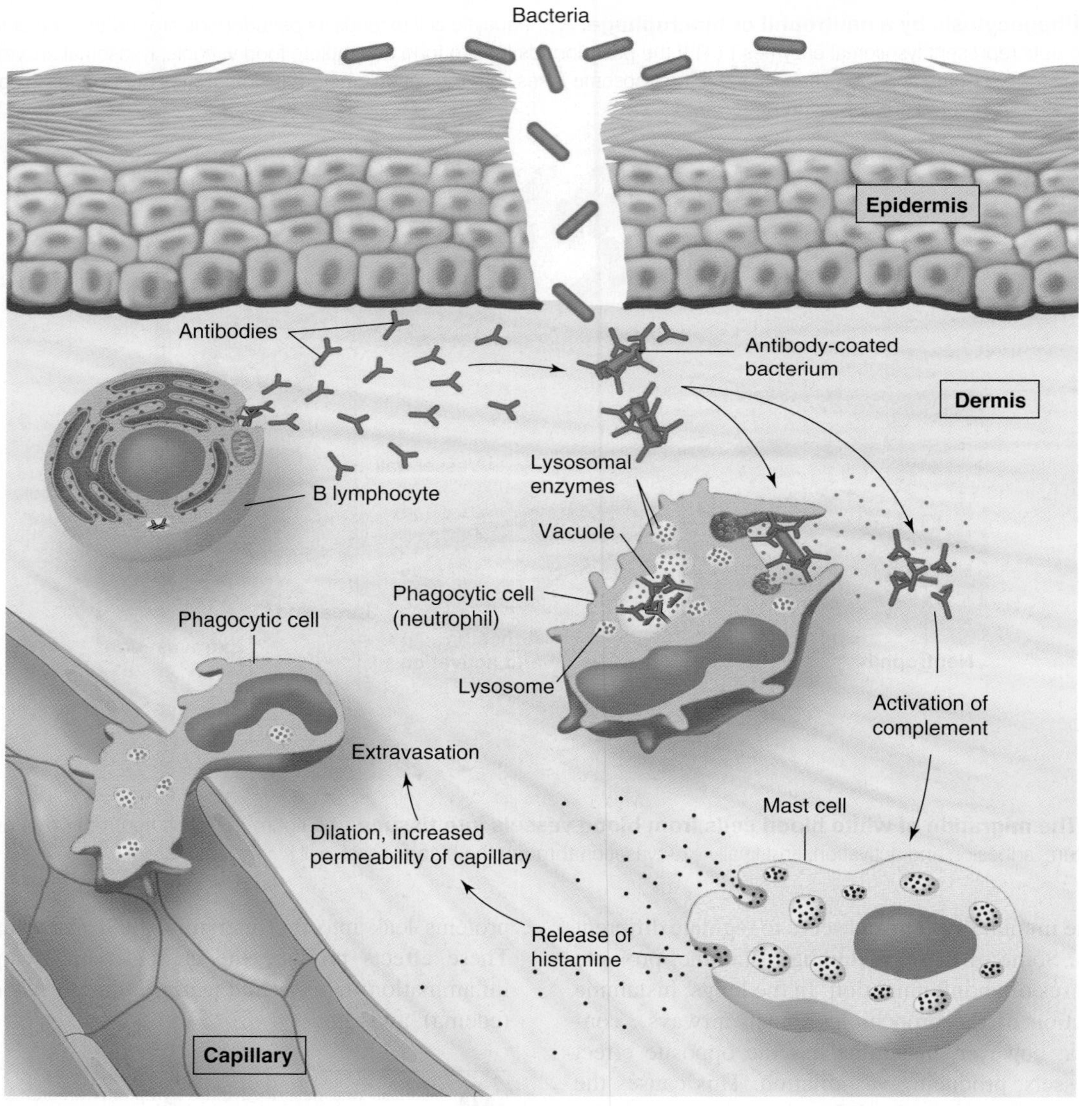

FIGURE 11.3 The events of a local inflammation. In this inflammatory reaction, antigens on the surface of bacteria are coated with antibodies and ingested by phagocytic cells. Symptoms of inflammation are produced by the release of lysosomal enzymes and by the secretion of histamine and other chemicals from tissue mast cells.

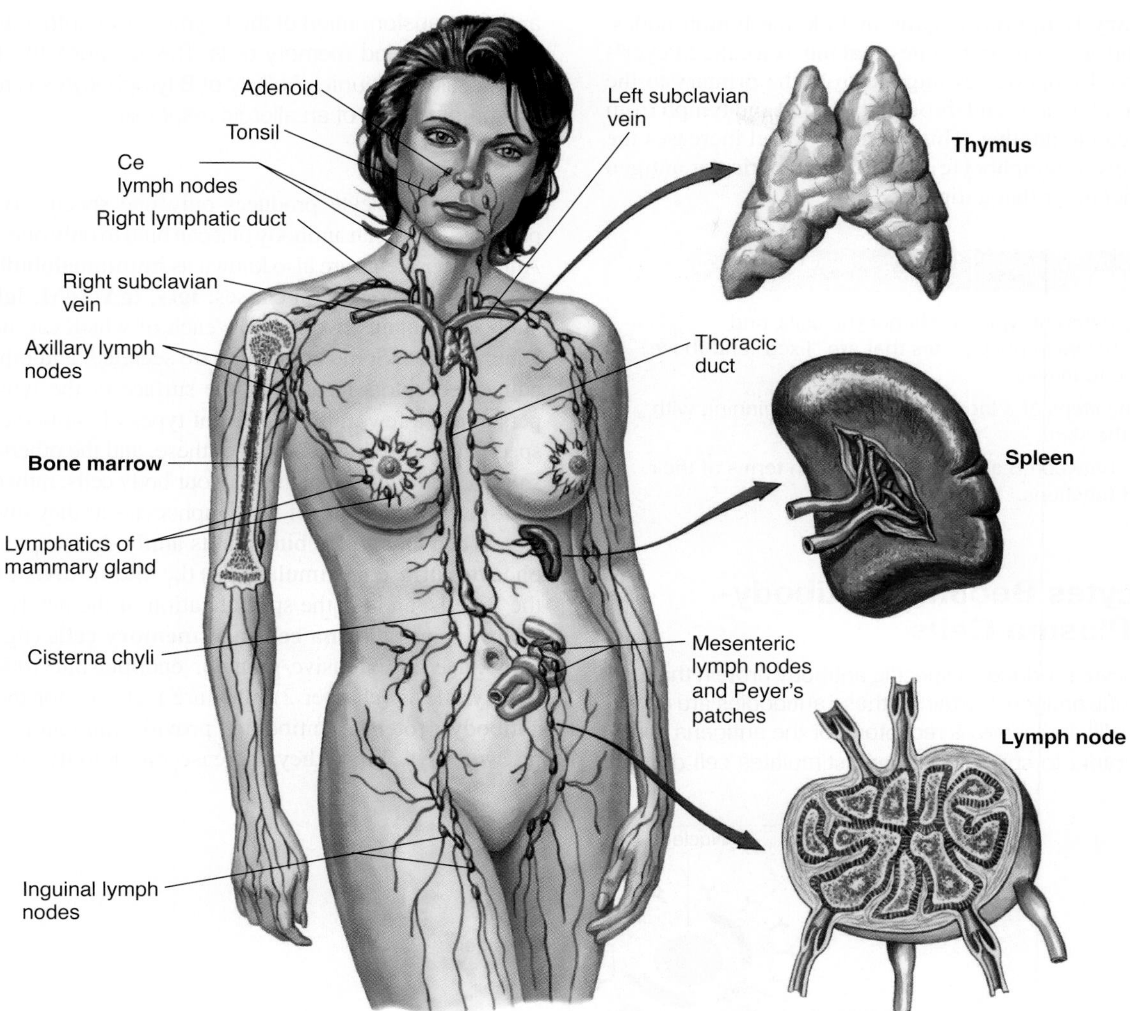

FIGURE 11.4 The location of lymph nodes along the lymphatic pathways. Lymph nodes are small bean-shaped bodies, enclosed within dense connective tissue capsules.

Lymphocytes, like the other formed elements of the blood, are produced in the myeloid tissue of the bone marrow (chapter 10). However, some of these lymphocytes can travel through the blood and seed the lymphoid tissue of the lymph nodes, thymus, and spleen (fig. 11.4), producing self-replacing lymphocyte colonies in these organs. The lymphocytes that seed the thymus become **T lymphocytes**, or **T cells**. The thymus, in turn, seeds other organs: about 65% to 85% of the lymphocytes in blood, and most of the lymphocytes in the *germinal centers* of the lymph nodes (where they "germinate" new lymphocytes by cell division), are T lymphocytes. Therefore, T cells are lymphocytes that either came from the thymus or had an ancestor that came from the thymus.

The lymphocytes that aren't T cells are **B lymphocytes**, or **B cells**. The "B" reminds us that these cells originated in the bone marrow. Because the B cells come from the bone marrow and the T cells come from the thymus, the bone marrow and thymus are known as the **primary lymphoid organs**. The B and T lymphocytes are indistinguishable under a microscope (without special techniques), but they function differently. The B lymphocytes produce antibodies, which are proteins that circulate in the blood and other body fluids. The B lymphocytes are thus said to provide **humoral immunity** ("humor" here is used in its ancient sense to mean a body fluid;[2] we still have the term *humidity* to indicate moisture). By contrast, T lymphocytes must come into actual physical contact with their victim cells in order to destroy them. The type of adaptive immunity provided by the T lymphocytes is thus known as **cell-mediated immunity**.

FYI [2]Hippocrates (460–370 B.C.) is generally credited with originating the concept that illness is caused by an imbalance in the four "humors," or body fluids, which also were believed to determine personality. These four humors were called *sanguine* (for blood, and indicating an amorous and courageous personality); *choleric* (for yellow bile, and indicating a quick, violent temper); *melancholic* (for black bile, and indicating someone who is pensive and depressive); and *phlegmatic* (for phlegm, and indicating a dull, unemotional personality). These terms are still used to describe personalities, and these concepts governed medical practice (leading to such techniques as bloodletting) for an amazingly long time. The four-humors theory of medicine ended only with the development of modern concepts of cellular pathology, pioneered by the German scientist Rudolf Virchow (1821–1902) and by the confirmation of the germ theory of disease and other discoveries by the great French scientist Louis Pasteur (1822–1895).

The **secondary lymphoid organs** include the lymph nodes, spleen, tonsils, and areas under the intestinal mucosa called Peyer's patches (fig. 11.4). Lymphocytes migrate from the primary to the secondary lymphoid organs, and through the blood and lymph from one lymphoid organ to another. This ceaseless travel increases the likelihood that a given lymphocyte specific for a particular antigen will be able to encounter that antigen.

CHECK POINT

1. Identify the different types of phagocytic cells, and distinguish between phagocytes that are "fixed" and those that are free to move.
2. Describe the steps of a local inflammation, beginning with a break in the skin.
3. Distinguish between B and T lymphocytes in terms of their origins and functions.

B Lymphocytes Become Antibody-Secreting Plasma Cells

Each B lymphocyte produces a specific antibody protein that can bind to a specific antigen. Certain of these antibodies are on the surface of the B cells and serve as receptors for the antigens. Binding of these receptors to specific antigens stimulates cell division and the transformation of the B lymphocyte into antibody-secreting plasma cells and memory cells. This provides the basis for active and passive immunity. Activity of B lymphocytes is also responsible for some aspects of an allergic response.

Each B lymphocyte produces only one specific type of antibody protein—that is, an antibody that can bind to only one specific antigen. Antibody proteins are also known as **immunoglobulins** (abbreviated *Ig*), and fall into several classes: **IgG**, **IgA**, **IgM**, **IgD**, and **IgE**. An antibody contains two regions, each of which can bind to one antigen molecule. Some antibodies are secreted, and others serve as the antigen receptors located on the surface of the lymphocytes. Each person has a few million different types of antibodies with different specificities. We inherit some of these, and the others we produce by *somatic mutations* (mutations in our body cells, rather than in the egg and sperm germ cells) of the lymphocytes as they divide.

When an antigen binds to its antibody receptor on the B lymphocyte surface, it stimulates (1) the mitotic division (chapter 2) of the B cells, and (2) the specialization of the newly formed B lymphocytes into **plasma cells** and **memory cells** (fig. 11.5). Plasma cells have an extensive granular endoplasmic reticulum for protein synthesis (chapter 2); they are factories for the production of antibody proteins. Antibodies provide immunological protection in two ways. First, they increase the activity of the phagocytic

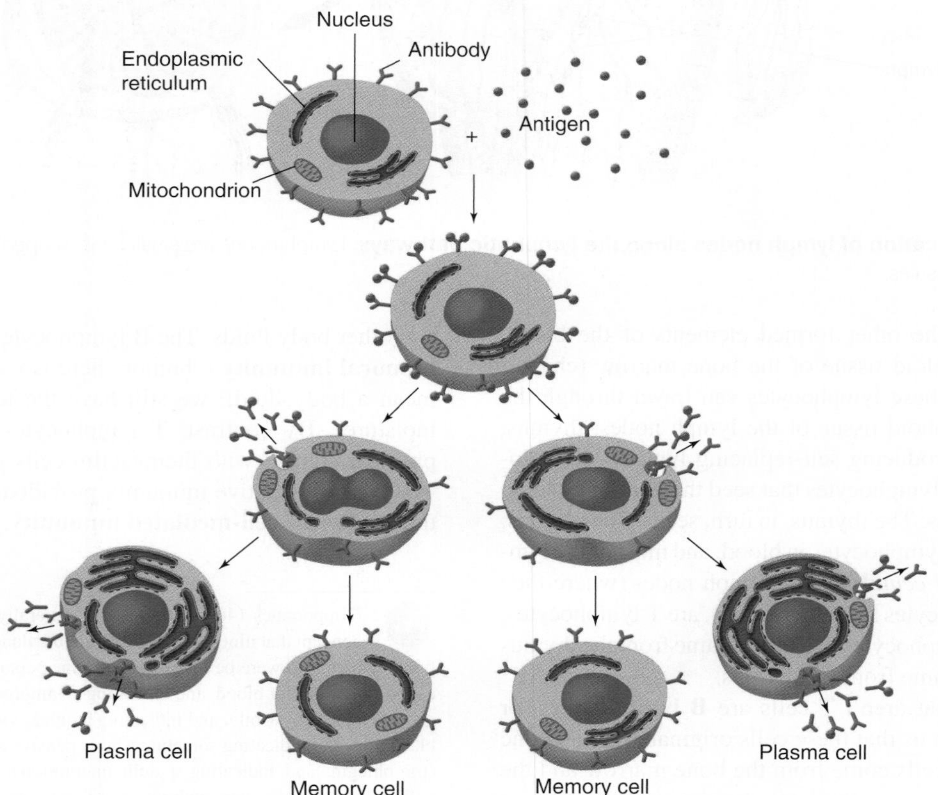

FIGURE 11.5 B lymphocytes are stimulated to become plasma cells and memory cells. B lymphocytes have antibodies on their surface that function as receptors for specific antigens. The interaction of antigens and antibodies on the surface stimulates cell division and the maturation of the B cells into memory cells and plasma cells. Plasma cells produce and secrete large amounts of the antibody.

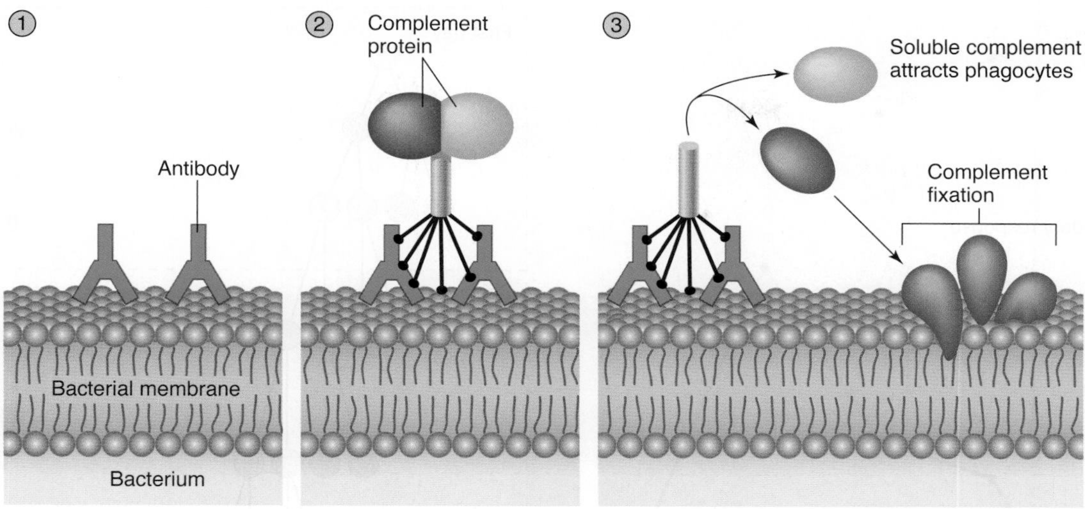

FIGURE 11.6 Complement fixation. The binding of antibodies to antigens on the surface of pathogens, such as bacteria, causes the activation of complement proteins in the plasma. Some of these activated complement proteins remain dissolved in the plasma and attract phagocytic cells. Others are inserted into the plasma membrane of the bacterial cell, in a process called complement fixation. This forms a membrane attack complex that helps destroy the bacterium.

cells: bacteria that are "buttered" with antibodies taste better to the phagocytic neutrophils and macrophages. Second, the binding of antibodies to antigens activates a group of plasma proteins known as the *complement* proteins.

A bacterial cell bears antigens on its surface that can become bound by antibodies. When this occurs, the binding of antigens to antibodies causes certain complement proteins in the plasma to become activated, which in turn causes other complement proteins to become inserted into the plasma membrane of the victim cell. This is called **complement fixation** (fig. 11.6). Some of the complement proteins form a **membrane attack complex**, which is a large pore that can kill the bacterial cell through the osmotic inflow of water. Notice that antibodies kill bacterial cells indirectly, either by promoting their phagocytosis by neutrophils and other phagocytic cells, or by activating complement proteins. You can think of antibodies as "kill signals" that flag antigen-bearing pathogenic cells for destruction by phagocytosis or complement fixation.

B Lymphocytes Provide Active and Passive Immunity

In the mid-eighteenth century, smallpox—a very *virulent* (disease-causing), often fatal disease—was widespread and greatly feared. About this time, some people in western Europe became aware that milkmaids who became ill with a related disease called cowpox were later immune to smallpox.[3] Today we know that the adaptive immune system that confers protection against cowpox cross-reacts with the antigens of smallpox to provide the immune protection.

FYI [3]An English physician named Edward Jenner (1749–1823) performed the first vaccinations based on this observation. In 1796 he inoculated an 8-year-old boy with a needle that he had previously rubbed in the pustules of a woman with cowpox, giving the boy this disease. After the boy recovered, Jenner next inoculated him with an otherwise deadly dose of smallpox, to which the boy (fortunately!) proved immune.

When a person is exposed to a particular pathogen for the first time, there is a latent period of 5 to 10 days before measurable amounts of specific antibodies appear in the blood. This sluggish **primary response** may not be enough to prevent a person from getting the disease. A subsequent exposure to that same antigen results in a **secondary response** (fig. 11.7), in which the antibody production is much more rapid, and maximum antibody concentrations in the blood are reached in less than 2 hours. This rapid rise in antibody production is usually able to prevent the disease. The induction of a secondary immune response thus provides what is known as **active immunity**. **Vaccinations** work by producing active immunity; a vaccination provides a first exposure to specific antigens, so that later exposure to the same or very similar antigens will evoke a secondary rather than a primary response. Of course, the antigens in the vaccination should have little or no virulence so that the person doesn't get the disease when the vaccination evokes the primary response.

The Clonal Selection Theory Accounts for Active Immunity

The immune system responds differently to the second exposure to an antigen than it did to the first, but how? It almost seems that the immune system learned from its previous exposure to a specific antigen, so it becomes better able to respond. This "learning" process responsible for the change from a primary to a secondary response (and thus for the effectiveness of vaccinations) is produced by a mechanism known as the **clonal selection theory**.

According to the clonal selection theory, each B lymphocyte is genetically able to produce a specific antibody against a specific antigen, even if it's never before encountered that antigen. This inherited ability to produce a particular antibody is reflected by antibodies on the lymphocyte surface that function as specific receptor proteins. When the lymphocyte is exposed to its specific antigen, the antigen-antibody reaction stimulates the lymphocyte to divide many times, until it produces a population of genetically identical cells—a *clone*.

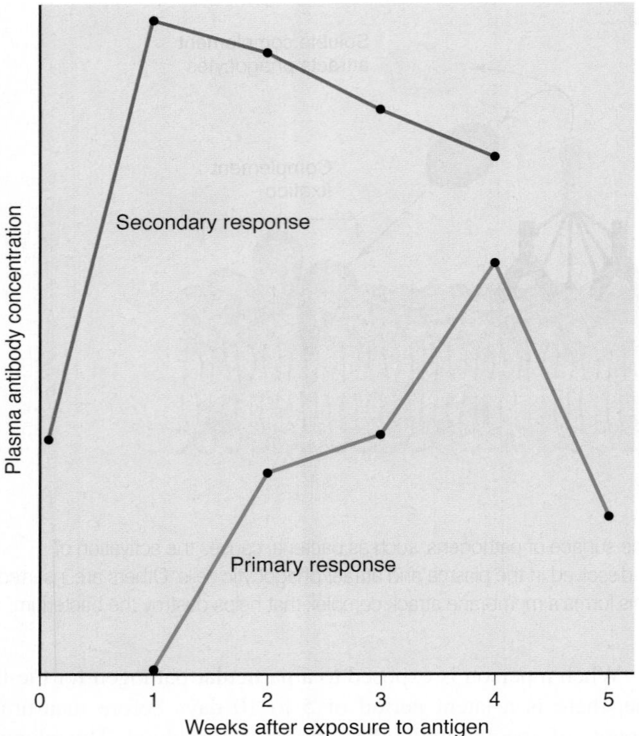

FIGURE 11.7 The primary and secondary immune responses. A comparison of antibody production in the primary response (upon first exposure to an antigen) to antibody production in the secondary response (upon subsequent exposure to the antigen). The more rapid production of antibodies in the secondary response is believed to be due to the development of lymphocyte clones produced during the primary response.

Some of these cells become plasma cells that secrete antibodies for the primary response; others are memory cells that can later become plasma cells and secrete antibodies for the secondary response if they are exposed to the same antigen (fig. 11.8).

Certain antigens stimulate T cells rather than B cells, and the T cells subsequently divide to produce a clone. Thus, T cells produce a secondary response when subsequently exposed to the same antigen, providing active immunity through the mechanisms described by the clonal selection theory. However, T cells don't secrete antibodies; their functions will be described in the next major section.

Clinical Investigation Clues

Remember that Helen got a vaccination for influenza (flu), and that last year's flu shot wouldn't protect her this year.

- What's in the flu vaccine to produce activity immunity, and how does the vaccine work to protect against the flu?
- What happens within Helen's body after she gets the vaccine that accounts for her immunity to that year's flu virus?
- Given that flu viruses differ in their antigens from year to year, why doesn't last year's vaccine work against this year's flu?

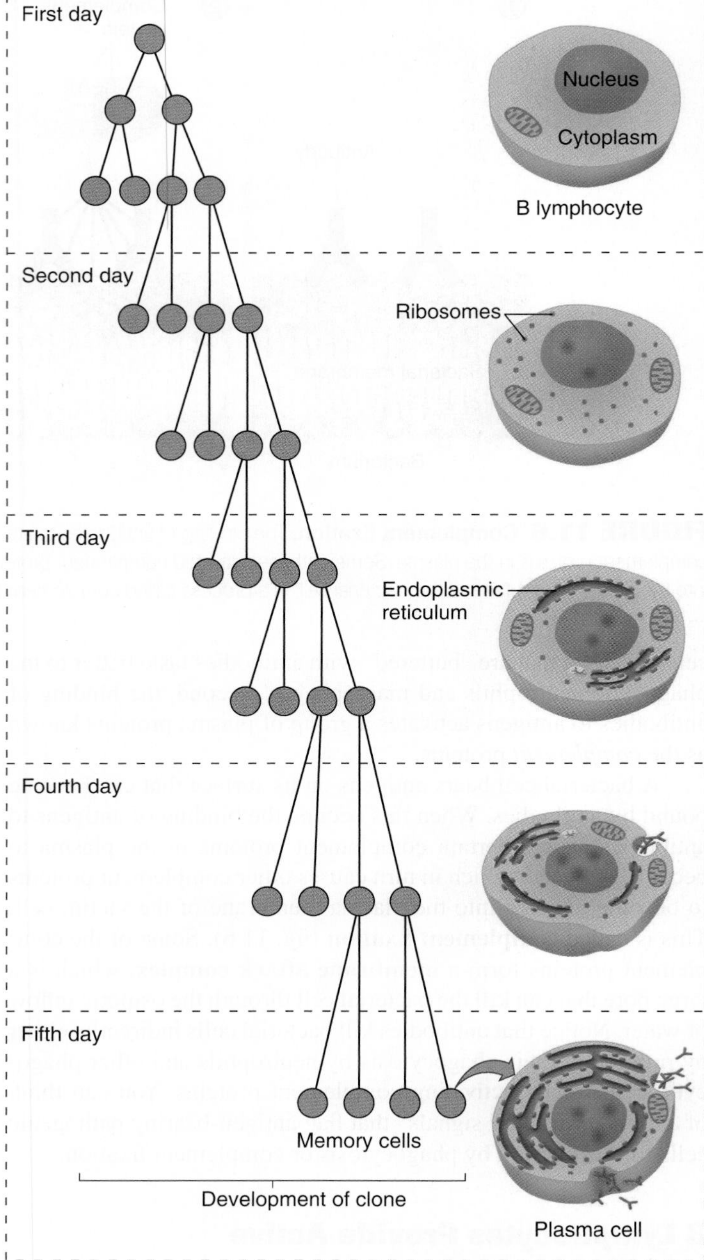

FIGURE 11.8 The clonal selection theory as applied to B lymphocytes. Most members of the B lymphocyte clone become memory cells, but some become antibody-secreting plasma cells.

Exogenous Antibodies Can Provide Passive Immunity

Passive immunity refers to the immune protection provided by exogenous antibodies—antibodies produced by another person or an animal. The person or animal providing the antibodies was exposed to the antigens, produced lymphocyte clones, and thus developed high blood concentrations of antibodies that provided them with active immunity. The antibodies in serum obtained from this blood can then be used for passive immunizations.

Passive immunizations are used to protect people who become exposed to extremely virulent infections or toxins, such as tetanus, hepatitis, rabies, and snake venom. In these cases, the affected person is injected with *antiserum* (serum containing antibodies), also called *antitoxin*, from an animal that had been previously exposed to the pathogen. Because the person never developed the lymphocyte clones and active immunity, another treatment with antiserum is needed if the person is exposed to the same antigens in the future.

A person's ability to develop active immunity—called **immunological competence**—doesn't begin to develop until about a month after birth. A fortunate result of this is that the fetus can't immunologically reject its mother. However, some antibodies from the mother can cross the placenta and enter the fetal blood, so that antibodies from the mother passively immunize the fetus. Because the fetus didn't develop the lymphocyte clones required to produce these antibodies, this passive immunization gradually disappears after birth. If the baby is breast-fed, it can receive supplementary passive immunization from the antibodies in its mother's milk, particularly in the *colostrum* (produced by the mammary glands for the first 2 to 3 days).

Clinical Investigation Clues

Remember that Helen required the injection of antitoxin to treat her snakebite, and that she would need another antitoxin treatment if she were bitten in the future.

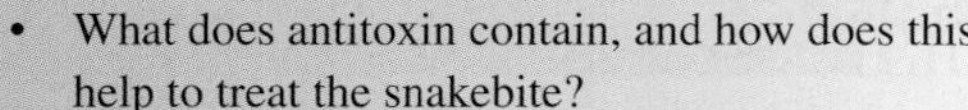

- What does antitoxin contain, and how does this help to treat the snakebite?

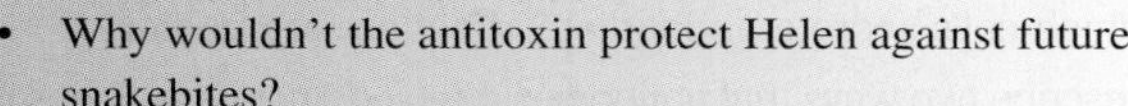

- Why wouldn't the antitoxin protect Helen against future snakebites?

B Lymphocytes Contribute to Allergy

The terms **allergy** and **hypersensitivity** are often used interchangeably, and refer to an abnormal immune response to particular antigens, which are known as *allergens* in these cases. There are two categories of allergy:

1. **Immediate hypersensitivity**, which is caused by an abnormal B cell response to an allergen. This produces symptoms within seconds or minutes.
2. **Delayed hypersensitivity**, which is caused by an abnormal T cell response (discussed in the next section). This produces symptoms within 24 to 72 hours after exposure to an allergen.

Immediate hypersensitivity can produce *allergic rhinitis* (chronic runny or stuffy nose), *allergic asthma*, *atopic dermatitis* (*urticaria*, or hives), and other conditions. These allergic responses differ from a normal immune response in the class of antibodies secreted by the B cells. In a normal secondary immune response, the B cells and plasma cells secrete antibodies of the immunoglobulin G (IgG) class. In an immediate hypersensitivity reaction, by contrast, the antibodies secreted are of the IgE class. The difference is that the IgG antibodies circulate in the blood and other body fluids, whereas antibodies of the IgE class attach to mast cells and basophils, which have membrane receptors for IgE. When these IgE antibodies, now attached to the mast cells and basophils, bind to their antigen (allergen), the mast cells and basophils are stimulated to release a number of chemicals that produce the allergic symptoms. These include *histamine* (fig. 11.9), *leukotrienes*, *prostaglandin D*, and other molecules.

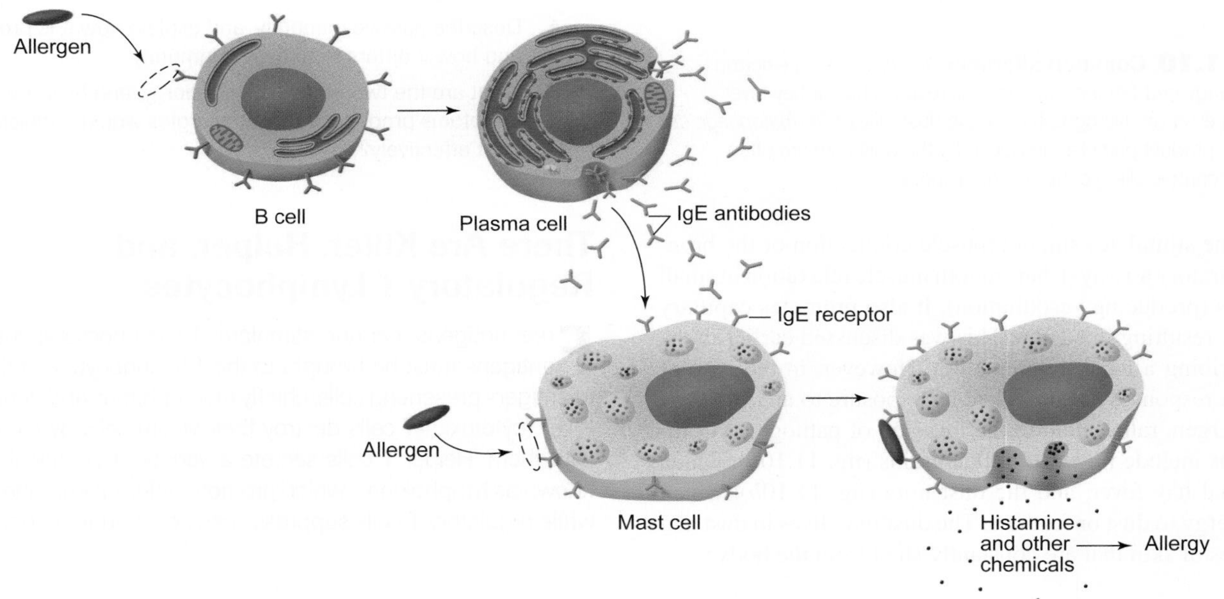

FIGURE 11.9 The mechanism of immediate hypersensitivity. Allergy (immediate hypersensitivity) is produced when antibodies of the IgE class attach to tissue mast cells. The combination of these antibodies with allergens (antigens that provoke an allergic reaction) causes the mast cell to secrete histamine and other chemicals that produce the symptoms of allergy.

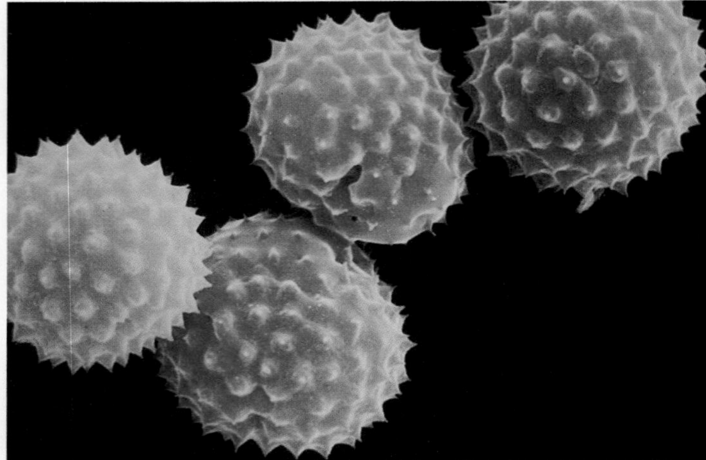

(a)

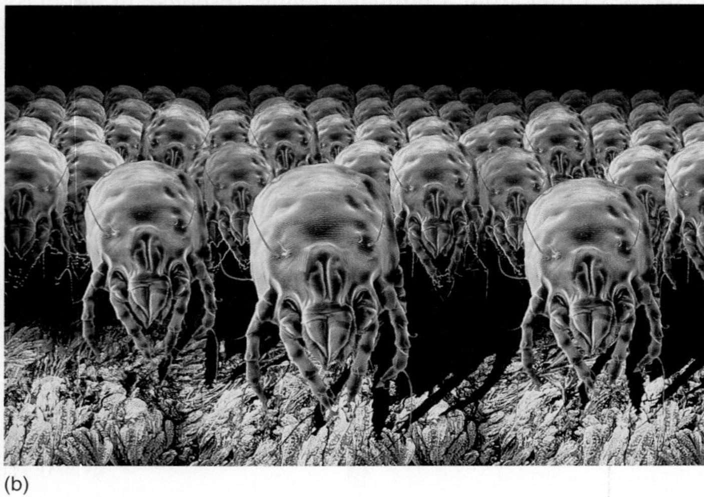

(b)

FIGURE 11.10 Common allergens. *(a)* A scanning electron micrograph of ragweed (*Ambrosia*), which is responsible for hay fever. *(b)* A scanning electron micrograph of house dust mites (*Dermatophagoides farinae*). Waste-product particles produced by the dust mite are often responsible for chronic allergic rhinitis and asthma.

Histamine stimulates smooth muscle contraction of the bronchioles (respiratory airways), but smooth muscle relaxation in small blood vessels (producing vasodilation); it also promotes capillary permeability, resulting in edema. This was discussed earlier in the section describing a local inflammation. However, in the case of allergy, these responses are provoked by exposure to an otherwise harmless allergen, rather than to the invasion of pathogens. Common allergens include ragweed pollen grains (fig. 11.10*a*), which cause seasonal hay fever, and the dust mite (fig. 11.10*b*), which causes an allergy to dust or feathers. The dust mite lives in dust and eats the scales of skin that are constantly shed from the body.[4]

FYI [4]Actually, most of the allergens are not in a dust mite's body but in its feces, which are tiny particles that can enter the nose much like pollen grains. Scientists estimate that there are more than 100,000 mite feces per gram of house dust! It's better not to think too much about this.

Delayed hypersensitivity includes **contact dermatitis**, provoked by poison oak, poison sumac, and poison ivy. Delayed hypersensitivity is caused by an abnormal T cell response, so antibodies aren't involved and antihistamines are ineffective. Corticosteroids (specifically, glucocorticoids such as hydrocortisone) are the only effective treatment, because they suppress the immune response (chapter 8). The skin tests for tuberculosis are also based on a delayed hypersensitivity reaction. If a person has been exposed to the tubercle bacillus and consequently has developed T cell clones, skin reactions appear within a few days after the tubercle antigens are rubbed into the skin with small needles (*tine* test) or are injected under the skin (*Mantoux* test).

CLINICAL INVESTIGATION CLUES

Remember that Helen sneezed, and the physician said that she probably had an allergy to pollen or dust. He prescribed an antihistamine for this allergy.

- What category of allergy does Helen's sneeze represent?
- Which cells are involved, and what do they do?
- What is the mechanism that produces the symptoms of this allergy?
- What aspects of an allergic reaction might antihistamines block?

CHECK POINT

1. Which cells secrete antibodies? What are antibodies?
2. Describe two ways that antibodies directed against antigens on the surface of bacteria help cause the bacteria's destruction.
3. Explain how the clonal selection theory helps explain active immunity.
4. Describe passive immunity, and explain how it is produced and how it differs from active immunity.
5. What are the two categories of allergy, and how are the symptoms produced? Which allergies would antihistamines treat effectively?

There Are Killer, Helper, and Regulatory T Lymphocytes

Free antigens cannot stimulate T lymphocytes; instead, the antigens must be brought to the T lymphocytes on the surface of antigen-presenting cells, chiefly macrophages and dendritic cells. Killer (cytotoxic) T cells destroy their victim cells by direct contact with them. Helper T cells secrete a variety of chemical cytokines, known as lymphokines, which promote different immune functions, while regulatory T cells suppress specific immune responses.

T lymphocytes (T cells), unlike the B lymphocytes, don't produce antibodies, and so they don't have antibody receptor proteins on their surfaces. The T cells do have receptor proteins for antigens, but these work differently than the receptors on B cells. The T cell

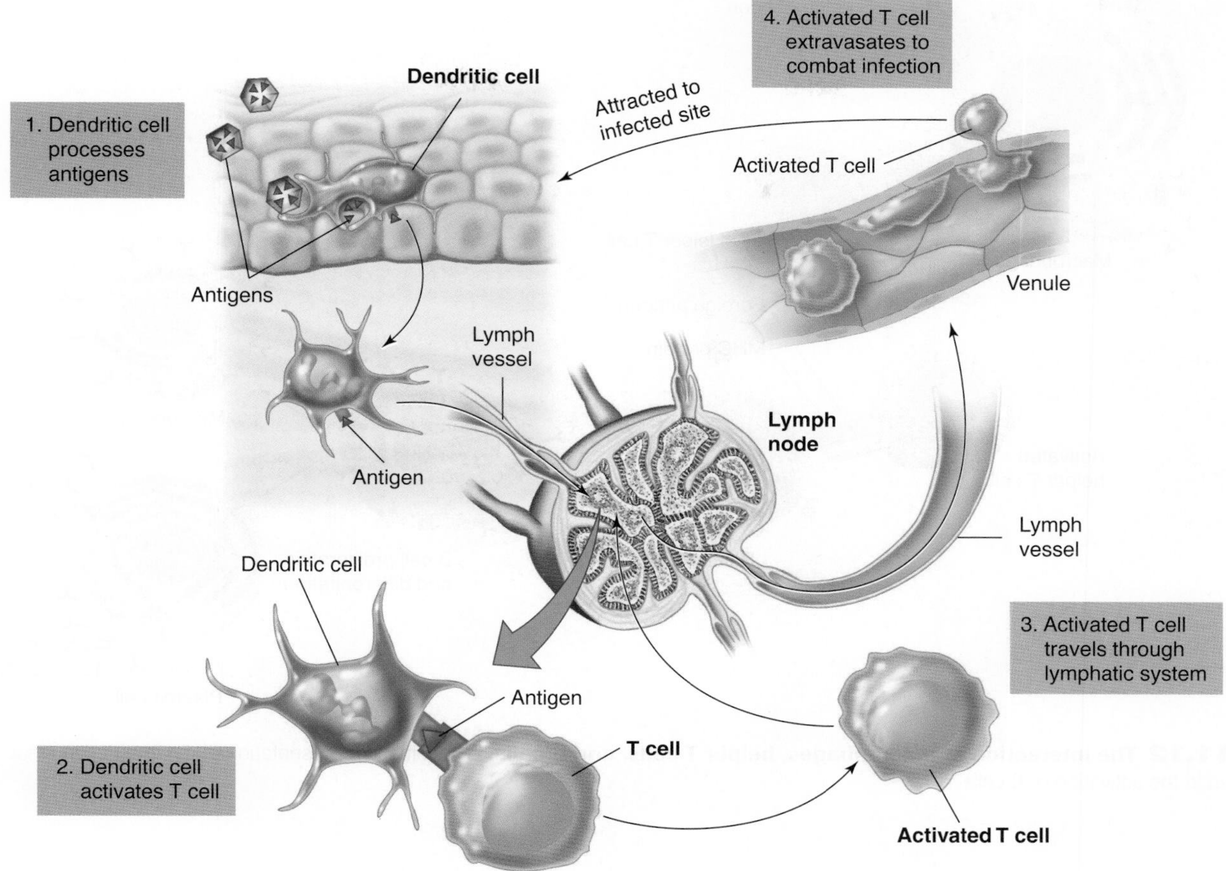

FIGURE 11.11 Migration of antigen-presenting dendritic cells to secondary lymphoid organs activates T cells. Once the T cells have been activated by antigens presented to them by the dendritic cells, the activated cells divide to produce a clone. Some of these cells then migrate from the lymphoid organ into the blood. Once in the blood, these activated T cells can home in on the site of the infection because of chemoattractant molecules produced during the inflammation.

receptors can't bind to free antigens. Instead, the antigens must be presented to the T cells by **antigen-presenting cells**, which are chiefly *macrophages* and **dendritic cells**. The antigen-presenting cells are especially concentrated at sites where pathogens might enter, such as the skin, intestinal mucosa, and lungs. For example, the epidermis contains a type of immature dendritic cells called **Langerhans cells**. A Langerhans cell engulfs antigens by phagocytosis and then moves them to its surface. In order to present these antigens to the correct T cell, the Langerhans cells must migrate through lymphatic vessels to the secondary lymphoid organs until it locates the T cell that is specific for that antigen (fig. 11.11).

In order for an antigen-presenting cell to stimulate a T cell, the antigen must be bound to a protein on the surface of the antigen-presenting cell. This protein is a member of a group of proteins called the **major histocompatability complex (MHC)**. Although this term relates to the requirement for the MHC proteins of donor and recipient tissues to match for an organ transplant, the function of MHC proteins within a person is to enable the T cell receptor to be activated by its specific antigen. For example, figure 11.12 shows a macrophage presenting an antigen, bound to an MHC protein, to a helper T cell (discussed shortly). The helper T cell is now activated, and figure 11.12 also shows that the activated helper T cell next interacts with a B cell, which also presents the antigen bound to an MHC protein. This interaction helps promote the immune function of that B cell. The general rule is that the antigen must be bound to MHC proteins on the surface of cells in order for any T cell to respond to the antigen.

Helper and Regulatory T Lymphocytes Modify Immune Responses

Helper T lymphocytes (helper T cells) aid the specific immune response of B lymphocytes as illustrated in figure 11.12. Helper T cells also stimulate the killer T lymphocytes through the secretion of chemical regulators known as **lymphokines**, which are the particular cytokine regulators secreted by lymphocytes (fig. 11.13).

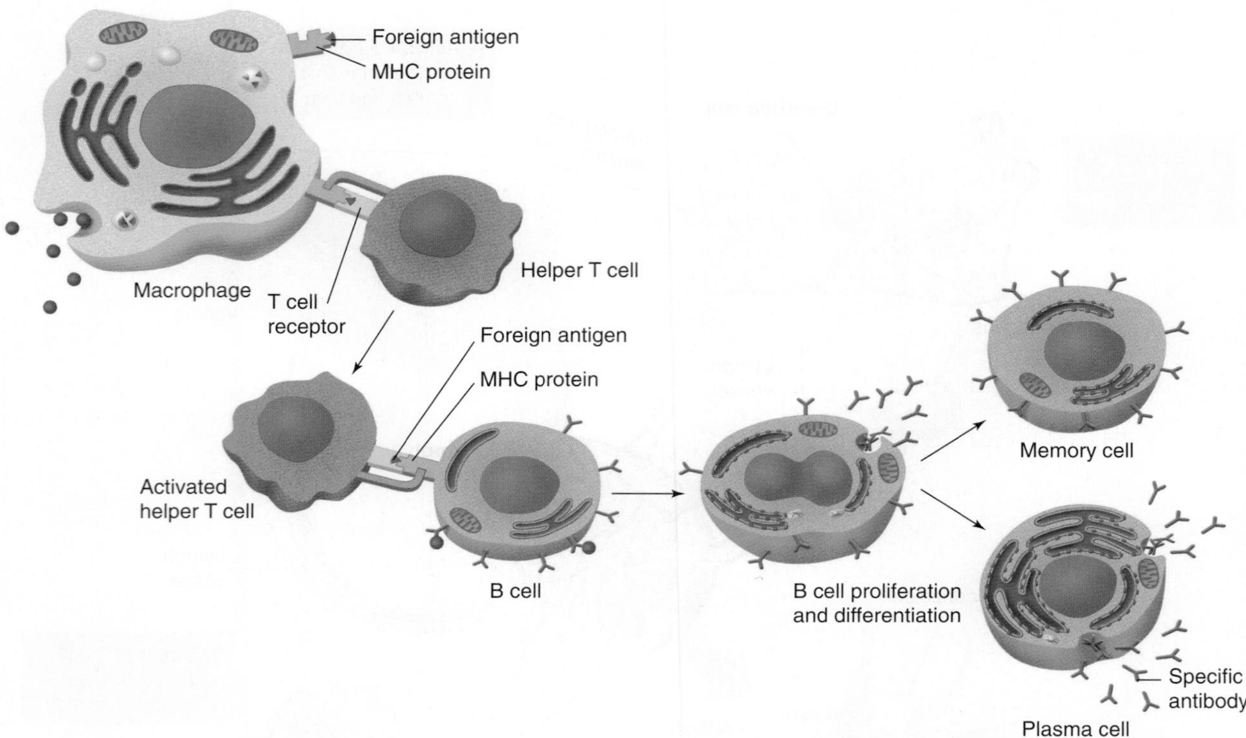

FIGURE 11.12 The interaction of macrophages, helper T cells, and B cells. A schematic representation of the interactions that can be involved in the activation of B cells.

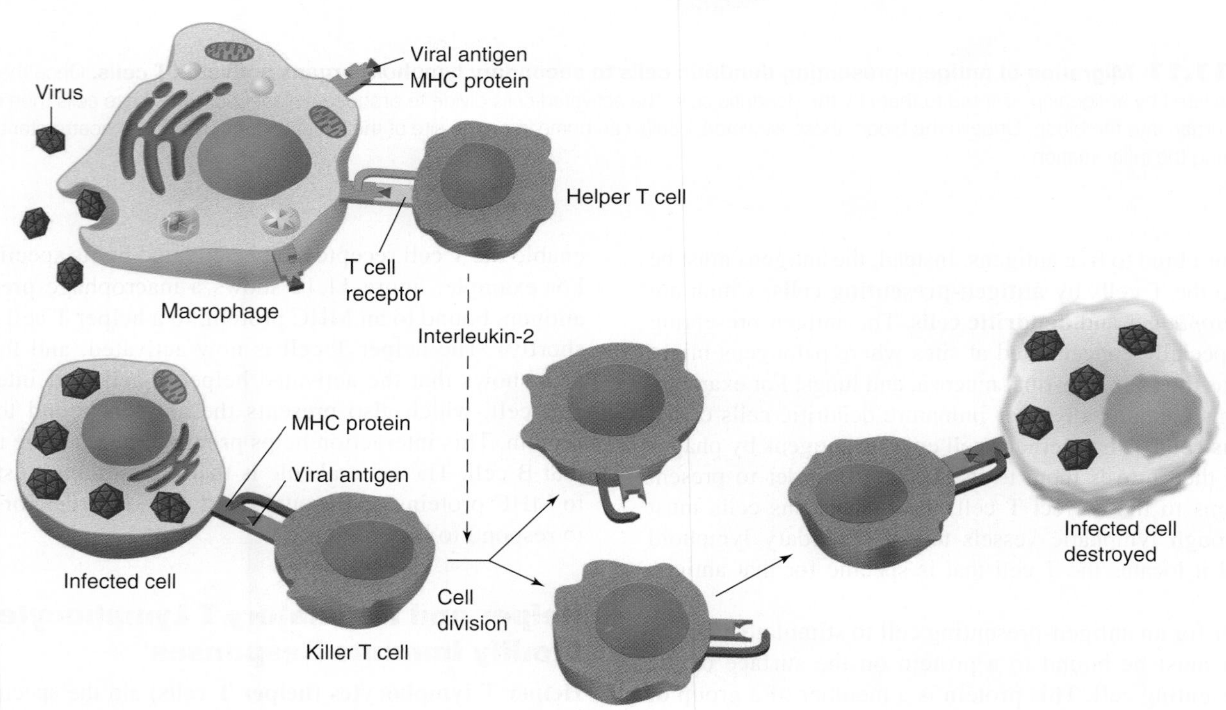

FIGURE 11.13 Helper T lymphocytes secrete lymphokines. Once the helper T lymphocyte in this figure has been activated by antigens presented to it by an antigen-presenting cell (here a macrophage), it secretes chemical regulators called lymphokines. The one shown here is interleukin-2, which stimulates the activation and cell division of killer T lymphocytes. These destroy their victim cell by direct contact (cell-mediated immunity).

A number of important lymphokines have been identified. Some have names, but others are called *interleukins*, followed by a number. For example:

1. *Interleukin-2* is released by helper T cells and is required for the activation of killer T cells, among other functions fig. 11.13). This is now used for some medical treatments.
2. *Interleukin-4* is secreted by helper T cells and is required for the cell division and clone formation of B cells.
3. *Granulocyte colony-stimulating factor* and *granulocyte-monocyte colony-stimulating factor* are lymphokines that promote the development of other leukocytes and are now available for medical treatments.

Clinical Applications

Acquired immune deficiency syndrome (AIDS) has killed approximately 25 million people worldwide. Today, more Americans have died of AIDS than were killed in World Wars I and II combined. AIDS is caused by the **human immunodeficiency virus (HIV)**, which specifically destroys the helper T lymphocytes. This results in decreased immunological function and greater susceptibility to infections, including *Pneumocystis carinii pneumonia.* Many people with AIDS also develop a previously rare form of cancer known as *Kaposi's sarcoma.* Drugs that act in a variety of ways to treat AIDS are currently available, and scientists are working diligently to discover new treatments, but there is currently no cure in sight.

Regulatory T lymphocytes (or **regulatory T cells**) were previously known as *suppressor T lymphocytes*. These provide a "brake" on the specific immune response (fig. 11.14) by inhibiting killer T lymphocytes and B lymphocytes. The mechanisms by which the regulatory T cells suppress specific immune responses are not completely understood, but by doing so they guard against diseases caused by the immune system, including allergy and autoimmune diseases (discussed in the "Physiology in Health and Disease" section). Thus, inadequate function of the regulatory T cells may partly explain why people get allergy and autoimmune diseases. On the other hand, the inappropriate functioning of regulatory T cells may also cause disease. This is suggested by observations that certain viral infections and cancers protect themselves from immune attack by recruiting regulatory T cells for their protection.

Killer T Lymphocytes Destroy by Cell-to-Cell Contact

Killer, or **cytotoxic**, **T lymphocytes (T cells)** identify antigens that are usually part of invading microorganisms, but sometimes can be part of the person's own body. For example, this occurs when a body cell has been infected by a virus and displays viral antigens on its surface (fig. 11.15), when a cell becomes transformed due to a malignancy (cancer), and when body cells display

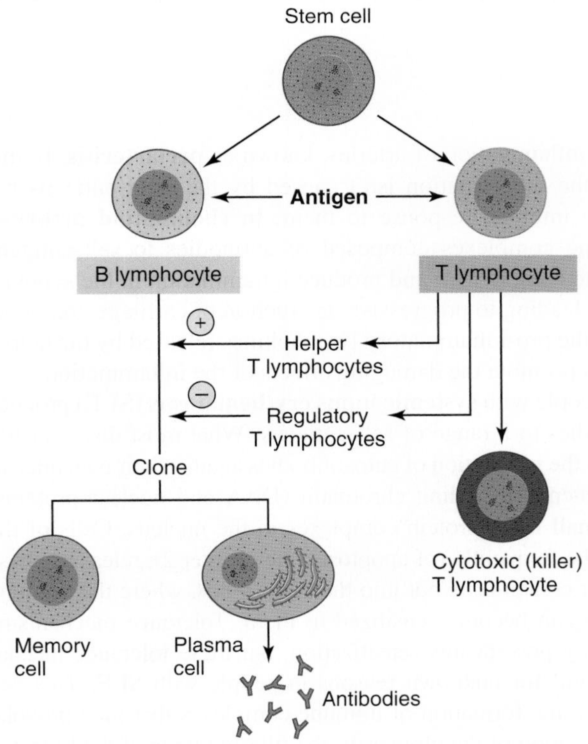

FIGURE 11.14 The effect of an antigen on B and T lymphocytes. A given antigen can stimulate the production of both B and T lymphocyte clones. The ability to produce B lymphocyte clones, however, is also influenced by the relative effects of helper and suppressor T lymphocytes.

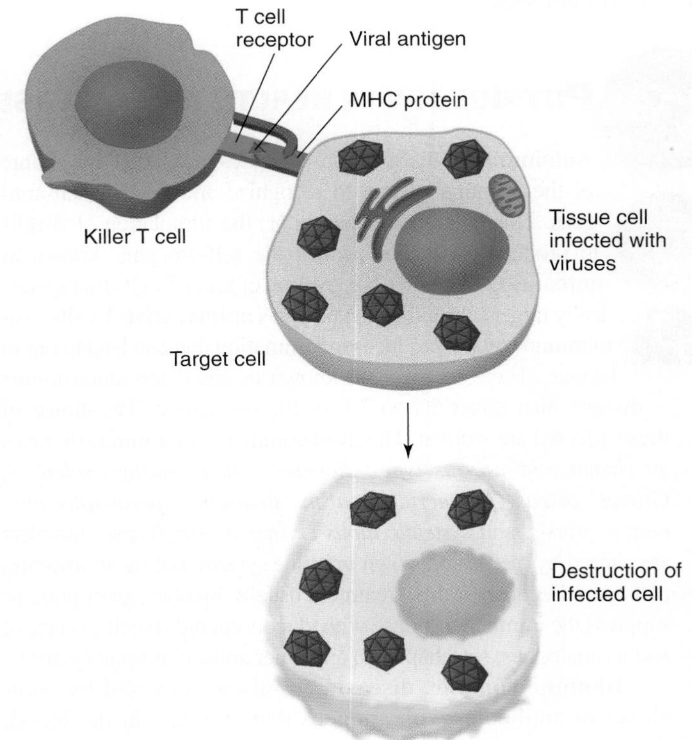

FIGURE 11.15 A killer T cell destroys an infected cell. In order for a killer T cell to destroy a cell infected with viruses, the T cell must interact with both the foreign antigen and the MHC protein on the surface of the infected cell.

normal "self" antigens that the immune system doesn't recognize and so doesn't tolerate. Killer T lympocytes defend against viral and fungal infections, and are also responsible for transplant rejection reactions and for the immunological surveillance against cancer. Although B cells fight most bacterial infections, some (such as the tubercle bacteria responsible for tuberculosis) are the targets of T cell attack.

B lymphocytes are like bombardiers, dropping bombs (antibodies) that can kill at a distance. The killer T lymphocytes are more like soldiers fighting hand-to-hand; they must be in actual physical contact with their victim cell. When this occurs, the killer T cells secrete molecules called *perforins* and enzymes called *granzymes*. The perforins enter the plasma membrane and form pores that kill the victim cell, similar to the way complement proteins form the membrane attack complex. The granzymes enter the victim cell and promote its destruction.

Natural killer (NK) cells, named because of their ability to kill tumor cells, are related to killer T lymphocytes but are considered part of the innate immune system. This is because, unlike the B and T lymphocytes that are part of the adaptive immune system, NK cells do not have receptors for specific antigens. Instead, the NK cells display an array of different receptors that allow them to attack malignant cells and to combat infections by different microorganisms. The NK cells destroy by cell-to-cell contact, and they secrete various cytokines that recruit the B and T lymphocytes. NK cells thus can provide a first line defense that is later backed up by the specific defenses provided by B and T lymphocytes.

CLINICAL APPLICATIONS

In response to *endotoxin*, a molecule released by bacteria, and cytokines including *interleukin-1* (released by helper T cells), macrophages are stimulated to produce *nitric oxide*. This molecule helps the macrophages destroy the bacteria, but nitric oxide released from macrophages also causes arterioles to dilate. If the amount of nitric oxide released is excessive, this can cause so much vasodilation (and lowered total peripheral resistance; see chapter 10) that blood pressure drops, producing the hypotension (low blood pressure) of **septic shock**. **Sepsis** is a leading cause of death in developed countries and kills more than 200,000 people yearly in the United States. Sepsis is caused by an infection that releases large amounts of endotoxin, which then stimulates macrophages to release pro-inflammatory cytokines and nitric oxide that promote a widespread inflammatory response and hypotension. This can result in tissue injury and organ failure.

CHECK POINT

1. What are the three major categories of T lymphocytes? How do the NK cells compare to the killer T lymphocytes?
2. Describe some of the ways that helper T lymphocytes enhance the response of the immune system to a pathogen.
3. Which types of diseases would be promoted by excessive function of regulatory T cells, and which types of diseases would be promoted by inadequate function of regulatory T cells?
4. Explain how killer T cells destroy their targeted victim cells.

PHYSIOLOGY IN HEALTH AND DISEASE

Autoimmune diseases are diseases produced by failure of the immune system to recognize and tolerate self-antigens. This failure can result in (1) the stimulation of B cells to produce antibodies against the self-antigens, known as **autoantibodies**; and (2) activation of killer T cells that specifically target self-antigens, known as **autoreactive T cells**. Autoimmune attack produces inflammation that can lead to organ damage. There are over 40 known or suspected autoimmune diseases that affect 5% to 7% of the population. Two-thirds of those affected are women. The most common autoimmune diseases are *rheumatoid arthritis*, *type 1 diabetes mellitus*, *multiple sclerosis*, *Graves' disease*, *glomerulonephritis*, *thyroiditis*, *pernicious anemia*, *psoriasis*, and *systemic lupus erythematosus*. These disorders are caused by the person's own immune system, not by an invading pathogen. Because of this, treatment usually involves attempting to suppress the immune response with glucocorticoids (such as cortisol and its analogues; see chapter 8) and other anti-inflammatory drugs.

Immune complex diseases are diseases caused by complexes of antibodies with antigens that circulate in the blood. When these enter organs, the antigen-antibody complexes can activate complement proteins and produce damage. Such immune complexes may result from the invasion of foreign antigens in bacteria, viruses, and parasites. For example, in *hepatitis B* the complexes of viral antigens and antibodies can cause inflammation of arteries, known as **periarteritis**. In this case, the inflammation isn't caused by the viral antigens but by the immune response to them. In **rheumatoid arthritis**, immune complexes composed of antibodies to self-antigens activate complement and produce inflammation in the synovial joints, leading to progressive destruction of cartilage and bone. Here, the pro-inflammatory lymphokines released by the helper T cells promote the damaging effects of the inflammation.

People with **systemic lupus erythematosus (SLE)** produce antibodies to a range of self-antigens. What most distinguishes SLE is the production of autoantibodies against their own nuclear constituents, including chromatin (DNA and nuclear proteins) and small RNA-protein complexes in the nucleus. Cells of the body die frequently (of apoptosis; see chapter 2), releasing these nuclear constituents out into the body fluids, where the immune system can become sensitized to them. Tolerance mechanisms normally prevent this sensitization, but these tolerance mechanisms fail for unknown reasons in people with SLE. This can result in the formation of immune complexes that may provoke inflammation of the glomeruli (the filtering units of the kidneys; see chapter 13), leading to **glomerulonephritis**. Multiple genes confer susceptibility to SLE, but interactions with the environment are also involved in initiating this disease. For example, exposure to sunlight, or various infections, may trigger SLE.

Physiology in Balance

Immune System

Working together with . . .

Respiratory System

Immune protection can go awry and produce asthma . . . p. 297

Circulatory System

Lymphatic vessels transport antigens and lymphocytes through the lymphoid organs . . . p. 240

Reproductive System

The placenta is an "immunologically privileged" site that prevents the maternal immune system from attacking fetal antigens . . . p. 373

Digestive System

Lymphatic nodules in the GI tract protect against pathogens that cross the mucosa . . . p. 348

Integumentary System

Langerhans cells in the skin present antigens to T lymphocytes . . . p. 157

Summary

The Immune System Defends Against Pathogens 280

- The innate immune system includes mechanisms that produce phagocytosis of invading pathogens (disease-causing agents) and that can result in fever.
- In a local inflammation, neutrophils arrive first, followed by monocytes that become transformed into macrophages.
- Mast cells release histamine, which causes constriction of bronchioles, dilation of blood vessels, and increased capillary permeability, resulting in edema.
- Fever can result from the release of endogenous pyrogen, a molecule produced by monocytes and macrophages.
- Cytokines are molecules released by cells of the immune system that regulate other cells of the immune system.
- B and T lymphocytes provide adaptive immunity, which is a defense against specific molecular targets, known as antigens.
- The bone marrow and thymus are the primary lymphoid organs, processing B and T cells, respectively; B lymphocytes provide humoral immunity by secreting antibodies, whereas T lymphocytes provide cell-mediated immunity.

B Lymphocytes Become Antibody-Secreting Plasma Cells 284

- Each B lymphocyte is genetically programmed to produce only one specific antibody, which it can secrete; the antibodies also serve as antigen receptor proteins on the surface of the B cells.
- Antibodies compose different classes of immunoglobulins: IgG, IgA, IgM, IgD, and IgE.
- When stimulated by its specific antigen, a B lymphocyte divides to produce memory cells and plasma cells, which are factories for the production and secretion of antibodies.
- Antibodies kill pathogenic cells bearing foreign antigens indirectly, by promoting phagocytosis and by activating complement; some complement proteins become fixed in the plasma membrane of the victim cell, forming a membrane attack complex that kills the cell.
- Upon first exposure to an antigen, a lymphocyte that is specific for it is stimulated to divide many times, producing a clone of genetically identical cells; antibodies are produced slowly in this primary response.
- Upon subsequent exposures to the same antigen, the lymphocyte clone can secrete a large amount of antibodies quickly; this is the secondary response, and is explained by the clonal selection theory.
- Vaccinations work by causing the development of lymphocyte clones and thus of active immunity.
- Injections of antiserum, or antitoxin, containing antibodies made by another animal, confer passive immunity.
- The immediate hypersensitivity type of allergic response occurs when B lymphocytes secrete antibodies of the IgE class; these attach to the surface of mast cells, promoting the release of histamine.
- Allergy that is delayed hypersensitivity is the result of an abnormal T cell response.

There Are Killer, Helper, and Regulatory T Lymphocytes 288

- T lymphocytes can't respond to free antigens; the antigens must be on the surface of antigen-presenting cells, chiefly macrophages and dendritic cells.
- In order for an antigen to be presented to the T cell receptor, the antigen must be bound to a member of the MHC (major histocompatability complex) group of proteins on the antigen-presenting cell.
- Helper T lymphocytes secrete a variety of lymphokines (the cytokines released by lymphocytes) that promote different aspects of the immune response.
- Regulatory T lymphocytes, previously called suppressor T lymphocytes, inhibit killer T and B cells, thereby keeping the immune system in check.
- Killer, or cytotoxic, T lymphocytes destroy cells invaded by viruses, cells that have become transformed in a cancer, and foreign transplanted tissue, as well as some bacteria.
- The killer T cells destroy their victim cells by direct cell-to-cell contact, where the killer T cell releases perforins and granzymes that kill the targeted cell.
- Natural killer (NK) cells are like killer T cells, but are part of the innate immune system.

Review Activities

Objective Questions: Test Your Knowledge

1. Which of the following is not a part of the innate immune system?
   a. Stomach acidity
   b. The function of neutrophils
   c. The function of lymphocytes
   d. A fever
2. Which of the following statements regarding antigens is true?
   a. They are always foreign molecules.
   b. They are produced by B lymphocytes.
   c. They are attacked by the innate immune system.
   d. They are targeted by the adaptive immune system.
3. Which of the following leukocytes can be transformed into macrophages during a local inflammation?
   a. Lymphocytes
   b. Monocytes
   c. Neutrophils
   d. Basophils
4. Extravasation refers to
   a. the squeezing of certain leukocytes through capillary channels to enter the interstitial fluid.
   b. the filtration of fluid out of capillary pores to result in a localized edema.
   c. the actions of histamine to promote vasodilation.
   d. the loss of blood in a hemorrhage.
5. Which of the following molecules can be classified as a cytokine?
   a. Endogenous pyrogen
   b. Endotoxin
   c. Acetylcholine
   d. Cortisol
6. Which of the following statements regarding humoral immunity is false?
   a. It is provided by B lymphocytes.
   b. It involves the secretion of antibodies.
   c. It's produced by cells processed in the bone marrow.
   d. It's provided by 65% to 85% of the lymphocytes in the blood.

7. Which of the following is not a secondary lymphoid organ?
   a. Lymph nodes
   b. Peyer's patches
   c. Thymus
   d. Tonsils
8. Which of the following statements about antibodies is false?
   a. Antibodies kill bacterial cells directly.
   b. Antibodies coat bacterial cells, increasing their phagocytosis.
   c. Antibodies activate complement proteins, which form a membrane attack complex.
   d. Antibodies are grouped into different classes of immunoglobulins.
9. Which of the following statements about a secondary immune response is false?
   a. It follows a primary immune response.
   b. It provides passive immunity.
   c. It results from the formation of lymphocyte clones.
   d. It causes more antibodies to be produced more quickly when exposed to a specific pathogen.
10. Antiserum, or antitoxin,
    a. is used for vaccinations.
    b. contains virulent pathogens.
    c. contains antibodies.
    d. produces lymphocyte clones.
11. According to the clonal selection theory,
    a. each lymphocyte is able to respond to only one particular antigen.
    b. each B lymphocyte is genetically able to produce only one specific antibody.
    c. exposure of a lymphocyte to its specific antigen causes it to proliferate.
    d. All of these statements are true.
12. When does immunological competence begin to develop?
    a. Before birth
    b. A month after birth
    c. A year after birth
    d. At puberty
13. The type of reaction that can be produced by poison ivy or poison oak is a(n)
    a. immediate hypersensitivity reaction caused by B cells.
    b. immediate hypersensitivity reaction caused by T cells.
    c. delayed hypersensitivity reaction caused by B cells.
    d. delayed hypersensitivity reaction caused by T cells.
14. In allergic rhinitis, the symptoms are caused by antibodies of the
    a. IgE class.
    b. IgG class.
    c. IgM class
    d. IgD class.
15. Which of the following statements about T cell stimulation is false?
    a. The antigen must be bound to an MHC protein.
    b. The antigen must be free in plasma or body fluids.
    c. The antigen must be presented on the surface of antigen-presenting cells.
    d. The antigen must be specific for the specific T cell receptor protein.
16. Which of the following statements regarding interleukin-2 is false?
    a. It's a cytokine.
    b. It's a lymphokine.
    c. It's secreted by B lymphocytes.
    d. It's needed for the activation of killer T lymphocytes.
17. Which of the following types of disease could be promoted by the inadequate action of regulatory T lymphocytes?
    a. Diseases caused by viruses
    b. Diseases caused by fungi
    c. Diseases caused by bacteria
    d. Diseases caused by the immune system (autoimmune diseases)
18. Perforins and granzymes are released by
    a. B lymphocytes.
    b. killer T lymphocytes.
    c. helper T lymphocytes.
    d. regulatory T lymphocytes.

Essay Questions 1: *Test Your Understanding*

1. Distinguish between the innate and adaptive immune systems.
2. List the types of phagocytic cells, and give examples of each type.
3. Step by step, describe a local inflammation, beginning when a break in the epidermis allows the entry of bacteria into the dermis.
4. What are cytokines, and what is their function in immunity?
5. Explain how a fever is produced.
6. Explain the meaning of the phrase "self vs. non-self" in immune function.
7. Distinguish between B and T lymphocytes in terms of their origins and general functions.
8. Identify the primary and secondary lymphoid organs.
9. Describe the relationship between B cells, plasma cells, and memory cells.
10. Explain two ways that antibodies indirectly promote the destruction of invading bacteria.
11. Distinguish between a primary and a secondary immune response.
12. Explain the clonal selection theory with regard to B lymphocytes, and explain how it accounts for active immunity.
13. Describe how passive immunizations are provided by a mother for her fetus and baby, and by the injection of an antitoxin.
14. Distinguish between immediate and delayed hypersensitivity, and explain how an immediate hypersensitivity reaction is produced in response to seasonal pollen.
15. Describe the requirements of the T cell receptors for the activation of T cells by specific antigens. How does this differ from the requirements for B cell activation?
16. Identify antigen-presenting cells, and describe how they help T lymphocytes become activated by antigens.
17. Identify two lymphokines, and explain how helper T lymphocytes aid the adaptive immune response by B cells and killer T cells.
18. What are MHC proteins, and what role do they play in the adaptive immune response?
19. Describe the role of regulatory T lymphocytes, and explain how diseases can be produced by either the inadequate function or the inappropriate activity of these cells.
20. Identify the role of killer T lymphocytes in the adaptive immune system, and explain how they kill their victim cells.

Essay Questions 2: *Test Your Analytical Ability*

1. Bacteria contain molecules that can stimulate the cells of the innate immune system. How does such stimulation differ from the way that the specific immune system is activated? Explain.
2. How does the adaptive immune system participate in a local inflammation?
3. Is a fever beneficial or not? Explain your answer.
4. Immunological competence, and the ability to distinguish "self" from "non-self," isn't developed in a fetus. Propose reasons why this would be beneficial. What compensations for this lack are available to the fetus?
5. Cytokines regulate specific cells of the immune system. What would you look for in a cell to determine if it was regulated by a particular cytokine? Explain.
6. What would happen if lymphocytes couldn't distinguish "self" from "non-self"? Suppose a T cell developed the ability to attack self-antigens in a particular organ. Propose two ways this attack could be prevented.
7. Mixing bacteria with purified antibodies against the bacterial antigens doesn't cause the bacteria's destruction. However, if plasma is also added, the antibodies cause the bacteria to die. Explain the mechanisms responsible for this observation.
8. Why do we give antitoxin after being bitten by a rabid dog or a poisonous snake, instead of generally vaccinating healthy people against future encounters with rabies or snakebite?
9. Do antihistamines cure allergy? Explain, in terms of the mechanisms of allergy and the actions of histamine.
10. Would antihistamines work in treating contact dermatitis? Explain. Would corticosteroids help to treat immediate as well as delayed hypersensitivity? Explain.
11. Cells that undergo malignant transformation in a cancer are supposed to be subjected to attack by killer T cells. How could the killer T cell recognize the transformed cells? What might prevent the killer T cell from attacking the transformed cells?
12. Interleukin-2 has been used in treating certain cancers. How might it help?
13. A suppression of helper T lymphocytes by HIV in AIDS is responsible for the appearance of "opportunistic infections." What do you suppose this term means, and how would suppression of helper T cells promote opportunistic infections?
14. NK cells are like killer T lymphocytes, but they are part of the innate immune system. Given this, propose how the NK cells could function as a first line of defense against cancer.

Web Activities

ARIS™ **For additional study tools, go to: www.aris.mhhe.com. Click on your course topic and then this textbook's author/title. Register once for a semester's worth of interactive activities and practice quizzing to help boost your grade!**

Respiratory System

CHAPTER 12

Chapter Outline

HOMEOSTASIS

Proper functioning of the respiratory system is a central requirement for life, and thus for all aspects of homeostasis. Breathing oxygenates the blood, and the hemoglobin in the red blood cells transports the oxygen to all body cells to maintain aerobic cell respiration. Without oxygen, the brain and heart can't long survive. Less obvious but almost as important, the process of breathing maintains homeostasis of the plasma carbon dioxide concentration. By maintaining homeostasis of the plasma carbon dioxide concentration, breathing also helps maintain homeostasis of the blood pH, because carbon dioxide forms carbonic acid in the blood. Indeed, breathing is adjusted by a negative feedback loop to maintain constancy of the plasma carbon dioxide concentration and blood pH, whereas the oxygen content of the blood has a lesser effect on breathing under usual conditions.

CLINICAL INVESTIGATION

But if homeostasis is not maintained . . .

William, a smoker who drove a delivery van in a large city, was in an accident in which a puncture wound to his chest caused collapse of a lung. This was treated during his hospital stay, where they performed various pulmonary function tests and blood gas measurements. The physician strongly urged William to stop smoking. "You have mild carbon monoxide poisoning," she said, "and although your vital capacity is currently in the normal range, you need to be careful. You don't want to get emphysema." She continued: "Other tests suggest you may have an obstructive disorder. Have you ever had asthma?" William said that he did, but that it was mild and he hadn't been taking anything for it. The physician gave him an inhaled drug that promoted dilation of the bronchioles (airways) whenever he had an asthma attack.

What caused the collapse of William's lung? How did he get mild carbon monoxide poisoning, and how was that detected? What is a vital capacity test, and what is its significance? What is emphysema, and why did the physician strongly advise William to stop smoking? What other pulmonary function test may have been performed that detected an obstructive disorder? What is asthma, and how does the prescribed medication treat this condition?

Breathing Replenishes Air for Gas Exchange in the Alveoli

Contraction and relaxation of muscles associated with the thoracic cavity change the volume of the lungs during breathing, allowing fresh air to be replenished in the alveoli, where gas exchange with the blood occurs. Air moves from higher to lower pressure, and so inspiration and expiration are influenced by changes in the pressure within the lungs, or intrapulmonary pressure. The ability of the lungs to inflate is also affected by their physical properties, including their elasticity and the surface tension within the alveoli. Pulmonary function tests assess ventilation (breathing) to allow the diagnosis of pulmonary disorders.

At first glance, you might expect a study of the physiology of the respiratory system to be short and relatively simple; we breathe in, we breathe out: how much more involved could it be? As you study this chapter, you'll see that there's quite a bit more to it. The lungs are amazing organs; they must be extremely thin-walled and highly vascular for gas exchange, and yet must be able to inflate and deflate under pressure without tearing. The oxygenation of the blood depends not only on pulmonary function, but also on the oxygen delivery system of the blood—the red blood cells and their content of hemoglobin. And so, oxygen transport by the blood is part of the story of the respiratory system. The other side of the gas-exchange coin is the transport of carbon dioxide by the blood to the lungs for elimination. And carbon dioxide transport is central to another very important subject in physiology and medicine—the homeostasis of the blood pH (also regulated by the kidneys, discussed in chapter 13). The study of respiratory physiology thus lays an important foundation for your knowledge pyramid in the physiology of several body systems.

Ventilation (breathing) brings air into and out of the lungs, so that **gas exchange** with the blood can occur. Gas exchange in the lungs refers to the net diffusion of oxygen from air to blood, and the net diffusion of carbon dioxide from blood to air. Because of this gas exchange, the blood in the pulmonary veins leaving the lungs has higher oxygen and lower carbon dioxide concentrations than the blood going to the lungs in the pulmonary arteries (chapter 10). In order for gas exchange to occur, fresh air must be channeled through the **conducting zone** of the respiratory system to the **respiratory zone**, which is the region of the lungs where gas exchange occurs.

The conducting zone consists of the **trachea** (windpipe), which branches to form ever-smaller **bronchi**, leading to numerous and very narrow **bronchioles** (fig. 12.1). The air passageways thus form a hollow tree, analogous to the arterial tree discussed in chapter 10. The conducting system ends at the *terminal bronchioles*, before channeling air into the respiratory zone. The respiratory zone consists of *respiratory bronchioles*, containing individual **alveoli** (singular, *alveolus*), and **alveolar sacs**, which are clusters of interconnected alveoli. Gas exchange occurs across the walls of the alveoli.

The numbers in parentheses in figure 12.1 indicate the extensive branchings of the bronchi and bronchioles. The trachea and the bronchi have supporting rings of cartilage, but the smaller bronchioles have walls of only smooth muscle. The bronchioles can thus constrict like arterioles, increasing the resistance to airflow analogous to the resistance function of arterioles to blood flow (chapter 10). This is an important process in the development of asthma (discussed later).

There are an estimated 300 million alveoli, providing an enormous surface area for the diffusion of gas molecules across the alveolar wall.[1] Considering this great surface area, and the extreme thinness of the alveolar wall (it's one cell thick), oxygen and carbon dioxide molecules can diffuse quickly across it. Only two thin cells separate the air in an alveolus from the blood in the surrounding capillaries: the squamous epithelial cell of the alveolar wall and the squamous capillary endothelial cell. This produces an average distance between air and blood of only about 2 μm. The blood capillaries completely surround the alveoli (fig. 12.2), so that gas exchange between the air and blood is normally very efficient.

Ventilation Results from Changes in Thoracic Volume

The *diaphragm* is a dome-shaped sheet of striated muscle that separates the body cavity into a lower abdominopelvic cavity and an upper **thoracic cavity** (or **thorax**). The central part of the thoracic cavity contains the heart, large blood vessels, trachea, esophagus, and thymus; the rest of the thoracic cavity contains the right and left lungs. Two wet epithelial membranes surround the structures in the central region of the thorax, called the *mediastinum*. The superficial membrane is the **parietal pleura**, which also lines the inside of the thoracic wall. The deeper

FYI [1]Although each alveolus is very tiny, the 300 million together provide a surface area of 60 to 80 square meters, or about 760 square feet. This is equivalent to the area of a courtyard 100 feet long by 76 feet wide! If alveoli were to coalesce, forming a smaller number of larger alveoli, the surface area for gas exchange would be reduced. Unfortunately, this does occur when a person has emphysema.

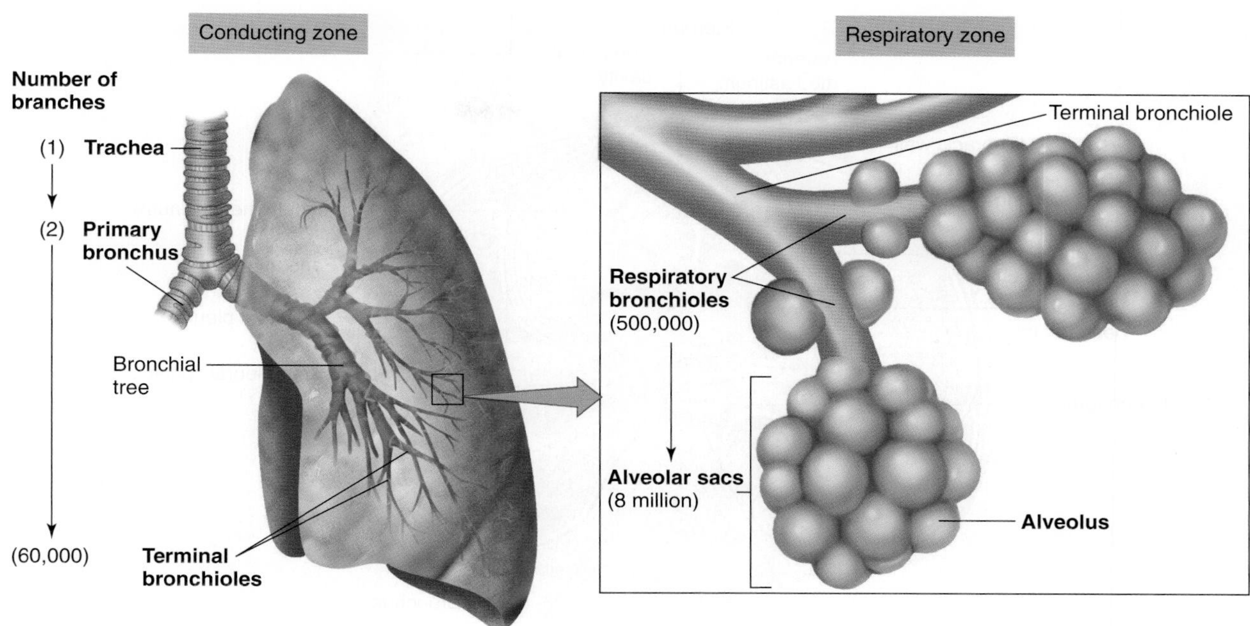

FIGURE 12.1 The conducting and respiratory zones of the respiratory system. The conducting zone consists of airways that conduct the air to the respiratory zone, which is the region where gas exchange occurs. The numbers of each member of the airways and the total number of alveolar sacs are shown in parentheses.

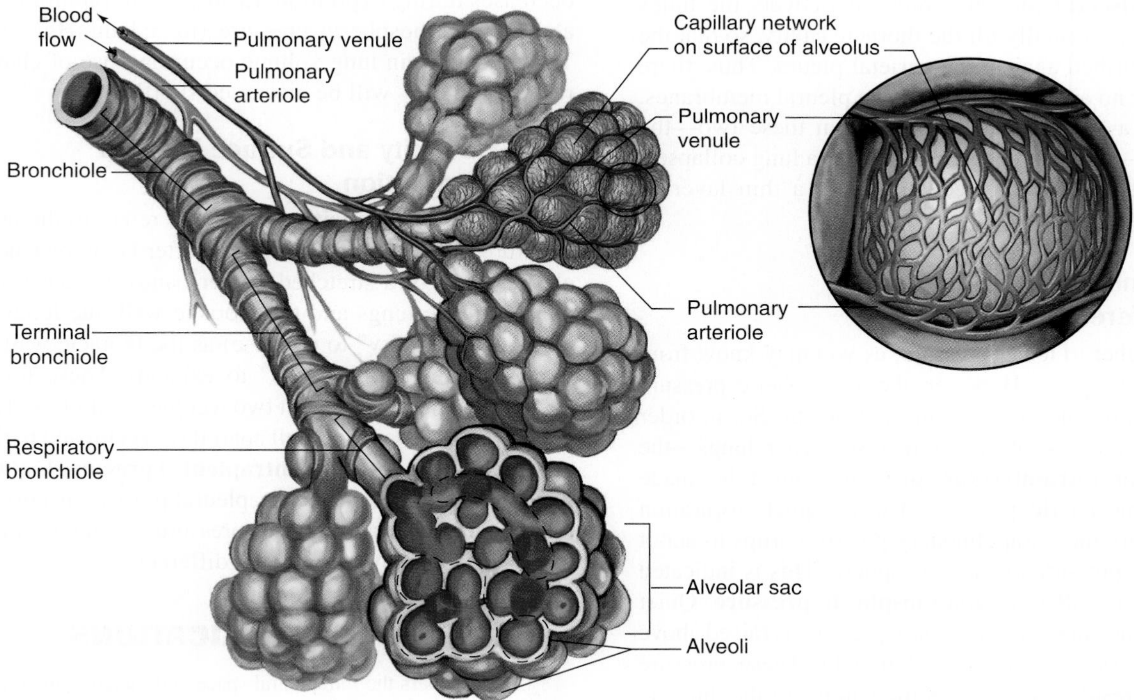

FIGURE 12.2 The relationship between alveoli and blood vessels. The extensive area of contact between the pulmonary capillaries and the alveoli allows for rapid exchange of gases between the air and blood.

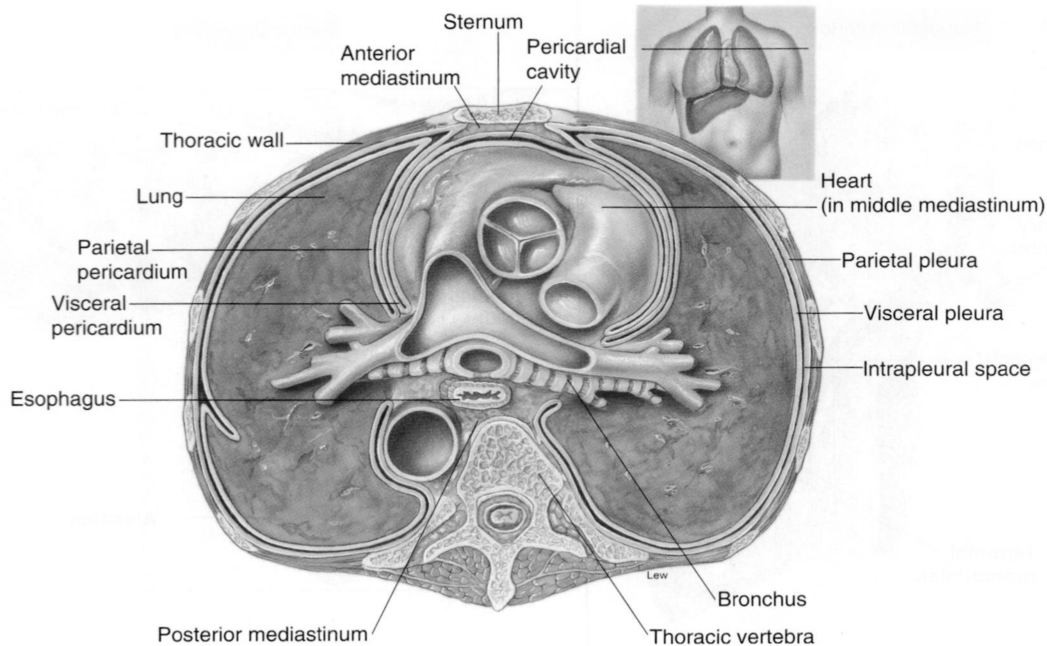

FIGURE 12.3 A cross section of the thoracic cavity. In addition to the lungs, the mediastinum and pleural membranes are visible. The parietal pleura is shown in green, and the visceral pleura in blue.

membrane is the **visceral pleura**, which also covers the lungs (fig. 12.3). The lungs normally fill the thoracic cavity, so that the visceral pleura is pushed against the parietal pleura. Thus, there is normally little or no air between these two pleural membranes. However, there is a "potential space" between these two—the *intrapleural space*—that can become real when a lung collapses. The intrapleural space normally contains only a thin layer of fluid, secreted by the parietal pleura.

Lungs Inflate and Deflate Because of Pressure Differences

Air moves from higher to lower pressure, as you may know from listening to weather reports. However, the atmospheric pressure is constant at this moment, before your next breath. So, in order for air to enter your lungs, the pressure inside your lungs—the **intrapulmonary**, or **intra-alveolar, pressure**—must be made lower than the atmospheric pressure. During quiet inspiration (unforced inhalation), the intrapulmonary pressure drops to about 3 mmHg below the pressure of the atmosphere. This is indicated as -3 mmHg, and is called a **subatmospheric pressure**. Quiet expiration occurs when the intrapulmonary pressure is raised above the atmospheric pressure by about $+3$ mmHg. These pressure changes occur as a result of changes in the volume of the thoracic cavity, as will be discussed shortly.

The intrapulmonary pressure changes result from changes in lung volume. This follows from **Boyle's law**, which states that the pressure of a given amount of gas (such as the air in the lungs) is inversely proportional to its volume. The lung volume increases during inspiration, causing the pressure within the lungs (the intrapulmonary pressure) to drop below the atmospheric pressure (to -3 mmHg, for example). Conversely, the lung volume decreases during expiration, raising the intrapulmonary pressure above the atmospheric pressure (to $+3$ mmHg, for example). These changes in lung volume occur because of changes in thoracic volume, as will be discussed shortly.

Lung Elasticity and Surface Tension Affect Ventilation

The lungs have **elasticity**, a term that refers to the tendency of a structure to return to its initial size after being distended (like the elastic recoil of a stretched rubber band). Because of the elastic tension of the lungs and the thoracic wall, the lungs pull in one direction (they "try" to collapse) as the thoracic wall pulls in the opposite direction (it "tries" to expand). These forces act like someone trying to pull apart two wet pieces of glass. The opposing elastic forces "trying" to pull apart the visceral and parietal pleurae produce a subatmospheric **intrapleural pressure** (between the two pleural membranes). This intrapleural pressure is normally always lower than the intrapulmonary pressure, so that each lung is stuck to the chest wall by the pressure difference.

CLINICAL APPLICATIONS

If air enters the intrapleural space and thereby raises the intrapleural pressure, the difference in pressure between the inside of the lungs (intrapulmonary pressure) and the outside of the lungs (intrapleural pressure) is abolished. As a result, the lung is no longer stuck to the thoracic wall; this is like releasing a stretched rubber band, and the lung's elastic recoil causes it to collapse. The condition of air entering the intrapleural space and causing the collapse of a lung is known as a **pneumothorax**. Fortunately, a pneumothorax usually causes only one lung to collapse, because each lung is contained in a separate pleural compartment.

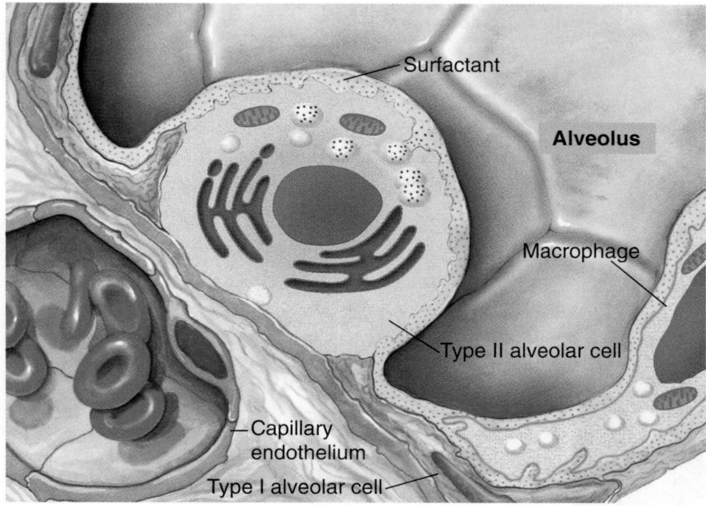

FIGURE 12.4 The production of pulmonary surfactant. Surfactant is produced by type II alveolar cells, and is a combination of particular proteins and phospholipids. The chief phospholipid surfactant is a derivative of lecithin.

CLINICAL INVESTIGATION CLUES

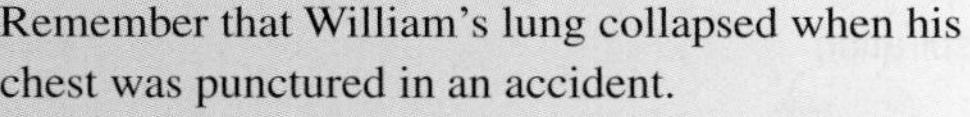

Remember that William's lung collapsed when his chest was punctured in an accident.

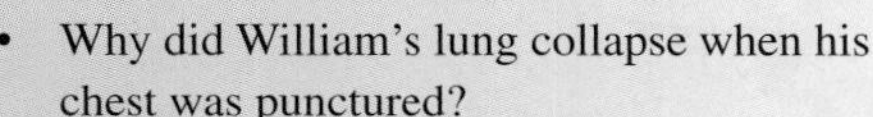

- Why did William's lung collapse when his chest was punctured?
- What is that condition called?
- How was William still able to breathe after his lung collapsed?

Compliance is defined as the change in the volume of a hollow organ (such as the lungs, urinary bladder, or blood vessels) per a given change in pressure. The compliance of the lungs relates to their ability to distend (stretch); the lungs are about 100 times more compliant (distensible) than a typical toy balloon. The physical properties of the lungs that act to resist distention (stretching) and lower lung compliance are its elasticity and surface tension. The alveolar wall is a slightly wet membrane, and water at a surface has **surface tension**—the attractive force, produced by hydrogen bonds (chapter 1), between water molecules at a fluid surface. The surface tension force in an alveolus acts to collapse the alveolus, raising the air pressure within it and forcing air out. This pressure is inversely proportional to the radius of the alveolus; in other words, the surface tension of smaller alveoli produces a higher pressure than that of larger alveoli.[2] This surface tension force should cause collapse of the lung alveoli, beginning with the smaller alveoli that blow their air into the larger ones.

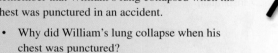

FYI [2]Here's an analogy that might be useful: imagine the water molecules in an alveolus as little people spread out in a circle, so they can just barely grasp each other's hands. As they pull on each other's hands, the force tends to make the circle smaller. Now that the circle's smaller, the little people can grab each other by the shoulders and pull harder, causing an even greater force acting to collapse the circle. The smaller the circle gets, the greater the force becomes.

Collapse of the alveoli does not normally occur, because the fluid in an alveolus contains a **surfactant** (fig. 12.4). A surfactant is a substance that lowers surface tension, and the pulmonary surfactant—a combination of particular proteins and phospholipids—is especially potent. Even better, the ability of pulmonary surfactant to lower the surface tension improves as the alveoli get smaller, as they do naturally during expiration. This prevents collapse of the alveoli during even a maximum expiration, so that there is a *residual volume* of air left in the lungs. Surfactant begins to be produced (by *type II alveolar cells*; fig. 12.4) in late fetal life.

CLINICAL APPLICATIONS

Because surfactant isn't produced until late fetal life, premature infants can be born with lungs lacking surfactant, a condition called **respiratory distress syndrome** (**RDS**). Considering that a full-term pregnancy lasts 37 to 42 weeks, RDS occurs in about 60% of babies born at less than 28 weeks, 30% of babies born at 28 to 34 weeks, and less than 5% of babies born after 34 weeks of gestation. Even under normal conditions, the first breath of a newborn is 15 to 20 times more difficult than afterward, because it has tiny alveoli with a high surface tension force to be overcome in order to inflate the alveoli. The effort required to inflate the lungs for a baby with RDS is much greater, and remains so. Fortunately, many babies with RDS can be saved by mechanical ventilators and by the use of exogenous surfactants, which are delivered into the baby's lungs by an endotracheal tube. The risk of RDS during pregnancy can be assessed by analysis of amniotic fluid (surrounding the fetus), and mothers can be given exogenous corticosteroids to accelerate the maturation of the fetus's lungs.

Muscle Contraction and Relaxation Change the Thoracic Volume

The volume of the thorax and lungs is increased during inspiration (inhalation) and decreased during expiration (exhalation). Changes in thoracic volume are produced by the contraction and relaxation of skeletal muscles. The chief muscle of ventilation is the **diaphragm**, which curves upward into the thoracic cavity when at rest (fig. 12.5). Between the bony portions of the ribs are the **external intercostal muscles** and the **internal intercostal muscles** (fig. 12.5).

An unforced (quiet) inspiration results primarily from contraction of the diaphragm, which lowers and flattens when it contracts (notice its change from fig. 12.6*a* to 12.6*b*). This increases thoracic volume in a vertical direction. Inspiration is also aided by contraction of the external intercostal muscles, which raise the ribs and increase thoracic volume laterally. During a forced inspiration, other skeletal muscles become involved: the *scalenes*, followed by the *pectoralis minor*, and then for an extreme effort, the *sternocleidomastoid muscles*. Because the lungs are stuck to the thoracic walls, their volume also increases, dropping the intrapulmonary pressure to subatmospheric levels (to about –3 mmHg for an unforced inspiration, and to more negative pressures for a forced inspiration). Air enters the lungs as a result, and inspiration stops when the intrapulmonary pressure is equal to the atmospheric pressure (when the difference in pressure is 0 mmHg).

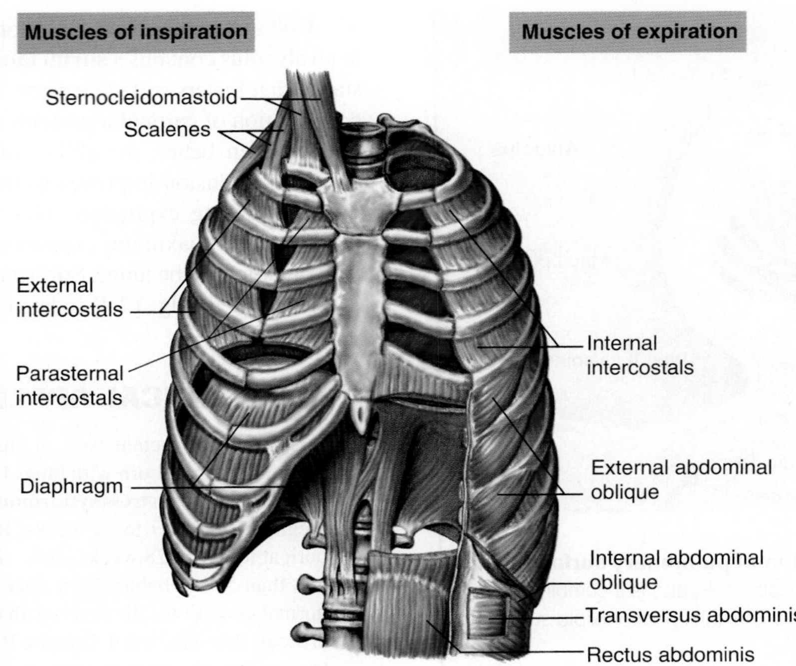

FIGURE 12.5 The muscles involved in breathing. The principal muscles of inspiration are shown on the left, and those of expiration are shown on the right.

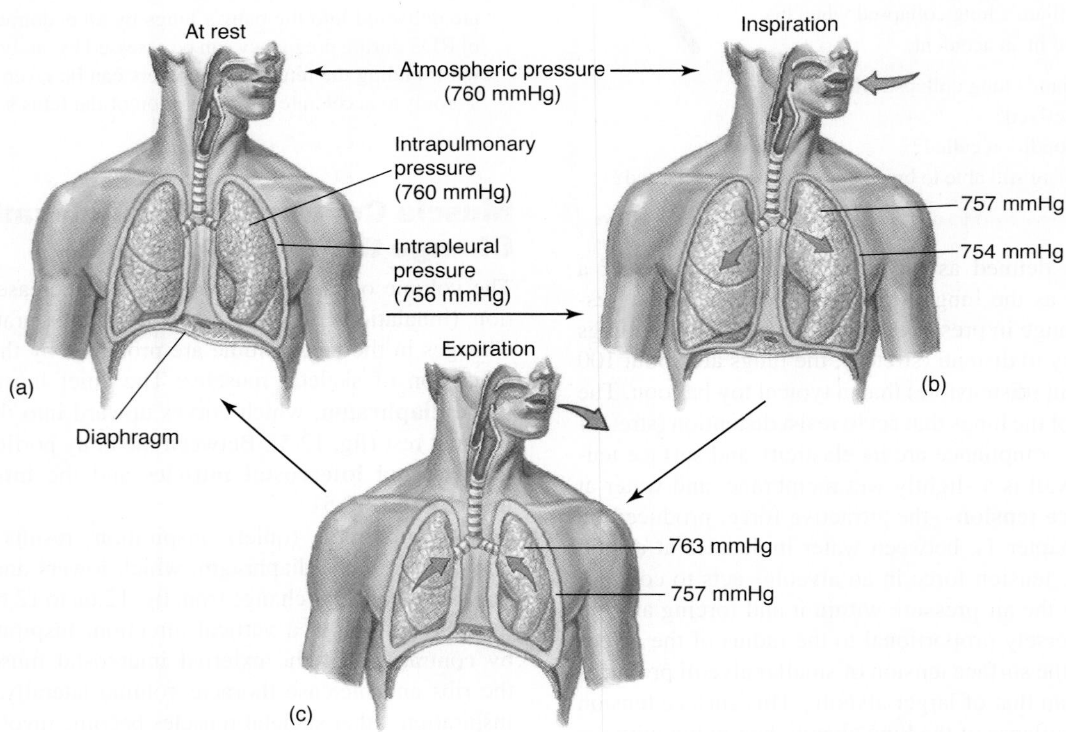

FIGURE 12.6 The breathing cycle. Pressures (at sea level) are shown (*a*) before inspiration, (*b*) during inspiration, and (*c*) during expiration. During inspiration, the intrapulmonary pressure is lower than the atmospheric pressure, and during expiration it is greater than the atmospheric pressure.

An unforced (quiet) expiration is a passive process. After becoming stretched by contractions of the diaphragm and thoracic muscles, the thorax and lungs recoil as a result of their elastic tension when the respiratory muscles relax. During a forced expiration, a further decrease in thoracic volume can be produced by contractions of the internal intercostal muscles, which depress the ribs. Contractions of abdominal muscles also help, because they press the abdominal organs up against the diaphragm, further

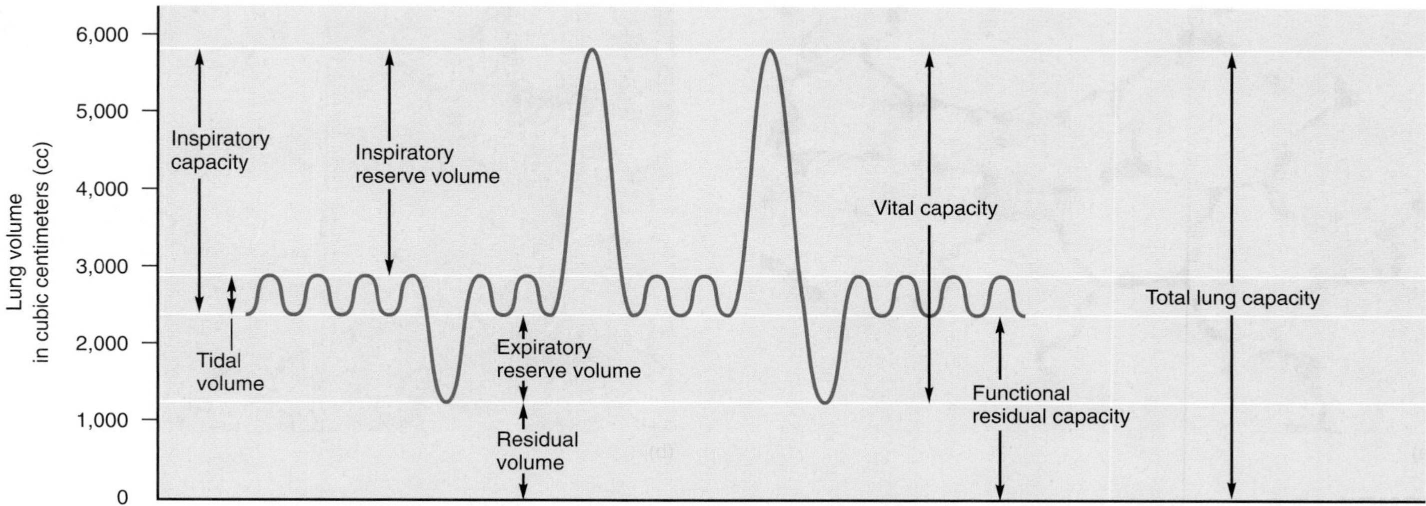

FIGURE 12.7 A spirogram showing lung volumes and capacities. A lung capacity is the sum of two or more lung volumes. The vital capacity, for example, is the sum of the tidal volume, the inspiratory reserve volume, and the expiratory reserve volume. Note that residual volume cannot be measured with a spirometer because it is air that cannot be exhaled. Therefore, the total lung capacity (the sum of the vital capacity and the residual volume) also cannot be measured with a spirometer.

decreasing thoracic volume. The decrease in thoracic and lung volume increases the intrapulmonary pressure above the atmospheric pressure, to about +3 mmHg in an unforced expiration and up to +20 mmHg or more in a forced expiration. Air is thus forced out of the lungs (fig. 12.6*c*) until, at the end of the expiration, the intrapulmonary pressure is again equal to the atmospheric pressure.

Pulmonary Function Can Be Tested by Spirometry

The health of the lungs is commonly tested by measuring the amount of air inhaled or exhaled into a measuring device called a *spirometer*, in a technique called **spirometry**. A graph of these breaths, called a *spirogram*, is illustrated in figure 12.7. Notice that some of the measurements are called "volumes" and others are called "capacities." A "capacity" is the sum of two or more "volumes."

The amount of air in an unforced breath is the **tidal volume**. Multiplying the tidal volume times the number of breaths per minute gives the **total minute volume**, a useful measurement that includes both the rate and depth of breathing. The maximum amount of air that can be expired after a maximum inspiration is the **vital capacity**. Since it's a "capacity," the vital capacity is the sum of the *inspiratory reserve volume*, *tidal volume*, and *expiratory reserve volume*. The expiratory reserve volume is the maximum amount of air that can be forcefully expired after an unforced expiration. This still leaves air left in the lungs that cannot be expired, called the *residual volume*. The sum of the expiratory reserve volume and the residual volume is the **functional residual capacity**. The vital capacity and the functional residual capacity are medically important measurements. The definitions of these and other lung volumes and capacities are provided in table 12.1.

TABLE 12.1 Terms Used to Describe Lung Volumes and Capacities

Term	Definition
Lung Volumes	The four nonoverlapping components of the total lung capacity
Tidal volume	The volume of gas inspired or expired in an unforced respiratory cycle
Inspiratory reserve volume	The maximum volume of gas that can be inspired during forced breathing in addition to tidal volume
Expiratory reserve volume	The maximum volume of gas that can be expired during forced breathing in addition to tidal volume
Residual volume	The volume of gas remaining in the lungs after a maximum expiration
Lung Capacities	Measurements that are the sum of two or more lung volumes
Total lung capacity	The total amount of gas in the lungs after a maximum inspiration
Vital capacity	The maximum amount of gas that can be expired after a maximum inspiration
Inspiratory capacity	The maximum amount of gas that can be inspired after a normal tidal expiration
Functional residual capacity	The amount of gas remaining in the lungs after a normal tidal expiration

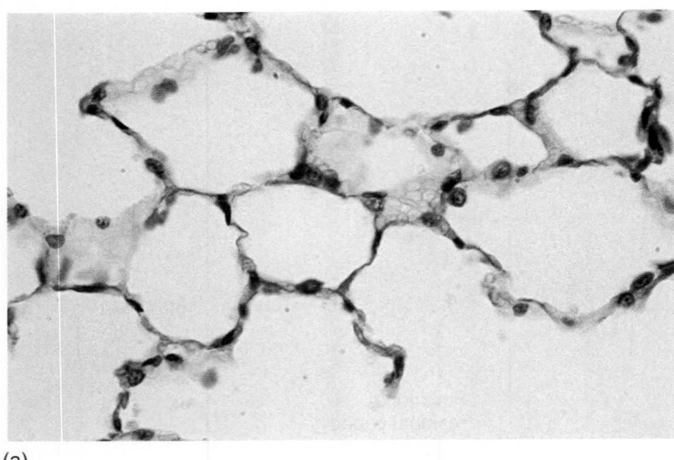

(a)

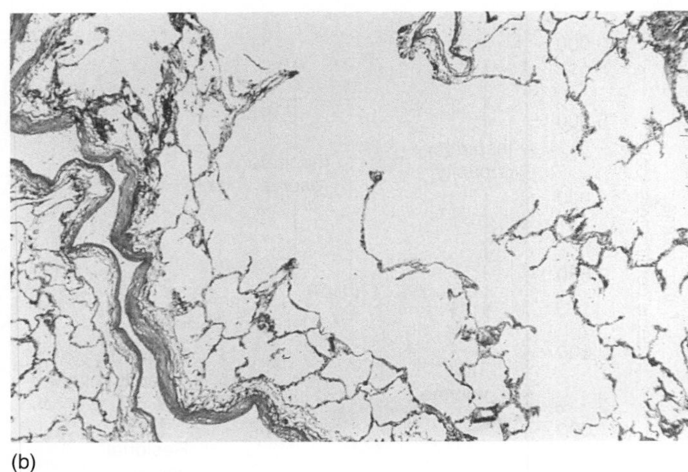

(b)

FIGURE 12.8 Emphysema destroys lung tissue. These are photomicrographs of tissue *(a)* from a normal lung and *(b)* from the lung of a person with emphysema. The destruction of lung tissue in emphysema results in fewer and larger alveoli.

An abnormally low vital capacity is produced by *pulmonary fibrosis*, in which collagen fibers reduce lung compliance, and by *emphysema*, in which alveoli are destroyed (fig. 12.8). Destruction of alveoli reduces the surface area for gas exchange, as described in the "Physiology in Health and Disease" section at the end of this chapter. Disorders that cause an abnormally low vital capacity are classified as **restrictive disorders**. By contrast, people with asthma have a normal vital capacity, but have increased airway resistance because of inflammation and smooth muscle constriction of the bronchioles. This makes it harder to breathe, which can be tested by measuring the time it takes to exhale a vital capacity.

Younger people can forcefully exhale about 80% or more of their vital capacity in the first second, a measurement called the **FEV_1** (FEV stands for the **forced expiratory volume**[3]). Disorders that cause an abnormally low FEV_1 are classified as **obstructive disorders**. A person with asthma, for example, may have an FEV_1 of, say, 60%, whereas another person of the same age has an FEV_1 of 80%. Asthma and acute bronchitis are purely obstructive disorders. In asthma, the bronchioles are "hyperresponsive," meaning that certain agents (such as allergens; chapter 11) stimulate bronchoconstriction. Because of this, asthma "attacks" are usually treated with inhaled bronchodilators that, like epinephrine, stimulate adrenergic receptors (chapter 6; also see the "Physiology in Health and Disease" section of this chapter).

By contrast, emphysema and other disorders classified as *chronic obstructive pulmonary disease* (*COPD*) are both restrictive (because of a low vital capacity) and obstructive (because of a low FEV_1). These diseases are progressive (they only get worse) and in most cases are caused by cigarette smoking.

FYI [3]There is a normal decline in the FEV_1 with age, but this decline appears to be accelerated by cigarette smoking. Smokers under the age of 35 who quit have improved lung function; those who quit after the age of 35 slow their age-related decline in FEV_1 to normal rates.

Clinical Investigation Clues

Remember that William was a smoker with a history of asthma, and that he had a normal vital capacity. However, another pulmonary function test gave evidence of an obstructive disorder; the physician prescribed an inhaled drug and strongly advised William to stop smoking.

- What is the vital capacity, and what disorders would be indicated if it were abnormal?
- What pulmonary diseases could William get if he didn't quit smoking?
- Which pulmonary function test indicates the presence of an obstructive disorder, and which obstructive disorder does William have?
- –What happens in asthma, and how does the inhaled drug work to treat it?

Check Point

1. Distinguish between the conducting and respiratory zones of the lungs, and explain how the structure of alveoli contributes to the efficiency of gas exchange.
2. Describe the location of the pleural membranes and the normal values for the intrapulmonary and intrapleural pressures; explain how a pneumothorax is produced.
3. Identify the nature and significance of pulmonary surfactant, and describe what happens in a premature infant when surfactant is lacking.
4. Explain how quiet inspiration and expiration are produced, giving values for the intrapulmonary pressures during the breathing cycle.
5. Define the vital capacity and the FEV_1, and explain how these values are used to test for restrictive and obstructive disorders.

Breathing Maintains Homeostasis of the Plasma O_2 and CO_2

The total pressure produced by the atmosphere is the sum of the pressures produced by each of its gases, which can be calculated as partial pressures. The partial pressure of O_2, the P_{O_2}, and of CO_2, the P_{CO_2}, can be measured in the blood to determine the function of the respiratory system, because normal breathing and gas exchange maintain homeostasis of these values. Breathing is adjusted to maintain homeostasis of plasma P_{O_2} and P_{CO_2} through chemoreceptor reflexes, which regulates the respiratory control centers in the brain stem.

The atmosphere is an ocean of gas that exerts pressure on all objects within it. This pressure, the **atmospheric pressure**, is measured with a mercury-filled barometer.[4] This device is a U-shaped tube partially filled with mercury (Hg), with one end open to the atmosphere and the other end sealed and under vacuum. The atmosphere pushes the mercury column up to a height dependent on the altitude. At sea level, the mercury column measures 760 mmHg, or 760 torr (fig. 12.9). This is also called *one atmosphere* pressure. The total atmospheric pressure is lower at higher altitudes (table 12.2).

According to **Dalton's law**, the total pressure of the atmosphere (or any gas mixture) is equal to the sum of the pressures that each gas in the mixture would exert independently. The pressure that each gas would exert independently is the **partial pressure** of that gas. The partial pressure of O_2 is indicated as P_{O_2}, the partial pressure of CO_2 is indicated as P_{CO_2}, the partial pressure of nitrogen (N_2) is indicated as P_{N_2}, and so on. If we consider only a dry atmosphere (so water vapor isn't a factor), the total atmospheric pressure would equal the $P_{N_2} + P_{O_2} + P_{CO_2}$.

Further, the partial pressure of each gas is proportional to the percentage of the gas in the total. Nitrogen, for example, composes 78% of the atmosphere, and so its partial pressure is easily calculated as 0.78×760 mmHg $= 593$ mmHg. Oxygen composes 21% of the air, and so its partial pressure is 0.21×760 mmHg $= 159$ mmHg. Since the total atmospheric pressure decreases with altitude, the partial pressures of each gas correspondingly decrease (table 12.2). Notice in table 12.2 that the P_{O_2} of the air in the alveoli is lower than in the surrounding atmosphere. This is partially due to gas exchange, which lowers the alveolar P_{O_2}, and partially due to the fact that there is a residual volume of air left after even a maximal expiration, so fresh air inspired at each breath must mix with the older air left over in the alveoli.

The P_{O_2} and P_{CO_2} of the Plasma Reflect the Amount Dissolved

We're familiar with our ability to dissolve solids, such as salt or sugar, in water. Although gases such as O_2 and CO_2 are invisible, they also can dissolve in water. According to **Henry's law**, the amount of a gas that can dissolve in water depends on:

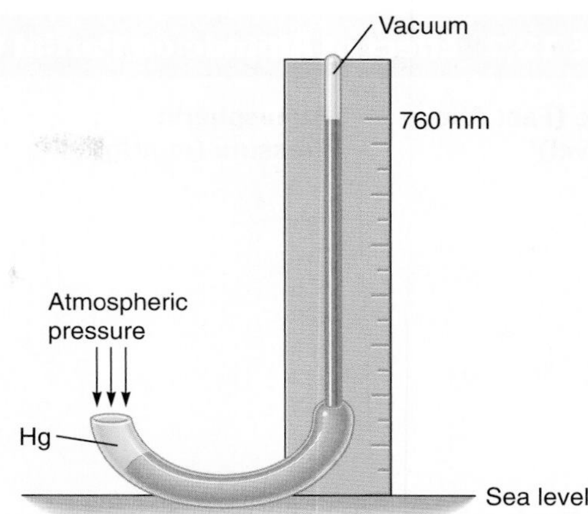

FIGURE 12.9 The measurement of atmospheric pressure. Atmospheric pressure at sea level can push a column of mercury to a height of 760 millimeters. This is also described as 760 torr, or one atmospheric pressure.

1. The temperature of the water; more of a gas can dissolve in colder water (if you heat up the water, gasses escape).
2. The solubility of the gas in water, a physical constant that reflects the properties of the particular gas; carbon dioxide is more soluble in water than oxygen, for example.
3. The partial pressure of the gas in the air above the water; more oxygen can be dissolved at a higher P_{O_2}, for example.

Because the solubility of oxygen (or carbon dioxide) is a physical constant, and our blood temperature is maintained relatively constant, the amount of oxygen (or carbon dioxide) dissolved in the blood is determined by their partial pressures in the air within the alveoli. If all of the oxygen is dissolved that's possible in water exposed to air with a particular P_{O_2}, we can say that the water now also has that same P_O (fig. 12.10*a*). This is useful because *the concentration of dissolved oxygen is directly proportional to the P_{O_2}*; for example, if the P_{O_2} were reduced by half, the oxygen concentration in the water would also be reduced by half.

Hospital personnel use this principle to calibrate blood gas machines, which can then be used to measure the P_{O_2} of blood (fig. 12.10*b*). The value of 100 mmHg in this figure is normal for arterial blood. These machines use electrodes that sense oxygen (and carbon dioxide) physically dissolved in the plasma; they don't detect the oxygen hidden in the red blood cells (which actually makes up most of the total oxygen in blood). So, you can think of the "P" in P_{O_2} and P_{CO_2} as standing for "plasma" oxygen and carbon dioxide, but realize that it actually stands for partial pressure.

Now, go back and examine the first column in table 12.2. Notice that the alveolar air has a P_{O_2} of 105 mmHg and the arterial blood has a value of 100 mmHg. Why the difference? Blood exposed to air with a P_{O_2} of 105 mmHg would have the same P_{O_2}, if gas exchange were perfect. The fact that the arterial blood instead has a P_{O_2} of 100 mmHg tells us that the gas exchange in

FYI [4]An Italian scientist, Evangelista Torricelli, invented the barometer in 1643. In his honor, atmospheric pressure is often measured in units of torr, which is equivalent to mmHg.

TABLE 12.2 Effect of Altitude on Partial Oxygen Pressure (P_{O_2})

Altitude (Feet Above Sea Level)*	Atmospheric Pressure (mmHg)	P_{O_2} in Air (mmHg)	P_{O_2} in Alveoli (mmHg)	P_{O_2} in Arterial Blood (mmHg)
0	760	159	105	100
2,000	707	148	97	92
4,000	656	137	90	85
6,000	609	127	84	79
8,000	564	118	79	74
10,000	523	109	74	69
20,000	349	73	40	35
30,000	226	47	21	19

* For reference, Pike's Peak (Colorado) is 14,110 feet; Mt. Whitney (California) is 14,505 feet; Mt. Logan (Canada) is 19,524 feet; Mt. McKinley (Alaska) is 20,320 feet; and Mt. Everest (Nepal and Tibet), the tallest mountain in the world, is 29,029 feet.

the lungs is extremely efficient, but it's not 100% efficient. If you again examine table 12.2, and compare arterial P_{O_2} values at different altitudes, you'll see that these values can fall from 100 mmHg at sea level to 69 mmHg at 10,000 feet, a mountain altitude commonly experienced in hiking or skiing.

Measurements of arterial blood P_{O_2} and P_{CO_2} are performed in hospitals to assess the effectiveness of pulmonary gas exchange. Arterial blood (from an artery of the systemic circulation; chapter 10), rather than a vein, is used for this because lung function modifies the blood returning to the left atrium and pumped from the left ventricle into the systemic arteries. Since a patient's room air has a normal P_{O_2} and P_{CO_2} for that altitude, the arterial values should also be in the normal range if pulmonary function is healthy. If the blood gases are abnormal, it may be caused by disorders of the lungs or pulmonary circulation, or by depressed breathing under anesthesia.

Breathing Maintains Homeostasis of Blood P_{O_2} and P_{CO_2}

We breathe to maintain a state of dynamic constancy—homeostasis—of the blood P_{O_2} and P_{CO_2} (and the blood pH, as will be described shortly). The concentrations of O_2 and CO_2 in the arterial blood are thus kept in their normal ranges because of gas exchange in the

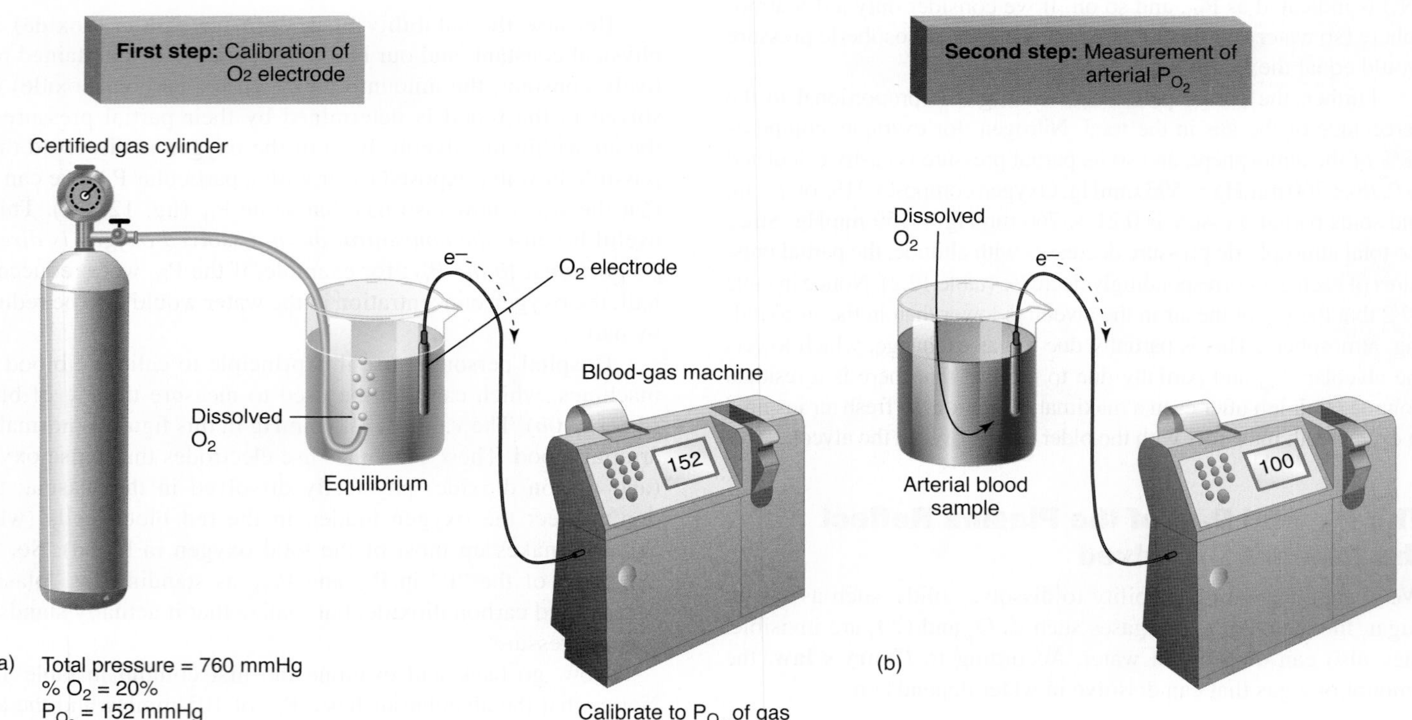

FIGURE 12.10 Blood gas measurements using the P_{O_2} electrode. (*a*) The electrical current generated by the oxygen electrode is calibrated so that the needle of the blood gas machine points to the P_{O_2} of the gas with which the fluid is in equilibrium. (*b*) Once standardized in this way, the electrode can be inserted into a fluid such as blood, and the P_{O_2} of this solution can be measured.

Clinical Applications

In a scuba dive, the gas pressure of the breathing tank increases by one atmosphere (760 mmHg) for every 33 feet below sea level. The partial pressures of the gases in the tank increase proportionately, increasing the concentrations of these gases in the blood. As the amount of dissolved nitrogen in the blood increases with increasing P_{N_2}, it can produce unusual effects in divers submerged for an hour or longer. This can produce **nitrogen narcosis** ("rapture of the deep"), resembling alcohol intoxication. If a diver ascends too rapidly, the decreasing P_{N_2} reduces the amount of N_2 that can remain dissolved, so that N_2 comes out of solution too quickly and forms tiny bubbles. The bubbles of N_2 gas can block small blood channels, producing muscle and joint pain as well as more serious effects. This condition is known as **decompression sickness**, commonly called "the bends." The primary treatment for this is **hyperbaric oxygen therapy**, in which a patient is given 100% oxygen gas at 2 to 3 atmospheres pressure. Hyperbaric oxygen therapy is also used to treat carbon monoxide poisoning, infections that could lead to gas gangrene, and other conditions.

lungs (fig. 12.11). Gas exchange also occurs in the tissues as a result of aerobic cell respiration, with a net diffusion of oxygen from the blood to the tissue cells, and a net diffusion of carbon dioxide from the tissues into the blood. As a result, the P_{O_2} is lower and the P_{CO_2} is higher in the venous than in the arterial blood. Because the venous blood gas values reflect tissue metabolism, arterial blood gas measurements are better indicators of lung function.

Sensors involved in the negative feedback control of breathing also monitor the blood gases in arterial blood. These are called *chemoreceptors*. There are two groups of chemoreceptors. The **peripheral chemoreceptors** are in the **aortic bodies**, located around the aortic arch, and in the **carotid bodies**, located in each common carotid artery near its branch forming the external and internal carotids (fig. 12.12). The other group of chemoreceptors is the **central chemoreceptors**, located in the medulla oblongata of the brain.

The peripheral chemoreceptors send sensory information via cranial nerves IX (glossopharyngeal) and X (vagus) to the **rhythmicity center** of the medulla oblongata. This is the neural center that controls automatic breathing. Neurons in the rhythmicity center have automatic, rhythmic activity to produce the breathing cycle, but the activity of these neurons is affected by sensory information from the chemoreceptors. The neurons of the rhythmicity center indirectly regulate motor neurons in the spinal cord that stimulate the respiratory muscles. For example, the phrenic nerve that stimulates the diaphragm has its cell bodies in the cervical region of the spinal cord.

The P_{CO_2} and pH of the Blood Stimulate Chemoreceptors

In **hypoventilation** (inadequate ventilation), the rate at which CO_2 is exhaled ("blown off") by the lungs is less than the rate at which it's produced by the tissue cells and enters the venous blood, causing a rise in the arterial P_{CO_2}. Conversely, **hyperventilation**

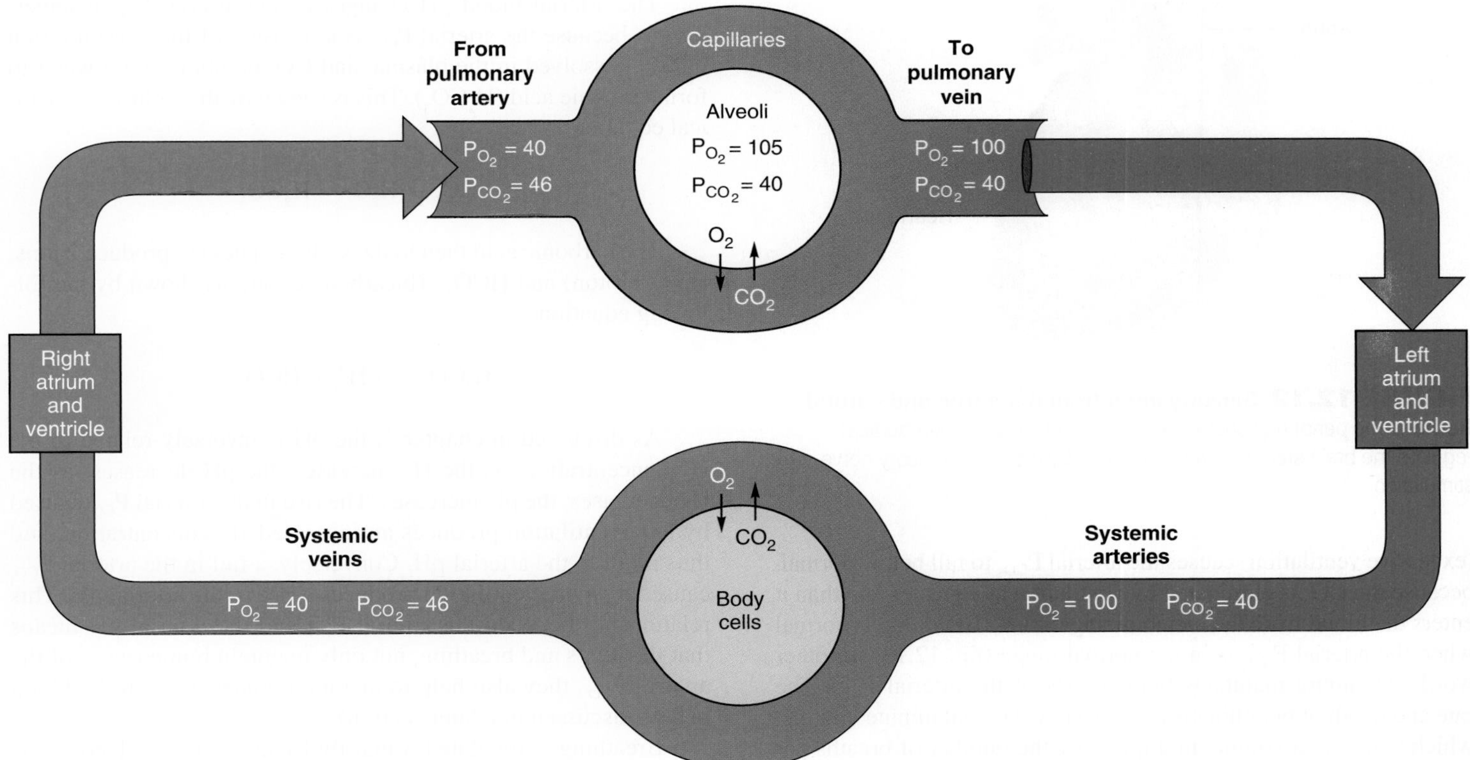

FIGURE 12.11 Partial pressures of gases in blood. The P_{O_2} and P_{CO_2} values of blood are a result of gas exchange in the lung alveoli and gas exchange between systemic capillaries and body cells.

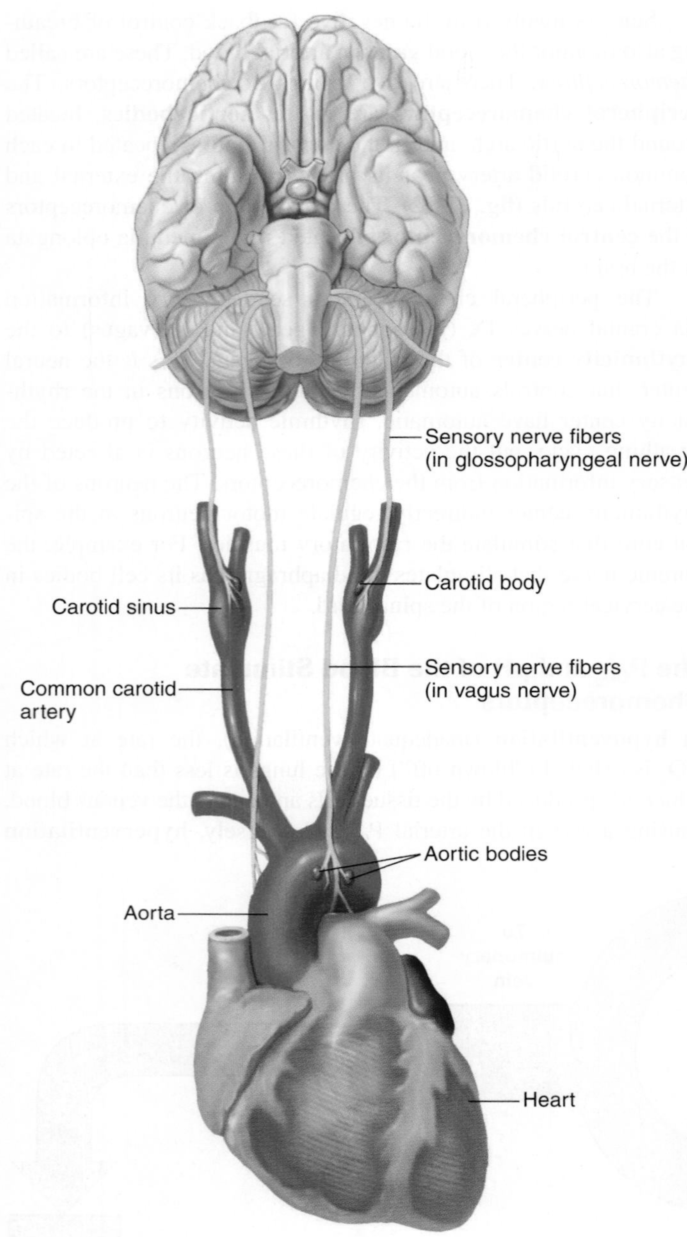

FIGURE 12.12 Sensory input from the aortic and carotid bodies. The peripheral chemoreceptors (aortic and carotid bodies) regulate the brain stem respiratory centers by means of sensory nerve stimulation.

(excessive ventilation) causes the arterial P_{CO_2} to fall below normal, because the CO_2 is eliminated by ventilation at a greater rate than it enters the blood from the metabolizing tissues. Breathing is normal when the arterial P_{CO_2} is in the normal range (fig. 12.13). In other words, breathing maintains homeostasis of the arterial P_{CO_2}. The rate and depth of breathing is measured by the total minute volume, which is the tidal volume multiplied by the number of breaths per minute. When a high total minute volume lowers the arterial P_{CO_2}, the person is hyperventilating. Conversely, when a low total minute volume raises the arterial P_{CO_2}, the person is hypoventilating.

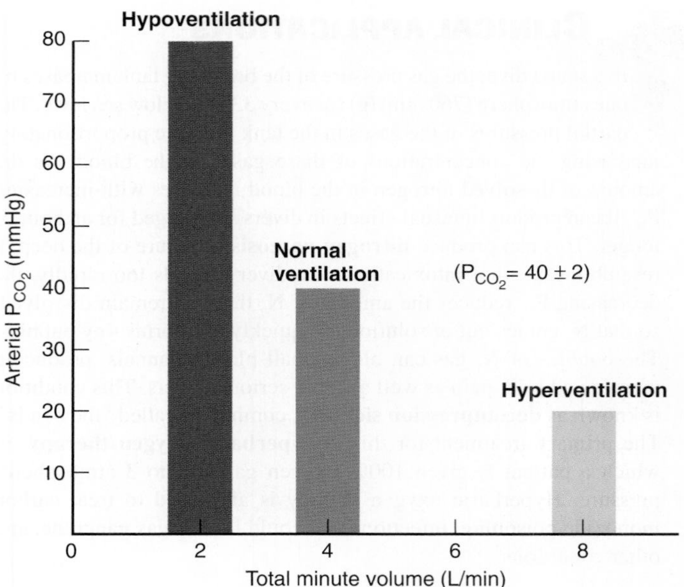

FIGURE 12.13 The relationship between total minute volume and arterial P_{CO_2}. Notice that these are inversely related: when the total minute volume increases by a factor of 2, the arterial P_{CO_2} decreases by half. The total minute volume measures breathing, and is equal to the amount of air in each breath (the tidal volume) multiplied by the number of breaths per minute. The P_{CO_2} measures the CO_2 concentration of arterial blood plasma.

The arterial blood pH changes as the arterial P_{CO_2} changes. This is because the arterial P_{CO_2} is a measure of the concentration of CO_2 dissolved in the plasma, and CO_2 combines with water to form carbonic acid (H_2CO_3). This is shown by the following chemical equation:

$$CO_2 + H_2O \rightarrow H_2CO_3$$

The carbonic acid then ionizes (dissociates) to produce 2 ions: H^+ (a proton) and HCO_3^- (bicarbonate ion), as shown by the following equation:

$$H_2CO_3^- \rightarrow H^+ + HCO_3^-$$

As discussed in chapter 1, the pH is inversely related to the H^+ concentration: as the H^+ increases, the pH decreases; as the H^+ decreases, the pH increases. The rise in the arterial P_{CO_2} caused by hypoventilation produces an increased H^+ concentration, and thus a fall in the arterial pH. Conversely, a fall in the arterial P_{CO_2} caused by hyperventilation produces a rise in the arterial pH. This relationship between the arterial P_{CO_2} and the arterial pH means that the lungs and breathing not only maintain homeostasis of the arterial P_{CO_2}, they also help to maintain homeostasis of the blood pH (as discussed in a later section).

Breathing is regulated primarily by the CO_2 dissolved in the plasma (as indicated by the P_{CO_2}). As CO_2 in the blood rises (due to hypoventilation), there is an associated fall in the arterial pH. This fall in blood pH stimulates the aortic and carotid bodies (the

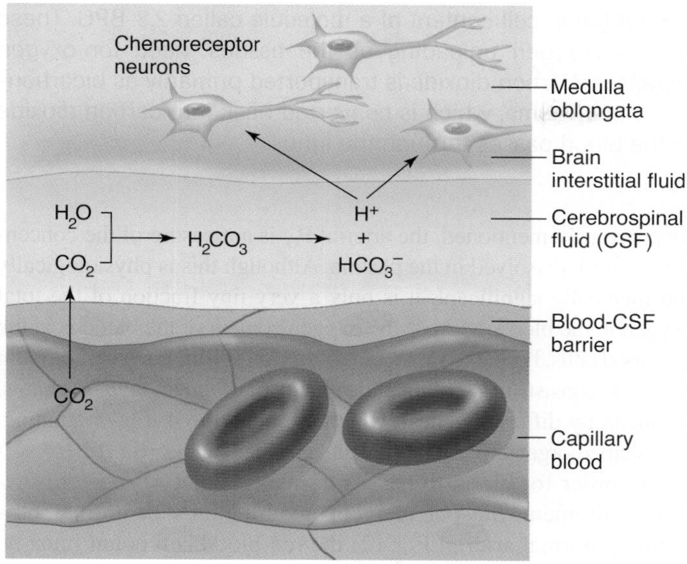

FIGURE 12.14 Stimulation of the central chemoreceptors. Carbon dioxide from the blood diffuses across the blood-brain barrier to form carbonic acid (H_2CO_3) in the cerebrospinal fluid and brain interstitial fluid. This lowers the pH in the brain and stimulates chemoreceptors in the medulla oblongata, which then stimulate the rhythmicity center in the medulla.

peripheral chemoreceptors) to cause an immediate increase in breathing. However, H^+ from the blood can't cross the blood-brain barrier to stimulate the central chemoreceptors. The central chemoreceptors in the medulla oblongata are different neurons from those in the rhythmicity center of the medulla. These neurons are stimulated by a fall in the pH of their surrounding brain interstitial fluid, which results from CO_2 diffusing across the blood-brain barrier (fig. 12.14).

Notice that the aortic and carotid bodies are directly stimulated by a decrease in the blood pH, and not by an increase in CO_2. This means that if any acids in the blood (such as lactic acid or ketone bodies) lower the blood pH, they will stimulate breathing. However, breathing is normally regulated by the changes in pH that result from changes in plasma CO_2, so that homeostasis of the arterial P_{CO_2} is maintained (see fig. 12.13).

The negative feedback control of breathing in response to a rise in arterial P_{CO_2} is illustrated in figure 12.15. The immediate increase in breathing caused by a rise in the arterial P_{CO_2} is produced by the peripheral chemoreceptors. However, the central chemoreceptors are responsible for 70% to 80% of the increased ventilation that occurs when the high arterial P_{CO_2} is sustained.[5]

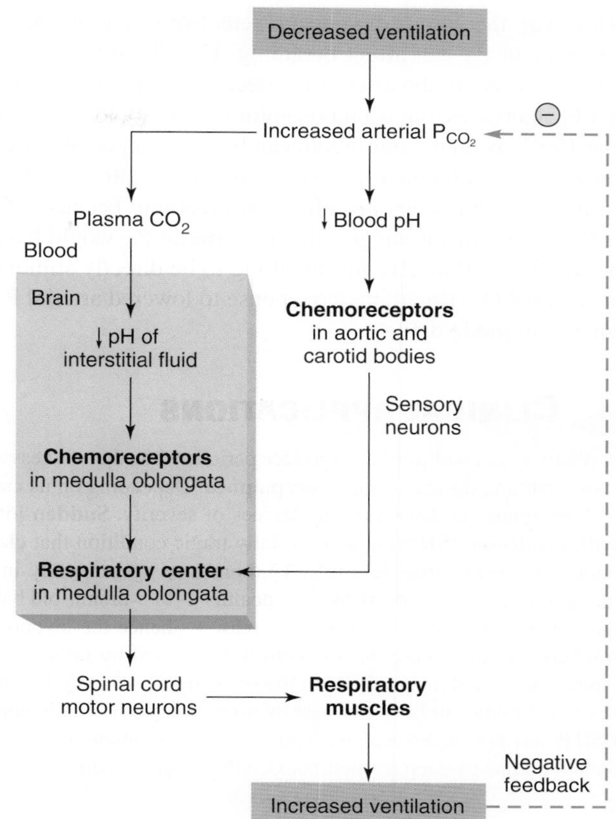

FIGURE 12.15 Chemoreceptor control of breathing. This figure depicts the negative feedback control of ventilation through changes in blood P_{CO_2} and pH. The blood-brain barrier, represented by the orange box, allows CO_2 to pass into the brain but prevents the passage of H^+.

CLINICAL APPLICATIONS

People who hyperventilate during psychological stress are sometimes told to breathe into a paper bag, so that they rebreathe their expired air, which is rich in CO_2. This procedure helps raise their arterial P_{CO_2} back up to the normal range. This is beneficial because low arterial P_{CO_2} causes a reflex cerebral vasoconstriction, which reduces blood flow and produces dizziness. Further, the reduced blood flow can cause a fall in pH of the brain interstitial fluid, stimulating the central chemoreceptors to cause more hyperventilation. Rebreathing into a paper bag can raise the arterial P_{CO_2} and help break this cycle.

A Low Arterial P_{O_2} Can Also Stimulate Breathing

The concentration of O_2 in the plasma (measured by the arterial P_{O_2}) has little effect on breathing at sea level or moderate altitudes. It's true that if you hyperventilate you can hold your breath longer, but that's not because you're significantly increasing your P_{O_2} (the lungs are already very efficient at oxygenating the blood), or because your blood has significantly more oxygen (normal breathing at sea level already loads about 97% of the hemoglobin with oxygen). Rather, your enhanced ability to hold your breath after hyperventilating is due to the lowering of arterial P_{CO_2}.

FYI [5]We can consciously override the negative feedback control of breathing, up to a point, when we decide to hyperventilate or when we hold our breath. However, the buildup of CO_2 will eventually stimulate breathing. So, when children try to hold their breath until they get their way, we can laugh—the chemoreceptor reflex will force them to breathe before any damage is done.

However, the carotid bodies are sensitive to the arterial P_{O_2}, and changes in P_{O_2} can affect breathing. Usually, this is an indirect effect—a rise in the arterial P_{O_2} decreases, and a lowering of arterial P_{O_2} increases, the chemoreceptor sensitivity to P_{CO_2}. If you breathe 100% oxygen, your breath can be held longer because the response to P_{CO_2} is blunted. If you go up to moderate altitudes, an increase in P_{CO_2} has a greater effect on breathing because of the lower P_{O_2}. Experiments suggest that the arterial P_{O_2} would have to fall below about 70 mmHg for breathing to be directly stimulated by the lowered O_2. Breathing in response to lowered arterial P_{O_2} is known as a **hypoxic drive**.

CLINICAL APPLICATIONS

A variety of conditions can produce periods when there is cessation of breathing during sleep, or **sleep apnea**. Depending on its cause, sleep apnea can have varying degrees of severity. **Sudden infant death syndrome (SIDS)** is an especially tragic condition that claims about 1 in 1,000 babies less than 12 months of age annually in the United States. Victims are apparently healthy 2- to 5-month-old babies who die in their sleep for no obvious reason—hence the layperson's term "crib death." These deaths seem to be caused by failure of the respiratory control mechanisms in the brain stem and/or by failure of the carotid bodies to be stimulated by arterial oxygen. The incidence of SIDS has declined since physicians began recommending that parents put infants to sleep on their backs rather than on their stomachs.

CHECK POINT

1. Explain how you calculate the partial pressure of a gas in a mixture, such as the atmosphere, and how that calculation changes as you go to a higher altitude.
2. Describe how the concentrations of dissolved gases in plasma are measured, and the physiological and medical significance of these measurements.
3. Explain how the pH of blood is affected by the concentration of dissolved CO_2, indicating the molecules and ions involved.
4. Define hypoventilation and hyperventilation, and explain how each affects the arterial P_{CO_2}.
5. Describe the negative feedback loop involving the chemoreceptor regulation of breathing.
6. Explain the contributions of the arterial P_{O_2} to the regulation of breathing.

The Blood Transports Oxygen and Carbon Dioxide

Hemoglobin in the red blood cells passing through the lungs binds to oxygen, so that the systemic arterial blood contains about 97% oxyhemoglobin. As this blood travels through the systemic capillaries, about one-fifth of its oxygen is unloaded to the tissues for cell respiration. The extent of this unloading depends on the pH and temperature of the tissue, and on the red blood cell content of a molecule called 2,3-BPG. These increase oxygen unloading as the tissues' need for oxygen increases. Carbon dioxide is transported primarily as bicarbonate in the plasma, which is converted back into carbon dioxide as the blood passes through the lungs.

As previously mentioned, the arterial P_{O_2} is a measure of the concentration of O_2 dissolved in the plasma. Although this is physiologically and medically significant, it is only a very tiny fraction of the total oxygen in the blood. Most of the oxygen content of the blood is in the red blood cells, bound to molecules of **hemoglobin**. Each hemoglobin molecule consists of 4 polypeptide chains (2 *alpha* and 2 *beta* chains, produced by different genes), which are bound to 4 iron-containing, disc-shaped organic pigment molecules called *hemes* (fig. 12.16).

In order for blood to have a normal content of oxygen, several requirements must be met: (1) breathing must be normal, producing a normal arterial P_{O_2}; (2) the red blood cell count must be in the normal range; (3) there must be a normal concentration of

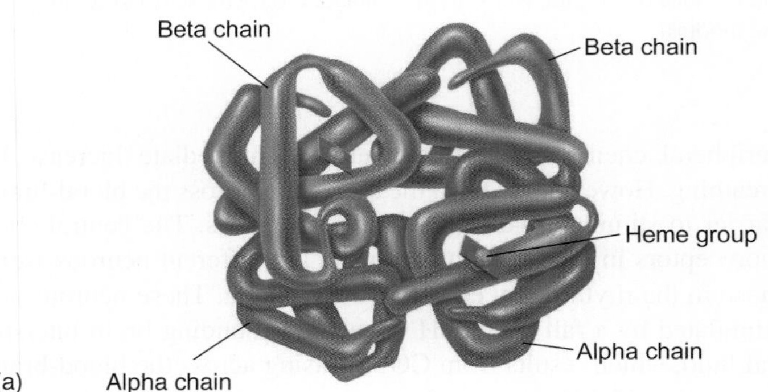

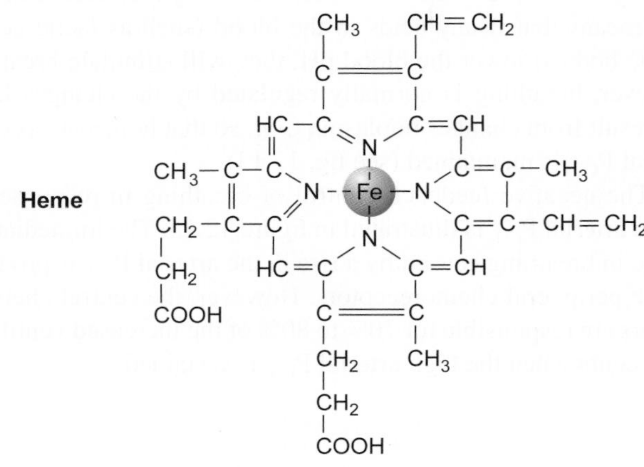

FIGURE 12.16 The structure of hemoglobin. (*a*) An illustration of the three-dimensional structure of hemoglobin in which the 2 alpha and 2 beta polypeptide chains are shown. The 4 heme groups are represented as flat structures with atoms of iron (*spheres*) in the centers. (*b*) The structural formula for heme.

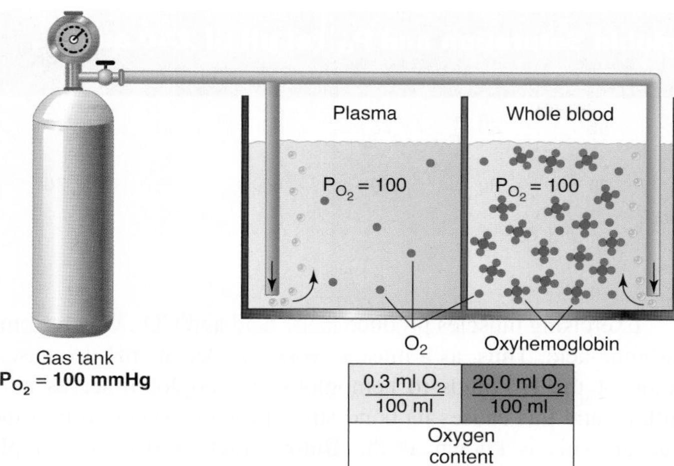

FIGURE 12.17 The oxygen content of blood. Plasma and whole blood that are brought into equilibrium with the same gas mixture have the same P_{O_2}, and thus the same number of dissolved oxygen molecules (shown as blue dots). The oxygen content of whole blood, however, is much higher than that of plasma because of the binding of oxygen to hemoglobin, which is normally only found in red blood cells (not shown).

hemoglobin; and (4) the hemoglobin must be in a normal form (not in the abnormal carboxyhemoglobin form discussed later). Under these conditions, arterial blood carries about 20 ml O_2 per 100 ml of blood (fig. 12.17). Figure 12.17 compares plasma (left) with whole blood (right). Notice that both are exposed to a gas mixture with a P_{O_2} of 100 mmHg, and so both the plasma and the whole blood have a P_{O_2} of 100 mmHg once all of the oxygen is dissolved. This corresponds to a plasma oxygen concentration of 0.3 ml O_2 per 100 ml blood in both right and left compartments. The far greater content of oxygen in whole blood is bound to hemoglobin, and so doesn't influence the P_{O_2} measurement.

Loading and Unloading of Oxygen Depend on the P_{O_2}

Normal heme iron is in the ferrous form (Fe^{2+}), which has an electron free to form a bond with an oxygen molecule (O_2). Because there are 4 heme groups per hemoglobin, each hemoglobin molecule can bond to 4 molecules of oxygen.[6] When this binding occurs, **deoxyhemoglobin** is converted into **oxyhemoglobin**. When oxyhemoglobin releases its oxygen to the tissues, it's converted back into deoxyhemoglobin. The binding and release of oxygen are known as the **loading** and **unloading** reactions, respectively; the loading reaction occurs in the lungs, and the unloading reaction occurs in the tissue capillaries. This is a reversible reaction, as shown by the following chemical equation:

$$\text{Deoxyhemoglobin} + O_2 \underset{\text{(tissues)}}{\overset{\text{(lungs)}}{\rightleftharpoons}} \text{Oxyhemoglobin}$$

Do you remember that the blood leaving the lungs has a P_{O_2} of about 100 mmHg (see fig. 12.11)? At this high a P_{O_2}, about 97% of the hemoglobin is loaded with oxygen; this is the *percent oxyhemoglobin*, also called the **oxyhemoglobin percent saturation**. This value is indicated in the leftmost column in table 12.3. When this blood travels through the tissue capillaries, the P_{O_2} decreases as the tissue cells consume oxygen in aerobic cell respiration. Scientists can measure the percent oxyhemoglobin at each value of P_{O_2}, as was done to construct table 12.3. Notice that the column labeled "Venous blood" has a P_{O_2} of 40 mmHg, and at that value, has a percent oxyhemoglobin of 75%. This means that 97% minus 75%, or 22% (a little over one-fifth), of the oxygen carried by the arterial blood is unloaded to the tissues. These are the normal values when a person is at rest.

The same information is presented in a graph (fig. 12.18), known as an *oxyhemoglobin dissociation curve*. We can use either the table or graph to determine what would happen to oxygen unloading during exercise. Suppose that you exercise, so that the P_{O_2} of the exercising muscles falls from 40 to 21 mmHg due to the increased metabolism. At a P_{O_2} of 21 mmHg, this blood contains 35% oxyhemoglobin (table 12.3). The arterial blood still has a P_{O_2} of 100 mmHg with a percent oxyhemoglobin of 97%, because increased breathing (*hyperpnea*) will maintain these values during normal levels of exercise. As the blood now travels through the capillaries in the exercising muscles, 97% minus 35%, or 62% of the oxygen carried by hemoglobin will be unloaded.

CLINICAL APPLICATIONS

Carboxyhemoglobin is hemoglobin with its heme groups bound to carbon monoxide instead of oxygen. Because the bond that hemoglobin forms with carbon monoxide is 210 times stronger than the bond it forms with oxygen, carbon monoxide tends to displace oxygen and remain bound as it goes through the circulation, thereby reducing the oxygen carried by the blood. Carboxyhemoglobin has a cranberry juice color, compared to the tomato juice color of oxyhemoglobin. The body actually produces a tiny amount of carbon monoxide, and so there is always some carboxyhemoglobin; however, according to federal standards, active nonsmokers should not have more than 1.5% carboxyhemoglobin. Higher percentages of carboxyhemoglobin constitute **carbon monoxide poisoning**. Nonsmokers living in certain smoggy cities can have 3% carboxyhemoglobin levels, whereas smokers there may have 10% or more carboxyhemoglobin. Much higher percentages can be produced by smoke inhalation or suicide attempts, and may cause death. People with carbon monoxide poisoning are often given hyperbaric oxygen therapy (previously described).

FYI [6]There are more than 280 million hemoglobin molecules in each red blood cell, so multiplying by 4 gives well over a billion molecules of oxygen carried by each red blood cell. Normal blood contains about 5 million red blood cells per cubic millimeter of blood—the size of a very small drop. We each have an average total of 5.5 L of blood. You can calculate the number of oxygen molecules if you like in a cubic millimeter of blood or for the total blood volume, but it's probably enough just to realize that these are very big numbers.

TABLE 12.3 Relationship Between Percent Oxyhemoglobin Saturation and P_{O_2} (at pH of 7.40 and Temperature of 37° C)

P_{O_2} (mmHg)	100	80	61	45	40	36	30	26	23	21	19
Percent Oxyhemoglobin	97	95	90	80	75	70	60	50	40	35	30
	Arterial Blood				Venous Blood						

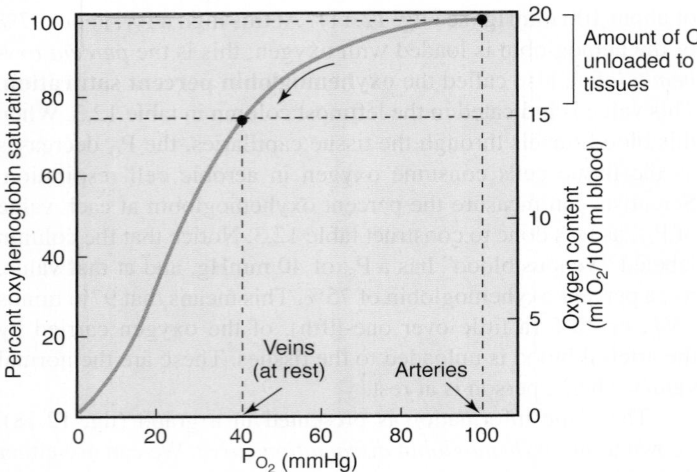

FIGURE 12.18 The oxyhemoglobin dissociaton curve. The percentage of oxyhemoglobin saturation and the blood oxygen content are shown at different values of P_{O_2}. Notice that the percent oxyhemoglobin decreases by about 22% as the blood passes through the tissue from arteries to veins, resulting in the unloading of a little less than 5 ml of O_2 per 100 ml of blood to the tissues.

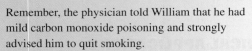

Clinical Investigation Clues

Remember, the physician told William that he had mild carbon monoxide poisoning and strongly advised him to quit smoking.

- What blood gas measurement revealed that William had carbon monoxide poisoning?
- How does carbon monoxide act as a poison?
- How was William exposed to carbon monoxide, and how would stopping smoking help?

Loading and Unloading of Oxygen Are Influenced by the pH and Temperature

When you exercise, the P_{O_2} of the exercising muscles decreases. An example of how this increases the percentage of oxygen unloading was presented earlier. However, the exercising muscles actually get more oxygen than would be predicted this way. This is because oxygen unloading to the tissues is also affected by the bond strength, or *affinity*, of hemoglobin for oxygen, which is subject to physiological regulation.

Exercising muscles produce lactic acid and CO_2, which forms carbonic acid. Thus, as a muscle works harder, its pH decreases. Some of the H^+ binds to hemoglobin (hemoglobin serves as a buffer), and this causes its bond strength with oxygen to become weaker. This is known as the **Bohr effect**. A decrease in pH (increase in H^+ concentration) slightly reduces the affinity of hemoglobin for oxygen, and thus the ability of hemoglobin to load with oxygen in the lungs. However, it more greatly reduces the bond strength for oxygen as the blood travels through the exercising muscles, promoting the unloading reaction. The net effect is that *a decrease in pH causes more oxygen to be unloaded to the tissues*. The opposite occurs if the pH were to rise; hemoglobin would load with slightly more oxygen in the lungs, but would unload less as it travels through the systemic capillaries.

This is shown graphically in figure 12.19 by the three oxyhemoglobin dissociation curves, constructed at three pH values. The middle graph was constructed at a pH of 7.4, which is the normal arterial pH. The graph to the right shows the increased oxygen unloading at a lower pH, and the one to the left shows the opposite effects at a higher pH. The S-shaped (sigmoidal) nature of these

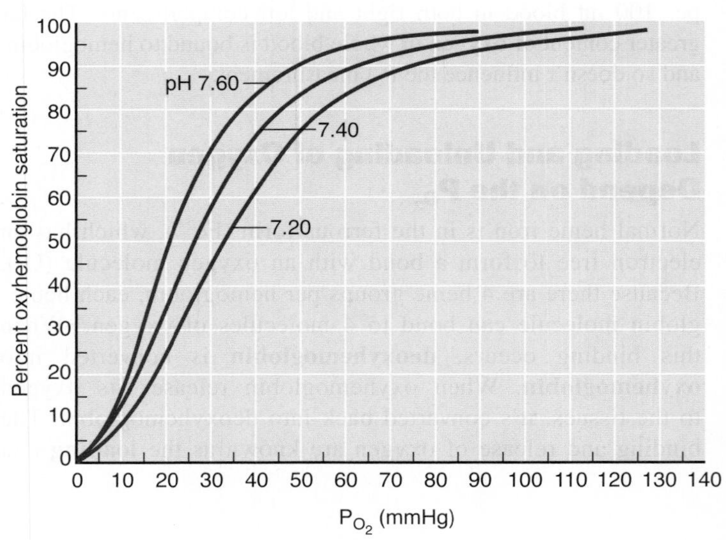

FIGURE 12.19 The effect of pH on the oxyhemoglobin dissociation curve. A decrease in blood pH (an increase in H^+ concentration) decreases the affinity of hemoglobin for oxygen at each P_{O_2} value, resulting in a "shift to the right" of the oxyhemoglobin dissociation curve. This is called the Bohr effect. A curve that is shifted to the right has a lower percent oxyhemoglobin saturation at each P_{O_2}.

curves indicates that changes in the lower range of P_{O_2} values (as occur in the systemic capillaries) greatly affect oxygen unloading, whereas changes in P_{O_2} at the higher ranges (as occur in the lungs) have a smaller effect on oxygen loading.

The affinity (bond strength) of hemoglobin for oxygen is also weakened by a higher temperature. The three curves shown in figure 12.19 were all constructed at the same temperature. If we were to construct three curves at the same pH but at three different temperatures, we would get graphs that looks much like figure 12.19. In summary, *an increase in temperature causes more oxygen to be unloaded to the tissues*. The exercising muscles are warmed by their faster metabolism, and so both the pH and temperature effects are operating to increase the unloading of oxygen as blood travels in the capillaries through these muscles.

Oxygen Unloading Is Increased by 2,3-BPG

Red blood cells produce **2,3-biphosphoglyceric acid (2,3-BPG)** as a side product of glycolysis. 2,3-BPG binds to deoxyhemoglobin, making it more stable. As a result, the unloading reaction is favored, where oxyhemoglobin is converted into deoxyhemoglobin plus oxygen. In summary, *an increase in 2,3-BPG increases the unloading of oxygen to the tissues*. You might thus predict that there would be an increased production of 2,3-BPG when there is less oxygen carried by the blood, so that the tissues would receive an adequate amount of oxygen. This does occur, because the production of 2,3-BPG is inhibited by oxyhemoglobin. As a result, a decrease in oxyhemoglobin causes an increase in 2,3-BPG, and thus in oxygen unloading.

This occurs normally as an adjustment to a high altitude, where the P_{O_2} of the air, and thus of the arterial blood, is lower than at sea level. Under these conditions, the hemoglobin will load a little less oxygen in the blood. As an example, going from sea level to an altitude of 8,000 feet decreases the arterial P_{O_2} from 100 to 74 mmHg (see table 12.2). Examining table 12.3, we can see that, at a P_{O_2} of 74, the percent oxyhemoglobin leaving the lungs would be about 93% (instead of the 97% at sea level). With less oxyhemoglobin, the red blood cells produce more 2,3-BPG. The greater amount of 2,3-BPG lowers the affinity of hemoglobin for oxygen, so that more of the oxygen carried by the blood is unloaded to the tissues (fig. 12.20). Longer-term adjustments to life at a high altitude include an elevated secretion of *erythropoietin* by the kidneys (chapter 10), which stimulates the bone marrow to increase its production of red blood cells and hemoglobin.

CLINICAL APPLICATIONS

The fetus has a different form of hemoglobin (called **hemoglobin F**) than its mother, who has adult hemoglobin (**hemoglobin A**). The reason that the fetus has a different form of hemoglobin than its mother is because hemoglobin F has a stronger affinity for oxygen than does hemoglobin A, causing oxygen to move from the mother's to the fetus's blood. Hemoglobin F has a higher affinity for oxygen because it can't bind to 2,3-BPG, which works to reduce the bond strength for oxygen in the mother's red blood cells. The fetus stops producing hemoglobin F and begins producing hemoglobin A at about week 38 of pregnancy. When the fetus switches from hemoglobin F to hemoglobin A, it destroys its old red blood cells and converts the heme groups into bile pigment, or *bilirubin* (chapter 14). Too much bilirubin can cause *jaundice* (a yellowing of the skin and mucus membranes), and jaundice produced for this reason is called **physiological neonatal jaundice**. Putting the babies under blue light converts the bilirubin into a more water-soluble derivative that they can excrete in the urine.

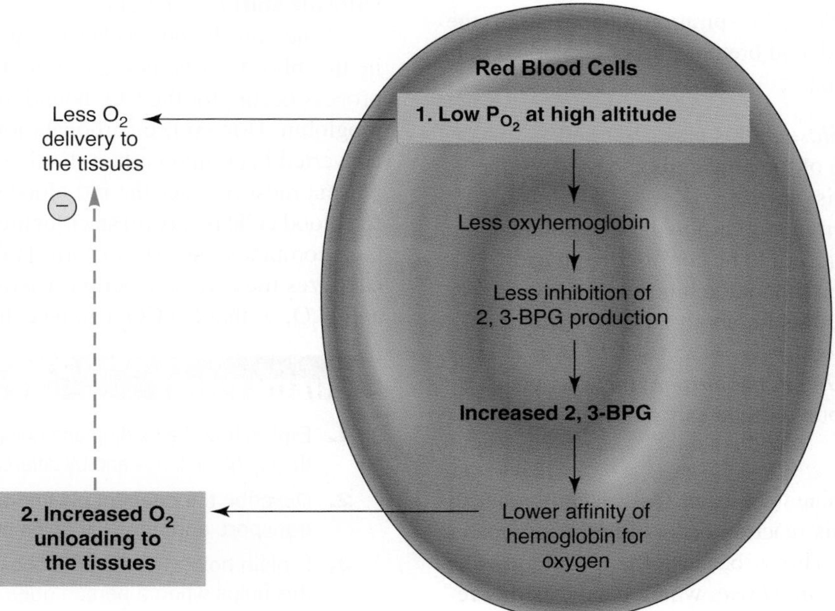

FIGURE 12.20 2,3-BPG promotes the unloading of oxygen to the tissues. Since production of 2,3-BPG is inhibited by oxyhemoglobin, a reduction in the red blood cell content of oxyhemoglobin (as occurs at the low P_{O_2} of high altitude) increases 2,3-BPG production. This lowers the affinity of hemoglobin for oxygen (decreases the bond strength), so that more oxygen can be unloaded. The dashed arrow and negative sign indicate the completion of a negative feedback loop.

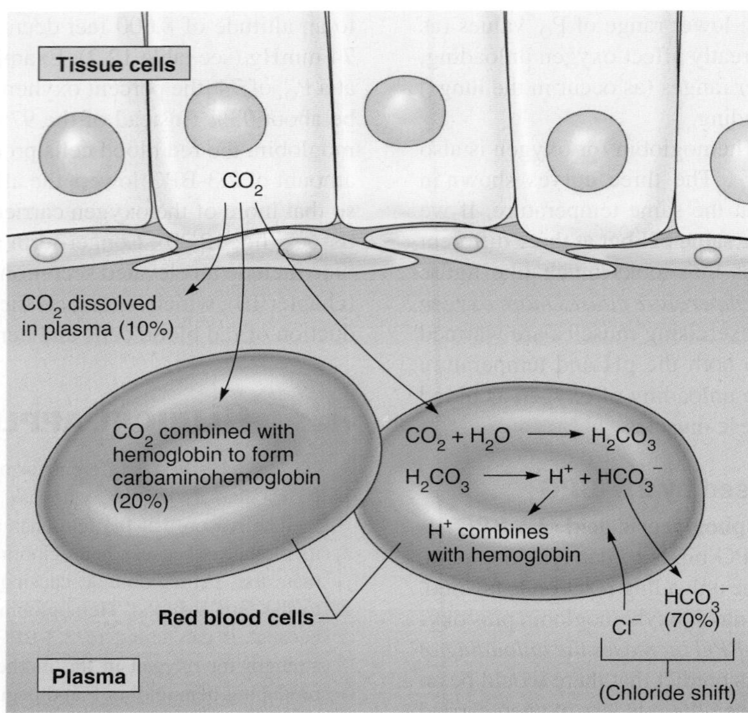

FIGURE 12.21 Carbon dioxide transport and the chloride shift. Carbon dioxide is transported in three forms: as dissolved CO_2 gas, attached to hemoglobin as carbaminohemoglobin, and as carbonic acid and bicarbonate. Percentages indicate the proportion of CO_2 in each of the forms. Notice that when bicarbonate (HCO_3^-) diffuses out of the red blood cells, Cl^- diffuses in to retain electrical neutrality. This exchange is the chloride shift.

Carbon Dioxide Is Transported Primarily as Bicarbonate

The blood transports carbon dioxide (CO_2) from the tissues, where the CO_2 is produced by aerobic cell respiration (chapter 2), to the lungs for elimination in the exhaled breath. Carbon dioxide is carried by the blood in three forms:

1. *As dissolved CO_2 in the plasma*; this produces the P_{CO_2} measurement, and about 10% of the total carbon dioxide in the blood is transported in this way.
2. *As carbaminohemoglobin*; this is carbon dioxide bound to an amino acid in hemoglobin (not to be confused with carboxyhemoglobin, where carbon monoxide binds to the iron in heme groups), and it accounts for about 20% of the total carbon dioxide in the blood.
3. *As bicarbonate ion (HCO_3^-) in the plasma*; this accounts for about 70% of the carbon dioxide carried by the blood (fig. 12.21).

Carbon dioxide can combine with water to form carbonic acid (H_2CO_3) in the plasma, but this reaction occurs to a much greater extent in the red blood cells. This is because the red blood cells contain the enzyme *carbonic anhydrase*, which catalyzes this reaction (fig. 12.21). The carbonic acid immediately dissociates to form H^+ and HCO_3^- (bicarbonate ion). Some of the H^+ binds to hemoglobin, and the rest diffuses out of the red blood cells into the plasma. There is also a concentration gradient for HCO_3^-, which diffuses out of the red blood cells. As the HCO_3^- diffuses out, an electrical gradient is established (because of the H^+ bound to hemoglobin) for chloride ion (Cl^-) to diffuse into the red blood cells. This exchange of Cl^- for HCO_3^- in the plasma is called the **chloride shift** (fig. 12.21).

When this blood reaches the lungs, some of the CO_2 dissolved in the plasma becomes gaseous and is exhaled, and a similar process occurs for the CO_2 bound to hemoglobin as carbaminohemoglobin. However, bicarbonate ion can't be exhaled; it must be converted back into carbon dioxide and water. The HCO_3^- in the plasma must re-enter the red blood cells, as Cl^- moves out of the red blood cells in a **reverse chloride shift** (fig. 12.22). The HCO_3^- then combines with H^+ to form H_2CO_3. Carbonic anhydrase now catalyzes the reverse reaction where H_2CO_3 is converted into CO_2 and H_2O, so that the CO_2 can be exhaled.

CHECK POINT

1. Explain how the loading and unloading reactions are affected by the P_{O_2} of the lungs and by different values of P_{O_2} in the tissues.
2. Describe the effect of pH and temperature on oxygen transport, and how this helps muscles during exercise.
3. Explain how 2,3-BPG affects oxygen transport, and how this helps when a person goes to a high altitude.
4. Describe the different forms in which carbon dioxide is transported by the blood.
5. Explain the events of the chloride shift and the reverse chloride shift.

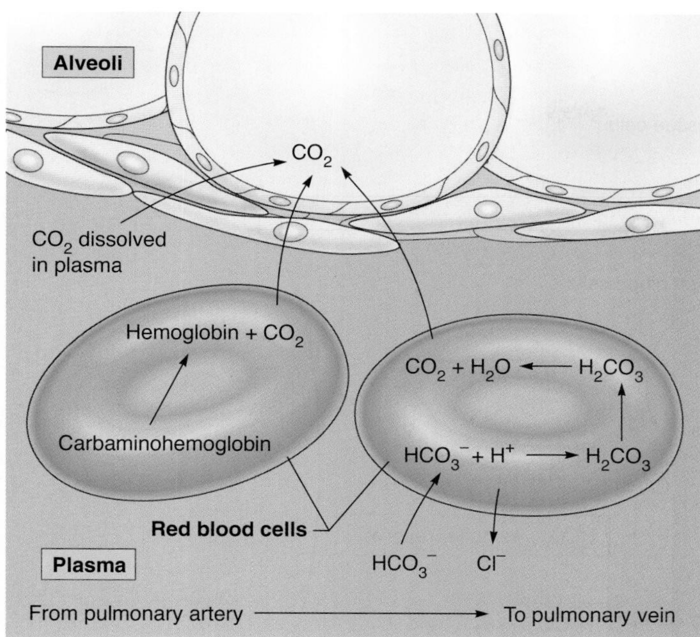

FIGURE 12.22 The reverse chloride shift in the lungs. Carbon dioxide is released from the blood as it travels through the pulmonary capillaries. A "reverse chloride shift" occurs during this time, and carbonic acid is transformed into CO_2 and H_2O. The CO_2 is eliminated in the exhaled air.

Breathing Helps Maintain Homeostasis of Blood pH

The arterial blood pH is maintained within an extremely narrow range by the lungs and kidneys. Breathing maintains a normal arterial P_{CO_2}, thereby ensuring that carbonic acid stays within the normal range. Other acids are nonvolatile (nongaseous), and cannot be exhaled. Instead, they're buffered by bicarbonate, and the maintenance of normal amounts of free bicarbonate is the responsibility of the kidneys.

Normal arterial blood has a pH of 7.40, within a range of about 7.35 to 7.45. This extremely narrow range tells us that homeostasis of blood pH, called **acid-base balance**, is very tightly controlled. Acid-base balance has two components, the *respiratory component* regulated by the lungs, and the *metabolic component* regulated by the free bicarbonate in the blood, which is in turn maintained by the kidneys.

The Lungs Are Responsible for Homeostasis of Carbonic Acid

We've already seen how breathing maintains homeostasis of the arterial P_{CO_2}. If we breathe too little (have too low a total minute volume), the concentration of CO_2 in the plasma will rise; this is *hypoventilation*. If we breathe too much (have too high a total minute volume), the concentration of CO_2 in the plasma will decrease; this is *hyperventilation*. The pH of arterial blood changes during hypoventilation and hyperventilation because of changes in the arterial P_{CO_2}. This is due to the formation of carbonic acid (H_2CO_3), which releases H^+ and lowers the pH as shown by the following equations:

$$CO_2 + H_2O \rightarrow H_2CO_3$$
$$H_2CO_3 \rightarrow H^+ + HCO_3^-$$

The regulation of acid-base balance by the lungs and breathing is known as the **respiratory component** of acid-base balance. If a person hypoventilates, so that the arterial P_{CO_2} is increased, there will be a buildup of carbonic acid that produces an increased plasma concentration of H^+. As a result, hypoventilation will cause the arterial pH to decrease, producing **respiratory acidosis**. "Respiratory" names the cause; "acidosis" refers to the lowering of the arterial pH below the normal range. This is different from "acidic," which refers to a pH below 7.000. For example, an arterial blood pH of 7.2 is acidotic but not acidic.

Conversely, if a person hyperventilates, too much CO_2 is "blown off" (the arterial P_{CO_2} will lower) and there will be less H^+ derived from carbonic acid in the blood. This raises the arterial pH, producing a **respiratory alkalosis** when the arterial pH becomes greater than 7.45. Breathing can help regulate the arterial pH because carbonic acid is converted into CO_2 gas in the lungs, as described in the previous section. Carbonic acid is thus a **volatile acid**, meaning that it can become gaseous and be eliminated in the exhaled breath.

Bicarbonate Buffers the Nonvolatile Acids

The blood contains several other acids besides carbonic acid. Lactic acid produced by exercising skeletal muscles, fatty acids released by the breakdown of fat in adipose tissue, and ketone bodies (four-carbon long acids, such as acetoacetic acid), derived from fatty acids in the liver (chapter 2), are all **nonvolatile acids** in the blood. These are always released into the blood at a certain rate, and can increase in the blood under certain conditions. For example, lactic acid is increased during exercise, and fatty acids and ketone bodies are increased whenever fat is broken down at a rapid rate (as in fasting, dieting, and uncontrolled type 1 diabetes mellitus; see chapter 8).

Being nonvolatile, these acids can't be eliminated by breathing. Instead, the bicarbonate (HCO_3^-) in the plasma must temporarily buffer the nonvolatile acids (fig. 12.23) until they can be eliminated by metabolism or other means. The nonvolatile acids release H^+ into the plasma, but this doesn't normally lower the blood pH because free bicarbonate in the plasma combines with the H^+ to form carbonic acid (H_2CO_3). This is indicated by the following formula:

$$HCO_3^- + H^+ \rightarrow H_2CO_3$$

This reaction lowers the concentration of H^+ in the plasma, thereby preventing the nonvolatile acids from causing acidosis. However, the amount of free bicarbonate is also reduced by this reaction. If the free bicarbonate disappears, the nonvolatile acids

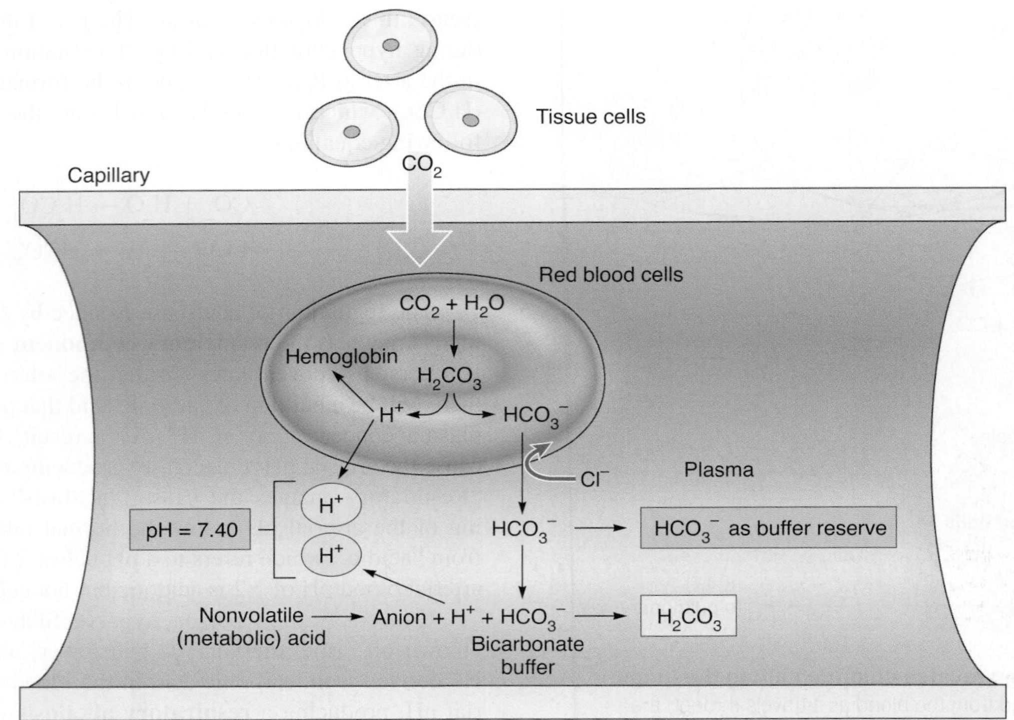

FIGURE 12.23 The effect of bicarbonate on blood pH. Bicarbonate released into the plasma from red blood cells buffers the H^+ produced by the ionization of nonvolatile acids (lactic acid, fatty acids, ketone bodies, and others). Binding of H^+ to hemoglobin also promotes the unloading of O_2.

will release unbuffered H^+, and the arterial pH will lower. When this lowers the blood pH below 7.35, the condition is described as **metabolic acidosis**. Metabolic acidosis doesn't normally occur, and so indicates pathological conditions. For example, it may be produced in uncontrolled type 1 diabetes mellitus, due to the overproduction of acidic ketone bodies (chapter 8). Metabolic acidosis may also be produced by excessive diarrhea, due to loss of bicarbonate (from pancreatic juice; chapter 14) that would otherwise be absorbed into the blood. This metabolic acidosis is produced because, with loss of bicarbonate, there is an increased ratio of H^+ to HCO_3^-.

Conversely, **metabolic alkalosis** occurs when there is an increased ratio of HCO_3^- to H^+ that produces an arterial pH greater than 7.45. This can be caused by taking too much exogenous bicarbonate (as in antacids), but is more commonly produced by excessive vomiting. This is because gastric juice is acidic (chapter 14), and vomiting loses the H^+ in gastric juice that otherwise would be returned to the blood. This raises the ratio of HCO_3^- to H^+, producing metabolic alkalosis. Metabolic acidosis and alkalosis constitute the **metabolic component** of the acid-base balance. Because the kidneys regulate the concentration of free bicarbonate in the plasma (chapter 13), the kidneys are responsible for maintaining homeostasis of the metabolic component of acid-base balance.

CLINICAL APPLICATIONS

Hospital personnel often deal with more complicated acid-base disturbances. A person with uncontrolled type 1 diabetes mellitus, for example, may produce too many ketone bodies and develop ketoacidosis (a type of metabolic acidosis). This lowers the arterial blood pH below 7.35, but not by as much as it might. Remember, the peripheral chemoreceptors (the aortic and carotid bodies) are stimulated by a fall in arterial pH. This causes breathing in excess of that required to maintain constancy of the arterial P_{CO_2}—in other words, it causes hyperventilation. Hyperventilation then lowers the arterial P_{CO_2} to produce a respiratory alkalosis. In summary, the person in this example would have a metabolic acidosis (ketoacidosis) that is partially compensated by a respiratory alkalosis.

CHECK POINT

1. Distinguish between the respiratory and metabolic components of acid-base balance.
2. Describe the respiratory component of acid-base balance, and how this relates to the regulation of breathing.
3. Describe the metabolic component of acid-base balance, and explain how metabolic acidosis and metabolic alkalosis may be produced.

Physiology in Health and Disease

People with pulmonary disorders frequently complain of **dyspnea**, which is a feeling of "shortness of breath." The dyspnea, wheezing, and other symptoms of **asthma** are produced by increased resistance to airflow through the bronchioles (asthma is an obstructive pulmonary disorder, as discussed previously). The increased resistance to air flow is caused by bronchoconstriction and inflammation that may be provoked by allergic reactions (chapter 11). Asthma may be treated on a sustained basis with glucocorticoid drugs (related to cortisol) that inhibit inflammation, thereby preventing or reducing the severity of "attacks." New drugs (such as Singulair) that block the action of leukotrienes, a type of regulatory fatty acid (related to prostaglandins) that promote asthma, are now also available for this purpose. Acute asthma attacks are commonly treated with inhaled drugs (such as Albuterol) that stimulate the β_2-adrenergic receptors (a type of receptor for epinephrine and norepinephrine; see chapter 6) that promote dilation of the bronchioles.

Alveolar tissue is destroyed in **emphysema**, resulting in fewer but larger alveoli (see fig. 12.8). The loss of alveoli reduces the ability of the bronchioles to remain open during expiration, causing *air trapping* during expiration when the bronchioles collapse. The most common cause of emphysema is cigarette smoking, which indirectly causes different protein-digesting enzymes to destroy the lung tissue. The loss of alveoli and air trapping reduces gas exchange, so that people with emphysema have difficulty in both oxygenating the blood and eliminating carbon dioxide. Because of this, people with emphysema must often breathe from an oxygen tank.

Chronic obstructive pulmonary disease (COPD) is characterized by chronic inflammation with narrowing of the airways and destruction of the alveolar walls. Included in the COPD category is *emphysema* and *chronic obstructive bronchiolitis*, which refers to fibrosis and obstruction of the bronchioles. The condition results in a faster age-related decline in the FEV_1 (discussed previously). COPD differs from asthma in that, unlike asthma, COPD is not reversible with the use of a bronchodilator such as Albuterol. Also unlike asthma, COPD is not helped much by inhaled glucocorticoids (drugs related to hydrocortisone). The vast majority of people with COPD are smokers, and stopping smoking once COPD has begun does not seem to stop its progression. In addition to the pulmonary problems directly caused by COPD, this condition increases the risk of pneumonia, pulmonary emboli (traveling blood clots), and heart failure. Patients with COPD may develop *cor pulmonale*—pulmonary hypertension with eventual failure of the right ventricle. COPD is now the fifth leading cause of death in the United States, and scientists have estimated that by 2020 it will become the third leading cause of death worldwide.

Physiology in Balance

Respiratory System

Working together with . . .

Cardiovascular System

The blood transports oxygen to all body cells . . . p. 240

Nervous System

The medulla oblongata contains a rhythmicity center that controls breathing, and sensory neurons from chemoreceptors also participate . . . p. 113

Endocrine System

Erythropoietin, secreted by the kidneys, stimulates the bone marrow to increase the red blood cell count . . . p. 186

Muscular System

The diaphragm and other muscles that insert on the thoracic cavity are needed for ventilation . . . p. 215

Urinary System

The kidneys filter and reabsorb bicarbonate to regulate the plasma bicarbonate . . . p. 324

Homeostasis Revisited

Topic	Set Point	Integrating Center	Sensors	Effectors
How homeostasis of arterial oxygen and carbon dioxide is maintained by the peripheral chemoreceptors when a person hypoventilates	Plasma pH	Respiratory control center in the medulla oblongata	Aortic and carotid bodies	Somatic motor neurons stimulate the diaphragm and other respiratory muscles to contract, causing breathing.
How homeostasis of arterial oxygen and carbon dioxide is maintained by the central chemoreceptors when a person hypoventilates	pH of brain interstitial fluid	Respiratory control center in the medulla oblongata	Central chemoreceptors in the medulla oblongata	Somatic motor neurons stimulate the diaphragm and other respiratory muscles to contract, causing breathing.
How the Bohr effect helps sustain aerobic respiration by exercising skeletal muscles	Oxygen available for aerobic respiration in muscle fibers.	Red blood cells	Hemoglobin molecules	The pH of exercising skeletal muscles decreases, reducing the affinity of hemoglobin for oxygen, thereby increasing oxygen unloading to the muscle fibers.
How the regulation of breathing operates to partially restore homeostasis of arterial pH when a person has a metabolic acidosis	Arterial pH	Respiratory control center in medulla oblongata	Aortic and carotid body chemoreceptors	Somatic motor neurons to respiratory muscles stimulate hyperventilation, which induces a respiratory alkalosis to partially offset the metabolic acidosis.
How 2,3-BPG in the RBCs helps sustain aerobic respiration when a person goes to a high altitude	Oxygen available for aerobic respiration.	Red blood cells	Enzyme that promotes production of 2,3-BPG	Reduced oxyhemoglobin stimulates the production of 2,3-BPG, which reduces the affinity of hemoglobin for oxygen, thereby increasing the unloading of oxygen to the tissues.

Summary

Breathing Replenishes Air for Gas Exchange in the Alveoli 298

- The lungs contain a conducting zone and a respiratory zone, where gas exchange occurs; gas exchange between the air and blood is produced by diffusion of oxygen and carbon dioxide across the walls of the alveoli.
- The lungs fill the thoracic cavity and are covered with a membrane called the visceral pleura, which is against the parietal pleura lining the wall of the thorax.
- The intrapulmonary pressure becomes subatmospheric during inhalation and greater than the atmospheric pressure during exhalation; the intrapleural pressure is normally less than the intrapulmonary pressure, so that the lungs are stuck to the chest wall.
- Lung compliance relates to its ability to become distended; this is opposed by the lung elasticity, which refers to its ability to recoil after being stretched, and by the surface tension within the alveoli.
- The surface tension within the alveoli tends to make them collapse; this is normally prevented by pulmonary surfactant, which lowers the surface tension.
- Inspiration (inhalation) is produced mostly by contraction of the diaphragm and external intercostal muscles, which expands the thoracic cavity and lungs; this decreases the intrapulmonary pressure below the atmospheric pressure.
- Quiet (unforced) expiration is produced by relaxation of the diaphragm and external intercostal muscles, which allows the elastic recoil of the thorax and lungs to decrease lung volume and increase the intrapulmonary pressure above the pressure of the atmosphere.
- Spirometry is a technique that measures lung volumes and capacities; a low vital capacity indicates restrictive lung disorders, including emphysema.
- Obstructive lung disorders, such as asthma, are detected by a low forced expiratory volume in 1 second (FEV_1); chronic obstructive pulmonary disease (COPD), including emphysema, is both obstructive and restrictive.

Breathing Maintains Homeostasis of the Plasma O_2 and CO_2 305

- The atmospheric pressure is the sum of the partial pressure of its individual gases; the partial pressure of a gas in a dry gas mixture is calculated by multiplying its percent composition by the total pressure of the gas.
- The atmospheric pressure is 760 mmHg (or torr) at sea level, which is also called 1 atmosphere pressure; this decreases with altitude.
- The concentration of a gas that can be dissolved in a fluid, such as plasma, is directly proportional to the partial pressure of the gas in contact with the fluid, and is measured by the partial pressure.
- Breathing maintains homeostasis of the partial pressures of carbon dioxide (P_{CO_2}) and oxygen (P_{O_2}) in the arterial blood.
- The arterial P_{CO_2} influences the arterial pH, because carbon dioxide combines with water to form carbonic acid, which releases H^+ into the plasma to lower its pH.
- The peripheral chemoreceptors, including the aortic and carotid bodies, are stimulated by a fall in arterial pH, and stimulate the rhythmicity center in the medulla oblongata to cause breathing.
- The central chemoreceptors in the medulla oblongata are stimulated by a fall in the pH of their surrounding interstitial fluid, caused by carbon dioxide diffusing across the blood-brain barrier.
- The chemoreceptor reflex functions to maintain a normal arterial P_{CO_2}, because this value varies inversely with the total minute volume.
- Hyperventilation occurs if a person breathes too much (has too high a total minute volume), so that the arterial P_{CO_2} falls and the arterial pH rises.
- Hypoventilation occurs if a person breathes too little (has too low a total minute volume), so that the arterial P_{CO_2} rises and the arterial pH falls.
- Breathing is not usually stimulated directly by a low arterial P_{O_2}; however, at moderate altitudes this can increase the chemoreceptor sensitivity to a rise in the arterial P_{CO_2}, and under more severe conditions can stimulate the carotid bodies directly, providing a "hypoxic drive" for breathing.

The Blood Transports Oxygen and Carbon Dioxide 310

- Deoxyhemoglobin loads with oxygen in the lungs to form oxyhemoglobin, which unloads its oxygen as the blood passes through the capillaries of the systemic circulation.
- At a normal P_{O_2} of 100 mmHg, arterial blood contains about 97% oxyhemoglobin; this is also called the oxyhemoglobin percent saturation.
- At rest, venous blood has a P_{O_2} of about 40 mmHg and a percent oxyhemoglobin of 75%, indicating that 22% (a little over one-fifth) of the oxyhemoglobin unloaded its oxygen to the tissues.
- During exercise, the P_{O_2} of the blood passing through the muscles is lowered because of increased metabolism, so that more of the oxyhemoglobin unloads its oxygen to the muscles.
- A decrease in pH causes more of the oxyhemoglobin to unload its oxygen to the tissues; this is known as the Bohr effect, and it operates to increase oxygen unloading to exercising muscles.
- An increase in temperature also weakens the bond strength (affinity) between hemoglobin and oxygen, so that more oxygen is unloaded to muscles warmed during exercise.
- 2,3-BPG (2,3-biphosphoglyceric acid) in the red blood cells weakens the affinity of hemoglobin for oxygen, and thus more oxygen is unloaded to the tissues; the production of 2,3-BPG is inhibited by oxyhemoglobin, so that a lowering of oxyhemoglobin causes more 2,3-BPG production.
- Because there is a lower P_{O_2} and percent oxyhemoglobin at a high altitude, there is an increased amount of 2,3-BPG in the red blood cells, and so a greater oxygen unloading to the tissues.
- Most of the CO_2 in the blood is carried as bicarbonate (HCO_3^-); this occurs when CO_2 and H_2O combine to form carbonic acid (H_2CO_3) in the red blood cells as they pass through the tissue capillaries, in a reaction catalyzed by carbonic anhydrase.
- The H_2CO_3 in the red blood cells dissociates to form HCO_3^- and H^+; the diffusion of HCO_3^- out of the red blood cells is accompanied by the inward diffusion of chloride (Cl^-), in an exchange called the chloride shift.
- Lesser amounts of CO_2 are carried dissolved in the plasma and bound to an amino acid in hemoglobin, forming carbaminohemoglobin; these release their CO_2 as the blood passes through the capillaries of the lungs.
- When blood passes through pulmonary capillaries, HCO_3^- in the plasma enters the red blood cells as Cl^- leaves, in a reverse chloride shift; this allows H_2CO_3 to be formed, which is then converted into CO_2 and H_2O so that the CO_2 can be exhaled.

Breathing Helps Maintain Homeostasis of the Blood pH 315

- Breathing regulates the arterial P_{CO_2}, and thus the amount of carbonic acid in the blood; this is called the respiratory component of acid-base balance.
- Hypoventilation raises the P_{CO_2} and thus lowers the pH to produce respiratory acidosis; hyperventilation lowers the P_{CO_2} and thus raises the pH to produce respiratory alkalosis.
- Carbonic acid is a volatile acid, and so can be regulated by breathing; lactic acid, fatty acids, and ketone bodies are nonvolatile acids, which must be buffered and then eliminated by metabolism and other means.
- Buffering of H^+ occurs when H^+ combines with HCO_3^- to form H_2CO_3 in the plasma.
- The metabolic component of acid-base balance refers to the buffering of nonvolatile acid by free bicarbonate in the plasma; metabolic acidosis occurs when these acids release too many H^+ and reduce the free bicarbonate buffer, or when bicarbonate buffer is lost due to excessive diarrhea.
- Metabolic alkalosis occurs when too much H^+ is lost, as in excessive vomiting, or when there is an abnormal increase in bicarbonate buffer (from exogenous sources).
- The kidneys regulate the availability of free bicarbonate in the plasma, and so are said to regulate the metabolic component of acid-base balance, while the lungs regulate the respiratory component.

Review Activities

Objective Questions: Test Your Knowledge

1. Which of the following is not a part of the conducting zone of the lungs?
   a. Trachea
   b. Terminal bronchioles
   c. Respiratory bronchioles
   d. Primary bronchi

2. Which of the following statements about gas exchange and alveoli is false?
   a. The distance separating the air and blood averages 100 μm.
   b. There are 2 cells separating the air and the blood.
   c. There are about 300 million alveoli.
   d. Gas exchange occurs by diffusion.

3. According to Boyle's law, the pressure of a gas
   a. decreases as its volume decreases.
   b. increases as its volume increases.
   c. decreases as its volume increases.
   d. is independent of its volume.
4. The intrapulmonary pressure is normally always
   a. greater than the atmospheric pressure.
   b. greater than the intrapleural pressure.
   c. less than the atmospheric pressure.
   d. less than the intrapleural pressure.
5. Which of the following statements about an intrapulmonary pressure of -3 mmHg is true?
   a. This indicates a pressure of 763 mmHg.
   b. This pressure causes air to leave the lungs.
   c. This pressure is produced when the lung volume decreases.
   d. This is the pressure at the beginning of a quiet (unforced) inspiration.
6. At the end of expiration, just before the next inspiration, the intrapulmonary pressure is
   a. 0 mmHg.
   b. -3 mmHg.
   c. $+3$ mmHg.
   d. variable, depending on the strength of the expiration.
7. Quiet (unforced) expiration is produced primarily by
   a. contraction of the internal intercostal muscles.
   b. contraction of the external intercostal muscles.
   c. contraction of the diaphragm.
   d. relaxation of muscles and elastic recoil of the thorax and lungs.
8. The surface tension inside of the alveoli
   a. increases the lung compliance.
   b. increases the lung elasticity.
   c. promotes collapse of the alveoli.
   d. produces a greater pressure in larger than in smaller alveoli.
9. Pulmonary surfactant consists of particular proteins combined with
   a. phospholipids.
   b. steroids.
   c. carbohydrates.
   d. nucleic acids.
10. The maximum amount of air that can be expired after a maximum inspiration is the
    a. tidal volume.
    b. vital capacity.
    c. functional residual capacity.
    d. total minute volume.
11. Which of the following statements about a person with emphysema is false?
    a. There is both a restrictive and obstructive pulmonary disease.
    b. There is an abnormal vital capacity but a normal FEV_1.
    c. The person has chronic obstructive pulmonary disease (COPD).
    d. The disease will get worse over time, even if the person quits smoking.
12. If a gas tank contains dry air with 50% oxygen at 1 atmosphere pressure, the partial pressure of oxygen in that tank is:
    a. 760 mmHg.
    b. 100 mmHg.
    c. 380 mmHg.
    d. 150 mmHg.
13. The P_{O_2} (partial pressure of oxygen) of arterial blood is directly proportional to the
    a. amount of oxygen contained in the red blood cells.
    b. total oxygen content of the blood.
    c. the concentration of oxygen dissolved in the plasma.
    d. solubility constant for oxygen.
14. Blood gas measurements to assess pulmonary function involve measuring the P_{O_2} and P_{CO_2} in a
    a. systemic artery.
    b. systemic vein.
    c. pulmonary artery.
    d. pulmonary vein.
15. In the normal control of breathing, the aortic and carotid bodies respond most directly to a
    a. rise in arterial pH.
    b. rise in arterial P_{CO_2}.
    c. fall in arterial P_{CO_2}.
    d. fall in arterial pH.
16. Which of the following statements about the central chemoreceptors is false?
    a. They are located in the medulla oblongata.
    b. They respond to a fall in the pH of brain interstitial fluid.
    c. They respond to H^+ crossing from the blood into the brain.
    d. They indirectly respond to a rise in CO_2 crossing the blood-brain barrier.
17. A person who is hyperventilating has a
    a. high arterial P_{CO_2}.
    b. low arterial P_{CO_2}.
    c. high arterial P_{O_2}.
    d. low arterial P_{O_2}.
18. At a blood P_{O_2} of 100 mmHg, which of the following occurs?
    a. Deoxyhemoglobin is converted into oxyhemoglobin in the lungs.
    b. Deoxyhemoglobin combines with oxygen in the systemic capillaries.
    c. Oxyhemoglobin unloads its oxygen in the systemic capillaries.
    d. Oxyhemoglobin is converted into deoxyhemoglobin in the lungs.
19. A percent oxyhemoglobin saturation of 97% is
    a. produced only be hyperventilating.
    b. a normal value for the systemic veins at rest.
    c. a normal value for the systemic arteries.
    d. produced only during exercise.
20. The unloading of oxygen to the tissue is increased by all of the following except
    a. a lowering of the P_{O_2} within the muscles that are exercising.
    b. an increased temperature in exercising muscles.
    c. an increased red blood cell content of 2,3-BPG.
    d. an increase in the pH within exercising muscles.
21. When a person goes to a high altitude, all of the following are true except
    a. There is a lower arterial P_{O_2} and percent oxyhemoglobin saturation.
    b. There is an increased production of 2,3-BPG.
    c. The affinity of hemoglobin for oxygen is increased as blood goes through the tissues.
    d. There is an elevated secretion of erythropoietin by the kidneys.

22. How is most of the carbon dioxide carried by the blood?
    a. Its dissolved in the plasma.
    b. Its bound to hemoglobin as carbaminohemoglobin.
    c. Its bound to hemoglobin as carboxyhemoglobin.
    d. Its converted into bicarbonate ion.

23. Which of the following statements about the events associated with the chloride shift is false?
    a. It occurs as blood passes through the pulmonary capillaries.
    b. Cl^- enters the red blood cells as HCO_3^- leaves.
    c. carbonic anhydrase converts CO_2 and H_2O into H_2CO_3.
    d. HCO_3^- and H^+ are formed within the red blood cells.

24. Which of the following statements about carbonic acid is false?
    a. It is formed in the tissue capillaries.
    b. It is increased when the arterial P_{CO_2} increases.
    c. It is eliminated by exhaling, and so is regulated by ventilation.
    d. It is a nonvolatile acid.

25. Which of the following statements about the metabolic component of acid-base balance is false?
    a. It is regulated by the kidneys.
    b. A metabolic acidosis begins when the blood pH falls below 7.000.
    c. Buffering of H^+ by bicarbonate forms H_2CO_3 in the plasma.
    d. Elimination of H^+ by excessive vomiting can cause a metabolic alkalosis.

Essay Questions 1: *Test Your Understanding*

1. Distinguish between the conducting and respiratory zones of the lungs, indicating the components of each.
2. Describe gas exchange in the lungs, explaining the anatomical reasons why it is so efficient.
3. Explain how the lungs become stuck to the chest wall.
4. Using Boyle's law, describe how air enters and leaves the lungs.
5. Explain how a pneumothorax is produced.
6. Describe the effect of surface tension on the lungs, and the role of pulmonary surfactant. Also, describe what happens when a premature baby is born without pulmonary surfactant.
7. Which muscles are used for quiet inspiration, and how do they cause a person to inhale? How is quiet expiration achieved? Describe the changing intrapulmonary pressures in the breathing cycle.
8. Distinguish between restrictive and obstructive pulmonary disorders, explaining how these affect pulmonary function tests.
9. Explain how the partial pressure of a gas is calculated, and how the partial pressure of the gas in a fluid is measured.
10. What is the physiological significance of arterial P_{O_2} and P_{CO_2} measurements? What might cause these measurements to become abnormal?
11. Explain how the atmospheric pressure, and partial pressure of oxygen, change as you go to a higher altitude.
12. Step by step, describe the chemoreceptor reflex control of breathing. Include both the peripheral and the central chemoreceptors in your answer.
13. Define hyperventilation, hypoventilation, and normal breathing with reference to the arterial P_{CO_2}.
14. Describe how, and when, breathing is affected by a low arterial P_{O_2}.
15. What are the loading and unloading reactions, and where in the body does each occur?
16. Compare the effects of P_{O_2} on the unloading reaction as blood passes through resting and exercising skeletal muscles.
17. Describe the Bohr effect, and the effect of temperature, on the transport of oxygen to muscles during exercise.
18. What causes the red blood cell production of 2,3-BPG to increase, how does this molecule affect hemoglobin, and what benefit does it confer on oxygen transport?
19. Describe the different forms in which carbon dioxide is transported by the blood. Include a step-by-step description of the chloride shift in your answer.
20. What occurs during the reverse chloride shift, and where in the body does this occur?
21. Distinguish between volatile and nonvolatile acids, and explain how the respiratory component of acid-base balance is regulated. Include an explanation of respiratory acidosis and alkalosis in your answer.
22. Describe the metabolic component of acid-base balance, and explain how it is regulated.

Essay Questions 2: *Test Your Analytical Ability*

1. What keeps the lungs stuck to the chest wall? Using the principles you describe, explain how a lung could collapse if the chest is punctured.
2. A person in a hospital who is being ventilated artificially from a gas tank inhales because of "positive pressure" from the tank. Explain what this might mean, and how it differs from the way we normally inhale.
3. The intrapulmonary pressure was described as being −3 mmHg at the start of a normal inspiration. Would it make any difference if this occurred at sea level or an altitude of 8,000 feet? Explain.
4. People with emphysema can have very tight thoracic muscles, gradually causing a "barrel chest," with bulging sternocleidomastoid muscles. What might cause these changes?
5. Explain the significance of surface tension forces in lung mechanics, and of pulmonary surfactant. What molecules comprise the surfactant, and what properties of these molecules allow them to have this function (hint: see chapter 1).
6. A person with emphysema has both restrictive and obstructive disease, and has an increased functional residual capacity. Explain these terms and the reasons why this statement is true.
7. Air that reaches the alveoli is saturated with water vapor, which dilutes the contributions of the other gases to the total pressure. Given that the partial pressure of water vapor is 47 mmHg, calculate the P_{O_2} of sea level air when water vapor pressure is considered.
8. What factors affect the concentration of oxygen dissolved in water? What will be the P_{O_2} of a bubbling brook at an altitude where the atmospheric pressure is one-third less than at sea level? Explain.
9. Given that champagne bubbles contain carbon dioxide, explain why (a) no bubbles are seen before the cork is popped; and (b) the cork of a champagne bottle pops out when it's loosened, after which time the bubbles appear. Then, use your answer to explain why divers get "the bends" (decompression sickness) if they ascend too rapidly.
10. Describe how the arterial P_{CO_2} is affected by breathing. Given that there is a few minutes reserve supply of oxygen in the blood (attached to hemoglobin in the red blood cells), suggest reasons why the chemoreceptor control of breathing is normally regulated by the P_{CO_2} rather than by the P_{O_2}.
11. What directly stimulates the peripheral and central chemoreceptors? How is that stimulus produced in each case? Would the peripheral and central chemoreceptors respond to an abnormally low blood pH produced by ketone bodies? Explain.
12. Suppose a person goes to a sufficiently high altitude to stimulate a hypoxic drive for breathing. Explain how this is produced, and what effect it would have on the arterial P_{CO_2} and pH.

13. Given your answer to the previous question, explain how this would affect (a) the total minute volume; and (b) the loading and unloading reactions of oxygen transport. What negative and positive effects would these changes have on a person at the high altitude?
14. When a person is exercising moderately, the increased total minute volume can be described more accurately as hyperpnea than as hyperventilation. Explain why this is so. What would happen if an exercising person did hyperventilate?
15. A person at very high altitudes can hyperventilate greatly. What produces this response, and how does it affect blood pH? Given this change, how would the loading and unloading reactions be affected? Explain.
16. What stimulates the production of 2,3-BPG? Given this, explain how 2,3-BPG can help partially compensate for anemia.
17. Explain the relationship between carbon dioxide transport and the homeostasis of blood pH. Include the regulation of both the respiratory and metabolic components of acid-base balance in your answer.
18. How does excessive vomiting, or taking too much exogenous bicarbonate as an antacid, cause metabolic alkalosis? How could the kidneys compensate for metabolic alkalosis? (Hint: The kidneys produce urine by filtering the blood plasma and then reabsorbing selected ions and molecules to varying degrees.)

Web Activities

ARIS™ For additional study tools, go to: www.aris.mhhe.com. Click on your course topic and then this textbook's author/title. Register once for a semester's worth of interactive activities and practice quizzing to help boost your grade!

13 CHAPTER

Urinary System

Chapter Outline

Homeostasis

The primary function of the kidneys is to maintain homeostasis of the blood volume (and thus the blood pressure) and of the chemical composition of the blood. By excreting nitrogen-containing waste products (urea and creatinine), the kidneys normally prevent their excessive accumulation in the blood. The kidneys reabsorb molecules (such as glucose) and ions (such as Na^+, K^+, Ca^{2+}, and others) needed in the blood, and thereby help maintain normal concentrations of these substances in the plasma. Kidney function is also needed for homeostasis of the blood pH, because H^+ is excreted in the urine while bicarbonate is normally reabsorbed.

Clinical Investigation

Helen had been losing sleep lately because of frequent urination, and felt more fatigued than usual. She also seemed to be constantly thirsty, even though she drank a lot of water. When she went on a camping trip and ran out of her usual supply of water, she became intensely thirsty and dizzy, whereas her friends suffered no ill effects with the same water intake. Her friends took her to a nearby hospital, where she was found to have low blood pressure (a mean arterial pressure under 80 mmHg) and elevated plasma creatinine. After giving her fluids and rest, a normal arterial blood pressure and plasma creatinine level were restored. A couple of weeks later, Helen provided her physician with a 24-hour urine sample. He told Helen that she produced a large volume (5 L/day) of urine that was abnormally dilute (about 100 mOsM). Helen asked if that meant that she was diabetic. The physician responded that her urine lacked glucose; however, she might have a certain type of diabetes, and he would like to measure a pituitary hormone to test a possible cause of her symptoms.

What might account for Helen's abnormally high volume of dilute urine? What type of diabetes does the physician suspect, and which hormone does he want to measure? How did this cause Helen to drink lots of water and yet still be thirsty and more fatigued than usual? Why was her plasma creatinine elevated after her camping trip when she was deprived of water and her blood pressure was low?

But if homeostasis is not maintained . . .

A virtual cadaver dissection experience

Urine Is Formed by Nephrons in the Kidneys

Urine formed by the kidneys is channeled through the ureters to the urinary bladder, and out of the urinary bladder in the urethra. The kidneys contain microscopic functional units, the nephrons, which produce urine by filtering plasma out of capillary beds known as glomeruli. The glomerular filtrate then enters the nephron tubules, which contain portions located in the renal cortex and renal medulla. Interactions between the tubules and surrounding blood vessels modify the glomerular filtrate to form the final urine.

The physiology of the kidneys involves processes studied in previous chapters: membrane transport processes (chapter 3), regulation by autonomic neurons (chapter 6) and hormones (chapter 8), capillary filtration and blood volume regulation (chapter 10), and others. However, knowledge of renal (kidney) physiology will also form a basis for future study of other systems, including the digestive system. This is because the kidneys and liver (part of the digestive system; chapter 14) both "clear" the blood of particular molecules—the kidneys do this by excreting these molecules in the urine, whereas the liver puts these molecules into the bile, so they can be excreted in the feces. The kidneys and liver are joined in this task by the lungs, which clear the blood of carbon dioxide in the exhaled breath (chapter 12). Thus, your study of renal physiology in this chapter will support and strengthen your understanding of other organ systems as you add yet another layer to your knowledge pyramid.

The **kidneys** are located posteriorly in the abdominal cavity, on either side of the vertebral column. Each kidney is about the size of a fist, and both produce urine that drains into long ducts called **ureters**, which then drain into the **urinary bladder**. Urine is excreted from the urinary bladder through a single duct, the **urethra**. The kidneys and their associated structures comprise the **urinary system** (fig. 13.1).

The urinary bladder has a muscular wall, the **detrusor muscle**. This muscle can generate action potentials in response to stretch, but it also receives parasympathetic axons, which release acetylcholine (ACh) to stimulate contraction and thus emptying of the bladder. Two sphincter muscles surrounding the urethra control the exit of urine from the bladder. The upper *internal urethral sphincter* is composed of smooth muscle; the lower *external urethral sphincter* is composed of striated muscle. The relaxation of these sphincters is required for urination, also called **micturition**.

Micturition is controlled by a reflex center in the sacral regions of the spinal cord. Filling of the urinary bladder activates stretch receptors, which send action potentials to the **micturition center**. The micturition center then activates parasympathetic neurons, which produce (1) contractions of the detrusor muscle; and (2) relaxation of the internal urethral sphincter. A person has

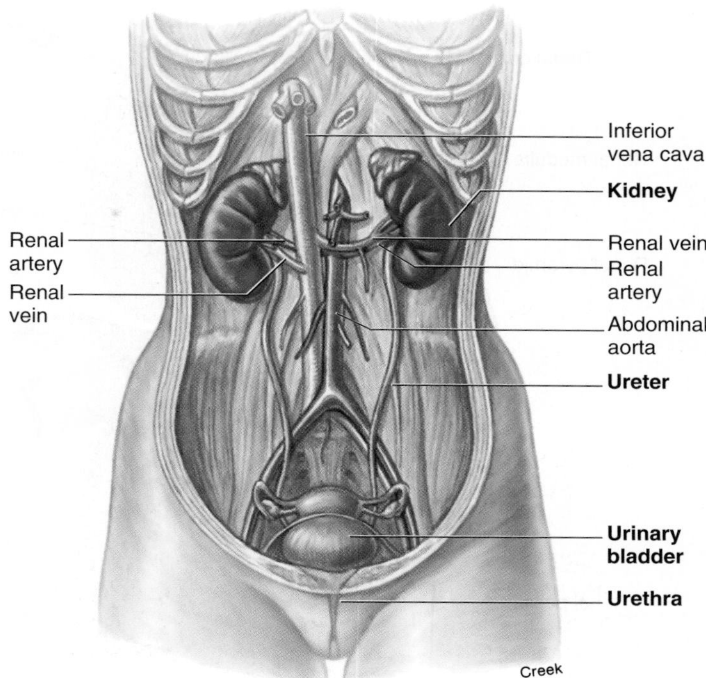

FIGURE 13.1 The organs of the urinary system. The urinary system of a female is shown; that of a male is the same, except that the urethra runs through the penis.

an urge to urinate at this point, but still retains conscious control over the external urethral sphincter. When the person consciously wants to urinate, descending motor tracts to the micturition center produces inhibition of the somatic motor neurons to the external urethral sphincter. This allows the external urethral sphincter to relax so that urination can occur.

Clinical Applications

Kidney stones are composed of crystals (of calcium oxalate, calcium phosphate, and other substances) and proteins that grow until they break loose and pass into the urine collection system. When a stone breaks loose and passes into a ureter, it produces steadily increasing pain, which can become so intense that the patient requires narcotic drugs. The calcium and other substances in kidney stones are normally present in urine, but they become supersaturated and crystalize to form stones for reasons not currently understood. The stones may be removed surgically or broken up by a noninvasive procedure called *shock-wave lithotripsy*.

Nephrons Consist of Tubules and Associated Blood Vessels

A section of a kidney shows two distinct regions: an outer **renal cortex**, and an inner **renal medulla** (fig. 13.2). The medulla contains **renal pyramids**, with their narrower portions projecting into cup-shaped structures called **calyces** (singular, *calyx*). The tip of each pyramid releases urine into a *minor calyx*, and several of

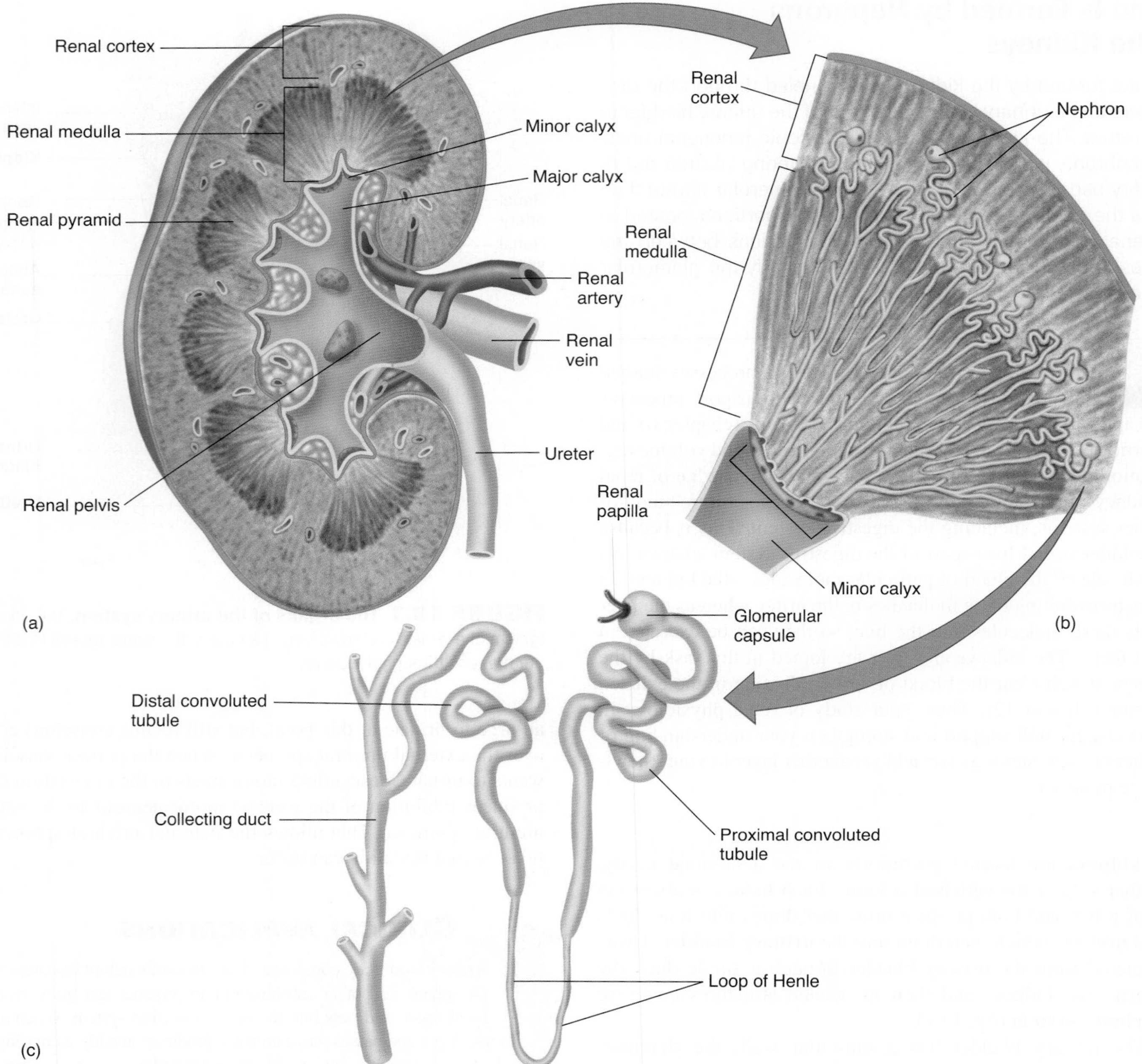

FIGURE 13.2 The structure of a kidney. The figure depicts (*a*) a coronal section of a kidney and (*b*) a magnified view of the contents of a renal pyramid. (*c*) A single nephron tubule, microscopic in actual size, is shown isolated.

these drain into a *major calyx.* The major calyces then merge to form a cavity in each kidney, the **renal pelvis**. This is the funnel-shaped origin of the ureter, so that urine can drain from the pelvis into the ureter of each kidney.

The renal cortex and medulla contain more than a million microscopic functioning units known as **nephrons**. Each nephron consists of small tubes, or **nephron tubules**, and their associated blood vessels. The nephron tubules receive a plasma filtrate (discussed shortly) and then modify this filtrate to form the final urine. Each tubule is highly twisted, or convoluted, and has several distinct parts (fig. 13.3):

1. The **glomerular** (or **Bowman's**) **capsule** is the bulbous beginning portion of the tubule that first receives the filtrate from plasma. All glomerular capsules are located in the renal cortex.
2. The **proximal convoluted tubule** (or just *proximal tubule*, for short) receives the filtrate from the glomerular capsule. The proximal tubules are also located in the renal cortex.
3. The **loop of Henle** (or **nephron loop**) receives fluid from the proximal tubule. The first portion of the loop, the **descending limb**, takes the fluid from the cortex into the medulla; the second portion of the loop, the **ascending limb**, returns to

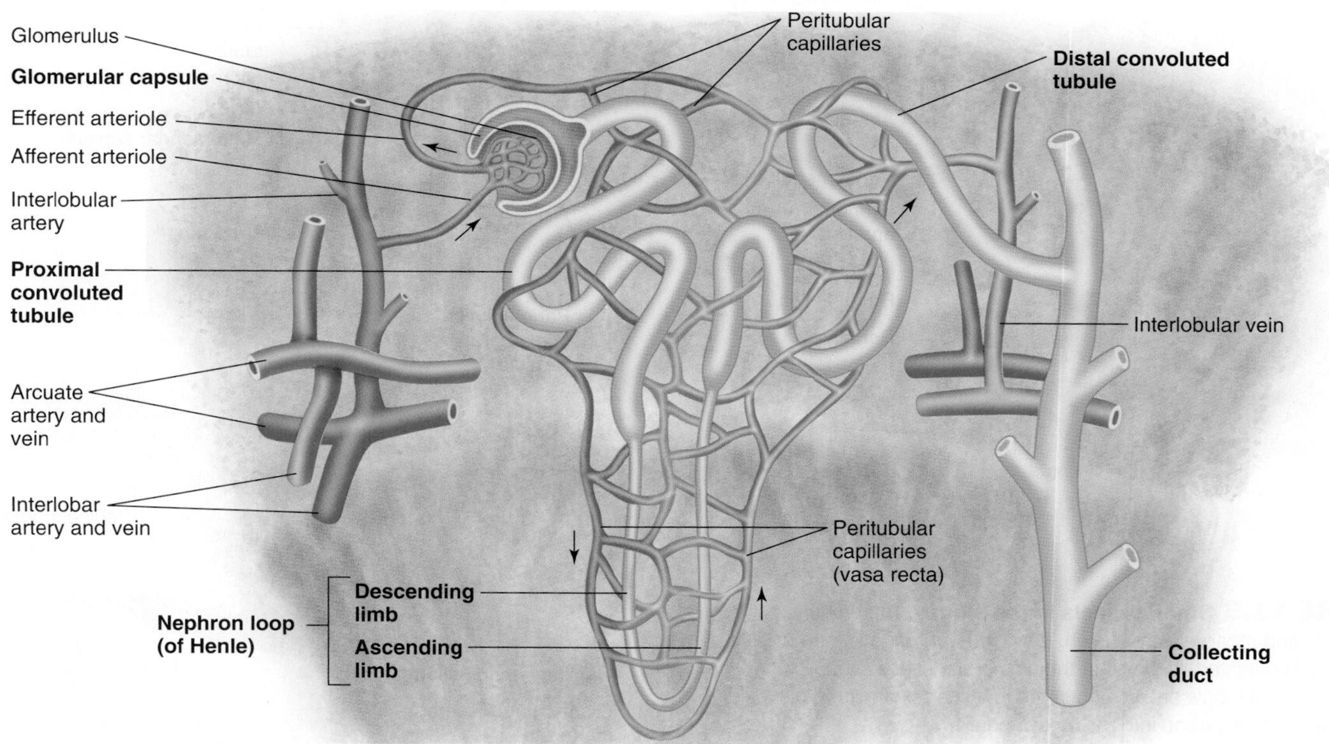

FIGURE 13.3 The nephron tubules and associated blood vessels. In this simplified illustration, the blood flow from a glomerulus to an efferent arteriole, to the peritubular capillaries, and to the venous drainage of the kidneys is indicated with arrows. The names for the different regions of the nephron tubules are indicated with boldface type.

the cortex. The loops of some nephrons extend deeper into the medulla than others (fig. 13.4). Also, the upper portions of the loop have thicker walls and different transport properties than the lower portions, and are sometimes referred to as the *thick segments*.

4. The **distal convoluted tubule** (or just *distal tubule*, for short) receives fluid from the ascending limb of the loop of Henle. The distal tubules are located in the renal cortex.
5. The **collecting duct** receives fluid from the distal convoluted tubules of several nephrons. Each collecting duct thus begins in the cortex—this is sometimes identified separately as the *cortical collecting duct*—and then descends into the medulla. The collecting ducts end at the tip of the renal pyramids, so that they can empty their urine into the calyces.

Nephron functions depend on interactions between the tubules and associated blood vessels. Arterial blood arrives in the *renal artery* and passes through branches (*interlobar* and *arcuate arteries*) to enter small *interlobular arteries* that extend into the renal cortex (fig. 13.5). From here, blood enters **afferent arterioles** that deliver the blood into tightly wound capillary beds, the **glomeruli** (singular, *glomerulus*). The vascular arrangement here is unique, because blood from the glomeruli doesn't pass into venules as it does elsewhere in the body. Instead, blood from the glomeruli next enters **efferent arterioles**. These vessels are structured more like arterioles than venules, and—like arterioles—they deliver blood

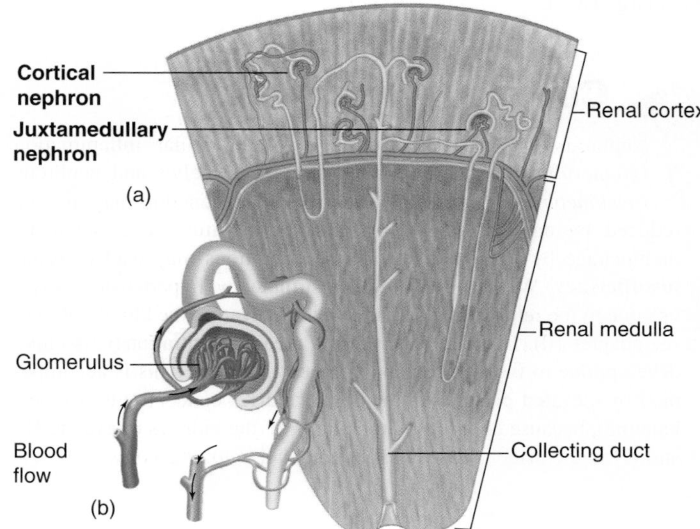

FIGURE 13.4 Two different nephron locations. (*a*) Cortical nephrons are located more superficially in the renal cortex, and their loops of Henle don't extend too deeply into the renal medulla. Juxtamedullary nephrons begin deeper in the cortex, near the medulla, and they have longer loops of Henle that extend deep into the renal medulla. (*b*) A glomerular capsule and early portion of the nephron tubule are illustrated with associated blood vessels.

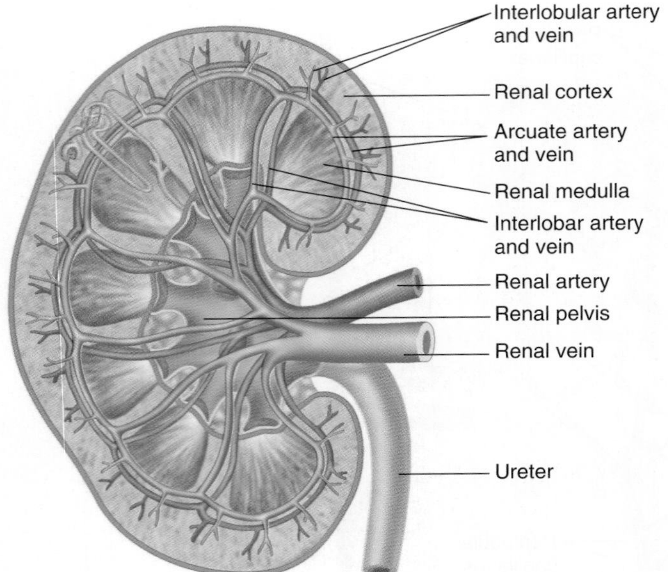

FIGURE 13.5 Blood vessels in the kidney. The branches of the renal artery and renal vein are shown. A single nephron, enlarged greatly for visibility, is depicted. Notice that the glomerular capsule, proximal tubule, and distal tubule are located in the renal cortex; the loop of Henle and collecting duct extend into the renal medulla.

into a bed of capillaries. These capillaries downstream from the glomerulus are called the **peritubular capillaries**, because they surround the tubules (see fig. 13.3). Blood then enters venules and veins that parallel the arteries, to exit the kidneys in the *renal veins* (fig. 13.5).

Clinical Applications

Nephrons may be destroyed due to glomerular inflammation (*glomerulonephritis*), infection of the renal pelvis and nephrons (*pyelonephritis*), or loss of a kidney. Renal function may also be reduced by damage caused by diabetes mellitus, arteriosclerosis, or blockage by kidney stones. These conditions may lead to **renal insufficiency**. When there is renal insufficiency, hypertension develops due to the retention of salt and water (elevating the blood volume; see chapter 10). **Uremia** (high plasma urea concentrations) also may develop, due to the reduced ability to excrete urea. This is accompanied by elevated plasma H^+ (acidosis) and K^+ concentration (hyperkalemia), because of the impaired ability of the kidneys to excrete H^+ and K^+ in the urine. These changes can lead to *uremic coma*.

Glomerular Filtrate Enters the Renal Tubules

The capillaries of the glomeruli have large pores, called *fenestrae* (fig. 13.6), which make them far more permeable to plasma fluid and dissolved solutes than typical capillaries, such as those in skeletal muscles. The fenestrae hold back the formed elements of the blood (blood cells and platelets), but are too large to prevent

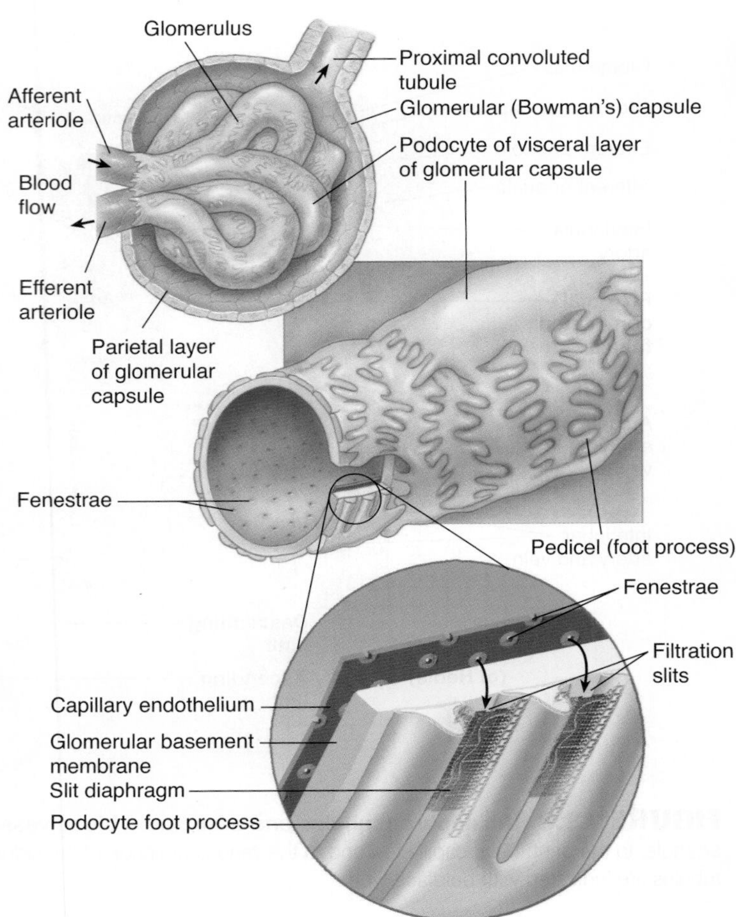

FIGURE 13.6 The structure of the glomerulus and capsule. An illustration of the relationship between the glomerular capillaries and the inner layer of the glomerular capsule. Notice that filtered molecules pass out of the fenestrae of the capillaries and through the filtration slits to enter the cavity of the capsule. Plasma proteins are excluded from the filtrate by the glomerular basement membrane and the slit diaphragm.

plasma proteins from entering the **glomerular filtrate**, the fluid that exits across the walls of the glomerular capillaries. Nevertheless, the glomerular filtrate (and urine) contains very little protein. This implies that some structure outside the fenestrae must provide a filtration barrier for proteins.

Just to the outside of the endothelial cells that comprise the walls of the glomerular capillaries is a **glomerular basement membrane** (fig. 13.6). This is a layer of glycoproteins that may provide a filtration barrier for proteins. The glomerulus and its basement membrane are next surrounded by extensions of the cells that compose the inner (visceral) layer of the glomerular capsule. These cells are called *podocytes*, and their tiny extensions (*pedicels*) wrap around the glomerular capillaries (fig. 13.6). The glomerular filtrate passes through the narrow slits between pedicels to enter the cavity of the glomerular capsule. Spanning the slits is a group of proteins that form the **slit diaphragm**, the structure that apparently prevents most plasma proteins from

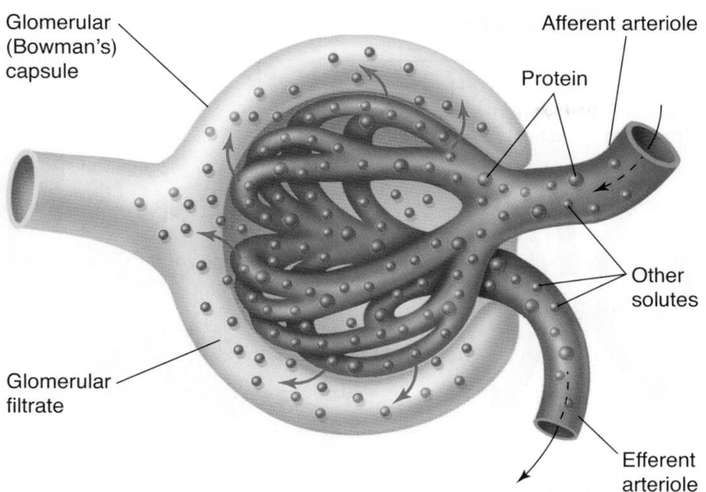

FIGURE 13.7 The formation of glomerular filtrate. Only a very small proportion of plasma proteins (*green spheres*) are filtered, but smaller plasma solutes (*purple spheres*) easily enter the glomerular ultrafiltrate. Arrows indicate the direction of filtration.

entering the filtrate. Figure 13.7 illustrates the selective filtration process, where glomerular filtrate is formed containing water and dissolved solutes, while the plasma proteins are retained in the capillaries.[1]

The forces that produce the glomerular filtrate are similar to those that act on other capillary beds to cause the formation of interstitial fluid (chapter 10; see fig. 10.31). The capillary blood pressure provides the force for filtering fluid out of the glomerulus, and this is opposed by the colloid osmotic pressure of the plasma proteins (drawing water into the capillaries by osmosis). The resulting net filtration pressure, together with the large size of the capillary fenestrae (pores), produces an extraordinarily large volume of filtrate. The **glomerular filtration rate** (**GFR**) is the volume of filtrate produced by both kidneys per minute, and averages 115 ml per minute in women and 125 ml per minute in men. If we take an average value of 120 ml per minute and multiply that by the number of minutes in a day, we get an incredible 180 L (about 45 gallons) of glomerular filtrate formed per day! Physiological mechanisms maintain a relatively constant GFR, while other mechanisms vary the final output of urine.

FYI [1]Here's an analogy that may be helpful: imagine that you have a wet sponge in your fist. Your fist is the wall of the glomerular capillaries, and the spaces between your fingers are the fenestrae. Perhaps you have some grease on the outside of your hand and between your fingers, to represent the basement membrane. Now, imagine that you stick your fist into a large, soft balloon, so that the wall of the balloon (representing the glomerular capsule) wraps around your fist. The part of the balloon in contact with your fist and surrounding it is the visceral layer of the glomerular capsule; the rest represents the parietal layer. Now, you must imagine something impossible for a balloon—that the balloon wall in contact with your fist has slits in it (with a molecular layer forming slit diaphragms). Water from the sponge passes between your fingers (fenestrae), through the grease (basement membrane), and then through the slits with their diaphragms to enter the interior of the balloon (the cavity of the glomerular capsule).

Because a person only has an average total blood volume of 5.5 L, most of the glomerular filtrate must be returned to the blood and recycled, a process known as *reabsorption*. We only excrete an average of about 1.5 L of urine per day out of the 180 L per day "filtered" (present in glomerular filtrate), although this varies under different conditions to maintain homeostasis of blood volume. In the face of a constant GFR, this variation in urine volume is adjusted by changes in the amount of salt and water reabsorbed (as described in the next section).

Clinical Applications

The term *dialysis* refers to the separation of molecules using their ability to diffuse across an artificial semipermeable membrane (chapter 3). This principle is used in the "artificial kidney" machine for **hemodialysis**. Like the walls of the glomerular capillaries, the artificial semipermeable membrane allows water and dissolved waste molecules (such as urea) to easily diffuse through the membrane pores, whereas plasma proteins are excluded by their larger size. However, unlike the tubules, the artificial membrane can't reabsorb Na^+, K^+, glucose, and other molecules needed in the blood. These substances are therefore included in the fluid around the dialysis membrane, so that there is no concentration gradient to cause their net diffusion out of the blood. Sometimes a patient's own peritoneal membranes (which line the abdominal cavity) are used as the dialysis membrane, in a technique called **continuous ambulatory peritoneal dialysis**.

Check Point

1. Describe the neural control of micturition (urination).
2. Describe the structure of the kidney, and the different regions of the nephron tubules. Identify which parts of the tubules are located in the renal cortex, and which in the renal medulla.
3. Identify the blood vessels associated with the nephron tubules, and describe the structure of the glomerulus and its associated capsule. Identify the filtration barrier that prevents proteins from entering the filtrate.
4. Identify the glomerular filtration rate (GFR), and describe how it relates to the volume of urine excreted per day.

The Kidneys Reabsorb Salt and Water from the Filtrate

Most of the salt and water in the glomerular filtrate are reabsorbed across the walls of the proximal tubules back into the blood. As the filtrate passes through the loop of Henle, transport processes cause Na^+ and Cl^-, as well as urea, to accumulate to a high concentration in the renal medulla. This provides an osmotic force for water reabsorption out of the collecting ducts, which are made permeable to water by antidiuretic hormone (ADH).

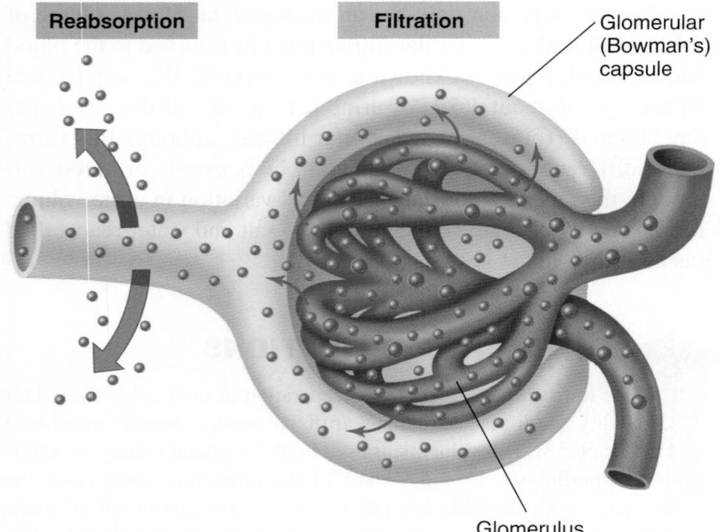

FIGURE 13.8 Filtration and reabsorption. Plasma water and its dissolved solutes (except proteins) enter the glomerular filtrate by filtration, but most of these filtered molecules are reabsorbed. The term *reabsorption* refers to the transport of molecules out of the tubular filtrate back into the blood.

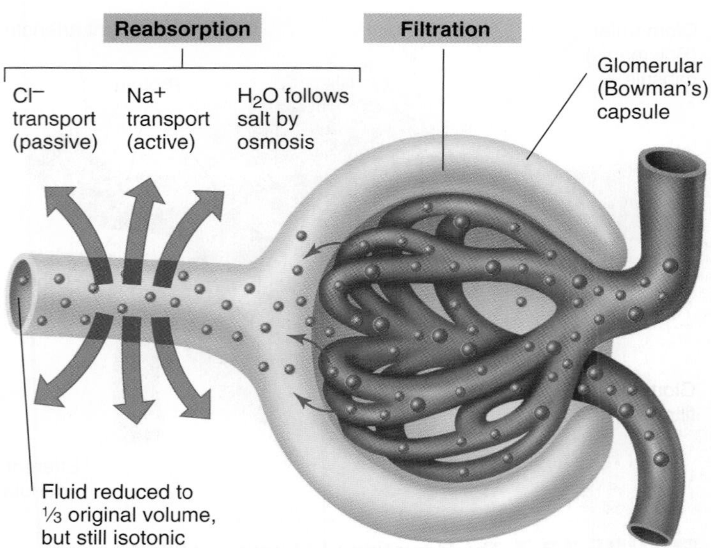

FIGURE 13.9 Salt and water reabsorption in the proximal tubule. Sodium is actively transported out of the filtrate and chloride follows passively by electrical attraction. Water follows the salt out of the tubular filtrate into the peritubular capillaries by osmosis.

Because urine is derived from blood plasma, increasing the volume of urine will lower the total blood volume, and thus the blood pressure. Conversely, a reduction in urine volume will help to increase the blood volume and pressure. The volume of urine produced by the kidneys must thus be carefully regulated to maintain homeostasis of the blood volume and pressure. To this end, we can excrete anywhere between a minimum of 400 ml of urine per day (the *obligatory water loss*) up to a theoretical maximum of about 23 L per day, depending on how much water we drink.

Reabsorption refers to the transport processes that allow substances in the glomerular filtrate to be returned to the vascular system (fig. 13.8). The cells of the tubule walls reabsorb solutes by either active or passive membrane transport (chapter 3), but water is reabsorbed only passively, by osmosis. This presents a problem: the glomerular filtrate is essentially *isotonic* to (has the same solute concentration as) the surrounding interstitial fluid of the renal cortex and the blood plasma. In order for water to be reabsorbed into the blood of the surrounding peritubular capillaries, a concentration (osmotic) gradient across the wall of the tubule must first be produced. This is accomplished in different ways by different segments of the nephron tubule.

The Proximal Tubule Reabsorbs Most of the Filtrate

Because the kidneys produce a huge volume of glomerular filtrate (180 L/day), most must immediately be reabsorbed to prevent an otherwise fatal drop in blood volume and pressure. In order for water to be drawn out of the proximal tubule by osmosis, an osmotic (concentration) gradient must first be produced. This involves transport processes across the plasma membrane of the epithelial cells that form the walls of the proximal tubules (fig. 13.9):

1. The plasma membrane actively transports Na^+ from the cytoplasm into the surrounding interstitial fluid and blood plasma. The Na^+/K^+ (ATPase) pumps responsible for this active transport (chapter 3) maintain a low Na^+ concentration in the cytoplasm. As a result, Na^+ continuously diffuses from the glomerular filtrate into the cytoplasm, and is then pumped into the surrounding interstitial fluid and plasma.
2. Chloride ion (Cl^-) follows the Na^+ passively by electrical attraction from the glomerular filtrate into the surrounding interstitial fluid and plasma. The net effect is an increase in the salt (NaCl) concentration of the interstitial fluid and the blood plasma of the peritubular capillaries.
3. The higher NaCl concentration of the fluid surrounding the proximal tubules serves as an osmotic gradient. Because the walls of the proximal tubules are permeable to water, water follows the NaCl into the interstitial fluid and then into the peritubular capillaries.

About 65% (almost two-thirds) of the glomerular filtrate is immediately reabsorbed across the walls of the proximal tubule. This reabsorption occurs constantly, regardless of how dehydrated or well hydrated a person may be, and consumes as much as 6% of the body's energy (calories) at rest. An additional smaller amount of water (about 20% of the filtrate) is also reabsorbed constantly across the descending limb of the loop of Henle (discussed in the next section), so that about 85% of the glomerular filtrate is always reabsorbed in the earlier parts of the nephrons. This still leaves 15% of the glomerular filtrate—27 L/day—to be reabsorbed in the later parts of the nephron (distal tubule and collecting duct). It is reabsorption in these later regions that is adjusted according to the body's state of hydration to maintain homeostasis of the blood volume.

The Loop of Henle Produces a Hypertonic Renal Medulla

The proximal tubule is freely permeable to water, allowing the osmotic movement of water to follow the transport of NaCl out of the filtrate. This allows equivalent amounts of salt and water to be removed from the filtrate, so that its volume is reduced but its concentration is unchanged. It was isotonic to blood plasma when it entered the glomerular capsule, and its still isotonic when it leaves the proximal tubule to enter the descending limb of the loop of Henle.

You may remember from chapter 3 that *osmolarity* is a measure of the total solute concentration of solutions, and isotonic solutions have a concentration of 300 *milliosmolar* (300 *mOsM*). Here we will simply use this number as a reference: solutions with a higher osmolarity are more concentrated (*hypertonic*), and those with a lower number are more dilute (*hypotonic*). The numbers shown in the loop of Henle in figure 13.10 indicate the changing osmolarity (concentration) of the filtrate as it travels through the loop. Notice that the osmolarity increases as the fluid descends from the cortex into the medulla, going from the isotonic 300 mOsM to the very hypertonic 1,400 mOsM at the bottom of the loop.

The increasing concentration of the filtrate in the descending limb is produced by the removal of water by osmosis into the surrounding peritubular capillaries. The water is drawn out of the descending limb because the interstitial fluid of the renal medulla is very hypertonic, and thus has a higher osmotic pressure. This is true for the following reasons:

1. The thick segment of the ascending limb actively pumps out NaCl into the surrounding interstitial fluid of the renal medulla (fig. 13.10). This is a different, and somewhat more complex, process than occurs in the proximal tubule, but the net effect is similar: salt is moved from the filtrate into the surrounding fluid.
2. The ascending limb of the loop is *not permeable to water*. As a result, the interstitial fluid becomes hypertonic while the filtrate that ascends into the distal tubule in the renal cortex becomes hypotonic (the osmolarity gets as low as 100 mOsM).
3. The blood vessels in the renal medulla also form loops (called the *vasa recta*), which allow them to carry away water but leave the NaCl to accumulate in the interstitial fluid of the medulla.

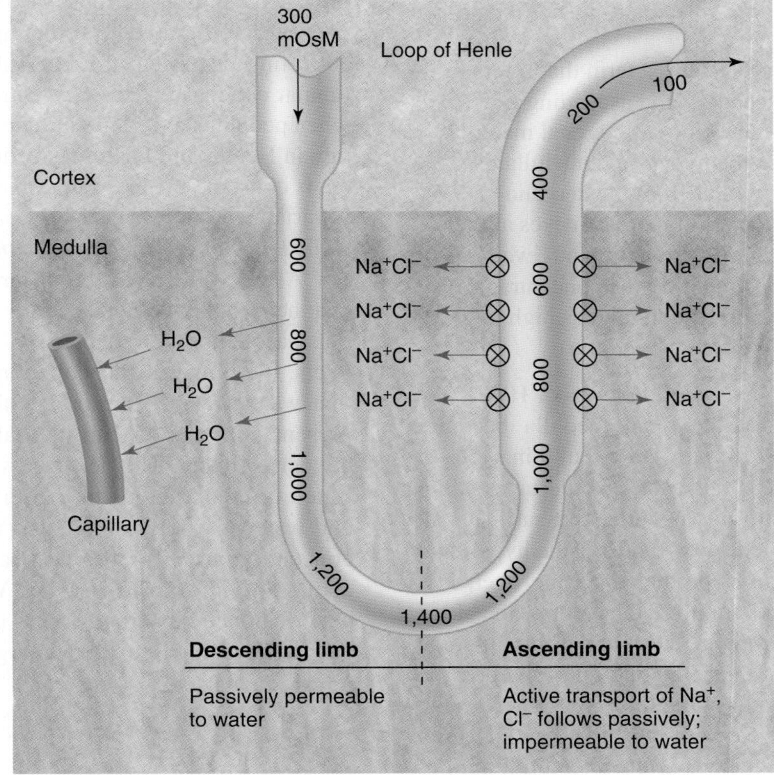

FIGURE 13.10 Transport processes in the loop of Henle. The ascending limb pumps NaCl into the interstitial fluid, making it hypertonic. This creates an osmotic gradient that draws water out of the descending limb to be reabsorbed into peritubular capillaries. Interactions between the descending and ascending limbs of the loop are described as a countercurrent multiplier system, creating a very hypertonic renal medulla. The deepest portion of the renal medulla has a concentration of 1,400 mOsM.

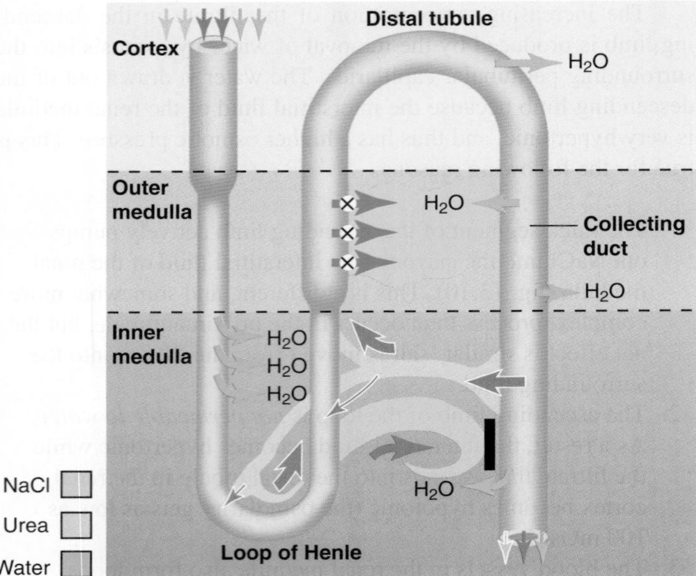

FIGURE 13.11 The role of urea in urine concentration. Urea diffuses out of the inner collecting duct and contributes significantly to the concentration of the interstitial fluid in the renal medulla. The active transport of Na^+ out of the ascending limbs also contributes to the hypertonicity of the medulla, so that water is reabsorbed by osmosis from the collecting ducts.

4. A **countercurrent multiplier system**, a term that refers to interactions between the descending and ascending limbs, operates to increase the concentration of the renal medulla. The ascending limb produces the hypertonic medulla by pumping out salt. This draws water from the descending limb, producing a hypertonic filtrate that arrives at the ascending limb, which then allows it to pump out even more NaCl. These interactions multiply the concentration of the interstitial fluid of the renal medulla to as high as 1,400 mOsM at the deepest part of the medulla.

Because of the transport processes in the loops of Henle (the countercurrent multiplier system), the renal medulla becomes increasingly salty and hypertonic. If you were to cut thin slices of a kidney and pop them into your mouth, they would taste saltier and saltier the deeper you cut into the medulla. But salt isn't all you'd taste (unfortunately): urea also contributes to the hypertonic concentration of the medulla. *Urea*, produced by the liver, is a waste product of the metabolism of amino acids. It enters the blood and is filtered out of the glomeruli. In the inner medulla, some urea diffuses through channels in the collecting ducts to enter the interstitial fluid. The ascending limb of the loop is also permeable to urea, so that the urea can recycle between the ascending limb and the collecting duct as it accumulates in the interstitial fluid of the medulla (fig. 13.11).

In summary, the transport processes of the loops of Henle accomplish three functions: (1) some water is reabsorbed across the walls of the descending limb; (2) the filtrate in the distal tubule in the renal cortex is made hypotonic (dilute, with a concentration as low as 100 mOsM); and (3) the interstitial fluid of the renal medulla is made very hypertonic (concentrated), with a concentration as high as 1,400 mOsM. This last function is the most important one, for reasons described in the next section.

Water Reabsorption from the Collecting Duct Requires ADH

Salt (NaCl) is pumped out of the ascending limbs of the loops and water is prevented from following it by osmosis; as a result, the fluid in the distal tubule is hypotonic, with a concentration of 100 mOsM. If we only needed to produce hypotonic urine—one that is dilute, with a high volume of water—we wouldn't need a loop of Henle, or a hypertonic renal medulla.[2] The hypertonic environment of the renal medulla provides the osmotic gradient for the reabsorption of water across the walls of the collecting ducts. This allows us to produce a small volume of very concentrated urine when we need to conserve water. The ability to conserve water and excrete concentrated urine is the "payoff" for the work performed by the loops of Henle.

The collecting ducts begin in the cortex and plunge into the medulla, all the way to the tips of the renal pyramids where they empty urine into the calyces. In the cortex, the collecting ducts receive hypotonic fluid from the distal tubules. In the medulla, the walls of the collecting ducts separate this dilute fluid from the surrounding hypertonic interstitial fluid. There is thus a steep concentration gradient promoting the osmosis of water from the collecting ducts into the interstitial fluid and then into the peritubular capillaries. This concentration gradient is constant, except when people take certain diuretic drugs (discussed in the "Physiology in Health and Disease" section). Although the osmotic gradient is constant, the reabsorption of water from the collecting ducts is variable. This is because *the permeability of the collecting duct walls to water is dependent on antidiuretic hormone (ADH).*

ADH stimulates the insertion of *aquaporin* channels—channels that specifically allow the diffusion of water across a plasma membrane (chapter 3)—into the walls of the collecting ducts. This makes the collecting duct permeable to water, so that water can be reabsorbed (fig. 13.12). When more ADH is present, more water can be reabsorbed across the walls of the collecting ducts. This leaves a smaller volume of more concentrated urine and helps to conserve water. When ADH secretion and water reabsorption are at a maximum, the urine will have a concentration of 1,400 mOsM and the person will excrete only about 400 ml of urine per day. When ADH is absent, very little water can be reabsorbed across the walls of the collecting ducts, and the person will excrete a large volume of dilute urine (with a

FYI [2]The bodies of freshwater bony fishes are more concentrated than their environment; they thus take in water by osmosis and must always excrete a large volume of dilute urine. Amphibians and reptiles lack loops of Henle, and so can't perform countercurrent multiplication to produce a hypertonic renal medulla. As a result, amphibians and reptiles can't produce urine that is any more concentrated than their blood plasma.

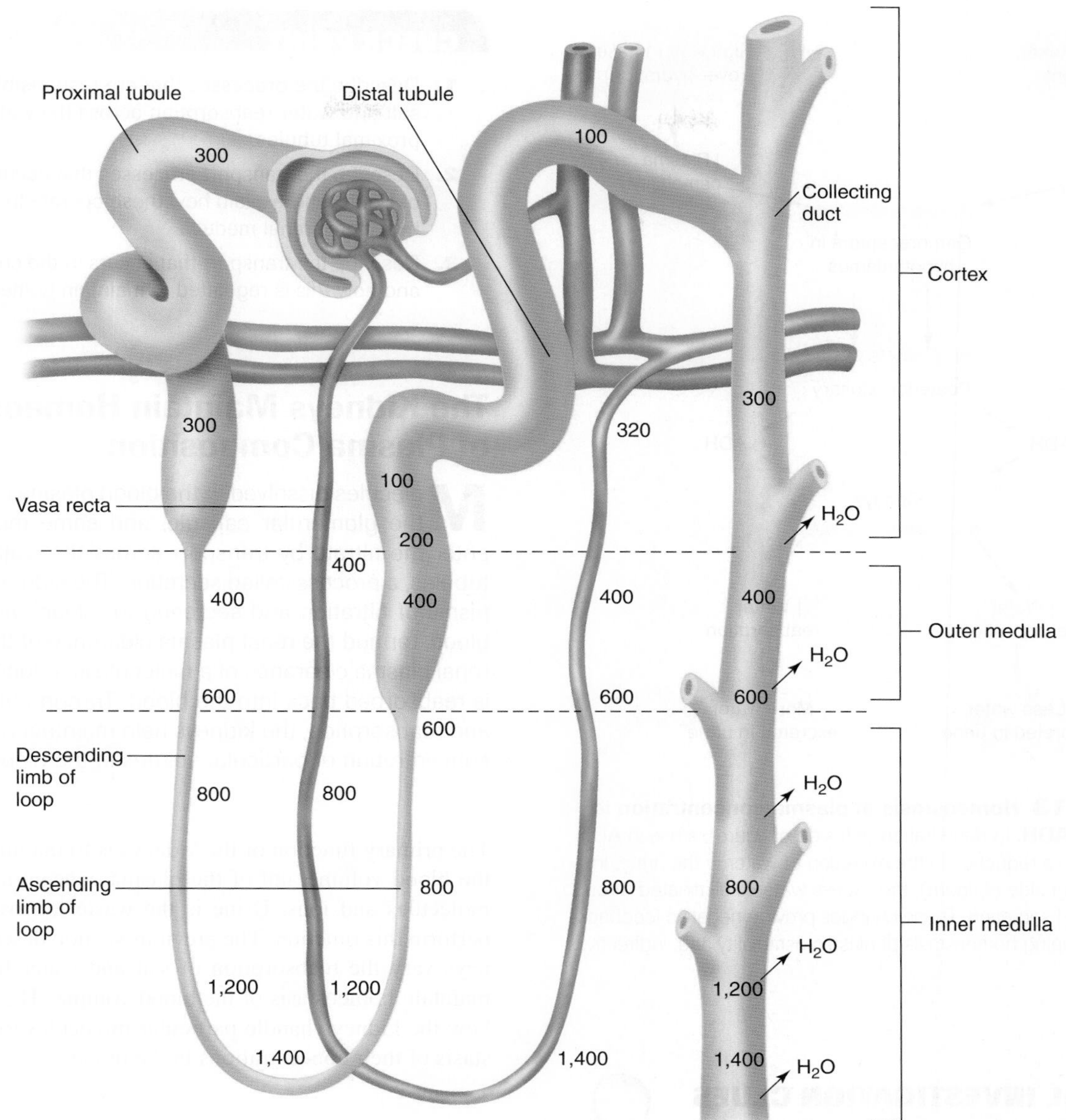

FIGURE 13.12 The osmolality of different regions of the kidney. The countercurrent multiplier system in the loop of Henle helps create a hypertonic renal medulla. Under the influence of antidiuretic hormone (ADH), the collecting duct becomes more permeable to water, and thus more water is drawn out by osmosis into the hypertonic renal medulla and peritubular capillaries. (Numbers indicate osmolarity.)

minimum concentration of 100 mOsM). Most of the time we're somewhere between these two extremes, reabsorbing an average of 99.2% of the original volume of glomerular filtrate and excreting 1.5 L of urine per day.

ADH is secreted by the posterior pituitary gland (chapter 8) when stimulated by **osmoreceptor** neurons in the hypothalamus. These neurons are activated by an increase in the plasma osmolarity, as may be produced by dehydration (or eating too much salt). Conversely, drinking too much water will dilute the blood and lower its osmolarity, so that the osmoreceptors are less stimulated. These changes influence the secretion of ADH accordingly, so that the kidneys will retain water or excrete more water as needed to maintain homeostasis of the plasma osmolarity (fig. 13.13).

Clinical Applications

Diabetes insipidus may be caused by (1) drinking too much water (*polydipsia*); (2) the inadequate secretion of ADH; or (3) inadequate ADH action due to a genetic defect in the ADH receptors or aquaporin channels. Without adequate ADH secretion or action, the collecting ducts are not very permeable to water. This results in the excretion of a large volume—greater than 3 L per day, and often 5 to 10 L per day—of dilute urine (with a concentration under 300 mOsM). Excretion of so much water can cause dehydration, which produces intense thirst. If the person drinks sufficient amounts of water, symptoms of dehydration are usually absent. However, a person with diabetes insipidus may have difficulty drinking enough to compensate for the large amounts of water lost in the urine.

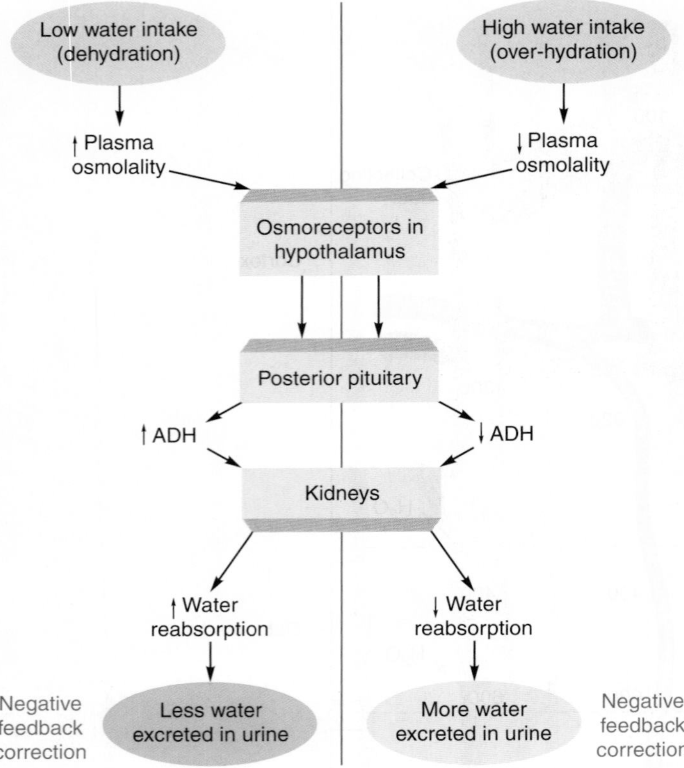

FIGURE 13.13 Homeostasis of plasma concentration is maintained by ADH. In dehydration (left side of figure), a rise in ADH secretion results in a reduction in the excretion of water in the urine. In overhydration (right side of figure), the excess water is eliminated through a decrease in ADH secretion. These changes provide negative feedback correction, maintaining homeostasis of plasma osmolality and, indirectly, blood volume.

Clinical Investigation Clues

Remember that Helen's 24-hour urine sample had an abnormally high volume and low osmolarity, and that she drank a lot of water. Also, the physician mentioned that she may have a form of diabetes, and wanted to measure a pituitary hormone.

- What form of diabetes does the physician suspect that Helen may have, and which hormone does he want to measure?
- Perhaps Helen simply drinks excessive amounts of water (has polydipsia); how would that influence her level of the pituitary hormone?
- Is polydipsia a likely primary cause of Helen's condition, given the observation that Helen became excessively dehydrated (compared to her friends), with low blood pressure, when she was deprived of her usual supply of water?

Check Point

1. Describe the processes that are responsible for salt and water reabsorption across the walls of the proximal tubules.
2. Describe the transport processes that occur in the loop of Henle, and explain how these operate to produce a hypertonic renal medulla.
3. Describe the transport that occurs in the collecting duct, and how this is regulated to maintain homeostasis.

The Kidneys Maintain Homeostasis of Plasma Composition

Molecules dissolved in the blood plasma can be filtered into the glomerular capsule, and some molecules can also enter the filtrate by transport across the walls of the nephron tubules, a process called secretion. The kidneys use the mechanisms of filtration and secretion to "clear" molecules from the blood, termed the renal plasma clearance of the molecules. The renal plasma clearance of a molecule is reduced if this molecule is reabsorbed back into the blood. Through filtration, secretion, and reabsorption, the kidneys help maintain homeostasis of the concentration of particular solutes in the blood plasma.

The primary function of the kidneys is to maintain homeostasis of the blood volume and of the plasma concentrations of particular molecules and ions. Urine is the waste left over as the kidneys perform this function. The previous section described how the kidneys vary the reabsorption of salt and water from the filtrate to maintain homeostasis of the blood volume. This section describes how the kidneys handle particular molecules to maintain homeostasis of their concentrations in the blood.

The Kidneys Clear Certain Molecules from the Plasma

The **renal plasma clearance** is the measure of the kidneys' ability to remove, or clear, particular molecules from the blood plasma. Molecules smaller than proteins in the plasma are easily "filtered," meaning that they easily enter the glomerular filtrate. If this is all that happens to these molecules, whatever was filtered will be excreted in the urine and thus cleared from the blood. In this case, the volume of blood cleared of that molecule per minute will be the same as the volume of blood filtered. Stated another way, the renal plasma clearance of that molecule will equal the *glomerular filtration rate* (*GFR*). Since the GFR is measured in ml per minute, the renal plasma clearance is also expressed in ml per minute. As described previously, the GFR averages about 120 ml/min.

However, if a molecule is *reabsorbed* to some degree, it will be transported out of the nephron tubule into the surrounding peritubular capillaries and returned to the blood (fig. 13.14). Thus, reabsorption decreases the renal plasma clearance of a

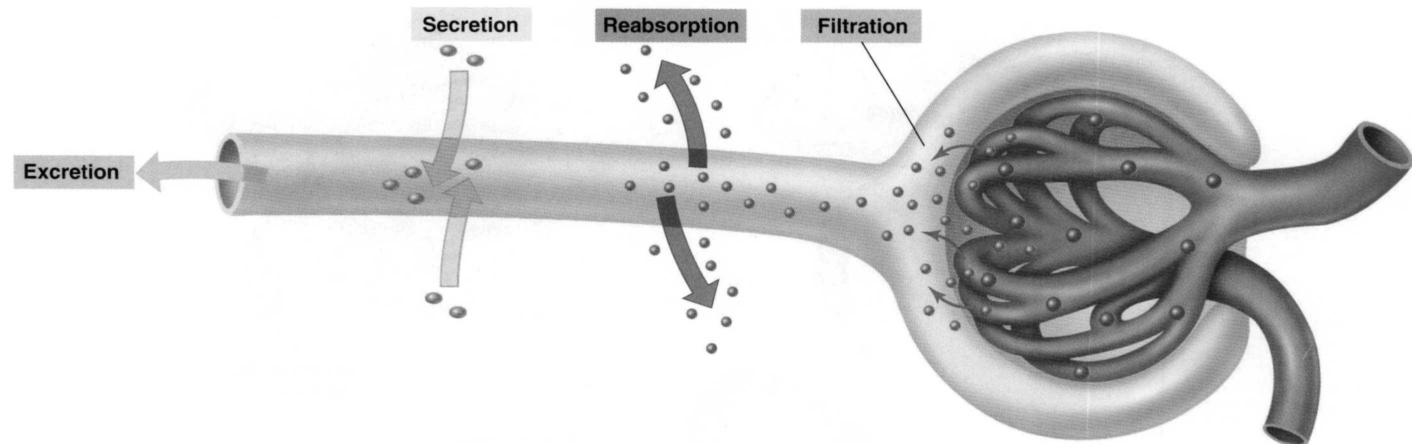

FIGURE 13.14 Secretion is the reverse of reabsorption. The term *secretion* refers to the active transport of substances from the peritubular capillaries into the tubular fluid. This transport is opposite in direction to that which occurs in reabsorption.

molecule. A molecule that is filtered but then reabsorbed will have a renal plasma clearance less than the GFR. If the molecule is completely reabsorbed, its renal plasma clearance will be zero.

The opposite of reabsorption is **secretion**: the transport of a molecule or ion from the peritubular blood, across the wall of the tubule, and into the fluid within the nephron (fig. 13.14). Secretion, like filtration, removes molecules from the blood plasma and adds them to the urine. Molecules that are both filtered and secreted have a renal plasma clearance greater than the GFR, because more than only the molecules filtered are cleared from the blood. There is a family of transporter proteins in the proximal tubule that secrete a large number of foreign molecules (such as antibiotics) that may have entered the blood for various reasons, allowing the kidneys to eliminate them at a rapid rate.

The Kidneys Handle Inulin, Glucose, and Penicillin Differently

Certain molecules can be used to illustrate how the processes of filtration, reabsorption, and secretion affect the renal plasma clearance. In summary:

1. Molecules that are filtered and reabsorbed have a renal plasma clearance less than the GFR. Examples include glucose and amino acids.
2. Molecules that are filtered and secreted have a renal plasma clearance greater than the GFR. Examples include penicillin and other foreign molecules.
3. Molecules that are filtered but neither reabsorbed nor secreted have a renal plasma clearance equal to the GFR. This is the case only of the plant molecule inulin, although the renal plasma clearance of creatinine approximates this condition.

Inulin Is Filtered, but Neither Reabsorbed Nor Secreted

Inulin is a plant product (a polysaccharide of fructose) that enters the glomerular filtrate but is neither reabsorbed nor secreted. Thus, its renal plasma clearance equals the GFR (fig. 13.15). Notice in figure 13.15 that the renal vein still contains inulin, but at a lower concentration than in the renal artery. The concentration of inulin is reduced in the blood going back to the general circulation because the amount that was contained in the filtered blood was eliminated in the urine. This means that the blood must make many passes through the kidneys to completely eliminate the inulin from the blood.

Clinical Applications

Measurements of the GFR are used clinically to assess kidney health. Most often, this involves measurements of the **creatinine** concentration in the blood and urine. Creatinine, a waste product derived from muscle creatine, enters the blood at a constant rate and is normally eliminated by the kidneys at a constant rate. The renal plasma clearance of creatinine is only slightly higher than the GFR, indicating that it is slightly secreted by the nephron tubules. Thus, the GFR can be measured to an approximate degree by the renal plasma clearance of creatinine. More often, a simple measurement of the plasma creatinine concentration can provide an index of the GFR and thus the health of kidney function.

Clinical Investigation Clues

Remember Helen's possible condition, and that she became intensely thirsty and dizzy when deprived of her usual supply of water on the camping trip. Also, under those conditions, Helen had a low blood pressure and an elevated plasma creatinine concentration.

- Could Helen have become dehydrated on her camping trip? How?
- How could dehydration account for her low blood pressure?
- What is the likely cause of the elevated plasma creatinine?
- Why did the plasma creatinine concentration return to normal after Helen was rehydrated and her normal blood pressure restored?

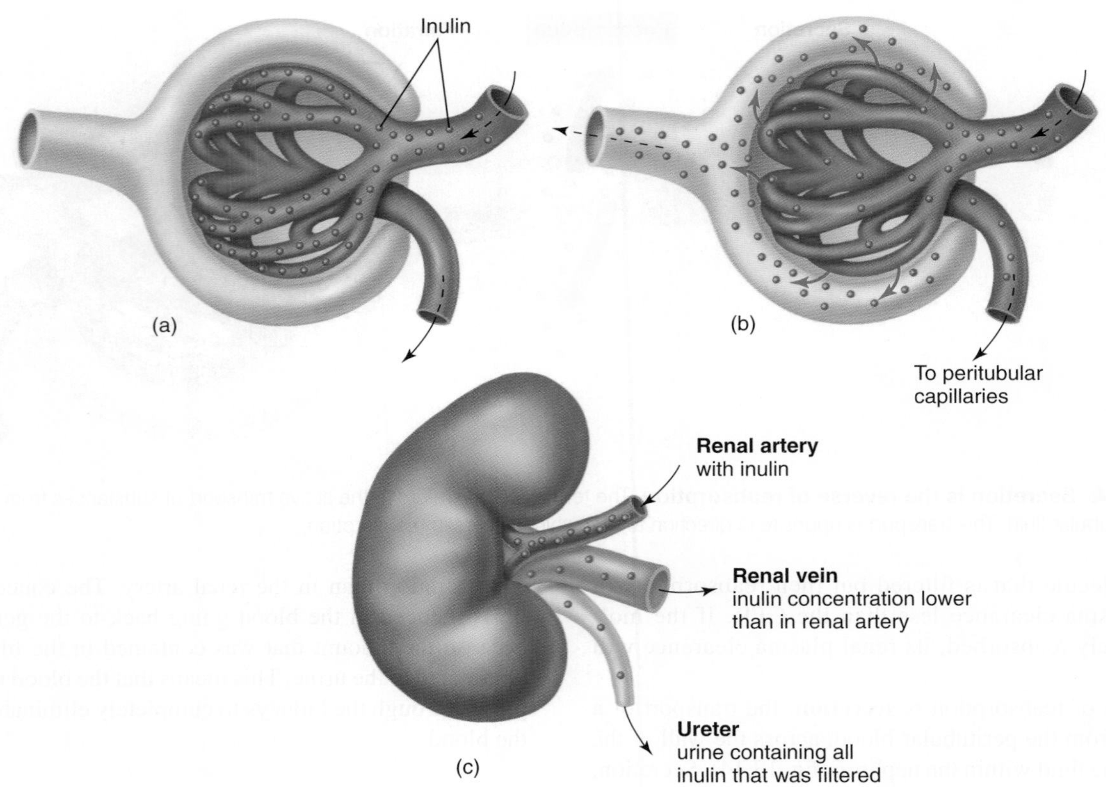

FIGURE 13.15 The renal clearance of inulin. (*a*) Inulin is present in the blood entering the glomeruli, and (*b*) some of this blood, together with its dissolved inulin, is filtered. All of this filtered inulin enters the urine, whereas most of the filtered water is returned to the vascular system (is reabsorbed). (*c*) The blood leaving the kidneys in the renal vein, therefore, contains less inulin than the blood that entered the kidneys in the renal artery. Since inulin is filtered but neither reabsorbed nor secreted, the inulin clearance rate equals the glomerular filtration rate (GFR).

Glucose Is Filtered but Normally Completely Reabsorbed

Glucose and amino acids are easily filtered out of the glomeruli, but are usually not present in the urine. This is because they are normally completely reabsorbed across the walls of the proximal tubules back into the blood (so that their renal plasma clearance is zero). Carrier proteins in the plasma membrane of the tubule epithelial cells are required for this reabsorption. As described in chapter 3, carrier-mediated transport has several characteristics, including the property of *saturation*. This means that if the concentration of transported molecules (such as glucose) is sufficiently high, all of the carrier proteins will be occupied. At this point, the rate of transport will reach a maximum, called the *transport maximum*.

The average transport maximum for glucose in the nephron tubules is 375 mg per minute. With a GFR of 120 ml/min, and a normal fasting plasma glucose concentration of 1 mg/ml, the tubules normally receive 120 mg per minute of glucose. This is well below the transport maximum, allowing all of the glucose to be reabsorbed. The blood glucose concentration would have to increase to more than 3 times the normal fasting level to saturate all of the carriers and exceed the transport maximum. Actually, some nephrons have a transport maximum below the average, so that their carriers become saturated at a lower blood glucose concentration. When the carriers are saturated, the glucose molecules that aren't transported and reabsorbed continue their journey through the tubules and "spill over" into the urine.

The plasma glucose concentration that must be reached for glucose to first appear in the urine is called the **renal plasma threshold** for glucose. This averages about 180–200 mg of glucose per 100 ml of blood, about twice the normal fasting levels. *Hyperglycemia*, an abnormally high fasting blood glucose concentration, will cause glucose to appear in the urine—a condition known as **glycosuria**—when the plasma glucose concentration is greater than the renal plasma threshold for glucose. This is the hallmark of *diabetes mellitus*, produced by the inadequate secretion and/or action of insulin (chapter 8).

CLINICAL INVESTIGATION CLUES

Remember that Helen produced copious amounts of dilute urine, and did not have glycosuria. Yet the physician said that she might have "a type of diabetes."

- Which type of diabetes is probably ruled out by the lack of glucose in the urine?
- What causes glucose to appear in the urine in people with that type of diabetes?
- People who "spill glucose" in the urine can produce large volumes of urine (have polyuria; see the "Physiology in Health and Disease" section at the end of this chapter). How would the osmolarity of the urine in this case differ from that of Helen's urine?

Penicillin and PAH Are Secreted by the Nephrons

Antibiotics, notably penicillin, are filtered and secreted: they are transported from the peritubular capillaries, across the walls of the proximal tubules, and into the filtrate. This allows the kidneys to quickly lower the blood concentration of penicillin, which could adversely reduce its effectiveness. For that reason, medications often combine penicillin with another molecule that binds to the same carrier proteins in the tubules, resulting in competition that decreases the rate of penicillin secretion and thus reduces its elimination in the urine.

PAH (**para-aminohippuric acid**) is a foreign (but harmless) molecule that can be infused into the blood to measure the *total renal blood flow*. This is because PAH is completely eliminated from the blood that enters the kidneys: whatever isn't filtered out of the glomeruli is removed from the peritubular capillaries by secretion across the tubule wall (fig. 13.16). Because all of the blood going to the kidneys is cleared of PAH, the renal plasma clearance of PAH is equal the total renal blood flow. This averages about 625 ml/min. Since the GFR averages 120 ml/min, this indicates that about 20% of the blood entering the kidneys is filtered out of the glomeruli. The remaining 80% of the renal blood flow passes into the efferent arterioles and peritubular capillaries, and so can only be cleared of molecules that are secreted.

CHECK POINT

1. What is the renal plasma clearance? Describe how the processes of filtration, reabsorption, and secretion affect the renal plasma clearance of a molecule.
2. How do the kidneys handle glucose? Explain how glycosuria can occur in a person with diabetes mellitus.
3. Describe how the renal plasma clearance of particular molecules can be used to measure the GFR and the total renal blood flow.
4. How do the kidneys handle penicillin, and what measures are taken to help keep penicillin in the blood for longer periods?

The Kidneys Maintain Homeostasis of Plasma Electrolytes and pH

The kidneys help maintain homeostasis of the plasma concentrations of many ions (electrolytes) by excreting them in the urine when their plasma concentration increases, and retaining them in the blood when their plasma concentration decreases. Through this means, the kidneys regulate the

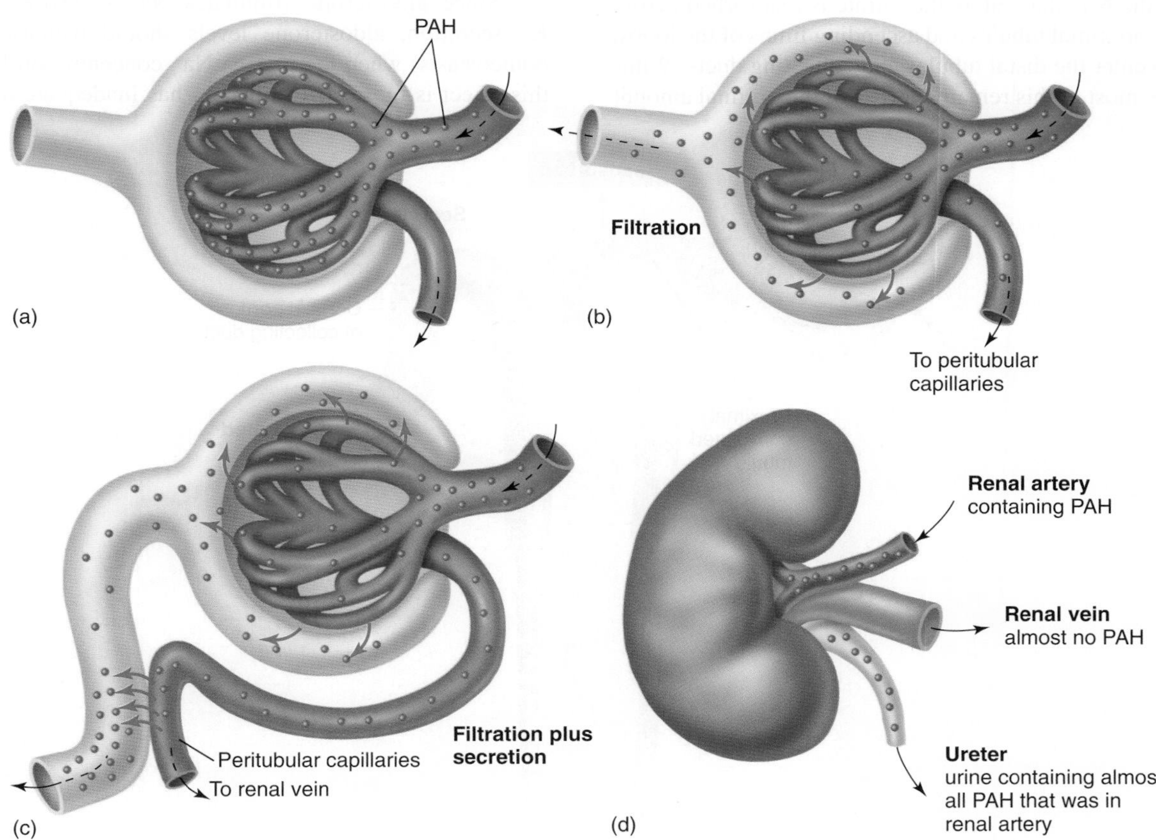

FIGURE 13.16 The renal clearance of PAH. Some of the para-aminohippuric acid (PAH) in glomerular blood (*a*) is filtered into the glomerular capsules (*b*). The PAH present in the unfiltered blood is secreted from the peritubular capillaries into the nephron (*c*), so that all of the blood leaving the kidneys is free of PAH (*d*). The clearance of PAH therefore equals the total renal blood flow.

concentration of blood bicarbonate, which is the major buffer in the plasma. By regulating the blood bicarbonate and excreting H^+ in the urine, the kidneys are responsible for the metabolic component of acid-base balance.

The kidneys regulate the concentrations of plasma electrolytes (ions), including sodium, potassium, chloride, calcium, phosphate, and bicarbonate ions. These ions are easily filtered (enter the glomerular filtrate), and some can also be secreted into the tubules; this allows the ions to be excreted in the urine when their blood concentrations increase. Conversely, each ion can be reabsorbed to varying degrees, depending on its concentration in the blood. The degree to which the renal tubules secrete or reabsorb an ion is adjusted to maintain homeostasis of the ion concentration in the blood plasma. These adjustments occur in response to particular hormones that act on the nephron. For example, parathyroid hormone (PTH, from the parathyroid glands; chapter 8) regulates how the nephron handles the calcium and phosphate ions; aldosterone, secreted by the adrenal cortex, regulates how the nephron handles Na^+, K^+, and H^+. You may recall from chapter 8 that aldosterone is classified as a mineralocorticoid, because it's a corticosteroid that regulates "mineral" (ion) balance.

Aldosterone Stimulates Na^+ Reabsorption and K^+ Secretion

About 90% of the Na^+ that enters the filtrate is reabsorbed across the walls of the proximal tubules and ascending limbs of the loops, leaving 10% to enter the distal tubules and collecting ducts. Without aldosterone, most of this remainder (8% of the original amount filtered) will be reabsorbed in the later regions of the nephron. This leaves 2% of the filtered Na^+ to be excreted in the urine, which is still a rather large amount (averaging 30 g/day). Aldosterone stimulates the last part of the distal tubule and **cortical collecting duct** (the portion of the collecting duct in the renal cortex) to reabsorb the remaining Na^+. Under maximum aldosterone stimulation, all of the Na^+ will be reabsorbed, so that none is excreted in the urine. The reabsorption of Na^+ is accompanied by Cl^- and water retention; you may remember that the aldosterone stimulation of salt and water retention was discussed in chapter 10 as an important regulator of the total blood volume, and thus the blood pressure.

About 90% of the K^+ that enters the glomerular filtrate is also reabsorbed across the walls of the proximal tubules (fig. 13.17). This reabsorption is constant and unregulated, leaving very little K^+ in the filtrate to be excreted in the urine. But what if you're fond of bananas? Bananas and some other fruits are rich sources of dietary potassium, and eating them will raise your plasma K^+ concentration. The only way that the additional K^+ can be excreted in the urine, after almost all of it has been reabsorbed, is for the K^+ to be later secreted into the tubules. This secretion of K^+ into the late distal tubule and cortical collecting duct (fig. 13.17) is stimulated by aldosterone. In a negative feedback loop, the adrenal cortex is stimulated to secrete aldosterone by a rise in the plasma concentration of K^+. Aldosterone then stimulates the secretion of K^+ into the cortical collecting ducts, thereby lowering the blood K^+ concentration to maintain homeostasis.

Since aldosterone stimulates Na^+ reabsorption as well as K^+ secretion, aldosterone levels should increase (to maintain homeostasis) when the plasma Na^+ concentration falls. However, this effect is indirect. If a person has inadequate dietary Na^+ (as

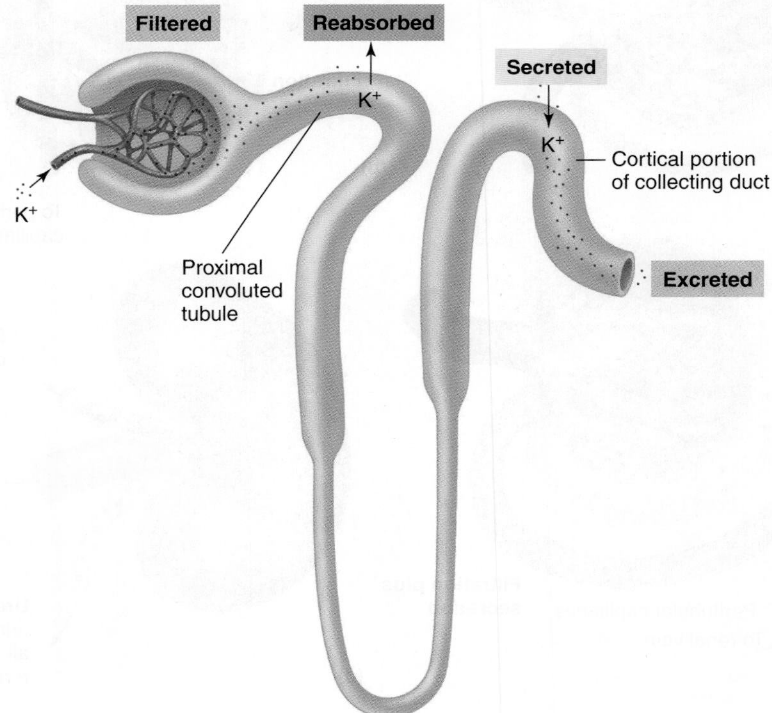

FIGURE 13.17 Potassium is reabsorbed and secreted. Potassium (K^+) is almost completely reabsorbed in the proximal tubule, but under aldosterone stimulation it is secreted into the cortical portion of the collecting duct. Almost all of the K^+ in urine is derived from secretion rather than from filtration.

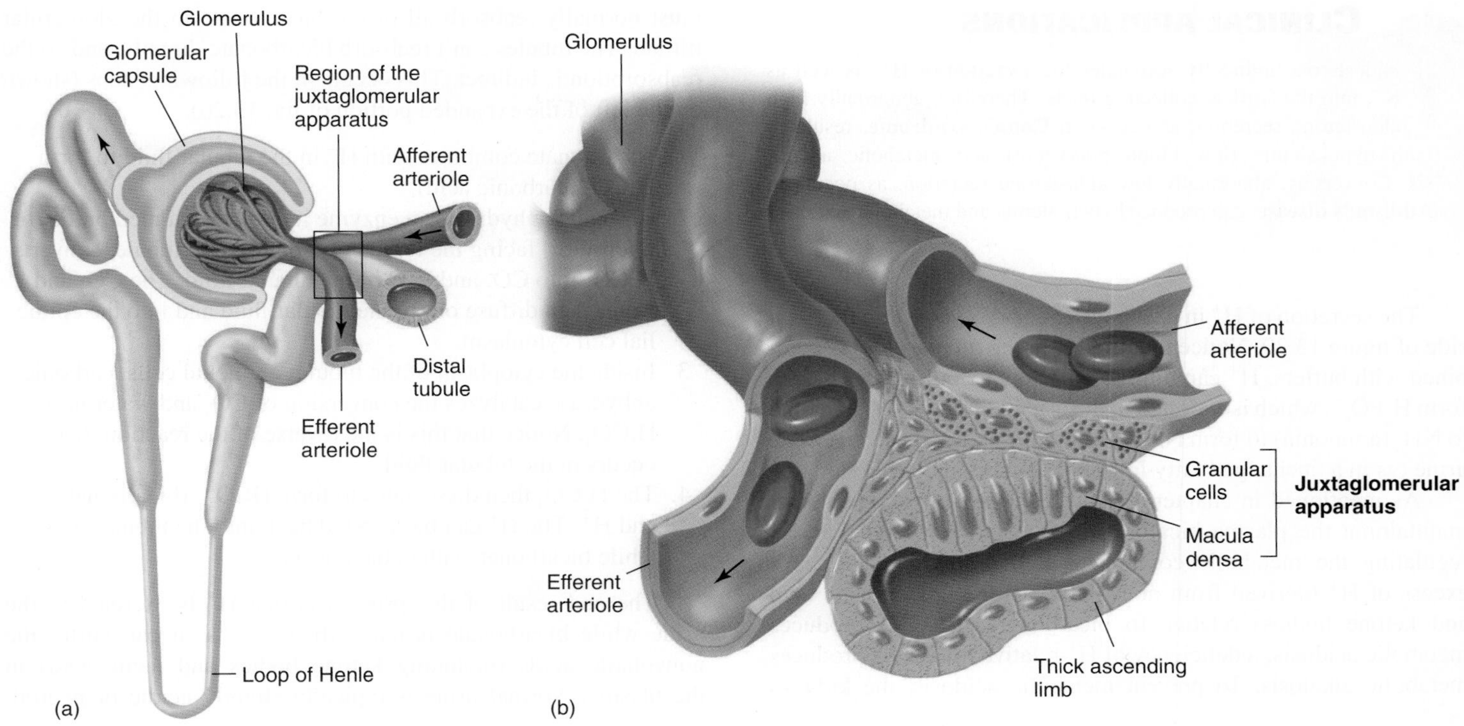

FIGURE 13.18 The juxtaglomerular apparatus. (*a*) The location of the juxtaglomerular apparatus. This structure includes the region of contact of the afferent arteriole with the last portion of the thick ascending limb of the loop. The afferent arterioles in this region contain granular cells that secrete renin, and the tubule cells in contact with the granular cells form an area called the macula densa, seen in (*b*).

salt), the blood volume and pressure will fall (chapter 10). This stimulates the **juxtaglomerular apparatus**, located where the afferent arteriole contacts the last portion of the ascending limb of the loop (fig. 13.18), to secrete the enzyme **renin**. Although this **renin-angiotensin-aldosterone system** was discussed in chapter 10, here is a brief review:

1. Renin catalyzes the conversion of a plasma protein, *angiotensinogen*, into *angiotensin I*.
2. Angiotensin I is inactive, but as it circulates in the blood it's converted by *angiotensin converting enzyme* (*ACE*) into **angiotensin II**.
3. Angiotensin II has numerous effects that raise the blood volume and pressure. It acts on arterioles to cause vasoconstriction, and it stimulates the adrenal cortex to secrete aldosterone.
4. The rise in aldosterone secretion stimulates increased Na^+ reabsorption in the late distal tubule and cortical collecting duct, so that more salt and water is retained in the blood.

The Kidneys Can Excrete H^+ and Reabsorb Bicarbonate

The reabsorption of Na^+ across the walls of the late distal tubule and cortical collecting duct creates an electrical attraction for K^+ or H^+ to be secreted (fig. 13.19). Acidosis (increased plasma H^+ concentration) increases the secretion of H^+ across the tubule, so that K^+ secretion is reduced. Conversely, alkalosis (lowered plasma H^+ concentration) reduces the secretion of H^+ and increases the secretion of K^+ into the filtrate. On the other hand, if a person has hyperkalemia (high plasma K^+ concentration), there is an increased K^+ secretion and a reduced secretion of H^+ into the tubular fluid. Hyperkalemia can thus cause an increase in the blood concentration of H^+ and acidosis.

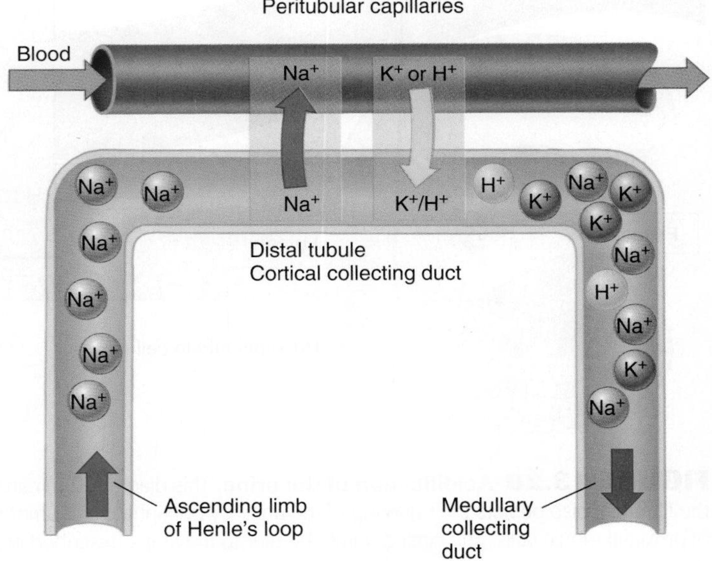

FIGURE 13.19 The reabsorption of Na^+ and secretion of K^+. In the distal tubule, K^+ and H^+ are secreted in response to the potential difference produced by the reabsorption of Na^+. High concentrations of H^+ may therefore decrease K^+ secretion, and vice versa.

Clinical Applications

Aldosterone indirectly stimulates the secretion of H^+, as well as K^+, into the cortical collecting ducts. Therefore, abnormally high aldosterone secretion, as occurs in **Conn's syndrome**, results in both hypokalemia (low blood potassium) and metabolic alkalosis. Conversely, abnormally low aldosterone secretion, as occurs in **Addison's disease**, can produce hyperkalemia and metabolic acidosis.

The secretion of H^+ in exchange for Na^+ is shown on the right side of figure 13.20. Notice that the H^+ in the tubular fluid is combined with buffers. H^+ can bind to a phosphate buffer, HPO_4^{-2}, to form $H_2PO_4^-$, which is excreted in the urine. Similarly, H^+ can bind to NH_3 (ammonia) to form NH_4^+ (ammonium ion); these give stale urine (as in a unchanged kitty litter box) its ammoniacal odor.

As mentioned in chapter 12, the kidneys are responsible for maintaining the plasma bicarbonate concentration, and thus for regulating the metabolic component of acid-base balance. An excess of H^+ (derived from nonvolatile acids such as lactic acid and ketone bodies) relative to bicarbonate (HCO_3^-) produces metabolic acidosis; a deficiency of H^+ relative to HCO_3^- produces metabolic alkalosis. To prevent metabolic acidosis, the kidneys must normally reabsorb all of the bicarbonate in the glomerular filtrate. The tubules can't reabsorb bicarbonate directly, and so the reabsorption is indirect. This occurs in the following steps (shown on the left of the expanded portion of fig. 13.20):

1. Bicarbonate combines with H^+ in the tubular fluid to form H_2CO_3 (carbonic acid).
2. Carbonic anhydrase, an enzyme in the plasma membrane (of microvilli) facing the lumen of the proximal tubule, converts H_2CO_3 into CO_2 and H_2O in the tubular fluid. The CO_2 and water then diffuse out of the tubular fluid and into the epithelial cell cytoplasm.
3. Inside the cytoplasm of the tubular epithelial cells, carbonic anhydrase catalyzes the conversion of CO_2 and water into H_2CO_3. Notice that this is the reverse of the reaction that occurs in the tubular fluid.
4. The H_2CO_3 then dissociates to form HCO_3^- (bicarbonate) and H^+. The H^+ can be secreted back into the tubular fluid, while bicarbonate enters the plasma.

The end result of this process is that H^+ is excreted in the urine while bicarbonate is reabsorbed, so that it can buffer the nonvolatile acids (including ketone bodies and lactic acid) in the plasma. Normal urine is typically slightly acidic or neutral,

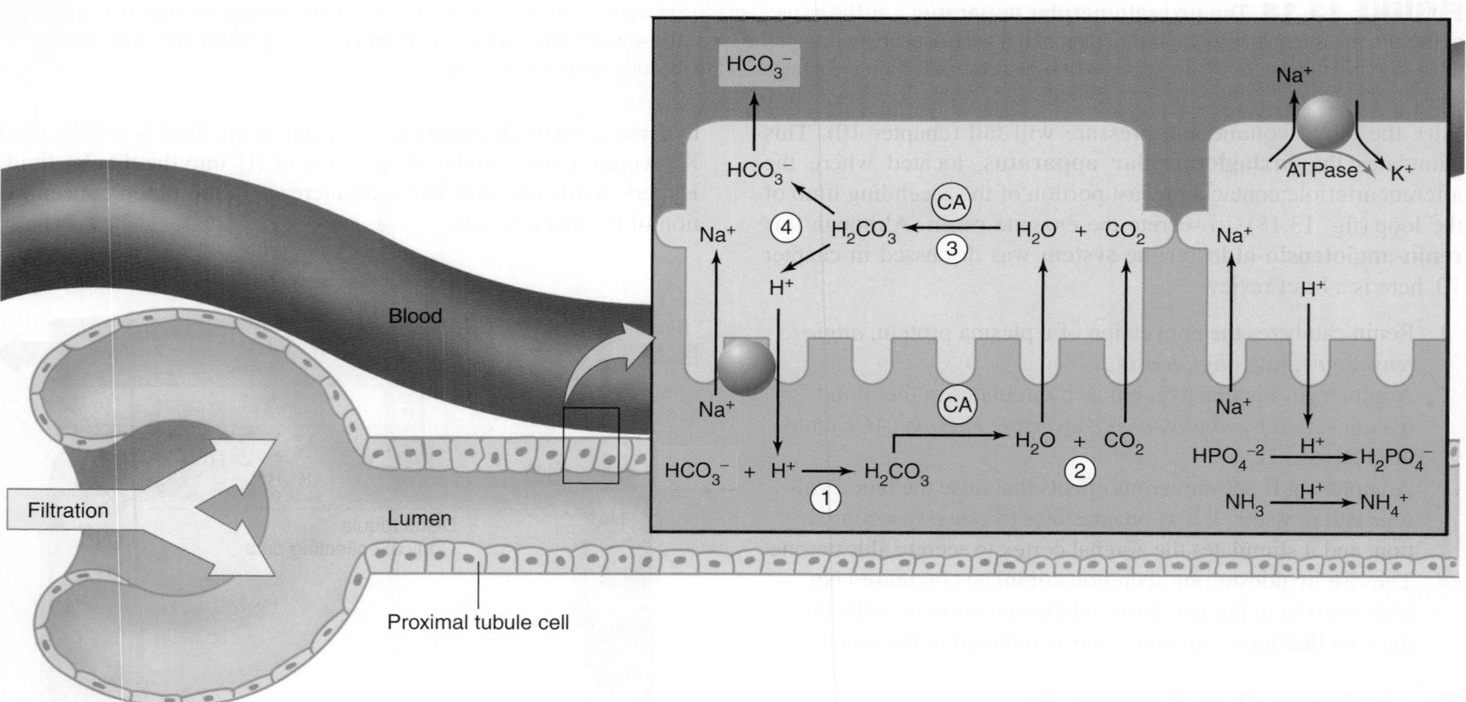

FIGURE 13.20 Acidification of the urine. This diagram summarizes how the urine becomes acidified and how bicarbonate is reabsorbed from the filtrate. It also depicts the buffering of the urine by phosphate and ammonium buffers. (CA = Carbonic anhydrase.) The inset depicts an expanded view of proximal tubule cells. The circled numbers refer to the steps described in the text.

usually with a pH lower than arterial blood (pH 7.35–7.45). The arterial pH is kept constant while the urine pH varies, depending on the body's production of the nonvolatile acids and on the acidity of foods and drinks. For example, drinking lemonade or coffee (which are acidic) will cause more H^+ to be excreted in the urine, which lowers the urine pH while maintaining homeostasis of the blood pH.

CHECK POINT

1. Describe how the nephron handles Na^+, and how Na^+ excretion and retention are regulated.
2. Describe how the nephron handles K^+, and how K^+ excretion and retention are regulated.
3. Explain how the plasma concentration of K^+ regulates aldosterone secretion, and how plasma Na^+ indirectly affects the renin-angiotensin-aldosterone system.
4. How does H^+ enter the urine, and in what forms is it excreted? How is the excretion of H^+ influenced by aldosterone and the plasma K^+ concentration?
5. Step by step, describe how bicarbonate is reabsorbed from the tubular fluid.

PHYSIOLOGY IN HEALTH AND DISEASE

A **diuretic** is a substance that increases urine volume. *Water* is the most common diuretic, acting to dilute the plasma (lower its osmolarity) and thereby reduce the stimulation of osmoreceptors in the hypothalamus. This lowers the secretion of ADH from the posterior pituitary, which reduces the permeability of the collecting ducts to water and causes *diuresis* (increased water excretion in the urine).

Osmotic diuretics are extra solutes in the tubular fluid. These increase the osmolarity of the fluid within the collecting ducts, so that the osmotic gradient (difference in concentration) between the tubular fluid and the interstitial fluid of the renal medulla is reduced. As a result, less water can be drawn out of the collecting ducts by osmosis, leaving more to be excreted in the urine. Glucose is an example of an endogenous molecule that can become an osmotic diuretic, if a person is hyperglycemic and the renal plasma threshold for glucose is exceeded. Because of this, a person with uncontrolled diabetes mellitus who "spills glucose" in the urine has *polyuria* (literally, "many urines") and can become dehydrated. Similarly, excessive production of ketone bodies (which can cause ketoacidosis; chapter 12) in uncontrolled type 1 diabetes mellitus results in *ketonuria*, and the extra ketone bodies in the tubular filtrate have an osmotic diuretic effect. A person on a strict weight-reducing diet, who has a rapid breakdown of fat and thus a high plasma level of ketone bodies (*ketosis*), can also have ketonuria. The resulting osmotic diuresis promotes dehydration, which is part of the reason dieters are advised to drink lots of water. *Mannitol* is an exogenous substance sometimes used clinically as an osmotic diuretic.

The most powerful clinical diuretics are the **loop diuretics**, including *furosemide* (*Lasix*). These inhibit as much as 25% of the salt transport out of the ascending limbs of the loops of Henle. Because of this, the interstitial fluid of the renal medulla is less concentrated (hypertonic), producing less of an osmotic gradient to draw water out of the collecting ducts. The **thiazide diuretics** (such as *hydrochlorothiazide*) inhibit up to 8% of the salt and water reabsorption by inhibiting Na^+ transport in the last part of the ascending limb and first part of the distal tubule, thereby reducing the osmotic gradient for water reabsorption. Although these are effective and commonly used diuretics, Lasix and hydrochlorothiazide have an undesirable side effect: they promote the excretion of K^+ in the urine, which lowers the plasma K^+ concentration (*hypokalemia*). Hypokalemia can cause neuromuscular disorders and ECG abnormalities. Because of this, people taking Lasix and hydrochlorothiazide should get their blood K^+ concentrations measured periodically, and must often take potassium supplements (in the form of KCl).

The hypokalemia in people taking Lasix or hydrochlorothiazide is caused by an increase in aldosterone-stimulated secretion of K^+ into the cortical collecting ducts. Because of this, some medications for the treatment of hypertension (high blood pressure; chapter 10) combine hydrochlorothiazide with one of the **potassium-sparing diuretics**. *Spironolactone* (such as *Aldactone*) diuretics block aldosterone action by competing for the aldosterone receptor proteins in the cells of the cortical collecting ducts. *Triamterene* (*Dyrenium*) is a potassium-sparing diuretic that acts more directly to block Na^+ reabsorption and K^+ secretion in the cortical collecting ducts. The diuretic actions of hydrochlorothiazide combined with the weaker diuretic but potassium-sparing actions of these drugs lower the blood volume, and thus the blood pressure, of people with hypertension.

Physiology in Balance

Urinary System

Working together with . . .

Cardiovascular System

The kidneys regulate the total blood volume and blood pressure . . . p. 240

Nervous System

Sympathetic neurons regulate blood flow through afferent arterioles, which helps maintain the GFR; parasympathetic neurons stimulate micturition . . . p. 86

Endocrine System

Antidiuretic hormone (vasopressin) stimulates water reabsorption and aldosterone stimulates salt and water reabsorption across the nephron tubules into the blood . . . p. 186

Muscular System

The smooth muscle of the urinary bladder (detrussor muscle) must contract to empty the bladder . . . p. 215

Respiratory System

The kidneys filter and reabsorb bicarbonate, while the lungs regulate the plasma CO_2 . . . p. 297

Homeostasis Revisited

Topic	Set Point	Integrating Center	Sensors	Effectors
How ADH acts on the renal nephrons to maintain homeostasis of the plasma osmolarity when a person is dehydrated	Plasma osmolarity	Hypothalamus	Osmoreceptors in the hypothalamus	Posterior pituitary secretes increased amounts of ADH, which increases the permeability of the collecting ducts to water, thereby promoting water reabsorption.
How transport processes in the nephrons that affect renal plasma clearance help maintain homeostasis of the plasma glucose concentration	Glucose flowing in the filtrate through the proximal tubules	Cells of the proximal tubules	Membrane transport carriers for glucose	Membrane transport carriers reabsorb glucose from the filtrate, preventing glucose from being excreted when the plasma glucose concentration is below a renal plasma threshold level.
How aldosterone acts on the nephrons to help maintain homeostasis of the plasma Na^+ concentration when there is inadequate dietary Na^+	Blood flow through the afferent arterioles	Juxtaglomerular apparatus	Juxtaglomerular apparatus	Renin secretion from the juxtaglomerular apparatus indirectly causes a rise in angiotensin II production, which stimulates aldosterone secretion from the adrenal cortex; aldosterone stimulates the cortical collecting ducts to reabsorb more Na^+, maintaining more Na^+ in the blood.
How aldosterone acts on the nephrons to help maintain homeostasis of the plasma K^+ concentration when the plasma K^+ rises	Plasma K^+ concentration	Adrenal cortex	Adrenal cortex	Aldosterone secreted by the adrenal cortex stimulates the cortical collecting ducts to secrete more K^+ into the filtrate, promoting its elimination from the blood.
How tubule transport processes for bicarbonate help maintain homeostasis of the plasma pH	Bicarbonate flowing in the filtrate through the proximal tubules	Cells of the proximal tubules	Carbonic anhydrase in the plasma membranes facing the filtrate	Carbonic anhydrase in the plasma membrane converts carbonic acid into CO_2 and H_2O; carbonic anhydrase in the cytoplasm converts CO_2 and H_2O into carbonic acid, allowing bicarbonate to enter the blood.

Summary

Urine Is Formed by Nephrons in the Kidneys 325

- The urine produced by the kidneys is channeled to the urinary bladder through the ureters, and then exits the urinary bladder in the urethra.
- Micturition, or urination, is controlled by the micturition center in the sacral region of the spinal cord, which stimulates contraction of the detrusor muscle of the bladder and relaxation of the internal urethral sphincter via activation of parasympathetic neurons.
- Each kidney contains more than a million nephrons; each nephron consists of tubules and associated blood vessels.
- The glomerular capsule, proximal tubule, distal tubule, and first portion of the collecting duct are located in the renal cortex; the descending and ascending limbs of the loop of Henle, and most of the collecting duct, are located in the renal medulla.
- Blood passes through an afferent arteriole to the glomerulus, and then from the glomerulus in an efferent arteriole; this

blood then enters peritubular capillaries, which surround the nephron tubules.
- Glomerular filtrate exits the glomerular capillaries through large channels called fenestrae; the filtrate then passes through a basement membrane and a slit diaphragm between pedicels, which are the tiny extensions of the inner (visceral) layer of the glomerular capsule.
- The glomerular filtration rate, or GFR, averages 120 ml/min, or 180 L/day; the filtrate is isotonic, but has a very low protein concentration.
- The filtration barrier that excludes proteins from the filtrate is the basement membrane and the slit diaphragm; the filtrate enters the glomerular capsule and then the proximal tubules.

The Kidneys Reabsorb Salt and Water from the Filtrate 329

- Reabsorption refers to the transport of substances from the glomerular filtrate, across the walls of the nephron tubules, and into the blood.
- Because there is such a large volume of filtrate produced per day, most must be reabsorbed; however, because the filtrate is isotonic to plasma, a concentration gradient must first be produced across the wall of the tubule for osmosis to occur.
- Approximately 65% of salt and water of the filtrate is constantly reabsorbed across the walls of the proximal tubules.
- The proximal tubules first actively transport Na^+, and Cl^- follows by electrical attraction; then water follows the NaCl by osmosis, causing NaCl and water to be reabsorbed into the peritubular capillaries.
- The tubular fluid remaining to enter the descending limb of the loop of Henle is reduced in volume but isotonic, with a concentration of 300 mOsM; this fluid descends into the renal medulla, which has a hypertonic interstitial fluid.
- The descending limb is permeable to water, but not to NaCl, so that water is drawn out by osmosis and reabsorbed into the peritubular capillaries.
- The ascending limb pumps NaCl into the surrounding interstitial fluid of the renal medulla, and loops of blood vessels (the vasa recta) keep the NaCl in the interstitial fluid; this helps to make the renal medulla hypertonic.
- Because the ascending limb isn't permeable to water, the tubular fluid that enters the distal tubule is hypotonic, with a concentration of 100 mOsM.
- The hypertonic renal medulla draws water out of the descending limb, which then supplies concentrated fluid to the ascending limb; the interactions between the two limbs form a countercurrent multiplier system.
- The countercurrent multiplier system of the loops of Henle make the interstitial fluid of the renal medulla very hypertonic, with a concentration as high as 1,400 mOsM in the deepest regions.
- Urea also contributes to the hypertonic renal medulla, because some urea can exit the collecting ducts and re-enter the ascending limbs; this allows urea to recycle and build up in the interstitial fluid of the renal medulla.
- The collecting ducts are permeable to water, and this permeability is adjusted by ADH (vasopressin), which stimulates the insertion of aquaporin channels into the plasma membranes in the collecting ducts.
- With more ADH, the collecting ducts are more permeable to water, and more water leaves the collecting ducts by osmosis and is reabsorbed into the blood; this occurs in response to dehydration, when osmoreceptors in the hypothalamus are stimulated.

The Kidneys Maintain Homeostasis of Plasma Composition 334

- The renal plasma clearance measures the kidney's ability to remove, or clear, molecules from the plasma; this is measured in ml/min, so that it can be compared to the GFR to ascertain how the kidneys "handle" a particular molecule.
- Secretion refers to transport of a molecule or ion from the tubular fluid, across the wall of the tubule, and into the blood; secretion, like filtration, rids the blood of molecules and thus increases their renal plasma clearance.
- If a molecule is filtered, but neither reabsorbed nor secreted, its renal plasma clearance will equal the GFR; inulin is the only molecule handled in this way, but the plasma creatinine clearance, or simply the plasma creatinine levels, are often used clinically as indicators of the health of the GFR.
- If a molecule is filtered and then reabsorbed, its renal plasma clearance will be less than the GFR; glucose is normally completely reabsorbed, because carrier proteins don't become saturated and reach their transport maximum until the plasma glucose concentration becomes at least twice as high as the normal fasting levels.
- Penicillin and some other antibiotics are secreted, and so their renal plasma clearance is greater than the GFR; PAH is a molecule that is completely secreted, and so its renal plasma clearance equals the total renal blood flow.

The Kidneys Maintain Homeostasis of Plasma Electrolytes and pH 337

- About 98% of Na^+ reabsorption doesn't require aldosterone stimulation; this includes the 90% that is reabsorbed in the proximal tubules, and most of the Na^+ that enters the distal tubules and collecting ducts.
- The remaining amount of Na^+ entering the late distal tubule and cortical collecting duct can't be reabsorbed without aldosterone stimulation; under maximum aldosterone stimulation, all of the Na^+ in the tubular fluid is reabsorbed.
- Almost all of the K^+ in the filtrate is reabsorbed in the proximal tubules, so that almost none of the filtered K^+ leaves in the urine; the only way that significant amounts of K^+ can be excreted is for the K^+ to be secreted into the late distal tubule and cortical collecting duct.
- Aldosterone secretion from the adrenal cortex is directly stimulated by a rise in the plasma K^+ concentration; aldosterone then stimulates the secretion of K^+ by the late distal tubule and cortical collecting duct into the tubular fluid and urine, maintaining homeostasis of the plasma K^+ concentration.
- Aldosterone secretion by the adrenal cortex is also stimulated by angiotensin II, which is formed as a consequence of the secretion of renin by the juxtaglomerular apparatus in the kidneys.
- The renin-angiotensin-aldosterone system is activated by a fall in the blood flow through the kidneys, and the action of aldosterone to stimulate reabsorption of Na^+ (together with Cl^- and water) helps to restore homeostasis of blood volume and pressure.
- The kidneys help to regulate the metabolic component of acid-base balance by excreting H^+ in the urine, where much of the H^+ combines with phosphate and ammonium buffers; the kidneys also reabsorb bicarbonate (HCO_3^-).
- The bicarbonate in the filtrate is reabsorbed indirectly, with the aid of carbonic anhydrase catalyzing the breakdown of carbonic acid in the filtrate and then the formation of carbonic acid in the cytoplasm of the tubule cells.

Review Activities

Objective Questions: Test Your Knowledge

1. Which of the following statements regarding the control of micturition is true?
   a. Sympathetic axons stimulate contraction of the detrusor muscle.
   b. Parasympathetic axons stimulate contraction of the internal urethral sphincter.
   c. Parasympathetic axons stimulate relaxation of the external urethral sphincter.
   d. The micturition center is located in the sacral region of the spinal cord.
2. The detrusor muscle is located in the
   a. urinary bladder.
   b. urethra.
   c. ureters.
   d. kidneys.
3. The part of the nephron that first receives a blood filtrate is the
   a. proximal tubule.
   b. distal tubule.
   c. glomerular capsule.
   d. collecting duct.
4. Which of the following parts of the nephron is not located entirely in the renal cortex?
   a. Glomerular capsule
   b. Collecting duct
   c. Proximal tubule
   d. Distal tubule
5. Which of the following statements about the renal pelvis is false?
   a. It is the origin of the ureter.
   b. It receives urine from the calyces.
   c. It contains renal pyramids.
   d. It is surrounded by the renal medulla.
6. Blood goes from the glomerulus directly into
   a. an efferent venule.
   b. an efferent arteriole.
   c. an interlobular vein.
   d. an interlobular artery.
7. Which of the following statements about plasma proteins is true?
   a. They are prevented from entering the filtrate by the basement membrane and slit diaphragm.
   b. They enter the filtrate and are excreted in the urine.
   c. They enter the filtrate and then are reabsorbed, so none are in the urine.
   d. They are prevented from entering the filtrate by the capillary fenestrae.
8. Which of the following statements regarding glomerular filtration is true?
   a. The net filtration pressure of the glomerular capillaries is unusually large.
   b. The capillary fenestrae are smaller than the typical capillary pores.
   c. All of the glomerular filtrate becomes urine.
   d. The glomerular filtration rate averages 120 ml/min, or 180 L/day.
9. Which of the following statements regarding the proximal tubule is false?
   a. The fluid entering the proximal tubule is isotonic.
   b. The fluid leaving the proximal tubule is isotonic.
   c. Active transport of water reduces the volume of the filtrate.
   d. Approximately two-thirds of the salt and water of the filtrate is reabsorbed here.

Match the nephron tubule region to its transport properties:

10. Proximal tubule — a. NaCl actively transported; impermeable to water
11. Descending limb — b. Permeable to water under ADH stimulation
12. Ascending limb — c. Constantly permeable to water
13. Collecting duct — d. Active transport of Na^+; Cl^- and water follow

14. Which of the following statements regarding the countercurrent multiplier system is false?
    a. This refers to interactions between the descending and ascending limbs of the loop of Henle.
    b. This increases the concentration of urea in the renal medulla.
    c. This helps to produce a very hypertonic renal medulla.
    d. Because of this, we can reabsorb water and excrete highly concentrated urine.
15. What is the concentration of tubular fluid in the distal tubule?
    a. 1,400 mOsM
    b. 300 mOsM
    c. 100 mOsM
    d. 10 mOsM
16. When a person produces the minimum volume of urine possible in a day, what would be its maximum concentration?
    a. 1,400 mOsM
    b. 300 mOsM
    c. 100 mOsM
    d. 10 mOsM
17. If a substance has a renal plasma clearance that is greater than the GFR, this means that this substance is
    a. filtered, but neither reabsorbed nor secreted.
    b. filtered and secreted.
    c. filtered and reabsorbed.
    d. neither filtered, secreted, nor reabsorbed.
18. Which of the following has a renal plasma clearance that is normally zero because it's completely reabsorbed?
    a. Protein
    b. Creatinine
    c. Urea
    d. Glucose
19. Which of the following statements is true regarding a substance that is filtered but neither reabsorbed nor secreted?
    a. It has a renal plasma clearance less than the GFR.
    b. It has a renal plasma clearance greater than the GFR.
    c. It is PAH.
    d. It is inulin.
20. Which of the following is both reabsorbed and secreted and has a renal plasma clearance that depends on aldosterone?
    a. Na^+
    b. K^+
    c. HCO_3^-
    d. Cl^-

21. The part of the nephron tubule that responds to aldosterone is the
    a. proximal tubule.
    b. thick segment of the ascending limb.
    c. cortical collecting duct.
    d. descending limb.

22. Which of the following is not stimulated by aldosterone?
    a. Na^+ reabsorption
    b. K^+ secretion
    c. H^+ secretion
    d. Na^+ secretion

23. Which of the following statements regarding the juxtaglomerular apparatus is false?
    a. It secretes renin into the blood.
    b. Its activation leads to the formation of angiotensin II.
    c. It's located where the efferent arteriole contacts the proximal tubule.
    d. It's stimulated by a fall in the blood flow through the kidney.

24. Which of the following statements about urine is false?
    a. H^+ combines with phosphate and ammonium buffers.
    b. Urine pH varies more than the arterial pH.
    c. Bicarbonate normally gives the urine an alkaline pH.
    d. Urine can have a pH less than 7.

25. Which of the following statements regarding bicarbonate is false?
    a. Bicarbonate enters the glomerular filtrate.
    b. Bicarbonate combines with H^+ in the filtrate to form H_2CO_3.
    c. Bicarbonate is transported out of the filtrate and into the cytoplasm of the tubular cells.
    d. The reabsorption of bicarbonate requires the action of carbonic anhydrase.

Essay Questions 1: *Test Your Understanding*

1. Describe the path of urine from the calyces to the outside of the body.
2. How is micturition (urination) regulated by the nervous system?
3. Draw a renal nephron and label its parts. Indicate which parts are located in the renal cortex and which parts are in the renal medulla.
4. Trace the course of arterial blood flow into the kidney, leading up to the glomerulus. Then trace the blood flow from the glomerulus to the renal vein.
5. Identify the structures that separate the blood plasma in the glomerular capillaries from the filtrate in the glomerular capsule. What are the forces involved in the production of glomerular filtrate?
6. Which components of blood enter the filtrate, and which do not? Identify the filtration barriers involved.
7. What is the average glomerular filtration rate (GFR)? How does this compare to the total blood volume, and what does that comparison imply? What is the obligatory water loss, and what is the theoretical maximum amount of urine that can be produced per day?
8. In a step-by-step manner, explain how NaCl and water are reabsorbed across the walls of the proximal tubules. Identify the percent of the filtrate volume reabsorbed at this region, and state if this is constant or variable.
9. Describe the membrane transport properties of the descending and ascending limbs of the loop of Henle. Explain how they interact as a countercurrent multiplier system.
10. What is the significance of the loop of Henle and the countercurrent multiplier system, in terms of kidney function and the maintenance of homeostasis?
11. Describe the permeability of the nephron tubule to urea and the movement of urea between different segments of the tubule. What function does this serve?
12. Describe how ADH (vasopressin) influences the permeability properties of the collecting duct. Explain how different levels of ADH secretion affect the urine volume and blood volume.
13. In a step-by-step manner, explain how urine volume and concentration are affected when a person drinks excessive amounts of water.
14. In a step-by-step manner, explain how urine volume and concentration are affected when a person is dehydrated.
15. What is the glomerular filtration rate (GFR), and how can it be measured by the renal plasma clearance of particular molecules?
16. Describe how the renal plasma clearance of substances is affected by reabsorption and secretion. Provide examples of substances that are reabsorbed or secreted.
17. Why is glucose normally absent from fasting urine? What must happen for glycosuria to occur?
18. What is creatinine, how do the kidneys handle it, and what is the significance of the plasma creatinine concentration?
19. Describe how the nephron tubule handles Na^+, and the role of aldosterone in the renal plasma clearance of Na^+.
20. Describe how the nephron tubule handles K^+, and the role of aldosterone in the renal plasma clearance of K^+.
21. Identify the location of the juxtaglomerular apparatus, and explain how its function relates to the secretion of aldosterone and homeostasis of blood volume.
22. Describe the interaction between K^+ and H^+ secretion and Na^+ reabsorption in the cortical collecting duct, and explain how these are affected by aldosterone.
23. How does the pH of urine compare with the pH of blood? Identify the buffers present in urine.
24. In a step-by-step manner, explain how the nephron tubule reabsorbs bicarbonate from the filtrate.

Essay Questions 2: *Test Your Analytical Ability*

1. Children too young to control when they urinate eventually develop voluntary control and get "potty trained." Which muscle do you think is involved in this training, and which branch of the nervous system? Explain.
2. Incontinence (inability to control urination) can be caused by an "overactive bladder." Which muscle is involved, and which branch of the nervous system? Propose a type of drug that would block neural stimulation of the bladder.
3. Explain how the glomerular filtration rate (GFR) could be affected by constriction and dilation of the afferent arterioles. How could this help to maintain a constant GFR, and why would constancy of the GFR be important?
4. Could you call the arrangement of vessels going to and from the glomerulus and peritubular capillaries a portal system? Explain, using the hypothalamo-hypophyseal portal system (chapter 8) for comparison.
5. Suppose a person had a mutation in a gene that coded for an important structural protein in the slit diaphragms. How would this influence the chemical composition of the urine? How could that affect the chemical composition of the blood, and what symptom might that produce (hint: see chapter 10)? Explain.
6. Although the GFR remains relatively constant, the volume of urine excreted per day can vary tremendously. What must change to produce this variation, and how does that variation help maintain homeostasis?

7. A drug that inhibited active transport in the proximal tubule would be rapidly fatal. Explain why this statement is true.
8. Because kangaroo rats have longer loops of Henle than we do, they can produce more concentrated urine. Explain, using a description of the transport properties of the loop and the countercurrent multiplier system.
9. Suppose that yesterday a person reabsorbed 99.2% of the original volume of glomerular filtrate, whereas today that person reabsorbed 98.2%. What probably produced this change in reabsorption? How did this change the urine volume from yesterday to today? Explain.
10. Alcohol inhibits ADH secretion. Given this, explain why alcoholic drinks are a poor choice for hydration, and why a person who drank a large volume of fluid containing alcohol on one day would be dehydrated on the next.
11. Suppose a drug inhibits all active transport throughout the nephron. This would be fatal, but for the sake of argument, suppose the person lived long enough to produce a urine sample that reflected the complete absence of active transport throughout the nephron. What would be the osmolarity of this urine? Explain.
12. Suppose a person is injected with a given amount of inulin. What is its renal plasma clearance, in ml/min? Given this number, and a total blood volume of 5 L, calculate how many passes of blood through the kidney would be required for the complete elimination of the inulin.
13. Suppose a person is injected with a given amount of PAH. What is its renal plasma clearance, in ml/min? Given that number, and a total blood volume of 5 L, calculate the percent of the total blood volume received by the kidneys.
14. The normal fasting blood glucose concentration varies from 70 to 110 mg/100 ml. Suppose a person's fasting blood sample tests hyperglycemic, but the urine test for glucose is negative. Explain how this could be possible.
15. Suppose a person with glomerulonephritis (inflammation of the glomeruli) has proteinuria and also shows edema. What is the relationship between these? (Hint: see chapter 10.)
16. A person given a drug that acts as a carbonic anhydrase inhibitor can develop metabolic acidosis. Explain how this can occur.
17. A person who takes a diuretic drug that blocks Na^+ reabsorption in the ascending limb or early distal tubule has more Na^+ entering the late distal tubule and cortical collecting duct. This person also has a tendency to lose excessive amounts of K^+ in the urine. Propose an explanation that relates these observations.

Web Activities

ARIS™ **For additional study tools, go to: www.aris.mhhe.com. Click on your course topic and then this textbook's author/title. Register once for a semester's worth of interactive activities and practice quizzing to help boost your grade!**

CHAPTER 14

Digestive System

Chapter Outline

Homeostasis

The body needs the subunits in food molecules for cell respiration, so that all of its cells have sufficient ATP to power the energy-requiring processes essential for life. These caloric (energy) needs may be temporarily provided by stored glycogen and fat, but homeostasis requires that food molecules derived from the GI tract replace these energy-storage molecules. In addition, food absorbed from the GI tract contains many essential molecules and ions required by body cells for their metabolism, and water absorbed from the GI tract is needed to replace lost water and thereby maintain homeostasis of blood volume and pressure. Proper functioning of the accessory digestive organs, including the liver, gallbladder, and pancreas, is required for digestive function and is needed to maintain homeostasis of the blood composition.

Clinical Investigation

But if homeostasis is not maintained . . .

Jim noticed that he developed stomach pains when he drank wine or coffee, and was puzzled when he felt a sharp pain below his right scapula (shoulder blade) whenever he ate peanut butter or bacon. He questioned his doctor about this, and during the examination his physician remarked that the whites of his eyes were yellowed, suggesting jaundice. Blood tests revealed an elevated level of conjugated bilirubin, but all other blood tests were normal, including tests for free bilirubin, urea and liver enzymes, and pancreatic amylase. However, he did have fat in his stools. His physician said that Jim probably had gallstones, and scheduled him for an x-ray. She advised Jim to avoid the foods that provoked his pains, and prescribed Prilosec for his stomach.

What caused Jim's stomach pains, and how would Prilosec help? What caused Jim's pain below his right scapula, and how is that related to gallstones? What is the relationship between gallstones and Jim's jaundice, and what do the results of his blood test reveal?

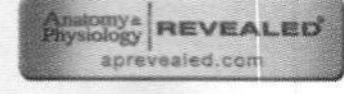

A virtual cadaver dissection experience

Regions of the Gastrointestinal Tract Have Different Specializations

The esophagus delivers food to the stomach, which secretes gastric juice containing hydrochloric acid (HCl) and a protein-digesting enzyme, pepsin. Portions of the stomach contents (chyme) are delivered to the small intestine, where most digestion and absorption of food occurs, and then to the large intestine, where water and electrolytes (ions) are absorbed. Nerves and hormones regulate gastric activity and the passage of chyme to the small intestine.

The digestive system absorbs molecules that originate from outside of the body (in food), so that they can enter the blood. Some of these molecules may leave the blood and enter the renal nephrons, but most will then be reabsorbed back into the blood. Substances that were ingested and not absorbed, and molecules produced by the body that enter the bile, are excreted in the feces. This excretion of waste products parallels the production of urine by the kidneys, and the elimination of carbon dioxide by the lungs. The liver, which is considered part of the digestive system, has several hundred functions that impact the physiology of all body systems. So, the construction of your knowledge pyramid is now well under way with the addion of digestive system physiology.

The **digestive system** includes the mouth, pharynx, stomach, small intestine, and large intestine, as well as the *accessory digestive organs*: the teeth and salivary glands in the head, and the liver, gallbladder, and pancreas in the abdominal cavity (fig. 14.1). The tubular portions of the digestive system—the esophagus, stomach, small intestine, and large intestine—are often termed the **digestive tract**, or **gastrointestinal (GI) tract**. The different regions and organs of the digestive system are specialized to perform different functions,[1] which include the following:

1. **Motility.** This refers to the movement of food through the esophagus to the stomach, the churning of the stomach contents, and then the movement of food through the small and large intestine.
2. **Secretion.** Digestive juices (gastric juice and pancreatic juice) are exocrine secretions, because they enter the lumen (cavity) of the GI tract. The stomach and small intestine also secrete several hormones that regulate digestive functions.
3. **Digestion.** This refers to the breakdown of food molecules into their smaller subunits—proteins into amino acids; starch and disaccharides into monosaccharides; and triglycerides (fats and oils) into fatty acids and glycerol. Complete digestion occurs in the small intestine.

FYI [1]In *Planaria* (a type of flatworm), the GI tract has only one opening—the mouth is also the anus. Each cell lining the GI tract of Planaria is exposed to food, the subunits produced by the digestion of food molecules, and wastes. By contrast, the GI tract of higher organisms (including us) allows one-way transport, from mouth to anus. This one-way transport permits our GI tract to have different specialized regions and function as a "dis-assembly line."

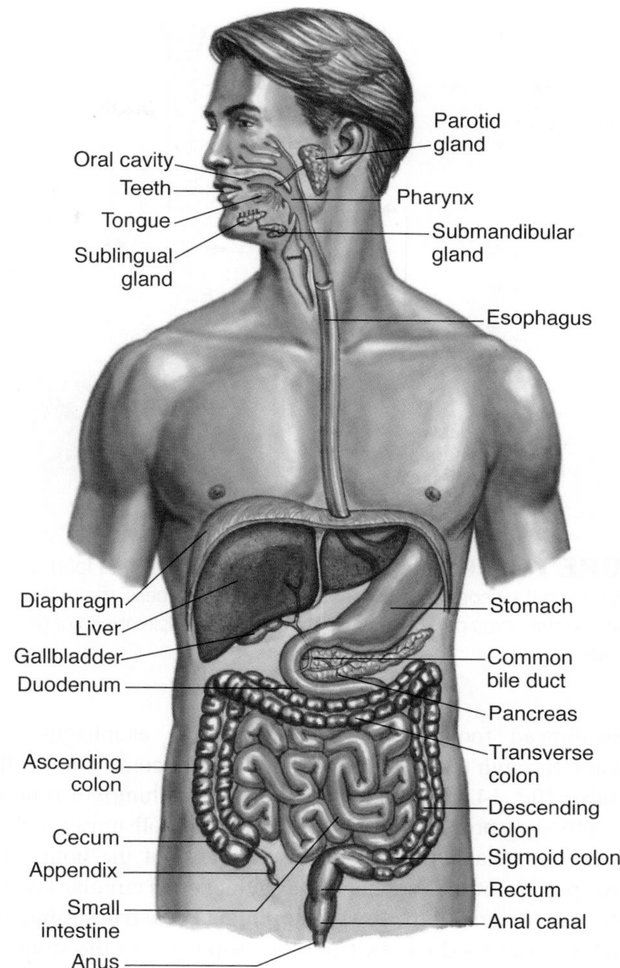

FIGURE 14.1 The organs of the digestive system. The digestive system includes the gastrointestinal tract and the accessory digestive organs.

4. **Absorption.** This refers to the transport of the subunit molecules from the lumen of the small intestine into the blood or lymph. Only molecules that are absorbed enter the body; wastes that remain in the intestinal lumen are outside of the body and are excreted in the feces. Most nutrient absorption occurs in the small intestine.
5. **Excretion.** Wastes that are not absorbed can be temporarily stored in the large intestine until they're excreted as feces.

The Esophagus Transports Food to the Stomach

The **esophagus** is a muscular tube, about 25 cm (10 in.) long, which joins the pharynx to the stomach. It's located posterior to the trachea in the thoracic cavity, and then passes through an opening in the diaphragm—the *esophageal hiatus*—to enter the abdominal cavity and join the stomach. The muscle type varies in different regions of the esophagus: the upper third contains skeletal muscle (and so its contractions are under voluntary control, to aid swallowing); the middle third has a mixture of skeletal and smooth muscle, and the last third contains only smooth muscle.

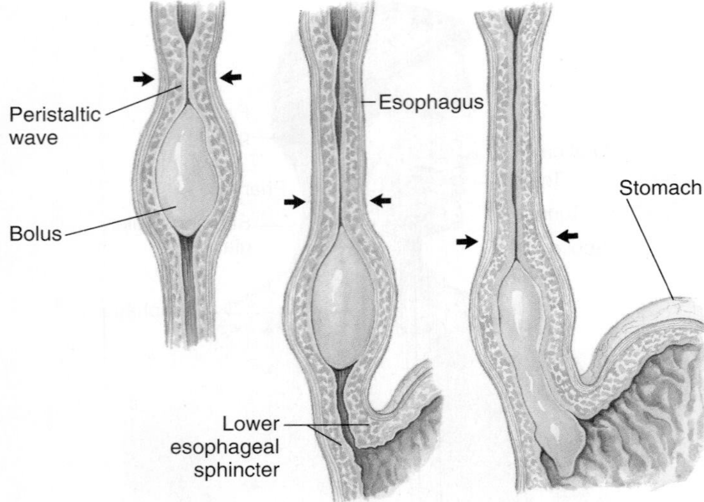

FIGURE 14.2 Peristalsis. A bolus (rounded mass) of food is pushed down the esophagus by peristalsis. In this process, the muscular wall behind the bolus contracts while the smooth muscle ahead of the bolus relaxes.

Swallowed food is pushed through the esophagus (and afterward, through the intestines) by a wavelike contraction called **peristalsis** (fig. 14.2). Movements of a *bolus* (a lump, or rounded mass) of food are produced by contractions of smooth muscle behind the bolus and relaxation of the muscle ahead of the bolus. The terminal portion of the esophagus has a thickened circular layer of smooth muscle termed the **lower esophageal (gastroesophageal) sphincter**. After food passes from the esophagus to the stomach, this sphincter layer closes, so that the stomach contents aren't normally pushed back up into the esophagus.

CLINICAL APPLICATIONS

The lower esophageal sphincter isn't a true sphincter muscle, and so it does at times permit the acidic contents of the stomach to enter the esophagus. This can create a burning sensation commonly called **heartburn**, although it really doesn't relate to the heart. This may be associated with a **hiatal hernia**, where a portion of the stomach protrudes up into the thoracic cavity through the esophageal hiatus in the diaphragm. The lower esophageal sphincter functions erratically in infants less than 1 year old, causing them to "spit up" following meals. Certain mammals, such as rodents, have a true lower esophageal sphincter that doesn't permit them to regurgitate. This is why poison is effective in killing rats and mice.

The Stomach Secretes Pepsin and Hydrochloric Acid (HCl)

The esophagus joins the stomach at its *cardiac region* (fig. 14.3); food from the esophagus mixes with **gastric juice**, the secretions of the stomach, to form a pasty material called **chyme**. Chyme passes into the *fundus* and *body* of the stomach to the *pyloric antrum,* and then must pass through the *pyloric sphincter* to leave the stomach and enter the first segment of the small intestine, the duodenum. Figure 14.3 also shows that the stomach has three layers of smooth muscles (*longitudinal*, *circular*, and *oblique smooth muscles*), and that the inner lining of the stomach has large folded ridges, the *gastric rugae*.

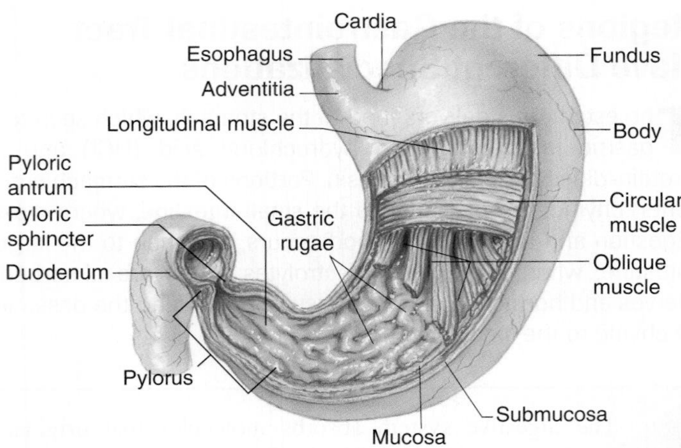

FIGURE 14.3 Regions and structures of the stomach. Notice that the pyloric region of the stomach includes the pyloric antrum (the wider portion of the pylorus) as well as the pyloric sphincter.

The GI tract, including the stomach, consists of four major tissue layers: the mucosa, submucosa, muscularis externa, and adventitia. The **mucosa** is the innermost layer, facing the lumen. Just to the outside of the mucosa is the **submucosa**, which is the major connective tissue layer. This is followed by the **muscularis externa**, consisting of two or three (in the stomach) layers of smooth muscles. A thin connective tissue layer, the **adventitia**, covers the GI tract. Figure 14.4 illustrates a microscopic view of the gastric mucosa and submucosa; notice that the mucosa

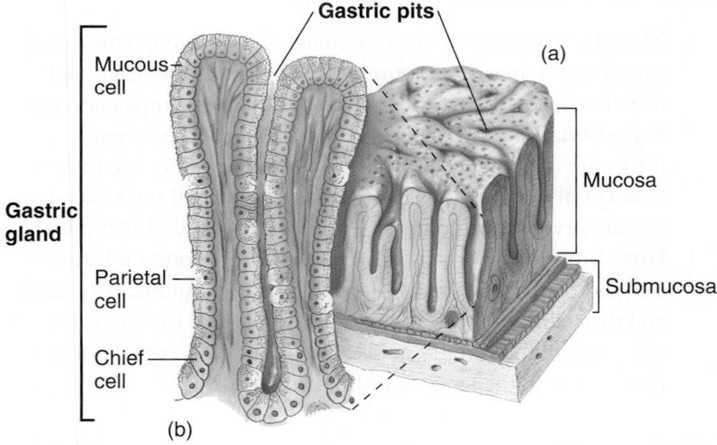

FIGURE 14.4 Gastric pits and gastric glands of the mucosa. (*a*) Gastric pits are the openings of the gastric glands. (*b*) Gastric glands consist of several types of cells (including mucous cells, chief cells, and parietal cells), each of which produces a specific secretion.

that covers the rugae itself forms tiny folds, the **gastric glands**. These are exocrine glands, because they have ducts that can channel the secretions of the cells along the ducts into the lumen of the stomach.

The gastric glands contain several cells that produce different secretions. These include:

1. **Goblet cells**, which secrete *mucus*.
2. **Parietal cells**, which secrete *hydrochloric acid (HCl)*.
3. **Chief cells**, which secrete *pepsinogen*, a less active precursor of the protein-digesting enzyme *pepsin*.
4. **Enterochromaffin-like (ECL) cells**, which secrete *histamine* and *5-hydroxytryptamine* (also called *serotonin*) as paracrine regulators (chapter 8).
5. **G cells**, which secrete the hormone *gastrin* into the blood.
6. Gastric cells that secrete *intrinsic factor*, a polypeptide required for the intestinal absorption of vitamin B_{12}.
7. Gastric cells that secrete *ghrelin*, a hormone that promotes hunger when the stomach is empty (discussed in a later section).

CLINICAL APPLICATIONS

The one function of the stomach that appears to be essential for life is the secretion of *intrinsic factor*. This polypeptide is needed for the absorption of vitamin B_{12} across the terminal portion of ileum, the last segment of the small intestine. Vitamin B_{12} is required for the maturation of red blood cells in the bone marrow. People with their stomachs surgically removed must have injections of vitamin B_{12}. Alternatively, they can take vitamin B_{12} orally if it's combined with intrinsic factor, to permit absorption. Without adequate vitamin B_{12}, a person develops **pernicious anemia**.

The parietal cells secrete H^+ (protons) into the stomach lumen by an active transport process, requiring carriers known as **H^+/K^+ ATPase pumps** (fig. 14.5). These transport H^+ uphill against a million-to-one concentration gradient, making gastric juice highly acidic with a pH of less than 2. Chloride ions (Cl^-) follow the H^+, so that the parietal cells secrete **hydrochloric acid (HCl)** into the gastric juice. The strong acidity of gastric juice serves three functions:

1. Proteins in food are denatured at the low pH. This means that their natural tertiary structure (shape; see chapter 1) is altered, making them more digestible.
2. Under acidic conditions, the less active enzyme pepsinogen, secreted by the chief cells, is converted into the fully active **pepsin** enzyme. This works because one weakly active pepsinogen can remove an amino acid sequence from a different pepsinogen, converting it into pepsin; once pepsin is formed, it can more effectively activate other pepsinogen molecules (fig. 14.6).
3. Pepsin has a pH optimum (see chapter 1) of about 2, meaning that it works best at this highly acidic pH.

The secretion of HCl by the parietal cells is stimulated by a variety of factors, including the hormone gastrin and the paracrine regulator histamine (fig. 14.7). **Gastrin** directly stimulates the parietal cells to secrete HCl, but its major action is to stimulate the ECL cells to secrete **histamine**. Histamine then binds to receptor proteins called the *H_2 histamine receptors* in the parietal cells to stimulate the secretion of HCl. The H_2 histamine receptors are different from those involved in allergic reactions (chapter 11), which are the H_1 histamine receptors blocked by such antihistamines as Benadryl.

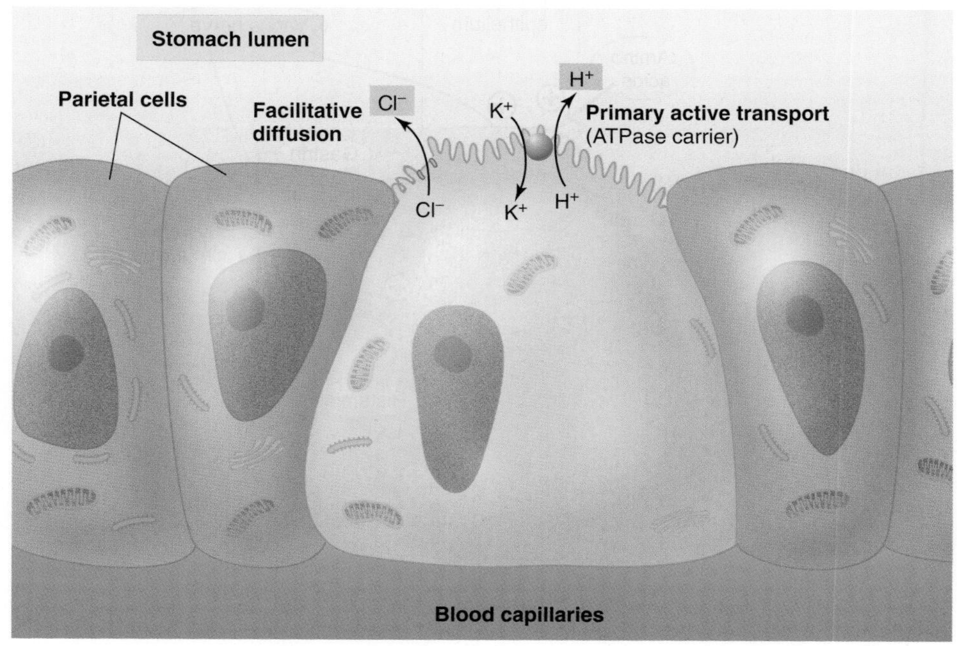

FIGURE 14.5 Secretion of gastric acid. Parietal cells in the gastric glands of the stomach secrete H^+ (protons) in exchange for K^+ by an active transport H^+/K^+ (ATPase) pump. Chloride ion (Cl^-) enters the stomach lumen by facilitative diffusion, using a carrier protein to move down its concentration gradient.

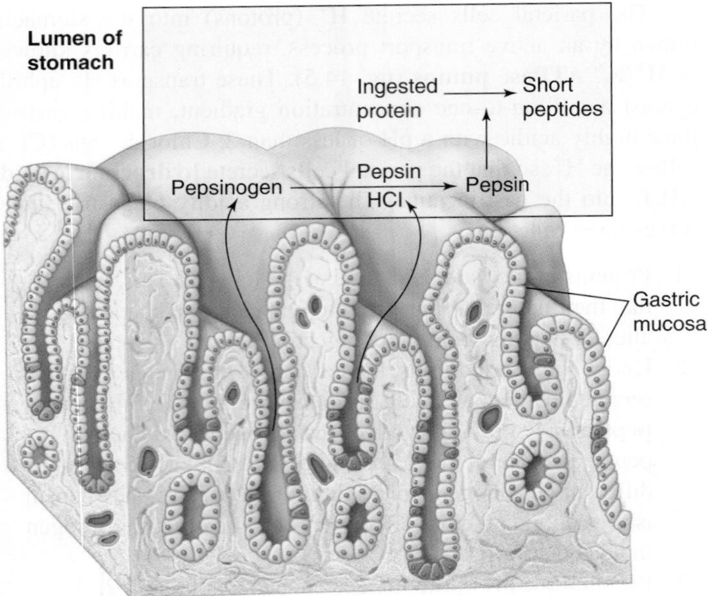

FIGURE 14.6 The activation of pepsin. The gastric mucosa secretes the inactive enzyme pepsinogen and hydrochloric acid (HCl). In the presence of HCl, the active enzyme pepsin is produced. Pepsin digests proteins into shorter polypeptides.

Pepsin is a protein-digesting enzyme, but proteins are only partially digested (converted into smaller polypeptide chains) in the stomach. There is minor digestion of lipids in the stomach, and only a negligible amount of starch digestion. (Starch digestion begins in the mouth due to an enzyme, salivary amylase, which is quickly inactivated in the acidic gastric juice.) So, the major digestive functions of the stomach are to (1) serve as a food reservoir when we eat large meals; (2) form chyme, containing partially digested proteins; and (3) send portions of chyme to the small intestine for further processing.

Clinical Applications

Gastroesophageal reflux disease (GERD) is a common disorder involving the reflux of acidic gastric juice into the esophagus. This is often treated with drugs that inhibit the H^+/K^+ pumps in the gastric mucosa. Such *proton pump inhibitors* include Prilosec and Prevacid. Because histamine (secreted from ECL cells) stimulates acid secretion, people with GERD or **peptic ulcers** (see the "Physiology in Health and Disease" section at the end of this chapter) may be treated with drugs that block histamine action. Such drugs, including Tagamet and Zantac, specifically block the H_2 histamine receptors in the parietal cells.

Clinical Investigation Clues

Remember that Jim had stomach pain when he drank wine and coffee, and his physician prescribed a proton pump inhibitor.

- Given that alcohol and caffeine stimulate acid secretion in the stomach, what may have caused Jim's pains? (Hint: see the "Physiology in Health and Disease" section at the end of this chapter.)
- What are "proton pumps," and why would a drug that inhibits their action help treat Jim's condition?

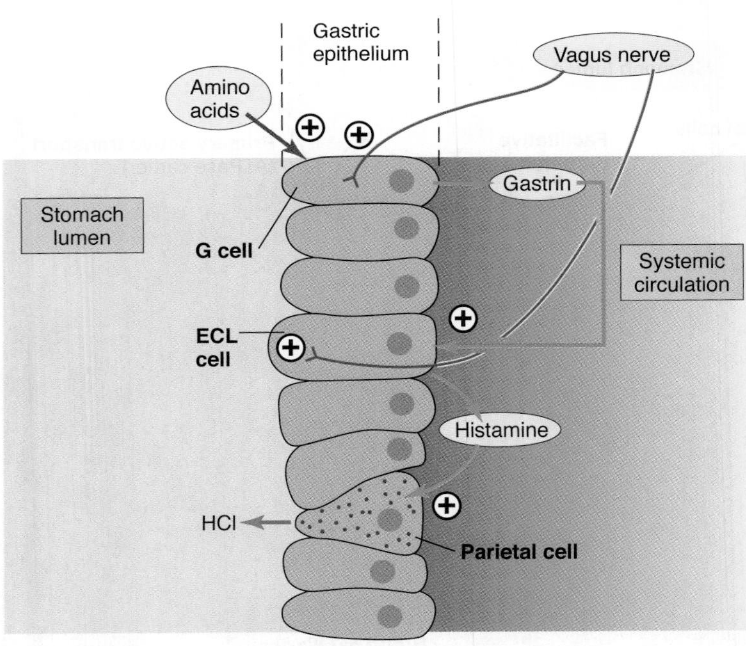

FIGURE 14.7 The regulation of gastric acid secretion. The presence of amino acids in the stomach lumen from partially digested proteins stimulates gastrin secretion. Gastrin secretion from G cells is also stimulated by vagus nerve activity. The secreted gastrin then acts as a hormone to stimulate histamine release from the ECL cells. The histamine, in turn, acts as a paracrine regulator to stimulate the parietal cells to secrete HCl. (⊕ = stimulation).

Gastric Function Is Regulated by Nerves and Hormones

To some extent, gastric motility and secretion are automatic. Contractions that push chyme through the pyloric sphincter can be initiated by neurons in the stomach wall that depolarize spontaneously. Also, partially digested food in the chyme can stimulate pepsinogen and HCl secretion independent of the effects of nerves and hormones. Superimposed on the intrinsic (automatic) regulation of the stomach is extrinsic regulation by nerves and hormones. Gastric regulation is divided into three phases: (1) the *cephalic phase*; (2) the *gastric phase*; and (3) the *intestinal phase*.

The **cephalic phase** refers to regulation of the stomach by the brain, which exerts control by means of the vagus nerves (fig. 14.7). This stimulates gastric secretions in anticipation of a meal, and can be activated by the thought, sight, or smell of food, and by other conditioned stimuli.[2] The vagus nerves stimulate the G cells to secrete gastrin, and the ECL cells to secrete histamine. Gastrin also stimulates the ECL cells to secrete histamine, which in turn stimulates the parietal cells to secrete HCl (fig. 14.7).

The arrival of food in the stomach stimulates the **gastric phase** of regulation. Intact food proteins have little effect, but when they are partially digested, the shorter polypeptides and amino acids stimulate pepsinogen and gastrin secretion. Gastrin then stimulates increased pepsinogen secretion from the chief cells; more significantly, it stimulates histamine secretion from the ECL cells. Histamine then stimulates the parietal cells to secrete HCl (fig. 14.7).

The **intestinal phase** of gastric regulation refers to the inhibition of the stomach that occurs when chyme enters the small intestine. The arrival of chyme into the duodenum stimulates sensory neurons, which evoke a neural reflex that inhibits gastric motility and secretion. The presence of fat in the chyme also stimulates the secretion of intestinal hormones that have an inhibitory effect on the stomach. The general term for such a hormone is **enterogastrone** (*entero* = intestine; *gastro* = stomach). There are several specific intestinal hormones that may have enterogastrone action, and the effect of these is to keep fatty chyme in the stomach for a longer period of time, allowing the intestine more time to process the fatty chyme.[3]

FYI [2]Conditioned reflexes were discovered by the Russian physiologist Ivan Petrovich Pavlov (1849–1936), who was awarded a Nobel Prize in Physiology or Medicine (in 1904) for his research on the digestive system. To his experimental animals (dogs), Pavlov famously paired the sight of food with another, conditioned stimulus (such as the ringing of a bell). After a number of trials, he discovered that the conditioned stimulus alone could evoke salivary and gastric secretions (in what he termed "psychic secretions"). This Pavlovian classical conditioning has become a fundamental concept in both physiology and psychology.

FYI [3]Because fat and oil in the chyme are the most potent stimuli for enterogastrone secretion, meals high in fat and oil trigger a more powerful inhibition of gastric motility and secretion. Thus, a meal of bacon and eggs goes through the GI tract slower than a carbohydrate breakfast such as oatmeal. This explains why we feel full for a longer time after fatty meals.

CHECK POINT

1. List and describe the functions of the digestive system.
2. Describe how food is transported along the esophagus, and how gastric juice is normally prevented from entering the esophagus.
3. List the cells of the gastric glands and the substances they secrete.
4. Explain the actions of intrinsic factor, ghrelin, gastrin, and histamine in stomach function.
5. Describe the regulation of gastric activity during the cephalic, gastric, and intestinal phases.

The Small and Large Intestine Have Different Structures and Functions

The first part of the small intestine, the duodenum, receives chyme from the stomach; it also receives bile from the liver and gallbladder, as well as pancreatic juice from the pancreas. Digestive enzymes in the pancreatic juice, together with enzymes stuck into the plasma membranes of cells lining the small intestine, digest food molecules into their subunits, which are then absorbed in the small intestine. The large intestine absorbs water, electrolytes, and some vitamins, and harbors a large population of bacteria that provide benefits for the body.

The **small intestine** (fig. 14.8) extends from the pyloric sphincter of the stomach to the *ileocecal valve*, which guards the entrance to the large intestine. It's called "small" only because its diameter is smaller than that of the large intestine; the small intestine is actually the longer of the two—3 meters (12 feet) in a living person, but twice that length in a cadaver because of relaxation of its smooth muscle. The first 20 to 30 cm (10 in.) is the **duodenum**; the next two-fifths is the **jejunum**, and the remainder is the **ileum**. Although the duodenum is the shortest segment of the small intestine, it's quite significant because (1) it's the first section to receive chyme from the stomach; (2) it receives bile from the common bile duct, which drains bile from the gallbladder and liver (see fig. 14.18); and (3) it receives pancreatic juice from the pancreatic duct (see fig. 14.18); pancreatic juice contains many digestive enzymes and bicarbonate.

Figure 14.8 (top) shows a cut-away view of the small intestine, with its *plicae circularis*. These are like the rugae of the stomach in that they're visible to the naked eye, but the plicae extend around the circumference of the small intestine. Figure 14.8 (bottom) illustrates a microscopic view of a section through the small intestine. Notice that the mucosa follows the pleats of the plicae and covers them with fingerlike projections called **villi** (singular, *villus*). The villi are microscopic projections of mucosa, containing a surface layer of simple columnar epithelium (chapter 2). Villi greatly increase the surface area of mucosa exposed to the chyme in the lumen.

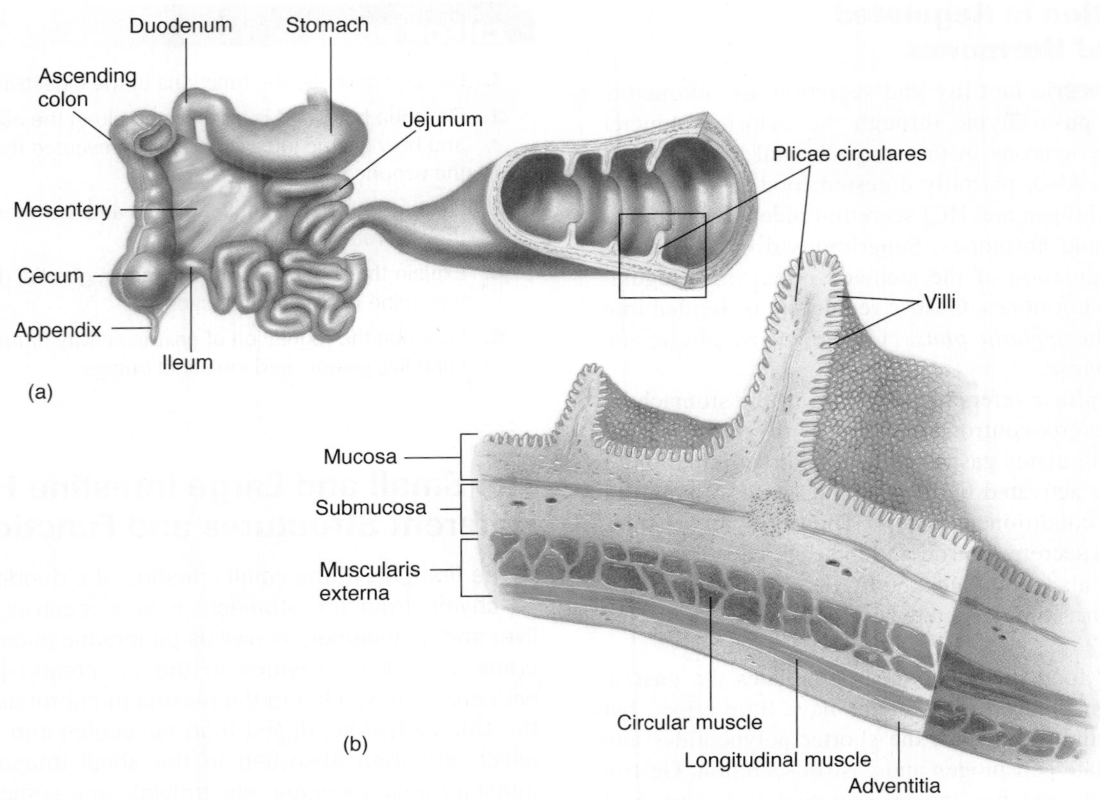

FIGURE 14.8 The small intestine. (*a*) The regions of the small intestine. (*b*) A section of the intestinal wall showing the tissue layers, plicae circulares, and villi.

A single villus is illustrated in figure 14.9. A simple columnar epithelium, containing mucus-secreting goblet cells, coats the villus, while its core is composed of connective tissue. Within the connective tissue of the villus is a network of blood capillaries and a single lymphatic capillary, known as a **lacteal**. The blood and lymphatic capillaries play an important role in the absorption of food molecules, as described in a later section. At the base of a villus, the epithelium pouches inward to form the **intestinal crypts**. These look something like the gastric glands, but the intestinal crypts don't secrete enzymes. Instead, they produce new epithelial cells by mitotic cell division. The new cells migrate out of the crypts and up to the tips of the villi, from which they are continuously shed.

The Small Intestine Digests and Absorbs Food

You can see the plicae circularis with your unaided eyes, but you need a microscope to see the intestinal villi. Going down even further in size, you need an electron microscope to see the extremely tiny **microvilli** (fig. 14.10). Microvilli are folds of plasma membrane on the apical (top) surface of the intestinal epithelial cells. Because they resemble the bristles of a brush, the microvilli on the apical surface of the villi are sometimes referred to as the *brush border*.

The microvilli provide a greatly increased surface area exposed to chyme in the intestinal lumen, improving the efficiency of digestion and absorption. A variety of digestive enzymes are embedded within the plasma membrane of the microvilli (fig. 14.11); these are collectively termed **brush border enzymes**. These enzymes complete the digestion of food molecules that was begun by other enzymes, such as pepsin in the stomach and the action of digestive enzymes in pancreatic juice (discussed later). For example, there are brush border enzymes that digest polypeptide chains and a variety of disaccharides (double sugars). Such brush border enzymes include *sucrase* (digesting sucrose, table sugar) and *lactase* (digesting lactose, milk sugar).

CLINICAL APPLICATIONS

The brush border enzyme *lactase* is required for the digestion of milk sugar, lactose. Lactase is present in all children under the age of 4, but becomes inactive to some degree in most adults of the world (people of Asian and African heritage are more often lactase deficient than Caucasians). Lactase deficiency causes **lactose intolerance**, in which undigested lactose in the intestine results in diarrhea, cramps, and gas. Yogurt is better tolerated than milk because yogurt contains lactase (produced by the yogurt bacteria), which becomes activated in the duodenum and digests lactose.

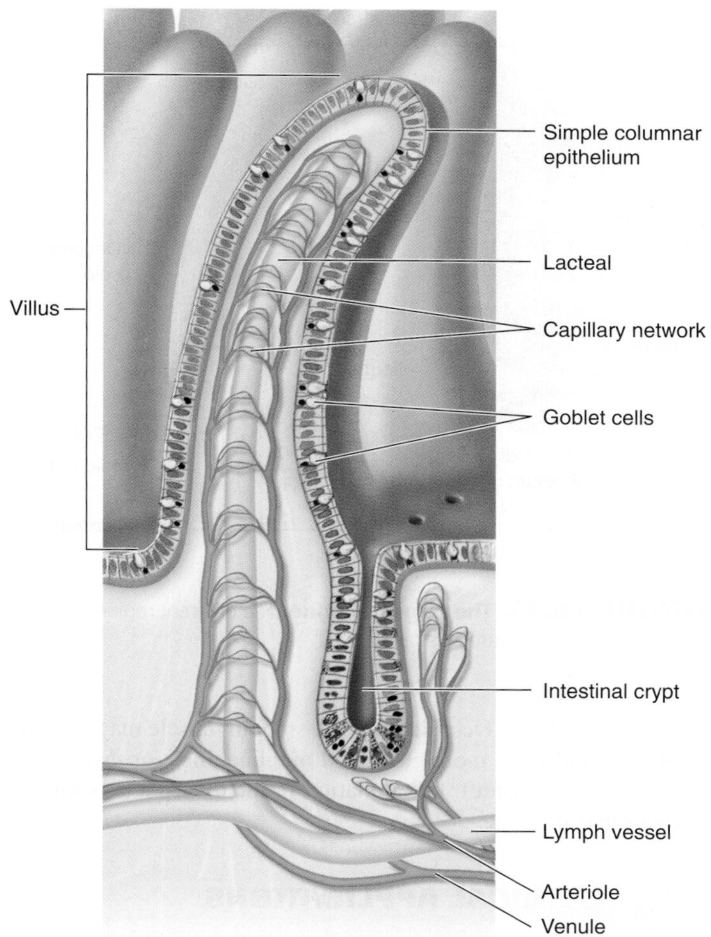

FIGURE 14.9 The structure of an intestinal villus. The figure also depicts an intestinal crypt, in which new epithelial cells are produced by mitosis.

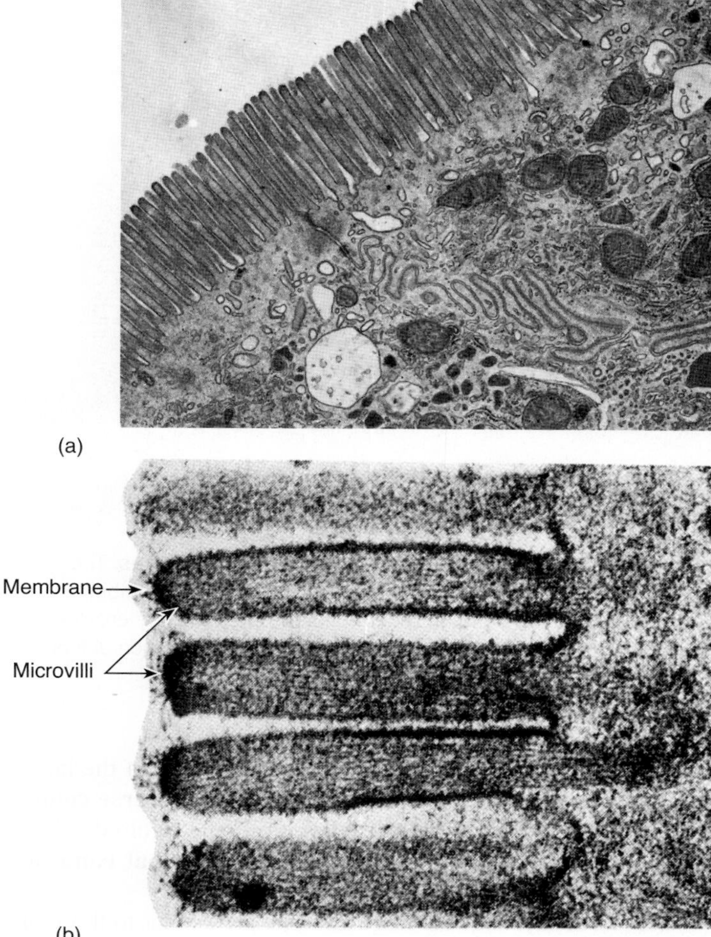

FIGURE 14.10 Electron micrographs of microvilli. Microvilli are evident at the apical surface of the columnar epithelial cells in the small intestine. These are seen here (*a*) at lower magnification and (*b*) at higher magnification. Microvilli increase the surface area for absorption and also have the brush border digestive enzymes embedded in their plasma membranes. From Keith R. Porter, D. H. Alpers, and D. Seetharan, "Pathophysiology of Diseases Involving Intestinal Brush-Border Proteins" in *New England Journal of Medicine*, Vol. 296, 1977, p. 1047, fig. 1.

Two major types of contractions occur in the small intestine: *peristalsis* and *segmentation*. Peristalsis in the small intestine is much weaker than in the esophagus and stomach. The major type of muscle contractions in the small intestine is **segmentation** (fig. 14.12). This refers to contractions that constrict the intestine at different segments, helping to more thoroughly mix the chyme. These segmental contractions occur more frequently at the proximal end (toward the stomach) than at the distal end (toward the anus) of the intestine, helping to produce the pressure head for motility of the chyme through the intestine.

Intestinal contractions occur automatically in response to pacemaker cells in the intestinal wall. The pacemaker cells depolarize spontaneously, much like the pacemaker cells of the heart (chapter 10), and lead to the production of action potentials and contraction of the smooth muscle cells. Parasympathetic axons, releasing acetylcholine (ACh), stimulate an increased rate of action potential production and thereby promote contractions and motility of the intestine.

Within the wall of the digestive tract are an estimated 100 million neurons, about the same number as in the spinal cord. These are located within the *submucosal plexus* (in the submucosa) and *myenteric plexus* (between smooth muscle layers in the muscularis externa), and form the **enteric nervous system** (sometimes also called the *enteric brain*). The enteric nervous system, like the CNS, contains interneurons, as well as different sensory and autonomic neurons. These neurons can coordinate complex reflex activities, including peristalsis and segmentation, without involving the spinal cord or brain.

The Large Intestine Absorbs Water, Electrolytes, and Some Vitamins

The **large intestine**, or **colon**, extends from the ileocecal valve to the anus. Chyme passes from the ileum of the small intestine, through the ileocecal valve, and into the **cecum**, which is a pouch at the origin of the large intestine (fig. 14.13). Waste

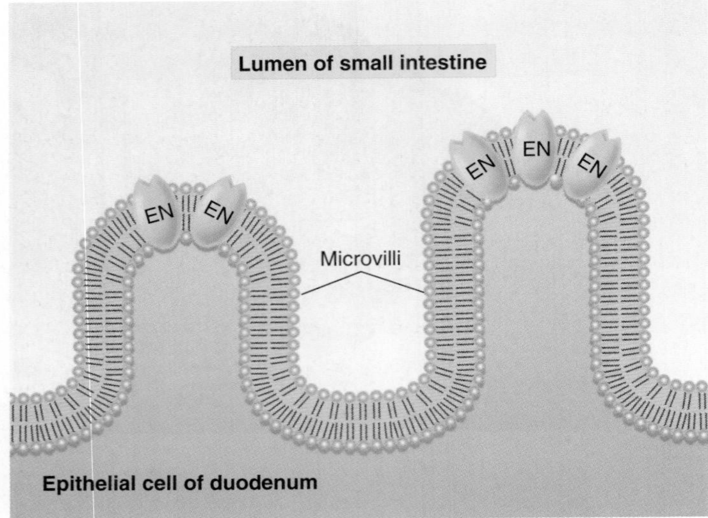

FIGURE 14.11 Location of brush border enzymes. The brush border enzymes (EN) are embedded in the plasma membrane of the microvilli in the small intestine. The active sites of these enzymes face the chyme in the lumen, helping to complete the digestion of food molecules.

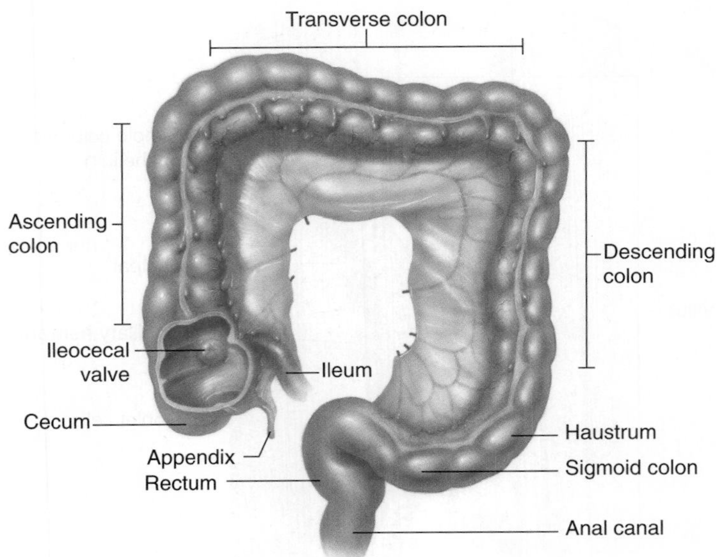

FIGURE 14.13 The large intestine. The different regions of the large intestine (colon) are illustrated.

material then passes through the following regions of the large intestine in sequence: the **ascending colon**, **transverse colon**, **descending colon**, and **sigmoid colon**. From the sigmoid colon, waste material passes through the **rectum** and **anal canal** to exit the body.

The tissue layers of the large intestine are similar to those of the small intestine, except that the mucosa of the large intestine doesn't have villi and microvilli (although it does have intestinal crypts). The outer wall of the large intestine forms pouches, or *haustra* (fig. 14.13). Occasionally, the smooth muscle may become weakened and form more elongated outpouchings, or *diverticula* (*divert* = turned aside). Inflammation in the diverticula is known as *diverticulitis*.

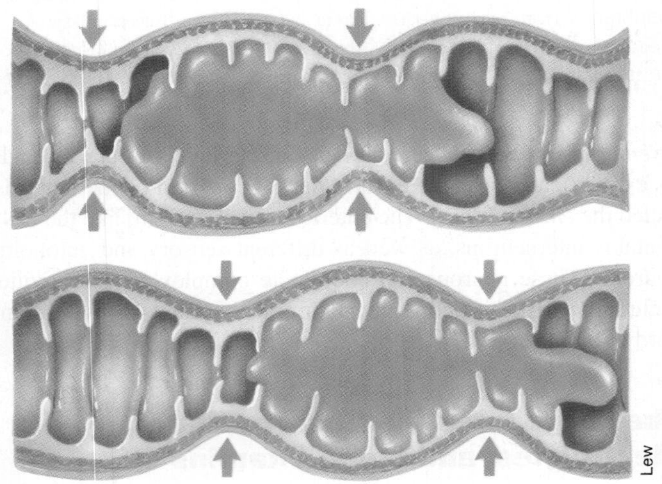

FIGURE 14.12 Segmentation of the small intestine. Simultaneous contractions of numerous segments of the intestine help to mix the chyme with digestive enzymes and mucus.

CLINICAL APPLICATIONS

The **appendix** is a short, thin outpouching from the cecum (fig. 14.13). It doesn't appear to have a digestive function, but contains numerous lymphatic nodules that may offer some immune protection. However, it's most famous because it can become dangerously inflamed, a condition called **appendicitis**. This is often detected in its later stages by pain in the lower right quadrant of the abdomen. If the appendix ruptures, the infection can spread throughout the abdominal cavity, causing inflammation of the *peritoneum* (the membranes that line the abdominal cavity and cover its organs), or *peritonitis*. This dangerous event can be prevented by an *appendectomy*, which is the surgical removal of the appendix.

The large intestine absorbs water, electrolytes (ions), several B vitamins, and vitamin K. The colon also harbors a large population of bacteria—collectively called the **intestinal microbiota** or **microflora**. How large? There are 10 times more bacterial cells in the colon (comprising more than 400 different species) than there are total human cells in the body! The intestinal microbiota provide their hosts (us) with several benefits:

1. They produce significant amounts of vitamin K and folic acid, which are then absorbed across the wall of the large intestine.
2. They produce short-chain fatty acids, which are used for energy by the epithelial cells of the colon and which aid the absorption of sodium, bicarbonate, magnesium, and iron in the large intestine.

3. They reduce the ability of pathogenic bacteria to cause disease, promoting anti-inflammatory processes that help protect the intestine from injury.

Clinical Applications

Diarrhea refers to excessive fluid excretion in the feces. In *cholera*, severe diarrhea and dehydration result from *enterotoxin* released by infecting bacteria. Enterotoxin stimulates active NaCl transport into the intestinal lumen, followed by the osmotic movement of water. In *celiac sprue*, susceptible people who eat *gluten* (proteins from grains such as wheat) develop diarrhea due to damage to the intestinal mucosa. People with *lactose intolerance* develop diarrhea because undigested lactose increases the osmolarity of the intestinal contents, with water following by osmosis.

Check Point

1. Describe the structure of the small intestine, identifying the parts that increase the surface area exposed to the chyme.
2. Identify the brush border, and the significance of brush border enzymes.
3. Describe the structure of the large intestine, and how it differs from the small intestine.
4. Explain the nature and significance of the intestinal microbiota.

The Liver, Gallbladder, and Pancreas Aid Digestion

The liver produces bile, which contains bile salts that emulsify fat in the small intestine. Bile also contains bilirubin (bile pigment), an excretory product derived from heme. The gallbladder stores and concentrates the bile. The exocrine tissue of the pancreas produces pancreatic juice, which contains many enzymes that are essential for digestion in the small intestine. Pancreatic juice also contains bicarbonate, which neutralizes the acidic chyme arriving from the stomach.

The **liver** is the largest internal organ, but in a sense it is only 1 to 2 cells thick. This is because the liver cells, or *hepatocytes*, are organized into *hepatic plates* that are only 1 to 2 cells thick. The hepatic plates are separated by large capillary spaces, called *sinusoids* (fig. 14.14). The sinusoids have big pores called *fenestrae*; this is like the glomerular capillaries (chapter 13), but unlike those, the hepatic (liver) sinusoids lack a basement membrane. Because of this, plasma proteins that carry fat and cholesterol can leave the sinusoids and enter the hepatocytes. The hepatic sinusoids also contain phagocytic Kupffer cells. These structural features of the liver permit it to very efficiently modify the blood passing through it.

Blood enters the liver from two sources. Arterial blood enters from the **hepatic artery**, and venous blood enters in the **hepatic portal vein**. The hepatic portal vein contains blood that had previously passed through the GI tract, and so it contains many of the products of digestion. This is a portal system because the sinusoids of the liver are a second capillary bed downstream from the capillaries in the GI tract (similar to the hypothalamo-hypophyseal portal system described in chapter 8).

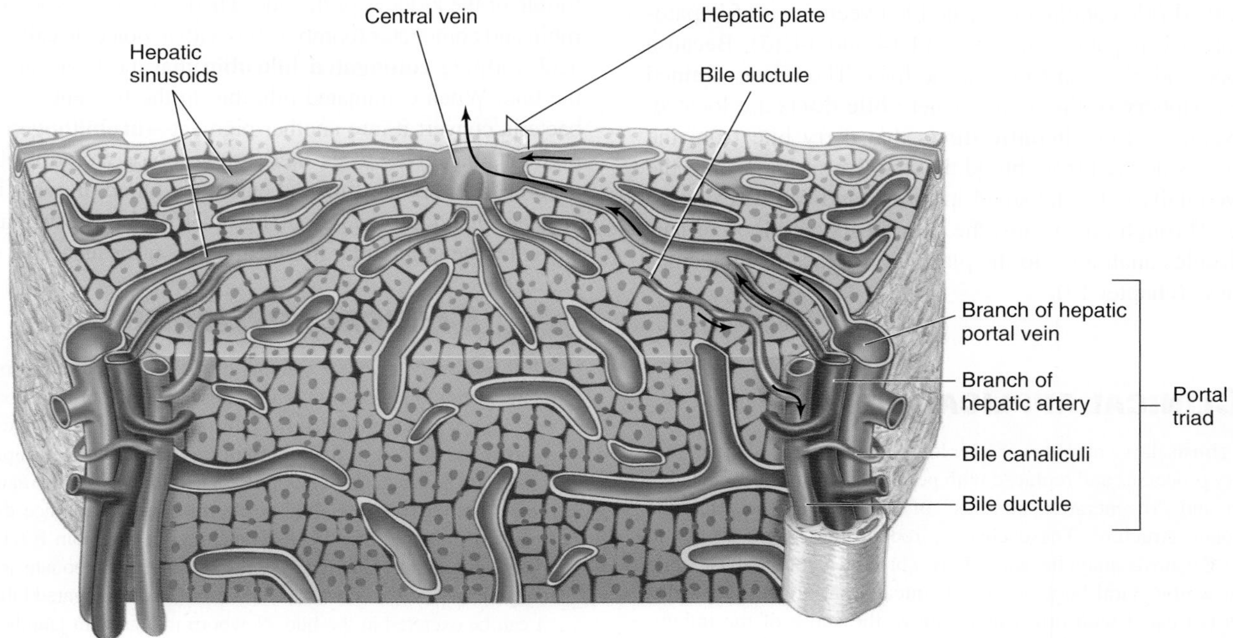

FIGURE 14.14 Microscopic structure of the liver. Blood enters a liver lobule through the vessels in a portal triad, passes through hepatic sinusoids, and leaves the lobule through a central vein. The central veins converge to form hepatic veins that transport venous blood from the liver.

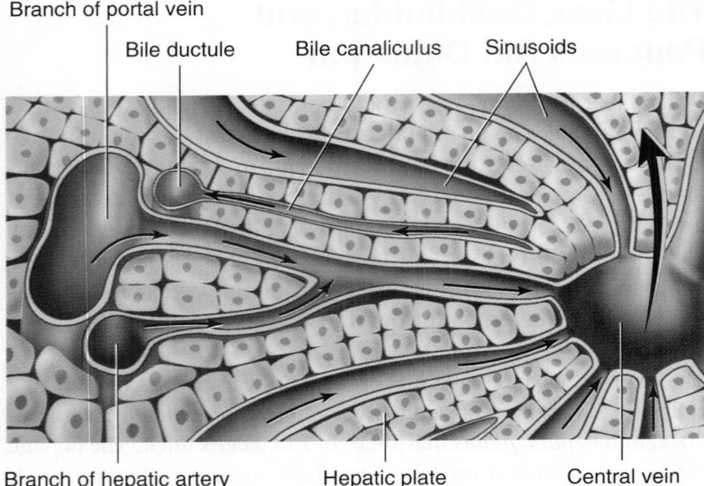

FIGURE 14.15 The flow of blood and bile in a liver lobule. Blood flows within sinusoids from a portal vein to the central vein (from the periphery to the center of a lobule). Bile flows within hepatic plates from the center to bile ductules at the periphery of a lobule.

The hepatic plates and sinusoids are organized into **liver lobules** (figs. 14.14 and 14.15). At the periphery of each lobule, opening into each sinusoid, are branches of the hepatic artery and the hepatic vein. Thus, these two bloods mix in the sinusoids as they flow toward the center of the lobule, to be drained out of a *central vein* (and eventually into the *hepatic vein*, returning blood to the heart). Thus, many molecules and ions absorbed in the intestine go through the liver first, before going to the heart and being distributed to the other organs in the body.

Bile is produced by the hepatocytes and secreted into thin channels, called *bile canaliculi*, located between rows of hepatocytes within each hepatic plate (figs. 14.14 and 14.15). Because of this, blood and bile don't mix in the liver. The bile is drained toward the periphery of the lobule, where **bile ducts** are located. These converge to form **hepatic ducts** that carry bile from the liver. Molecules derived from blood that the liver secretes into the bile will eventually enter the small intestine, and can be excreted in the feces. Through this means, the liver clears the blood of particular molecules analogous to the plasma clearance of molecules by the kidneys (chapter 13).

Clinical Applications

In **cirrhosis**, large numbers of liver lobules are destroyed by inflammatory processes and replaced with permanent, scarlike connective tissue and "regenerative nodules" of hepatocytes that lack the normal plate structure. These changes reduce the liver's ability to function. Cirrhosis may be caused by chronic alcohol abuse, bile duct obstruction, viral hepatitis, or chemicals that attack liver cells. Cirrhosis is treated with drugs that remove the cause of the inflammation (such as antiviral drugs, if the cause is viral hepatitis) and with anti-inflammatory drugs.

The Liver Has Many Diverse Functions

The liver is the first organ to receive the absorbed molecules from the intestine, has intimate contact with the blood, and can remove molecules from the blood by secreting them into the bile. Combined with the large number of different enzymes present in hepatocytes, these associations allow the liver to perform a great many functions that impact all of the body systems. Here is a partial list of general hepatic functions:

1. *Production and secretion of bile*. Bile salts emulsify fat, aiding fat digestion; bilirubin is an excretory form of heme, allowing its elimination from the blood.
2. *Detoxication of the blood*. Phagocytosis by Kupffer cells, and enzymatic alteration of molecules within the hepatocytes, allow the liver to remove toxins from the blood. In the process, urea, uric acid, and other wastes are produced that are less toxic than their parent molecules.
3. *Regulation of blood glucose*. The liver removes excess glucose from the blood and stores it as glycogen. Also, the liver can secrete glucose into the blood, which it derives from glycogen and from other molecules.
4. *Metabolism of lipids*. The liver excretes cholesterol from the blood by converting it into bile acids. The liver also produces ketone bodies from free fatty acids.
5. *Synthesis of plasma proteins*. The liver produces albumin (the plasma protein most responsible for the colloid osmotic pressure of the blood; chapter 10), carrier proteins for nonpolar hormones and lipids, and clotting factors.

Bilirubin (bile pigment) is produced in the liver, spleen, and bone marrow from the heme groups (minus the iron) of hemoglobin. Bilirubin is not very water soluble, and so circulates in the blood bound to proteins. This prevents it from entering either the glomerular filtrate of the kidneys or the bile. The liver removes some of the bilirubin and conjugates (combines) it with a molecule called *glucuronic* acid, forming **conjugated bilirubin** that the liver can secrete into the bile. When conjugated bilirubin in the bile enters the intestine, bacteria convert it into another pigment—**urobilinogen**. This is the pigment that gives feces their brown color. Some urobilinogen is absorbed by the intestine and enters the blood, in what is known as an *enterohepatic circulation* (fig. 14.16). As a result, urobilinogen can enter the glomerular filtrate and give urine an amber color.

Clinical Applications

Jaundice is a yellow staining of the tissues produced by high blood concentrations of bilirubin. A high concentration of conjugated bilirubin may be produced by gallstones, which block bile excretion. A high concentration of free bilirubin can be caused by a rapid rate of red blood cell destruction, as in *physiological jaundice of the newborn*. In this condition, otherwise healthy newborns develop jaundice due to a rapid breakdown of red blood cells containing hemoglobin F (chapter 12). In premature infants, jaundice may result from an inadequate amount of liver enzymes needed to convert free bilirubin into conjugated bilirubin, so that it can be excreted in the bile. Newborn infants with jaundice are usually treated with blue light, which converts free bilirubin into a more polar form that can dissolve in plasma and thus be excreted in the bile and urine.

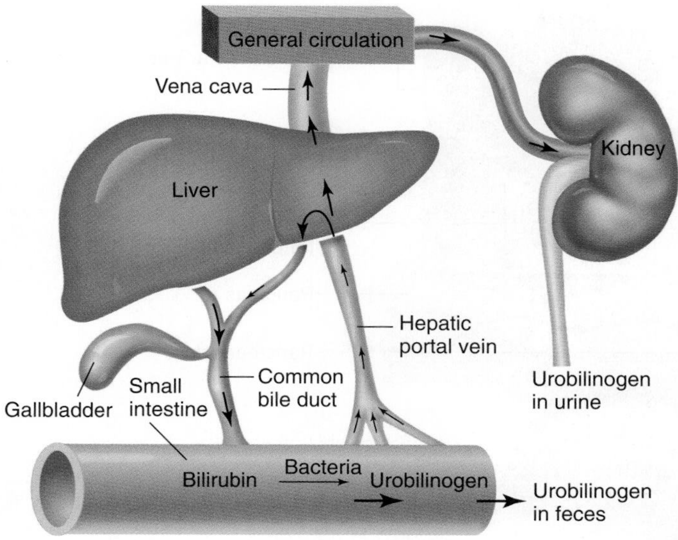

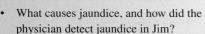

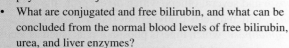

FIGURE 14.16 The enterohepatic circulation of urobilinogen. Bacteria in the intestine convert bilirubin (bile pigment) into urobilinogen. Some of this pigment leaves the body in the feces; some is absorbed by the intestine and is recycled through the liver. A portion of the urobilinogen that is absorbed enters the general circulation and is filtered by the kidneys into the urine.

(a) Cholic acid structure: OH, CH_3, CH–CH_2–CH_2—COOH, HO, OH

Cholic acid (a bile acid)

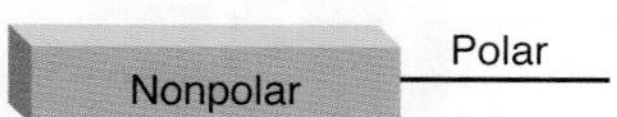

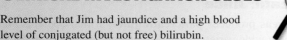

(b) Simplified representation of bile acid

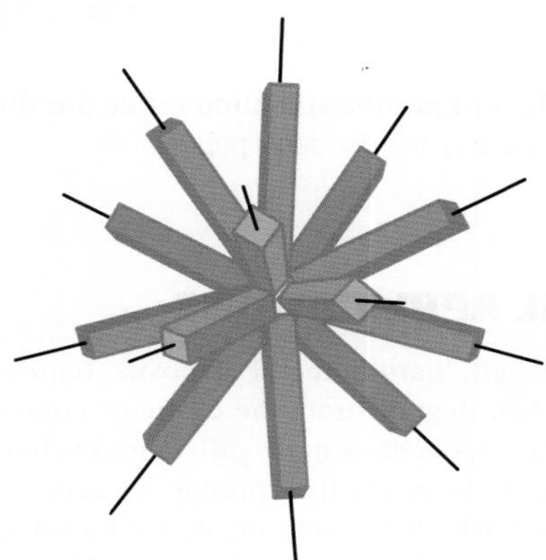

(c) **Micelle** of bile acids

FIGURE 14.17 Bile acids form micelles. (*a*) Cholic acid, a bile acid. (*b*) A simplified representation of a bile acid, emphasizing that part of the molecule is polar, but most is nonpolar. (*c*) Bile acids in water aggregate to form associations called micelles. Cholesterol and lecithin, being nonpolar, can enter the micelles. The bile acids in the micelles serve to emulsify triglycerides (fats and oils) in the chyme.

CLINICAL INVESTIGATION CLUES

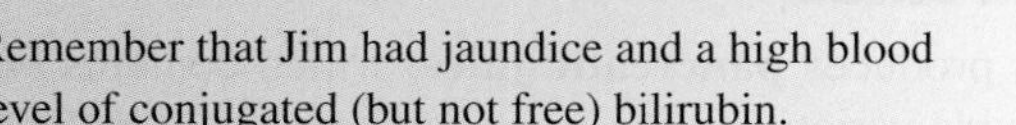

Remember that Jim had jaundice and a high blood level of conjugated (but not free) bilirubin.

- What causes jaundice, and how did the physician detect jaundice in Jim?
- What are conjugated and free bilirubin, and what can be concluded from the normal blood levels of free bilirubin, urea, and liver enzymes?
- What is the possible relationship between gallstones and a high blood level of conjugated bilirubin?

The liver converts cholesterol into **bile acids**, which have polar hydroxyl (OH; see chapter 1) groups. This serves two purposes: (1) it produces the bile acids, which aid digestion; and (2) it's the major way cholesterol is eliminated from the body. Bile acids bound to the amino acids glycine and taurine form **bile salts**. These molecules huddle together to form aggregates known as **micelles** (fig. 14.17). Nonpolar lipids including cholesterol and lecithin enter these micelles, and the dual nature of the bile acids—part polar and part nonpolar—allows them to emulsify fat and oil (triglycerides). **Emulsification** refers to the breaking up of large fat (or oil) globules into smaller droplets, creating a finer suspension of fat in water. This provides more surface area of fat and oil for attack by the fat-digesting enzymes (lipases), which are dissolved in the water surrounding each emulsified droplet. It is the lipase enzymes, not the bile acids, that digest (hydrolyze) the triglycerides.

Bile is stored and concentrated in the **gallbladder**, a saclike organ attached to the inferior surface of the liver. The gallbladder receives bile from the liver by way of the *bile ducts*, *hepatic duct*, and *cystic duct* (fig. 14.18). When the gallbladder fills with bile, it expands to the size and shape of a small pear. Contraction of the smooth muscle in the gallbladder ejects the bile through the cystic duct into the *common bile duct*, which transports it to the duodenum. When the small intestine is empty, a sphincter muscle at the end of the common bile duct closes, forcing bile through the cystic duct into the gallbladder.

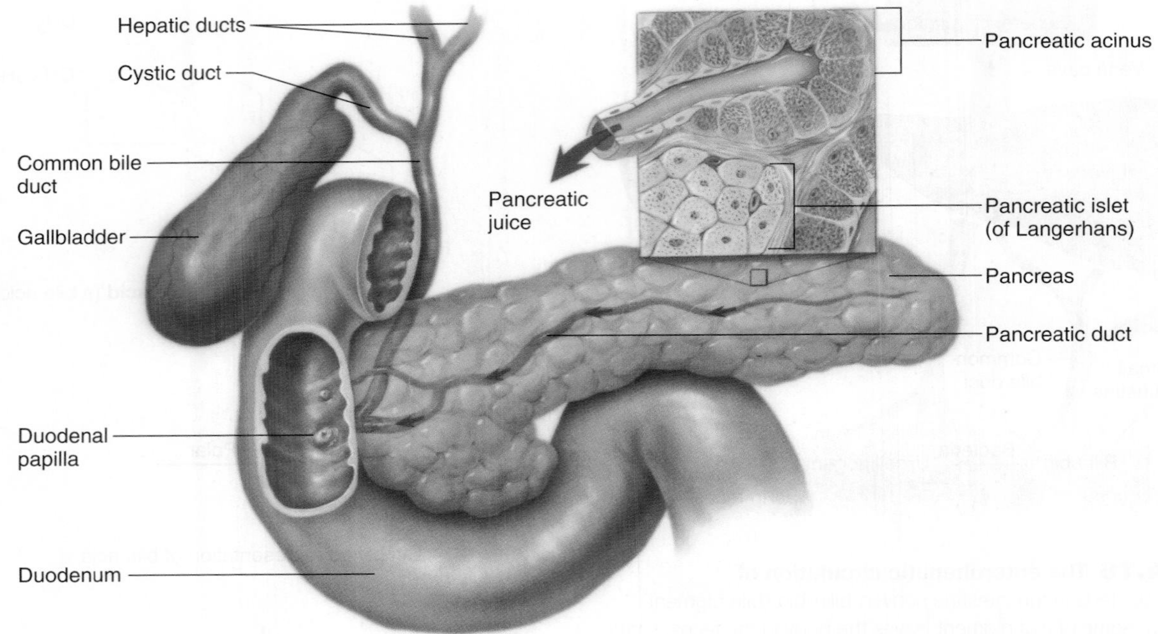

FIGURE 14.18 Bile and pancreatic juice enter the duodenum. The pancreatic duct joins the common bile duct as they empty into the duodenum at a common opening, the duodenal papilla.

CLINICAL APPLICATIONS

Gallstones are small, hard mineral deposits (*calculi*) that can cause pain when they obstruct the cystic or common bile ducts. The major component of most gallstones is cholesterol. In order for gallstones to form, the liver produces a supersaturated solution of cholesterol, which becomes crystalized when exposed to excessive mucus secreted by the gallbladder. The combination of these crystals and the mucus form a sludge that impedes the emptying of the gallbladder. Sometimes the gallstones can be dissolved by taking bile acids orally; sometimes, they can be shattered by high-energy waves given to a patient in a water bath, a technique called *extracorporeal shock-wave lithotripsy*. However, the most effective treatment is surgical removal of the gallbladder.

CLINICAL INVESTIGATION CLUES

Remember that Jim has fat in his stools, jaundice, and a high blood level of conjugated bilirubin. Based on this, his physician stated that he probably has gallstones.

- What are gallstones and where are they found?
- How could gallstones cause an elevated blood level of conjugated bilirubin and jaundice?

The Exocrine Pancreas Produces Pancreatic Juice

The **pancreas** produces **pancreatic juice**, which contains *bicarbonate* and a wide variety of digestive enzymes. These include:

1. *Pancreatic amylase*, which digests starch
2. *Trypsin* and several others, which digest proteins
3. *Lipase*, which digests triglycerides (fats and oils)
4. *Phospholipase*, which digests phospholipids (such as lecithin)
5. *Ribonuclease* and *deoxyribonuclease*, which digests nucleic acids

Pancreatic juice is an exocrine product released into the pancreatic duct, which empties into the duodenum together with the common bile duct at the *duodenal papilla* (fig. 14.18). The exocrine tissue is organized into **pancreatic acini** (fig. 14.19), composing the majority of pancreatic tissue. You may recall from chapter 8 that the pancreas is also an endocrine gland, because it contains the *islets of Langerhans* (fig. 14.19) that secrete the hormones insulin and glucagon.

Figure 14.19 indicates *zymogen granules* in the acinar cells. A *zymogen* is an enzyme with little or no activity until it becomes activated. The pancreatic enzymes are produced in this form, to protect the pancreatic tissue from self-digestion. Once in the small intestine, the inactive form of trypsin (called *trypsinogen*) is converted into the active trypsin by a brush border enzyme. Brush border enzymes, you may recall, are stuck into the plasma membranes of the microvilli in the small intestine. Once trypsin is activated, it then activates the other pancreatic enzymes (fig. 14.20).

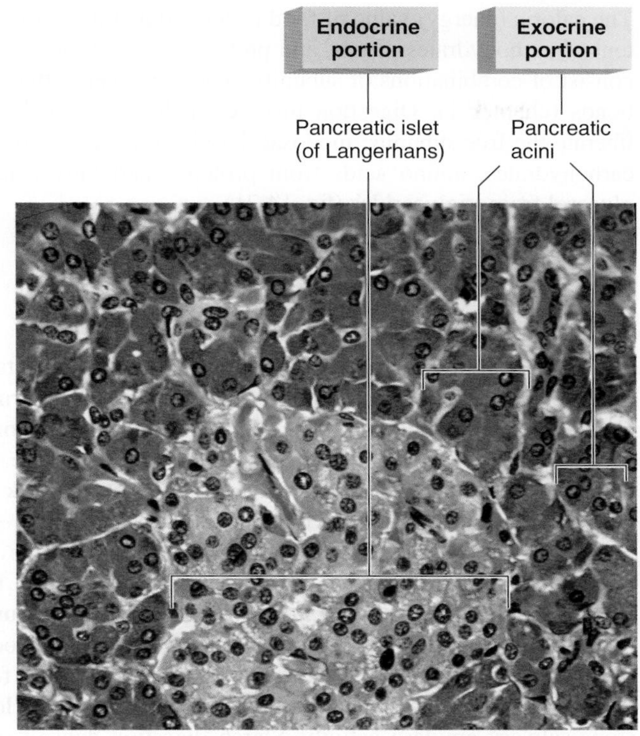

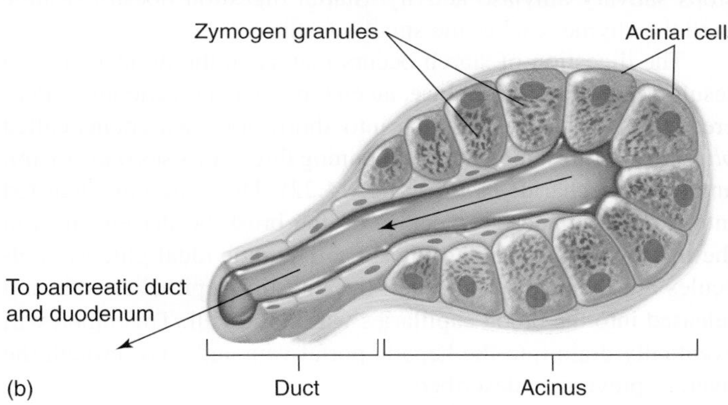

FIGURE 14.19 The pancreas is both an exocrine and an endocrine gland. (*a*) A photomicrograph of the endocrine and exocrine portions of the pancreas. (*b*) An illustration depicting the exocrine pancreatic acini, where the acinar cells produce inactive enzymes stored in zymogen granules. The inactive enzymes are secreted by way of a duct system into the duodenum.

Clinical Investigation Clues

Remember that Jim had a normal level of pancreatic amylase in the blood, a test for inflammation of the pancreas.

- What is pancreatic amylase, and what is its function?
- Where is pancreatic amylase normally active, and how does it become activated?

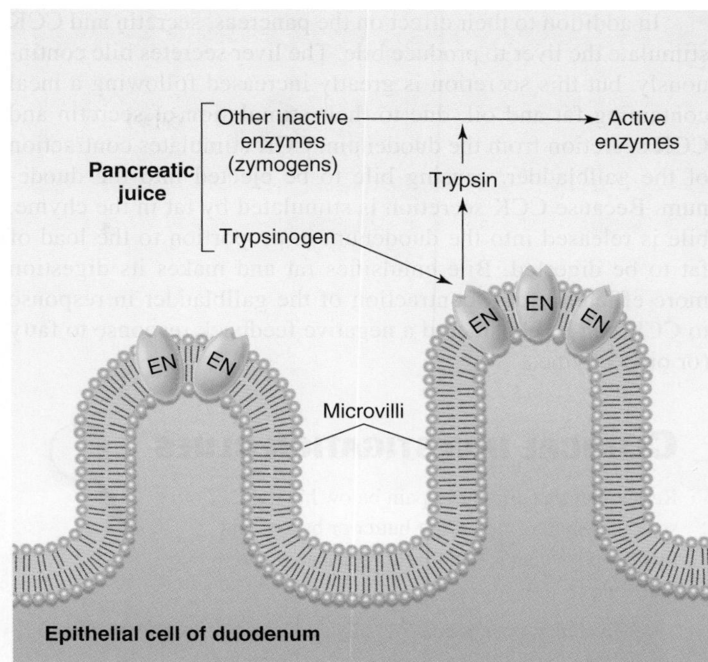

FIGURE 14.20 The activation of pancreatic juice enzymes. The pancreatic protein-digesting enzyme trypsin is secreted in an inactive form known as trypsinogen. This inactive enzyme (zymogen) is activated by a brush border enzyme, enterokinase (EN), located in the cell membrane of microvilli. Active trypsin in turn activates other zymogens in pancreatic juice.

Intestinal Hormones Stimulate Secretion of Pancreatic Juice and Bile

When chyme is pushed through the pyloric sphincter into the duodenum, it triggers the secretion of enterogastrone, the general term for hormones that inhibit the stomach (the intestinal phase of gastric regulation, previously discussed). This reduces the entry of new chyme, allowing the small intestine time to digest the load of chyme it just received. At the same time, the small intestine secretes other hormones that stimulate the release of pancreatic juice and bile to mix with the chyme in the duodenum.

When acidic chyme enters the duodenum and the pH falls below 4.5, the duodenum is stimulated to release **secretin**. Secretin is a hormone that stimulates the pancreas to secrete bicarbonate (HCO_3^-) into the pancreatic juice. When pancreatic juice arrives in the duodenum and mixes with the acidic chyme, the bicarbonate neutralizes the chyme (actually making it somewhat alkaline) as a negative feedback response to the low pH. In addition, the fat and protein content of the chyme stimulate the small intestine to secrete another hormone, **cholecystokinin (CCK)**. CCK stimulates the pancreas to produce and secrete digestive enzymes, such as lipase, amylase, and trypsin. Because lipase digests fat, a molecule that stimulates CCK secretion, this effect of CCK constitutes a negative feedback response to fatty chyme. Thus, two hormones (secretin and CCK) stimulate the pancreas to produce the two different components (bicarbonate and digestive enzymes) of pancreatic juice.

In addition to their effect on the pancreas, secretin and CCK stimulate the liver to produce bile. The liver secretes bile continuously, but this secretion is greatly increased following a meal containing fat and oil, due to their stimulation of secretin and CCK secretion from the duodenum. CCK stimulates contraction of the gallbladder, causing bile to be ejected into the duodenum. Because CCK secretion is stimulated by fat in the chyme, bile is released into the duodenum in proportion to the load of fat to be digested. Bile emulsifies fat and makes its digestion more efficient; thus, contraction of the gallbladder in response to CCK can be considered a negative feedback response to fatty (or oily) chyme.

Clinical Investigation Clues

Remember that Jim had a pain below his right scapula when he ate peanut butter or bacon, and that his physician stated that Jim probably had gallstones.

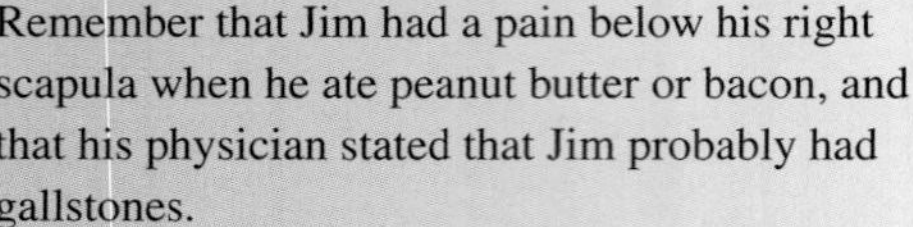

- Given that peanut butter is oily and bacon fatty, how would eating these foods influence the gallbladder?
- Given that the presence of gallstones can evoke a referred pain (chapter 7) below the right scapula, how did eating these foods cause Jim's pain?

Check Point

1. Describe the microscopic structure of the liver, explaining the pathway for blood and bile through the liver lobules.
2. Describe some of the different functions of the liver.
3. Identify the composition and functions of bile and pancreatic juice.
4. Explain how the secretions of bile and pancreatic juice are regulated.

Food Molecules Are Digested by Hydrolysis Reactions and Absorbed

Enzymes in the digestive system hydrolyze food molecules into their subunits: monosaccharides, amino acids, and so on. Monosaccharides and amino acids are transported into blood capillaries within the villi; absorbed fat is released into the lacteals and transported by the lymphatic system. Hunger is controlled by the hypothalamus and influenced by hormones from the GI tract and adipose tissue.

The caloric (energy) value of food is derived mainly from its content of carbohydrates, lipids, and proteins. These food molecules consist of combinations of subunits joined together with covalent bonds (chapter 1). Digestion involves hydrolysis reactions that liberate the free subunits: monosaccharides such as glucose from carbohydrates, amino acids from proteins, and fatty acids and glycerol from triglycerides (fig. 14.21).

Carbohydrates Are Digested into Monosaccharides

The most common form of carbohydrates in our food is starch, a polysaccharide of glucose subunits joined together to form long chains with occasional branches (fig. 14.21). The most common food sugars are disaccharides, including sucrose (table sugar, consisting of glucose bonded to fructose) and lactose (milk sugar, consisting of glucose bonded to galactose).

The digestion of starch begins in the mouth with the action of an enzyme in saliva, **salivary amylase**. This enzyme breaks the bonds (by hydrolysis reactions) between glucose subunits in the starch molecules, but most people don't chew their food long enough for this digestion to proceed very far. When the food is swallowed and mixes in the stomach with gastric juice, the low pH stops salivary amylase activity. Starch digestion doesn't resume until the chyme reaches the small intestine.

The digestion of starch occurs mainly in the duodenum as a result of **pancreatic amylase**, an enzyme in pancreatic juice. Pancreatic amylase digests starch into short, branched chains called *oligosaccharides*; molecules containing three glucoses (*maltriose*); and the disaccharide *maltose* (fig. 14.22). These are next digested into separate glucose molecules by the brush border enzymes in the microvilli of the small intestine. The individual glucose molecules are then moved through the intestinal epithelial cells and released into the blood capillaries within the villi. This blood will eventually drain into the hepatic portal vein and pass through the liver, as previously described.

Proteins Are Digested into Amino Acids

The digestion of proteins begins in the stomach, with the action of pepsin. Pepsin can liberate some amino acids, but its primary effect is to digest large proteins into shorter polypeptide chains. This helps make the chyme more homogeneous, but it isn't required for complete protein digestion, which occurs in the small intestine. Most protein digestion occurs in the duodenum and jejunum, largely due to trypsin and other protein-digesting enzymes in pancreatic juice.

As a result of the action of these pancreatic juice enzymes, proteins are digested into amino acids, dipeptides, and tripeptides (consisting, respectively, of 2 or 3 amino acids bonded together).

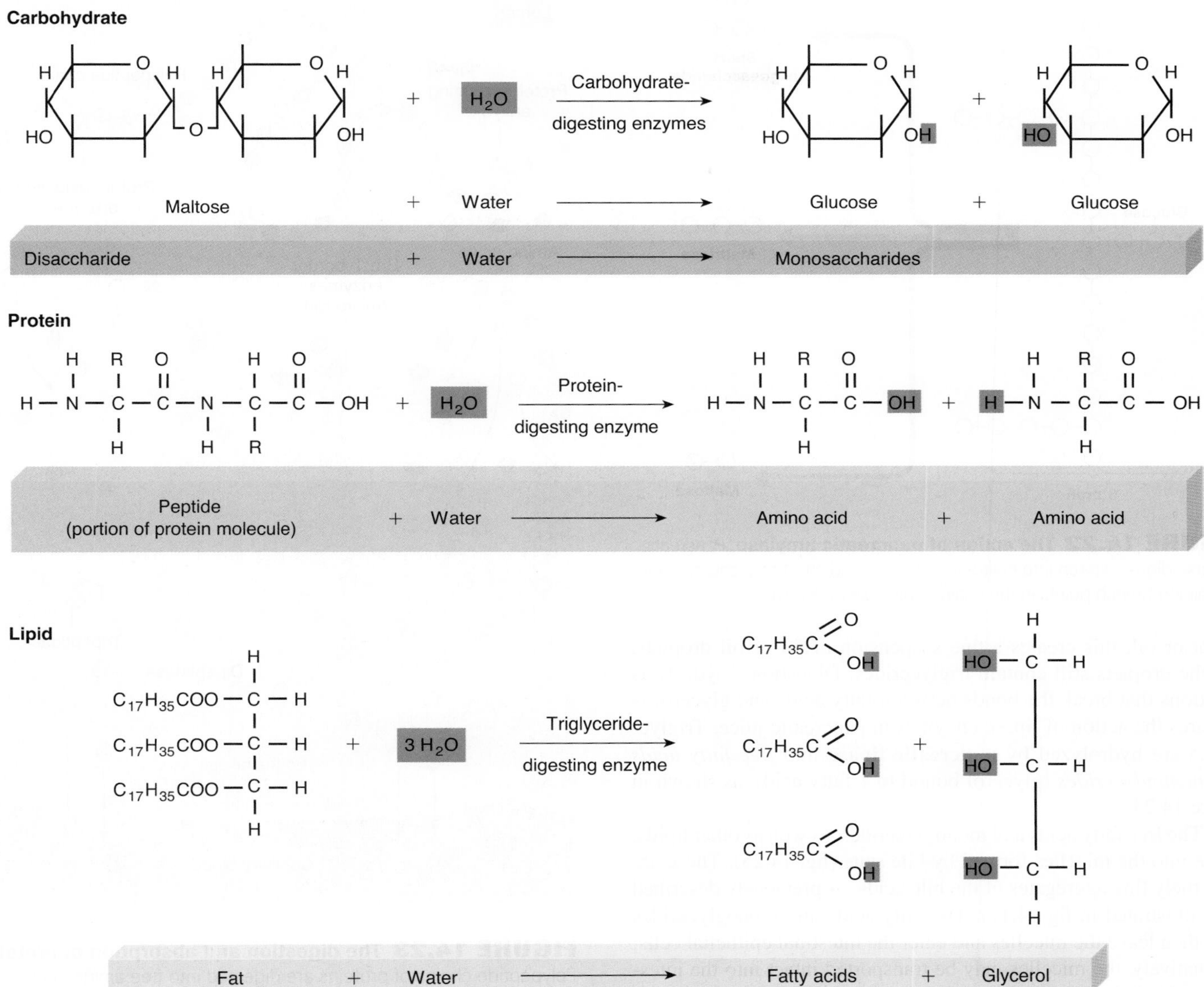

FIGURE 14.21 The digestion of food molecules through hydrolysis reactions. These reactions ultimately release the subunit molecules of each food category.

The free amino acids are transported through the intestinal epithelial cells and released into the blood capillaries of the villi. The dipeptides and tripeptides are transported into the epithelial cell cytoplasm, and then digested into separate amino acids within the cells (fig. 14.23). These amino acids enter the capillary blood, which eventually drains into the hepatic portal vein and passes through the liver.

Triglycerides Are Transported by the Lymphatic System

Triglycerides (fat and oil) are digested almost exclusively in the small intestine. When the fatty chyme reaches the duodenum, the small intestine secretes the hormones CCK and secretin, triggering the release of bile and pancreatic juice. Bile emulsifies the globules

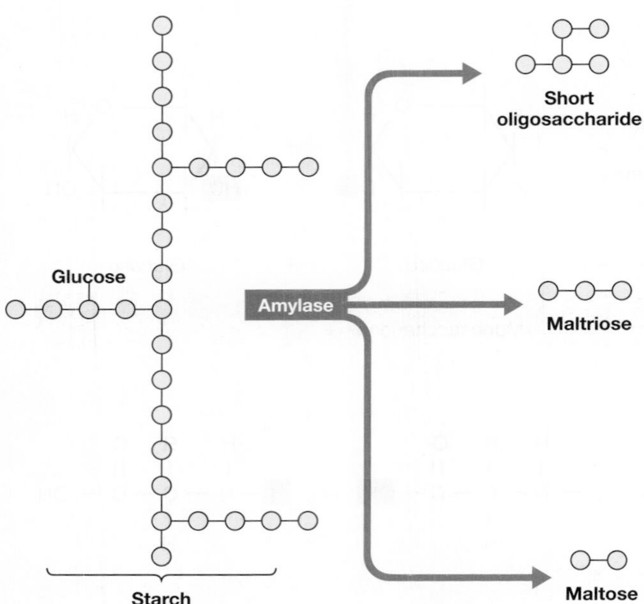

FIGURE 14.22 The action of pancreatic amylase. Pancreatic amylase digests starch into maltose, maltriose, and short oligosaccharides containing branch points in the chain of glucose molecules.

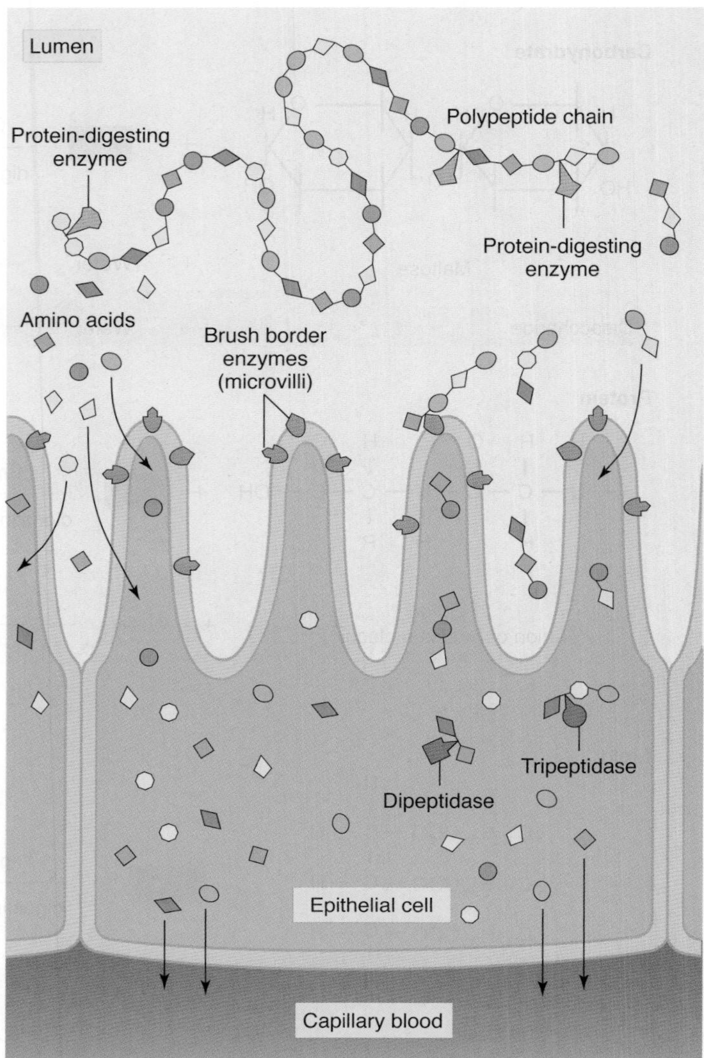

FIGURE 14.23 The digestion and absorption of proteins. Polypeptide chains of proteins are digested into free amino acids, peptides, and tripeptides by the action of pancreatic juice enzymes and brush border enzymes. The amino acids, dipeptides, and tripeptides enter duodenal epithelial cells. Dipeptides and tripeptides are hydrolyzed into free amino acids within the epithelial cells, and these products are secreted into capillaries that carry them to the hepatic portal vein.

of fat or oil; this creates a fine suspension of fat or oil droplets, but the droplets still contain triglycerides. Digestion—hydrolysis reactions that break the bonds between fatty acids and glycerol—requires the action of lipase enzymes in pancreatic juice. Triglycerides are hydrolyzed by **pancreatic lipase** into *free fatty acids* and *monoglycerides* (glycerol bound to 1 fatty acid), as shown in figure 14.24.

The free fatty acids and monoglycerides, as well as other lipids, move into the micelles formed by bile salts (fig. 14.25). These are extremely tiny aggregates of the bile acids, as previously described (and illustrated in fig 14.17). The fatty acids and monoglycerides may then leave the micelles and enter the intestinal epithelial cells; alternatively, the micelles may be transported intact into the intestinal epithelial cells. Either way, the free fatty acids and monoglycerides inside the cytoplasm are put back together again (through dehydration synthesis reactions; see chapter 1) to re-form triglycerides (fig. 14.26). Then, the newly formed triglycerides (together with cholesterol) are bound to carrier proteins, forming tiny particles known as **chylomicrons**. The chylomicrons are released from the epithelial cells into the lacteals, the lymphatic capillaries within the villi (fig. 14.26).

Lymphatic vessels transport the chylomicrons into the *thoracic duct*, which ultimately drains into the *left subclavian vein* (chapter 10; see fig. 10.32). Thus, the triglycerides in the chylomicrons enter the general circulation without first passing through the liver, unlike the monosaccharides and amino acids from digested carbohydrates and proteins. As the chylomicrons pass through the blood capillaries in muscles and adipose tissue, an enzyme called *lipoprotein lipase* hydrolyzes the triglycerides. This releases free fatty acids and glycerol that can be used for cell respiration by the skeletal muscles, or for the synthesis of new fat molecules in the adipose cells.

CLINICAL INVESTIGATION CLUES

Remember that Jim had fatty stools and his physician told him that he probably had gallstones.

- How does bile participate in the digestion of fat?
- How could gallstones cause undigested fat to appear in the stools?

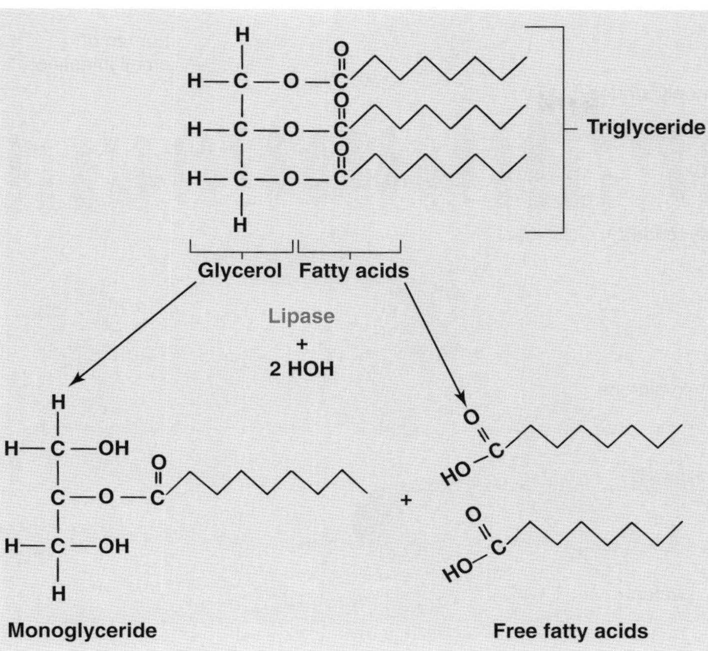

FIGURE 14.24 The digestion of triglycerides. Pancreatic lipase digests fat (triglycerides) by cleaving off the first and third fatty acids. This produces free fatty acids and monoglycerides. Sawtooth lines indicate hydrocarbon chains in the fatty acids.

The Hypothalamus Regulates Hunger and Metabolic Rate

The body obtains energy from the digestion and absorption of nutrients; if more caloric energy is consumed than is immediately required, the excess energy is stored for later use. We "bank" excess energy in the form of fat in adipose tissue, with lesser amounts stored as glycogen in the muscles and liver. These reserves are tapped when the body's caloric expenditure exceeds its intake from food. The caloric expenditure of the body has three components:

1. **Basal metabolic rate (BMR)** is the energy expenditure of a relaxed, resting person at a comfortable temperature. This accounts for about 60% of the total calorie expenditure of an average adult.
2. **Adaptive thermogenesis** is the heat energy (calories) expended to (a) maintain homeostasis of body temperature, and (b) digest and absorb food.
3. **Physical activity** increases the metabolic rate of skeletal muscles, which can significantly raise the body's metabolic rate, depending on the type and intensity of the exercise.

We maintain homeostasis of body weight by eating enough on a sustained basis to compensate for the calories we burn in these activities. Dieters have personal experience with the difficulty of losing weight, which can be attributed to the physiological mechanisms that fight to maintain body weight homeostasis. These mechanisms regulate hunger, and thus food intake, as well as the metabolic rate.

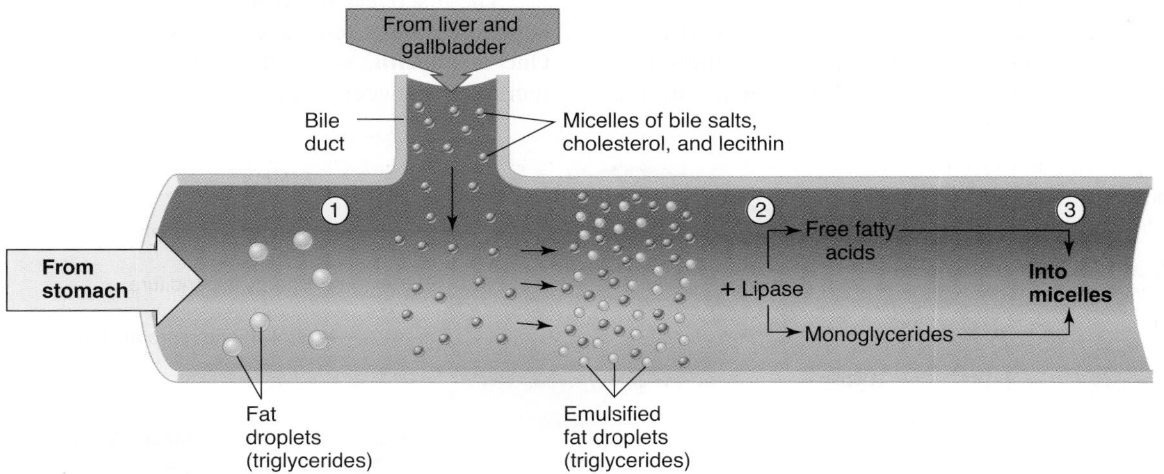

FIGURE 14.25 Fat digestion and emulsification. The three steps indicate the fate of fat in the small intestine. The digestion of fat (triglycerides) releases fatty acids and monoglycerides, which become associated with micelles of bile salts secreted by the liver.

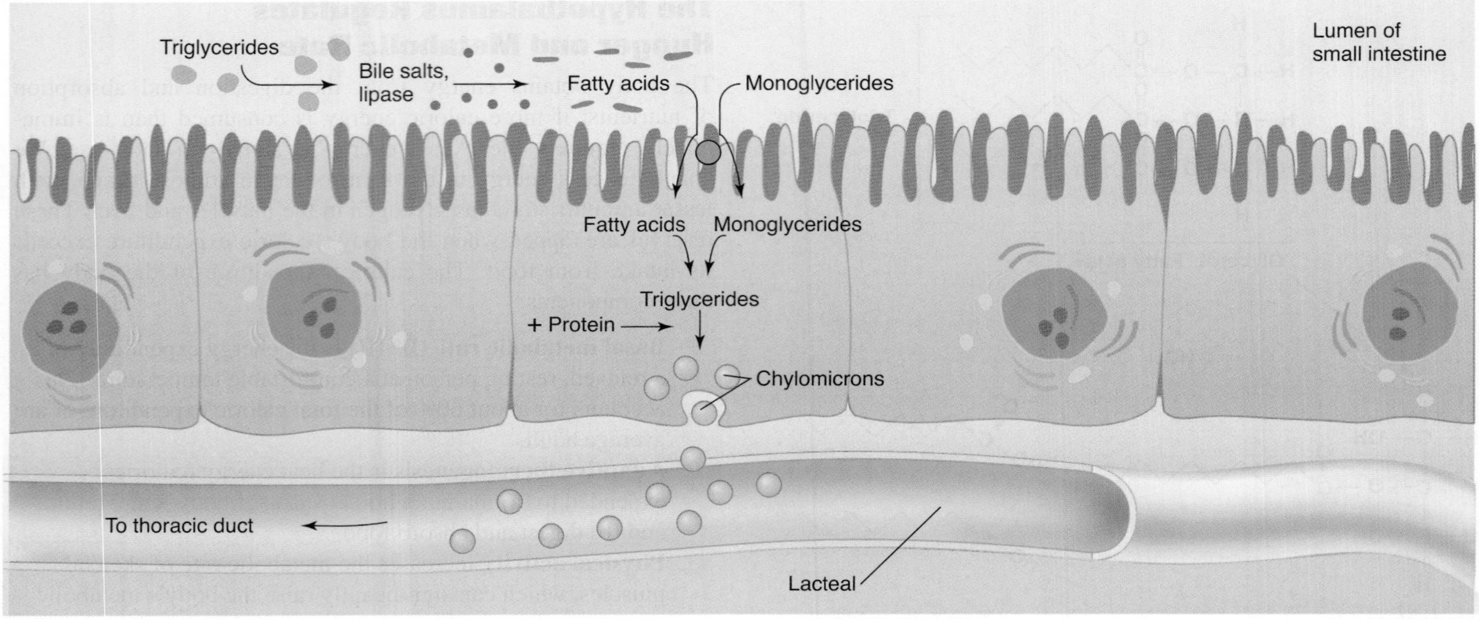

FIGURE 14.26 The absorption of fat. Fatty acids and monoglycerides from the micelles within the small intestine are absorbed by epithelial cells and converted intracellularly into triglycerides. These are then combined with protein to form chylomicrons, which enter the lymphatic vessels (lacteals) of the villi. These lymphatic vessels transport the chylomicrons to the thoracic duct, which empties them into the venous blood (of the left subclavian vein).

The **arcuate nucleus** in the hypothalamus is believed to be the hunger center of the brain, sending axons to other brain regions involved in hunger and eating behavior. There appear to be two groups of neurons in the arcuate center that regulate hunger. One group releases polypeptide neurotransmitters that act to decrease hunger; the other releases different polypeptides that stimulate hunger. The activity of these two groups of neurons is influenced by other brain areas; in this way, the smell or appearance of food, or even talking about it, can increase appetite. In addition, hunger is influenced by chemical signals (hormones) released by the GI tract and the adipose tissue.

The discovery of GI tract hormones that influence hunger is ongoing, but a few have already been identified (fig. 14.27). One is **ghrelin**, secreted by the stomach. Ghrelin secretion increases between meals when the stomach is empty, and

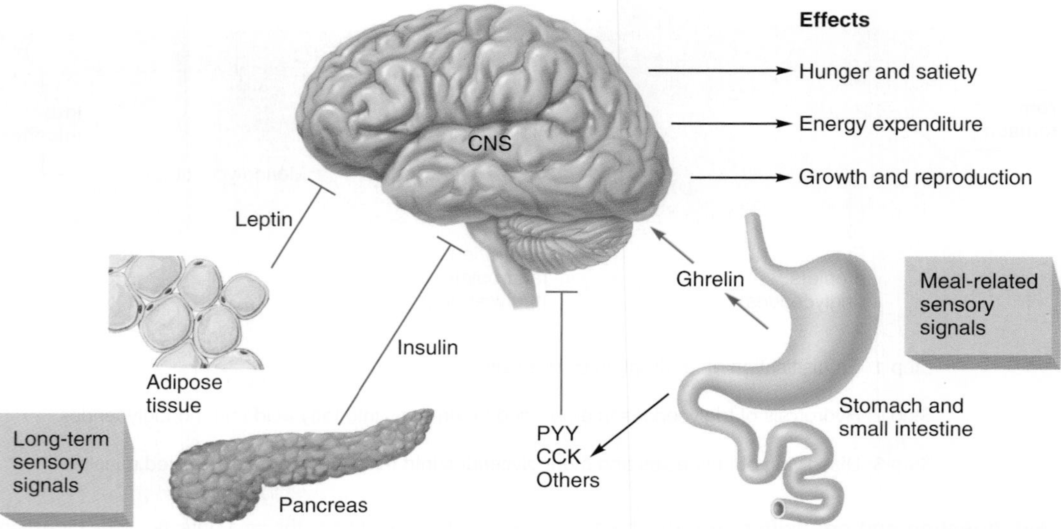

FIGURE 14.27 Hormonal signals that regulate feeding and energy expenditures. The inhibitory sensory signals are shown in red, and the stimulatory signal (ghrelin) in green. The central nervous system (CNS) integrates this sensory information with other information (smell, taste, and psychological factors) to help regulate hunger and satiety, energy expenditures, as well as growth and reproduction.

stimulates hunger. The small intestine secretes a different hormone, *cholecystokinin* (*CCK*), after a meal; in addition to its previously described effects, CCK reduces appetite. Thus, ghrelin and CCK act antagonistically to regulate hunger on a short-term basis, just before and after meals. Another recently discovered intestinal hormone, **polypeptide Y-Y (PYY)**, seems to suppress hunger on a more intermediate term, perhaps helping to space meals.

The adipose tissue secretes hormones that act as *satiety factors* (satiety refers to a feeling of fullness, suppressing appetite), which regulate hunger on a more long-term basis than the hormones secreted by the GI tract. One well-established satiety factor secreted by adipose tissue is **leptin**. The secretion of leptin increases as the adipose cells store increasing amounts of fat. Leptin then acts on the hypothalamus to suppress appetite and increase the metabolic rate. These actions reduce the intake of calories while raising caloric expenditure, thereby promoting the loss of fat in a negative feedback manner.

Conversely, when fat is broken down (as in a diet), the adipose cells secrete less leptin. A reduced secretion of leptin increases appetite and decreases metabolic rate. These actions of leptin and other chemical signals from adipose tissue help explain why it can be so difficult to lose weight—the body fights to maintain homeostasis of its *adiposity* (amount of stored fat). Insulin, secreted by the β-cells in the islets of Langerhans, is also believed to function as a satiety factor (fig. 14.27).

CHECK POINT

1. Identify the enzymes involved in carbohydrate digestion and their locations. Describe how monosaccharides are absorbed.
2. Identify the enzymes involved in protein digestion and their locations. Describe how amino acids are absorbed.
3. Describe the roles of bile and pancreatic juice in fat digestion, and explain how fat is absorbed and used by the body.
4. List and describe the three components of the body's caloric expenditures.
5. Identify the hormones of the GI tract and adipose tissue that influence hunger and satiety, and explain their actions.

PHYSIOLOGY IN HEALTH AND DISEASE

Peptic ulcers are erosions of the mucosa layer of the stomach or duodenum, produced by the action of hydrochloric acid (HCl). The stomach is normally protected from the damaging effects of HCl because (1) it has an *adherent layer of mucus* (stuck to the apical surface of the epithelial cells) containing alkaline bicarbonate (HCO_3^-); (2) there are tight junctions between adjacent epithelial cells, protecting the underlying layers from the HCl in gastric juice; and (3) epithelial cell division is very rapid, replacing the entire epithelium every 3 days. The duodenum is likewise normally protected from the acidity of chyme because (1) it also has an adherent layer of alkaline mucus, and glands in the submucosa (Brünner's glands) that secrete an alkaline mucus; and (2) alkaline pancreatic juice is released into the duodenum in response to the arrival of acidic chyme, as previously described.

When the stomach's barriers to self-destruction are broken down, acid can leak through the mucosa to the submucosa and cause inflammation, known as **acute gastritis**. The inflammation is worsened by histamine released from mast cells (chapter 11), which stimulates the parietal cells to secrete even more HCl. This is why drugs that block the H_2 histamine receptor (such as Tagamet and Zantac) are used to treat gastritis. Peptic ulcers of the stomach or duodenum can result if the secretion of acid is too great, the production of bicarbonate in the adherent layer of mucus is deficient, or some other aspects of the barriers to the damaging effects of HCl are deficient.

The 2005 Nobel Prize in Physiology or Medicine was awarded to two scientists who discovered that *Helicobacter pylori*, a common bacterium present in almost half of all adults, is the cause of most cases of peptic ulcers of the stomach and duodenum. As a result of this discovery, peptic ulcers can now be treated with a proton pump inhibitor (a drug such as Prilosec, which inhibits the K^+/H^+ pumps in the parietal cells of the stomach) combined with two different antibiotics to suppress the *H. pylori* infection.

People with gastritis or peptic ulcers should avoid substances that stimulate acid secretion, such as coffee and alcohol, and often must take antacids (such as Tums). Also, they can take H_2 histamine receptor blockers (such as Zantac) or proton pump inhibitors (such as Prilosec). There are tests now available that can determine if *H. pylori* is present, and if so antibiotics such as *amoxicillin* and *clarithromycin* may also be prescribed to treat the gastritis or peptic ulcers.

Physiology in Balance

Digestive System

Working together with . . .

Cardiovascular System

The hepatic portal vein transports digestion products to the liver . . . p. 240

Nervous System

Parasympathetic neurons regulate gastric secretions during the cephalic phase, and stimulate other aspects of digestive system function . . . p. 135

Endocrine System

The stomach secretes the hormone gastrin, and the small intestine secretes the hormones secretin and CCK to regulate digestive functions . . . p. 186

Muscular System

Smooth muscles produce peristaltic movements and segmentation in the GI tract . . . p. 215

Respiratory System

The respiratory system provides oxygen, required for the aerobic respiration of molecules obtained from the GI tract . . . p. 297

Homeostasis Revisited

Topic	Set Point	Integrating Center	Sensors	Effectors
How food proteins in the stomach stimulate the gastric phase of stomach regulation	The protein composition of chyme in the stomach	Parietal cells of the gastric mucosa	G cells of the gastric mucosa	Partially digested protein stimulates the G cells to secrete gastrin, which stimulates the ECL cells to secrete histamine; histamine then stimulates the parietal cells to secrete HCl.
How acidic chyme entering the duodenum affects the function of the pancreas	pH of chyme in the duodenum	Cells of the duodenum that produce secretin	Cells of the duodenum that produce secretin	Secretin is secreted and stimulates the pancreas to produce and secrete bicarbonate into the pancreatic juice, helping to neutralize the acidic chyme that entered the duodenum.
How fat in the chyme entering the duodenum affects the function of the pancreas	Fat content of chyme in the duodenum	Cells of the duodenum that produce cholecystokinin (CCK)	Cells of the duodenum that produce cholecystokinin (CCK)	CCK, secreted by the duodenum, stimulates the pancreas to produce and secrete lipase (and other enzymes), promoting fat digestion.
How fat in the chyme entering the duodenum affects the function of the gallbladder	Fat content of chyme in the duodenum	Cells of the duodenum that produce cholecystokinin (CCK)	Cells of the duodenum that produce cholecystokinin (CCK)	CCK, secreted by the duodenum, stimulates the contractions of the gallbladder, promoting the entry of bile into the duodenum; bile emulsifies fat and aids its digestion.
How changes in the amount of fat stored in the adipose tissue helps regulate hunger	Amount of fat stored in the adipose tissue	Adipocytes	Adipocytes	Adipocytes secrete leptin (and other hormones); leptin suppresses hunger when the amount of stored fat increases.

Summary

Regions of the Gastrointestinal Tract Have Different Specializations 349

- The esophagus moves food toward the stomach by a wavelike contraction known as peristalsis.
- The lower esophageal sphincter is normally closed, preventing acidic gastric juice from being pushed through into the esophagus.
- The gastric mucosa contains gastric glands, with parietal cells that secrete hydrochloric acid (HCl); chief cells that secrete pepsinogen; ECL cells that secrete histamine; G cells that secrete gastrin; and cells that secrete intrinsic factor (for vitamin B_{12} absorption) and other substances.
- The parietal cells secrete H^+ (protons) into the gastric lumen using a H^+/K^+ ATPase active transport pump.
- The HCl in gastric juice helps activate pepsin, a protein-digesting enzyme that partially digests food proteins into shorter polypeptide chains.
- In the cephalic phase, the brain stimulates secretion via the vagus nerves, which release ACh to stimulate gastric secretion directly, and also indirectly via gastrin and histamine.
- In the gastric phase of gastric regulation, partially digested proteins stimulate pepsinogen and gastrin secretion; gastrin stimulates histamine secretion from the ECL cells, and the histamine stimulates the parietal cells to secrete HCl.
- In the intestinal phase, the arrival of chyme in the duodenum stimulates a neural reflex that inhibits gastric activity; also, fat in the chyme stimulates the duodenum to secrete an enterogastrone hormone, which inhibits stomach activity.

The Small and Large Intestine Have Different Structures and Functions 353

- The mucosa of the small intestine has microscopic, fingerlike projections called villi, which increase the surface area for digestion and absorption.

- The central core of each villus has connective tissue with blood capillaries and a lymphatic capillary, called a lacteal; there are also intestinal crypts, where new epithelial cells are produced by mitosis.
- The plasma membrane on the apical (top) surface of each epithelial cell on a villus is folded to produce microvilli; embedded in the microvilli are various brush border enzymes.
- Brush border enzymes complete the digestion of food molecules, and include such enzymes as lactase, which digests lactose (milk sugar).
- Peristalsis and segmentation, the simultaneous contractions of different intestinal segments to help mix the chyme, are coordinated by neurons in the GI tract that form an enteric nervous system, so that these reflexes can occur without the involvement of the CNS.
- The large intestine absorbs many ions (electrolytes) and water, as well as certain vitamins that may be produced by intestinal bacteria.
- The large population of intestinal bacteria—called the intestinal microbiota—serves many functions, including the production of vitamin K, folic acid, and short-chain fatty acids; they also help protect the body against pathogenic (disease-causing) bacteria and promote anti-inflammatory processes.

The Liver, Gallbladder, and Pancreas Aid Digestion 357

- The liver contains microscopic lobules, which consist of hepatic plates (1 to 2 cells thick) separated by wide, porous capillaries called sinusoids.
- Small branches of the hepatic artery and hepatic portal vein empty blood into the sinusoids, and the bloods mix as they flow toward the center of the hepatic lobule, to be drained by central veins.
- The hepatic portal vein delivers blood that has already passed through the capillaries in the GI tract, and so the liver is the first organ to get most of the products of digestion.
- The hepatocytes (liver cells) produce bile and secrete it into bile canaliculi between adjacent rows of cells in each hepatic plate; bile drains into hepatic ducts that carry the bile from the liver.
- The liver has many diverse functions, including detoxication of the blood, production of bile, regulation of blood glucose, metabolism of lipids, and the synthesis of many of the plasma proteins.
- Bilirubin (bile pigment) is formed from heme groups (minus the iron) derived from hemoglobin; the liver conjugates (combines) the free bilirubin with other molecules, forming conjugated bilirubin.
- The liver secretes conjugated bilirubin in the bile; bacteria in the intestine convert it into urobilinogen, and some of this is absorbed and travels back to the liver in the hepatic portal vein (in an enterohepatic circulation).
- Bile acids are derivatives of cholesterol that have polar groups, allowing them to form aggregations called micelles and to emulsify fat; emulsification refers to the breaking up of larger globules of triglycerides into smaller droplets, producing a finer suspension of triglycerides in water.
- Pancreatic acini form exocrine units that secrete pancreatic juice into the pancreatic duct, emptying into the duodenum.
- Pancreatic juice contains bicarbonate and many enzymes, which are produced in an inactive form (as zymogens) that are activated within the small intestine by a brush border enzyme.
- The low pH of acidic chyme entering the duodenum stimulates the small intestine to release secretin, a hormone that stimulates bicarbonate production in the pancreas.
- The fat and protein content of chyme in the duodenum stimulates the small intestine to secrete cholecystokinin (CCK), which stimulates the pancreas to produce the digestive enzymes of pancreatic juice.
- When fatty chyme arrives in the duodenum, it stimulates the secretion of CCK and secretin, which promote bile production in the liver; CCK also stimulates contraction of the gallbladder, causing the ejection of bile into the duodenum.

Food Molecules Are Digested by Hydrolysis Reactions and Absorbed 362

- Digestion reactions involve hydrolysis, in which the covalent bonds joining subunit molecules together are broken.
- Starch digestion begins in the mouth by the action of salivary amylase, but this doesn't proceed very far; it continues in the small intestine under the action of pancreatic amylase.
- Complete digestion of starch and disaccharides into monosaccharides requires the action of brush border enzymes, including lactase (digesting lactose) and sucrase (digesting sucrose).
- Protein digestion begins in the stomach, with the action of pepsin; pepsin cleaves large proteins into shorter polypeptide chains.
- Protein digestion continues in the small intestine with the action of trypsin and other protein-digesting enzymes in pancreatic juice.
- Monosaccharides and amino acids are absorbed across the epithelial cells of the small intestine into the blood capillaries in the villi, so that they eventually travel in the hepatic portal vein to the liver.
- Triglycerides (fat and oil) in the chyme are emulsified by bile salts and then hydrolyzed by pancreatic lipase into free fatty acids and monoglycerides.
- The fatty acids and monoglycerides enter the micelles and then pass into the cytoplasm of the epithelial cells; once in the cytoplasm, they are put together again (through dehydration synthesis) to make triglycerides.
- The newly formed triglycerides are bound to carrier proteins to form tiny particles called chylomicrons, which are released into the lacteals within the villi; they travel through the lymphatic system and eventually enter the blood at the left subclavian vein.
- The expenditure of calories can be divided into three components: the basal metabolic rate; adaptive thermogenesis (including calories spent in temperature regulation and digestion); and physical activity.
- The arcuate nucleus in the hypothalamus is the center for the control of hunger; it is influenced by higher brain centers and hormones secreted by the GI tract and adipose tissue.
- The hormones ghrelin (from the stomach) and the intestinal hormones CCK and PYY act antagonistically to regulate hunger on a short-term basis.
- Leptin, a satiety factor from the adipose tissue, and insulin from the pancreas, may act on the hypothalamus to reduce hunger on a long-term basis as adipose tissue grows, helping to maintain homeostasis of body weight.

Review Activities

Objective Questions: Test Your Knowledge

1. The movement of the subunits of food molecules across the wall of the digestive tract into the blood is known as
   a. excretion
   b. digestion
   c. absorption
   d. motility

2. In peristaltic contractions of the esophagus,
   a. smooth muscle ahead of the bolus contracts while smooth muscle behind it relaxes.
   b. smooth muscle behind the bolus contracts while smooth muscle ahead of it relaxes.
   c. the lower esophageal sphincter contracts while smooth muscle elsewhere relaxes.
   d. the entire smooth muscle of the esophagus contracts simultaneously.

Match the cell type of the gastric glands with its secretion:

3. Parietal cells	a. histamine
4. G cells	b. hydrochloric acid
5. ECL cells	c. pepsinogen
6. Chief cells	d. gastrin

7. The pH of gastric juice is
   a. greater than 10.
   b. about 7.
   c. between 5 and 7.
   d. less than 2.
8. Which of the following statements about pepsin is false?
   a. It completely digests proteins into separate amino acids.
   b. It is formed in the stomach lumen.
   c. It has a pH optimum of about 2.
   d. It is produced when 1 pepsinogen molecule removes an amino acid sequence from another pepsinogen molecule.
9. Which of the following about proton pumps is true?
   a. They are carrier proteins located in the duodenum.
   b. They are protein channels that allow passive transport in the stomach.
   c. They are H^+/K^+ ATPase active transport carriers in the stomach.
   d. They are Na^+/K^+ ATPase active transport carriers in the colon.
10. Hydrochloric acid secretion in the stomach is most directly stimulated by
    a. pepsinogen.
    b. gastrin.
    c. histamine.
    d. acetylcholine.

Match the phase of gastric regulation with the most correct description:

11. Cephalic phase	a. stimulated by partially digested food proteins
12. Gastric phase	b. inhibited by histamine secreted from mast cells
13. Intestinal phase	c. stimulated by the vagus nerves
	d. inhibited by enterogastrone and neural reflexes

14. Which of the following statements about brush border enzymes is true?
    a. They are located in the lacteals.
    b. They are secreted by the gastric glands.
    c. They are secreted by the intestinal crypts.
    d. They are stuck into the plasma membrane of microvilli.

Match the digestive enzyme with its correct description:

15. Lipase	a. produced by both salivary glands and the pancreas
16. Lactase	b. produced by gastric glands
17. Amylase	c. is a brush border enzyme
	d. produced by the pancreas

18. Which of the following statements about the enteric nervous system is false?
    a. It contains interneurons, like the CNS.
    b. It must relay information to the CNS for reflex contractions, including peristalsis.
    c. It has about the same number of neurons as the spinal cord.
    d. It contains sensory and autonomic neurons.
19. The first organ to receive monosaccharides and amino acids from the small intestine is the
    a. liver.
    b. pancreas.
    c. heart.
    d. brain.
20. Which of the following statements regarding bile pigment is true?
    a. It's produced only in the liver.
    b. The liver converts conjugated bilirubin into free bilirubin.
    c. The liver converts conjugated bilirubin into urobilinogen.
    d. Bilirubin is derived from heme, minus the iron.
21. Which of the following statements regarding the emulsification of fat by bile is true?
    a. Emulsification is produced by the action of bile salts in the colon.
    b. Triglycerides are hydrolyzed into their subunits during emulsification.
    c. Larger fat globules are broken up into a finer suspension.
    d. Bile acids are derived from lecithin.
22. Which of the following statements regarding pancreatic juice is false?
    a. Its produced by the pancreatic acini.
    b. Trypsinogen is activated by a brush border enzyme in the small intestine.
    c. Pancreatic juice enzymes are produced as zymogens.
    d. Its enzymes digest disaccharides into monosaccharides.
23. Which of the following statements regarding secretin and CCK is false?
    a. Secretin primarily stimulates pancreatic enzyme production.
    b. Secretin secretion is stimulated by acidic chyme in the duodenum.
    c. CCK secretion is most stimulated by fatty chyme in the duodenum.
    d. CCK stimulates contraction of the gallbladder.
24. Which of the following statements regarding chylomicrons is true?
    a. They consist of a carrier protein bound to fatty acids and monoglycerides.
    b. They are absorbed from the intestinal lumen into the epithelial cells.
    c. They transport triglycerides from the intestine through the lymphatic system.
    d. They enter the blood capillaries in the villi, and then the hepatic portal vein.

25. The hormone leptin is secreted
    a. by the stomach and causes increased appetite.
    b. by the small intestine and reduces appetite.
    c. by the adipose tissue and reduces appetite.
    d. in increased amounts when a person loses weight.

Essay Questions 1: *Test Your Understanding*

1. Describe peristalsis in the esophagus, and explain why gastric juice doesn't normally enter the esophagus.
2. Identify the cells of the gastric glands and their secretions, and explain how hydrochloric acid is secreted.
3. List the benefits provided by the strong acidity of gastric juice, and explain the role of histamine in the regulation of HCl secretion.
4. Identify the three phases of gastric regulation, and explain the events that occur in each phase. Also, explain why a fatty or oily meal stays longer in the stomach than a strictly carbohydrate meal.
5. What is the duodenum, and what features make it unique?
6. Describe the structure of villi and microvilli, as well as the intestinal crypts, and explain their functions.
7. Identify the location and nature of the brush border enzymes, and explain their significance.
8. Describe the nature and significance of the enteric nervous system.
9. What is the intestinal microbiota, and what is its physiological significance?
10. Describe the structure of a liver lobule, and trace the path of blood and bile through the liver.
11. Identify some of the functions of the liver.
12. Describe the formation and fate of bilirubin, in all of its forms.
13. Describe the origin of the bile acids and explain their functions.
14. What are zymogens, and how are the zymogens in pancreatic juice activated?
15. Identify the stimuli for secretin and CCK secretion, and describe how they affect the liver, gallbladder, and pancreas.
16. Describe how food starch is digested and absorbed.
17. Describe how food protein is digested and absorbed.
18. Describe the digestion and absorption of fat and oil, and explain how this differs from the digestion and absorption of carbohydrates and proteins.
19. Describe the different components of the metabolic rate, and explain how they can vary.
20. Explain how hormones from the GI tract and adipose tissue affect the hunger center in the hypothalamus and the metabolic rate.

Essay Questions 2: *Test Your Analytical Ability*

1. When a person has a hiatal hernia (where the stomach protrudes through the esophageal hiatus of the diaphragm), there is likelihood of severe heartburn due to gastroesophageal reflux. Provide a possible explanation for this association.
2. The esophagus, small intestine, and large intestine have two layers of smooth muscle in the muscularis externa, whereas the stomach has three layers. What difference in the function of these regions might this difference in structure serve?
3. Tagamet is an antihistamine drug used for gastritis and heartburn, whereas Benadryl is an antihistamine drug used to treat allergy. Explain how these two drugs work, and how they differ in their actions.
4. Describe how a "proton pump inhibitor" would help treat gastritis and heartburn, and compare its effect on gastric juice to the action of a histamine receptor blocker.
5. Your mouth can water and your stomach gurgles when you look and smell good food. Explain the mechanisms responsible.
6. A cheeseburger makes you feel full for a longer time than a hamburger of the same size without the cheese. Propose a physiological explanation for this difference.
7. Explain how microvilli in the small intestine benefit both digestion and absorption.
8. Explain the cause of lactose intolerance, and propose how special milk called Lactaid may get around this problem.
9. What is the enteric nervous system, and what benefits do you think it provides?
10. People who take too many antibiotics may develop inflammatory bowel disease—chronic inflammation of the intestine. Propose an explanation for this association.
11. "Bacteria only cause disease." Evaluate that statement in view of the intestinal microbiota.
12. In a step-by-step manner, explain why feces have a brown color. What color do you think feces would have if the intestinal microbiota were destroyed? Explain.
13. Some drugs are less potent when taken orally (as pills) than when taken by other means (such as by skin patch, nasal spray, inhalation, or injection). Explain several possible reasons for this.
14. Explain how jaundice caused by excessive red blood cell destruction could be distinguished from jaundice caused by gallstones. How could liver disease cause jaundice?
15. Suppose a person had the gallbladder surgically removed because of gallstones. How would this affect fat digestion and absorption? Explain.
16. Suppose a person had the stomach removed (had a gastrectomy) and subsequently developed pernicious anemia. Explain how these two observations could be related.
17. Vitamin K is needed for the liver's production of some clotting factors. Explain how a person who took excessive amounts of antibiotics could develop a prolonged clotting time.
18. A person is told to provide a fasting sample of blood plasma for laboratory testing. The technician looks at the sample of plasma and says, "This is turbid [cloudy]; you ate something fatty or oily before coming in." Provide a step-by-step explanation of how eating caused turbidity of the plasma.
19. Use the steps involved in protein digestion and absorption to explain why polypeptide hormones such as insulin and growth hormone cannot effectively be taken in the form of pills.
20. With regard to the effects of GI tract and adipose tissue hormones on the arcuate nucleus of the hypothalamus, explain how we get hungry between meals, and how it becomes increasingly harder to lose weight the further along a person gets in a diet.

Web Activities

For additional study tools, go to: www.aris.mhhe.com. Click on your course topic and then this textbook's author/title. Register once for a semester's worth of interactive activities and practice quizzing to help boost your grade!

CHAPTER 15

Male and Female Reproductive Systems

HOMEOSTASIS

The theme of the reproductive system, unlike that of the other body systems, isn't homeostasis but rather the perpetuation of the species. However, the health of the reproductive system depends on the health of the other body systems and on their ability to maintain homeostasis. For example: the abnormal function of many endocrine glands often results in abnormal reproductive function; changes in the nervous system associated with stress can interfere with both female and male reproduction; and the homeostasis of body weight, acting through hormone secretion from adipocytes, affects the reproductive system. Thus, a healthy body that maintains homeostasis is required for optimal function of the reproductive system.

CLINICAL INVESTIGATION

But if homeostasis is not maintained . . .

Michelle went to the physician complaining that she had been missing her period for a few months, and although she had been trying to get pregnant, her pregnancy tests were negative. He inquired about her low body weight, and Michelle answered that she didn't eat much, was under great stress at work, and tried to relax by running long distances daily. A laboratory test revealed that Michelle's husband David had a low sperm count, and when the physician later saw David he found him to be well masculinized and very muscular. David said he worked out at the gym and took anabolic steroids. All of the laboratory tests for Michelle were normal, and the physician advised her to eat more and try to gain weight, and to take steps to reduce her stress and her level of exercise to try to regain her period. He told David to stop using anabolic steroids, because they were likely responsible for his low sperm count.

What did the physician think is the likely cause of Michelle's missed period? How does pregnancy cause a missed period, and how does a pregnancy test work? What possible cause of a low sperm count was ruled out by the observation that David was well masculinized? What are anabolic steroids, and how could taking them cause David's low sperm count?

Chapter Outline

A virtual cadaver dissection experience

The Male Reproductive System Depends on the Function of Testes

The testes are the primary sex organs in males, secreting testosterone and producing sperm, the male gametes. These functions depend on stimulation by the anterior pituitary gland, through its secretion of the gonadotropic hormones FSH and LH. The testes consist of two compartments: interstitial Leydig cells, which secrete testosterone under LH stimulation, and the seminiferous tubules, which produce sperm by meiotic cell division.

You're now near the top of the knowledge pyramid you've been constructing during your course in human physiology. An understanding of the reproductive system is particularly dependent on your previous background in the physiology of the nervous and endocrine systems, but it also relies in part on an understanding of the cardiovascular, respiratory, urinary, and digestive systems, among others. However, this is a "fundamentals" textbook for an introductory physiology course, and so even at this point in your study there is still quite a bit more to learn. New material can be added to your existing framework of knowledge with relative ease; you've already done the "heavy lifting" of building a strong foundation for future growth in your coursework and careers.

In sexual reproduction, **gametes** (*sperm* and *egg cells*, or *ova*) are produced within the **gonads**, the primary sexual organs—**testes** (singular, *testis*) in males and **ovaries** (singular, *ovary*) in females. Gametes are *haploid* cells, produced by *meiosis* of diploid parent cells (described in Appendix 4). Because each gamete has 23 chromosomes, the *diploid* cell—the **zygote**—formed at fertilization contains 46 chromosomes. The first 22 pairs of chromosomes are the *autosomal chromosomes*, whereas the 23rd pair is the *sex chromosomes*. In a female, these consist of two X chromosomes; in a male, there is one X and one Y chromosome per diploid cell[1] (chapter 2). Although the Y chromosome is short (fig. 15.1), it codes for the **testis-determining factor** (**TDF**; fig. 15.2) that causes early embryonic gonads to become testes. When the Y chromosome and TDF are absent, the embryonic gonads become ovaries. The gene coding for the TDF has been isolated and named *SRY* (for "sex determining region Y").

The Anterior Pituitary Gland Regulates the Testes

The testes are composed of two compartments, or parts: the **seminiferous tubules** and the **interstitial tissue**, which contains **Leydig cells** (fig. 15.3). The seminiferous tubules are hollow tubes, and account for 90% of the weight of an adult testis. The interstitial tissue is a web of loose connective tissue between the tubules and contains the endocrine Leydig cells. The Leydig cells produce *androgens* (male hormones), mainly **testosterone**.

The anterior pituitary gland secretes two gonadotropic hormones: FSH (follicle-stimulating hormone) and LH (luteinizing hormone). As explained in chapter 8, the names of these two hormones describe their action in females, but the same names are

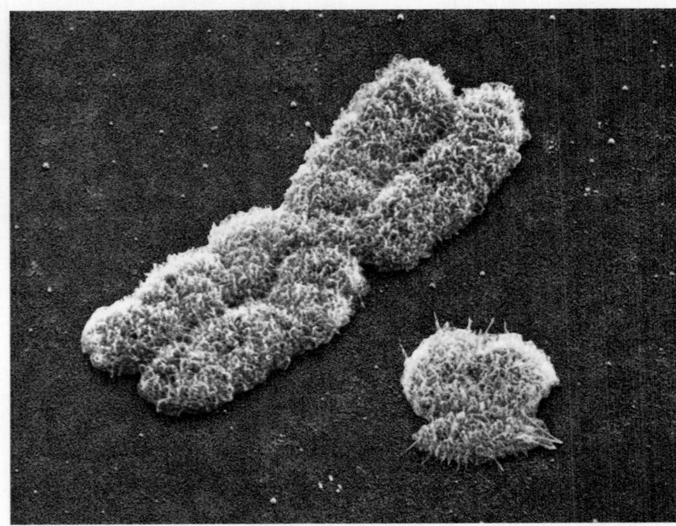

FIGURE 15.1 The human X and Y chromosomes. Magnified about 10,000×, these images reveal that the X and Y chromosomes are dramatically different in size and shape. Despite the small size and simple appearance of the Y chromosome, recent studies have uncovered its surprising complexity and sophistication.

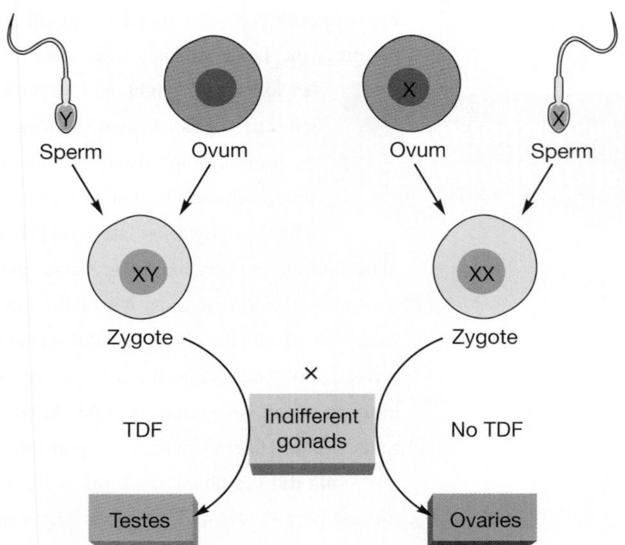

FIGURE 15.2 The chromosomal sex and the development of embryonic gonads. The very early embryo has "indifferent gonads" that can develop into either testes or ovaries. The testis-determining factor (TDF) is a gene located on the Y chromosome. In the absence of TDF, ovaries will develop.

FYI [1]The X chromosome has recently been shown to have 1,098 genes. Surprisingly many of these X-linked genes are responsible for certain diseases—there are presently 168 diseases known to be caused by mutations in 113 X-linked genes. These diseases are more common in males, who have only one X chromosome and so can't carry these genes in a recessive state. In contrast to its larger sister, the poor little Y chromosome has only about 80 genes.

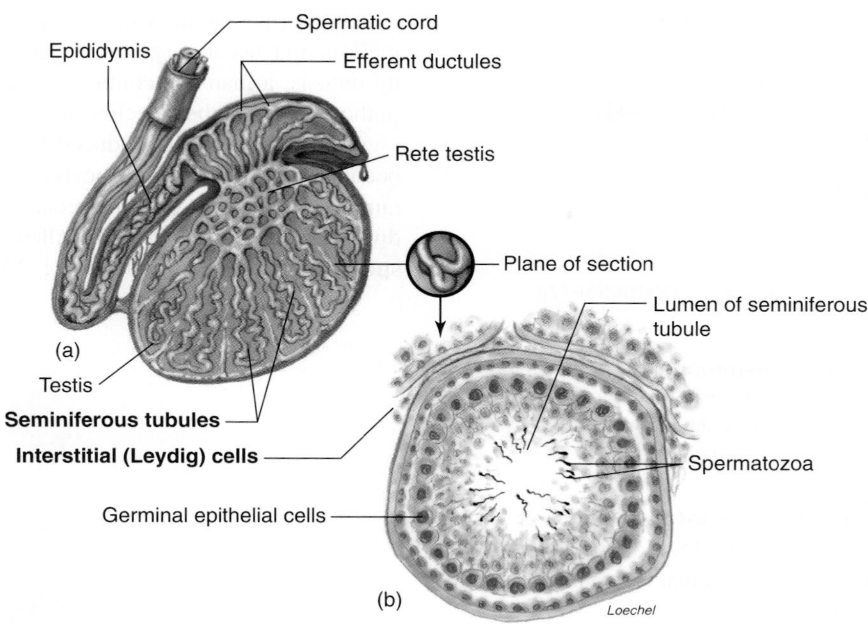

FIGURE 15.3 Microscopic structure of a testis. The seminiferous tubules contain a germinal epithelium, which undergoes meiosis to produce spermatozoa. The interstitial (Leydig) cells secrete the hormone testosterone. (*a*) The testis is illustrated in sagittal section; and (*b*) a cross section of a seminiferous tubule is depicted.

commonly used for these hormones in males to avoid confusion. FSH receptor proteins are located only in the seminiferous tubules, suggesting that these tubules are the targets of FSH action in men. Receptors for LH are exclusively located in the Leydig cells, suggesting that the Leydig cells are the targets of LH action in men.

The hypothalamus, by secreting gonadotropin-releasing hormone (GnRH), stimulates the anterior pituitary gland to secrete FSH and LH (as described in chapter 8). When stimulated by LH, the Leydig cells secrete testosterone. Testosterone then exerts negative feedback inhibition on LH secretion from the anterior pituitary, and on GnRH secretion from the hypothalamus (fig. 15.4). However, scientists found that testosterone itself doesn't fully inhibit FSH secretion. The seminiferous tubules secrete a polypeptide hormone called **inhibin** that specifically inhibits FSH secretion. Thus, in a pet that has been neutered or spayed (castrated, with its gonads removed), the pituitary secretion of LH rises because of lack of testosterone, and the secretion of FSH rises because of lack of inhibin.

Figure 15.4 presents a symmetrical picture: LH stimulates the Leydig cells, which secrete something (testosterone) that keeps LH levels from rising; FSH stimulates the seminiferous tubules, which secretes something (inhibin) that prevents FSH from rising. However, this symmetrical picture is a little misleading, because there is interaction between the two compartments of the testes. For example, testosterone stimulates the seminiferous tubules to produce sperm.

Another complexity of the male reproductive system is that testosterone may not be the active hormone inside of its target cells. In chapter 8, you learned that thyroxine (T_4) is a prehormone that must be changed within its target cells into the active form, triiodothyronine (T_3), before it can have an effect. Similarly, testosterone is actually a prehormone within many of its target cells. For example, to stimulate the cells of the prostate gland (an accessory male reproductive organ), testosterone must first be changed into **dihydrotestosterone (DHT)** by the *enzyme 5α-reductase* (fig. 15.5). In some other target cells, such as neurons in the hypothalamus responsible for the negative feedback effects of testosterone, the

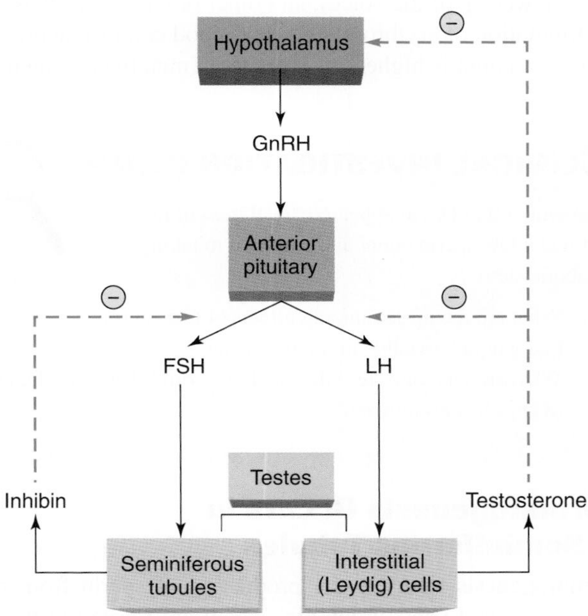

FIGURE 15.4 The anterior pituitary and testes. The seminiferous tubules are the targets of FSH action; the interstitial (Leydig) cells are targets of LH action. Testosterone secreted by the Leydig cells inhibits LH secretion; inhibin secreted by the tubules inhibits FSH secretion.

FIGURE 15.5 Derivatives of testosterone. Testosterone secreted by the interstitial (Leydig) cells of the testes can be converted into active metabolites in the brain and other target organs. These active metabolites include DHT and estradiol.

testosterone must first be converted into **estradiol-17β** (or just *estradiol*) by the enzyme *aromatase* (fig. 15.5). Estradiol is the major female hormone, but here you can see that it also plays a role in the male reproductive system.

Testosterone and its derivatives have numerous effects on a man's body. Among these, androgens stimulate:

1. *Spermatogenesis*. Testosterone acts on the seminiferous tubules to stimulate the production of sperm.
2. *Secondary sexual characteristics*. This includes the growth and maintenance of the sex accessory organs (prostate and seminal vesicles); growth of the penis; growth of facial and axillary (underarm) hair; and body growth at puberty.
3. *Anabolism*. Because of the stimulation of protein synthesis by testosterone and its derivatives, these hormones promote the growth of the musculoskeletal system, the larynx (causing lowering of the voice), and other organs, as well as the stimulation of erythropoiesis (red blood cell production; the rbc count is higher, on average, in men than women).

CLINICAL INVESTIGATION CLUES

Remember that David appeared "well masculinized" but had a low sperm count and admitted to taking anabolic steroids.

- What are the masculinizing effects of testosterone that would be visible to the physician?
- What are anabolic steroids, and how would that cause David to appear masculinized?

Spermatogenesis Occurs in the Seminiferous Tubules

Spermatogenesis refers to the production of sperm from parent stem cells, called **spermatogonia**, which are located at the outer region of the seminiferous tubules. The spermatogonia and the cells derived from them compose the **germinal epithelium** of the seminiferous tubules (they "germinate" sperm). Spermatogonia can duplicate themselves by mitosis (chapter 2); some of these cells remain spermatogonia, while others become specialized to undergo meiosis and become sperm. Because spermatogonia are renewed by mitosis, a man can continue to produce sperm throughout life (although the numbers decline at different rates among the elderly).

Some of the cells produced by the mitosis of spermatogonia become **primary spermatocytes**, which are diploid cells (containing 46 chromosomes) specialized to undergo meiotic cell division. At the first meiotic cell division, 2 haploid **secondary spermatocytes** are produced (fig. 15.6). Although each of these

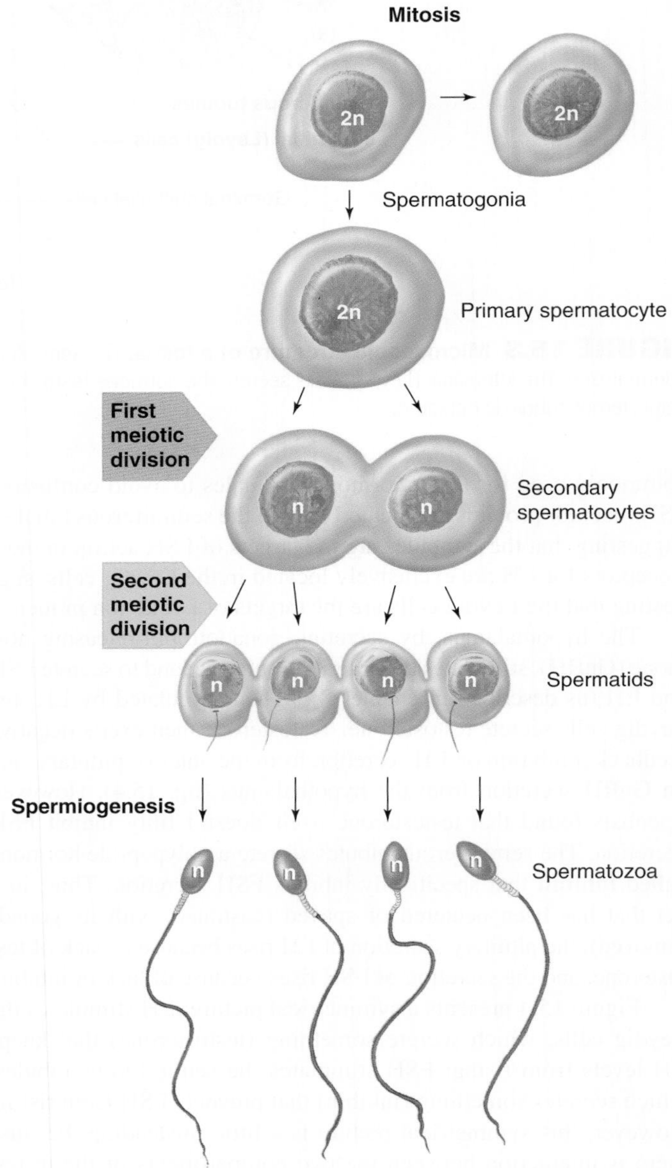

FIGURE 15.6 Spermatogenesis. Spermatogonia undergo mitotic division in which they replace themselves and produce a daughter cell that will undergo meiotic division. This cell is called a primary spermatocyte. Upon completion of the first meiotic division, the daughter cells are called secondary spermatocytes. Each of these completes a second meiotic division to form spermatids. Notice that the 4 spermatids produced by the meiosis of a primary spermatocyte are interconnected. Each spermatid forms a mature spermatozoon.

cells is haploid (with 23 chromosomes), each of their chromosomes contains 2 duplicate *chromatids*, which are separated into different cells by the second meiotic division (Appendix 4). At the end of the second meiotic division, there are 4 haploid **spermatids**. These are near the lumen of the seminiferous tubule, and have interconnected cytoplasm (fig. 15.6). The cytoplasm is removed as the cells become more specialized, eventually forming 4 **spermatozoa** (or just *sperm*) that are released into the lumen. At this point, the spermatozoa are nonmotile; they can't use their flagella to swim. The sperm are moved through to the *rete testis*, so that they can leave the testis and enter the head (first part) of the **epididymis** (see fig. 15.3). The sperm mature and become motile as they pass through the epididymis.

The walls of the seminiferous tubules also contain nongerminal cells (those that don't form sperm), known as **Sertoli cells** (also called *nurse cells*, or *sustentacular cells*). Each Sertoli cell extends from the basement membrane surrounding the tubule to the lumen, and each has tight junctions with the next. Because of this, the Sertoli cells form a continuous barrier around the circumference of the tubule. Although each Sertoli cell has tight junctions with the others around it, these junctions are broken around the germ cells. Thus, the spermatogonia, primary and secondary spermatocytes, and spermatids are surrounded by Sertoli cell cytoplasm (fig. 15.7). The tight junctions between adjacent Sertoli cells must continuously break ahead of the developing sperm and re-form behind it as the germinal cells move towards the lumen. The Sertoli cells perform a number of functions. They:

1. *Supply the developing sperm with needed molecules.* For example, the X chromosome of a developing germ cell is inactive, and so the Sertoli cells may supply molecules coded by the X chromosome.
2. *Make the seminiferous tubules an immunologically privileged site.* The Sertoli cells must normally protect the developing sperm from attack by the man's own immune system, because the secondary spermatozoa and spermatids are the only haploid cells in his body and would otherwise be immunologically rejected.
3. *Eliminate the cytoplasm from spermatids, converting them into spermatozoa.* This process, called *spermiogenesis*, is illustrated in figure 15.8.
4. *Produce androgen-binding protein (ABP) and inhibin.* The Sertoli cells secrete ABP into the lumen, and its ability to bind androgens may help concentrate testosterone in the tubules and thereby promote its stimulation of spermatogenesis. Inhibin exerts negative feedback inhibition of FSH secretion.

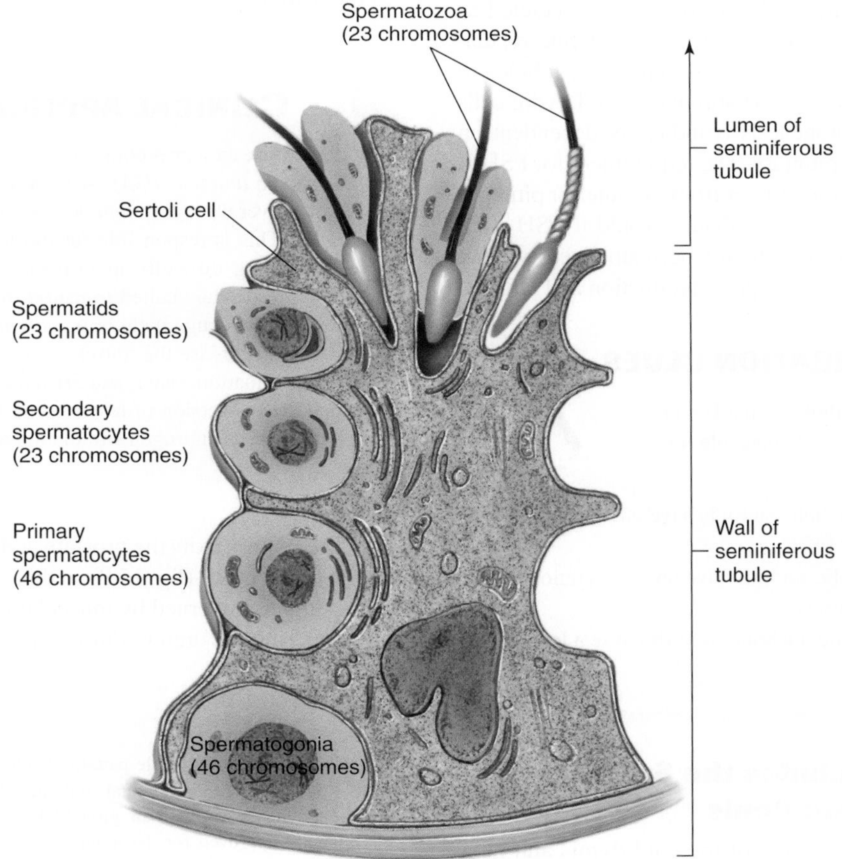

FIGURE 15.7 Sertoli cells nurse developing sperm. The Sertoli cell extends across the width of the tubule and surrounds each developing germ cell. It is believed that one of the functions of Sertoli cells is to provide substances needed by the cells undergoing meiosis.

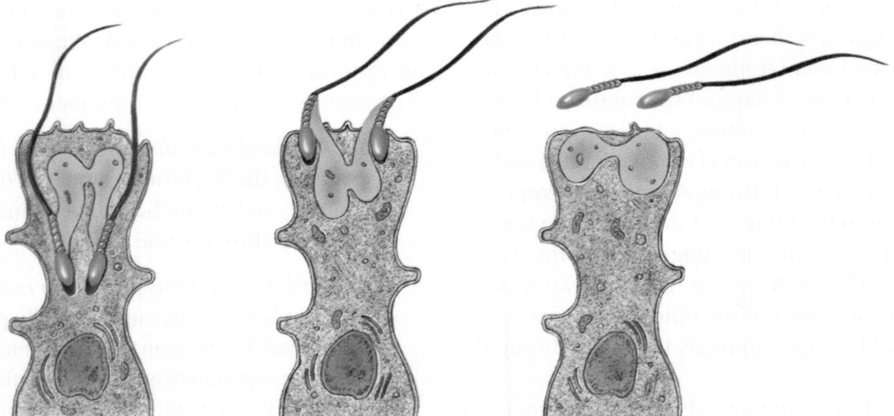

FIGURE 15.8 The processing of spermatids into spermatozoa. As the spermatids develop into spermatozoa, most of their cytoplasm is pinched off as residual bodies and ingested by the surrounding Sertoli cell cytoplasm.

5. *Contain FSH receptor proteins.* The target cells of FSH are the Sertoli cells; thus, any action of FSH in the testes must be produced through its stimulation of the Sertoli cells. These actions of FSH include the production of ABP and inhibin, and the enhancement of spermatogenesis.

Spermatogenesis is dependent on testosterone stimulation. This stimulation starts spermatogenesis at puberty, and is needed to maintain spermatogenesis in the adult. Testosterone is secreted as a hormone into the blood; it can also work as a paracrine regulator, traveling from the Leydig cells to the seminiferous tubules to stimulate spermatogenesis. Because LH stimulates the Leydig cells to secrete testosterone, spermatogenesis is indirectly dependent on LH secretion from the anterior pituitary. The requirement for FSH is less clear. At puberty, the secretion of FSH from the anterior pituitary allows spermatogenesis to occur earlier than it would if FSH were absent. In adults, spermatogenesis can still be maintained without FSH, but FSH is needed for optimum sperm production and fertility.

CLINICAL INVESTIGATION CLUES

Remember that the physician thought that David's use of anabolic steroids could be responsible for his low sperm count.

- What effect would the anabolic steroids have on David's levels of LH and FSH?
- What effect would anabolic steroids have on the secretion of testosterone by David's testes?
- By what mechanisms could anabolic steroids cause a low sperm count?

The Male System Includes the Sex Accessory Organs and Penis

Spermatozoa leave the tail (last part) of the epididymis and next enter the **ductus (vas) deferens**, a tube that leaves the *scrotum* and through the *inguinal canal* into the pelvic cavity (fig. 15.9). The paired **seminal vesicles** then secrete a fluid that composes about 60% of the volume of the semen and is rich in fructose, which serves as an energy source for the sperm. After this point, the ductus deferens becomes the **ejaculatory duct**. The ejaculatory duct is short (about 2 cm) because it enters the **prostate gland** and merges with the prostatic **urethra**. There are numerous pores in the walls of the prostatic urethra, and the fluid secreted by the prostate adds citric acid, calcium, and coagulation proteins to the semen.

CLINICAL APPLICATIONS

Prostate cancer is commonly tested using a blood test for *prostate-specific antigen* (*PSA*). A more common disorder, affecting most men over 60 to different degrees, is **benign prostatic hyperplasia (BPH)**. This is responsible for most cases of bladder outlet obstruction, causing difficulty in urination. BPH treatment may involve a surgical procedure called *transurethral resection* (*TUR*), or the use of drugs. These drugs include α_1*-adrenergic receptor blockers* (chapter 6), which decrease the muscle tone of the prostate and bladder neck, making urination easier, and 5α*-reductase inhibitors*. The latter drugs block the conversion of testosterone into dihydrotestosterone (DHT), which reduces androgen stimulation and thus the size of the prostate.[2]

The urethra exits the prostate and then runs through the ventral region of the **penis** (fig. 15.9) so that it can discharge either urine or semen; this is governed by muscular contractions that prevent urine from entering the urethra during ejaculation. The urethra is located

FYI [2]Patients with male-pattern baldness (*androgenetic alopecia*) who were undergoing clinical trials of a 5α-reductase inhibitor drug to treat BPH reported unexpected hair growth! Finasteride (trade named Propecia) was subsequently approved for the treatment of both conditions, and can cause hair growth in many men with baldness of the vertex (crown of the scalp), which is apparently promoted by dihydrotestosterone (DHT). Pregnant women or women who might get pregnant shouldn't take finasteride, because the normal development of a male embryo's external genitalia also requires stimulation by DHT.

within a column of *erectile tissue*—spongy tissue that can fill with blood to make the penis turgid during an erection. This is the *corpus spongiosum*; in addition, there are two columns of erectile tissue dorsally in the penis, called the *corpora cavernosa* (fig. 15.10). Erection is produced by parasympathetic nerve-induced vasodilation that allows blood to enter the corpora cavernosa. The major neurotransmitter producing this effect is nitric oxide (NO), which causes relaxation of the vascular smooth muscle. As the penis becomes turgid, the venous blood flow out of the penis is partially blocked, further aiding the erection.

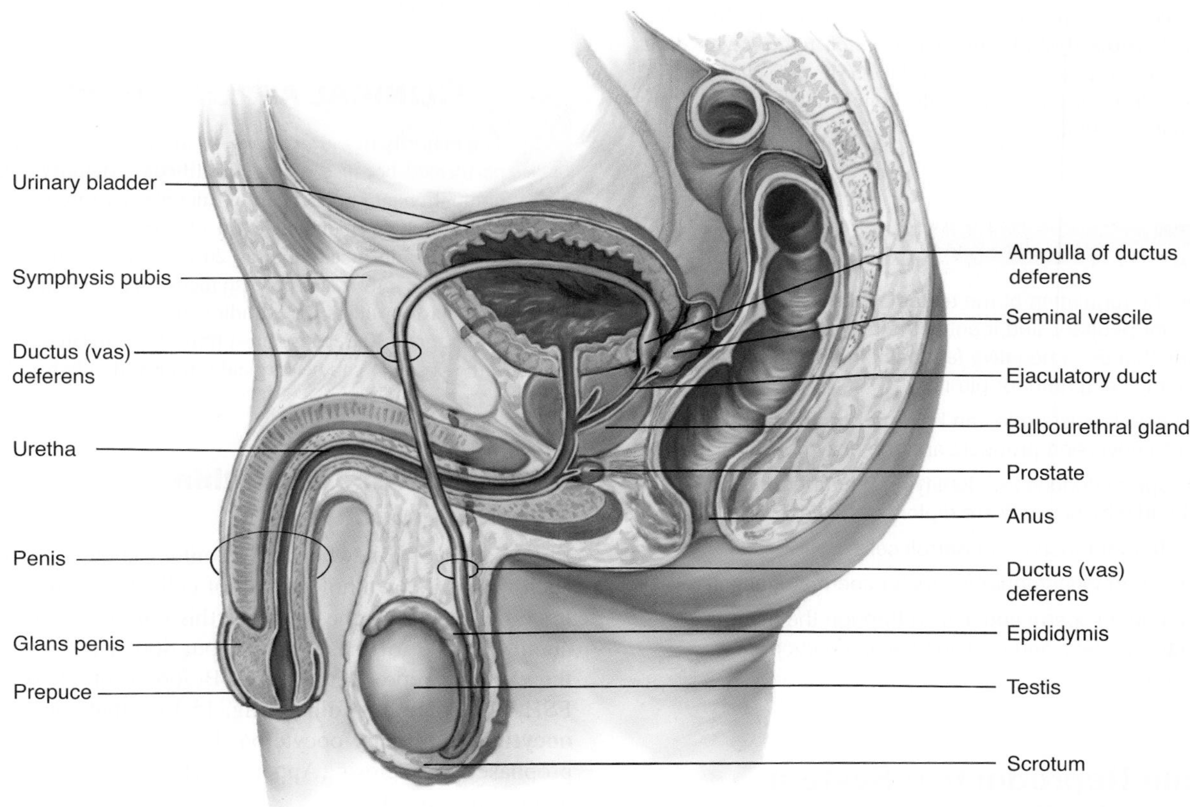

FIGURE 15.9 The organs of the male reproductive system. The male organs are seen here in a sagittal view.

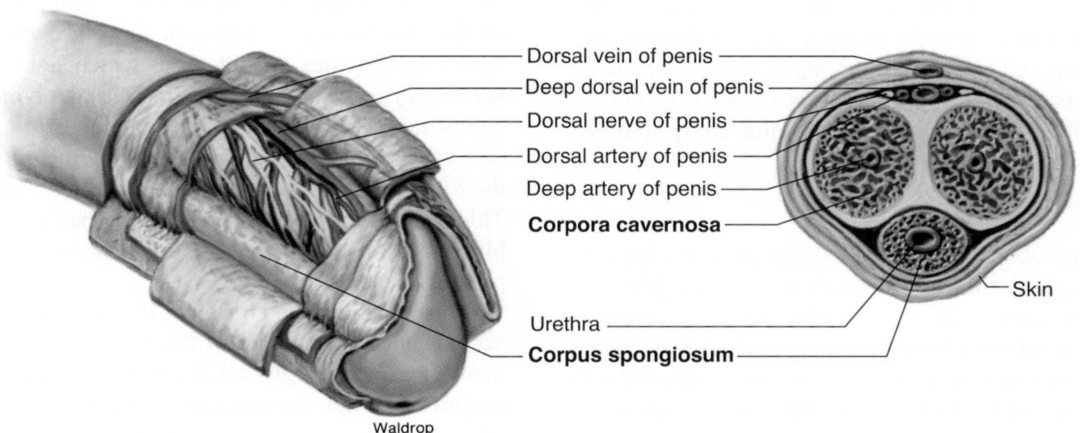

FIGURE 15.10 The structure of the penis. The attachment, blood and nerve supply, and arrangement of the erectile tissue are shown in both longitudinal and cross section.

Clinical Applications

Nitric oxide, released in the penis in response to parasympathetic nerve activation, enters the smooth muscle cells in the arterioles and stimulates the production of a second messenger, *cyclic guanosine monophosphate* (*cGMP*). The cGMP causes the smooth muscle cells to relax and the vessels to dilate, so that more blood can flow to the corpora cavernosa and produce erection. A particular *cGMP phosphodiesterase* enzyme then breaks down cGMP, ending the erection. **Erectile dysfunction** is now often treated with drugs such as *sildenafil* (Viagra), which block the cGMP phosphodiesterase enzyme. These drugs increase the cellular concentration of cGMP and thereby promote erection.

Check Point

1. Describe the regulation of the testes by the hypothalamus and anterior pituitary, and identify the testicular hormones that exert negative feedback inhibition on the hypothalamus and anterior pituitary.
2. Describe how testosterone can be a prehormone, and identify its conversion products and the enzymes involved.
3. Describe spermatogenesis, identifying the different cells involved and whether they are diploid or haploid.
4. List the different functions of Sertoli cells.
5. Explain the hormonal requirements for spermatogenesis.
6. Trace the path of sperm and semen through the male reproductive system, and explain how an erection is produced.

The Female Reproductive System Depends on the Function of Ovaries

The ova, or oocytes (egg cells), are released from the ovaries at ovulation, and travel through the uterine (fallopian) tubes toward the uterus. The ova don't complete meiosis until they're fertilized, and if that occurs the embryo implants into the inner lining, or endometrium, of the uterus. Each ovum develops within its own ovarian follicle, and the development of the ova and follicles is accompanied by cyclic changes in ovarian hormone secretion. These cyclic hormonal changes cause the buildup and shedding of the endometrium, among other body changes that occur during the menstrual cycle.

The two ovaries are about the size and shape of large almonds. The **uterine (fallopian) tubes** have extensions, called *fimbriae*, that partially cover each ovary (fig. 15.11). When an *ovum*, or *oocyte* (egg cell), is released from an ovary—in a process called **ovulation**—it's drawn into the uterine tube by the beating of cilia on the epithelial lining of the uterine tube. The lumen of each uterine tube is continuous with the **uterus**, a muscular, pear-shaped organ held in place within the pelvic cavity by ligaments. The uterus consists of three layers: an outer layer of connective tissue, the **perimetrium**; a thick, middle layer of smooth muscle, the **myometrium**; and an inner stratified epithelium, the **endometrium** (fig. 15.11). The uterus narrows to form the *cervix* (= neck), which opens to the **vagina** (fig. 15.12). The opening of the vagina is covered by inner **labia minora** and outer **labia majora**. The **clitoris**, a small structure composed mainly of erectile tissue (related to the corpora cavernosa of the penis), is located at the anterior margin of the labia minora.

Clinical Applications

The majority of *hysterectomies* (surgical removal of the uterus) are performed because of **uterine fibroids (leiomyomas)**. These are nonmalignant (noncancerous) neoplasms (growths) in the uterus that also include abundant extracellular matrix. Fibroids can be as small as 10 mm or as large as 20 cm, and produce such symptoms as pelvic discomfort and profuse menstrual bleeding. Uterine fibroids have receptor proteins for estradiol and progesterone, which can stimulate their growth. Because most fibroids are located within the uterine wall, they usually can be surgically removed only by a hysterectomy.

Oogenesis Occurs Within Ovarian Follicles

The ovaries of a newborn girl contain about 2 million oocytes, each within its own hollow ball of cells, the **ovarian follicles**. By the time the girl reaches puberty, this number has been reduced to about 400,000; of these, only about 400 will be ovulated during the woman's reproductive years. Before a follicle is stimulated by FSH, it's a **primary follicle** (fig. 15.13*a*) that contains a **primary oocyte**. The primary oocyte has begun meiosis, but is arrested at prophase I of meiosis (Appendix 4), and so is still diploid (with 46 chromosomes).

During a monthly cycle (discussed in the next section), FSH stimulates some of the follicles in one of the ovaries to begin growing and developing. This includes cell division in the follicle cells, producing **granulosa cells** that surround the oocyte and begin to fill the follicle. The granulosa cells are endocrine, secreting an increasing amount of *estradiol* (the major estrogen) at this stage of the monthly cycle. Some of these growing follicles develop fluid-filled cavities (*vesicles*), at which point they're known as **secondary follicles** (fig. 15.13*a*). During a typical monthly cycle, one of these secondary follicles continues to grow as its vesicles fuse to form a single fluid-filled cavity, the *antrum*. At this stage, the follicle is known as a **graafian (mature) follicle** (fig. 15.13*b*). This follicle will grow so large that it becomes like a thin-walled blister on the surface of the ovary, ready to expel its oocyte in the process of ovulation.

The change from a primary to a secondary ovarian follicle is accompanied by changes of the oocyte it contains. The development of a mature egg cell, ready to become fertilized, is called **oogenesis**. When the primary oocyte completes its first meiotic division, it produces 2 cells, each of which contains a haploid number (23) of chromosomes. Each of the chromosomes at this point consists of 2 duplicate strands, the *chromatids*. However, these

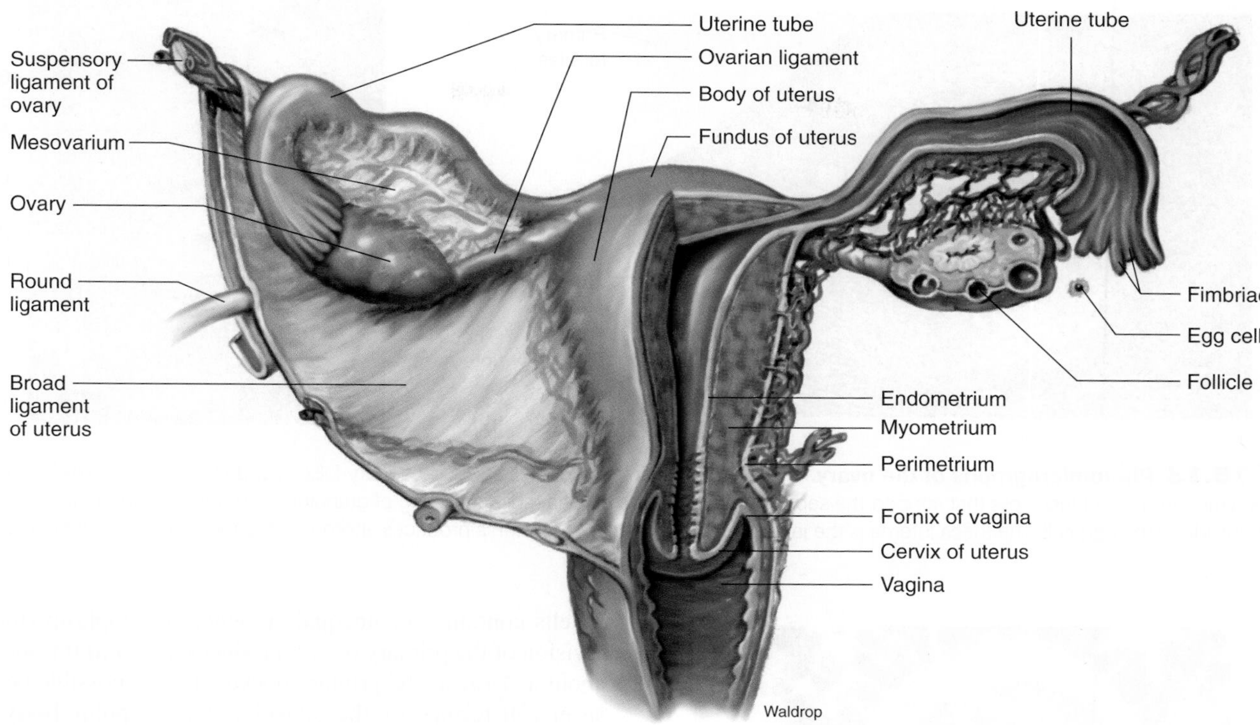

FIGURE 15.11 The uterus, uterine tubes, and ovaries. The supporting ligaments can also be seen in this posterior view.

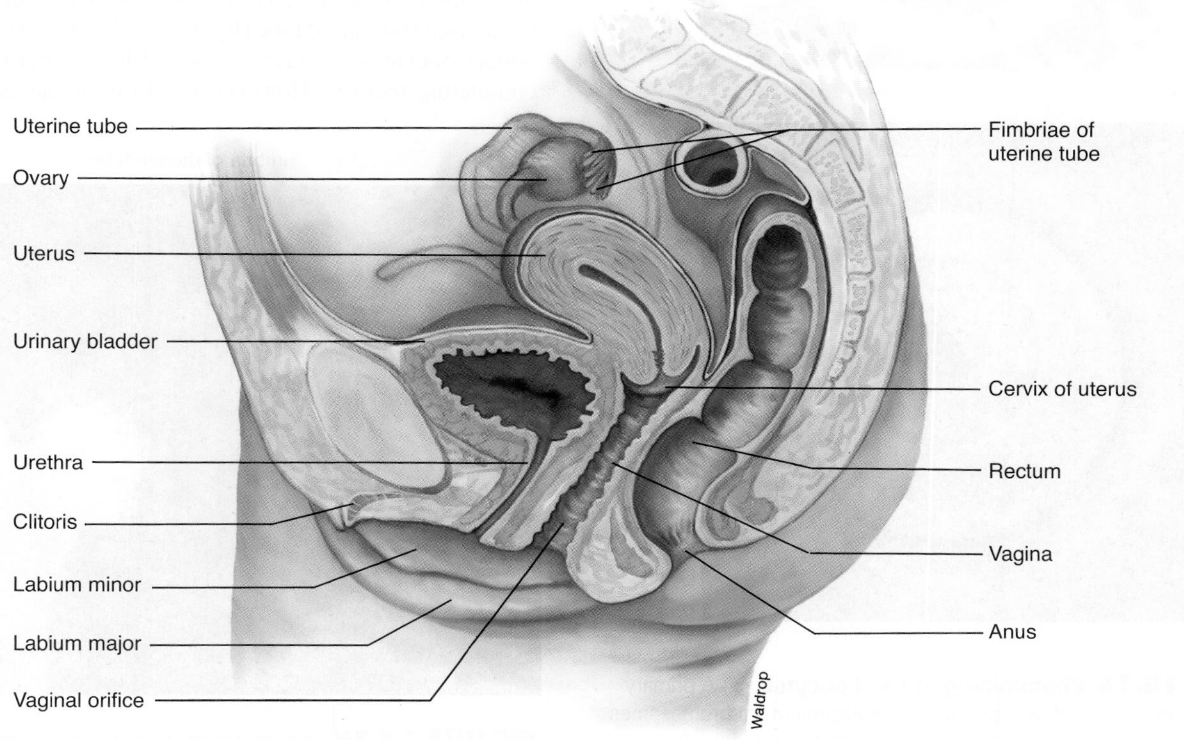

FIGURE 15.12 The organs of the female reproductive system. These are shown in sagittal section.

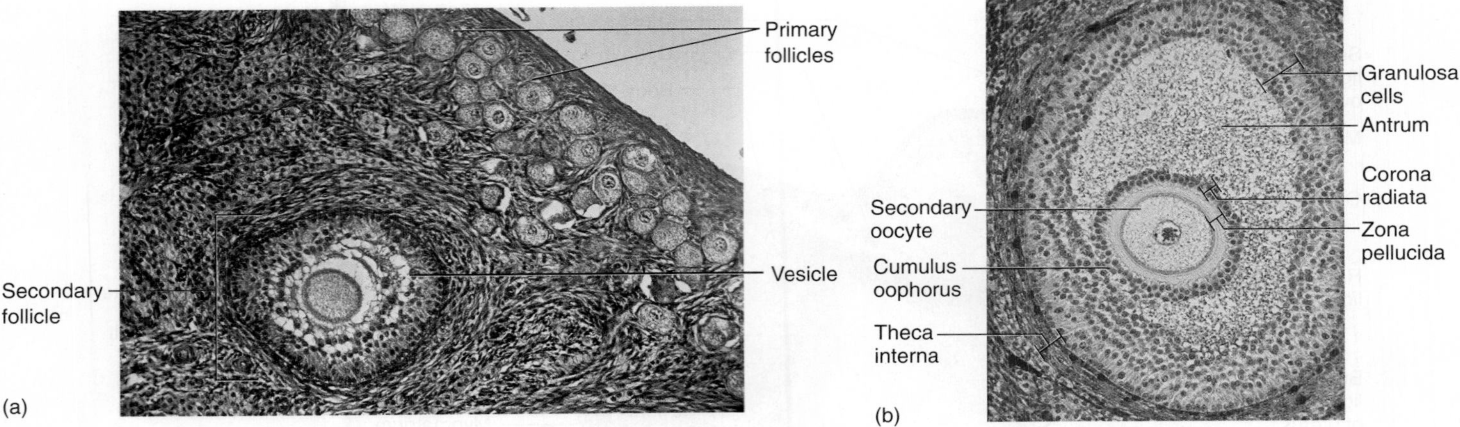

FIGURE 15.13 Photomicrographs of the ovary. These show (*a*) primary follicles and 1 secondary follicle; and (*b*) a graafian follicle. The cumulus oophorous is a mound of granulosa cells that support the secondary oocyte. The corona radiata is a ring of granulosa cells that surround the zona pellucida (clear zone) outside of the egg cell. The theca interna is the inner wall of the ovarian follicle; this produces androgen that the granulosa cells convert into estradiol.

2 cells contain very unequal amounts of cytoplasm (fig. 15.14); division of the primary oocyte produces a **secondary oocyte** that's about as large as the primary oocyte. This is possible because the sister cell formed by the division, called a **polar body**, is quite small and eventually disintegrates. The secondary oocyte is contained within the mature graafian follicle, and it is the egg cell that will be released at ovulation.

The secondary oocyte starts its second meiotic division, but meiosis freezes at metaphase II. Thus, a secondary oocyte, arrested at metaphase II, is released from the ovary at ovulation and enters the uterine (fallopian) tube (fig. 15.15). If it isn't fertilized, the secondary oocyte disintegrates about 24 hours after ovulation, never completing meiosis. However, fertilization causes the secondary

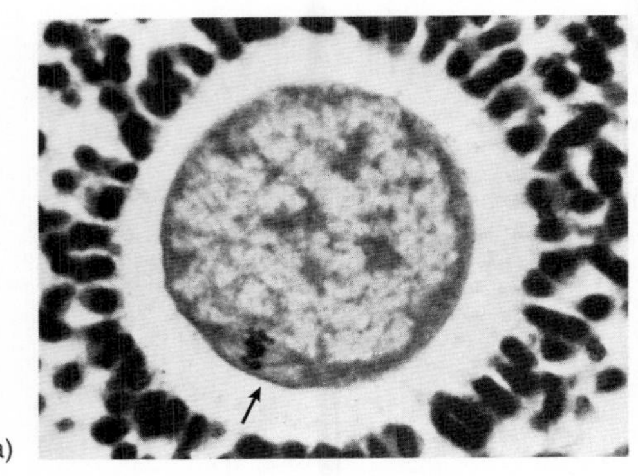

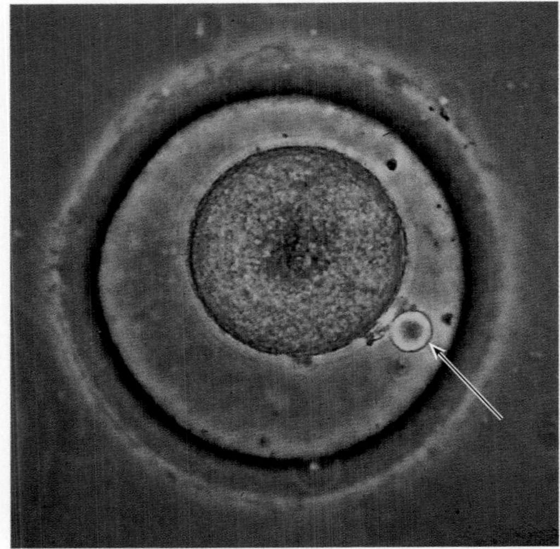

FIGURE 15.14 Photomicrographs of oocytes. (*a*) A primary oocyte at a metaphase I of meiosis. Notice the alignment of chromosomes (*arrow*). (*b*) A human secondary oocyte formed at the end of the first meiotic division. Also shown is the first polar body (*arrow*).

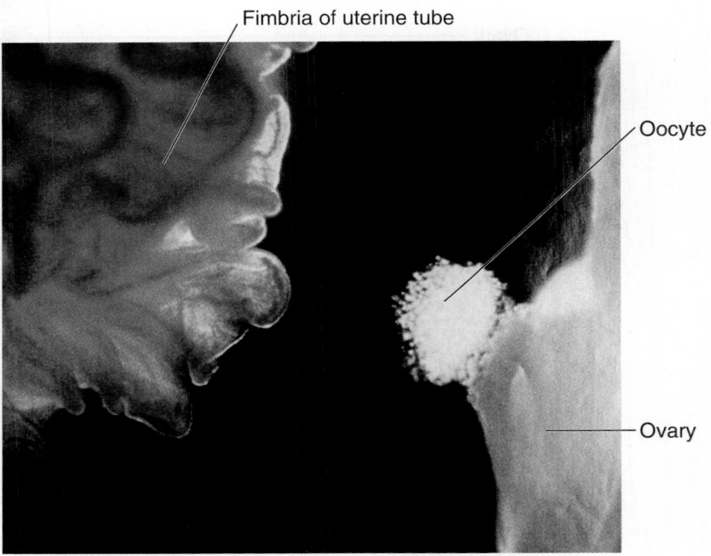

FIGURE 15.15 Ovulation from a human ovary. Notice the cloud of fluid and granulosa cells surrounding the ovulated oocyte.

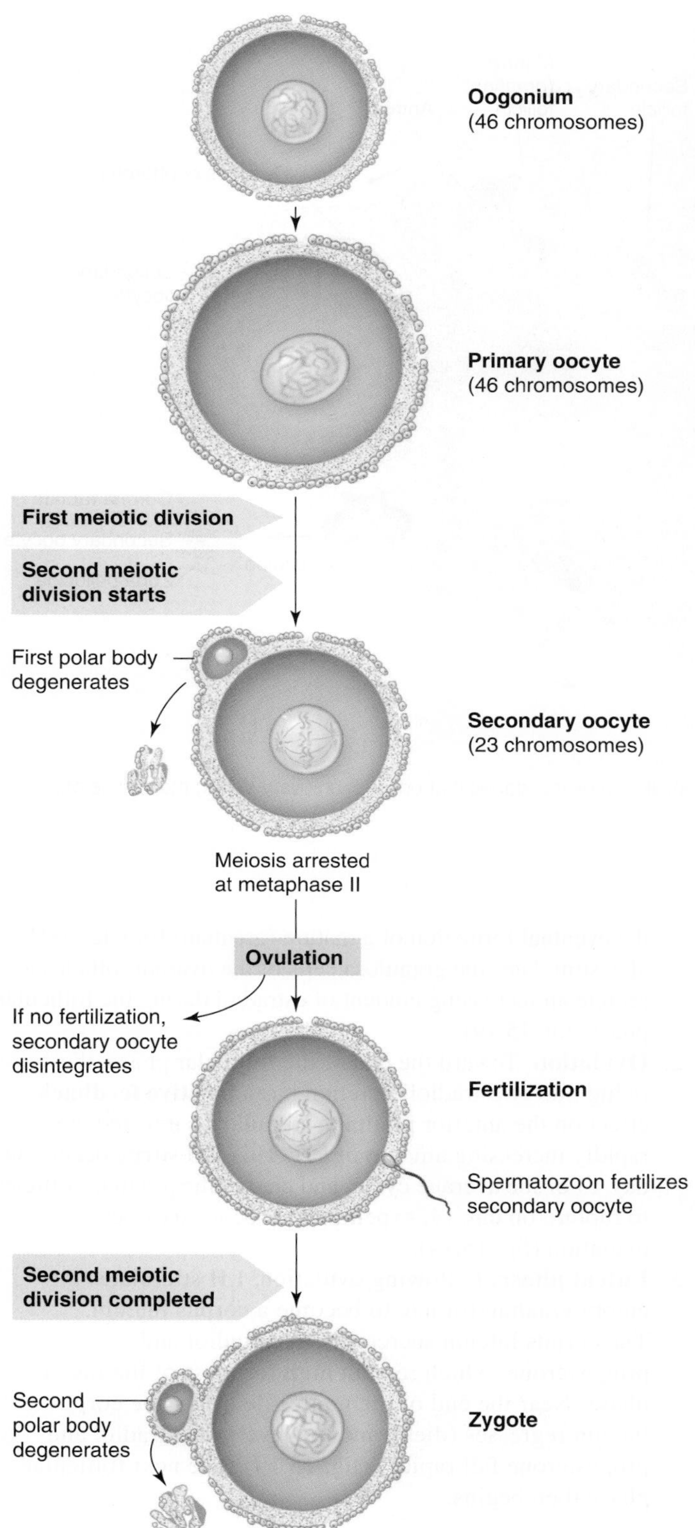

FIGURE 15.16 Oogenesis. During meiosis, each primary oocyte produces a single haploid gamete. If the secondary oocyte is fertilized, it forms a second polar body and its nucleus fuses with that of the sperm cell to become a zygote.

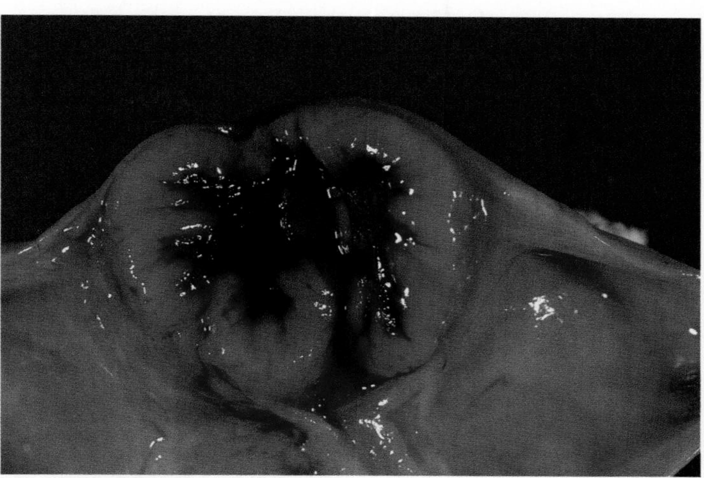

FIGURE 15.17 A corpus luteum in a human ovary. This structure is formed from the empty graafian follicle following ovulation.

oocyte to complete the second meiotic division (fig. 15.16). Like the first division, the cytoplasm is divided very unequally, producing one large cell and another polar body. Each of these contains the haploid number of chromosomes (23) as the chromatids are separated. These chromosomes in the large cell are now joined with the haploid number of chromosomes from the spermatozoan, producing a zygote.[3]

After the graafian follicle has ruptured at ovulation, LH (luteinizing hormone) stimulates it to become a different structure, new to this cycle: a **corpus luteum** (fig. 15.17). The corpus luteum secretes estradiol, like the granulosa cells of growing ovarian follicles. Additionally, the corpus luteum secretes a new ovarian steroid hormone, **progesterone**. Thus, progesterone is absent prior to ovulation, and is present in a woman's blood only after the corpus luteum has formed. The cyclic changes in the ovarian follicles (fig. 15.18), accompanied by cyclic changes in ovarian hormone secretion, produce the changes characteristic of the menstrual cycle.

The Menstrual Cycle Is Caused by Cyclic Secretion of Hormones

Menstrual cycles repeat at about 1-month intervals (*menstru* = monthly). Bleeding occurs when the inner two-thirds of the endometrium—the *stratum functionale*—is shed with accompanying bleeding. The stratum functionale is built up by mitotic cell division during the cycle, and is shed (leaving only the

FYI [3]Why is division of the cytoplasm so unequal, so that meiosis produces only 1 full cell (and 2 polar bodies)? It's because sperm are swimming packets of chromosomes; they don't contribute cytoplasm or mitochondria at fertilization. Because of this, the unfertilized secondary oocyte needs to be as large as possible for the sake of the zygote. Actually, the oocyte is so large it can be seen without a microscope; it's about the size of the period at the end of this sentence.

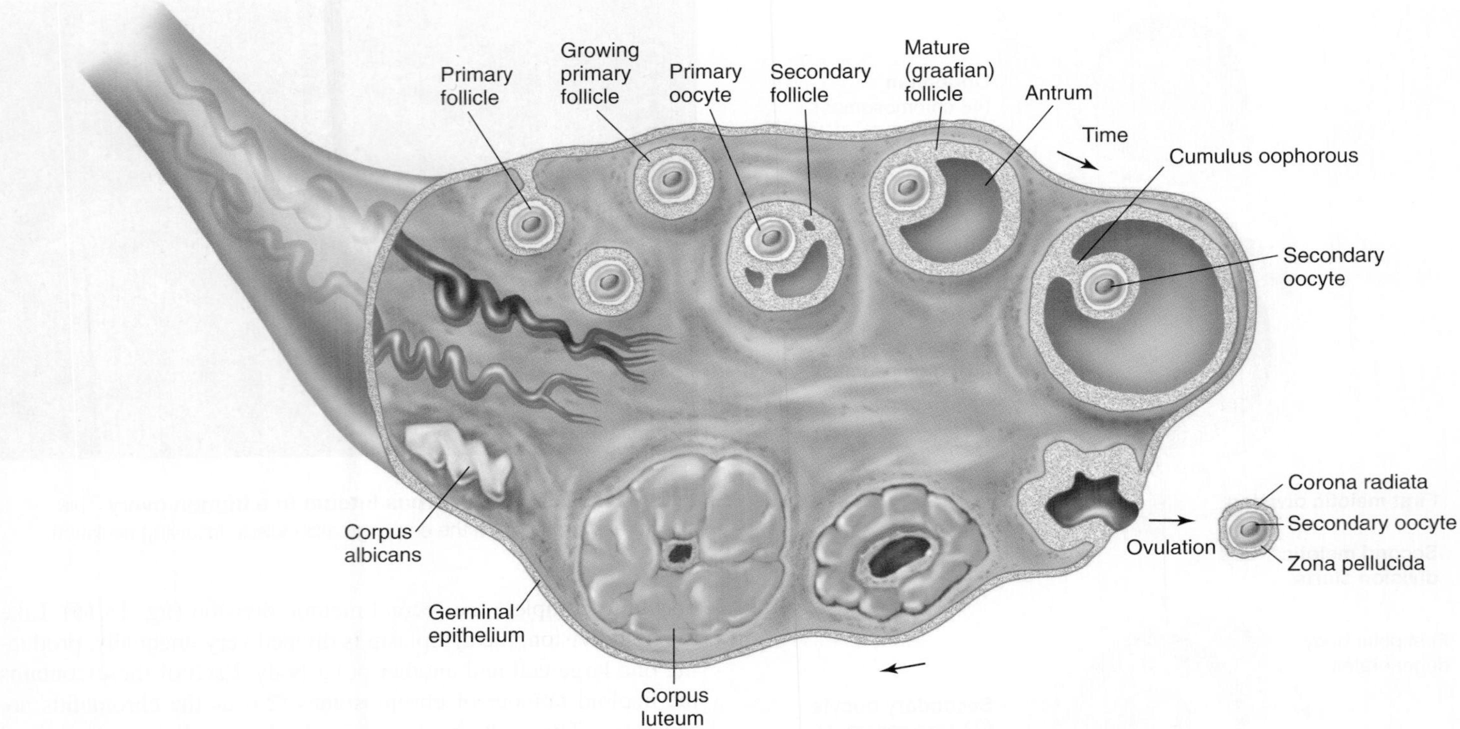

FIGURE 15.18 Stages of ovum and follicle development. This diagram illustrates the stages that occur in an ovary during the course of a monthly cycle. The arrows indicate changes with time.

bottom layers of the endometrium, the *stratum basale*) at the beginning of each nonfertile cycle. Bleeding results from breaking of unique *spiral arteries* that invade the built-up endometrium during the cycle.[4]

Because it's a cycle, the nonfertile menstrual cycle doesn't really have a beginning or end. However, for the sake of description, it's said to begin with the first day of menstruation (day 1). The average cycle lasts 28 days, although this varies substantially. In terms of the effects of pituitary gonadotropins on the ovarian follicles, the cycle can be described by the following phases:

1. **Follicular phase**. This lasts from day 1 to about day 14, the day of ovulation in the average 28-day cycle. During the follicular phase, FSH stimulates the growth of follicles and the eventual formation of a mature (graafian) follicle. FSH also stimulates the granulosa cells of the ovarian follicles to secrete an increasing amount of estradiol during the follicular phase (fig. 15.19).
2. **Ovulation**. Toward the end of the follicular phase, the rising tide of estradiol secretion has a **positive feedback** effect on the anterior pituitary, stimulating it to secrete a rapidly increasing amount of LH. This **LH surge** occurs on day 13 of the average cycle, and causes the graafian follicle to rupture on day 14, expelling the secondary oocyte in ovulation (fig. 15.19).
3. **Luteal phase**. Following ovulation, LH stimulates the empty graafian follicle to become a corpus luteum. The corpus luteum secretes both estradiol and progesterone, which remain high for most of the luteal phase. Near the end of the nonfertile cycle, the corpus luteum regresses (dies), and the levels of estradiol and progesterone fall rapidly (fig. 15.19). The next follicular phase then begins.

Notice that there is a positive feedback effect of estradiol on LH secretion at about the midpoint of the cycle, just prior to ovulation. This is the only example of a positive feedback control of the anterior pituitary we've encountered. Considering that the LH

FYI [4]Only humans, apes, and Old-World monkeys (that is, all primates except South American monkeys) have menstrual cycles, because only they bleed when they shed the stratum functionale of the endometrium at the beginning of a cycle, when estradiol secretion is low. Your pet dog or cat may bleed during her cycle, but the bleeding occurs more midcycle, either just before or after ovulation, and is not associated with shedding of this endometrial layer. All female mammals that don't have menstrual cycles instead have *estrous cycles*, so-called because they go into heat ("estrus") around the time they ovulate, when estradiol secretion is high.

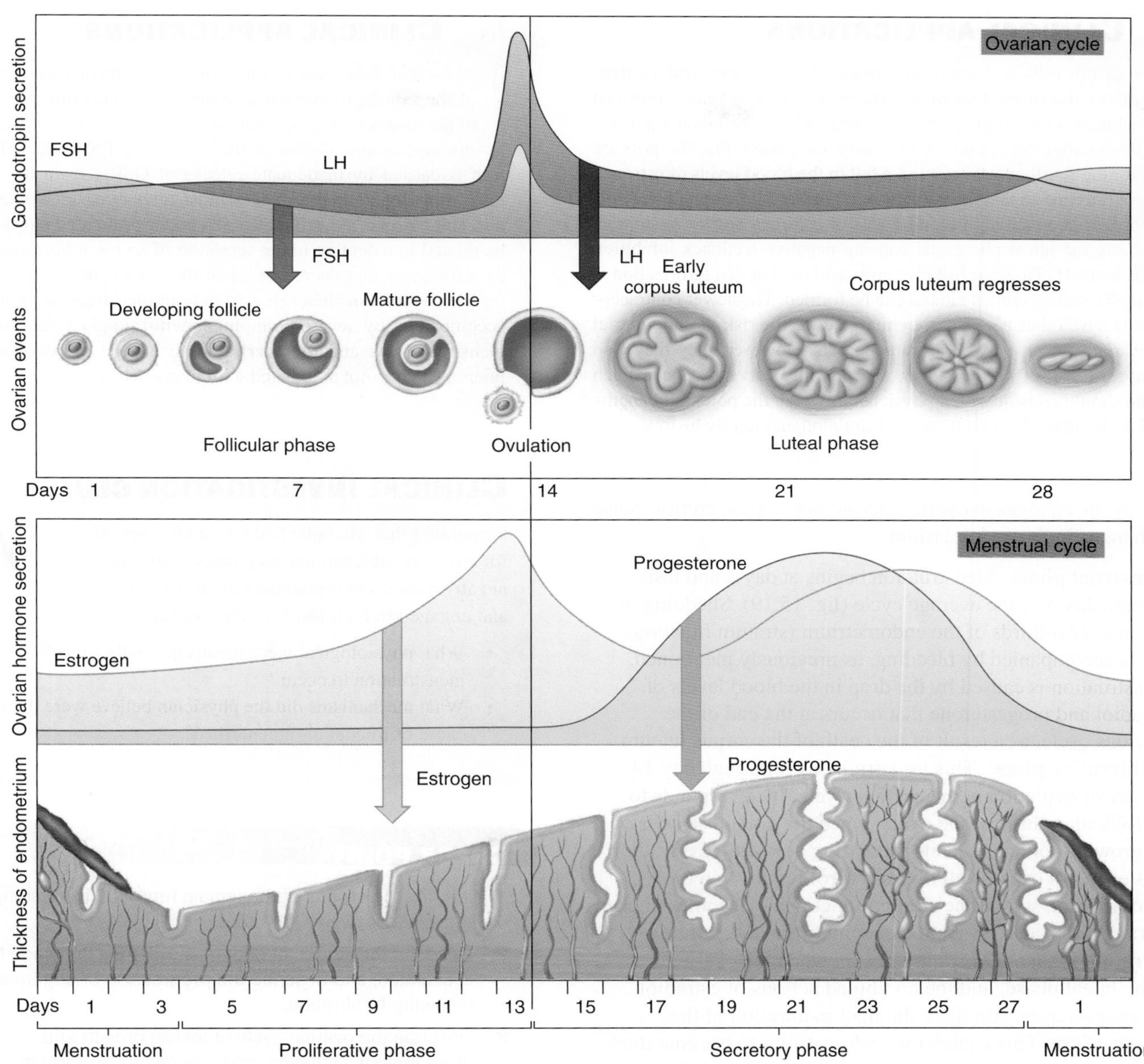

FIGURE 15.19 The cycle of ovulation and menstruation. The downward arrows indicate the effects of the hormones.

surge triggers ovulation, it's important for the LH surge to be timed correctly, when the follicle and oocyte are fully mature and ready to ovulate. The positive feedback effect of estradiol on LH secretion at that time is a signal sent from the ovary to the anterior pituitary, ensuring that the LH surge and ovulation are timed correctly.[5]

FYI [5]Cats and rabbits are reflex ovulators. If they don't have sexual intercourse, they don't ovulate. Whenever they do have coitus (sexual intercourse), a reflex surge in LH secretion is triggered, causing ovulation. Thus, there is no "safe period" for a cat or rabbit, which explains their legendary fertility.

Immediately after ovulation (during the luteal phase), estradiol—now joined by progesterone—from the corpus luteum has the opposite effect. These hormones (separately and together) have a negative feedback effect on FSH and LH. This is the more familiar effect: estradiol and progesterone inhibit GnRH secretion from the hypothalamus, and gonadotropin secretion from the anterior pituitary. The low levels of FSH and LH during the luteal phase of the cycle serve a very useful function—they prevent new follicles from growing and ovulating after one ovulation has occurred. This is a natural birth control mechanism, preventing successive ovulations from occurring in one cycle.

Clinical Applications

About 60 million women worldwide currently use **oral contraceptives** (**birth control pills**). These contain a synthetic estrogen combined with synthetic progesterone, which are taken each day for 3 weeks after the last day of the menstrual period. Placebo pills are taken for the fourth week, to cause a fall in the blood levels of estrogen and progesterone so that menstruation can occur. The birth control pills immediately produce high blood levels of estrogen and progesterone, mimicking the luteal phase and causing negative feedback inhibition of FSH and LH. Thus, no follicles grow and ovulate (so fertilization is prevented), and no corpus luteum can be formed. The newer contraceptive pills have other benefits: they may reduce the risk of endometrial and ovarian cancer, as well as osteoporosis. However, they may also increase the risk of breast cancer, and possibly cervical cancer. Each woman should consult with a physician to weigh the potential benefits and risks in light of her own medical situation and family history.

The cyclic changes in ovarian steroid hormone secretion cause cyclic changes in the endometrium:

1. **Menstrual phase**. Menstruation begins at day 1 and lasts through day 4 of the average cycle (fig. 15.19). Shedding of the inner two-thirds of the endometrium (stratum functionale) is accompanied by bleeding, as previously mentioned. Menstruation is caused by the drop in the blood levels of estradiol and progesterone that occurs at the end of the previous cycle, as a result of the death of the corpus luteum.
2. **Proliferative phase**. This lasts from day 4 through day 14, the day of ovulation in the average cycle. It corresponds to the follicular phase of the ovary, when the granulosa cells of the growing follicles secrete increasing amounts of estradiol. This estradiol causes cell division (proliferation) in the endometrium, so that the stratum functionale becomes thicker. Spiral arteries also invade the endometrium at this time.
3. **Secretory phase**. This phase occurs when the ovaries are in the luteal phase, and the combined actions of estradiol and progesterone stimulate the final maturation of the endometrium. This causes the endometrium to become thick, highly vascular, spongy in appearance, and rich in glycogen. In short, it becomes a suitable place for an embryo to implant if fertilization has occurred.

Near the end of the luteal phase of the ovarian cycle, when the endometrium is in its secretory phase, the corpus luteum dies. It could be sustained by luteinizing hormone (LH), but—through its secretion of estradiol and progesterone—the corpus luteum suppresses LH secretion by negative feedback. So, in a sense, the corpus luteum commits suicide. Death of the corpus luteum results in a fall in blood levels of estradiol and progesterone, which produces menstruation and the beginning of the next cycle. If the oocyte is fertilized these events must stop: something must save the corpus luteum, thereby preventing menstruation and the loss of the embryo. The embryo saves itself; it secretes a hormone that sustains its mother's corpus luteum, as described in the next section.

Clinical Applications

Abnormal menstruations are among the most common disorders of the female reproductive system. The term **amenorrhea** refers to the absence of menstruation. *Functional amenorrhea* is caused by inadequate stimulation of the ovaries by FSH and LH, which in turn is caused by inadequate release of GnRH from the hypothalamus. Functional amenorrhea is most often seen in women who are thin and athletic, as well as women under prolonged stress. This may be related to a decline in the secretion of *leptin*, a hormone produced by adipocytes that decreases when the adipocytes get smaller (chapter 14). **Dysmenorrhea** refers to painful menstruation, which may be accompanied by severe cramping; **menorrhagia** refers to excessive menstrual flow; and **metrorrhagia** to uterine bleeding at irregular intervals that is not associated with menstruation.

Clinical Investigation Clues

Remember that Michelle had missed her period for a few months, but her pregnancy tests were negative; also, she was underweight, stressed, and engaged in high levels of physical activity.

- What physiological mechanisms normally cause menstruation to occur?
- What mechanisms did the physician believe were the likely cause of Michelle's amenorrhea?

Check Point

1. Identify the organs of the female reproductive system, and the tissue layers of the uterus.
2. Describe the stages of oogenesis, identify the stage that is ovulated, and describe the progression of oogenesis following fertilization.
3. Describe the ovarian cycle of follicle growth and development, ovulation, and corpus luteum formation, and the hormones secreted at the different phases of the ovarian cycle.
4. Describe the positive feedback and negative feedback effects of ovarian hormones on the secretion of anterior pituitary gonadotropins, and the physiological significance of these effects.
5. Identify the phases of the endometrium during the menstrual cycle, and explain how these phases are related to the ovarian cycle and the secretion of ovarian hormones.
6. Explain the events at the end of a nonfertile cycle that lead to menstruation.

Gestation, Parturition, and Lactation Are Female Reproductive Functions

The embryo prevents its mother from menstruating by secreting hCG, a hormone that maintains the corpus luteum. Embryonic and fetal development, or gestation, is accompanied by the development of the placenta. The placenta serves such functions as gas exchange, detoxication, elimination of waste products, and the secretion of a number of hormones. The placenta is even believed to set the time of childbirth, or parturition, through its hormonal secretions. After parturition, prolactin and oxytocin from the mother's pituitary promote lactation, the ability of the mother to nurse the baby.

During sexual intercourse, a man ejaculates an average of about 300 million sperm into the vagina; of these, only about 100 sperm survive to enter each uterine tube. And of those, only 1 has a chance of fertilizing the egg cell, an event that occurs within the uterine tube. When a secondary oocyte is released at ovulation and enters the uterine tube, two layers surround its plasma membrane: (1) a clear layer of glycoproteins, the **zona pellucida**; and (2) a layer of granulosa cells, the **corona radiata**. In order for a spermatozoan to fertilize the oocyte, it must first tunnel through these layers. This is made possible by digestive enzymes in a cap—the **acrosome**—that crowns the head of the sperm (fig. 15.20), which is a nucleus containing highly condensed chromosomes. When the sperm encounters the corona radiata and zona pellucida, an *acrosomal reaction* occurs in which the cap develops openings that release the digestive enzymes (fig. 15.21).

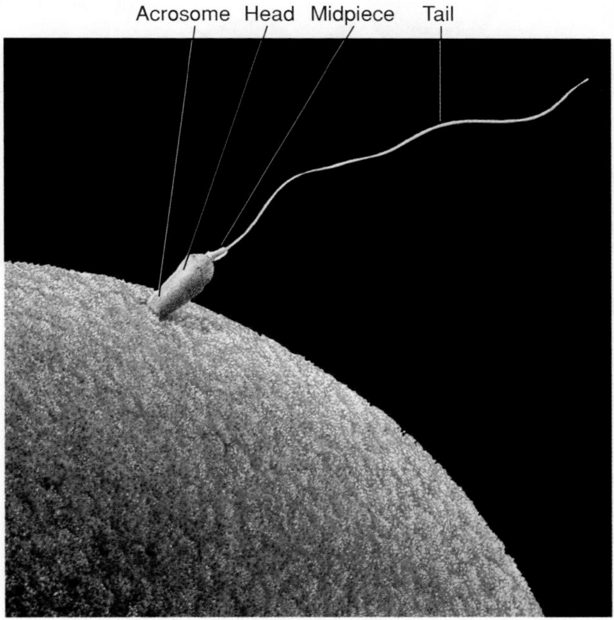

FIGURE 15.20 The parts of a spermatozoan. This is a scanning electron micrograph showing a spermatozoan in contact with a secondary oocyte.

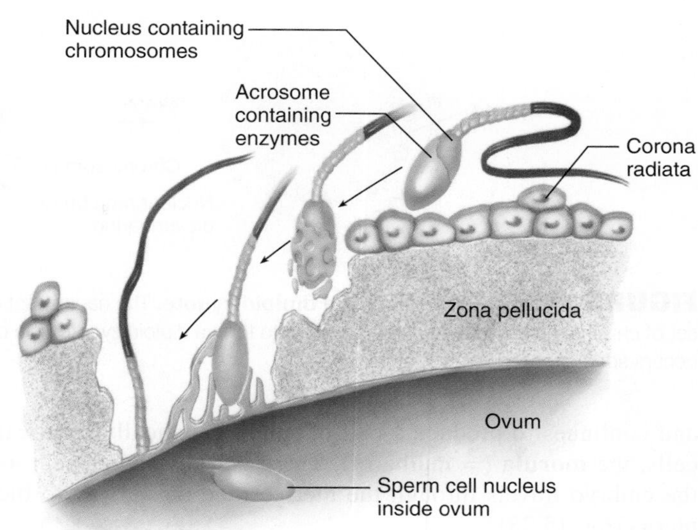

FIGURE 15.21 Fertilization. When the sperm reaches the corona radiata and zona pellucida surrounding the oocyte, the acrosome ruptures and releases digestive enzymes. This allows the sperm to tunnel its way to the oocyte. At fertilization, the plasma membrane of the sperm becomes continuous with the plasma membrane of the oocyte, and the sperm nucleus moves into the oocyte cytoplasm.

Clinical Applications

The techniques of ***in vitro* fertilization** are sometimes used to produce pregnancies. Physicians can estimate the time of ovulation by waiting 36 to 38 hours after the LH surge, and then can collect the secondary oocyte by aspiration. Alternatively, the woman can be treated with powerful FSH-like hormones that cause the development of multiple follicles, and the preovulatory oocytes can be collected surgically. The oocytes and sperm may be placed together in a petri dish for 2 or 3 days for fertilization, or the physician may use a microscope to inject a sperm through the zona pellucida directly into the cytoplasm of the oocyte, a technique called *intracytoplasmic sperm injection* (*ICSI*). A number of embryos may be produced at the same time, and the surplus frozen in liquid nitrogen. Three or more embryos, at the 4-cell stage, are usually transferred at the same time into the woman's uterus. The likelihood of a successful implantation is low, and the procedure is expensive, but it may be the only way that some women can become pregnant.

Meiosis of the ovulated secondary oocyte is arrested at metaphase II, but when the oocyte becomes fertilized it completes its division, forming another polar body and a fertilized egg cell, or zygote. The zygote is diploid with 23 chromosomes from the mother's oocyte and 23 from the father's sperm (fig. 15.22). It soon divides by mitosis, forming 2 identical cells, each of which is about half the size of the zygote. Mitotic division of the diploid zygote produces diploid daughter cells,

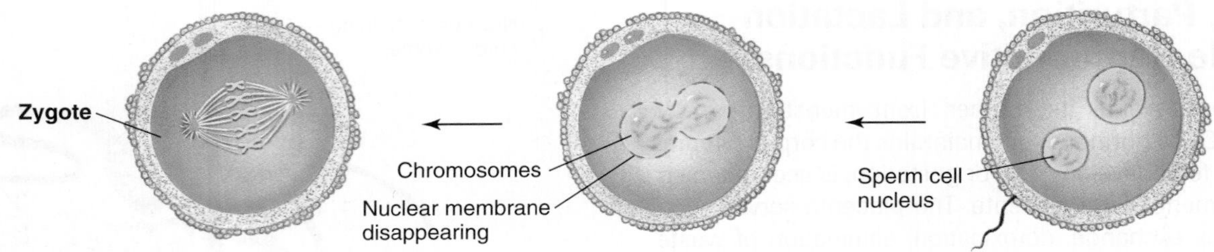

FIGURE 15.22 Formation of a diploid zygote. The haploid set of chromosomes from the mother's secondary oocyte combines with the haploid set of chromosomes from the father's sperm to form a diploid zygote. The chromosomes of the zygote are shown lining up for mitotic cell division, which occurs soon after the zygote has formed.

and continues to produce a ball of small, genetically identical cells, the **morula** (= mulberry). These cell divisions occur as the embryo travels through the uterine tube on its way to the uterus (fig. 15.23).

The embryo reaches the uterus about 3 days after ovulation, and contains 32 to 64 cells by the fourth day following fertilization. The embryo remains unattached to the endometrium for the next 2 days, during which time it changes into a **blastocyst** (fig. 15.23). The blastocyst consists of two parts: (1) the **inner cell mass**, which will become the fetus; and (2) a surrounding membrane, the **chorion**, which will become part of the placenta. The cells that form the chorion are called *trophoblast cells*. On the sixth day following fertilization, the blastocyst attaches to the endometrium and starts the process of **implantation** (fig. 15.24). By the seventh day, the blastocyst is completely buried in the endometrium.

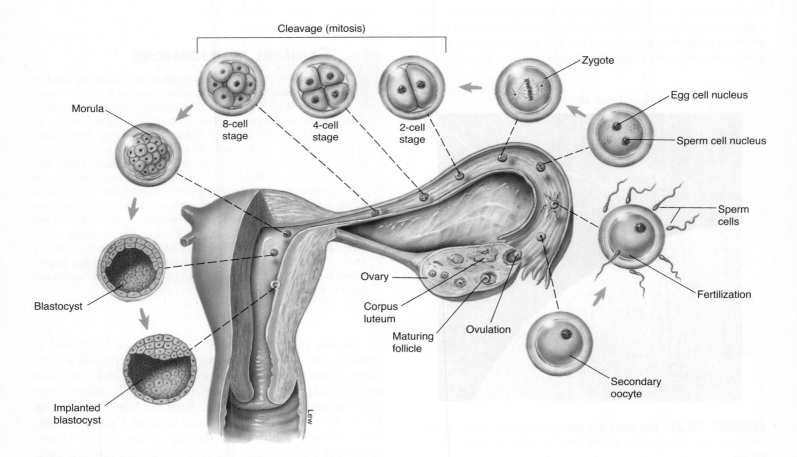

FIGURE 15.23 Fertilization, cleavage, and the formation of a blastocyst. A diagram showing the ovarian cycle, fertilization, and the events of the first week following fertilization. Implantation of the blastocyst begins between the fifth and seventh day and is generally complete by the tenth day.

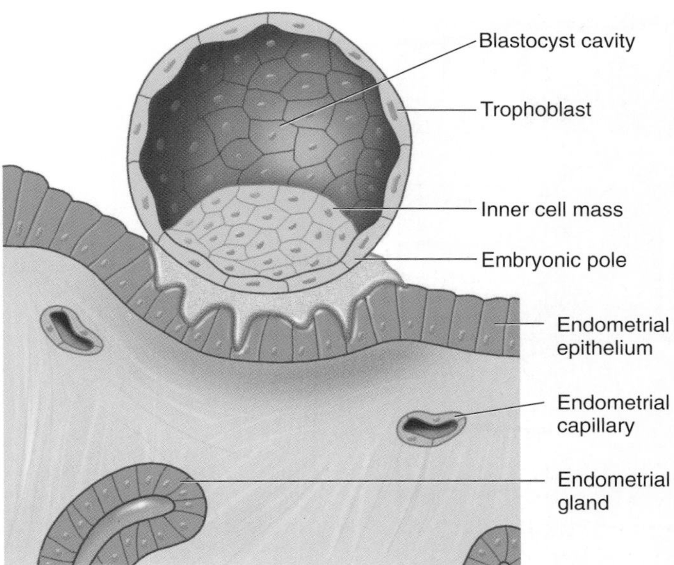

FIGURE 15.24 Implantation of the blastocyst. A blastocyst attached to the endometrium at about the sixth day following fertilization is illustrated. It will become completely buried in the endometrium by the tenth day.

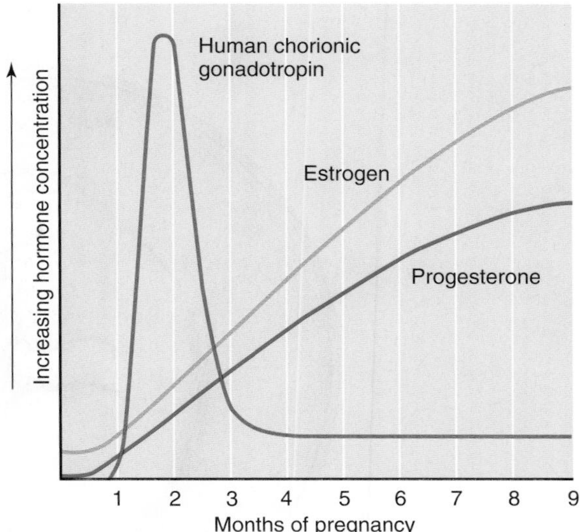

FIGURE 15.25 The secretion of human chorionic gonadotropin (hCG). This hormone is secreted by trophoblast cells during the first trimester of pregnancy, and it maintains the mother's corpus luteum for the first 5½ weeks. After that time, the placenta becomes the major sex-hormone-producing gland, secreting increasing amounts of estrogen and progesterone throughout pregnancy.

The trophoblast cells of the chorion secrete a hormone, **human chorionic gonadotropin (hCG)**, which acts like LH; it prevents the death of the corpus luteum that otherwise would have occurred at the end of a nonfertile cycle. Because this, the corpus luteum continues to secrete estradiol and progesterone (fig. 15.25), preventing menstruation. Since hCG, unlike LH, is not secreted by the anterior pituitary in response to GnRH from the hypothalamus, the secretion of hCG isn't inhibited by high blood levels of estradiol and progesterone. Despite this, the secretion of hCG declines and the corpus luteum regresses at about the fifth to sixth week of pregnancy (fig. 15.25), for reasons that aren't completely understood. The blood levels of estrogen and progesterone nevertheless continue to increase throughout the pregnancy (fig. 15.25), because after the first few weeks the major source of these hormones is the placenta, rather than the mother's ovaries. These high blood levels of estrogen and progesterone prevent menstruation during pregnancy, among their other effects.

Clinical Applications

Because hCG is secreted by the cells of the chorionic membrane of the embryo, and not by the mother's endocrine glands, all **pregnancy tests** assay (test) for hCG in urine or blood. Modern pregnancy tests detect the beta subunit of hCG (one of two different polypeptide chains that comprise the protein), which is unique to hCG and provides the least amount of cross-reaction with related hormones. Pregnancy tests use *monoclonal antibodies* (produced by lymphocyte clones; see chapter 11), which are specific for the beta subunit of hCG and are produced by animals such as rabbits injected with hCG. Home pregnancy tests, using monoclonal antibodies that react with hCG in urine, are generally accurate in the week following the first missed menstrual period.

Clinical Investigation Clues

Remember that Michelle missed her period for a few months, but her pregnancy tests were negative.

- What physiological mechanisms prevent menstruation during pregnancy?
- How does a pregnancy test detect pregnancy?
- Why didn't Michelle menstruate, and what relationship did that have to her difficulty in becoming pregnant?

The Placenta Performs Functions Essential for Pregnancy

As the blastocyst implants, the chorion develops *chorionic villi*, projections that invade blood-filled cavities in the endometrium that surrounds the embryo. The chorionic villi form a leafy-appearing structure, the **chorion frondosum** (*frond* = leaf). Remember, the chorion and its derivatives are genetically the same as the embryo and fetus, because they're derived from the zygote. Meanwhile, the surrounding endometrium undergoes cellular growth, accumulation of glycogen, and other changes known as the *decidual reaction*. This forms a specialized region of the endometrium called the **decidua basalis**. The **placenta** (fig. 15.26) is the combination of the chorion frondosum (fetal tissue) and decidua basalis (maternal tissue).

The disc-shaped human placenta is continuous at its outer surface with the smooth part of the chorion, which bulges into the uterine cavity. Immediately beneath the chorion is the *amnion*, which surrounds the entire embryo. The embryo (and later the fetus), together with its umbilical cord, is located within the fluid-filled **amniotic sac** (fig. 15.26). Amniotic fluid is isotonic and

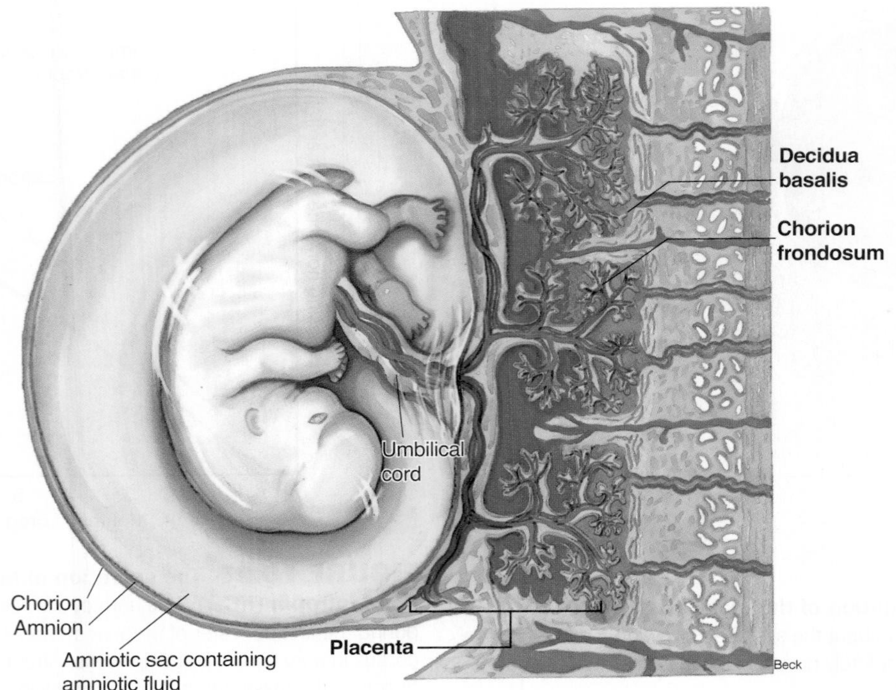

FIGURE 15.26 The amniotic sac and placenta. Blood from the embryo is carried to and from the chorion frondosum by umbilical arteries and veins. The maternal tissue between the chorionic villi is known as the decidua basalis; this tissue, together with the chorionic villi, forms the functioning placenta. The fetus lies within the fluid-filled amniotic sac.

contains cells sloughed off the fetus, placenta, and amniotic sac. Because all of these cells are derived from the zygote, they are genetically the same, and can be aspirated and used for genetic testing in a technique called *amniocentesis*.

Blood from the fetus is sent to the chorion frondosum of the placenta in 2 **umbilical arteries**; blood returns from the placenta to the fetus in a single **umbilical vein** (fig. 15.27). The blood sent to the placenta in the umbilical arteries is low in oxygen, because oxygen can't be obtained from the fetus's lungs by the pulmonary circulation. That's why the umbilical arteries are color-coded blue in figure 15.27. Maternal arteries to the decidua basalis of the endometrium deliver oxygen to the pools of blood that surround the chorionic villi. As a result of gas exchange, the umbilical vein returns oxygenated blood (shown by the red color in fig. 15.27) to the fetus. This blood is the only fully oxygenated blood in the fetus, because it mixes with blood low in oxygen when the umbilical vein joins the inferior vena cava (chapter 10).

In addition to performing gas exchange like the lungs, the placenta performs a detoxication function like the liver, converting many potentially harmful molecules into less toxic forms. The placenta also behaves like two different endocrine glands, secreting sex steroids like those from the gonads as well as polypeptide hormones that resemble some from the anterior pituitary. The placental hormones include:

1. **Estrogens**. The placenta takes an androgen called *DHEA* (*dehydroepiandrosterone*), secreted by the fetus's adrenal cortex, and converts it into estrogens, chiefly *estriol*. This estrogen affects both the mother and fetus in many ways during the course of the pregnancy.
2. **Progesterone**. The placenta secretes much more progesterone than is secreted during the luteal phase of a nonfertile cycle. This works together with placental estrogens to maintain the endometrium in a developed state, promote development of the mammary glands, and produce other effects during the pregnancy.

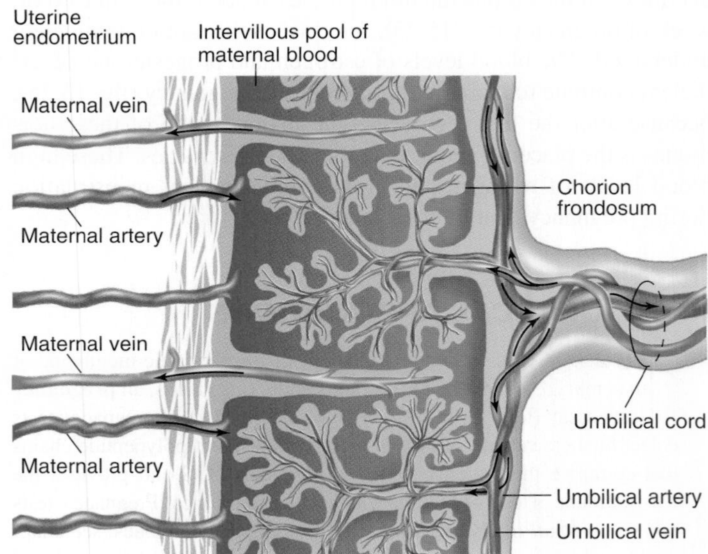

FIGURE 15.27 The circulation of blood within the placenta. Maternal blood is delivered to and drained from the spaces between the chorionic villi. Fetal blood is brought to blood vessels within the villi by branches of the umbilical artery and is drained by branches of the umbilical vein.

3. **Chorionic gonadotropin (hCG)**. This is similar to LH in its effects, helping to maintain the mother's corpus luteum during the first 5½ weeks of pregnancy.
4. **Chorionic somatomammotropin (hCS)**. This is similar in its action to 2 anterior pituitary hormones: growth hormone and prolactin.
5. **Placental corticotrophin-releasing hormone (placental CRH)**. Secretion of this hormone is believed to be essential for timing the onset of labor (parturition) in humans.

The mechanisms responsible for initiating **parturition** (childbirth) in humans are not completely understood. The placenta produces *CRH* (corticotrophin-releasing hormone) only in primates, and only in humans and the great apes does the secretion of CRH rise rapidly during pregnancy. Many scientists now believe that the rate of increase in the blood levels of CRH is the most important determinant of when parturition will occur. CRH stimulates ACTH secretion from the anterior pituitary, and thus cortisol secretion from the adrenal cortex in both the fetus and the mother. Cortisol then stimulates the placenta to secrete more CRH in a positive feedback loop. Cortisol also promotes the maturation of the fetal lungs (stimulating surfactant production; see chapter 12) and other tissues.

ACTH, secreted in response to CRH, also stimulates the fetal adrenal cortex to secrete DHEA. This is a weak androgen that is rapidly converted by the placenta into estrogens. Estrogens and progesterone (from the placenta) help maintain the pregnancy and prepare the mammary glands for lactation following delivery of the baby. In most mammals, a drop in progesterone causes parturition, but in humans the progesterone levels don't fall at the onset of labor. Scientists believe that instead the uterus may have mechanisms that interfere with its ability to respond to progesterone as parturition approaches.

During the course of the pregnancy, the uterine smooth muscle (the myometrium) undergoes changes that make it contract more strongly, and contract in response to stretch. These changes include the development of gap junctions between smooth muscle cells, so that they can contract in unison to expel the fetus for delivery. Once parturition begins, contraction of the myometrium at delivery is stimulated by (1) **oxytocin**, secreted by the posterior pituitary; and by (2) **prostaglandins**, which are regulatory fatty acids produced within the uterus. Indeed, labor can be induced by the injection of oxytocin, or sometimes by the insertion of a suppository of prostaglandins in the vagina.

Lactation Requires the Action of a Number of Hormones

Each **mammary gland** is composed of 15 to 20 *lobes* (fig. 15.28), divided by adipose tissue. The adipose tissue determines the size and shape of the breast, but not the woman's ability to nurse. Each lobe is composed of *lobules*, which contain glandular *alveoli* that secrete milk in a lactating woman. Milk is secreted into *mammary*

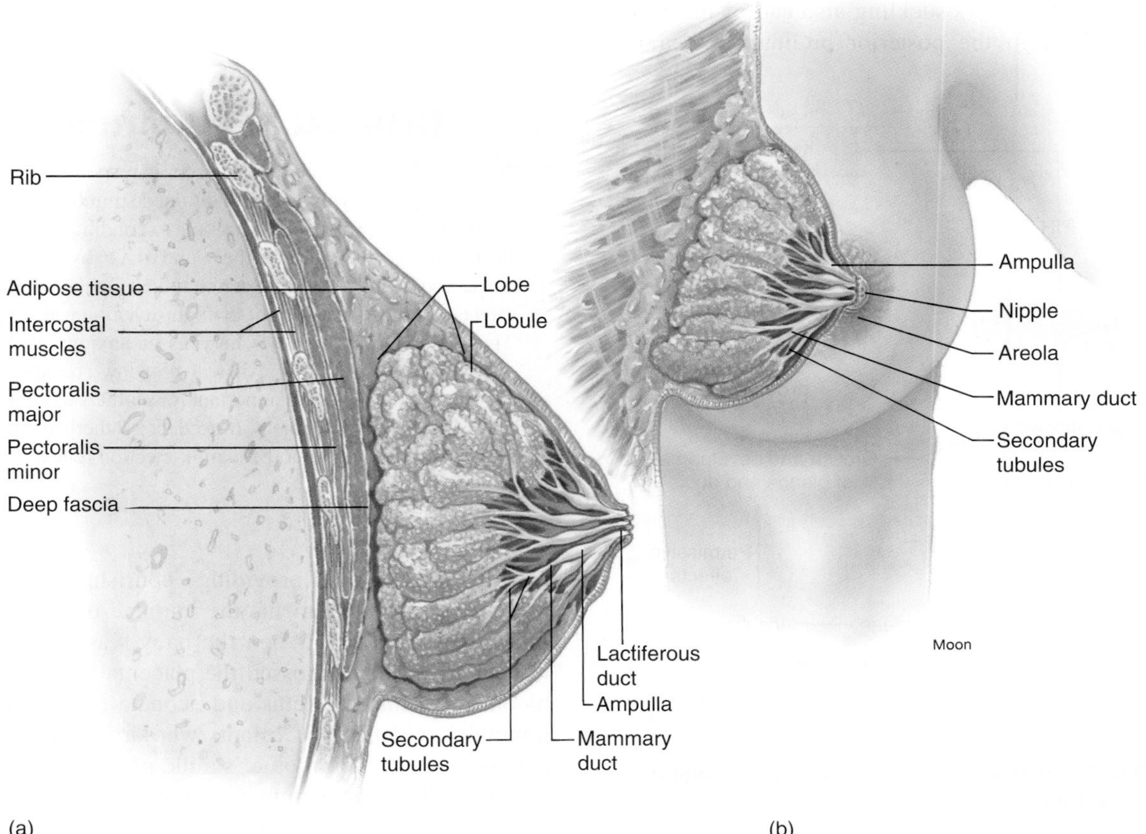

FIGURE 15.28 The structure of the breast and mammary glands. (*a*) A sagittal section and (*b*) an anterior view partially sectioned.

ducts, which converge to form *lactiferous ducts* that drain at the nipple. The lumen of each lactiferous duct is expanded to form an *ampulla*, just beneath the surface of the nipple, where milk accumulates during nursing.

The mammary glands grow and develop during the pregnancy: estrogen stimulates proliferation of the tubules and ducts, and progesterone stimulates development of the mammary alveoli (fig. 15.29). As illustrated in figure 15.29, the permissive action of several other hormones (insulin, cortisol, and thyroxine) is needed for the mammary glands to respond properly to estrogen and progesterone stimulation. The very high blood level of estrogen in a pregnant woman has another important effect on lactation: it stimulates the hypothalamus to secrete *prolactin-inhibiting hormone* (*PIH*). PIH inhibits the anterior pituitary from secreting **prolactin** (fig. 15.29), the hormone that stimulates milk production by the mammary glands. Thus, during the course of pregnancy, the high blood levels of estrogen help prepare the mammary glands for lactation but prevent prolactin secretion and action.

After the placenta is expelled as the *afterbirth*, the blood concentration of estrogen declines sharply, causing the hypothalamus to stop its secretion of PIH. As a result, the anterior pituitary now secretes a large amount of prolactin, which stimulates the production of milk proteins by the mammary glands. The act of nursing (breast-feeding) helps maintain high levels of prolactin secretion through a *neuroendocrine reflex*. In this reflex, the baby's suckling stimulates sensory endings in the breast, which relay impulses to the hypothalamus that inhibit the secretion of PIH. The baby's suckling also causes the reflex secretion of *oxytocin* from the posterior pituitary. When it's secreted in a lactating woman in response to suckling, oxytocin stimulates contraction of the lactiferous ducts, resulting in the **milk-ejection reflex**, or **milk letdown** (fig. 15.30). Oxytocin promotes contraction of the uterus at the same time that it stimulates the milk-ejection reflex.

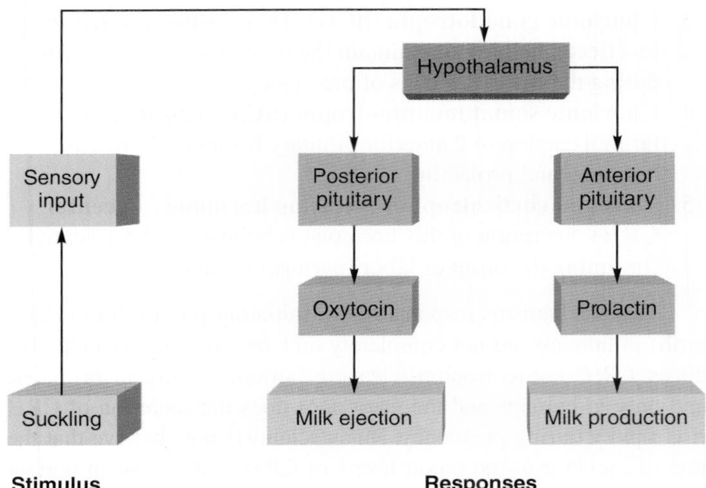

FIGURE 15.30 Milk production and the milk-ejection reflex. Lactation occurs in two stages: milk production (stimulated by prolactin) and milk ejection (stimulated by oxytocin). The stimulus of sucking triggers a neuroendocrine reflex that results in increased secretion of oxytocin and prolactin.

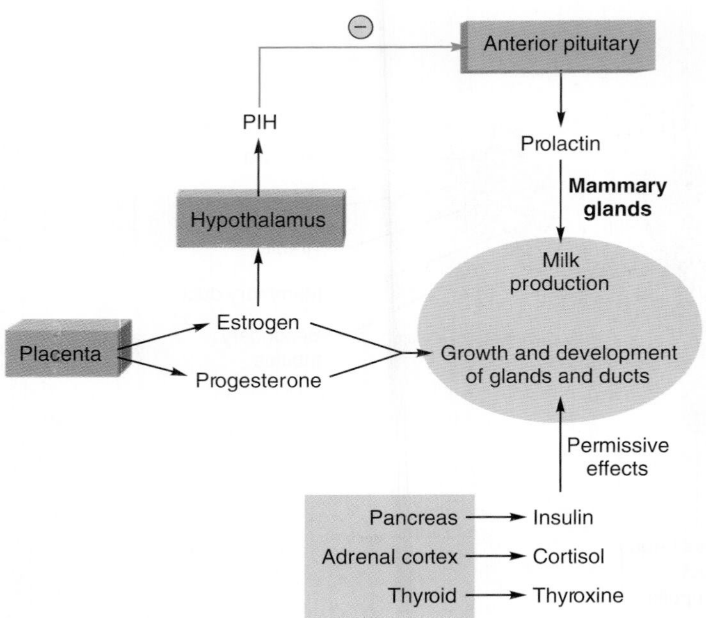

FIGURE 15.29 The hormonal control of mammary gland development and lactation. Notice that milk production is prevented during pregnancy by estrogen inhibition of prolactin secretion. This inhibition is accomplished by the stimulation of PIH (prolactin-inhibiting hormone) secretion from the hypothalamus.

Clinical Applications

The milk-ejection reflex can occur in response to conditioned stimuli (those paired with the unconditioned stimulus of the baby suckling). In this way, milk letdown becomes a conditioned response (discussed in chapter 14); for example, the sound of a baby crying can elicit oxytocin secretion and the milk-ejection reflex. On the other hand, adrenergic effects produced in the fight-or-flight reaction can suppress this reflex. Thus, if a woman becomes nervous or anxious while breast-feeding, she will produce milk, but it will not flow (because milk letdown is prevented). It's therefore important for mothers to nurse their babies in a quiet and calm environment. If needed, synthetic oxytocin can be given as a nasal spray to promote the milk-ejection reflex.

In addition to providing nourishment for the baby, breast-feeding supplements the immune protection given to the infant by its mother. When the baby was a fetus, some of the mother's antibodies crossed the placenta and entered the fetus's blood, providing the fetus and neonate (newborn) with passive immunity (chapter 11). Infants who are breast-fed also receive antibodies from their mother's milk, providing additional passive immunity. The first mammary gland secretion, called *colostrum*, is especially rich in antibodies. This helps protect the baby until it's old enough to produce its own antibodies and develop its own active immunity.

The End Is Also the Beginning

We shall not cease from exploration
And the end of all our exploring
Will be to arrive where we started
And know the place for the first time.

T. S. Eliot[1]

And so we reach the end of this textbook, at the beginning of a new life. The baby may grow to adulthood and become interested in how the body functions, and may want a career in the health professions. If so, an introductory human physiology course and textbook might be in the child's future. There will certainly be many changes between then and now, because physiology is a rapidly advancing science. I hope that this textbook will provide you with a solid basis for your future coursework and careers, and for a lifetime of further learning.

CHECK POINT

1. Describe the events that occur at fertilization, and how the embryo changes as it travels through the uterine tube.
2. Identify the parts of the blastocyst and their functions.
3. Identify the parts of the placenta and their origins. List some of the functions of the placenta.
4. Describe the hormones secreted by the placenta, and their functions.
5. Describe how the mammary glands are affected during a pregnancy, and explain the roles of prolactin and oxytocin in lactation.

PHYSIOLOGY IN HEALTH AND DISEASE

Regenerative medicine is the name for the emerging field of medical treatment using stem cells. This field is still in its infancy and has engendered ethical controversies that affect its progress. This is because, although stem cell therapies hold great promise for the treatment of diseases that are presently intractable, the most promising stem cells are those obtained from surplus embryos banked at fertility clinics. Scientific advances may change this medical field at its roots, and so the terms of the ethical debate may also change in the future.

The zygote and its early cleavage cells are **totipotent**—each cell could form the entire embryo if it were implanted into a uterus. An adult cell is highly differentiated (specialized), but has the same genes as the zygote from which it was derived by mitosis. If the nucleus of an adult cell is transplanted into the cytoplasm of a secondary oocyte—a technique called **somatic cell nuclear transfer**—the DNA can become reprogrammed to be totipotent. The use of such a cell for human reproduction, termed **reproductive cloning**, is condemned by almost everyone. However, the cell can also be used for **therapeutic cloning**, a technique that allows the reprogrammed cell to grow *in vitro* up to the blastocyst stage, at which time cells from the inner cell mass are isolated.

The cells derived *in vitro* from the inner cell mass are called **embryonic stem (ES) cells**. These are described as **pluripotent**, because they can form all of the body tissues but not the trophoblast cells of the chorion. Thus, ES cells aren't totipotent and can't be implanted into a uterus for reproductive cloning. Techniques may be developed that change cultured ES cells into neurons for the treatment of Alzheimer's disease, Parkinson's disease, spinal cord injuries and others; into beta cells of the pancreatic islets for the treatment of type 1 diabetes; and into other tissues. If a patient's cell provided the nucleus for the somatic cell nuclear transfer and the specialized tissues that resulted, the patient's immune system wouldn't reject the transplanted tissues. However, stem cell therapies may present other problems, such as the possibility that undifferentiated cells could form tumors.

Adult stem cells are found in many different protected locations in the body. For example, there are neural stem cells in particular regions of the brain; intestinal epithelial stem cells in the intestinal crypts; stem cells for the formed elements of the blood in the bone marrow; and others. These are described as **multipotent**, because they can give rise to some differentiated cell types, but not to the full range of different tissues that pluripotent ES cells can produce. The multipotent adult stem cells normally produce only a limited number of cell types; neural stem cells form neurons and neuroglia, for example. There have been scientific reports of exceptions—for example, of neural stem cells that have been induced to differentiate into blood and muscle cells—but these reports are highly controversial.

Scientific advances may discover techniques for turning multipotent adult stem cells into pluripotent cells, similar to ES cells (there are now preliminary reports that this has been done). Or other discoveries may be made that alter the nature of stem cell research and regenerative medicine therapies. If so, this may satisfy the present ethical criticism of stem cell research, but may open other, and perhaps presently unforeseen, areas of controversy. As a student of human physiology and as a future health professional, its particularly important for you to stay informed as this story unfolds.

FYI [1]T. S. Eliot, "Little Gidding," last stanza, Collected Poems, 1909–1962, p. 208 (1963).

Physiology in Balance

Male and Female Reproductive Systems

Working together with . . .

Cardiovascular System

The fetal blood becomes oxygenated in the placenta . . . p. 240

Nervous System

Parasympathetic neurons release nitric oxide, which causes vasodilation in the penis, engorging the erectile tissue and producing an erection . . . p. 135

Endocrine System

Estradiol and progesterone from the ovaries stimulate thickening of the endometrium, and the withdrawal of these hormones produces menstruation . . . p. 186

Muscular System

Smooth muscle in the uterus gains gap junctions and produces strong contraction in response to oxytocin . . . p. 215

Respiratory System

Cortisol, secreted by the adrenal cortex, stimulates the fetal lungs to produce surfactant . . . p. 297

Summary

The Male Reproductive System Depends on the Function of Testes 374

- Testes form in the embryo because of a gene on the Y chromosome; ovaries form in the absence of this gene.
- The anterior pituitary secretes LH, which stimulates the interstitial Leydig cells in the testes, and FSH, which stimulates the seminiferous tubules of the testes.
- The Leydig cells secrete testosterone, which exerts negative feedback inhibition of LH secretion; the seminiferous tubules secrete inhibin, which inhibits FSH secretion.
- Testosterone must be converted into other molecules within many target cells in order to regulate them; these active derivatives include dihydrotestosterone (DHT) and estradiol.
- Testosterone and its derivatives stimulate spermatogenesis, promote secondary male sexual characteristics, and stimulate anabolism.
- Spermatogonia are diploid parent cells that give rise to primary spermatocytes, which undergo meiosis in the process of spermatogenesis.
- Haploid secondary spermatocytes are formed after the first meiotic division; spermatids are formed after the second meiotic division, and are then converted into spermatozoa.
- Sertoli cells are nongerminal cells in the seminiferous tubules that are the targets of FSH action; they are needed to aid the developing germ cells and to eliminate cytoplasm from the spermatids, converting them into spermatozoa.
- Sertoli cells make the seminiferous tubules an immunologically privileged site, protecting the sperm from the man's immune system; they also produce inhibin and ABP (androgen-binding protein).
- The seminal vesicles and prostate add fluid and needed molecules to the semen.
- Erection of the penis is produced by the engorgement of the erectile tissue with blood, as a result of vasodilation of penile blood vessels produced by the action of nitric oxide.

The Female Reproductive System Depends on the Function of Ovaries 380

- An egg cell (ovum or oocyte) is released from an ovary at ovulation, and enters a uterine (fallopian) tube and then the uterus; the epithelial lining of the uterus is the endometrium, and the smooth muscle wall is the myometrium.
- Before an ovarian follicle is stimulated by FSH, it's a primary follicle containing a primary oocyte; FSH stimulates a follicle to grow and develop fluid-filled cavities, when it becomes a secondary follicle.
- FSH also stimulates granulosa cells within the ovarian follicles to secrete increasing amounts of estradiol, the major estrogen.
- During a particular cycle, 1 follicle will grow very large, develop a fluid-filled single cavity, the antrum, and become a mature, or graafian, follicle.
- The primary oocyte within a stimulated follicle finishes the first meiotic division, forming a secondary oocyte and a small polar body, which disintegrates.
- The secondary oocyte is arrested at metaphase II of meiosis; when a graafian follicle ovulates, this is the egg cell that is released from the ovary.
- Ovulation is triggered by a surge in LH secretion, which is caused by a positive feedback effect of estradiol secreted by the granulosa cells of the growing follicles.
- The secondary oocyte enters a fallopian tube and disintegrates after about 24 hours if it isn't fertilized; if it's fertilized, it will complete the meiotic division, forming a single large cell and a second small polar body.
- Fertilization occurs in a uterine tube, and the fertilized oocyte is called a zygote.
- The empty graafian follicle after ovulation is changed by LH stimulation into a corpus luteum, which secretes progesterone as well as estradiol.
- In a female cycle prior to ovulation, the ovaries are in the follicular phase; following ovulation, the ovaries are in the luteal phase.
- During the follicular phase of the ovaries, the rising secretion of estradiol stimulates the proliferative phase of the endometrium.
- During the luteal phase of the ovaries, progesterone and estradiol stimulate the secretory phase of the endometrium.
- Estradiol and progesterone exert negative feedback inhibition of FSH and LH during the luteal phase, and the suppression of LH causes death of the corpus luteum near the end of the luteal phase.
- The menstrual phase, in which the stratum functionale of the endometrium is shed with bleeding, results from the drop in estradiol and progesterone caused by death of the corpus luteum near the end of the previous cycle.

Gestation, Parturition, and Lactation Are Female Reproductive Functions 387

- The ovulated oocyte is surrounded by a clear zona pellucida, and a layer of granulosa cells called a corona radiata; the sperm tunnels its way through these by releasing digestive enzymes from its cap, the acrosome.
- The zygote immediately starts to undergo mitosis, or cleavage, forming 2 identical diploid cells, which also undergo mitosis; this forms a ball of small cells, the morula, which enters the uterus at about the third day after conception.
- In the next 2 days, the embryo becomes a blastocyst that consists of two parts: an inner cell mass that will become the embryo, and a layer called the chorion that will contribute to the placenta.
- The chorion secretes a hormone, human chorionic gonadotropin (hCG), which acts like LH to maintain the mother's corpus luteum; this keeps estradiol and progesterone secretion high, so that menstruation is prevented.
- The placenta consists of the chorion frondosum and a specialized part of the endometrium, the decidua basalis.
- Umbilical arteries deliver oxygen-poor blood from the fetus to the placenta, which oxygenates the blood that returns to the fetus in the umbilical vein.
- The placenta secretes estrogens (chiefly estriol) and progesterone, as well as hormones that resemble some secreted by the anterior pituitary.
- Corticotropin-releasing hormone (CRH) is secreted by the placenta in increasing amounts during gestation, and is believed to be the primary factor timing the onset of parturition (childbirth).
- Contractions of the myometrium during labor are stimulated by oxytocin from the posterior pituitary, and by prostaglandins produced within the uterus.
- The high levels of estrogen and progesterone secreted by the placenta promote the development of the mammary glands during a pregnancy; however, high estrogen prevents lactation because it stimulates the hypothalamus to produce prolactin-inhibiting hormone (PIH).
- The stimulus of a baby suckling activates neuroendocrine reflexes that stimulate prolactin secretion from the anterior pituitary and oxytoxin secretion from the posterior pituitary.
- In a lactating woman, oxytocin stimulates the milk-ejection reflex, or milk letdown.

Review Activities

Objective Questions: Test Your Knowledge

1. Which of the following statements about the formation of the testes is true?
   a. There must be only one X chromosome.
   b. Testes form due to a gene on the X chromosome.
   c. Testes form due to a gene on the Y chromosome.
   d. Formation of testes requires a gene on the X and a gene on the Y chromosome.

2. Inhibin is a hormone
   a. secreted by the Leydig cells that inhibits LH secretion.
   b. secreted by the seminiferous tubules that inhibits LH secretion.
   c. secreted by the Leydig cells that inhibits FSH secretion.
   d. secreted by the seminiferous tubules that inhibits FSH secretion.

3. The enzyme 5α-reductase
   a. converts testosterone into dihydrotestosterone (DHT).
   b. converts testosterone into estradiol.
   c. converts testosterone into progesterone.
   d. converts testosterone into dehydroepiandrosterone (DHEA).

4. Which of the following hormones stimulates spermatogenesis most directly and powerfully?
   a. FSH
   b. Testosterone
   c. LH
   d. Estradiol

5. Which of the following hormones most stimulates anabolism?
   a. Progesterone
   b. Cortisol
   c. Testosterone
   d. Estradiol

6. Which of the following cells is haploid, but with 2 chromatids per chromosome?
   a. Spermatogonia
   b. Primary spermatocytes
   c. Secondary spermatocytes
   d. Spermatids

7. These germ cells have their cytoplasm removed through phagocytosis by Sertoli cells:
   a. Spermatogonia
   b. Primary spermatocytes
   c. Secondary spermatocytes
   d. Spermatids

8. Which of the following is not a function of Sertoli cells?
   a. Production of testosterone
   b. Production of inhibin
   c. Production of androgen-binding protein (ABP)
   d. Protection of the developing germ cells from the man's immune system

9. About 60% of the volume of the semen is produced by the
   a. seminiferous tubules.
   b. seminal vesicles.
   c. epididymis.
   d. prostate.

10. Engorgement of the corpora cavernosa with blood during an erection of the penis result primarily from the action of
    a. acetylcholine.
    b. norepinephrine.
    c. testosterone.
    d. nitric oxide.

11. Which of the following statements about the female reproductive system is false?
    a. The uterus is lined by an epithelium that contains cilia at its surface.
    b. The stratum functionale of the endometrium is shed at menstruation.
    c. Only certain primates have menstrual cycles.
    d. At puberty, there are about 400,000 follicles in the ovaries.

12. FSH stimulates a rising secretion of estradiol from
    a. primary oocytes.
    b. secondary oocytes.
    c. granulosa cells.
    d. corpora lutea.

13. Which cells are formed in oogenesis by the end of the first meiotic division?
    a. 2 secondary oocytes
    b. 1 secondary oocyte and 1 polar body
    c. 1 secondary oocyte and 2 polar bodies
    d. 1 secondary oocyte and 1 primary oocyte

14. The egg cell that's ovulated is a
    a. secondary oocyte that's completed meiosis.
    b. secondary oocyte that's arrested at prophase I of meiosis.
    c. secondary oocyte that's arrested at telophase II of meiosis.
    d. secondary oocyte that's arrested at metaphase II of meiosis.

15. Which of the following statements regarding a corpus luteum is false?
    a. It's formed following ovulation.
    b. It's formed from the ovulated graafian follicle.
    c. It persists from one cycle to the next.
    d. It depends on LH stimulation.

16. Which of the following statements regarding progesterone is false?
    a. Little or no progesterone is secreted prior to ovulation.
    b. Granulosa cells of the follicles secrete progesterone.
    c. Progesterone secretion is characteristic of the luteal phase of the cycle.
    d. Progesterone secretion falls at the end of the luteal phase of the cycle.

Match the phase of the ovarian cycle with its description:

17. Follicular phase	a. Progesterone secretion rises.
18. Ovulation	b. Estradiol secretion rises.
19. Luteal phase	c. Estradiol secretion rises and progesterone falls.
	d. LH secretion rises sharply just prior to this.

20. Estradiol and progesterone exert negative feedback inhibition of FSH and LH secretion during
    a. the follicular phase.
    b. ovulation.
    c. the luteal phase.
    d. the proliferative phase.

21. Menstruation is caused by
    a. a rise in estradiol and progesterone levels.
    b. a fall in estradiol and progesterone levels.
    c. a rise in estradiol and a fall in progesterone levels.
    d. a fall in estradiol and a rise in progesterone levels.
22. Digestive enzymes that help the sperm to reach the egg cell's plasma membrane are located in the
    a. corona radiata.
    b. zona pellucida.
    c. cumulus oophorous.
    d. acrosome.
23. Following fertilization,
    a. the zygote immediately implants into the uterine tube.
    b. the zygote becomes a morula when it reaches the uterus.
    c. the zygote divides by meiosis.
    d. the morula stage embryo implants into the endometrium.
24. Which of the following statements regarding human chorionic gonadotropin (hCG) is false?
    a. It's secreted by the inner cell mass of the blastocyst.
    b. It acts like LH.
    c. It maintains the corpus luteum.
    d. It's normally absent if a woman isn't pregnant.
25. Which of the following statements regarding the placenta is true?
    a. It secretes large amounts of a hormone that acts like FSH.
    b. It's composed of the chorion frondosum and the decidua basalis.
    c. It makes the blood in the umbilical arteries fully oxygenated.
    d. It acts like the kidneys and produces urine.
26. Parturition in humans is believed to be initiated by
    a. ACTH from the fetal anterior pituitary.
    b. cortisol from the fetal adrenal cortex.
    c. rising CRH secreted by the placenta.
    d. falling progesterone secreted by the placenta.
27. Contractions of the myometrium during labor and the milk-ejection reflex are stimulated by
    a. prolactin from the anterior pituitary.
    b. oxytocin from the anterior pituitary.
    c. prolactin from the posterior pituitary.
    d. oxytocin from the posterior pituitary.

Essay Questions 1: *Test Your Understanding*

1. Explain the requirements for an embryo to develop either testes or ovaries.
2. Identify the two compartments of the testes, and explain how they are regulated by the pituitary gonadotropins.
3. Describe the negative feedback regulation of the gonadotropins in males.
4. How does testosterone function as a prehormone? What are its major derivatives, and which enzymes produce them?
5. Describe the germ cells formed at each step of spermatogenesis. Identify the hormonal requirements for spermatogenesis.
6. Describe the structure and location of Sertoli cells, and list their functions.
7. List, in sequence, the path of sperm from the testes out of the body. In which location do sperm become motile? What do the seminal vesicles and prostate contribute to the semen?
8. Describe the erectile tissue of the penis, and the mechanisms that produce an erection.
9. Describe the changes that occur in ovarian follicles leading up to ovulation.
10. Describe oogenesis up to the time of ovulation.
11. How and when is a corpus luteum formed, and what is its function?
12. Describe the phases in the ovaries during a nonfertile female cycle, identifying the effects of the gonadotropic hormones on the ovary and the hormones that the ovaries secrete in each phase.
13. How and when do the ovaries exert a positive feedback effect on the anterior pituitary, and what function is served by this?
14. How and when do the ovaries exert a negative feedback effect on the anterior pituitary, and what function is served by this?
15. Identify the phases of the endometrium during the female cycle.
16. Step by step, explain how menstruation is produced when the oocyte isn't fertilized.
17. What is the acrosome, and what is its function?
18. Describe the completion of oogenesis and the formation of a zygote following fertilization.
19. Identify the parts of a blastocyst and what these parts form. Describe when and where a blastocyst is produced and what normally happens to it.
20. What is hCG, and what is its function? Use this information to explain why a woman usually misses her period if she's pregnant.
21. Identify the fetal and maternal components of the placenta, and describe some of its functions.
22. Identify the hormones secreted by the placenta and describe some of their functions.
23. What is believed to be the major factor timing the onset of labor? Identify the factors that stimulate uterine contractions during labor.
24. Describe the hormonal requirements for mammary gland development, and explain why a woman doesn't lactate while she is pregnant.
25. What stimulates milk production and milk letdown in a nursing mother? Describe the reflexes that allow a nursing infant to obtain milk.

Essay Questions 2: *Test Your Analytical Ability*

1. Suppose an error occurred in mitotic cell division in an embryo, resulting in the cells of the early embryo having the genotype XXY (with an extra sex chromosome, for a total of 47). Would this embryo develop testes or ovaries? Explain.
2. Suppose you have your pet male dog neutered (castrated). What would happen to its blood levels of FSH and LH? Now suppose you inject it with daily amounts of testosterone: what would happen to its blood levels of FSH and LH? Explain.
3. Given that dihydrotestosterone (DHT) is responsible for the development of male-pattern baldness, what might happen to a man with this condition who takes a drug that inhibits 5α-reductase? What other effects might this drug have? Explain.
4. The theca interna of ovarian follicles produce testosterone, but the granulosa cells are rich in aromatase enzyme. Explain the significance of these observations.
5. Why and how is LH secretion required for spermatogenesis? What is the significance of FSH secretion for spermatogenesis? How might this effect be exerted, given that FSH acts on the Sertoli cells, not on the germinal cells?
6. When a man gets a vasectomy (in which sperm transport along the vas deferens is blocked in some way), his immune system may eventually become sensitized to his own sperm and attack them. How are sperm normally protected from this, and what may have occurred to allow immunological attack?
7. A man who has been taking anabolic steroids gets a physical exam, and the physician notices that the testes seem abnormally small. What might account for this?

8. Suppose a new drug that blocks the ability of FSH to activate its receptors is developed for male contraception. How would such a drug affect spermatogenesis? Would it abolish fertility? Explain.
9. How does Viagra treat erectile dysfunction? How might this drug's effects interact dangerously with those of nitroglycerin, a vasodilator used to treat angina pectoris (chapter 10)?
10. Meiosis in oogenesis forms only 1 cell, whereas meiosis in spermatogenesis results in 4 sperm cells. Explain this difference, and what advantages this difference may provide.
11. A virgin rabbit has never ovulated, nor developed a corpus luteum. How could injection of an extract from a pregnant woman's urine cause the rabbit to ovulate? How could you tell she ovulated, if you were to look at her ovaries?
12. How do birth control pills work? How can women taking birth control pills menstruate?
13. Suppose you were to inject a man with hCG. What would be its effect, if any? Explain.
14. Explain how the corpus luteum can be said to commit suicide at the end of a nonfertile cycle.
15. People sometimes say that the birth control pill "tricks your brain into thinking you're pregnant." Explain the physiological basis for this statement, comparing a woman taking birth control pills with a pregnant woman.
16. Although cells obtained by amniocentesis may derive from the placenta, they can be used to do genetic analysis of the fetus. Explain why this is true.
17. The fetal adrenal cortex has a unique zone (absent after birth) that secretes dehydroepiandrosterone (DHEA), a weak androgen. During pregnancy, the placenta secretes increasing amounts of estriol, an estrogen. Explain the relationship between these two statements.
18. Pregnant women are sometimes given injections of oxytocin to induce labor. Explain why this works best if the woman is ready to begin parturition.
19. The fetus is never exposed to as high an oxygen concentration as it will experience after it's born and begins breathing. Explain why this statement is true.
20. Explain how the sound of a baby crying can cause a lactating mother's milk to flow.

Web Activities

ARIS™ **For additional study tools, go to: www.aris.mhhe.com. Click on your course topic and then this textbook's author/title. Register once for a semester's worth of interactive activities and practice quizzing to help boost your grade!**

Appendix 1

Calculations of Concentrations

The simplest way to express the concentration of a solution is by indicating the weight of the solute—in milligrams (mg), for example—per given volume of solution, such as per 100 milliliters (mL). Thus, a solution that contains 1.17 g of sodium chloride (NaCl) per 100 mL can be expressed as 1.17 g/100 mL, or 1.17 g/dL (deciliter, which is one-tenth of a liter, or 100 mL). A solution containing 3.6 g of glucose per 100 mL is expressed as 3.6 g/100 mL or 3.6 g/dL.

But how does the number of solute molecules in 1.17 g of NaCl compare with the number of solute molecules in 3.6 g of glucose? There's no way to know, just given these weights. However, if we know how much 1 molecule weighs compared to the other, we can find out. First, we need to know the molecular weight of the solute molecules. This can be determined by adding the atomic weights of each element in the molecule. For example:

	Sodium chloride (NaCl)	*Glucose* ($C_6H_{12}O_6$)
Atomic weights:	Na = 23.0	C_6 $12 \times 6 = 72$
	Cl = 35.5	H_{12} $1 \times 12 = 12$
		O_6 $16 \times 6 = 96$
Molecular weights:	58.5	180

Second, we need to know that the molecular weight in grams—called a **mole**—of different compounds contains the same number of molecules (this is known as *Avogadro's number*—6.02×10^{23}). Thus, 58.5 g of NaCl and 180 g of glucose contain equal numbers of molecules. **Molarity (M)** is a unit of concentration indicating moles per liter of solution. If we dissolve 1 mole of solute in 1 liter of solution, we have a 1 molar (1 M) solution. Getting back to the two solutions previously described, we can determine their relative numbers of solute molecules per liter by calculating their molar concentrations. Remember that 1 mole of NaCl weighs 58.5 g and 1 mole of glucose weighs 180 g.

$$\frac{1.17\text{ g NaCl}}{100\text{ mL}} \times \frac{1\text{ mole}}{58.5\text{ g}} \times \frac{1{,}000\text{ mL}}{1\text{ L}} = 0.20\text{ M}$$

and

$$\frac{3.6\text{ g glucose}}{100\text{ mL}} \times \frac{1\text{ mole}}{180\text{ g}} \times \frac{1{,}000\text{ mL}}{1\text{ L}} = 0.20\text{ M}$$

These calculations reveal that the two solutions have the same number of solute molecules per liter of solution. However, there is a difference between these two solutions, because NaCl is an ionic compound, and each NaCl molecule in solution dissociates to form equal numbers of Na^+ and Cl^- ions. Thus, a 1 M NaCl solution has 1 M Na^+ and 1 M Cl^- ions. These produce a combined solute concentration of 2.0 osmolar, as described in chapter 2. The 0.2 M NaCl solution in this example has 0.2 M Na^+ and 0.2 M Cl^-, giving an osmolarity of 0.4 OsM or 400 millisosmolar (mOsM). By contrast, the glucose solution has a concentration of 0.2 OsM, or 200 mOsM, because glucose is not an ionic compound.

The concentrations of ions are sometimes expressed as **milliequivalents per liter (mEq/L).** A milliequivalent is a millimole of an ion multiplied by its number of charges. Thus, because the solution in the previous example contains 0.2 M Na^+, or 200 millimolar (mM) of Na^+, the concentration of Na^+ can also be expressed as $200 \times 1 = 200$ mEq/L of Na^+. The sign of the charges doesn't affect the calculation, so Cl^- here would also have a concentration of $200 \times 1 = 200$ mEq/L of Cl^-. If a different solution had a concentration of 200 mM Ca^{2+}, this concentration could be expressed as $200 \times 2 = 400$ mEq/L of Ca^{2+}. Clinical laboratory measurements of Na^+, Cl^-, HCO_3^-, Ca^{2+}, and other ions are frequently indicated in milliequivalent per liter concentrations.

Appendix 2

Steps of Glycolysis and the Krebs Cycle

The individual steps of glycolysis are given in figure A.1. Notice that glucose is converted into its isomer fructose at step 2, and phosphate groups from 2 ATP are added to form fructose 1,6-biphosphate by step 3. These 2 ATP represent an energy "investment." The 6-carbon-long fructose is hydrolyzed into two 3-carbon-long molecules (3-phosphoglyceraldehyde) at step 4, and 2 ATP are produced at step 6, making up the 2 ATP that were used previously. At step 9, 2 more ATP are produced, so that glycolysis provides a net gain of 2 ATP (the sum of –2 ATP in the early steps and +4 ATP in the later steps). Two NAD are converted into NADH + H^+ at step 5, but these aren't used for ATP production in glycolysis—that process requires the electron transport system and oxidative phosphorylation of aerobic respiration (chapter 2).

The individual steps of the Krebs cycle are indicated in figure A.2. The Krebs cycle begins in step 1, when a 2-carbon-long molecule (the acetyl group of acetyl CoA) joins with a 4-carbon long molecule (oxaloacetic acid) to form a 6-carbon-long molecule (citric acid). Enzymes then produce sequential changes that eventually reproduce an oxaloacetic acid (step 8). As the cycle turns, 1 ATP is produced at step 5. Also, 3 NAD are converted into NADH + H^+, and 1 FAD is converted into $FADH_2$, using hydrogen atoms from the intermediate molecules of the Krebs cycle. The electrons from each hydrogen atom used to form NADH and $FADH_2$ will eventually be donated to the electron transport system, and the transport of electrons will be used to power the formation of ATP by oxidative phosphorylation (chapter 2).

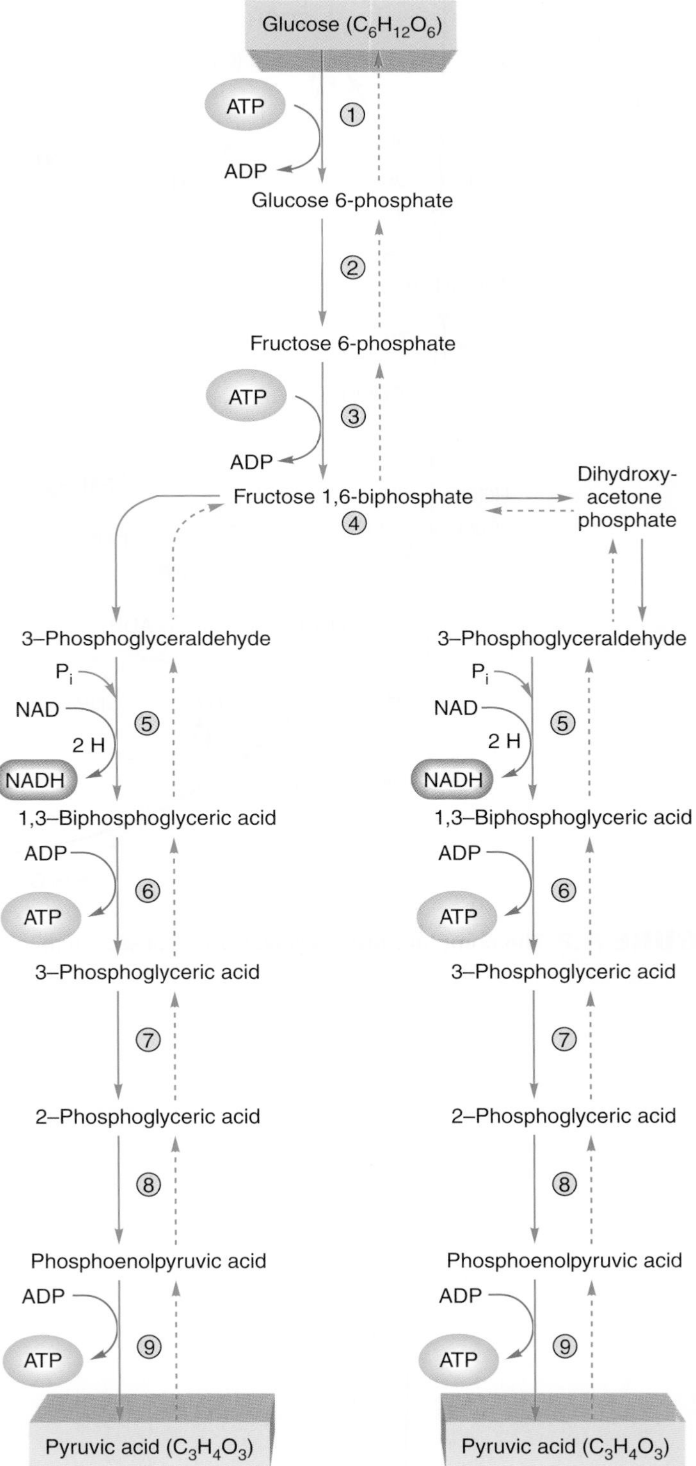

FIGURE A.1 Glycolysis. In glycolysis, 1 glucose is converted into 2 pyruvic acids in nine separate steps. In addition to 2 pyruvic acids, the products of glycolysis include 2 NADH and 4 ATP. Since 2 ATP were used at the beginning, however, the net gain is 2 ATP per glucose. Dashed arrows indicate reverse reactions that may occur under other conditions.

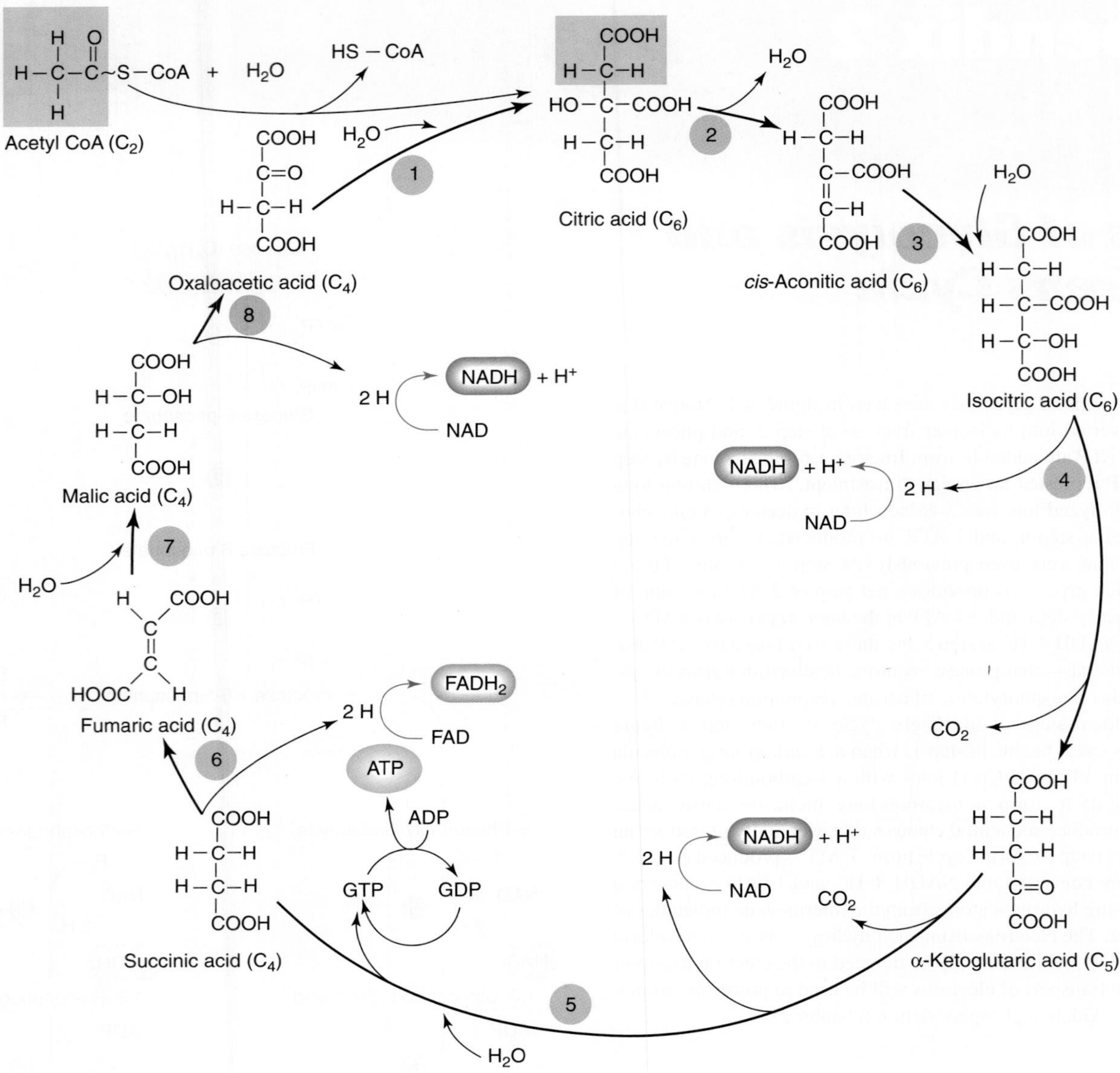

FIGURE A.2 The complete Krebs cycle. Notice that each "turn" of the cycle produces 1 ATP, 3 NADH, and 1 $FADH_2$.

Appendix 3

Answers to Objective Questions: Test Your Knowledge

Chapter 1

1. c 2. d 3. d 4. c 5. a 6. b 7. d 8. c 9. b 10. a 11. d 12. c 13. b 14. c 15. d 16. a 17. d 18. c

Chapter 2

1. c 2. a 3. c 4. b 5. c 6. a 7. d 8. b 9. d 10. a 11. b 12. d 13. d 14. b 15. c 16. a 17. c 18. b 19. c 20. a

Chapter 3

1. a 2. c 3. b 4. d 5. c 6. a 7. d 8. c 9. b 10. a 11. c 12. d 13. d 14. b 15. a 16. d 17. b 18. c 19. d 20. a 21. c

Chapter 4

1. c 2. d 3. b 4. a 5. b 6. a 7. d 8. c 9. a 10. d 11. a 12. b 13. d 14. c 15. b 16. d 17. a 18. c 19. a 20. d 21. d 22. b

Chapter 5

1. d 2. b 3. c 4. a 5. b 6. c 7. d 8. d 9. c 10. d 11. a 12. c 13. b 14. a 15. d 16. b 17. b 18. d

Chapter 6

1. d 2. c 3. a 4. b 5. c 6. b 7. c 8. a 9. a 10. c 11. b 12. d 13. c 14. d 15. a 16. d 17. b 18. c 19. a 20. d 21. b

Chapter 7

1. b 2. d 3. c 4. d 5. a 6. c 7. c 8. c 9. a 10. b 11. a 12. c 13. b 14. d 15. d 16. c 17. c 18. c 19. a 20. c 21. b 22. a 23. d 24. c 25. c 26. a 27. b 28. c 29. b

Chapter 8

1. a 2. d 3. c 4. b 5. d 6. a 7. b 8. d 9. a 10. b 11. d 12. a 13. d 14. c 15. d 16. d 17. b 18. c 19. a 20. b 21. d 22. a 23. c 24. c 25. c 26. a 27. c 28. d 29. b 30. d

Chapter 9

1. c 2. b 3. d 4. a 5. c 6. a 7. b 8. d 9. d 10. b 11. c 12. a 13. c 14. a 15. c 16. d 17. a 18. b 19. c 20. b 21. a 22. d 23. c 24. b 25. c

Chapter 10

1. a 2. c 3. d 4. c 5. d 6. a 7. b 8. c 9. c 10. d 11. d 12. b 13. c 14. b 15. c 16. d 17. a 18. c 19. a 20. d 21. c 22. b 23. a 24. b 25. a 26. d 27. d 28. a 29. d 30. b

Chapter 11

1. c 2. d 3. b 4. a 5. a 6. d 7. b 8. a 9. b 10. c 11. d 12. c 13. d 14. a 15. b 16. c 17. d 18. b

Chapter 12

1. c 2. a 3. c 4. b 5. d 6. a 7. d 8. c 9. a 10. b 11. b 12. c 13. c 14. a 15. d 16. c 17. b 18. a 19. c 20. d 21. c 22. d 23. a 24. d 25. b

Chapter 13

1. d 2. a 3. c 4. b 5. c 6. b 7. a 8. d 9. c 10. d 11. c 12. a 13. b 14. b 15. c 16. a 17. b 18. d 19. d 20. b 21. c 22. d 23. c 24. c 25. c

Chapter 14

1. c
2. b
3. b
4. d
5. a
6. c
7. d
8. a
9. c
10. c
11. c
12. a
13. d
14. d
15. d
16. c
17. a
18. b
19. a
20. d
21. c
22. d
23. a
24. c
25. c

Chapter 15

1. c
2. d
3. a
4. b
5. c
6. c
7. d
8. a
9. b
10. d
11. a
12. c
13. b
14. d
15. c
16. b
17. b
18. d
19. a
20. c
21. b
22. d
23. b
24. a
25. b
26. c
27. d

Appendix 4

Meiosis

When a cell is going to divide, either by mitosis or meiosis, the DNA is replicated (forming chromatids) and the chromosomes become shorter and thicker. At this point the cell has forty-six chromosomes, each of which consists of two duplicate chromatids.

The short, thick chromosomes seen at the end of the G_2 phase can be matched as pairs, the members of each pair appearing to be structurally identical. These matched chromosomes are called **homologous chromosomes**. One member of each homologous pair is derived from a chromosome inherited from the father, and the other member is a copy of one of the chromosomes inherited from the mother. Homologous chromosomes do not have identical DNA base sequences; one member of the pair may code for blue eyes, for example, and the other for brown eyes. There are twenty-two homologous pairs of *autosomal chromosomes* and one pair of *sex chromosomes*, described as X and Y. Females have two X chromosomes, whereas males have one X and one Y chromosome (fig. A.3).

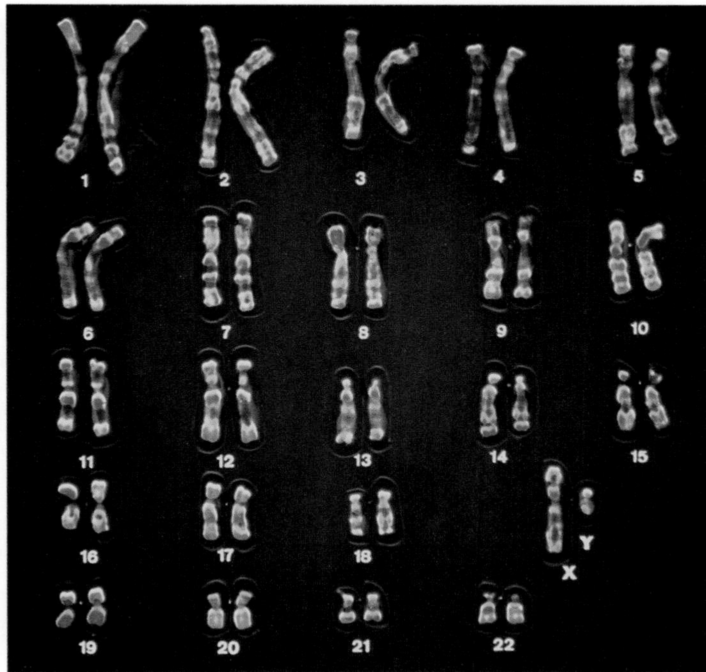

FIGURE A.3 A karyotype, in which chromosomes are arranged in homologous pairs. A false-color light micrograph of chromosomes from a male arranged in numbered homologous pairs, from the largest to the smallest.

Meiosis, which has two divisional sequences, is a special type of cell division that occurs only in the gonads (testes and ovaries), where it is used only in the production of gametes—sperm cells and ova. In the first division of meiosis, the homologous chromosomes line up side by side, rather than single file, along the equator of the cell. The spindle fibers then pull one member of a homologous pair to one pole of the cell, and the other member of the pair to the other pole. Each of the two daughter cells thus acquires only one chromosome from each of the twenty-three homologous pairs contained in the parent. The daughter cells, in other words, contain twenty-three rather than forty-six chromosomes. For this reason, meiosis (from the Greek *meion* = less) is also known as **reduction division**.

At the end of this cell division, each daughter cell contains twenty-three chromosomes—but *each of these consists of two chromatids*. (Since the two chromatids per chromosome are identical, this does not make forty-six chromosomes; there are still only twenty-three *different* chromosomes per cell at this point.) The chromatids are separated by a second meiotic division. Each of the daughter cells from the first cell division itself divides, with the duplicate chromatids going to each of two new daughter cells. A grand total of four daughter cells can thus be produced from the meiotic cell division of one parent cell. This occurs in the testes, where one parent cell produces four sperm cells. In the ovaries, one parent cell also produces four daughter cells, but three of these die and only one progresses to become a mature egg cell.

The stages of meiosis are subdivided according to whether they occur in the first or the second meiotic cell division. These stages are designated as prophase I, metaphase I, anaphase I, telophase I; and then prophase II, metaphase II, anaphase II, and telophase II (table A.1 and fig. A.4).

The reduction of the chromosome number from forty-six to twenty-three is obviously necessary for sexual reproduction, where the sex cells join and add their content of chromosomes together to produce a new individual. The significance of meiosis, however, goes beyond the reduction of chromosome number. At metaphase I, the pairs of homologous chromosomes can line up with either member facing a given pole of the cell. (Recall that each member of a homologous pair came from a different parent.) Maternal and paternal members of homologous pairs are thus randomly shuffled. Hence, when the first meiotic division occurs, each daughter cell will obtain a complement of twenty-three chromosomes that are randomly derived from the maternal or paternal contribution to the homologous pairs of chromosomes of the parent cell.

In addition to this "shuffling of the deck" of chromosomes, exchanges of parts of homologous chromosomes can occur at prophase I. That is, pieces of one chromosome of a homologous pair can be exchanged with the other homologous chromosome in a process called *crossing-over* (fig. A.5). These events together result in **genetic recombination** and ensure that the gametes produced by meiosis are genetically unique. This provides additional genetic diversity for organisms that reproduce sexually, and genetic diversity is needed to promote survival of species over evolutionary time.

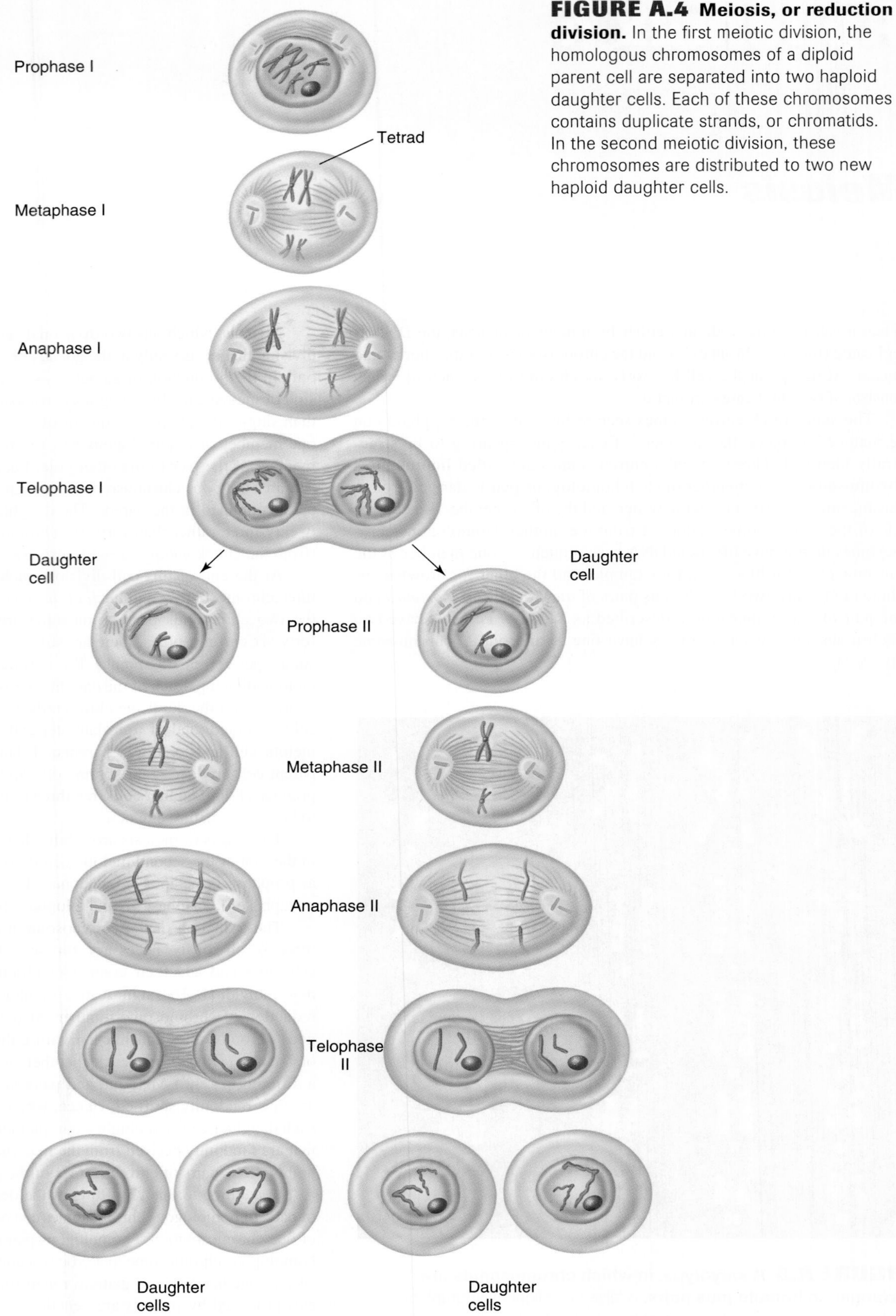

FIGURE A.4 Meiosis, or reduction division. In the first meiotic division, the homologous chromosomes of a diploid parent cell are separated into two haploid daughter cells. Each of these chromosomes contains duplicate strands, or chromatids. In the second meiotic division, these chromosomes are distributed to two new haploid daughter cells.

TABLE A.1 Stages of Meiosis

Stage	Events
First Meiotic Division	
Prophase I	Chromosomes appear double-stranded. Each strand, called a chromatid, contains duplicate DNA joined together by a structure known as a centromere. Homologous chromosomes pair up side by side.
Metaphase I	Homologous chromosome pairs line up at equator. Spindle apparatus is complete.
Anaphase I	Homologous chromosomes separate; the two members of a homologous pair move to opposite poles.
Telophase I	Cytoplasm divides to produce two haploid cells.
Second Meiotic Division	
Prophase II	Chromosomes appear, each containing two chromatids.
Metaphase II	Chromosomes line up single file along equator as spindle formation is completed.
Anaphase II	Centromeres split and chromatids move to opposite poles.
Telophase II	Cytoplasm divides to produce two haploid cells from each of the haploid cells formed at telophase I.

(a) First meiotic prophase

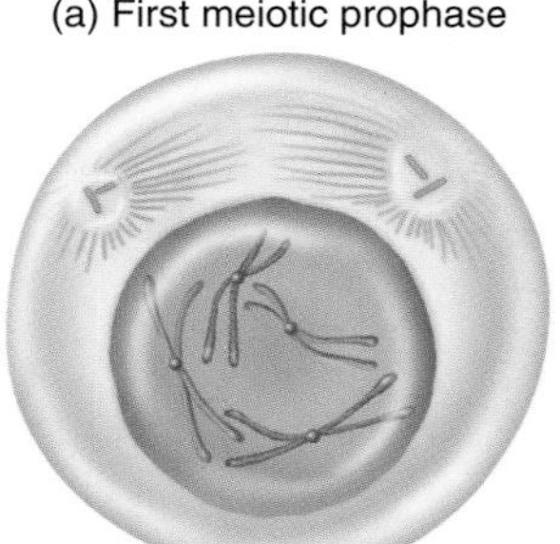

Chromosomes pairing

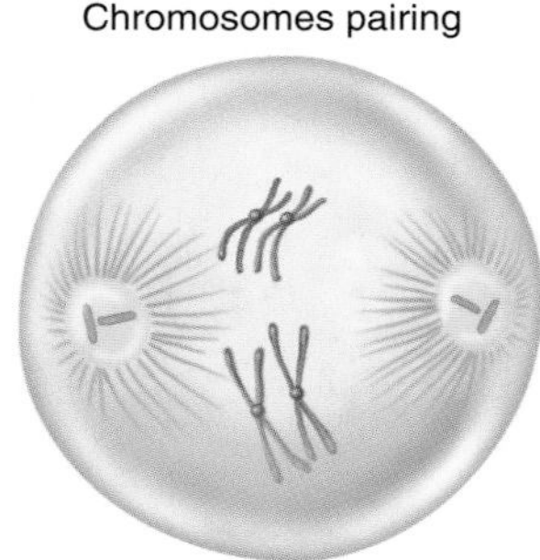

Chromosomes crossing-over

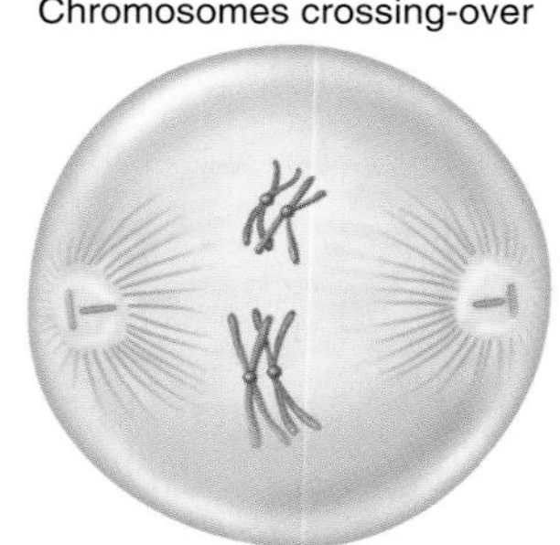

(b) Crossing-over

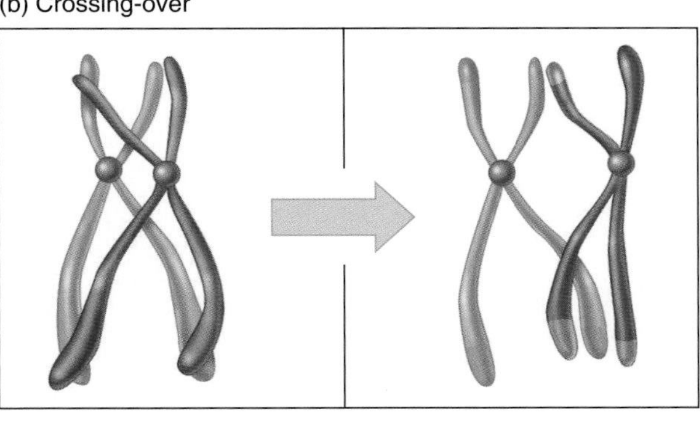

FIGURE A.5 Crossing-over. (*a*) Genetic variation results from the crossing-over of tetrads, which occurs during the first meiotic prophase. (*b*) A diagram depicting the recombination of chromosomes that occurs as a result of crossing-over.

Appendix 5

Correlation Guide to Laboratory Manual

A Laboratory Guide to Human Physiology, Concepts and Clinical Applications, by Stuart Fox, is a stand-alone lab manual that provides support for the concepts and applications discussed in this text. Many figures in the laboratory guide are black-and-white versions of those in this text, which helps to reinforce the correlation between lecture and laboratory. Fundamental physiology concepts discussed in the text are also covered in the manual within the context of laboratory exercises. In the laboratory setting, some of the practical applications of these concepts can be explored as analytical and quantitative skills are developed. Further, the review activities at the end of each exercise serve as useful tools for studying the concepts for both lab and lecture.

The following is a guide to the correlation between topics covered in this textbook and in the laboratory exercises:

Chapters 1 and 2 of Fundamentals of Human Physiology

Laboratory Guide Exercises in Chemical and Cellular Concepts.

Exercise 1.1 Use of a microscope; estimation of size of microscopic objects; use of the metric system; unit conversion by dimensional analysis; review of mitosis and meiosis.

Exercise 1.2 The characteristics of the primary tissues and their microscopic appearance.

Exercise 1.3 The nature of homeostasis and the characteristics of negative feedback loops, as demonstrated in a water bath and in resting pulse rates. Calculations of ranges, averages, and set points are performed.

Exercise 2.1 The nature of carbohydrates, proteins, and lipids is discussed as plasma glucose, protein, and cholesterol concentration are measured. Calculations based on ratios are required.

Exercise 2.2 The differences between amino acids are discussed in the context of thin-layer chromatography.

Exercise 2.3 The different classes of the plasma proteins are discussed in the context of plasma electrophoresis.

Exercise 2.4 Characteristics of enzymes are discussed as the activities of selected enzymes are measured and calculated.

Exercise 2.5 Genetic information and its translation into enzyme proteins is discussed in the context of inborn errors in metabolism, some of which are tested in students' urine.

Chapter 3 of Fundamentals of Human Physiology

Laboratory Guide Exercise in Membrane Transport.

Exercise 2.6 Diffusion, osmosis, dialysis, and tonicity are discussed. Students observe osmosis and the effects of different saline solutions on red blood cells, and learn to calculate osmolarity.

Chapters 4, 5, and 6 of Fundamentals of Human Physiology

Laboratory Guide Exercises in the Nervous System.

Exercise 3.1 The membrane potential and action potential are discussed in the context of action potential recording from a frog sciatic nerve.

Exercise 3.2 The release of neurotransmitters at synapses, and synaptic potentials (EPSPs and IPSPs) are discussed in the performance of electroencephalograph recordings.

Exercise 3.3 Spinal nerves, sensory and somatic motor neurons, spinal reflexes, and spinal cord tracts are discussed as students perform the knee-jerk and other reflex tests.

Chapter 7 of Fundamentals of Human Physiology

Laboratory Guide Exercises in Sensory Physiology.

Exercise 3.4 Student exercises review cutaneous sensations and referred pain, while the neural pathways of sensation and sensory representation in the postcentral gyrus are discussed.

Exercise 3.5 The structure and function of the eye is described as students perform refractive tests, use the ophthalmoscope, test for the pupillary reflex and blind spot, and perform other exercises examining vision.

Exercise 3.6 The structure and function of the ear is discussed as students perform laboratory tests for conductive deafness.

Exercise 3.7 The vestibular apparatus is described while the semicircular canals are stimulated and students observe vestibular nystagmus.

Exercise 3.8 The physiology of taste is explained as students try to map their taste sensations on their tongues.

Chapter 8 of Fundamentals of Human Physiology

Laboratory Guide Exercises in Endocrine Physiology.

Exercise 4.1 The structure and function of different endocrine glands is described as students observe these glands under a microscope.

Exercise 4.2 The categories and chemical structures of steroid hormones are explored and students perform thin-layer chromatography of the steroid hormones.

Exercise 4.3 The physiology of the islets of Langerhans, the action of insulin, and diabetes mellitus are described in this exercise on insulin shock in a small fish.

Chapter 9 of Fundamentals of Human Physiology

Laboratory Guide Exercises in Muscle Physiology.

Exercise 5.1 The neural control of skeletal muscle contraction, including excitation-contraction coupling, is discussed in a laboratory exercise involving stimulation of a frog nerve and muscle.

Exercise 5.2 Twitch, summation, tetanus and fatigue are described and demonstrated in an exercise involving either frog muscles or students' forearms.

Exercise 5.3 The concepts of motor units, antagonistic muscles, and concentric/eccentric contractions are discussed as students record electromyograms (EMGs).

Chapters 10 and 11 of Fundamentals of Human Physiology

Laboratory Guide Exercises in Blood and Circulation.

Exercise 6.1 Students perform a red blood cell count, measure hemoglobin concentration and hematocrit, and perform calculations involving these measurements while learning about the role of hemoglobin in oxygen transport.

Exercise 6.2 Students perform white blood cell counts and differential counts while learning about the different leukocytes and their roles in immunity.

Exercise 6.3 The ABO and Rh antigen systems are discussed as students perform blood typing.

Exercise 6.4 The process of blood clotting is discussed and clotting tests are performed.

Exercise 7.1 The physiology of the heart and its regulation is demonstrated and discussed as the actions of epinephrine, Ca^{2+}, K^+, and various drugs are applied to a frog heart.

Exercise 7.2 Students record each other's electrocardiogram (ECG) as the electrical properties of the heart is discussed together with important arrhythmias.

Exercise 7.3 The effects of exercise on the ECG is determined in the context of the neural regulation of the heart.

Exercise 7.4 The pattern of electrical conduction in the heart is presented as students determine the mean electrical axis of the heart.

Exercise 7.5 The cardiac cycle is followed by listening to the heart sounds, and the correlation between the electrical events in the heart and the action of its valves is described.

Exercise 7.6 The determinants of arterial blood pressure and how blood pressure is regulated are discussed as students learn the origin of the Korotkoff sounds and how to measure blood pressure.

Exercise 7.7 Aerobic capacity and lactate threshold are discussed as students take a measurement of their cardiovascular fitness.

Chapter 12 of Fundamentals of Human Physiology

Laboratory Guide Exercises in Respiratory Physiology.

Exercise 8.1 Ventilation and the mechanics of breathing are discussed as students measure their pulmonary function and learn about obstructive and restrictive pulmonary diseases.

Exercise 8.2 The neural regulation of breathing and the chemoreceptor reflex are explored as students determine the effects of exercise on the respiratory system.

Exercise 8.3 Oxygen transport by the blood is discussed in the context of measurements of oxyhemoglobin saturation.

Exercise 8.4 The regulation of acid-base balance in the blood by breathing is explored as students measure pH and demonstrate increased carbon dioxide production during exercise.

Chapter 13 of Fundamentals of Human Physiology

Laboratory Guide Exercises in the Urinary System.

Exercise 9.1 Regulation of kidney function by ADH and the renin-angiotensin-aldosterone system is discussed, and milliequivalents are calculated as students take measurements of their urine before and after water loading.

Exercise 9.2 The concept of renal plasma clearance is explained and students obtain a measure of their renal plasma clearance of urea.

Exercise 9.3 Students learn the nature of various constituents of the urine as they perform a clinical urinalysis of their own urine.

Chapter 14 of Fundamentals of Human Physiology

Laboratory Guide Exercises in the Digestive System.

Exercise 10.1 The structure and function of different regions of the GI tract, as well as the liver and pancreas, is explained and explored with the use of a microscope.

Exercise 10.2 Students test the digestive action of salivary amylase, pepsin, and pancreatic lipase, as well as the action of bile salts, as the actions of these digestive enzymes are described.

Exercise 10.3 The nature of the BMR and methods for its calculation, body composition analysis and calculation of the body mass index, and basic concepts of nutrition are described as students maintain a 3-day dietary record and analyze their data.

Chapter 15 of Fundamentals of Human Physiology

Laboratory Guide Exercises in the Reproductive System.

Exercise 11.1 The hormonal regulation of the menstrual and estrus cycles is described as students perform and analyze a vaginal smear of a rat.

Exercise 11.2 The secretion and action of hCG are described in the context of performing pregnancy tests of students' urine.

Exercise 11.3 A discussion of dominance-recessive relationships and patterns of heredity provide the framework for students' tests for sickle-cell anemia, PTC tasting, and color blindness.

Glossary

Keys to Pronunciation

Most of the words in this glossary are followed by a phonetic spelling that serves as a guide to pronunciation. The phonetic spellings reflect standard scientific usage and can be interpreted easily following a few basic rules.

1. Any unmarked vowel that ends a syllable or that stands alone as a syllable has the long sound. For example, *ba, ma,* and *na* rhyme with *fay; be, de,* and *we* rhyme with *fee; bi, di,* and *pi* rhyme with *sigh; bo, do,* and *mo* rhyme with *go.* Any unmarked vowel that is followed by a consonant has the short sound (for example, the vowel sounds in *hat, met, pit, not,* and *but*).
2. If a long vowel appears in the middle of a syllable (followed by a consonant), it is marked with a macron (¯). Similarly, if a vowel stands alone or ends a syllable but should have short sound, it is marked with a breve (˘).
3. Syllables that are emphasized are indicated by stress marks. A single stress mark (′) indicates the primary emphasis; a secondary emphasis is indicated by a double stress mark (″).

A

ABO system The most common system of classification for red blood cell antigens. On the basis of antigens on the red blood cell surface, individuals can be type A, type B, type AB, or type O.

absorption *(ab-sorp′shun)* The transport of molecules across epithelial membranes into the body fluids.

accommodation *(ă-kom″ŏ-da′shun)* Adjustment; specifically, the process whereby the focal length of the eye is changed by automatic adjustment of the curvature of the lens to bring images of objects from various distances into focus on the retina.

acetylcholine *(ă-sēt″l-ko′lēn)* (**ACh**) An acetic acid ester of choline—a substance that functions as a neurotransmitter chemical in somatic motor nerve and parasympathetic nerve fibers.

acetylcholinesterase *(ă-sēt″l-ko″lĭ-nes′tĕ-rās)* (AChE) An enzyme in the membrane of postsynaptic cells that catalyzes the conversion of ACh into choline and acetic acid. This enzymatic reaction inactivates the neurotransmitter.

acid A molecule that donates free protons (H^+) to a solution.

acidic solution A solution with a pH of less than 7.000.

acromegaly *(ak″ro-meg′ă-le)* A condition caused by hypersecretion of growth hormone from the pituitary after maturity and characterized by enlargement of the extremities, such as the nose, jaws, fingers, and toes.

ACTH Adrenocorticotropic *(ă-dre″no-kor″-ă-ko-trop′ik)* **hormone** A hormone secreted by the anterior pituitary that stimulates the adrenal cortex.

actin *(ak′tin)* A structural protein of muscle that, along with myosin, is responsible for muscle contraction.

action potential An all-or-none electrical event in an axon or muscle fiber in which the polarity of the membrane potential is rapidly reversed and reestablished.

active immunity The ability to mount a specific immune response against antigens. Active immunity involves the development of lymphocyte clones that are able to recognize specific antigens.

active transport The movement of molecules or ions across the cell membranes of epithelial cells by membrane carriers. An expenditure of cellular energy (ATP) is required.

adaptive immunity Also called specific immunity, this refers to the immune response to specific antigens. Adaptive immunity is provided by the functions of B and T lymphocytes.

adaptive thermogenesis The production of body heat required to help maintain homeostasis of body temperature and to digest and absorb food.

adenohypophysis *(ad″n-o-hi-pof′ĭ-sis)* The anterior, glandular lobe of the pituitary gland that secretes FSH (follicle-stimulating hormone), LH (luteinizing hormone), ACTH (adrenocorticotropic hormone), TSH (thyroid-stimulating hormone), GH (growth hormone), and prolactin. Secretions of the anterior pituitary are controlled by hormones secreted by the hypothalamus.

ADH Antidiuretic *(an″te-di″yŭ-ret′ik)* **hormone** Also known as *vasopressin.* A hormone produced by the hypothalamus and released from the posterior pituitary. It acts on the kidneys to promote water reabsorption, thus decreasing the urine volume.

adrenal cortex *(ă-dre′nal kor′teks)* The outer part of the adrenal gland. Derived from embryonic mesoderm, the adrenal cortex secretes corticosteroid hormones, including aldosterone and hydrocortisone.

adrenal medulla *(mĕdul′ă)* The inner part of the adrenal gland. Derived from embryonic postganglionic sympathetic neurons, the adrenal medulla secretes catecholamine hormones—epinephrine and (to a lesser degree) norepinephrine.

adrenergic *(ad″rĕ-ner′jik)* Denoting the actions of epinephrine, norepinephrine, or other molecules with similar activity (as in *adrenergic receptor* and *adrenergic stimulation*).

adrenergic receptors Receptor proteins within the plasma membranes of the cells of smooth muscles, cardiac muscles, or glands that respond to epinephrine and norepinephrine. There are different alpha (α) and beta (β) adrenergic receptors.

adult stem cells Stem cells found in some organs of adults that serve to replenish the more specialized cells of the organs. Adult stem cells are described as multipotent, because they form a variety of specialized cell types, but not normally all of the types that pluripotent embryonic stem cells are capable of forming.

aerobic *(ă-ro′bik)* **capacity** The ability of an organ to utilize oxygen and respire aerobically to meet its energy needs.

aerobic respiration The metabolic pathway that requires oxygen as a final electron acceptor, and that breaks down organic molecules into carbon dioxide and water as energy is provided for the production of ATP.

afferent *(af′er-ent)* Conveying or transmitting inward, toward a center. Afferent neurons, for example, conduct impulses toward the central nervous system; afferent arterioles carry blood toward the glomerulus.

agglutinate (*ă-gloot′n-āt*) A clumping of cells (usually erythrocytes) as a result of specific chemical interaction between surface antigens and antibodies.

albumin (*al-byoo′min*) A water-soluble protein produced in the liver; the major component of the plasma proteins.

aldosterone (*al-dos′ter-ōn*) The principal corticosteroid hormone involved in the regulation of electrolyte balance (mineralocorticoid).

alkaline solution A solution with a pH greater than 7.000.

alkalosis (*al″kă-lo′sis*) An abnormally high alkalinity of the blood and body fluids (blood pH > 7.45).

allergy (*al′er-je*) A state of hypersensitivity caused by exposure to allergens. It results in the liberation of histamine and other molecules with histamine-like effects.

all-or-none law The statement that a given response will be produced to its maximum extent in response to any stimulus equal to or greater than a threshold value. Action potentials obey an all-or-none law.

alpha motoneuron (*al′fă mo″tŏ-noor′on*) The type of somatic motor neuron that stimulates extrafusal skeletal muscle fibers.

alveoli (*al-ve′ŏ-li*); sing., *alveolus* Small, saclike dilations (as in *lung alveoli*).

amylase (*am′il-ās*) A digestive enzyme that hydrolyzes the bonds between glucose subunits in starch and glycogen. Salivary amylase is found in saliva and pancreatic amylase is found in pancreatic juice.

an- (Gk.) Without, not.

anabolic steroids (*an″ă-bol′ik ster′oidz*) Steroids with androgen-like stimulatory effects on protein synthesis.

anabolism (*ă-nab′ŏ-liz″em*) Chemical reactions within cells that result in the production of larger molecules from smaller ones; specifically, the synthesis of protein, glycogen, and fat.

anaerobic respiration (*an-ă-ro′bik res″pĭ-ra′shun*) A form of cell respiration involving the conversion of glucose to lactic acid in which energy is obtained without the use of molecular oxygen.

anaerobic threshold The maximum rate of oxygen consumption that can be attained before a significant amount of lactic acid is produced by the exercising skeletal muscles through anaerobic respiration. This generally occurs when about 60% of the person's total maximal oxygen uptake has been reached.

androgen (*an′drŏ-jen*) A steroid hormone that controls the development and maintenance of masculine characteristics; primarily testosterone secreted by the testes, although weaker androgens are secreted by the adrenal cortex.

anemia (*ă-ne′me-ă*) An abnormal reduction in the red blood cell count, hemoglobin concentration, or hematocrit, or any combination of these measurements. This condition is associated with a decreased ability of the blood to carry oxygen.

angina pectoris (*an-ji′nă pek′tŏ-ris*) A thoracic pain, often referred to the left pectoral and arm area, caused by myocardial ischemia.

angiotensin-converting enzyme (ACE) The enzyme that converts angiotensin I into the much more biologically active molecule, angiotensin II.

angiotensin II (*an″je-o-ten′sin*) An 8-amino-acid polypeptide formed from angiotensin I (a 10-amino-acid precursor), which in turn is formed from the cleavage of a protein (angiotensinogen) by the action of renin, an enzyme secreted by the kidneys. Angiotensin II is a powerful vasoconstrictor and a stimulator of aldosterone secretion from the adrenal cortex.

anion (*an′i-on*) An ion that is negatively charged, such as chloride, bicarbonate, or phosphate.

anterior pituitary (*pĭ-too′ĭ-ter-e*) *See* adenohypophysis.

antibodies (*an′tĭ-bod″ēz*) Immunoglobulin proteins secreted by B lymphocytes that have been transformed into plasma cells. Antibodies are responsible for humoral immunity. Their synthesis is induced by specific antigens, and they combine with these specific antigens but not with unrelated antigens.

anticoagulant (*an″te-ko-ag′yŭ-lant*) A substance that inhibits blood clotting.

anticodon (*an″te-ko′don*) A base triplet provided by 3 nucleotides within a loop of transfer RNA that is complementary in its base-pairing properties to a triplet (the codon in mRNA). The matching of codon to anticodon provides the mechanism for translation of the genetic code into a specific sequence of amino acids.

antigen (*an′tĭ-jen*) A molecule able to induce the production of antibodies and to react in a specific manner with antibodies.

apnea (*ap′ne-ă*) The temporary cessation of breathing.

apoptosis (*ap″ŏ-to′sis*) Cellular death in which the cells show characteristic histological changes. It occurs as part of programmed cell death and other events in which cell death is a physiological response.

aquaporins (*ă-kwă-por′inz*) The protein channels in a cell (plasma) membrane that permit osmosis to occur across the membrane. In certain tissues, particularly the collecting ducts of the kidney, aquaporins are inserted into the cell membrane in response to stimulation by antidiuretic hormone.

aqueous humor (*a′kwe-us*) A fluid produced by the ciliary body that fills the anterior and posterior chambers of the eye.

artery (*ar′tĕ-re*) A vessel that carries blood away from the heart.

astigmatism (*ă-stig′mă-tiz″em*) Unequal curvature of the refractive surfaces of the eye (cornea and/or lens), so that light that enters the eye along certain meridians does not focus on the retina.

atherosclerosis (*ath″ĕ-ro-sklĕ-ro′sis*) A common type of arteriosclerosis in which raised areas, or plaques, within the tunica interna of medium and large arteries are formed from smooth muscle cells, cholesterol, and other lipids. These plaques occlude the arteries and serve as sites for the formation of thrombi.

atomic number A whole number representing the number of positively charged protons in the nucleus of an atom.

ATP Adenosine triphosphate (*ăden′ŏ-sēn tri-fos′fāt*). The universal energy carrier of the cell.

atrioventricular node (*a″tre-o-ven-trik′yŭ-lar nōd*) A specialized mass of conducting tissue located in the right atrium near the junction of the interventricular septum. It transmits the impulse into the bundle of His; also called the *AV node.*

atrioventricular valves One-way valves located between the atria and ventricles. The AV valve on the right side of the heart is the tricuspid, and the AV valve on the left side is the bicuspid, or mitral, valve.

atrophy (*at′rŏ fe*) A gradual wasting away, or decrease in mass and size of an organ; the opposite of hypertrophy.

autoantibody (*aw″to-an′tĭ-bod″e*) An antibody that is formed in response to, and that reacts with, molecules that are part of one's own body.

autonomic (*aw″tŏ-nom′ik*) **nervous system (ANS)** The part of the nervous system that involves control of smooth muscle, cardiac muscle, and glands. The autonomic nervous system is subdivided into the sympathetic and parasympathetic divisions.

autoregulation (*aw″to-reg′yŭ-la′shun*) The ability of an organ to intrinsically modify the degree of constriction or dilation of its small arteries and arterioles, and thus to regulate the rate of its own blood flow. Autoregulation may occur through myogenic or metabolic mechanisms.

autosomal chromosomes (*aw″to-so′mal kro′mŏ-sōmz*) The paired chromosomes; those other than the sex chromosomes.

axon (*ak′son*) The process of a nerve cell that conducts impulses away from the cell body.

B

baroreceptors (*bar″o-re-sep′torz*) Receptors for arterial blood pressure located in the aortic arch and the carotid sinuses.

Barr body A microscopic structure in the cell nucleus produced from an inactive X chromosome in females

basal ganglia (*ba′sal gang′gle-ă*) Gray matter, or nuclei, within the cerebral hemispheres, forming the corpus striatum, amygdaloid nucleus, and claustrum.

basal metabolic (*ba′sal met″ă-bol′ik*) **rate (BMR)** The rate of metabolism (expressed as oxygen consumption or heat production) under resting or basal conditions 8 to 12 hours after eating.

base A molecule or ion that removes free protons (H^+) from a solution.

basophil (*ba′sŏ-fil*) The rarest type of leukocyte; a granular leukocyte with an affinity for blue stain in the standard staining procedure.

bile (*bīl*) Fluid produced by the liver and stored in the gallbladder that contains bile salts, bile pigments, cholesterol, and other molecules. The bile is secreted into the small intestine.

bile salts Salts of derivatives of cholesterol in bile that are polar on one end and nonpolar on the other end of the molecule. Bile salts have detergent or surfactant effects and act to emulsify fat in the lumen of the small intestine.

bilirubin (*bil″ĭ-roo′bin*) Bile pigment derived from the breakdown of the heme portion of hemoglobin.

blastocyst (*blas′tŏ-sist*) The stage of early embryonic development that consists of an inner cell mass, which will become the embryo, and surrounding trophoblast cells, which will form part of the placenta. This is the form of the embryo that implants in the endometrium of the uterus beginning at about the fifth day following fertilization.

bleaching reaction The photodissociation of rhodopsin, which occurs when light strikes the retina. This leads to electrochemical changes in associated neurons that convey the sense of vision.

blood-brain barrier The structures and cells that selectively prevent particular molecules in the plasma from entering the central nervous system.

blood plasma The liquid portion of the blood, as distinct from the formed elements of the blood (red and white blood cells and blood platelets).

B lymphocytes (*lim′fŏ-sīts*) Lymphocytes that can be transformed by antigens into plasma cells that secrete antibodies (and are thus responsible for humoral immunity). The *B* stands for *bursa equivalent,* which is believed to be the bone marrow.

Bohr effect The effect of blood pH on the dissociation of oxyhemoglobin. Dissociation is promoted by a decrease in the pH.

Boyle's law The statement that the pressure of a given quantity of a gas is inversely proportional to its volume.

bradycardia (*brad″ĭ-kar′de-ă*) A slow cardiac rate; less than 60 beats per minute.

Broca's area The area of the cerebral cortex, usually located in the left hemisphere, that controls the motor ability to speak. Damage to Broca's area diminishes the ability to speak but not language comprehension.

bronchiole (*brong′ke-ōl*) The smallest of the air passages in the lungs, containing smooth muscle and cuboidal epithelial cells.

brush border enzymes Digestive enzymes that are located in the cell membrane of the microvilli of intestinal epithelial cells.

buffer A molecule that serves to prevent large changes in pH by either combining with H^+ or by releasing H^+ into solution.

bundle of His (*hiss*) A band of rapidly conducting cardiac fibers originating in the AV node and extending down the atrioventricular septum to the apex of the heart. This tissue conducts action potentials from the atria into the ventricles.

C

calorie (*kal′ŏ-re*) A unit of heat equal to the amount of heat needed to raise the temperature of 1 gram of water by 1° C.

cAMP cyclic adenosine monophosphate (*ă-den′ŏ-sēn mon″o-fos′fāt*) A second messenger in the action of many hormones, including catecholamine, polypeptide, and glycoprotein hormones. It serves to mediate the effects of these hormones on their target cells.

cancer A tumor characterized by abnormally rapid cell division and the loss of specialized tissue characteristics. This term usually refers to malignant tumors.

capillary (*kap′ĭlar″e*) The smallest vessel in the vascular system. Capillary walls are only 1 cell thick, and all exchanges of molecules between the blood and tissue fluid occur across the capillary wall.

cardiac control center The nucleus in the medulla oblongata that controls the cardiac rate. This control is exerted through the autonomic nervous system.

cardiac cycle The repeating pattern of the heart's contractions and relaxations, including atrial systole and diastole and ventricular systole and diastole.

cardiac (*kar′de-ak*) **muscle** Muscle of the heart, consisting of striated muscle cells. These cells are interconnected, forming a mass called the myocardium.

cardiac output The volume of blood pumped by either the right or the left ventricle each minute.

carrier-mediated transport The transport of molecules or ions across a cell membrane by means of specific protein carriers. It includes both facilitated diffusion and active transport.

carrier protein A protein within the plasma membrane that promotes either the facilitated diffusion or active transport of molecules across the plasma membrane; it has the properties of specificity and saturation.

catabolism (*kă-tab′ŏ-liz-em*) Chemical reactions in a cell whereby larger, more complex molecules are converted into smaller molecules.

catecholamines (*kat″ĕ-kol′ă-mēnz*) A group of molecules that includes epinephrine, norepinephrine, L-dopa, and related molecules. The effects of catecholamines are similar to those produced by activation of the sympathetic nervous system.

cations (*kat′i-ions*) Positively charged ions, such as sodium, potassium, calcium, and magnesium.

cell-mediated immunity Immunological defense provided by T cell lymphocytes that come into close proximity with their victim cells (as opposed to humoral immunity provided by the secretion of antibodies by plasma cells).

cellular respiration (*sel′yŭ-lar res″pĭ-ra′shun*) The energy-releasing metabolic pathways in a cell that oxidize organic molecules such as glucose and fatty acids.

cerebellum (*ser″ĕ-bel′um*) A part of the metencephalon of the brain that serves as a major center of control in the extrapyramidal motor system.

cerebral lateralization (*ser′ĕ-bral lat″er-al-ĭ-za′shun*) The specialization of function of each cerebral hemisphere. Language ability, for example, is lateralized to the left hemisphere in most people.

channel protein A protein that spans the thickness of the plasma membrane and provides an aqueous channel for the passage of specific ions and water molecules.

chemically regulated channel (ligand-regulated channel) A protein channel through the plasma membrane that is generally closed until a neurotransmitter binds to its receptor protein. In some cases, the chemically regulated channel is also the receptor protein; in other cases, the chemically regulated channel and the receptor protein are separate molecules.

chemoreceptors In the regulation of breathing, there are peripheral chemoreceptors—the aortic and carotid bodies—and central chemoreceptors, in the medulla oblongata. The peripheral chemoreceptors directly respond to the pH of blood, and the central chemoreceptors respond to the pH of brain interstitial fluid.

chloride (*klor′īd*) **shift** The diffusion of Cl^- into red blood cells as HCO_3^- diffuses out of the cells. This occurs in tissue capillaries as a result of the production of carbonic acid from carbon dioxide.

cholecystokinin (*ko″lĭ-sis″to-ki′nin*) **(CCK)** A hormone secreted by the duodenum that acts to stimulate contraction of the gallbladder and to promote the secretion of pancreatic juice.

cholinergic (*ko″lĭ-ner′jik*) Denoting nerve endings that, when stimulated, release acetylcholine as a neurotransmitter, such as those of the parasympathetic system.

chromatin (*kro′mă-tin*) Threadlike structures in the cell nucleus consisting primarily of DNA and protein. They represent the extended form of chromosomes during interphase.

chromosome (*kro′mŏ-sōm*) A structure in the cell nucleus, containing DNA and associated proteins, as well as RNA, that is made according to the genetic instructions in the DNA. Chromosomes are in a compact form during cell division; hence, they become visible as discrete structures in the light microscope at this time.

chylomicron (*ki″lo-mi′kron*) A particle of lipids and protein secreted by the intestinal epithelial cells into the lymph and transported by the lymphatic system to the blood.

chyme (*kīm*) A mixture of partially digested food and digestive juices that passes from the pylorus of the stomach into the duodenum.

cilia (*sil′e-ă*); sing., *cilium* Tiny hairlike processes extending from the cell surface that beat in a coordinated fashion.

circadian (*ser″kă-de′an*) **rhythms** Physiological changes that repeat at approximately 24-hour periods. They are often synchronized to changes in the external environment, such as the day-night cycles.

cirrhosis (*sĭ-ro′sis*) Liver disease characterized by the loss of normal microscopic structure, which is replaced by fibrosis and nodular regeneration.

clonal (*klōn′al*) **selection theory** The theory in immunology that active immunity is produced by the development of clones of lymphocytes able to respond to a particular antigen.

clone (*klōn*) **1.** A group of cells derived from a single parent cell by mitotic cell division; since reproduction is asexual, the descendants of the parent cell are genetically identical. **2.** A term used to refer to cells as separate individuals (as in white blood cells) rather than as part of a growing organ.

CNS Central nervous system That part of the nervous system consisting of the brain and spinal cord.

cochlea (*kok′le-ă*) The organ of hearing in the inner ear where nerve impulses are generated in response to sound waves.

codon (*ko′don*) The sequence of 3 nucleotide bases in mRNA that specifies a given amino acid and determines the position of that amino acid in a polypeptide chain through complementary base pairing with an anticodon in transfer RNA.

colloid osmotic (*kol′oid oz-mot′ik*) **pressure** Osmotic pressure exerted by plasma proteins that are present as a colloidal suspension; also called *oncotic pressure.*

complement fixation The insertion of particular activated complement proteins from the blood plasma into the plasma membrane of cells targeted by the immune system for destruction.

compliance (*kom-pli′ans*) **1.** A measure of the ease with which a structure such as the lung expands under pressure. **2.** A measure of the change in volume as a function of pressure changes.

concentric contraction An isotonic muscle contraction in which the muscle shortens when it contracts. Distinguished from an eccentric muscle contraction, in which a load on the muscle causes it to lengthen despite its contraction.

conducting zone The structures and airways that transmit inspired air into the respiratory zone of the lungs, where gas exchange occurs. The conducting zone includes such structures as the trachea, bronchi, and larger bronchioles.

cone Photoreceptor in the retina of the eye that provides color vision and high visual acuity.

connective tissue One of the four primary tissues, characterized by an abundance of extracellular material.

cornea (*kor′ne-ă*) The transparent structure forming the anterior part of the connective tissue covering of the eye.

corpora quadrigemina (*kor′por-ă kwad″rĭ-jem′ĭ-na*) A region of the mesencephalon consisting of the superior and inferior colliculi. The superior colliculi are centers for the control of visual reflexes; the inferior colliculi are centers for the control of auditory reflexes.

corpus callosum (*kor′pus kă-lo′sum*) A large transverse tract of nerve fibers connecting the cerebral hemispheres.

corpus luteum The yellow body within an ovary that forms from the ruptured follicle after it has released its egg cell at ovulation. The corpus luteum secretes estradiol and progesterone.

corticosteroid (*kor″tĭ-ko-ster′oid*) Any of a class of steroid hormones of the adrenal cortex, consisting of glucocorticoids (such as hydrocortisone) and mineralocorticoids (such as aldosterone).

cotransport Also called *coupled transport* or *secondary active transport.* Carrier-mediated transport in which a single carrier transports an ion (e.g., Na^+) down its concentration gradient while transporting a specific molecule (e.g., glucose) against its concentration gradient. The hydrolysis of ATP is indirectly required for cotransport because it is needed to maintain the steep concentration gradient of the ion.

countercurrent multiplier system The interaction that occurs between the descending limb and the ascending limb of the loop of Henle in the kidney. This interaction results in the multiplication of the solute concentration in the interstitial fluid of the renal medulla.

cretinism (*krēt′n-iz″em*) A condition caused by insufficient thyroid secretion during prenatal development or the years of early childhood. It results in stunted growth and inadequate mental development.

curare (*koo-ră-re*) A chemical derived from plant sources that causes flaccid paralysis by blocking ACh receptor proteins in muscle cell membranes.

Cushing's syndrome Symptoms caused by hypersecretion of adrenal steroid hormones as a result of tumors of the adrenal cortex or ACTH-secreting tumors of the anterior pituitary.

cytokine (*si′to-kīn*) An autocrine or paracrine regulator secreted by various tissues.

cytoplasm (*si′tŏ-plaz″em*) The semifluid part of the cell between the cell membrane and the nucleus, exclusive of membrane-bound organelles. It contains many enzymes and structural proteins.

cytoskeleton (*si″to-skel′ĕ-ton*) A latticework of structural proteins in the cytoplasm arranged in the form of microfilaments and microtubules.

D

Dalton's law The statement that the total pressure of a gas mixture is equal to the sum that each individual gas in the mixture would exert independently. The part contributed by each gas is known as the partial pressure of the gas.

dark adaptation The ability of the eyes to increase their sensitivity to low light levels over a period of time. Part of this adaptation involves increased amounts of visual pigment in the photoreceptors.

declarative memory The memory of factual information. This can be contrasted with nondeclarative memory, which is the memory of perceptual and motor skills.

delayed hypersensitivity An allergic response in which the onset of symptoms may not occur until 2 or 3 days after exposure to an antigen. Produced by T cells, it is a type of cell-mediated immunity.

dendrite (*den′drīt*) A relatively short, highly branched neural process that carries electrical activity to the cell body.

dendritic (*den-drit′ik*) **cells** The most potent antigen-presenting cells for the activation of helper T lymphocytes. The dendritic cells originate in the bone marrow and migrate through the blood and lymph to lymphoid organs and to nonlymphoid organs such as the lungs and skin.

deoxyhemoglobin (*de-ok″se-he″mō-glo′bin*) The form of hemoglobin in which the heme groups are in the normal reduced form but are not bound to a gas. Deoxyhemoglobin is produced when oxyhemoglobin releases oxygen.

depolarization (*de-po″lar-ĭ-za′shun*) The loss of membrane polarity in which the inside of the cell membrane becomes less negative in comparison to the outside of the membrane. The term is also used to indicate the reversal of membrane polarity that occurs during the production of action potentials in nerve and muscle cells.

diabetes insipidus (*di″ă-be′tēz in-sip′ĭ-dus*) A condition in which inadequate amounts of antidiuretic hormone (ADH) are secreted by the posterior pituitary. It results in inadequate reabsorption of water by the kidney tubules, and thus in the excretion of a large volume of dilute urine.

diabetes mellitus (*mĕ-li′tus*) The appearance of glucose in the urine due to the presence of high plasma glucose concentrations, even in the fasting state. This disease is caused by either a lack of sufficient insulin secretion or by inadequate responsiveness of the target tissues to the effects of insulin.

dialysis (*di-al′ĭ-sis*) A method of removing unwanted elements from the blood by selective diffusion through a porous membrane.

diastole (*di-as′tŏ-le*) The phase of relaxation in which the heart fills with blood. Unless accompanied by the modifier *atrial,* diastole refers to the resting phase of the ventricles.

diastolic (*di″ă-stol′ik*) **blood pressure** The minimum pressure in the arteries that is produced during the phase of diastole of the heart. It is indicated by the last sound of Korotkoff when taking a blood pressure measurement.

differentiation The process by which unspecialized cells become increasingly more specialized in terms of both structure and function.

diffusion (*dĭ-fyoo′zhun*) The net movement of molecules or ions from regions of higher to regions of lower concentration.

digestion The process of converting food into molecules that can be absorbed through the intestine into the blood.

dihydrotestosterone (DHT) The hormone, formed by the enzyme 5 α-reductase from testosterone within some target organs, which produces the androgenic effect of testosterone within those organs.

1,25-dihydroxyvitamin (*di″hi-drok″se-vi′tă-min*) **D_3** The active form of vitamin D produced within the body by hydroxylation reactions in the liver and kidneys of vitamin D formed by the skin. This is a hormone that promotes the intestinal absorption of Ca^{2+}.

disaccharide (*di-sak′ă-rīd*) Any of a class of double sugars; carbohydrates that yield 2 simple sugars, or monosaccharides, upon hydrolysis.

diuretic (*di″yŭ-ret′ik*) A substance that increases the rate of urine production, thereby lowering the blood volume.

DNA Deoxyribonucleic (*de-ok″se-ri″bo-noo-kle′ik*) acid A nucleic acid composed of nucleotide bases and deoxyribose sugar that contains the genetic code.

dopamine (*do′pă-mēn*) A type of neurotransmitter in the central nervous system; it is also the precursor of norepinephrine, another neurotransmitter molecule.

dyspnea (*disp-ne′ă*) Subjective difficulty in breathing.

E

eccentric (*ek-sen′trik*) **contraction** A muscle contraction in which the muscle lengthens despite its contraction, due to a greater external stretching force applied to it. The contraction in this case can serve a shock absorbing function, for example, when the quadriceps muscles of the leg contract eccentrically upon landing when a person jumps from a height.

ECG Electrocardiogram (*ĕ-lek″tro-kar′de-ŏ-gram*) (also abbreviated EKG) A recording of electrical currents produced by the heart.

edema (*ĕ-de′mă*) Swelling resulting from an increase in tissue fluid.

EEG **electroencephalogram** (*ĕ-lek″tro-en-sef′ă-lŏ-gram*) A recording of the electrical activity of the brain from electrodes placed on the scalp.

effector (*ĕ-fek′tor*) **organs** A collective term for muscles and glands that are activated by motor neurons.

efferent (*ef′er-ent*) Conveying or transporting something away from a central location. Efferent nerve fibers conduct impulses away from the central nervous system, for example, and efferent arterioles transport blood away from the glomerulus.

elasticity (*ĕ″las-tis′ĭ-te*) The tendency of a structure to recoil to its initial dimensions after being distended (stretched).

electrocardiogram (ECG or EKG) The recording of the cyclic electrical activity of the heart, consisting of the P wave, QRS complex, and T wave.

electron-transport system (ETS) The system of molecules in the cristae of mitochondria that accepts electrons from NADH and $FADH_2$ and ultimately passes them to atoms of oxygen.

elephantiasis (*el″ĕ-fan-ti′ă-sis*) A disease in which the larvae of a nematode worm block lymphatic drainage and produce edema. The lower areas of the body can become enormously swollen as a result.

embryonic stem cells Also called ES cells, these are the cells of the inner cell mass of a blastocyst. Embryonic stem cells are pluripotent, and so are potentially capable of differentiating into all tissue types except the trophoblast cells of a placenta.

emphysema (*em″fĭ-se′mă em″fĭ-ze′mă*) A lung disease in which alveoli are destroyed and the remaining alveoli become larger. It results in decreased vital capacity and increased airway resistance.

emulsification (*ĕ-mul″sĭ-fĭ-ka′shun*) The process of producing an emulsion or fine suspension. In the small intestine, fat globules are emulsified by the detergent action of bile.

end-diastolic (*di″ă-stol′ik*) **volume** The volume of blood in each ventricle at the end of diastole, immediately before the ventricles contract at systole.

endocrine (*en′dŏ-krin*) **glands** Glands that secrete hormones into the circulation rather than into a duct; also called *ductless glands.*

endocytosis (*en″do-si-to′sis*) The cellular uptake of particles that are too large to cross the cell membrane. This occurs by invagination of the cell membrane until a membrane-enclosed vesicle is pinched off within the cytoplasm.

endogenous (*en-doj′ĕ-nus*) Denoting a product or process arising from within the body (as opposed to exogenous products or influences, which arise from external sources).

endometrium (*en″do-me′tre-um*) The mucous membrane of the uterus, the thickness and structure of which vary with the phases of the menstrual cycle.

endoplasmic reticulum (*en-do-plaz′mik rĕ-tik′yŭ-lum*) An extensive system of membrane-enclosed cavities within the cytoplasm of the cell. Those with ribosomes on their surface are called rough endoplasmic reticulum and participate in protein synthesis.

enteric (*en-ter′ik*) A term referring to the intestine.

enterochromaffin (*en″ter-o-kro″maf′in*) **-like (ECL) cells** Cells of the gastric epithelium that secrete histamine. The ECL cells are stimulated by the hormone gastrin and by the vagus nerve; the histamine from ECL cells, in turn, stimulates gastric acid secretion from the parietal cells.

enterogastrone A generic name for a hormone secreted from the small intestine that inhibits the emptying of chyme from the stomach into the duodenum during the intestinal phase of gastric regulation.

enzyme (*en′zim*) A protein catalyst that increases the rate of specific chemical reactions.

epi- (Gk.) Upon, over, outer.

epidermis (*ep″ĭ-der′mis*) The stratified squamous epithelium of the skin, the outer layer of which is dead and filled with keratin.

epididymis (*ep″ĭ-did′ĭ-mis*); pl., *epididymides* A tubelike structure outside the testes. Sperm pass from the seminiferous tubules into the head of the epididymis and then pass from the tail of the epididymis to the ductus (vas) deferens. The sperm mature, becoming motile, as they pass through the epididymis.

epinephrine (*ep″ĭ-nef′rin*) A catecholamine hormone secreted by the adrenal medulla in response to sympathetic nerve stimulation. It acts together with norepinephrine released from sympathetic nerve endings to prepare the organism for "fight or flight"; also known as *adrenaline.*

EPSP excitatory postsynaptic (*pōst″sĭ-nap′tik*) **potential** A graded depolarization of a postsynaptic membrane in response to stimulation by a neurotransmitter chemical. EPSPs can be summated, but they can be transmitted only over short distances; they can stimulate the production of action potentials when a threshold level of depolarization is attained.

erythroblastosis fetalis (*ĕ-rith″ro-blas-to′sis fe-tal′is*) Hemolytic anemia in an Rh-positive newborn caused by maternal antibodies against the Rh factor that have crossed the placenta.

erythrocyte (*ĕ-rith′rŏ-sīt*) A red blood cell. Erythrocytes are the formed elements of blood that contain hemoglobin and transport oxygen.

erythropoietin (*ĕ-rith″ro-poi′ĕ-tin*) A hormone secreted by the kidneys that stimulates the bone marrow to produce red blood cells.

estradiol (*es″tră-di′ol*) **-17 β** The major estrogen (female sex steroid hormone) secreted by the ovaries.

excitation-contraction coupling The means by which electrical excitation of a muscle results in muscle contraction. This coupling is achieved by Ca^{2+}, which enters the muscle cell cytoplasm in response to electrical excitation and which stimulates the events culminating in contraction.

exocrine (*ek′sŏ-krin*) **gland** A gland that discharges its secretion through a duct to the outside of an epithelial membrane.

exocytosis (*ek″so-si-to′sis*) The process of cellular secretion in which the secretory products are contained within a membrane-enclosed vesicle. The vesicle fuses with the cell membrane so that the lumen of the vesicle is open to the extracellular environment.

extrapyramidal (*ek″stră-pĭ-ram′ĭ-dl*) **tracts** Neural pathways that are situated outside of, or that are "independent of," pyramidal tracts. The major extrapyramidal tract is the reticulospinal tract, which originates in the reticular formation of the brain stem and receives excitatory and inhibitory input from both the cerebrum and the cerebellum. The extrapyramidal tracts are thus influenced by activity in the brain involving many synapses, and they appear to be required for fine control of voluntary movements.

extravasation Also called diapedesis, this refers to the ability of white blood cells to leave blood vessels and enter the surrounding connective tissues.

F

facilitated (*fă-sil′ĭ-ta″tid*) **diffusion** The carrier-mediated transport of molecules through the cell membrane along the direction of their concentration gradients. It does not require the expenditure of metabolic energy.

fibrillation (*fib″rĭ-la′shun*) A condition of cardiac muscle characterized electrically by random and continuously changing patterns of electrical activity and resulting in the inability of the myocardium to contract as a unit and pump blood. It can be fatal if it occurs in the ventricles.

fibrin (*fi′brin*) The insoluble protein formed from fibrinogen by the enzymatic action of thrombin during the process of blood clot formation.

fovea centralis (*fo′ve-ă sen-tra′lis*) A tiny pit in the macula lutea of the retina that contains slim, elongated cones. It provides the highest visual acuity (clearest vision).

Frank–Starling Law of the Heart The statement describing the relationship between end-diastolic volume and stroke volume of the heart. A greater amount of blood in a ventricle prior to contraction results in greater stretch of the myocardium, and by this means produces a contraction of greater strength.

FSH Follicle-stimulating hormone One of the two gonadotropic hormones secreted by the anterior pituitary. In females, FSH stimulates the development of the ovarian follicles; in males, it stimulates the production of sperm in the seminiferous tubules.

G

GABA gamma-aminobutyric (*gam″ă-ă-me″no-byoo-tir′ik*) **acid** An amino acid believed to function as an inhibitory neurotransmitter in the central nervous system.

gamete (*gam′ēt*) Collective term for haploid germ cells: sperm and ova.

gamma motoneuron (*gam′ă mo″tŏ-noor′on*) The type of somatic motor neuron that stimulates intrafusal fibers within the muscle spindles.

ganglion (*gang′gle-on*) A grouping of nerve cell bodies located outside the brain and spinal cord.

gap junctions Specialized regions of fusion between the cell membranes of two adjacent cells that permit the diffusion of ions and small molecules from one cell to the next. These regions serve as electrical synapses in certain areas, such as in cardiac muscle.

gas exchange The diffusion of oxygen and carbon dioxide down their concentration gradients that occurs between pulmonary capillaries and alveoli, and between systemic capillaries and the surrounding tissue cells.

gastric juice The secretions of the gastric mucosa. Gastric juice contains water, hydrochloric acid, and pepsinogen as major components.

gastrin (*gas′trin*) A hormone secreted by the stomach that stimulates the gastric secretion of hydrochloric acid and pepsin.

general adaptation syndrome (GAS) The activation of the hypothalamus–anterior pituitary–adrenal cortex axis in response to nonspecific stressors. In humans, this results in an elevation in cortisol secretion.

genetic transcription The process by which RNA is produced with a sequence of nucleotide bases that is complementary to a region of DNA.

genome (*je′nom*) All of the genes of an individual or in a particular species.

ghrelin (*gre′lin*) A hormone that stimulates hunger. It is secreted by the stomach in increased amounts when the stomach is empty.

gigantism (*ji-gan′tiz″em*) Abnormal body growth due to the excessive secretion of growth hormone.

glomerular filtrate Fluid filtered through the glomerular capillaries into the glomerular (Bowman's) capsule of the kidney tubules.

glomerular (*glo-mer′yŭ-lar*) **filtration rate (GFR)** The volume of blood plasma filtered out of the glomeruli of both kidneys each minute. The GFR is measured by the renal plasma clearance of inulin.

glomeruli (*glo-mer′yŭ-li*) The tufts of capillaries in the kidneys that filter fluid into the kidney tubules.

glomerulonephritis (*glo-mer″yŭ-lo-nĕ-fri′tis*) Inflammation of the renal glomeruli; associated with fluid retention, edema, hypertension, and the appearance of protein in the urine.

glucagon (*gloo′că-gon*) A polypeptide hormone secreted by the alpha cells of the islets of Langerhans in the pancreas that acts to promote glycogenolysis and raise the blood glucose levels.

glucocorticoid (*gloo″ko-kor′tĭ-koid*) Any of a class of steroid hormones secreted by the adrenal cortex (corticosteroids) that affects the metabolism of glucose, protein, and fat. These hormones also have anti-inflammatory and immunosuppressive effects. The major glucocorticoid in humans is hydrocortisone (cortisol).

gluconeogenesis (*gloo″ko-ne″ŏ-jen′ĭ-sis*) The formation of glucose from noncarbohydrate molecules, such as amino acids and lactic acid.

GLUT An acronym for *glucose transporters.* GLUT proteins promote the facilitated diffusion of glucose into cells. One isoform of GLUT, designated GLUT4, is inserted into the cell membranes of muscle and adipose cells in response to insulin stimulation and exercise.

glutamate (*gloo′tă-māt*) The ionized form of glutamic acid, an amino acid that serves as the major excitatory neurotransmitter of the CNS. *Glutamate* and *glutamic acid* are terms that can be used interchangeably.

glycogen (*glī'kŏ-jen*) A polysaccharide of glucose—also called *animal starch*—produced primarily in the liver and skeletal muscles. Similar to plant starch in composition, glycogen contains more highly branched chains of glucose subunits than does plant starch.

glycogenesis (*glī"kŏ-jen'ĭ-sis*) The formation of glycogen from glucose.

glycogenolysis (*glī"ko-jĕ-nol'ĭ-sis*) The hydrolysis of glycogen to glucose-1-phosphate, which can be converted to glucose-6-phosphate. The glucose-6-phosphate then may be oxidized via glycolysis or (in the liver) converted to free glucose.

glycolysis (*glī"kol'ĭ-sis*) The metabolic pathway that converts glucose to pyruvic acid. The final products are two molecules of pyruvic acid and two molecules of reduced NAD, with a net gain of two ATP molecules. In anaerobic respiration, the reduced NAD is oxidized by the conversion of pyruvic acid to lactic acid. In aerobic respiration, pyruvic acid enters the Krebs cycle in mitochondria, and reduced NAD is ultimately oxidized by oxygen to yield water.

glycosuria (*glī"kŏ-soor'e-ă*) The excretion of an abnormal amount of glucose in the urine (urine normally contains only trace amounts of glucose).

Golgi (*gol'je*) **apparatus** A network of stacked, flattened membranous sacs within the cytoplasm of cells. Its major function is to concentrate and package proteins within vesicles that bud off from it.

Golgi tendon organ A tension receptor in the tendons of muscles that becomes activated by the pull exerted by a muscle on its tendons; also called a *neurotendinous receptor.*

gonad (*go'nad*) A collective term for testes and ovaries.

gonadotropic (*go"nad-ŏ-trop'ik*) **hormones** Hormones of the anterior pituitary that stimulate gonadal function—the formation of gametes and secretion of sex steroids. The two gonadotropins are FSH (follicle-stimulating hormone) and LH (luteinizing hormone), which are essentially the same in males and females.

G-protein An association of three membrane-associated protein subunits, designated alpha, beta, and gamma, that is regulated by guanosine nucleotides (GDP and GTP). The G-protein subunits dissociate in response to a membrane signal and, in turn, activate other proteins in the cell.

graafian (*graf'e-an*) **follicle** A mature ovarian follicle, containing a single fluid-filled cavity, with the ovum located toward one side of the follicle and perched on top of a hill of granulosa cells.

granular leukocytes (*loo'kŏ-sīts*) Leukocytes with granules in the cytoplasm. On the basis of the staining properties of the granules, these cells are of three types: neutrophils, eosinophils, and basophils.

Graves' disease A hyperthyroid condition believed to be caused by excessive stimulation of the thyroid gland by autoantibodies. It is associated with exophthalmos (bulging eyes), high pulse rate, high metabolic rate, and other symptoms of hyperthyroidism.

gray matter The part of the central nervous system that contains neuron cell bodies and dendrites but few myelinated axons. It forms the cortex of the cerebrum, cerebral nuclei, and the central region of the spinal cord.

growth hormone (GH) A hormone secreted by the anterior pituitary that stimulates growth of the skeleton and soft tissues during the growing years and that influences the metabolism of protein, carbohydrate, and fat throughout life.

gyrus (*ji'rus*) A fold or convolution in the cerebrum.

H

hCG **human chorionic gonadotropin** (*kor'e-on-ik gon-ad"ŏ-tro'pin)* A hormone secreted by the embryo that has LH-like actions and that is required for maintenance of the mother's corpus luteum for the first 10 weeks of pregnancy.

heart murmur An abnormal heart sound caused by an abnormal flow of blood in the heart. Murmurs are due to structural defects, usually of the valves or septum.

heart sounds The sounds produced by closing of the AV valves of the heart during systole (the first sound) and by closing of the semilunar valves of the aorta and pulmonary trunk during diastole (the second sound).

helper T cells A subpopulation of T cells (lymphocytes) that help stimulate antibody production of B lymphocytes by antigens.

hematocrit (*he-mat'ŏ-krit*) The ratio of packed red blood cells to total blood volume in a centrifuged sample of blood, expressed as a percentage.

hemoglobin (*he'mŏ-glo"bin*) The combination of heme pigment and protein within red blood cells that acts to transport oxygen and (to a lesser degree) carbon dioxide. Hemoglobin also serves as a weak buffer within red blood cells.

Henry's law The statement that the concentration of gas dissolved in a fluid is directly proportional to the partial pressure of that gas.

hepatic (*hĕ-pat'ik*) Pertaining to the liver.

hiatal hernia (*hi-a'tal her'ne-ă*) A protrusion of an abdominal structure through the esophageal hiatus of the diaphragm into the thoracic cavity.

high-density lipoproteins (*lip"o-pro'te-inz*) (**HDLs**) Combinations of lipids and proteins that migrate rapidly to the bottom of a test tube during centrifugation. HDLs are carrier proteins that are believed to transport cholesterol away from blood vessels to the liver, and thus to offer some protection from atherosclerosis.

histamine (*his'tă-mēn*) A compound secreted by tissue mast cells and other connective tissue cells that stimulates vasodilation and increases capillary permeability. It is responsible for many of the symptoms of inflammation and allergy.

homeostasis (*ho"me-o-sta'sis*) The dynamic constancy of the internal environment, the maintenance of which is the principal function of physiological regulatory mechanisms. The concept of homeostasis provides a framework for understanding most physiological processes.

hormone (*hor'mōn*) A regulatory chemical produced in an endocrine gland that is secreted into the blood and carried to target cells that respond to the hormone by an alteration in their metabolism.

humoral immunity (*hyoo'-mor-al ĭ-myoo'nī-te*) The form of acquired immunity in which antibody molecules are secreted in response to antigenic stimulation (as opposed to cell-mediated immunity).

hydrocortisone (*hi"drŏ-kor'tĭ-sōn*) The principal corticosteroid hormone secreted by the adrenal cortex, with glucocorticoid action; also called *cortisol.*

hydrophilic (*hi"drŏ-fil'ik*) Denoting a substance that readily absorbs water; literally, "water loving."

hydrophobic (*hi"drŏ-fo'bik*) Denoting a substance that repels, and that is repelled by, water; literally, "water fearing."

hyperbaric (*hi"per-bar'ik*) **oxygen** Oxygen gas present at greater than atmospheric pressure.

hyperkalemia (*hi"per-kă-le'me-ă*) An abnormally high concentration of potassium in the blood.

hyperopia (*hi"per-o'pe-ă*) A refractive disorder in which rays of light are brought to a focus behind the retina as a result of the eyeball being too short; also called *farsightedness.*

hyperplasia (*hi"per-pla'ze-ă*) An increase in organ size because of an increase in the number of cells as a result of mitotic cell division.

hyperpolarization (*hi"per-po"lar-ĭ-za'shun*) An increase in the negativity of the inside of a cell membrane with respect to the resting membrane potential.

hypersensitivity (*hi"per-sen"sĭ-tiv'ĭ-te*) Another name for *allergy*; an abnormal immune response that may be immediate (due to antibodies of the IgE class) or delayed (due to cell-mediated immunity).

hypertension (*hi"per-ten'shun*) High blood pressure. Classified as either primary, or essential, hypertension of unknown cause or secondary hypertension that develops as a result of other, known disease processes.

hypertonic (*hi"per-ton'ik*) Denoting a solution with a greater solute concentration, and thus a greater osmotic pressure, than plasma.

hypertrophy (*hi-per′trŏ-fe*) Growth of an organ because of an increase in the size of its cells.

hypodermis (*hi″pŏ-der′mis*) A layer of fat beneath the dermis of the skin.

hypotension (*hi″po-ten′shun*) Abnormally low blood pressure.

hypothalamo-hypophyseal (*hi″po-thă-lam′o-hi″po-fĭ-se′al*) **portal system** A vascular system that transports releasing and inhibiting hormones from the hypothalamus to the anterior pituitary.

hypothalamo-hypophyseal tract The tract of nerve fibers (axons) that transports antidiuretic hormone and oxytocin from the hypothalamus to the posterior pituitary.

hypothalamus (*hi″po-thal′ă-mus*) An area of the brain lying below the thalamus and above the pituitary gland. The hypothalamus regulates the pituitary gland and contributes to the regulation of the autonomic nervous system, among its many functions.

I

immediate hypersensitivity Hypersensitivity (allergy) that is mediated by antibodies of the IgE class and that results in the release of histamine and related compounds from tissue cells.

immunoglobulins (*im″yŭ-no-glob′yŭ-linz*) Subclasses of the gamma globulin fraction of plasma proteins that have antibody functions, providing humoral immunity.

implantation (*im″plan-ta′shun*) The process by which a blastocyst attaches itself to and penetrates the endometrium of the uterus.

inhibin (*in-hib′in*) Believed to be a water-soluble hormone secreted by the seminiferous tubules of the testes that specifically exerts negative feedback control of FSH secretion from the anterior pituitary.

innate immunity Also called nonspecific immunity, this refers to the aspects of immune attack that are not targeted against specific antigens. Innate immunity includes phagocytosis, fever, protection afforded by epithelial membranes and gastric acid, and attack by the complement system.

insulin (*in′sŭ-lin*) A polypeptide hormone secreted by the beta cells of the islets of Langerhans in the pancreas that promotes the anabolism of carbohydrates, fat, and protein. Insulin acts to promote the cellular uptake of blood glucose and, therefore, to lower the blood glucose concentration; insulin deficiency produces hyperglycemia and diabetes mellitus.

intestinal microbiota Also called the intestinal microflora, this refers to the normal population of commensal bacteria that reside in the intestine. The intestinal microbiota provide certain vitamins and help strengthen the immune system.

intrapleural pressure The pressure outside of the lungs, in the cavity between the visceral and parietal pleural membranes of the thoracic cavity. This is normally lower than the intrapulmonary (intra-alveolar) pressure.

intrapulmonary pressure Also called the intra-alveolar pressure, this is the pressure of the air within the lungs. It is normally greater than the intrapleural pressure outside of the lungs in the thoracic cavity.

inulin (*in′yŭ-lin*) A polysaccharide of fructose, produced by certain plants, that is filtered by the human kidneys but neither reabsorbed nor secreted. The clearance rate of injected inulin is thus used to measure the glomerular filtration rate.

in vitro (*in ve′tro*) Occurring outside the body, in a test tube or other artificial environment.

IPSP inhibitory postsynaptic potential A hyperpolarization of the postsynaptic membrane in response to a particular neurotransmitter chemical, which makes it more difficult for the postsynaptic cell to attain the threshold level of depolarization required to produce action potentials. IPSPs are responsible for postsynaptic inhibition.

ischemia (*ĭ-ske′me-ă*) A rate of blood flow to an organ that is inadequate to supply sufficient oxygen and maintain aerobic respiration in that organ.

islets of Langerhans (*ī′letz of lang′er-hanz*) Encapsulated groupings of endocrine cells within the exocrine tissue of the pancreas, including alpha cells that secrete glucagon and beta cells that secrete insulin; also called *pancreatic islets.*

isometric (*i″sŏ-met′rik*) **contraction** Muscle contraction in which there is no appreciable shortening of the muscle.

isotonic (*i″sŏ-ton′ik*) **contraction** Muscle contraction in which the muscle shortens in length and maintains approximately the same amount of tension throughout the shortening process.

isotonic solution A solution having the same total solute concentration, osmolality, and osmotic pressure as the solution with which it is compared; a solution with the same solute concentration and osmotic pressure as plasma.

J

jaundice (*jawn′dis*) A condition characterized by high blood bilirubin levels and staining of the tissues with bilirubin, which imparts a yellow color to the skin and mucous membranes.

juxtaglomerular (*juk″stă-glo-mer′yŭ-lar*) **apparatus** A renal structure in which regions of the nephron tubule and afferent arteriole are in contact with each other. Cells in the afferent arteriole of the juxtaglomerular apparatus secrete the enzyme renin into the blood, which activates the renin-angiotensin system.

K

ketogenesis (*ke″to-jen′ĭ-sis*) The production of ketone bodies.

kilocalorie (*kil′ŏ-kal″ŏ-re*) A unit of measurement equal to 1,000 calories, which are units of heat. (A kilocalorie is the amount of heat required to raise the temperature of 1 kilogram of water 1° C.) In nutrition, the kilocalorie is called a big calorie (Calorie).

Krebs (*krebz*) **cycle** A cyclic metabolic pathway in the matrix of mitochondria by which the acetic acid part of acetyl CoA is oxidized and substrates provided for reactions that are coupled to the formation of ATP.

L

lactic acid fermentation The metabolic pathway that includes glycolysis, the conversion of glucose into pyruvic acid, and then the conversion of pyruvic acid into lactic acid.

lactate threshold A measurement of the intensity of exercise. It is the percentage of a person's maximal oxygen uptake at which a rise in blood lactate levels occurs. The average lactate threshold occurs when exercise is performed at 50% to 70% of the maximal oxygen uptake (aerobic capacity).

lactose (*lak′tōs*) Milk sugar; a disaccharide of glucose and galactose.

lactose intolerance The inability of many adults to digest lactose because of a deficiency of the enzyme lactase.

law of complementary base pairing The law that states that, in DNA, adenine can form hydrogen bonds only with thymine (and vice versa), and guanine can form hydrogen bonds only with cytosine (and vice versa) in a double helix. Adenine in DNA bonds only to uracil in the RNA.

law of specific nerve energies This law states that a person will perceive the modality of sensation characteristic of a particular sensory receptor.

length-tension relationship In a striated muscle, the relationship between the resting length of the muscle fiber and the strength of its contraction, as measured by the tension it produces. There is an ideal resting length, which produces the strongest contraction of the muscle fibers.

leptin (*lep′tin*) A hormone secreted by adipose tissue that acts as a satiety factor to reduce appetite. It also increases the body's caloric expenditure.

leukocyte (*loo′kŏ-sīt*) A white blood cell.

Leydig (*lī′dig*) **cells** The interstitial cells of the testes that serve an endocrine function by secreting testosterone and other androgenic hormones.

limbic (*lim′bik*) **system** A group of brain structures, including the hippocampus, cingulate gyrus, dentate gyrus, and amygdala. The limbic system appears to be important in memory, the control of autonomic function, and some aspects of emotion and behavior.

lipid (*lip′id*) An organic molecule that is nonpolar, and thus insoluble in water. Lipids include triglycerides, steroids, and phospholipids.

lipolysis (*li-pol′ĭ-sis*) The hydrolysis of triglycerides into free fatty acids and glycerol.

low-density lipoproteins (*lip″o-pro′te-inz*) **(LDLs)** Plasma proteins that transport triglycerides and cholesterol to the arteries. LDLs are believed to contribute to arteriosclerosis.

lower motor neuron The motor neuron that has its cell body in the gray matter of the spinal cord and that contributes axons to peripheral nerves. This neuron innervates muscles and glands.

lung surfactant (*sur-fak′tant*) A mixture of lipoproteins (containing phospholipids) secreted by type II alveolar cells into the alveoli of the lungs. It lowers surface tension and prevents collapse of the lungs.

luteinizing (*loo′te-ĭ-ni″zing*) **hormone (LH)** A gonadotropic hormone secreted by the anterior pituitary. In a female, LH stimulates ovulation and the development of a corpus luteum; in a male, it stimulates the Leydig cells to secrete androgens.

lymph (*limf*) A fluid derived from tissue fluid that flows through lymphatic vessels, returning to the venous bloodstream.

lymphocyte (*lim′fŏ-sīt*) A type of mononuclear leukocyte; the cell responsible for humoral and cell-mediated immunity.

lymphokine (*lim′fŏ-kīn*) Any of a group of chemicals released from T cells that contribute to cell-mediated immunity.

lysosome (*li′sŏ-sŏm*) An organelle containing digestive enzymes that is responsible for intracellular digestion.

M

mast cell A type of connective tissue cell that produces and secretes histamine and heparin.

maximal oxygen uptake The maximum rate of oxygen consumption by the body per unit time during heavy exercise. Also called the *aerobic capacity,* the maximal oxygen uptake is commonly indicated with the symbol $\dot{V}_{O_2}$ max.

mean arterial pressure An adjusted average of the systolic and diastolic blood pressures. It averages about 100 mmHg in the systemic circulation and 10 mmHg in the pulmonary circulation.

mechanoreceptor (*mek″ă-no-re-sep′tor*) A sensory receptor that is stimulated by mechanical means. Mechanoreceptors include stretch receptors, hair cells in the inner ear, and pressure receptors.

medulla oblongata (*mĕ-dul′ă ob″long-gătă*) A part of the brain stem that contains neural centers for the control of breathing and for regulation of the cardiovascular system via autonomic nerves.

melatonin (*mel″ă-to′nin*) A hormone secreted by the pineal gland that produces darkening of the skin in lower animals and that may contribute to the regulation of gonadal function in mammals. Secretion follows a circadian rhythm and peaks at night.

membrane potential The potential difference or voltage that exists between the two sides of a cell membrane. It exists in all cells but is capable of being changed by excitable cells (neurons and muscle cells).

membranous labyrinth (*mem′bră-nus lab′ĭ-rinth*) A system of communicating sacs and ducts within the bony labyrinth of the inner ear.

Ménière's (*mān-yarz′*) **disease** Deafness, tinnitus, and vertigo resulting from a disease of the labyrinth.

menstrual (*men′stroo-al*) **cycle** The cyclic changes in the ovaries and endometrium of the uterus that lasts about a month. It is accompanied by shedding of the endometrium, with bleeding, and occurs only in humans and the higher primates.

menstruation (*men″stroo-a′shun*) Shedding of the outer two-thirds of the endometrium with accompanying bleeding as a result of a lowering of estrogen secretion by the ovaries at the end of the monthly cycle. The first day of menstruation is taken as day 1 of the menstrual cycle.

messenger RNA (mRNA) A type of RNA that contains a base sequence complementary to a part of the DNA that specifies the synthesis of a particular protein.

metabolic acidosis (*as″ĭ-do′sis*) **and alkalosis** (*al″kă-lo′sis*) Abnormal changes in arterial blood pH due to changes in nonvolatile acid concentration (for example, changes in lactic acid or ketone body concentrations) or to changes in blood bicarbonate concentration.

metabolism (*mĕ-tab′ŏ-liz-em*) All of the chemical reactions in the body. It includes those that result in energy storage (anabolism) and those that result in the liberation of energy (catabolism).

micelle (*mi-sel′*) A colloidal particle formed by the aggregation of numerous molecules.

micro- (L.) Small; also, 1-millionth.

microvilli (*mi″kro-vil′i*) Tiny fingerlike projections of a cell membrane. They occur on the apical (lumenal) surface of the cells of the small intestine and in the renal tubules.

micturition (*mik″tŭ-rish′un*) Urination.

milk-ejection reflex Also called milk letdown, this term refers to the contractions of the mammary ducts that propels milk from the mammary glands toward the nipple. The milk ejection reflex occurs in response to oxytocin secreted from the posterior pituitary.

mineralocorticoid (*min″er-al-o-kor′tĭ-koid*) Any of a class of steroid hormones of the adrenal cortex (corticosteroids) that regulate electrolyte balance.

mitosis (*mi-to′sis*) Cell division in which the 2 daughter cells receive the same number of chromosomes as the parent cell (both daughters and parent are diploid).

molar (*mo′lar*) Pertaining to the number of moles of solute per liter of solution.

mole (*mōl*) The number of grams of a chemical that is equal to its formula weight (atomic weight for an element or molecular weight for a compound).

monoamine (*mon″o-am′ēn*) Any of a class of neurotransmitter molecules containing one amino group. Examples are serotonin, dopamine, and norepinephrine.

monoamine oxidase (*mon′o-am′ēn ok′sĭ-dās*) **(MAO)** An enzyme that degrades monoamine neurotransmitters within presynaptic axon endings. Drugs that inhibit the action of this enzyme thus potentiate the pathways that use monoamines as neurotransmitters.

monocyte (*mon′o-sīt*) A mononuclear, nongranular, phagocytic leukocyte that can be transformed into a macrophage.

mononuclear phagocyte (*fag′ŏ-sīt*) **system** A term used to describe monocytes and tissue macrophages.

monosaccharide (*mon″ŏ-sak′ă-rīd*) The monomer of the more complex carbohydrates. Examples of monomers are glucose, fructose, and galactose. Also called a *simple sugar.*

monosynaptic muscle stretch reflex The reflex contraction of a skeletal muscle stimulated by stretching of the muscle. This activates the stretch receptors, the muscle spindles. Each sensory neuron from the spindles makes a synapse in the spinal cord with a motor neuron, so that only one synapse is crossed within the CNS.

motor neuron (*noor′on*) An efferent neuron that conducts action potentials away from the central nervous system to effector organs (muscles and glands). It forms the ventral roots of spinal nerves.

motor unit A lower motor neuron and all of the skeletal muscle fibers stimulated by branches of its axon. Larger motor units (more muscle fibers per neuron) produce more force when the unit is activated, but smaller motor units afford a finer degree of neural control over muscle contraction.

muscarinic ACh receptors (*mus″kă-rin′ik ACh re-sep′torz*) Receptors for acetylcholine that are stimulated by postganglionic parasympathetic neurons. Their name is derived from the fact that they are also stimulated by the chemical muscarine, derived from a mushroom.

muscle spindle (*mus′el spin′d′l*) A sensory organ within skeletal muscle that is composed of intrafusal fibers. It is sensitive to muscle stretch and provides a length detector within muscles.

myelin (*mī'ĕ-lin*) **sheath** A sheath surrounding axons formed from the cell membrane of Schwann cells in the peripheral nervous system and from oligodendrocytes in the central nervous system.

myocardial infarction (MI) (*mi"ŏ-kar'de-al in-fark'shun*) An area of necrotic tissue in the myocardium that is filled in by scar (connective) tissue.

myofibril (*mi"ŏ-fi'bril*) A subunit of striated muscle fiber that consists of successive sarcomeres. Myofibrils run parallel to the long axis of the muscle fiber, and the pattern of their filaments provides the striations characteristic of striated muscle cells.

myoglobin (*mi"ŏ-glo'bin*) A molecule composed of globin protein and heme pigment. It is related to hemoglobin but contains only one subunit (instead of the four in hemoglobin). Myoglobin is found in striated muscles, wherein it serves to store oxygen.

myoneural (*mi"ŏ-noor'al*) **junction** A synapse between a motor neuron and the muscle cell that it innervates; also called the *neuromuscular junction.*

myopia (*mi-o'pe-ă*) A condition of the eyes in which light is brought to a focus in front of the retina because the eye is too long; also called *nearsightedness.*

myosin (*mi'ŏ-sin*) The protein that forms the A bands of striated muscle cells. Together with the protein actin, myosin provides the basis for muscle contraction.

N

necrosis (*nĕ-kro'sis*) Cellular death within tissues and organs as a result of pathological conditions. Necrosis differs histologically from the physiological cell death of apoptosis.

negative feedback inhibition When referring to the interaction between the anterior pituitary and one of its target glands, this refers to an inhibitory effect produced by the target gland's hormones on the secretion of the corresponding anterior pituitary hormones.

negative feedback loop A response mechanism that serves to maintain a state of internal constancy, or homeostasis. Effectors are activated by changes in the internal environment, and the inhibitory actions of the effectors serve to counteract these changes and maintain a state of balance.

nephron (*nef'ron*) The functional unit of the kidneys, consisting of a system of renal tubules and a vascular component that includes capillaries of the glomerulus and the peritubular capillaries.

nerve A collection of motor axons and sensory dendrites in the peripheral nervous system.

neurilemma (*noor"ĭ-lem'ă*) The sheath of Schwann and its surrounding basement membrane that encircles nerve fibers in the peripheral nervous system.

neuroglia (*noo-rog'le-ă*) The supporting cells of the central nervous system that aid the functions of neurons. In addition to providing support, they participate in the metabolic and bioelectrical processes of the nervous system, also called *glial cells.*

neurohypophysis (*noor"o-hi-pof'ĭ-sis*) The posterior part of the pituitary gland that is derived from the brain. It releases vasopression (ADH) and oxytocin, both of which are produced in the hypothalamus.

neuron (*noor'on*) A nerve cell, consisting of a cell body that contains the nucleus; short branching processes called dendrites that carry electrical charges to the cell body; and a single fiber, or axon, that conducts nerve impulses away from the cell body.

neurotransmitter (*noor"o-trans'mit-er*) A chemical contained in synaptic vesicles in nerve endings that is released into the synaptic cleft, where it causes the production of either excitatory or inhibitory postsynaptic potentials.

neutron (*noo'tron*) An electrically neutral particle that exists together with positively charged protons in the nucleus of atoms.

neutral solution A solution with a pH of 7.000, that of pure water.

nicotinic ACh receptors (*nik"ŏ-tin'ik ACh re-sep'torz*) Receptors for acetylcholine located in the autonomic ganglia and in neuromuscular junctions. Their name is derived from the fact that they can also be stimulated by nicotine, derived from the tobacco plant.

nitrogenous base One of a group of molecules in DNA and RNA that includes adenine, guanine, cytosine, and thymine (if DNA) or uracil (if RNA).

nociceptor (*no"sĭ-sep'tor*) A receptor for pain that is stimulated by tissue damage.

nodes of Ranvier (*ran've-a*) Gaps in the myelin sheath of myelinated axons, located approximately 1 mm apart. Action potentials are produced only at the nodes of Ranvier in myelinated axons.

norepinephrine (*nor"ep-ĭ-nef'rin*) A catecholamine released as a neurotransmitter from postganglionic sympathetic nerve endings and as a hormone (together with epinephrine) by the adrenal medulla; also called *noradrenaline.*

nuclear receptors Receptors that bind to both a regulatory ligand (such as a hormone) and to DNA. The nuclear receptors, when activated by their ligands, regulate genetic expression (RNA synthesis).

nucleotide (*noo'kle-ŏ-tīd*) The subunit of DNA and RNA macromolecules. Each nucleotide is composed of a nitrogenous base (adenine, guanine, cytosine, and thymine or uracil); a sugar (deoxyribose or ribose); and a phosphate group.

nucleus, cell The organelle, surrounded by a double saclike membrane called the nuclear envelope (nuclear membrane), that contains the DNA and genetic information of the cell.

O

oligodendrocyte (*ol"ĭ-go-den'drŏ-sīt*) A type of glial cell that forms myelin sheaths around axons in the central nervous system.

oncogene (*on'kŏ-jēn*) A gene that contributes to cancer. Oncogenes are believed to be abnormal forms of genes that participate in normal cellular regulation.

oogenesis (*o"ŏ-jen'ĕ-sis*) The formation of ova in the ovaries.

optic (*op'tik*) **disc** The area of the retina where axons from ganglion cells gather to form the optic nerve and where blood vessels enter and leave the eye. It corresponds to the blind spot in the visual field caused by the absence of photoreceptors.

organ A structure in the body composed of two or more primary tissues that performs a specific function.

organelle (*or"gă-nel'*) A structure within cells that performs specialized tasks. Organelles include mitochondria, the Golgi apparatus, endoplasmic reticulum, nuclei, and lysosomes. The term is also used for some structures not enclosed by a membrane, such as ribosomes and centrioles.

organ of Corti (*kor'te*) The structure within the cochlea that constitutes the functional unit of hearing. It consists of hair cells and supporting cells on the basilar membrane that help to transduce sound waves into nerve impulses; also called the *spiral organ.*

osmolarity A measure of the concentration of solutes in a solution that indicates the total molarity of all of the solutes per liter of solution.

osmoreceptor (*oz"mŏ-re-cep'tor*) A sensory neuron that responds to changes in the osmotic pressure of the surrounding fluid.

osmosis (*oz-mo'sis*) The passage of solvent (water) from a more dilute to a more concentrated solution through a membrane that is more permeable to water than to the solute.

osmotic (*oz-mot'ik*) **pressure** A measure of the tendency for a solution to gain water by osmosis when separated by a membrane from pure water. Directly related to the osmolality of the solution, it is the pressure required to just prevent osmosis.

osteoblast (*os'te-ŏ-blast*) A bone-forming cell.

osteoclast (*os'te-ŏ-klast*) A cell that resorbs bone by promoting the dissolution of calcium phosphate crystals.

osteoporosis (*os"te-o-pŏ-ro'sis*) Demineralization of bone, seen most commonly in post-menopausal women and patients who are inactive or paralyzed. It may be accompanied by pain, loss of stature, and other deformities and fractures.

ovarian follicles The structures within the ovary within which egg cells (oocytes) develop. Granulosa cells of the ovarian follicles secrete the major estrogen, estradiol, in response to FSH stimulation.

ovary (*o'vă-re*) The gonad of a female that produces ova and secretes female sex steroids.

ovulation (*ov-yŭ-la'shun*) The extrusion of a secondary oocyte from the ovary.

oxygen (*ok'sĭ-jen*) **debt** The extra amount of oxygen required by the body after exercise to metabolize lactic acid and to supply the higher metabolic rate of muscles warmed during exercise.

oxyhemoglobin (*ok"se-he"mŏ-glo'bin*) A compound formed by the bonding of molecular oxygen with hemoglobin.

oxyhemoglobin percent saturation The ratio, expressed as a percentage, of the amount of oxyhemoglobin compared to the total amount of hemoglobin in blood.

oxidative phosphorylation The process by which mitochondria add phosphate groups to ADP to form (by phosphorylation) ATP, using energy provided by the electron transport system (ETS). This is the major process that produces ATP in the cell.

oxytocin (*ok"sĭ-to'sin*) One of the two hormones produced in the hypothalamus and released from the posterior pituitary (the other being vasopressin). Oxytocin stimulates the contraction of uterine smooth muscles and promotes milk letdown in females.

P

pacemaker potential The spontaneous depolarization of the heart's pacemaker (the SA node) that produces action potentials that result in contractions of the atria and ventricles.

PAH para-aminohippuric (*par'ă-ă-me'no-hi-pyoor"ik*) **acid** A substance used to measure total renal plasma flow because its clearance rate is equal to the total rate of plasma flow to the kidneys. PAH is filtered and secreted by the renal nephrons but not reabsorbed.

pancreatic (*pan"kre-at'ik*) **islets** *See* islets of Langerhans.

pancreatic juice The secretions of the pancreas that are transported by the pancreatic duct to the duodenum. Pancreatic juice contains bicarbonate and the digestive enzymes trypsin, lipase, and amylase.

paracrine (*par'ă-krin*) **regulator** A regulatory molecule produced within one tissue that acts on a different tissue of the same organ. For example, the endothelium of blood vessels secretes a number of paracrine regulators that act on the smooth muscle layer of the vessels to cause vasoconstriction or vasodilation.

parasympathetic (*par"ă-sim"pă-thet'ik*) Pertaining to the craniosacral division of the autonomic nervous system.

parathyroid (*par"ă-thi'roid*) **hormone (PTH)** A polypeptide hormone secreted by the parathyroid glands. PTH acts to raise the blood Ca^{2+} levels primarily by stimulating resorption of bone.

Parkinson's disease A tremor of the resting muscles and other symptoms caused by inadequate dopamine-producing neurons in the basal nuclei of the cerebrum. Also called *paralysis agitans.*

parturition (*par"tyoo-rish'un*) The process of giving birth; childbirth.

passive immunity Specific immunity granted by the administration of antibodies made by another organism.

passive transport The transport of molecules and ions across a plasma membrane from higher to lower concentrations; does not require ATP.

pathogen (*path'ŏ-jen*) Any disease-producing microorganism or substance.

pepsin (*pep'sin*) The protein-digesting enzyme secreted in gastric juice.

peptic ulcer (*pep'tik ul'ser*) An injury to the mucosa of the esophagus, stomach, or small intestine caused by the breakdown of gastric barriers to self-digestion or by excessive amounts of gastric acid.

peripheral resistance The resistance to blood flow through the arterial system. Peripheral resistance is largely a function of the radius of small arteries and arterioles. The resistance to blood flow is proportional to the fourth power of the radius of the vessel.

peristalsis (*per"ĭ-stal'sis*) Waves of smooth muscle contraction in smooth muscles of the tubular digestive tract. It involves circular and longitudinal muscle fibers at successive locations along the tract and serves to propel the contents of the tract in one direction.

phagocytosis (*fag"ŏ-si-to'sis*) Cellular eating; the ability of some cells (such as white blood cells) to engulf large particles (such as bacteria) and digest these particles by merging the food vacuole in which they are contained with a lysosome containing digestive enzymes.

phosphocreatine (*fos'-fo-kre'ă-tin*) An organic phosphate molecule in muscle cells that serves as a source of high-energy phosphate for the synthesis of ATP; also called *phosphocreatine.*

phosphodiesterase (*fos"fo-di-es'ter-ās*) An enzyme that cleaves cyclic AMP into inactive products, thus inhibiting the action of cyclic AMP as a second messenger.

photoreceptors (*fo"to-re-sep'torz*) Sensory cells (rods and cones) that respond electrically to light. They are located in the retina of the eyes.

pineal (*pin'e-al*) **gland** A gland within the brain that secretes the hormone melatonin. It is affected by sensory input from the photoreceptors of the eyes.

pinocytosis (*pin"ŏ-si-to'sis*) Cell drinking; invagination of the cell membrane to form narrow channels that pinch off into vacuoles. This permits cellular intake of extracellular fluid and dissolved molecules.

pituitary (*pĭ-too'ĭ-ter-e*) **gland** Also called the *hypophysis.* A small endocrine gland joined to the hypothalamus at the base of the brain. The pituitary gland is functionally divided into anterior and posterior portions. The anterior pituitary secretes ACTH, TSH, FSH, LH, growth hormone, and prolactin. The posterior pituitary releases oxytocin and antidiuretic hormone (ADH), which are produced by the hypothalamus.

placenta Formed from both maternal tissue (the decidua basalis) and fetal tissue (the chorion frondosum), the placenta nourishes the developing embryo and fetus, provides it with oxygen, eliminates some waste products, secretes a variety of hormones, and performs other functions vital to fetal development and parturition.

plasma (*plaz'mă*) The fluid portion of the blood. Unlike serum (which lacks fibrinogen), plasma is capable of forming insoluble fibrin threads when in contact with test tubes.

plasma cells Cells derived from B lymphocytes that produce and secrete large amounts of antibodies. They are responsible for humoral immunity.

plasmalemma (*plaz"mă-lem'ă*) The cell membrane; an alternate term for the selectively permeable membrane that encloses the cytoplasm of a cell.

platelet (*plāt'let*) A disc-shaped structure, 2 to 4 micrometers in diameter, derived from bone marrow cells called megakaryocytes. Platelets circulate in the blood and participate (together with fibrin) in forming blood clots.

pluripotent (*ploo-rip'ŏ-tent*) A term used to describe the ability of early embryonic cells to specialize to produce all tissues except the trophoblast cells of the placenta.

pneumothorax (*noo"mo-thor'aks*) An abnormal condition in which air enters the intrapleural space, either through an open chest wound or from a tear in the lungs. This can lead to the collapse of a lung (atelectasis).

PNS The peripheral nervous system, including nerves and ganglia.

polar body A small daughter cell formed by meiosis that degenerates in the process of oocyte production.

polar molecule A molecule in which the shared electrons are not evenly distributed, so that one side of the molecule is negatively (or positively) charged in comparison with the other side. Polar molecules are soluble in polar solvents such as water.

polypeptide (*pol"e-pep'tīd*) A chain of amino acids connected by covalent bonds called peptide bonds. A very large polypeptide is called a protein.

polypeptide Y-Y (PYY) This is a hormone secreted by the small intestine that suppresses appetite on an intermediate basis, helping to space meals.

polysaccharide (*pol″e-sak′ă-rīd*) A carbohydrate formed by covalent bonding of numerous monosaccharides. Examples are glycogen and starch.

positive feedback A response mechanism that results in the amplification of an initial change. Positive feedback results in avalanche-like effects, as occur in the formation of a blood clot or in the production of the LH surge by the stimulatory effect of estrogen.

posterior pituitary *See* neurohypophysis.

postganglionic neurons Autonomic neurons that have cell bodies in autonomic ganglia and axons that innervate smooth muscles, cardiac muscles, or glands.

postsynaptic (*pŏst″sĭ-nap′tik*) **inhibition** The inhibition of a postsynaptic neuron by axon endings that release a neurotransmitter that induces hyperpolarization (inhibitory postsynaptic potentials).

potential (*pŏ-ten′shal*) **difference** In biology, the difference in charge between two solutions separated by a membrane. The potential difference is measured in voltage.

preganglionic neurons Autonomic neurons that have cell bodies in the CNS and axons that synapse within autonomic ganglia.

prehormone (*pre-hor′mōn*) An inactive form of a hormone secreted by an endocrine gland. The prehormone is converted within its target cells to the active form of the hormone.

progesterone (*pro-jes′tĕ-rōn*) A steroid hormone secreted by the corpus luteum of the ovaries and by the placenta. Secretion of progesterone during the luteal phase of the menstrual cycle promotes the final maturation of the endometrium.

prolactin (*pro-lak′tin*) A hormone secreted by the anterior pituitary that stimulates lactation (acting together with other hormones) in the postpartum female. It may also participate (along with the gonadotropins) in regulating gonadal function in some mammals.

proprioceptor (*pro″pre-o-sep″tor*) A sensory receptor that provides information about body position and movement. Examples are receptors in muscles, tendons, and joints and in the semicircular canals of the inner ear.

prostaglandin (*pros′tă-glan′din*) Any of a family of fatty acids that serve numerous autocrine regulatory functions, including the stimulation of uterine contractions and of gastric acid secretion and the promotion of inflammation.

protein (*pro′te-in*) The class of organic molecules composed of large polypeptides, in which over a hundred amino acids are bonded together by peptide bonds.

proteome (*pro′te-om*) All of the different proteins produced by a genome.

proton (*pro′ton*) A unit of positive charge in the nucleus of atoms.

pseudopod (*soo″dŏ-pod*) A footlike extension of the cytoplasm that enables some cells (with amoeboid motion) to move across a substrate. Pseudopods also are used to surround food particles in the process of phagocytosis.

pulmonary (*pul′mŏ-ner″e*) **circulation** The part of the vascular system that includes the pulmonary arteries and pulmonary veins. It transports blood from the right ventricle of the heart through the lungs, and then back to the left atrium of the heart.

pupil The opening at the center of the iris of the eye.

Purkinje (*pur-kin′je*) **fibers** Specialized conducting tissue in the ventricles of the heart that carry impulses from the bundle of His to the myocardium of the ventricles.

pyramidal (*pĭ-ram′ĭ-dal*) **tracts** Motor tracts that descend without synaptic interruption from the cerebrum to the spinal cord, where they synapse either directly or indirectly (via spinal interneurons) with the lower motor neurons of the spinal cord; also called *corticospinal tracts.*

pyrogen (*pi′rŏ-jen*) A fever-producing substance.

Q

QRS complex The principal deflection of an electrocardiogram, produced by depolarization of the ventricles.

R

reabsorption (*re″ab-sorp′shun*) The transport of a substance from the lumen of the renal nephron into the peritubular capillaries.

receptor potential Also called the generator potential, this is a depolarization produced in a receptor cell in response to a sensory stimulus.

receptor protein A specific protein in either the plasma membrane, the cytoplasm, or the nucleus of a cell that binds to a particular regulatory molecule; this specific binding is required for the function of the regulatory molecule.

reciprocal innervation (*rĭ-sip′rŏ-kal in″er-va′shun*) The process whereby the motor neurons to an antagonistic muscle are inhibited when the motor neurons to an agonist muscle are stimulated. In this way, for example, the extensor muscle of the elbow joint is inhibited when the flexor muscles of this joint are stimulated to contract.

recruitment (*rĭ-kroot′ment*) In terms of muscle contraction, the successive stimulation of more and larger motor units in order to produce increasing strengths of muscle contraction.

referred pain Pain originating in deep, visceral organs that is perceived to be coming from particular body surface locations. This is believed to be caused by sensory neurons from both locations synapsing on the same interneurons in the same spinal cord level.

refractory (*re-frak′tŏ-re*) **period** The period of time during which a region of axon or muscle cell membrane cannot be stimulated to produce an action potential (absolute refractory period), or when it can be stimulated only by a very strong stimulus (relative refractory period).

releasing hormones Polypeptide hormones secreted by neurons in the hypothalamus that travel in the hypothalamo-hypophyseal portal system to the anterior pituitary and stimulate the anterior pituitary to secrete specific hormones.

REM sleep The stage of sleep in which dreaming occurs. It is associated with rapid eye movements (REMs). REM sleep occurs 3 to 4 times each night and lasts from a few minutes to over an hour.

renal (*re′nal*) Pertaining to the kidneys.

renal plasma clearance The volume of plasma from which a particular solute is cleared each minute by the excretion of that solute in the urine. If there is no reabsorption or secretion of that solute by the nephron tubules, the renal plasma clearance is equal to the glomerular filtration rate.

renal plasma threshold The minimum plasma concentration of a substance required for that substance to be excreted in the urine. At a plasma concentration lower than this, all of the substance will be reabsorbed from the filtrate so that none appears in the urine.

renal pyramid (*pĭ′ră-mid*) One of a number of cone-shaped tissue masses that compose the renal medulla.

renin (*re′nin*) An enzyme secreted into the blood by the juxtaglomerular apparatus of the kidneys. Renin catalyzes the conversion of angiotensinogen into angiotensin II.

renin-angiotensin-aldosterone system The relationship between the secretion of renin by the juxtaglomerular apparatus of the kidneys, the production of angiotensin II, and the stimulation of aldosterone secretion from the adrenal cortex.

rennin (*ren′in*) A digestive enzyme secreted in the gastric juice of infants that catalyzes the digestion of the milk protein casein.

repolarization (*re-po″lar-ĭ-za′shun*) The reestablishment of the resting membrane potential after depolarization has occurred.

respiratory acidosis (*rĭ-spīr′ă-tor-e as″ĭ-do′sis*) A lowering of the blood pH to below 7.35 as a result of the accumulation of CO_2 caused by hypoventilation.

respiratory alkalosis (*al″kă-lo′sis*) A rise in blood pH to above 7.45 as a result of the excessive elimination of blood CO_2 caused by hyperventilation.

respiratory center The nucleus in the medulla oblongata that controls the reflex regulation of ventilation (breathing).

respiratory distress syndrome (RDS) A lung disease of the newborn, most frequently occurring in premature infants, that is caused by abnormally high alveolar surface tension as a result of a deficiency in lung surfactant; also called *hyaline membrane disease.*

respiratory zone The region of the lungs in which gas exchange between the inspired air and pulmonary blood occurs. It includes the respiratory bronchioles, in which individual alveoli are found, and the terminal alveoli.

resting potential The potential difference across a cell membrane when the cell is in an unstimulated state. The resting potential is always negatively charged on the inside of the membrane compared to the outside.

reticular (*rĕ-tik′yŭ-lar*) **activating system (RAS)** A complex network of nuclei and fiber tracts within the brain stem that produces nonspecific arousal of the cerebrum to incoming sensory information. The RAS thus maintains a state of alert consciousness and must be depressed during sleep.

retina (*ret′-ĭ-nă*) The layer of the eye that contains neurons and photoreceptors (rods and cones).

retinal (retinaldehyde, retinene) The photopigment, bound to the protein opsin, that forms rhodopsin in the photoreceptors of the eyes.

rhodopsin (*ro-dop′sin*) Visual purple. A pigment in rod cells that undergoes a photochemical dissociation in response to light and, in so doing, stimulates electrical activity in the photoreceptors.

rhythmicity center The area of the medulla oblongata that controls the rhythmic pattern of inspiration and expiration.

RNA Ribonucleic (*ri″bo-noo-kle′ik*) acid. A nucleic acid consisting of the nitrogenous bases adenine, guanine, cytosine, and uracil; the sugar ribose; and phosphate groups. There are three types of RNA found in cytoplasm: messenger RNA (mRNA), transfer RNA (tRNA), and ribosomal RNA (rRNA).

rods One of the two categories of photoreceptors (the other being cones) in the retina of the eye. Rods are responsible for black-and-white vision under low illumination.

S

saltatory (*sal′tă-tor-e*) **conduction** The rapid passage of action potentials from one node of Ranvier to another in myelinated axons.

sarcolemma (*sar″cŏ-lem′ă*) The cell membrane of striated muscle cells.

sarcomere (*sar′kŏ-mēr*) The structural subunit of a myofibril in a striated muscle; equal to the distance between two successive Z lines.

sarcoplasmic reticulum (SR) (*sar″kŏ-plaz′mik rĕ-tik′yŭ-lum*) The smooth or agranular endoplasmic reticulum of striated muscle cells. It surrounds each myofibril and serves to store Ca^{2+} when the muscle is at rest.

Schwann (*shvan*) **cell** A supporting cell of the peripheral nervous system that forms sheaths around peripheral nerve fibers. Schwann cells also direct regeneration of peripheral nerve fibers to their target cells.

sclera (*skler′ă*) The tough white outer coat of the eyeball, continuous anteriorly with the clear cornea.

second messenger A molecule or ion whose concentration within a target cell is increased by the action of a regulator molecule (e.g., a hormone or neurotransmitter) so as to stimulate the metabolism of that target cell in a way characteristic of the actions of the regulator molecule—that is, in a way that mediates the intracellular effects of the regulator molecule.

secretin (*sĕ-kre′tin*) A polypeptide hormone secreted by the small intestine in response to acidity of the intestinal lumen. Along with cholecystokinin, secretin stimulates the secretion of pancreatic juice into the small intestine.

secretion (*sĕ-kre′shun*) **renal** The transport of a substance from the blood through the wall of the nephron tubule into the urine.

segmentation The contractions of the smooth muscle in the intestine at different segments, serving to mix the chyme and move it distally.

semicircular canals Three canals of the bony labyrinth that contain endolymph, which is continuous with the endolymph of the membranous labyrinth of the cochlea. The semicircular canals provide a sense of equilibrium.

semilunar (*sem″e-loo′nar*) **valves** The valve flaps of the aorta and pulmonary artery at their juncture with the ventricles.

seminiferous tubules (*sem″ĭ-nif′er-us too′byoolz*) The tubules within the testes that produce spermatozoa by meiotic division of their germinal epithelium.

sensory neuron (*noor′on*) An afferent neuron that conducts impulses from peripheral sensory organs into the central nervous system.

series elastic component The tendons and other connective tissues of a muscle that are stretched by contraction of a muscle. They have elasticity, and their recoil helps the muscle to lengthen when it relaxes.

serotonin (*ser″ŏ-to′nin*) Monoamine neurotransmitter, chemically known as 5-hydroxytryptamine, derived from the amino acid L-tryptophan. Serotonin released at synapses in the brain have been associated with the regulation of mood and behavior, appetite, and cerebral circulation.

Sertoli (*ser-to′le*) **cells** Nongerminal supporting cells in the seminiferous tubules. Sertoli cells envelop spermatids and appear to participate in the transformation of spermatids into spermatozoa; also called *sustentacular cells.*

serum (*ser′um*) The fluid squeezed out of a clot as it retracts; supernatant when a sample of blood clots in a test tube and is centrifuged. Serum is plasma from which fibrinogen and other clotting proteins have been removed as a result of clotting.

set point The level of a particular measurement of the internal environment that homeostatic mechanisms operate to maintain.

shock As it relates to the cardiovascular system, a rapid, uncontrolled fall in blood pressure, which in some cases becomes irreversible and leads to death.

sinoatrial (*si″no-a′tre-al*) **node** A mass of specialized cardiac tissue in the wall of the right atrium that initiates the cardiac cycle; the SA node, also called the *pacemaker.*

skeletal muscle pump A term used with reference to the effect of skeletal muscle contraction on the flow of blood in veins. As the muscles contract, they squeeze the veins, and in this way help move the blood toward the heart.

sleep apnea A temporary cessation of breathing during sleep, usually lasting for several seconds.

sliding filament theory The theory that the thick and thin filaments of a myofibril slide past each other during muscle contraction, decreasing the length of the sarcomeres but maintaining their own initial length.

smooth muscle A specialized type of nonstriated muscle tissue composed of fusiform single-nucleated fibers. It contracts in an involuntary, rhythmic fashion in the walls of visceral organs.

somatesthetic (*so″mat-es-thet′ek*) **sensations** Sensations arising from cutaneous, muscle, tendon, and joint receptors. These sensations project to the postcentral gyrus of the cerebral cortex.

somatic (*so-mat′ik*) **motor neuron** A motor neuron in the spinal cord that innervates skeletal muscles. Somatic motor neurons are categorized as alpha and gamma motoneurons.

somatostatin (*so″mă-tŏ-stat′n*) A polypeptide produced in the hypothalamus that acts to inhibit the secretion of growth hormone from the anterior pituitary. Somatostatin is also produced in the islets of Langerhans of the pancreas, but its function there has not been established.

sounds of Korotkoff (*kŏ-rot′kof*) The sounds heard when blood pressure measurements are taken. These sounds are produced by the turbulent flow of blood through an artery that has been partially constricted by a pressure cuff.

spatial summation The adding together of excitatory and inhibitory postsynaptic potentials (EPSPs and IPSPs) produced in different locations of the dendrites and cell bodies of postsynaptic neurons.

spermatocyte (*sper-mat′ŏ-sīt*) A diploid cell of the seminiferous tubules in the testes that divides by meiosis to produce spermatids.

spermatogenesis (*sper″mă-to-jen′ĭ-sis*) The formation of spermatozoa, including meiosis and maturational processes in the seminiferous tubules.

spermatozoon (*sper″mă-to-zo′on*) pl., *spermatozoa* or, loosely, *sperm.* A mature sperm cell, formed from a spermatid.

sphygmomanometer (*sfig″mo-mă-nom′ĭ-ter*) A manometer (pressure transducer) used to measure the blood pressure.

spirometry The measurements of pulmonary volumes and capacities. These measurements help to assess pulmonary function.

stem cells Cells that are relatively undifferentiated (unspecialized) and able to divide and produce different specialized cells.

striated (*stri′āt-ed*) **muscle** Skeletal and cardiac muscle, the cells of which exhibit cross banding, or striations, because of the arrangement of thin and thick filaments.

stroke volume The amount of blood ejected from each ventricle at each heartbeat.

sub- (L.) Under, below.

summation (*sŭ-ma′shun*) In neural physiology, the additive effects of graded synaptic potentials. In muscle physiology, the additive effects of contractions of different muscle fibers.

suprachiasmatic (*soo″pră-ki″az-mot′ik*) **nucleus (SCN)** The primary center for the regulation of circadian rhythms. Located in the hypothalamus, the SCN is believed to regulate circadian rhythms by means of its stimulation of melatonin secretion from the pineal gland.

surfactant (*sur-fak′tant*) In the lungs, a mixture of phospholipids and proteins produced by alveolar cells that reduces the surface tension of the alveoli and contributes to the elastic properties of the lungs.

synapse (*sin′aps*) The junction across which a nerve impulse is transmitted from an axon terminal to a neuron, a muscle cell, or a gland cell either directly or indirectly (via the release of chemical neurotransmitters).

systemic (*sis-tem′ik*) **circulation** The circulation that carries oxygenated blood from the left ventricle via arteries to the tissue cells and that carries blood depleted of oxygen via veins to the right atrium; the general circulation, as compared to the pulmonary circulation.

systole (*sis′-tŏ-le*) The phase of contraction in the cardiac cycle. Used alone, this term refers to contraction of the ventricles; the term *atrial systole* refers to contraction of the atria.

T

tachycardia (*tak″ĭ-kar′de-ă*) An excessively rapid heart rate, usually applied to rates in excess of 100 beats per minute (in contrast to bradycardia, in which the heart rate is very slow—below 60 beats per minute).

target organ The organ that is specifically affected by the action of a hormone or other regulatory process.

T cell A type of lymphocyte that provides cell-mediated immunity (in contrast to B lymphocytes, which provide humoral immunity through the secretion of antibodies). There are three subpopulations of T cells: cytotoxic (killer), helper, and suppressor.

testes (*tes′tēz*); sing; *testis* Male gonads. Testes are also known as *testicles.*

testis-determining factor (TDF) The product of a gene located on the short arm of the Y chromosome that causes the indeterminate embryonic gonads to develop into testes.

testosterone (*tes-tos′tĕ-rōn*) The major androgenic steroid secreted by the Leydig cells of the testes after puberty.

tetanus (*tet′n-us*) A sustained muscle contraction. Produced by the summation of muscle twitches, incomplete tetanus occurs when there are small jerks between the twitches, and complete tetanus occurs when a smooth muscle contraction is produced by a sufficiently rapid summation of twitches.

theophylline (*the-of′ĭ-lin*) A drug found in certain tea leaves that promotes dilation of the bronchioles by increasing the intracellular concentration of cyclic AMP (cAMP) in the smooth muscle cells. This effect is due to inhibition of the enzyme phosphodiesterase, which breaks down cAMP.

thorax (*thor′aks*) The part of the body cavity above the diaphragm; the chest.

threshold The minimum stimulus that just produces a response.

thrombin (*throm′bin*) A protein formed in blood plasma during clotting that enzymatically converts the soluble protein fibrinogen into insoluble fibrin.

thrombocyte (*throm′bŏ-sīt*) A blood platelet; a disc-shaped structure in blood that participates in clot formation.

thyroglobulin (*thi-ro-glob′yŭ-lin*) An iodine-containing protein in the colloid of the thyroid follicles that serves as a precursor for the thyroid hormones.

thyroxine (*thi-rok′sin*) Also called *tetraiodothyronine,* or T_4. The major hormone secreted by the thyroid gland. It regulates the basal metabolic rate and stimulates protein synthesis in many organs. A deficiency of this hormone in early childhood produces cretinism.

total minute volume The product of tidal volume (ml per breath) and ventilation rate (breaths per minute).

total peripheral resistance The resistance to blood flow through the arteries of the body. An uncompensated increase in total peripheral resistance increases the arterial blood pressure and reduces the output of ventricle.

totipotent (*tōtip′ŏ-tent*) The ability of a cell to differentiate into all tissue types, and thus to form a new organism when appropriately stimulated and placed in the correct environment (a uterus).

tracts A collection of axons within the central nervous system that forms the white matter of the CNS.

triiodothyronine (*tri″i-ŏ′dŏ-thi′ro-nēn*) (**T_3**) A hormone secreted in small amounts by the thyroid; the active hormone in target cells formed from thyroxine.

tropomyosin (*tro″pŏ-mi′ŏ-sin*) A filamentous protein that attaches to actin in the thin filaments. Together with another protein called troponin, it acts to inhibit and regulate the attachment of myosin cross bridges to actin.

troponin (*tro′pŏ-nin*) A protein found in the thin filaments of the sarcomeres of skeletal muscle. A subunit of troponin binds to Ca^{2+}, and as a result causes tropomyosin to change position in the thin filament.

TSH **Thyroid-stimulating hormone** Also called *thyrotropin (thi″rŏ-tro′pin)* A hormone secreted by the anterior pituitary that stimulates the thyroid gland.

twitch A rapid contraction and relaxation of a muscle fiber or a group of muscle fibers.

U

universal donor A person with blood type O, who is able to donate blood to people with other blood types in emergency blood transfusions.

universal recipient A person with blood type AB, who can receive blood of any type in emergency transfusions.

uremia (*yoo-re′me-ă*) The retention of urea and other products of protein catabolism as a result of inadequate kidney function.

urobilinogen (*yoo″rŏ-bi-lin′ŏ-jen*) A compound formed from bilirubin in the intestine. Some is excreted in the feces and some is absorbed and enters the enterohepatic circulation, where it may be excreted either in the bile or in the urine.

V

vaccination (*vaks″sĭ-na′shun*) The clinical induction of active immunity by introducing antigens into the body so that the immune system becomes sensitized to them. The immune system will mount a secondary response to those antigens upon subsequent exposures.

vagina (*vă-ji′nă*) The tubular organ in the female leading from the external opening of the vulva to the cervix of the uterus.

vagus (*va′gus*) **nerve** The tenth cranial nerve, composed of sensory dendrites from visceral organs and preganglionic parasympathetic nerve fibers. The vagus is the major parasympathetic nerve in the body.

Valsalva's (*val-sal'vaz*) **maneuver** Exhalation against a closed glottis, so that intrathoracic pressure rises to the point that the veins returning blood to the heart are partially constricted. This produces circulatory and blood pressure changes that could be dangerous.

vasoconstriction (*va"zo-kon-strik'shun*) A narrowing of the lumen of blood vessels as a result of contraction of the smooth muscles in their walls.

vasodilation (*va"zo-di-la'shun*) A widening of the lumen of blood vessels as a result of relaxation of the smooth muscles in their walls.

vasomotor center The nucleus in the medulla oblongata that regulates blood vessel constriction and dilation, thereby affecting blood pressure and flow.

vasopressin (*va"zo-pres'in*) Another name for antidiuretic hormone (ADH), released from the posterior pituitary. The name *vasopressin* is derived from the fact that this hormone can stimulate constriction of blood vessels.

vein A blood vessel that returns blood to the heart.

ventilation (*ven"tĭ-la'shun*) Breathing; the process of moving air into and out of the lungs.

venous return The return of blood in veins to the atria. In the systemic circulation, this is aided by the action of skeletal muscle pumps.

vertigo (*ver'tĭ-go*) A sensation of whirling motion, either of oneself or of external objects; dizziness, or loss of equilibrium.

vestibular (*vĕ-stib'yŭ-lar*) **apparatus** The parts of the inner ear, including the semicircular canals, utricle, and saccule, that function to provide a sense of equilibrium.

villi (*vil'i*) Fingerlike folds of the mucosa of the small intestine.

virulent (*vir'yŭ-lent*) Pathogenic, or able to cause disease.

vital capacity The maximum amount of air that can be forcibly expired after a maximal inspiration.

voltage-regulated channel (voltage-gated channel) A protein channel through the plasma membrane that is generally closed at the resting membrane potential but opens in response to a threshold level of depolarization; primarily located in the axons of neurons and in muscle cells.

W

Wernicke's area The area of the cerebral cortex, usually located in the left hemisphere, required for language comprehension. Damage to Wernicke's area can result in speech that is fluid but nonsensical.

white matter The portion of the central nervous system composed primarily of myelinated fiber tracts. This forms the region deep to the cerebral cortex in the brain and the outer portion of the spinal cord.

Z

zygote (*zi'gōt*) A fertilized ovum.

Photo Credits

Appendix

Appendix Photo #1: SPL/Photo Researchers, Inc.

Chapter 1

Opener: Royalty Free/Corbis; p. 23(skier) (snowboarder) (kayaker): Royalty Free/Corbis; (yoga): Image Source/Jupiter Images; (skateboarder): TRBfoto/ Getty Images; (rock climber): Royalty Free/Corbis

Chapter 2

Opener: Royalty Free/Corbis; 2.3: © Dr. Dennis Kunkel/Phototake; 2.4a: Custom Medical Stock Photo, Inc.; 2.4b: © Biophoto Associates/Photo Researchers, Inc.; 2.6a & 2.7a: Courtesy Keith R. Porter; 2.8a: Courtesy of Mark S. Ladinsky and Kathryn E. Howell, University of Colorado; 2.16a–e, 2.31–2.34: © Ed Reschke; 2.35a,b: © Ray Simons/Photo Researchers; 2.35c, 2.36a: © Ed Reschke; 2.39: © The McGraw-Hill Companies, Inc./Photo by Dr. Alvin Telser; 2.40a & 2.41a: © Ed Reschke; p. 56: Digital Vision/Getty Images; p. 57 (biker): Royalty Free/Corbis; (skate boarder): TRBfoto/Getty Images (kayaker), (rock climber): Royalty Free/Corbis; (balance beam): © Digital Vision; (roller blader): Royalty Free/Corbis; (yoga): Image Source/Jupiter Images

Chapter 3

Opener: Karl Weatherly/Getty Images; 3.1a,b: © Jeon/Visuals Unlimited; p. 80: Digital Vision/Getty Images; p. 81(biker): Karl Weatherly/Getty Images; (skate boarder): TRBfoto/Getty Images; (kayaker): © Royalty Free/Corbis; (rock climber): Royalty Free/Corbis; (snowboarder): © Royalty Free/Corbis; (yoga): Image Source/Jupiter Images

Chapter 4

Opener: Ryan McVay/Getty Images; 4.7(both): From H. Webster, The Vertebrate Peripheral Nervous System, John Hubbard (ed.), © 1974 Plenuim Publishing Corporation; p. 107: Digital Vision/Getty Images; p. 108 (ice skater): Ryan McVay/Getty Images; (skate boarder): TRBfoto/Getty Images; (kayaker), (rock climber), (snowboarder): © Royalty Free/Corbis; (yoga): Image Source/Jupiter Images

Chapter 5

Opener: © Royalty Free/Corbis; 5.3b: © The McGraw-Hill Companies, Inc./Karl Rubin, Photographer; 5.5: From W.T. Carpenter and R.W. Buchanan, "Medical Progress: Schizophrenia" in New England Journal of Medicine, 330:685, 1994, fig 1A. Copyright ©1994 Massachusetts Medical Society. All rights reserved; 5.19a,b: Reprinted Figure 2 (1st and 3rd panels) with permission from R.J. Dolan, SCIENCE 298:1191–1194. Copyright 2002 AAAS; p. 130: Digital Vision/Getty Images; p. 131 (ice skater): © Royalty Free/Corbis; (skate boarder): TRBfoto/Getty Images; (kayaker): © Royalty Free/Corbis; (hand stand): © Digital Vision; (snowboarder), (rock climber): © Royalty Free/Corbis

Chapter 6

Opener: Royalty Free/Corbis; p. 151: Digital Vision/ Getty Images; p. 152 (woman gymnast): Royalty Free/Corbis; (skate boarder): TRBfoto/Getty Images; (kayaker), (snowboarder): © Royalty Free/Corbis; (balance beam): © PhotoAlto/Alamy; (rock climber): Royalty Free/Corbis; (yoga): Image Source/Jupiter Images; p. 153: Royalty Free/Corbis

Chapter 7

Opener: © PhotoAlto/Alamy; 7.10a: Courtesy of Dean E. Hillman; 7.20a: © Thomas Sims; 7.22b: Courtesy of P.N. Farnsworth, University of Medicine and Dentistry; 7.26b: © Ominkron/Photo Researchers; p. 180: Digital Vision/Getty Images; p. 181 (balance beam): © PhotoAlto/Alamy; (skate boarder): TRBfoto/Getty Images; (kayaker), (snowboarder): © Royalty Free/Corbis; (ice skater): Ryan McVay/Getty Images; (rock climber): Royalty Free/Corbis; (ballet couple): © Digital Vision

Chapter 8

Opener: TRBfoto/Getty Images; 8.1b: © Ed Reschke; 8.14: © Martin M. Rotker; 8.17: © L.V. Bergman/ The Bergman Collection; 8.21a,b: Reproduced from J. Bone Miner Res 1986; 1: 15–21 with permission of the American Society for Bone and Mineral Research; p. 208: Digital Vision/Getty Images; p. 209 (skateboarder): TRBfoto/Getty Images; (roller blader): Royalty Free/Corbis; (yoga): Image Source/ Jupiter Images; (snowboarder), (tandem bicycle), (rock climber): © Royalty Free/Corbis; (ballet couple): © Digital Vision

Chapter 9

Opener: © Digital Vision; 9.2: © Ed Reschke; 9.4a & 9.7a: © Dr. H.E. Huxley; 9.20: © Ed Reschke; p. 234: Digital Vision/Getty Images; p. 235: (hand stand): © Digital Vision; (gymnast): © Royalty Free/Corbis; (yoga): Image Source/Jupiter Images; (snowboarder): © Royalty Free/Corbis; (balance beam): © PhotoAlto/ Alamy; (rock climber): © Royalty Free/Corbis

Chapter 10

Opener: © Royalty Free/Corbis; 10.4: Nature Volume 413, Issue 6855, 4 October 2001, cover. Photo by Dr. John W. Weisel, University of Pennsylvania; 10.6: © Stuart Fox; 10.34: From E.K. Markell and M. Vogue, Medical Parasitology, 6th edition, 1986, W.B. Saunders Company; 10.39a,b: © Niels A. Lassen, Coppenhagen, Denmark; 10.41: © Biophoto Associates/Photo Researchers; p. 271: Digital Vision/ Getty Images; p. 272 (snow boarder): © Royalty Free/Corbis; (kayaker): © Royalty Free/Corbis; (skate boarder): TRBfoto/Getty Images; (rock climber): © Royalty Free/Corbis; (hand stand): © Digital Vision; (yoga): Image Source/Jupiter Images

Chapter 11

Opener: © Royalty Free/Corbis; 11.10a: © Stan Flegler/Visuals Unlimited; 11.10b: © Dennis Kunkel/ Phototake; p. 292: Digital Vision/Getty Images; p. 293 (tandem bicycle), (kayaker), (snow boarder): © Royalty Free/Corbis; (ballet couple): © Digital Vision; (rock climber): Royalty Free/Corbis; (balance beam): © PhotoAlto/Alamy

Chapter 12

Opener: © Royalty Free/Corbis; 12.8a,b: © R. Calentine/Visuals Unlimited; p. 317: Digital Vision/Getty Images; p. 318 (kayaker), (snowboarder), (roller blader): Royalty Free/Corbis; (skate boarder): TRBfoto/Getty Images; (hand stand): © Digital Vision; (yoga): Image Source/Jupiter Images

Chapter 13

Opener: Image Source/Jupiter Images; p. 341: Digital Vision/Getty Images; p. 342 (yoga): Image Source/Jupiter Images; (snowboarder): © Royalty Free/Corbis; (ice skater): Ryan McVay/Getty Images; (skate boarder): TRBfoto/Getty Images; (hand stand): © Digital Vision; (kayaker): © Royalty Free/Corbis

Chapter 14

Opener: Royalty Free/Corbis; 14.10a,b: From Keith R. Porter, D.H. Alpers, and D. Seetharan,

"Pathophysiology of Diseases Involving Intestinal Brush-Border Proteins" in New England Journal of Medicine, Vol. 296, 1977, p. 1047, fig. 1. Copyright ©1977 Massachusetts Medical Society. All rights reserved; 14.19a: © Ed Reschke; p. 367: Digital Vision/Getty Images; p. 368 (rock climber), (snowboarder), (gymnast): Royalty Free/Corbis; (skate boarder): TRBfoto/Getty Images; (hand stand): © Digital Vision; (kayaker): © Royalty Free/Corbis

Chapter 15

Opener: © Digital Vision; 15.1: Science Photo Library/Photo Researchers, Inc.; 15.13a,b: © Ed Reschke; 15.14a,b: From Richard J. Blandau, A Textbook of Histology, 10th edition. © W.B. Saunders; 15.15: © Landrum B. Shettles, MD; 15.17: © Martin M. Rotker; 15.20: © F. Leroy/SPL/Photo Researchers; p. 393: Digital Vision/Getty Images; p. 394 (ballet couple): © Digital Vision; (snowboarder), (gymnast): Royalty Free/Corbis; (skate boarder): TRBfoto/Getty Images; (hand stand): © Digital Vision; (kayaker): © Royalty Free/Corbis

Index

Note: Page references followed by the letters *f* and *t* indicate figures and tables, respectively.

A

B

D

E

F

G

H

I

M

N

O

P

Q

R

S

T

U

V

W

X

Y

Z